TRANSPORT IN THE URBAN ENVIRONMENT

Published by

The Institution of Highways & Transportation

ISBN 0 902933 21 3
© The Institution of Highways & Transportation, 6 Endsleigh Street, London WC1H ODZ
Published June 1997
Printed in England by Mayhew McCrimmon Printers Ltd, Essex

Transport In The Urban Environment

Structure of Contents

Foreword

by The Rt Hon Dr Gavin Strang MP, Minister for Transport

Our towns and cities are a major focus of activity for almost all of us. Nearly 90% of us live, and a similar proportion of the working population work, in urban areas. We go shopping and take part in many leisure activities there. We wish to be able to travel in safety, we wish to travel without unreasonable delay, and to go about our daily activities in unpolluted and attractive environments. In turn, industry and commerce rely on efficient communications.

Our towns and cities deserve transport systems which provide the genuine choice necessary to meet our varied transport needs. Transport planning must address the variety of purposes for which journeys are made and supply the most appropriate means to carry out these journeys. Whilst it will be essential for some journeys to be made by car or by lorry, others can be made by public transport, by walking or by cycling. We need to give full consideration to the mobility needs of disabled people. An effective and integrated transport policy, linked to the planning process, is essential to provide a proper balance between all modes of transport and safeguard our environment for the future.

I applaud the efforts of The Institution of Highways & Transportation in bringing forward this timely and substantial publication. Its many contributors have brought a wealth of expertise to its production. It provides a clear view of the range of techniques which can be used to make sustainable and integrated transport systems a reality in our towns and cities. I am sure *Transport in the Urban Environment* will be an invaluable reference source for everyone seeking to make the towns and cities we travel in safer, cleaner, thriving and attractive places.

The Rt Hon Dr Gavin Strang MP, Minister for Transport

Introduction

By the President of The Institution of Highways & Transportation

In 1987, the IHT, in association with the Department of Transport, published the technical manual, *Roads and Traffic in Urban Areas*, which soon became better known as RTUA or the 'Brown Book'. It quickly proved to be an indispensable source of information for all practitioners in highway and traffic engineering, transportation and town planning, as well as others involved in urban planning and development.

Since then, the emergence of sustainable development as a key objective of transport policy, both within the UK and internationally, has precipitated radical changes in thinking about urban transport systems which are progressively influencing their design and operation. Earlier assumptions of providing for unrestrained growth have been replaced by a developing consensus in favour of demand management and a fully integrated approach to travel across all modes. The emphasis has moved towards journey–time reliability rather than speed of travel and towards the pursuit of overall environmental quality. These trends, supported by new legislation, new technology and improved standards of design have required fundamental changes to the original publication and, significantly, a change of title, to reflect these trends.

This new publication, entitled *Transport in the Urban Environment*, is intended to promote the design and management of transport infrastructure and systems of operation, consistent with the principles of sustainable development and integrated with broader urban environmental objectives. Greater importance has been placed on the roles of public transport, cycling and walking, with guidance on increasing their relative priority.

The manual provides a comprehensive guide to every aspect of urban transport and will I am sure, like its predecessor, rapidly become an indispensable 'reference of first resort'. Although the pace of change is likely to continue, for example in the funding and management arrangements for the road network, the adoption of sustainable development principles should ensure that the manual remains a source of best practice for many years.

I would like to thank all those involved in the production of the manual, particularly members of the Steering Group, under the Chairmanship of Professor Peter Hills OBE, and all the contributing authors. Thanks are also due to the Department of Transport, the Institution of Civil Engineers and the County Surveyors' Society, for financing the project and also to the numerous individuals and bodies who responded to consultation drafts. I am especially indebted to the Managing Editor, Bill Mackay, who bore the brunt of the editing work, and to Wendy Hooper, Executive Editor to IHT, for arranging and overseeing the publication.

On behalf of the Institution, I am pleased to commend *Transport in the Urban Environment* to all with a professional interest in the future of our urban areas and the quality of their transport systems.

Michael Kendrick OBE
President
June 1997

Acknowledgements

The drafting and production of this manual, over the two–year period from April 1995 to March 1997, has involved the co–operation and unstinting goodwill of many people and the financial backing of several organisations. The IHT is grateful to all those people and to those who provided sources of data and information necessary to compile this volume. Particular thanks are due to the main contributors identified below.

Steering Group

Professor Peter Hills OBE	University of Newcastle upon Tyne (Chairman)
Derek Palmer	The Institution of Highways & Transportation (Secretary)
Professor Richard Allsop OBE	University College London
Chris Carter	Independent Consultant
David Green	Babtie Group
Chris Jeffrey	Gateshead Metropolitan Borough Council
Alan Lovell	West Sussex County Council/CSS
Terry Mulroy	Transportation Planning (International)/ICE
Mike Talbot	Department of Transport

Managing Editor

Bill Mackay	Independent Consultant (formerly of JMP Consultants)

Executive Editor

Mrs Wendy Hooper	IHT Publications

Contributing Authors (in alphabetical order)

Stephen Alexander	Hampshire County Council
Elizabeth Ampt	Steer, Davies and Gleave
Gordon Baker	SIAS Ltd
Peter D Bayless	Hampshire County Council
Peter de la Bertauche	Surrey County Council
Howard Blessington	Oscar Faber
Ian Braddock	Oscar Faber
Fiona Brown	Halcrow Fox
Richard Brown	Oscar Faber
Edward Chorlton	Devon County Council
Jim Coates CB	Independent Consultant
Dr Denvil Coombe	The MVA Consultancy
Alison Cox	Oscar Faber
Ellis Dallal	Woking Borough Council
Dr David Davies	Independent Consultant
Colin Eastman	Birmingham City Council
Callum M Findlay	Surrey County Council
James Fletcher	Independent Consultant
Keith Gardner	London Planning Advisory Committee
Kevin Gardner	LT Buses
David R Gilson	Leeds City Council
Mervyn Hallworth	Leeds City Council
David H Hodgkinson	Sevenoaks District Council

Neil Houldsworth	Leeds City Council
David Jeffrey	Atkins Wootton Jeffreys
Professor Peter Jones	University of Westminster
Alex Macauley	The City of Edinburgh Council
Professor Peter Mackie	University of Leeds
Professor Keith Madelin OBE	University of Birmingham
John Manson	Devon County Council
Michael Mason	Oscar Faber
Don McIntyre	Freight Transport Association
John Nicholson	Basingstoke and Deane Borough Council
Ken Oastler	Corporation of London
Derek Palmer	The Institution of Highways & Transportation
John Phillips	Independent Consultant
Dr Stephen Potter	Open University
Dr Tony Rhodes	University of Newcastle upon Tyne
Martin G Richards OBE	The MVA Group
Malcolm Roberts	Colin Buchanan and Partners
Ian Routledge	Oscar Faber
Barbara Sabey ISO	Road Safety Consultant
Philip W Schnepp	Babtie Group
Juliet Soloman	University of Westminster
David E Tarrant	Hampshire County Council
Richard Taylor	Devon County Council
John D Wallis	Leeds City Council
Tony Young	Independent Consultant

The Department of Transport, The HighwaysAgency, The Department of the Environment, The Scottish Office, The Welsh Office, and The Department of the Environment (Northern Ireland), have also contributed greatly to the document through the willingness of many of its officials to comment in detail on the draft chapters as they were produced. A few were also Contributing Authors. Although individuals are not named, nonetheless, the Institution is most grateful for their contributions.

Sponsors

The Institution would like to express its gratitude to three organisations who have sponsored the production of this Manual through their substantial financial contributions:

Department of Transport

CSS (formerly the County Surveyors' Society)

Institution of Civil Engineers' Research Fund

Preface

Purpose and Scope

The intention in publishing this Manual, as with the brown book *Roads and Traffic in Urban Areas* which it has superseded, is to provide 'a reference of first resort' for all those engaged professionally or otherwise in the fields of urban transportation planning, traffic and environmental management, traffic safety and highway design. Contained within a single volume (or disc), this manual is a comprehensive guide to almost all aspects of transport in urban areas and every attempt has been made to present current best practice. It provides an easy–to–read guide, with an excellent list of references and further reading, which allows ready access to more detailed and technical advice and guidance. While changes will inevitably occur, for example in traffic regulations or the arrangements for funding transport investments, this document will continue to be a source of best practice for many years.

Definitions

In the preparation of RTUA, many minds were exercised in trying to pin down the most appropriate definition of 'an urban area'. Candidate definitions ranged from abstruse considerations such as the sense of place or of neighbourhoods and the overlapping catchments for different services to more mundane and practical considerations such as the designated boundaries of local authorities or even of areas subject to a certain speed–limit or equipped with street–lights with a certain spacing. The change in title, which now makes no explicit reference to urban areas, has led to a more relaxed view of the need for precise definitions. Certainly, no attempt has been made here to define what constitutes "the urban environment".

Suffice it to say that the scope of this volume is intended to cover all contiguous built–up areas, of any size, from a large village or small self–contained town up to the largest conurbation, but that the advice on policy and best practice will vary considerably with the scale, structure, intensity and balance of the land–uses in the area concerned.

Another definitional problem centres on the distinction to be made between *transport* and *transportation*. Some would argue that the words are synonymous and others that transport is the 'proper' English word and transportation its variant, based upon American usage. Both, in fact, are to be found in the Oxford English Dictionary, where the suffix '–ation' refers 'to the system of'. This implies that transportation deals with the whole system of determining why, where and when people and goods are conveyed, whereas transport is the means or act of conveyance. An excellent analogy is the distinction between *taxation*, as the system of determining the amounts of money that must be paid how and by whom, and *tax* which represents the actual payment. With this distinction in mind, the words transportation and transport are both used in the volume, each in its appropriate context.

Structure and Content

The volume is divided into six parts.

PART I (*Issues, Responsibilities and Principles*) provides an overview of the key issues, including the intrinsic importance of travel–demand and transport–supply. Sustainable development issues are discussed, as well as the importance of developing sustainable urban transport policies. Past travel–patterns and future trends are described, as is the recent growth in the negative impacts of traffic growth. Transport policy is covered, focusing on objective–led planning and the role of targets. The roles, responsibilities and powers of both central and local government are described, including issues relating to Transport Policies and Programmes, funding, the Transport and Works Act and trunk road planning. Unlike the previous edition, *Transport in the Urban Environment* also looks to the future of urban transportation, reflecting on the possible impact of changes in society, such as the growth of tele–working and tele–shopping, as well as changes in household size and location.

PART II (*The Transportation Planning Process*) describes the context for urban transportation plans. The components of transport policy are described, in detail, as are the different methods of data collection. Urban transportation plans must be based upon an understanding of travellers' responses to different potential policies; so, one chapter is devoted to the role of modelling in estimating travellers' responses. Policies cannot be adopted in urban areas without proper appraisal of their economic and environmental effects, which are the subject of another chapter. However, for measures to be introduced successfully, public acceptance and approval are required, so one chapter concentrates upon methods of involving the public.

PART III, the largest part, deals with issues relating to *Traffic, Safety and Environmental Management*. This part covers general issues, such as managing the use of the road system, town centres, procedures for implementing traffic management measures and enforcement. It also covers information for transport users, road safety, environmental management, technology for network management, car parking, traffic calming and the control of speed, as well as 'demand management' ie traffic restraint. A separate chapter is devoted to each different mode: pedestrians; cycling; measures to assist public transport; and the management of HGVs.

PART IV deals with the *Highway and Traffic Considerations for Development*. Chapters cover the transport aspects of new developments and of development control, development–related parking, residential developments and non–residential parking. As well as covering the legislative framework of such matters, this part considers design issues and the planning and provision for different road–users.

PART V considers in detail *The Development and Design of Highways and other Infrastructure Schemes*. Chapters cover design considerations, alternative concepts of road capacity, as well as the procedures for the planning and approval of transport infrastructure schemes. Detailed design guidance is provided on fixed track systems, highway links, priority junctions, roundabouts, traffic signal control, co–ordinated signal systems, signalised gyratories and grade–separated junctions. Other chapters deal with the highway in cross–section, as well as the general consideration of junction design.

Finally, PART VI reports on variations applying to *Northern Ireland, Scotland and Wales*. There is also a glossary of initials and acronyms and a general index to the whole volume.

Limitations

In view of its wide subject coverage, this Manual cannot provide the level of detailed advice required, for all planning and design purposes, across all of the subjects with which it deals. This is recognised by the inclusion, at the end of each chapter, of a broad range of references and sources from which further information can be sought.

Diagrams, Figures, Illustrations and Photographs

References to diagrams, figures, illustrations and photographs are given in the comprehensive index at the end. Where appropriate, sources are given with the items and the IHT is most grateful for these contributions. The cover was designed by David Sexton and the photographs for the Part dividers were supplied by David Nicholls and Derek Palmer.

Currency of the advice given

The material compiled for this Manual was assembled between 1995 and 1997 and every attempt has been made to reflect the latest position, at the date of publication, in terms of legislation, practice and research. Inevitably, as time passes, legislation is changed, new ideas and policies are advanced and new techniques develop, in some fields faster than in others. Whilst every endeavour was made to include the latest legislative position and references, up to 31 October 1996, later information has been included wherever possible.

Government Regional Offices:

Advice and guidance on transportation, development and planning matters can often be obtained, in the first instance, by enquiry directed to the relevant Government Regional Office. These are listed below.

Government Office for the North East
Stanegate House
2 Groat Market
Newcastle upon Tyne NE1 4YN
Tel: 0191–201–3300
Fax: 0191–202–3744

Government Office for the North West
Sunley Tower
Piccadilly Plaza
Manchester M1 4BE
Tel: 0161–952–4000
Fax: 0161–952–4255

Government Office for Merseyside
Cunard Building
Pier Head
Water Street
Liverpool L3 1QB
Tel: 0151–224–6300
Fax: 0151–224–6470

Government Office for Yorkshire and the Humber
PO Box 213
City House
New Station Street
Leeds LS1 4US
Tel: 0113–280–0600
Fax: 0113–283–6394

Government Office for the East Midlands
The Belgrave Centre
Stanley Place
Talbot Street
Nottingham NG1 5GG
Tel: 0115–971–9971
Fax: 0115–971–2404

Government Office for the West Midlands
77 Paradise Circus
Queensway
Birmingham B1 2DT
Tel: 0121–212–5000
Fax: 0121–212–5071

Government Office for the South West
4th Floor
The Pithay
Bristol BS1 2PB
Tel: 0117–900–1700
Fax: 0117–900–1915

Government Office for the Eastern Region
Enterprise House
Vision Park
Chivers Way
Histon
Cambridge CB4 4ZR.
Tel: 01223–202000
Fax: 01234–796341

Government Office for the South East
Bridge House
1 Walnut Tree Close
Guildford GU1 4GA
Tel: 01483–882255
Fax: 01483–882259

Government Office for London
Riverwalk House
157–161 Millbank
London SW1P 4RR
Tel: 0171–217–3456
Fax: 0171–217–3450

Documents produced by the European Commission can be obtained directly from:

Commission for the European Communities Publications
Rue de la Loi 200
25, 6th Floor
1049 Bruxelles
Belgium
Tel: 00–32–2–295–5491
Fax: 00–32–2–296–0740
Email:http:\\europa.eu.int

Alternatively, application can be made to the **CEC offices in the UK:**

Office in the UK
Jean Monnet House
8 Storey's Gate
London SW1P 3AT
Tel: 0171–973–1992
Fax: 0171–973–1900

Office in Northern Ireland
Windsor House
9–15, Bedford Street
Belfast BT2 7EG
Tel: 01232–240708
Fax: 01232–248241

Office in Scotland
9 Alva Street
Edinburgh EH2 4PH
Tel: 0131–225–2058
Fax: 0131–226–4105

Office in Wales
4 Cathedral Road
Cardiff CF1 9SG
Tel: 01222–371631
Fax: 01222–395489

Orders for IHT publications should be addressed to:

The Institution of Highways & Transportation
6 Endsleigh Street
London WC1H ODZ
Tel: 0171–387 2525
Fax: 0171–387–2808
E–mail: publicat@iht.org

Further information about other IHT publications is available on the Internet:
http://www.public@iht.org

Stationery Office publications (including published legislation) can be obtained from:

The Stationery Office
Orders Department
PO Box 276
London SW8 5DT
Tel: 0171–873–0011
Fax: 0171–873–8200
E–mail: books.orders@theSO.co.uk

or from Stationery Office bookshops and approved agents

Transport Research Laboratory publications are available from:

TRL Library Service
Transport Research Laboratory
Old Wokingham Road
Crowthorne
Berks RG45 6AU
Tel: 01344–770783/4
Fax: 01344–770193
E–mail: sandrao@lib.trl.co.uk

List of Contents

Chapter 16   Road Safety   209

Chapter 17   Environmental Management   223

Chapter 18   Technology for Network Management   235

Issues, Responsibilities and Principles

Chapter 1 Transport in the Urban Environment

1.1 Importance of Travel and Transport

Travel is an essential part of daily life in the modern world and one that is valued highly. The average household in the UK now spends approximately 15% of its total expenditure on travel and 84% of that is on car–ownership and use. This is about the same proportion as is spent on housing and only slightly less than that spent on food, thus reflecting the high priority attached to personal mobility. Throughout the twentieth century, the growth of the national economy has been closely associated with the increasing movement of people and goods, particularly by road. Over the last 50 years, the growth in road traffic has, generally, mirrored the rise in Gross Domestic Product (GDP) and real personal income.

On average, people in Great Britain spend about an hour a day travelling, which represents about six percent of their waking hours: with similar proportions observed in Germany and some other European countries (Jones *et al*, 1993). Despite large increases in traffic levels, perhaps surprisingly, the total number of trips per person per day and the time spent travelling have increased very little (see Chapter 2). Traffic growth has resulted from a switch to car–use and away from public transport, walking and cycling, coupled with an increase in the average distance travelled per trip.

Most people need to travel in order to reach places of work or education, to purchase goods and services and to take part in a wide variety of social and leisure activities outside their homes, as detailed in Chapter 2. While some journeys can be tiring and stressful, and more are becoming so, many people still say that much of their travelling, particularly when driving, is an enjoyable experience (Lex, 1995).

Nationally, in 1994, 68% of households owned at least one car and, of these, 23% owned two or more. Rates of car ownership are inversely related to population density, being highest in rural areas, where 81% of

Photograph 1.1: A typical urban street with many functions.

Photograph 1.2: Servicing frontage shops from the kerbside.

households own one or more cars, and lowest in the conurbations (59%).

Around three percent of the workforce in Great Britain is employed in the vehicle building and servicing industry and it is estimated that about 15% of GDP is related to the construction, maintenance and use of motor vehicles. In 1995/96, approximately four percent of total government expenditure (c. £10bn) was used to develop, maintain and operate surface transport systems; of this, 60% was spent directly on road construction and maintenance. Total revenue from the use of road vehicles, in the form of vehicle excise duty and fuel tax, was approximately £19.5bn per year, in 1995, at 1995 prices. A similar amount is invested each year by private individuals and firms in new cars.

The movement of goods, mainly by road, is a vital part of the production, consumption and export processes of the economy and almost all activities carried out at different commercial and industrial land-uses rely, to some extent, on access by road and servicing by road-based vehicles. In addition, social activities and personal aspirations are now heavily geared towards road transport, particularly private cars. Both factors combine to make it difficult to encourage the use of alternative modes.

1.2 Consequences of Traffic Growth

Traffic growth, in the past, has been directly associated with increased economic activity and

higher living standards and many individuals have benefited from lifestyles built around use of the car. However, the aggregate effects of continually increasing traffic levels in some areas are now causing serious economic, environmental and social problems. The projected increases in traffic growth (see Chapter 2), if unrestrained, will accentuate many of these problems and the prospects of this are causing considerable concern among both professionals and the public. Furthermore, the strength of the links between economic growth and increases in road traffic is being questioned.

In October 1994, the Royal Commission on Environmental Pollution published its 18th Report on Transport and the Environment (RCEP, 1994). This raised questions over the future direction of transport policy in the UK, given the increasing impact that the forecast growth in car-use was likely to have upon the environment. Partly in response, the Government initiated a debate (DOT, 1995a) to highlight key issues and to assess the public's reaction to policy options. In April 1996, the Government published the first green paper on transport for almost 20 years, Transport: the Way Forward (DOT, 1996a) [Sa]. This represented the Government's formal response to the issues raised by the transport debate [NIa].

Economic Costs

The main negative effects of vehicular traffic levels in Great Britain, on the performance of the network and on the surrounding areas, can be summarised under the following headings:

❑ 'traffic congestion' – resulting in delays to road vehicles and passengers. A total congestion cost of £19bn per year has been estimated as the value of time and material resources wasted in traffic congestion, virtually all in urban areas. However, the assumptions required and the absence of an agreed 'base' level makes such estimates subject to considerable uncertainty (Newbery, 1995);

❑ 'traffic accidents' – estimated to cost £13.3bn per year (DOT, 1996b). This estimate, however, is based upon individuals' willingness to pay (ex ante) for marginal risk-reduction. The corresponding (ex post) cost, based upon a gross output calculation would be only about five billion pounds per year;

❑ 'air pollution' – resulting in various respiratory problems and in a deterioration of building facades. The annual national health cost has been estimated to be £2.3bn per year and, for example, Oxford City Council has estimated that the cost of cleaning the facades of historic buildings in the city centre alone have amounted to £150,000 per year (Jarman, 1994);

❑ 'traffic noise' – causing nuisance to people both on the street and in adjacent buildings. The

Photograph 1.3: Congestion – vehicular and pedestrian.

economic cost of this has been estimated at around £3 bn per year (RCEP, 1994);

❑ 'severance of neighbourhoods and visual intrusion' – these have not been quantified nationally but they can have significant adverse social and psychological impacts; and

❑ 'loss of land, due to road–building' – in the six year period, 1985 to 1990, approximately 0.1% of the total land area of England was taken for road–building (two–thirds of this being rural land) (DOE, 1996). The total area of land occupied by roadspace of all kinds, however, is less than 1.5% of the land–area of the UK.

Some of these effects, notably congestion, impact directly upon vehicle occupants. Those resulting in increased business costs are passed on to consumers and others, such as adverse effects on health, affect people more widely. Health concerns may go wider still, as some doctors attribute the recorded increases in obesity and heart disease, in part, to greater reliance on cars and the reduction in trips on foot and by bicycle.

Public concern

These negative effects of road traffic receive considerable media attention and are of increasing concern both to the general public and politically. Typically, 80%–90% of the adult population of the UK regard traffic congestion, air pollution from traffic and traffic accidents as 'very' or 'fairly' serious national problems – a view which is unaffected by whether people live in urban or rural environments. Surveys by the Department of the Environment have recorded an increase in the proportion of the public being 'very worried' about air pollution from traffic, up from 23% in 1986 to 40% in 1993.

The severity of these problems is often perceived to be less locally than nationally, with fewest people being affected in rural areas and most in the conurbations with other urban areas lying in between. Table 1.1 shows that the greatest disparity between residents in rural areas and conurbations is in relation to air pollution.

1.3 Sustainable Development Issues

In addition to the localised impacts of traffic growth, there is increasing concern about the more global impacts of the production, use and disposal of motor vehicles, in view of the worldwide efforts to develop more sustainable economic and social systems. At the United Nations Conference on Environment and Development at Rio de Janiero, June 1992, the UK government committed this country to seeking more sustainable economic development.

The most widely accepted definition of 'sustainable development' is contained in the report of the World Commission on Environment and Development (WCED, 1987), where it is defined as: "Development that meets the needs of the present, without compromising the ability of future generations to meet their own needs."

Although the practical interpretation of this definition is still subject to considerable debate, the road traffic system as a whole consumes large

Aspects of public concern	Conurbation Residents	Other Urban Residents	Rural Residents	All Respondents
Traffic congestion	65	47	41	50
Air pollution	65	47	31	45
Traffic noise	45	32	25	32
Traffic accidents	42	32	26	33
Loss of land	30	21	20	23

Table 1.1: Percentage of respondents believing each issue to be a 'very' or 'fairly' serious problem locally. Source: Jones and Haigh, (1994).

Photograph 1.4: Pedestrian priority.

quantities of non–renewable resources, mainly as fuel, but also in the construction and maintenance of vehicles and road networks and produces a wide range of noxious and non–degradable outputs. There is disagreement as to how far current rates of global resource depletion and emissions are sustainable but there is little dispute that, if European levels of car–ownership and use were to extend throughout the 'Third World', the situation would certainly not be sustainable from a global point of view.

Photograph 1.5: Bus priority by re–allocating urban road–space.

One of the major features of sustainability, that has prompted a re–appraisal of transport policy, has been concern about the production of carbon dioxide emissions and the implications for global warming. As part of the European Union, the UK government is committed to stabilising national carbon dioxide emissions by the year 2000 at its 1990 level (see Chapter 17). Transport currently accounts for 25% of UK emissions and, unlike most other sectors, this is rising in line with the increase in traffic. Road transport accounts for 87% of the total transport carbon dioxide emissions, rail and shipping five percent each and air transport three percent (OST, 1995).

Concern for more sustainable transport is resulting in pressure on the motor vehicle industry to produce more fuel–efficient vehicles, so as to generate less carbon dioxide, and to design systems so that the materials from these vehicles can be recycled at the end of their useful life. However, while industry has developed cars that, for a given engine size, have become more fuel efficient, the average engine size of new cars purchased by consumers has become larger, thereby partly offsetting this increase in fuel efficiency.

Carbon dioxide emissions contribute to global warming, regardless of the location or time of day at which they are generated, thus forcing a wider re-appraisal of transport policy than would be triggered by the need simply to tackle localised congestion or air quality problems.

NB: A 'Position Paper' by The Institution of Highways & Transportation provides information and recommendations on Road Transport, The Environment and Sustainable Development (IHT 1996a).

1.4 Objectives–Led Transportation Strategies

Growing concern about the local and global effects of current road traffic levels, and a realisation that projected traffic growth could not be accommodated physically in most urban areas, has led to many local authorities developing comprehensive transportation strategies based on agreed local objectives [NIa].

Defining Objectives

The objectives–led approach has been encouraged by the Department of Transport, who require this as the starting point for the development of the so–called 'package' bids by local highway authorities in their annual Transport Policy and Programme (TPP) submissions (DOT, 1995b) [NIb] [Wa]. Higher–level objectives of local transport strategies now typically include:
❑ reductions in traffic congestion;
❑ improvements in air quality;
❑ reductions in traffic accidents;
❑ reductions in (growth of) carbon dioxide emissions; as well as
   ❑ enhancing economic vitality and viability;
   ❑ improving the visual quality of the urban environment;
   ❑ increasing the availability of alternative modes to the car; and
   ❑ improving provision for disabled and other vulnerable road–users.

There is much confusion over the distinction between goals, objectives, policy measures (ie solutions), performance indicators and targets. The Department of Transport is particularly concerned that, in developing local transport packages, objectives should not be couched in terms of implementing a particular solution (eg 'to build a light rail system') but, instead, should describe desired outcomes in terms that are as independent as possible of the means of achieving them (DOT, 1995b).

Monitoring Progress

Quantitative measures of each objective can be helpful, so that progress in achieving objectives can be assessed. These can range from the objective measurement of physical quantities, such as traffic volumes and exhaust emissions, to the subjective measurement of public satisfaction. Some local authorities go further and set specific targets for improvements in performance levels, as recommended by the Royal Commission on Environmental Pollution (RCEP, 1994).

Care is needed in defining objectives and devising appropriate measurements for monitoring purposes. Given the broad aim of reducing the negative impacts of traffic growth, setting objectives and monitoring progress can be carried out at three levels:
❑ 'primary measurements' – recording directly the changes in a specific traffic impact where a policy–objective has been set, such as reducing the number and severity of road traffic accidents;
❑ 'secondary measurements' – recording changes in the total volume, or mix, of road traffic or in the rate of traffic growth, as a proxy for monitoring the achievement of several primary objectives, such as reductions in traffic noise, severance and road congestion; and
❑ 'tertiary measurements' – aimed at measuring changes in behaviour, such as reductions in average car trip–length or changes in modal split, that in turn affect the overall volumes of road traffic, which in their turn impact on the primary objectives.

Secondary and tertiary measurements are often easier to make and, when improvements are recorded, several policy–objectives may be being achieved at the same time. However, links between secondary and tertiary measurements and primary objectives may change over time and need to be kept under review. For example, with the increase in the numbers of catalytic convertors in the car stock, there is a no longer a linear relationship between traffic levels and many forms of air pollution.

Further details on setting objectives and monitoring the effectiveness of policy instruments are provided in the IHT Guidelines on Developing Urban Transport Strategies (IHT, 1996b).

1.5 Developing Urban Transport Policy Packages to Achieve Sustainable Development

Typical Urban Packages

Through the 'package' approach [NIc], local authorities are encouraged to develop a set of transportation and land–use policies which aim to enhance the economic vitality and viability of towns, while at the same time reducing, or at least containing, the negative environmental and economic effects associated with heavy concentrations of road traffic. Most of this effort is focused on managing urban car–use, although some authorities take into account ways of tackling urban goods movement as well.

Car–related problems are dealt with by implementing a combination of 'carrots' (improvements to the modal alternative of public transport, walking and cycling) and 'sticks' (measures to restrain or discourage less essential car–use, through time or cost penalties). In combination, these actions are designed to shift the balance of advantage away from private cars for certain types of journey (see Section 2.7 and Chapter 21 on Demand Management).

These measures are usually supported by complementary land–use policies, by providing better information to travellers, such as warning of problems on the road network and giving details of public transport alternatives (see Chapter 15), and by public awareness campaigns designed to encourage more responsible use of cars (see Chapter 10).

Future Developments
Some increase in urban road capacity is likely to result from the combination of new road construction, route–guidance and network management technologies. However, local authorities are increasingly active in the way they manage road capacity, by deciding the levels of vehicular traffic to be allowed on different parts of the road network, in order to reduce traffic congestion and the various harmful environmental impacts, and by allocating limited urban roadspace among competing groups of road–users.

The implementation of sustainable transportation policies in more congested areas requires a significant shift away from private cars in favour of public transport, cycling and walking. In some cases, in the interests of regenerating local shops and services, it may also reverse the previous trend of limiting kerbside parking and loading, which favoured 'through' traffic at the expense of local business activity.

To be successful, this shift in the balance between the provision for cars and other transport modes needs to be matched by a parallel shift in public opinion. There are signs that this is happening, where national surveys reveal diminishing support for major new road–building and increasing support for improved modal alternatives and even for car–restraint policies (TSG, 1996). Given the public awareness campaigns being run by many local authorities, this shift in public opinion is likely to continue, although the pace of change in travel behaviour might be quite slow.

Most restraint measures have, historically, taken the form of physical capacity constraints on the road network, often relying on congestion itself as a restraint mechanism. Alternatively, parking controls have been used to limit the overall number of spaces or to influence the mix of types of vehicles parked in an area. However, parking controls do not affect 'through' traffic and some forms of parking, particularly privately owned non–residential (PNR) spaces, are not amenable to public control under current legislation.

Advances in transport telematics offer the possibility of restraining moving traffic, either by introducing electronic road–use pricing or by limiting access to certain areas to vehicles of a particular type, such as buses or delivery vehicles or vehicles which are displaying an appropriate electronic tag, such as local residents or Orange Badge holders. As the situation develops, it is likely that local authorities will move away from relying on a single solution, such as pricing, to restrain moving traffic towards the more pluralistic situation that has evolved in relation to parking controls. Here, local objectives are met by prioritising the use of limited parking spaces through a combination of physical, regulatory and pricing measures.

1.6 Relationship to Policies of Safety and Sustainability

When a road link or junction is subjected to traffic flow which exceeds its capacity, congestion ensues. Speeds drop, delays are caused to vehicles, journey times become unpredictable, fuel is wasted by queueing traffic and vehicle exhausts pollute the air to a greater extent than would be the case with the same volume of smoothly–flowing traffic. However, it is not always the case that increasing the capacity of a congested link also improves its safety record. For example, the severity of accidents tends to worsen with increased traffic speed. Safety audits of individual schemes are, therefore, needed to predict their consequences for accidents (see Chapter 16).

Likewise, it is not automatic that increasing the capacity of an urban link provides benefits in terms of sustainability. When the capacity of a congested link is increased, there will be immediate reductions in overall fuel consumption and pollution from vehicle exhausts. However, in conditions where traffic demand has been suppressed by congestion, the extra capacity may lead to traffic growth, with consequent increases in fuel usage and pollution. If, on the other hand, the extra capacity is provided as part of an area–wide plan to strengthen the road hierarchy, then the benefits of diverting traffic away from residential environments onto the improved link may be realised without encouraging further traffic growth.

1.7 World and European Perspectives

Many countries, elsewhere in Europe and in other parts of the world, face problems similar to those found in the UK and there is much useful experience that can be drawn upon (eg Jones, 1993: OECD, 1994 and 1996; Pharoah et al, 1995 and Tolley, 1990). Also, a number of libraries maintain good sources of overseas publications which are usually available via e–mail, hard copy or for inspection for a fee.

An awareness of thinking at the European Commission is useful for providing background on existing and proposed legislation at the European level. Future developments in legislation and standards affecting the environmental impact of transport, in such areas as emissions and modal change for passengers and freight, are likely to be stimulated by the Commission. European legislation is likely to have a significant influence on urban transport, possibly even at scheme level if, for example, the relative importance of different objectives for pedestrianised or traffic–calmed areas is changed.

Information on European initiatives can be obtained directly from the European Commission in Brussels or from the UK Office of the Commission. It is important to realise that approximately two–thirds of the twenty–three Directorates are concerned with transport in one way or another. Apart from DGVII – Transport, the more involved ones for transport practitioners are DGII – Economic and Financial Affairs, DGVIII – Development, DGXII – Science, Research and Development, DGXIII – Telecommunications (for transport telematics), DGXVI – Regional Policies and DGXVI – Energy. Many of these Directorates have research and development programmes in which transport plays a key part and some funding may be available to assist in experimental or innovative schemes.

Specific areas of the United Kingdom may be eligible for European grant–aid, in particular circumstances. The best known of these are regional development plans to assist peripheral areas, such as Northern Ireland, declining industrial areas and rural areas.

The UK Office of the Commission can assist in providing information, as can Government departments, including the DOT, DOE and DTI [Sb]. The European Investment Bank (EIB) also provides assistance on grant–eligibility.

Some European cities have invested heavily in rail–based public transport systems and, where the city centre remains a strong economic force and parking controls have been in place for a long period (eg as in Munich, Vienna and Zurich), the use of these services is high and patronage has been rising for many years. In such cities, public transport trips may account for between 25% and 35% of all trips made by city residents. The proportion of trips made by car in these cities has remained stable over many years – or, in some cases, has declined – despite very high levels of household car–ownership. However, such cities generally have higher residential densities than in the UK and it is notable that their public transport systems require substantial public subsidies, both for investment and for operation.

In smaller cities, improved public transport services have been provided by investing in a modern, high quality, bus fleet, with segregated rights–of–way at congested points in the road network and real–time information systems to keep passengers informed of current operating conditions. In larger cities, with rail–based networks, more emphasis is now placed on providing high quality feeder and orbital bus routes, with the same livery and image of quality and reliability as the core rail networks.

Studies have shown that public transport improvements, such as these, can lead to significant and sustained increases in patronage but there is little evidence of decline in traffic on parallel roads. In order to achieve traffic reductions, some of the capacity of the road network needs to be reallocated to other road–users. However, in areas where the road network is heavily congested, traffic levels may stabilise, with the additional journeys, resulting from more economic activity, being accommodated on the public transport network.

In cities that have invested in substantial bicycle networks (eg Amsterdam and Hannover) around 15% of residents' trips are now made by that mode – a doubling, in the case of Hannover, over a 15 year period. In residential areas, there is a move away from providing separate cycle–lanes to reducing traffic speeds, through traffic–calming measures and cyclists sharing space with other road traffic. Most European cities have experienced a decline in walking trips and efforts are now being made to improve facilities for pedestrians with the aim of reversing this trend.

Forms of traffic restraint vary from one area to another. In northwest Europe, parking controls have been the primary means of traffic restraint (coupled with local pedestrianisation schemes); in Italy and Greece, various access controls have been widely introduced to limit traffic in city centres (commonly reducing traffic levels by 10% – 20%). Singapore has been operating an Area Licensing road–use pricing scheme in its city centre for over twenty years, which, coupled with parking restraint and public transport improvements, has reduced road traffic by around 40%. More recently, Norway has introduced toll rings, in several cities, but these were aimed primarily at raising revenue for transport infrastructure and have had little impact on traffic levels (which they were not designed to reduce), although there is some evidence of time–of–day switching.

1.8 References

DOE (1996)	'Indicators of Sustainable Development for the United Kingdom', Stationery Office.
DOT (1995a)	'Transport: The Way Ahead', DOT.
DOT (1995b)	'Transport Policies and Programmes Submissions for 1996 – 1997; Supplementary Guidance Notes on the Package Approach', DOT.
DOT (1996a)	'Transport: The Way Forward', Stationery Office.
DOT (1996b)	'1996 Highways Economic Note 1', DOT.
IHT (1996a)	'Road Transport, The Environment and Sustainable Development', The Institution of Highways & Transportation.
IHT (1996b)	'Guidelines on Developing Urban Transport Strategies', The Institution of Highways & Transportation.
Jarman M (1994)	'Valuing Wider Environmental Benefits from an Urban Traffic Restraint Package', Transportation Planning Systems, 2 (2) pp 23 – 37.
Jones P (1993)	'Study of Policies in Overseas Cities for Traffic and Transport', Report to the Department of Transport by the Transport Studies Group, University of Westminster.
Jones P, Bovey P, Orfeuil JP, and Salomon I (1993)	'Transport Policy: the European Laboratory' in Salomon I, Bovey P and Orfeuil JP (eds) 'A Billion Trips a Day: Tradition and Transition in European Travel Patterns', Kluwer.
Jones P and Haigh D (1994)	'Reducing Traffic Growth by Changing Attitudes and Behaviour', PTRC European Transport Forum.
Lex (1995)	'What Drives the Motorist?', Lex Report onMotoring, MORI.
Newbery DM (1995)	'Reforming Road Taxation', Report for the Automobile Association.
OECD (1994)	'Congestion Control and Demand Management', Organisation for Economic Co–operation and Development, Paris.
OECD (1996)	'Sustainable Transport in Central and Eastern European Cities', Organisation for Economic Co–operation and Development, Paris.
OST (1995)	'Transport: Some Issues in Sustainability', Office of Science and Technology.
Pharoah T and Apel D (1995)	'Transport Concepts in European Cities' Avebury, Aldershot.
RCEP (1994)	Royal Commission on Environmental Pollution 18th Report October 1994 ' Transport and the Environment', Stationery Office.
Tolley R (1990) (ed)	'The Greening of Urban Transport', Belhaven Press.
TSG (1996)	'Public Attitudes to Transport Policy and the Environment: An in–depth exploratory study.' Summary Report to the Department of Transport, University of Westminster, Transport Studies Group, April.

WCED (1987)   'Our Common Future', World
                     Commission on Environment and
                     Development, Oxford University
                     Press.

1.9 Further Information

ICE (1996)   'Sustainability and Acceptability
                     in Infrastructure Development',
                     Thomas Telford, London.

OECD (1995)   'Urban Travel and Sustainable
                     Development', European
                     Conference of Ministers of
                     Transport.

OECD (1996)   'Daily Mobility: Can it be
                     Reduced or Transferred to Other
                     Modes?', European Conference of
                     Ministers of Transport, Round
                     Table 102, Paris, May 1996.

Chapter 2 Travel Patterns in Urban Areas

2.1 Introduction

Travel is an integral part of the social and commercial fabric of urban communities. As a result, it is vital to understand both the make–up of the demand for movement and the patterns of travel behaviour, for the overall success of the multi–modal transport systems common to all urban areas. Increasingly, transportation planning is moving away from merely satisfying demand towards managing or influencing demand and the resulting patterns of travel (see Chapter 21).

2.2 Trends in the Patterns of Travel

Vehicles and Traffic

The increase in vehicular travel that has taken place between the early 1950s and the mid–1990s has had a fundamental effect on the economy of the UK and the sorts of lives people lead. Over that period, the number of motor vehicles in the UK grew from four million in 1950, of which about half were private cars,

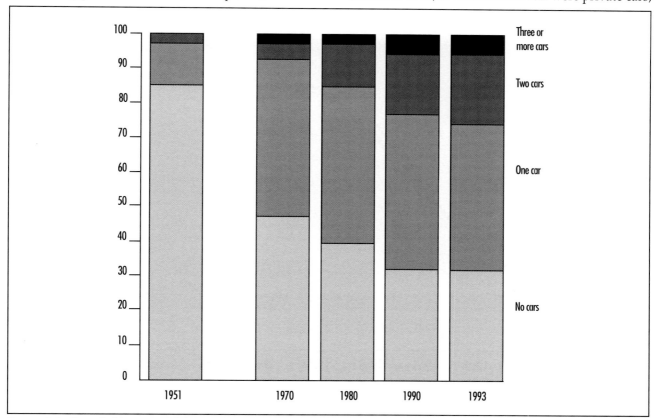

Figure 2.1: Household car availability (%).   Source: DOT (1995a).

Category of Vehicle	1950	1960	1970	1980	1990	1995
Private Cars	2.0	4.9	10.0	14.7	20.2	21.4
Goods Vehicles (>3.5T GVW)	0.4	0.5	0.5	0.5	0.5	0.4
Public Transport	0.1	0.1	0.1	0.1	0.1	0.1
Motor Cycles/Mopeds	0.6	1.6	0.9	1.4	0.8	0.6
Agricultural Vehicles	0.3	0.4	0.4	0.4	0.4	0.3
Other (incl light goods)	0.6	1.0	1.6	2.1	2.7	2.6
Total	**4.0**	**8.5**	**13.5**	**19.2**	**24.7**	**25.4**

Table 2.1: Motor vehicles licensed in the UK (millions).   Source: DOT (1995a) and (1995b).

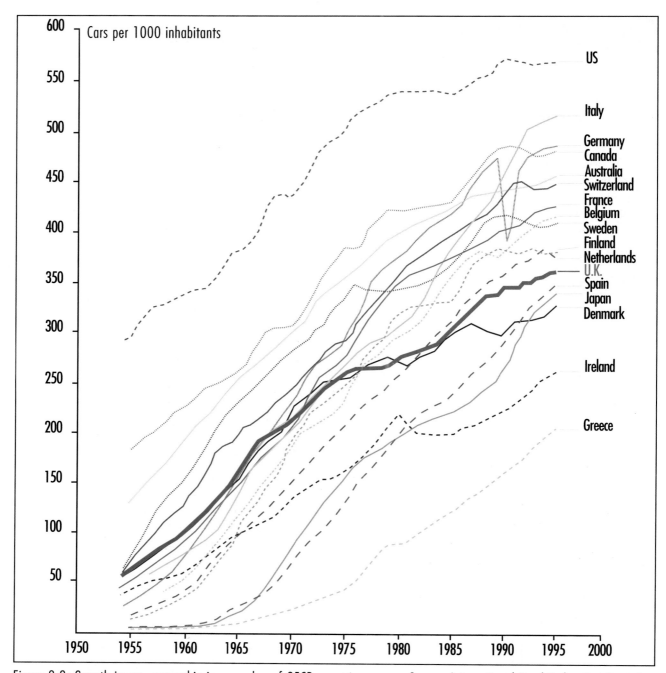

Figure 2.2: Growth in car–ownership in a number of OECD countries.   Source: International Road Federation Statistics.

to over 25 million in 1995, 83% of which were cars (see Table 2.1).

In the early fifties, only 14% of households had a car. By 1970, just over half of households (52%) in the UK had access to a car and, by 1993, 68% of households had the regular use of one or more cars (see Figure 2.1). In 1996, more than 2.0 m new cars were bought in the UK.

By way of comparison, it is interesting to examine Figure 2.2 which demonstrates the strong relationship between real income per head and car–ownership.

This accounts for car–ownership levels in the UK still being well below those in richer countries, including the majority of European Union member–States. This demonstrates the potential that still exists for car–ownership in the UK to increase, even if present levels appear to be high. For example, if the UK had the same car–ownership level in 1995 as Italy and Germany did (about 500 cars per 1000 inhabitants), there would have been 7.5 million more cars on the roads.

Not surprisingly, the relentless increase in car–ownership in the UK has been reflected in corresponding increases in road traffic. In 1952, the

TRANSPORT IN THE URBAN ENVIRONMENT

Type of Traffic		1952	1984	1994	1994 Growth Index (1952=1.0)
Passenger Vehicles:	Cars and Taxis	30.6	244.0	345.1	11.3
	Motor Cycles etc.	6.0	8.1	4.1	0.7
	Large Buses/Coaches	4.2	3.9	4.7	1.1
All Passenger Motor Vehicles		40.8	256.0	353.9	8.7
Bicycles (estimated)		22.9	6.4	4.4	0.2
All Passenger Vehicles (including bicycles)		63.7	262.4	358.3	5.6
Freight Vehicles:	Light Vans	8.7	24.5	38.5	4.4
	Goods Vehicles	11.3	22.6	29.6	2.6
All Motor Vehicles		60.8	303.1	422.0	6.9

Table 2.2: Road traffic (billion vehicle–kilometres) in 1952, 1984 and 1994. Source: DOT (1995a).

Mode of Travel	1952	1984	1994	1994 Growth Index (1952=1.0)
Buses & Coaches	92	49	43	0.5
Cars and Vans	58	441	596	10.3
Motor Cycles	7	8	4	0.6
Pedal Cycles	23	6	5	0.2
All Road Modes	180	504	648	3.6
Rail–based modes	39	37	35	0.9
Aircraft	0.2	4	5	25.0
Pedestrians	n/a	18	16	–

Table 2.3: Personal Travel (billion passenger–kilometres) in 1952, 1984 and 1994. Source: DOT (1995a) and Macafee *et al* (1996).

Freight Transport Mode	1952	1984	1994	1994 Growth Index (1952=1.0)
(a) Goods Lifted				
Road vehicles	861	1400	1689	2.0
Railways	289	79	97	0.3
Waterways	50	140	140	2.8
Pipelines	2	88	125	62.5
All Freight lifted	1202	1707	2051	1.7
(b) Goods Moved				
Road vehicles	31	100	144	4.6
Railways	37	13	13	0.4
Waterways	20	60	52	2.6
Pipelines	0.2	10	12	60.0
All Freight moved	88	183	221	2.5

Table 2.4: Goods lifted (million tonnes) and goods moved (billion tonne–kilometres) in 1952, 1984 and 1994. Source: DOT (1995a).

Transport Mode	Greater London (%)	Major Cities and Metropolitan Counties* (%)	Other Large Urban Areas Population over 100,000 (%)	Medium Urban Areas Population 25,000 to 100,000 (%)	Small Urban Areas Population 3,000 to 25,000 (%)
Walk	34	30	31	28	29
Car/Van	47	54	58	62	62
Public Transport	17	14	8	7	5
Other	2	2	3	3	4
Total	100	100	100	100	100
Average no. of journeys per person per year	1,025	1,006	1,077	1,055	1,071

*West Midlands, Greater Manchester, West Yorkshire, Greater Glasgow, Merseyside, South Yorkshire and Tyne and Wear.

Table 2.5: Modal split (%) of journeys according to size of urban area.   Source: National Travel Survey 1993/95.

Transport Mode	Greater London		Major Cities and Metropolitan Counties		Other Large Urban Areas, Population over 100,000		Medium Urban Areas Population 25,000 to 100,000		Small Urban Areas Population 3,000 to 25,000		Average for all Urban Areas	
	miles	(%)	miles	(%)	miles	(%)	miles	(%)	miles	(%)	miles	(%)
Walk	280	(3)	280	(3)	300	(3)	290	(3)	260	(2)	280	(3)
Car/Van	5,540	(68)	6,560	(76)	8,090	(81)	8,420	(82)	10,400	(84)	7,960	(80)
Public Transport	2,110	(26)	1,600	(18)	1,250	(13)	1,100	(11)	1,170	(10)	1,410	(14)
Other	230	(3)	250	(3)	340	(3)	370	(4)	440	(4)	330	(3)
Total	8,160	100	8,690	100	9,980	100	10,180	100	12,270	100	9,980	100

Table 2.6 Average number of miles (and %) travelled per person per year by type of area in 1993/95.

Source: National Travel Survey 1993/95.

4.5 million vehicles then on the roads travelled just over 60 billion vehicle–kilometres. By 1994, the number of vehicles had risen nearly six–fold, to 25.2 million, and motor traffic some seven–fold to 422 billion vehicle–kilometres (see Table 2.2).

Travel and Traffic Growth
The increase in passenger vehicle traffic, between 1952 and 1994, was two and a half times that of road goods vehicles. Passenger motor vehicle traffic increased almost ninefold, with a 38% increase in the 10 years 1984–94, and car traffic increased over elevenfold, including a 41% growth in the 10 years to 1994.

Personal Travel
Total passenger–kilometres rose by a smaller proportion than traffic (see Table 2.3). The 5.6 times increase in traffic by road passenger vehicles from 1952 was associated with only a 3.6 times increase in the volume of personal travel, principally because of

the decline in the use of non–car modes.

The larger increase in vehicular traffic, relative to passenger–kilometres, reflects the fact that growth was in the use of cars rather than public transport modes. Even within the car–use category, traffic volumes grew faster than personal travel by car, because of a decline in average vehicle occupancy.

Freight Movement
Freight transport exhibited a similar trend, with the growth in vehicular traffic rising faster than the amount of goods carried. Growth and changes in the economy of the UK, and in company logistics, mean that goods are being transported more often and further. Tables 2.4 and 2.2 show that the tonnage of goods moved by road in Britain in 1994 was twice that in 1952, while the vehicle–kilometres travelled by goods vehicles increased by 2.6 times.

The average length of haul for different types of lorry

Type of highway	Car/taxis	motor cycles	bus/ coaches	light vans	goods vehicles	bicycles	Total (%)
Motorways	76.6	0.3	0.8	8.2	14.1	n/a	100.0
Built–up major roads	82.5	1.2	1.6	9.0	4.9	0.8	100.0
Non–built–up major roads	80.0	0.8	0.8	9.2	9.0	0.2	100.0
Built–up minor roads	83.1	1.2	1.6	8.7	2.9	2.5	100.0
Non–built–up minor roads	81.8	1.1	0.7	10.1	4.6	1.7	100.0
All roads	81.0	1.0	1.1	9.0	6.9	1.0	100.0

Table 2.7: Average road traffic composition (%) in 1994.   Source: based on DOT (1995a) and unpublished tables.

has remained more or less the same over the years but the commercial drive to improve efficiency has led to increased use of heavier vehicles. These vehicles make longer journeys, which is reflected in the growth of HGV traffic. The greater carrying–capacity and average pay–load factor of these vehicles has meant that the traffic growth is significantly less than the growth in tonne–kilometres.

Influence of City–size

Passenger travel patterns in urban areas vary according to the size of settlement. Travel can be measured either by the number of journeys or by volume. Both measures are relevant to a basic understanding of travel demand. In terms of journeys, there is little difference in the number of trips undertaken by people in urban areas of different sizes (see Table 2.5). The 1993–95 National Travel Survey (Macafee et al, 1996) recorded an average ranging between 1006 and 1077 trips per annum per person, for a range of urban areas from 3000 people, right up to Greater London, with a population of 6.8 million people.

The National Travel Survey (NTS) is a household travel survey covering Great Britain. From 1965–1986, it was undertaken periodically but, from 1989, it became a continuous survey with any three years' data providing a representative sample for the whole of Britain. An analysis of traffic growth derived from the 1993/95 NTS shows that modal split varies with the size of settlement average (Macafee et al, 1996). The use of cars is highest, as might be expected, by residents of small towns and public transport use is greatest by people living in the large cities.While the difference in public transport use, ranging from 17% of trips in London down to five percent in small settlements, might be expected, the large proportion of journeys undertaken on foot varies relatively little

with settlement size and confirms the importance of pedestrian access in all types of urban area.

When travel is measured in terms of distance, the importance of modes with longer average– journey lengths, such as trains, is increased but car–use predominates nevertheless (see Table 2.6).

In terms of the overall distances people travel, there is a large range with respect to the size of settlement, from an average of around 8160 miles in London and the major cities to over 12000 miles travelled per person per year in the smallest urban areas.

2.3 Urban Traffic

Composition of Traffic

Table 2.7 shows the average mix of traffic in 1994, by percent, on different types of road in Great Britain, excluding pedestrians using the adjacent footways.

Taking an average across all road types in Great Britain in 1994, the mix of road traffic comprised 81% private cars and taxis, nine percent light vans, seven per cent other goods vehicles, one per cent buses, one per cent motorcycles and one percent pedal cycles. When account is taken of the relative effect of different types of vehicle in the traffic stream, measured in terms of passenger car units (p.c.u), then the preponderance of private cars and taxis falls to 71%.

On British motorways, the proportion of goods vehicles (14.1%) is double the average for all roads (6.9%), with a corresponding drop in figures for all other categories of road traffic, notably cars down to 77%. Goods vehicles are relatively much less common on minor roads. Proportions of bus/coach traffic are highest in built–up areas, being 1.6% of traffic flow on both major and minor roads, and cycling is much more common on minor roads, being 2.5% and 1.7% of traffic flow in built–up and non–built up areas, respectively.

In city centres, the traffic mix is usually rather different. In central London, for example, traffic composition is typically 75% cars and taxis, 16% light and other goods vehicles, four per cent motorcycles, three percent buses and coaches and two percent bicycles.

Other Road–Users

Traffic flow figures describe the volume of vehicles moving through the road network. In urban areas in particular, it is important to recognise that there are other groups of road–users, who also claim use of the carriageway, and these include pedestrians and stationary vehicles that are either parking or loading. In traditional high streets, loading provision is important in maintaining the competitiveness of local businesses. Also, since the average car spends over 96% of the time parked, parked vehicles are an important 'consumer' of space and the control of parking both on– and off–street is an important policy issue (see Chapter 19).

In residential areas, many households rely on the provision of on–street parking spaces in order to store their cars. As yet unpublished NTS figures suggest that, nationally, 22% of all cars are parked on–street at night. In conurbations, such as London, with denser housing and fewer off–street facilities, the figure rises to 44%. This proportion is likely to increase significantly in the future as car–ownership rises, particularly among lower income groups occupying properties less able to accommodate cars off–street.

2.4 Traffic Growth

Historical Traffic Growth

Table 2.2 showed that total vehicle–kilometres on the road network in Great Britain increased by a factor of seven between 1952 and 1994, with car traffic increasing elevenfold. However, growth rates have shown marked differences by type of road. Figure 2.3 shows the pattern of growth in traffic, by different classes of road, between 1984 and 1994. Overall, traffic grew by 39%, with over double this rate on motorways (84%). Note the difference in the relationship between built–up roads (those with a speed limit of under 40 miles/h) and non–built up roads. On non–built-up major roads traffic grew at about the national average (41% over 10 years) whereas on minor roads it grew by only five percent. Conversely, traffic on built–up major roads grew at

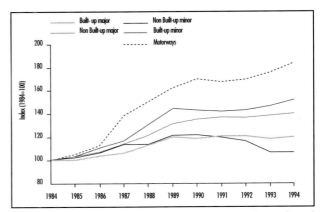

Figure 2.3: Growth of motor vehicle traffic by class of road 1984–1994.

half the national average (20%) but that using minor roads grew faster than the average rate (52%). This is probably attributable to an increase in 'rat–running' and the increased use of cars for local trips which make little use of the main road network.

Lane and Judd (1992) examined peak traffic growth–rates in urban areas of different size between 1986 and 1991. They found some evidence that these growth–rates were constrained by the available urban road capacity.

Reasons for Traffic Growth

The volume of traffic on the road network in a given time–period is determined by the number of vehicle–trips and the average length of each trip; for example, road traffic could increase by 100% either as a result of the present trips doubling in length or by twice the number of vehicle trips being made on the network at current trip–lengths. In practice, traffic growth reflects a combination of both these tendencies.

Comparing NTS data for 1975/76 and 1993/95 shows an overall increase in total distance travelled of 54%. This can be attributed mainly to:
- ❑ increased population ....four percent:
- ❑ more journeys ....13%; and
- ❑ longer journeys ....32%.

A comparison of data for 1965 and 1985 (May, 1993) concluded that: '...one of the major causes of traffic growth has been the encouragement of longer journeys'. In London, growth was only 11%, but three–quarters of the growth was attributed to increased trip–length and only one–quarter to increased numbers of trips.

Growth in car traffic can be attributed to a number of factors, including increasing disposable incomes, rising car–ownership, dispersion of land–uses and increasing traffic speeds outside urban areas. Growth in numbers of goods vehicles can be related to increases in GDP, although increased lorry size has enabled more goods to be transported per vehicle.

Traffic Forecasts

In 1989, the Department of Transport published new National Road Traffic Forecasts (NRTF). These predicted higher rates of traffic growth than previous estimates, suggesting a growth of between 24% (low) and 47% (high) between 1988 and the year 2000, increasing to 83% and 142%, respectively, by the year 2025. These forecasts caused concern within many local highway authorities and triggered a debate about the feasibility of providing for this scale of future traffic growth in many parts of the country.

The latest forecasts, re–indexed to 1994, suggest an average traffic growth across the road network of between 58% and 92% by the year 2025 (DOT, 1995a). The figures for cars and taxis are slightly lower, ranging between a 57% and an 87% increase, over the 1994 base, by the year 2025.

A review of the forecasts from a number of recent urban studies (May, 1993), in Birmingham, Edinburgh and London, found predicted increases in annual car–kms of around 1.4% per year over the next twenty years (ie a total 32% increase) being about two–thirds of the average national growth–rate. The study by Lane and Judd (1992) also found urban peak traffic growth between 1986 and 1991 to be well below the forecast increases in car–based work–trips for that period. Taken together, these studies suggest that the forecast average increases in traffic growth are unlikely to be realised in the larger urban areas.

2.5 Components of Change

Journey lengths and journey–numbers.

Four major elements make up changes in the overall volume of traffic, as follows:
- ❑ the number of trips per unit of time;
- ❑ the average length of trip;
- ❑ the mode of transport used; and
- ❑ the size and occupancy of the vehicles.

For both passenger and freight traffic, increasing average journey–length has been an important element in the growth of traffic volumes. For passenger travel, the average number of journeys of all lengths that people have made increased by 17% between 1975/76 and 1993/95. However, the distance travelled over the period increased by 38%, mostly due to a 24% increase in average journey–length, from 8.0 to 10.0 kilometres (5.0 to 6.2 miles).

A picture thus emerges of the growth of overall personal travel being generated mainly by people travelling further rather than more often. This is explored in more detail in Table 2.8 which shows the distribution of all journeys by length. In fact, the growth in trips is markedly greater for those in households with cars than those without.

Whilst the rise in journey numbers for the longest journeys (100 miles and over) only went up from 0.4% to 0.6%, it accounted for 30% of the overall growth in mileage from 1975/76 to 1993/95. The next biggest contributor was the next size category (50 to 100 miles) with a 20% share.

There was a 37% increase in journeys of length 2 miles to under five miles, but this only boosted overall mileage by about five percent. Of course, in terms of urban mileage, the large increase in medium–length journeys is of special interest, as high fuel consumption and high emissions of pollutants are features of short and medium–length car journeys.

One curious feature has been the fall in journeys of one mile to two miles and the rise in journeys of under one mile. Despite the slight under–recording of short walk–journeys, this probably reflects the fact that the number of short walk–journeys has changed little over the years, whilst walk–journeys of one mile or more have fallen sharply – probably mostly replaced by longer car–journeys.

Comparison of changes in journey patterns between the 1985/86 and 1993/95 National Travel Surveys provides an indication of the relative roles of journey numbers and journey distances (see Tables 2.9 and 2.10) and trip–purpose (see Table 2.11) over this period. These years are used because, for methodological reasons, comparisons are more reliable than for other years.

Between 1985/86 and 1993/95, there was very little change in either numbers of journeys or average journey–length by London residents (see Tables 2.9, 2.10). By contrast, in other urban areas (continuously built–up settlements of over 3000 people), there was an increase in the total number of journeys of all lengths of between two percent and five percent (see Table 2.10). If journeys of only one mile or over are considered, the increases are between six percent and 13%, reflecting the lengthening of journeys noted earlier. When distances travelled are considered (see Table 2.9), the absolute distances in 1993/95 increased with the smaller size of settlement. Residents in small settlements now travel over 40% more miles per year than those in large cities.

Journey Length	Under 1 mile	1 mile to < 2 miles	2 miles to < 5 miles	5 miles to < 10 miles	10 miles to < 25 miles	25 miles to < 50 miles	50 miles to < 100 miles	100 miles and over	All Journeys	
Year	(%)	(%)	(%)	(%)	(%)	(%)	(%)	(%)	(%)	average no. per person per year
1975/76	29.9	19.9	24.8	13.8	8.2	2.1	0.8	0.4	100.0*	935
1985/86	32.7	18.2	24.4	13.0	8.2	2.2	0.9	0.4	100.0	1024
1989/91	29.2	17.9	25.6	13.9	9.2	2.5	1.1	0.6	100.0	1091
1993/95	29.5	16.4	25.6	14.4	9.6	2.7	1.1	0.6	100.0*	1052
rise in journey nos. from 1975/76 to 1993/95	+31	−13	+37	+22	+24	+9	+5	+3	−	+118 journeys
rise in mileage from 1975/76 to 1993/95	+12	−13	+100	+143	+371	+309	+333	+546	−	+1801 miles

N.B.*Statistical 'rounding' explains why these do not add up to exactly 100%

Table 2.8: Distributions of journey–length from 1975/76 to 1993/95.   Source: National Travel Surveys.

Years	Greater London		Major Cities and Metropolitan Counties		Other Large Urban Areas Population over 100,000		Medium Urban Areas Population 25,000 to 100,000		Small Urban Areas Population 3,000 to 25,000		Average for all	
	journeys	miles	journeys	miles	journeys	miles	journeys	miles	journeys	miles	journeys	miles
1985/86	692	8,148	630	6438	714	8213	700	8247	672	9409	685	8198
1993/95	676	8,163	714	8,690	760	9,978	751	10,180	744	12,270	734	9,978
% change	−2	0	+13	+35	+6	+21	+7	+23	+11	+30	+7	+22

Table 2.9 Average number of journeys per person per year by type of area in 1985/86 and 1993/95. (NB Excluding journeys under one mile).   Source: National Travel Surveys.

Years	Greater London	Major Cities and Metropolitan Counties	Other Large Urban Areas Population over 100,000	Medium Urban Areas Population 25,000 to 100,000	Small Urban Areas Population 3,000 to 25,000	All urban areas
1985/86	1,047	958	1,054	1,038	1,036	1,030
1993/95	1,025	1,006	1,077	1,055	1,071	1,051
% change	−2	+5	+2	+2	+3	+2

Table 2.10 Average number of journeys per person per year by type of area in 1985/86 and 1993/95. (NB Including journeys under 1 mile).   Source: National Travel Surveys.

Trip Purpose	Settlement Size											Average for all Urban Areas	
	Greater London		Major Cities and Metropolitan Counties		Other Large Urban Areas Population > 100,000		Medium Urban Areas Population 25,000 to 100,000		Small Urban Areas Population 3,000 to 25,000					
	miles	change	miles	change	miles	change	miles	change	miles	change			miles	change
Commuting	1611	−383	1566	+208	1991	+450	1844	+62	2483	+562			1929	+229
On Business	777	+50	702	+101	989	+115	1197	+404	1379	+497			1026	+228
Education and education escort	455	+185	308	+94	359	+50	367	+76	477	+139			390	+98
Other personal business and escort	1038	+74	1005	+305	1232	+209	1283	+330	1539	+408			1234	+260
Shopping	830	+114	1039	+281	1101	+250	1210	+184	1636	+483			1178	+254
Social/ Sport/ Entertainment	2339	−155	2446	+553	2829	+400	2796	+468	3178	+498			2752	+363
Holiday/ day trip/ other	1114	+129	1624	+710	1478	+291	1483	+409	1579	+277			1470	+347
Total all purposes	8164	+14	8690	+2252	9979	+1765	10180	+1933	12271	+286			19979	+1779

Table 2.11 Mileage per person per year by journey purpose and type of area in 1993/95 and change in mileage since 1985/86. Source: National Travel Surveys.

In 1985/86, the pattern was less consistent, particularly in Metropolitan areas, where a lower base meant higher proportional increases in mileage (35%) than for other areas (21% – 30%). It should be noted, of course, that much of the travel by residents from smaller settlements will be in rural hinterland areas.

Journey purpose

Distance travelled per person has grown in about all trip purposes between 1985/86 and 1993/5 (see Table 2.11). Commuting continues to account for more mileage than any other single type of journey, although it is contracting in importance as the number of full time workers continues to fall and working patterns change (Macafee *et al*, 1996). It increased overall by 12% during this period, whereas education trips increased by 25%, business and shopping both by 22% and holiday/day trips by 26%. For most trip-purposes, the distances travelled in 1993/5 increased with smaller settlement size. The main exception is in London where commuting, business and education mileage are greater than those in the Metropolitan areas. In the smallest urban areas, commuting distances are 59% greater than in Metropolitan areas, business mileage 95% and education 55% greater. Shopping mileage is about double that in London.

Over the eight-year period from 1985/6, the increases in mileage for commuting, business, shopping and personal business were all greatest in the smallest urban areas, although there was no consistent pattern across the different sized areas. For education trips, the largest increase was in London and for sport/entertainment and holiday/day-trip in Metropolitan areas. Some of the reasons for these changes are probably to be found in changes in land-use, with greater residential and employment dispersal, changes in shopping patterns and in education provision and choice.

The strength of the growth in the 'commuting' and 'business' categories in the smaller urban areas probably reflects the shift in population to such settlements, as a result of a trend towards longer distance commuting, which may also explain the high growth in distances travelled on social journeys. Other factors may be related more directly to the increasing cost and poor provision of public transport.

Geographical and Temporal Distribution of Travel

From these historical data on travel trends, a picture emerges of not just an increase in mobility but also a greater dispersal of travel. Travel-patterns are becoming less concentrated upon the major trip-purposes and, partly as a consequence of this, also becoming more dispersed in place and in time. There was a slight spreading of peak-hour trips, although the strong modal shift towards cars for work and school journeys means that the volume of peak-hour motorised traffic grew, despite the wider dispersal of travel (see Photograph 2.1).

There were also changes in the geographical pattern of journeys. Within urban areas, the old pattern of predominantly radial trips, in and out of town centres, shifted towards one in which peripheral and cross-town trips feature more prominently. This

Photograph 2.1: Peak-hour traffic flow into London — A41 near Hendon

reflects changes in land-use patterns and the distribution of employment and facilities, such as shopping and leisure. This dispersal includes trips to new urban fringe or out-of-town developments and to other adjacent towns and is reflected in the surge in numbers of the middle-distance trips observed earlier in this chapter.

Overall, therefore, growth in travel has involved, basically, a constant number of journeys becoming more dispersed, geographically, in time and between purposes and concentrated more upon the mode of transport that serves such dispersed travel-patterns best – namely, the private car.

2.6 Influences on Travel Trends

Travel-Generating Land-Uses

Changes in personal travel-patterns in urban areas derive largely from changes that take place in society and in the economy. Core factors are the rising levels of car (and driving-licence) ownership, associated with growth in real income and the declining real costs of car- ownership and use. This growth is also associated with the gradually increased amount of travel. Local shops are being replaced by hypermarkets, retail warehouses and shopping centres towards the edges of town; hospitals and schools are tending to get larger and more remote; workplace are less centralised; and so journeys of all kinds are increasing in length. Better roads and more comfortable vehicles have also contributed towards these trends.

Settlement Patterns and Population Shifts

Coupled with the greater dispersal of facilities has been a trend to a more dispersed settlement pattern. Since the early 1900s, there has been a dispersal of population away from major cities to the suburbs. This was initially made possible through developments in public transport, trains for the middle classes and trams and buses for others. The spread of car-ownership accelerated population dispersal, so that, from the 1970s, suburbanisation did not merely involve decentralisation of people to a city's fringes but also well beyond. Another generation of technology is taking this a stage further. As information technologies loosen the physical ties between home and work place, more people can now choose to live in even more remote locations. Telecommuting may reduce journey-to-work travel but, as noted in Section 2.1, small urban settlements are very travel-intensive and rural areas even more so. Thus, a cut in the number of work journeys could be more than offset by additional trips (and therefore traffic) generated by other travel-demands.

At the regional and national levels, these long–term shifts in population have led to a decline in the resident populations of large cities and conurbations, with growth occurring mainly in free–standing towns, in the smaller cities (under 200,000 population) and in the semi–urbanised country fringe areas. This trend has been particularly strong in southern England, with a band of high population growth–areas from Cornwall across to East Anglia, along the Channel coast and in the Welsh borders. Retirement migration has represented a major element of growth in some of the more peripheral areas. Nevertheless, the decentralisation of economically–active population from larger cities to small urban areas and rural locations has been a major trend of the last two decades. Between 1971 and 1991, the population of the major cities in Britain declined by 1.80 million, while that of non–metropolitan Britain rose by 3.44 million (Champion, 1993). The fastest rates of population loss by the major cities were in the 1970s but subsequent inner city development has slowed their rate of loss. The populations of Greater London and the principal cities in the UK, after declining for many years, are now close to being stable.

These population shifts have compounded the trends towards higher car–use, for there are now more people in the types of settlement, particularly the smallest urban areas, where motorised travel is highest.

Travel–Generating Social Changes

Lower density developments have been reinforced by lower densities of occupation, as average household size has declined. This is associated with long–term demographic trends, including fewer children per household, more young adults setting up home on their own rather than living with their parents and more divorces and separations, as well as longer life–expectancy, which leads to an increasing proportion of one or two elderly–person households.

The growth of double income families frequently leads to compromise decisions on home/workplace location, which can increase overall travelling requirements. Recent changes in the employment and housing markets also indirectly generate more travel. For many, the job market has become more unstable, with contract work replacing what were previously permanent posts. The 1990s recession in the housing market made selling and moving more difficult, so many people are now prepared to commute long distances to take up a new job rather than move house. This has contributed towards the growth in long distance commuting, which has largely replaced short distance migration to take up a new job.

Another social trend that has important indirect impacts on travel generation is the greater choice in public services, such as schools and health. In the past, schools and health facilities were mostly provided on the basis of local catchment areas. With more freedom to choose over a large geographical range, many people are willing to travel much further by car to what used to be very localised facilities. This will have contributed to the sharp increase in escort trips and education trips by car, noted earlier.

Overall, the roots of the growth of mobility are to be found in these structural changes in society, the transport implications of which are sometimes not immediately apparent. However indirect, these have important links into the physical factors that are known to influence travel–patterns, such as the distribution of land–uses, and all tend to encourage further growth in car–ownership and use.

2.7 Influencing Future Travel Trends

The structural changes that lie at the root of the growth in mobility, and vehicular traffic in particular, are a self–reinforcing process that has, for several decades, increased the amount of travel that people undertake. It is a process that is far from complete, as projections for future levels of traffic and travel indicate (see Section 2.2), and many of the contributing factors appear to be deeply entrenched in the nation's way of life.

A combination of congestion, the costs of accommodating the growth in vehicular traffic, the effect of traffic on the quality of life and, increasingly, concerns about the environmental effects of traffic all make the accommodation of an indefinite continuation of these past traffic growth trends undesirable, particularly in the urban environment. Already, in the larger urban settlements, congestion has begun to contain traffic growth, albeit at great economic cost. As noted by the IHT (1992), congestion is likely: "...to occur in more (urban) areas and for longer periods of the day".

If congestion is itself not to be the only moderating influence on demand, then a range of other policies must be deployed to manage the overall level of travel–demand, the peak–period use of the road network and traffic levels on particular streets (see Chapter 21). The various components of transport policy are set out in Chapter 6. Since unrestrained traffic growth cannot reasonably be accommodated in most medium–sized and larger urban areas – there is an inescapable need to 'manage' the demand for road space, so as to prevent some of the more serious environmental impacts from taking place.

The means to achieve these objectives are the main thrust of Parts 2 and 3 of this book.

2.8 Information Needs

An understanding of the key factors behind the generation of the demand for travel becomes increasingly important when planning future accessibility and mobility. People's propensity to travel under different circumstances is not sufficiently understood.

Managing the demand for traffic growth likewise requires an understanding of the key components of change and the underlying choice factors. This permits identification of where transport demand management policies can be most effectively applied. Equally, it is important to identify where policy can and cannot make a difference, which traffic generating elements have to be accepted, such as growth in double income households, and which are amenable to influence. It is also important to identify where other areas of policy have significant indirect transport effects. This is accepted in respect of planning and land–use policies but not so much for issues such as health, education and welfare or fiscal policies.

The change from a policy of building substantial amounts of new road space in response to demand to one of seeking to manage key elements in the generation of that demand requires a change in the type of information needed to implement a transportation strategy. The information needs of traditional, demand–led, transport policies were relatively simple and involved measuring, for example, traffic flows, origins and destinations, vehicle types etc. These are still needed but there is also a need for statistics of understanding and not merely those of measurement.

This implies more emphasis on 'softer', more socially–based methodologies, which require the use of techniques and measurements that have previously been only on the fringes of transportation studies. These are behavioural methodologies that assess the public acceptance (or otherwise) of demand management measures, of the effects of different combinations of measure and of the effects over time and in different places and circumstances (see Chapter 7). Equally, there is a need for transportation modelling to provide a more complete and dynamic understanding of travel and traffic generation (see Chapter 8).

Local authorities possess large amounts of factual information on traffic counts, traffic speed surveys, parking supply figures, public transport supply and accidents. Many local authorities have good land–use data and some have conducted attitudinal and behavioural surveys that could have relevance for managing traffic demand. However, there is little indication of the existence of up–to–date local household travel surveys, which are required for a more comprehensive analysis of local travel behaviour.

2.9 References

Champion A (1993)	'A decade of regional and local population change', Town and Country Planning, March 1993.
DOT (1995a)	'Transport Statistics Great Britain 1995', Stationery Office
DOT (1995b)	'New motor vehicle registrations', Stationery Office.
IHT (1992)	'Position Paper Traffic Congestion in Urban Areas', The Institution of Highways & Transportation.
Lane R and Judd M (1992)	'Peak Hour Traffic Growth', TRICS Conference, 1992.
Macafee, K et al (1996)	'National Travel Survey 1993/5 Transport Statistics Report', Stationery Office.
May A (1993)	Transport Policy and Management Chapter 17 in D Banister and K Button (Eds) 'Transport, the Environment and Sustainable Development', E & FN Spon, London.

2.10 Further Information

DOE	'Northern Ireland Transport Statistics' DOE Belfast [Sa].
DOE/DOT (1994)	Planning Policy Guidance Note (PPG 13) 'Transport', Stationery Office [Sb] [Wa].
DOT (Annual)	'National Travel Survey Report', Stationery Office.
DOT (Annual)	'Transport Statistics Great Britain', Stationery Office.

DOT
(1971,1981&1991):   'London Area Transport Survey', London Research Centre.

Hughes P (1993)   'Personal Travel and the Greenhouse Effect' Earthscan, London.

Jones P and
Haigh D (1994)   'Reducing Traffic Growth by Changing Attitudes and Behaviour', PTRC European Transport Forum, September 1994.

Newbury DM
(1995)   'Reforming Road Taxation', Report to the Automobile Association, September 1995.

Oak Ridge
National
Laboratory (1995)   Energy Transportation Data Book, Vols 15 and 16, US Department of Energy.

Potter S et al (1996) 'Vital Travel Statistics', Landor Publications London.

Potter S (1996)   'The Passenger Trip Length Surge', Transport Planning Systems.

RCEP (1994)   Royal Commission on Environmental Pollution, Eighteenth Report, October 1994. 'Transport and the Environment', Stationery Office.

SACTRA (1994)   Standing Advisary Committee on Trunk Road Assessment December 1994 'Trunk Roads and the Congestion of Traffic', Stationery Office.

Transport
Statistics
Users Group
(1995)   'Sources of Transport Statistics', Information Research Network, London

Chapter 3 Transport Policy

3.1 Introduction

It is over 30 years since the Buchanan report, Traffic in Towns (Buchanan *et al*, 1963) highlighted the challenges that unrestrained growth in personal travel would present and, today, society faces many major problems in meeting the seemingly insatiable demand for travel. The end of the 1980s and early 1990s have seen a period of considerable change for urban transport policy. The publication of the National Road Traffic Forecasts (NRTF) in 1989 (DOT, 1989a) sparked the debate about the inability to supply enough urban roadspace to meet the predicted demand and heralded an era in which demand management is now clearly central to transportation planning (see Chapter 21). Issues of sustainability have also focused attention on encouraging the use of environmentally less harmful and economically more efficient means of transport (DOE, 1994). The early 1990s saw the main local authority associations producing transport policy documents (AMA, 1990; ACC, 1991 and ADC, 1991) and the Rees Jeffreys Road Fund sponsored the report, Transport: the New Realism, (Goodwin *et al*, 1991). Similar concerns about the direction of transport policy were raised by the professional institutions (eg IHT, 1992). In 1994, the Departments of the Environment and Transport jointly produced National Planning Policy Guidance (PPG13) on Transport (DOE/DOT, 1994) [Sa], once again stressing the need to integrate land–use and transportation planning [NIa]. There has also been the rise of the new generation of Integrated Transport Strategies (May, 1991; and Steer, 1995). Finally, in the mid 1990s, there was the national debate on transport policy initiated by the Department of Transport's discussion document, Transport: The Way Ahead (DOT, 1995c), culminating in the publication of the green paper Transport – the Way Forward (DOT, 1996a) [Sb], setting out the Government's position [NIb].

3.2 Formulating a Transportation Strategy

The Institution of Highways & Transportation's document, Guidelines on Developing Urban Transport Strategies (IHT, 1996) points out that there is a widely accepted view that the main stages of the process of formulating a transportation strategy for an urban area should include:

❑ agreement on a set of objectives which the strategy should seek to satisfy;
❑ analysis of present and future problems on the transport system;
❑ exploration of potential solutions for solving the problems and meeting the objectives;
❑ assessment of the ideas, seeking combinations which perform better as a whole than the sum of the individual components; and
❑ selection and phasing of the preferred transportation strategy, taking account of the views of the public and of transport providers.

The Department of Transport (DOT) now encourages local authorities to submit bids for funding through the 'package' approach [NIc] [Wa]. The intention is both to encourage a 'balanced' approach to the provision of urban transport, with highway construction, traffic management, public transport and demand management each playing an appropriate and complementary role, and to allow local authorities greater discretion on what elements to include in a package. The current DOT view is that a package should consist of five elements, which come in a logical order: the current situation, problems and issues; objectives; strategy; proposed schemes and measures; and intended outcome (DOT, 1995b; and GOL, 1996a). Here, the term 'package' covers the entire submission made by a local authority to the DOT for funding of elements of its transportation strategy. As can be seen from the DOT's guidance, the transportation strategy is a central requirement of a Package Bid. The Department sees the distinction between the elements of a Package Bid, as follows: "Broadly, objectives are what is desired in wider terms; a strategy is the way this will be achieved; schemes and measures are the physical means of doing it; and the outcome is what the package area will look like at the end" (DOT 1995b; and GOL 1996a).

The Transportation Strategy Development Process

Chapter 2 of the IHT Guidelines (IHT, 1996) details 26 steps, or stages, that provide a logical process for formulating a transportation strategy for an urban area. These steps are shown in Figure 3.1. The steps are, on the whole, sequential, although there is feedback between several elements. The strategy should cover the actions that are necessary, to develop an urban area's transport system, for a

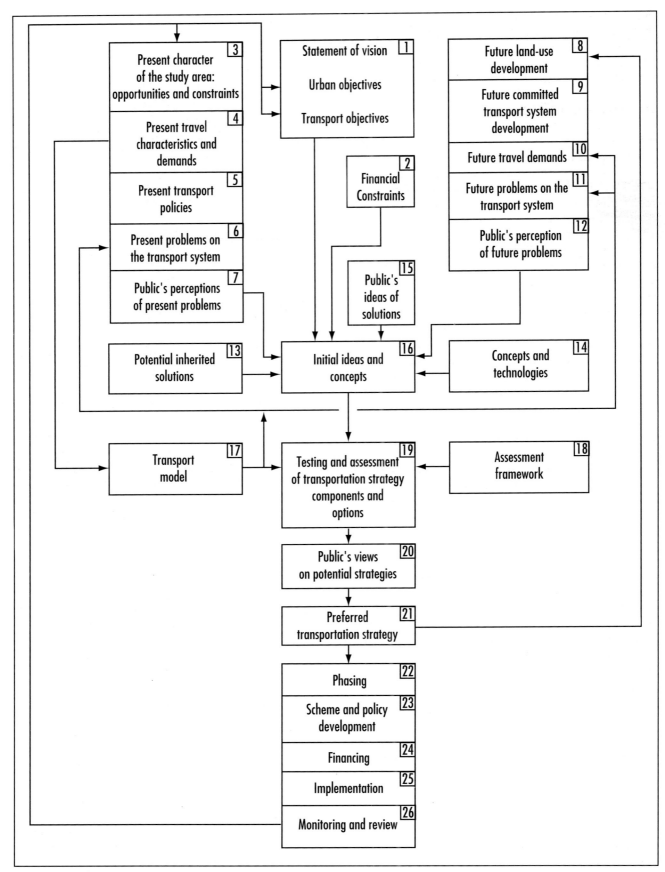

Figure 3.1: The transportation strategy development process.

period of ten to 20 years into the future. This means that the implementation of the individual components that make up the strategy will need to be phased over time and that the strategy should be flexible enough to meet changing circumstances.

Two complementary approaches to urban strategy development are the so-called 'top down' approach, which examines, at a broad scale, ways of achieving objectives through the integration of strategy components, and the so-called 'bottom up' approach, in which particular attention is paid to the detailed analysis of problems and their solutions (Coombe, 1991).

The main differences between the current series of integrated transport studies and the land-use/transportation studies of the 1960s have been identified (May, 1991) as:
❑ the emphasis on a vision, which provides a context for developing transportation policy-objectives;
❑ the appreciation that financial targets, or constraints, need to be identified at the outset;
❑ the treatment of a wide range of transportation policy (and, potentially, land-use policy) instruments;
❑ the use of 'cartoon' strategies, to test the full range of instruments and to identify the interactions between them;
❑ the emphasis on strategies which maximise the potential for synergy between transportation policy instruments;
❑ the treatment of these strategies as frameworks to facilitate action rather than as fixed blueprints;
❑ the use of multi-criteria evaluation methods, with reliance both on model output and professional judgement;
❑ the testing of robustness, as a way of treating uncertainty;
❑ the speed with which studies have to be conducted; and
❑ the development of strategic sketch-planning models.

The most notable features of the current integrated transport studies (Steer, 1995) are:
❑ multi-client commissioning body;
❑ explicit attention given to objectives, broadly defined and embracing a vision of the area under study;
❑ a medium- to long-term focus; and
❑ inclusion of all parts of the transport system but an appraisal of strategies rather than of individual schemes.

Finally, this approach has identified several policy issues of considerable importance (May, 1991), such as:
❑ the need to consider policies which encourage a reduction in average trip-length, without necessarily reducing trip-making;
❑ the need to accept that synergy between transportation policy instruments is achievable, provided that the search for possible instruments is drawn more widely than simply infrastructure provision;
❑ the importance of pricing measures, in particular, as regulators of demand which, virtually alone among policy instruments, can have area-wide effects;
❑ the need to develop the concept of road-user charging (so far untried in the UK), particularly for major conurbations such as London but, in due course, elsewhere also;
❑ the need to accept a common basis for evaluation of all elements of transportation policy; and
❑ the importance of providing finance, on a consistent basis, for different types of measure and removing undue constraints on its availability.

3.3 Levels of Policy Formulation

This section covers the main levels of policy formulation of relevance to transportation policy in the UK. For further reference, Section 3.5 provides an overview of the Statutory Planning Framework and Chapter 8 of the IHT Guidelines on Developing Urban Transport Strategies (IHT, 1996) explains the planning process and the hierarchy of development plans in the UK.

International/European Level
European Union (EU) policy and legislation is having a growing influence on transport policy in the UK (Harman, 1995; and Miles, 1995). The EU's Common Transport Policy places a lot of emphasis on environmental matters. Major objectives for the Common Transport Policy (Miles, 1995) are:
❑ the continued reinforcement and proper functioning of the internal market;
❑ the development of coherent, integrated, transport systems for the Community as a whole, using the best available technology;
❑ the development of transport infrastructure to reduce disparities between the regions of the Community, especially land-locked and peripheral regions;
❑ the adoption of programmes that can contribute to the solution of major environmental problems; and
❑ the promotion of safety, social measures and links to third world countries.

The Common Transport Policy defines the EU's long–term aim as to provide 'sustainable mobility' and calls for the adoption of the right balance of policies, favouring 'coherent integrated transport systems' and measures to ensure that the development of transport systems contributes to a sustainable pattern of development, by respecting the environment and contributing to the solution of major environmental problems. This supports growing international concerns, highlighted by the United Nations Earth Summit Conference held at Rio de Janeiro in 1992, which has resulted in the UK Government's sustainable development strategy (DOE, 1994). The United Nations Economic Commission for Europe (UN–ECE) also has responsibility for certain elements of transport policy, including co–ordinating signing throughout Europe.

National Level

The UK Government's transport policies are set out in the green paper Transport – The Way Forward (DOT, 1996a) (see Section 6.1) [NIb] [Sb]. The Department of Transport produces annual guidance to local authorities on their Transport Policies and Programmes (TPPs) [Wb], through Local Authority Circulars (DOT, 1995a) and Supplementary Guidance Notes on the Package Approach (DOT, 1995b and GOL, 1996a) [NIc]. The Department of the Environment (DOE) gives national planning policy guidance (PPGs) [Sc] on the operation of the development plan system [NIa] [We]. Those of most relevance to urban transportation planning are:

❑ PPG12 on Development Plans and Regional Planning Guidance [Sa], which gives guidance on the form and content of development plans (DOE, 1992) – the Government expects full national coverage of up–to–date Development Plans by 1997;

❑ PPG13 on Transport [Sa], which was jointly produced with the Department of Transport, provides guidance on how local authorities should integrate land–use and transportation planning, with the aim of reducing the need to travel, particularly by car (DOE/DOT, 1994); and

❑ PPG6 on 'Town Centres and Retail Developments' [Sd], which gives guidance on the planning of all aspects of development in town centres, as well as retail development (DOE, 1996a). In particular, the document emphasises the importance of a coherent town centre parking strategy and promotes the idea of accessibility to retail developments by a choice of transport modes.

The Environment Act 1995 (HMG, 1995) requires that local authorities review air quality in their area against standards and targets set by the Government in the National Strategy [NId]. The draft United Kingdom National Air Quality Strategy, to be finalised after consultation, will set these standards, objectives and timetables for the eight pollutants of most concern (DOE, 1996b).

In transport, as in other sectors, a distinction needs to be drawn between measures where action is appropriate at a national level, such as the setting of vehicle–emission and fuel–quality standards, with enforcement procedures, and measures more suitable for inclusion in local Air Quality Management Area Action Plans, dealing with general pollution from traffic.

The Government is currently preparing national guidance on traffic management and air quality, which will give advice to local authorities on the development of an Air Quality Management: Area Action Plan (DOE/DOT, 1996). Responsibility for producing such plans rests with Unitary authorities and District councils. Unitary authorities should integrate their Air Quality Management Area Action Plan with their transport plan. District Councils will have to negotiate with the relevant County Council, as the Highway Authority for the road network, to develop transport plans. Early discussion of underlying principles and ongoing co–operation between the two bodies will be essential (see Chapter 17).

Regional Level

The Government has set up integrated Government Offices for the Regions, which cover the responsibilities of the Departments of the Environment, Employment and Education, Trade and Industry and Transport [NIe] [Wd]. These bring together land–use and transportation planning, with wider regeneration and business competitiveness issues. Of most relevance is the role of Regional Planning Guidance, which develops national planning policy at a regional level and sets a context for the development of urban transportation strategies.

Local Level

The effective implementation of urban transportation strategies should be within the context of the statutory development plan [NIf]. There are three types of development plan.

❑ 'Structure plans', which provide the overall strategy and policy framework for land–use and transport in county areas. These are currently prepared by county councils, although the re–organisation of local government in England will lead, in some places, to structure plans in future being prepared jointly by counties and

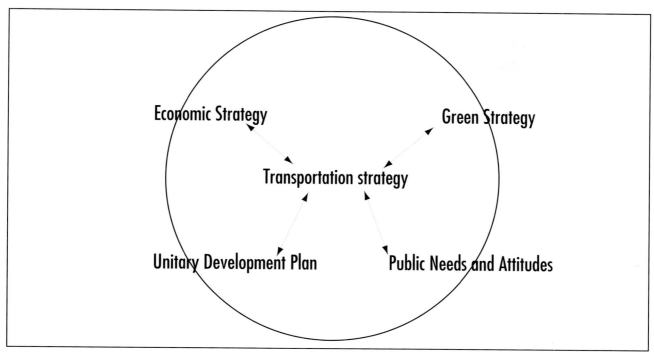

Figure 3.2: Leeds Transportation Strategy Context.

Unitary authorities.

❑ 'Local plans', which provide detailed local policies and proposals for implementation of the Structure Plan strategy according to local circumstances. These are prepared by district authorities and must be in general conformity with the Structure Plan for the area.

❑ 'Unitary development plans' (UDPs), which cover the roles of both structure and local plans, with Part 1 setting out the key strategic policy framework for the area and Part 2 providing the detailed local policies and proposals for implementation. These are currently prepared by the London Boroughs and Metropolitan Districts and will, in future, also be prepared by the new Unitary Authorities.

The other key document, linked to the development plan, is the Transport Policies and Programmes (TPP) document [Se], which is submitted annually by all highway authorities, which include the County and Unitary authorities, in support of their bids for funding capital projects [NIc] [We]. The introduction of the Package Approach to local transport funding in 1993 has encouraged a re-emphasis, within the TPP process, on producing an integrated transport system.

3.4 Transport Policy–Objectives in Context

A starting point for the formulation of an urban transportation strategy is a statement of the vision to which the city or town aspires. Such statements are usually about the desired attributes of the city or town, as a whole, and tend not to mention transport specifically but they do provide the broad goals to which transport policies should contribute and, thus, the context for any transportation strategy. In particular, they are driven by issues of quality–of–life, economic regeneration and the viability and vitality of city and town centres. These broad vision statements are important in reflecting the aspirations of local communities as to how they want their areas to develop.

One of the earliest examples of a vision statement is the Fourfold Vision specified by the London Planning Advisory Committee (May, 1991). This set four themes which were seen as being of equal importance and which were used to establish a mutually–supportive policy framework. The four themes were:

❑ London as a civilised city, offering a high quality environment for all Londoners;
❑ London as a world centre of international trade and business;
❑ London as a city of opportunities for all; and
❑ London as a city of stable and secure residential neighbourhoods, capable of sustained community development.

Similarly, the Birmingham Integrated Transport Study (Jones et al, 1990) started with a vision of:

❑ Birmingham having a national and international standing, equivalent to that of other European

TRANSPORT IN THE URBAN ENVIRONMENT

29

provincial capitals;
❑ Birmingham maintaining its special and high level role as a regional centre; and
❑ Birmingham providing a social and cultural environment, in which its diverse groups of residents could each play a satisfying and distinctive part.

The City of Leeds has also formulated a vision statement, which states that "Leeds has the potential to become a leading European City, with increasing prosperity, an enviable environment and a city centre to be proud of..." (Steer, 1995). The transportation strategy was developed from a public consultation programme called 'Think Leeds/Talk Transport' and from two parallel exercises producing a *green* strategy and an *economic* strategy for the city, as well as being linked into the Unitary Development Plan. This demonstrates the need to establish local ownership and identity of the overall vision and the strategies to achieve it. The relationships between these elements are shown in Figure 3.2 (Steer, 1995).

The value in developing such broad vision statements is that they focus attention first on the issues of 'What sort of city or town do you want?'; which leads to the question 'How best can transport help to realise that vision?' The answers to this latter question then help to specify the higher-level transportation policy-objectives.

It is important to be clear as to the objectives of an urban transportation strategy as a basis for identifying problems and justifying solutions. Without such definitions, real problems may be overlooked and inappropriate solutions pursued. Statements which indicate a preferred strategy (such as the use of demand management) should not be presented as objectives in their own right. Such statements may lead to other effective solutions being overlooked; they also fail to provide a justification for the measures proposed.

Objectives, once established, serve several functions:
❑ they help to identify the problems to be overcome, both now and in the future;
❑ they can be used to design schemes and strategies and provide guidance on the types of solution that might be appropriate and the locations in which they are needed (see Chapter 6);
❑ they act as constraints, clarifying what should be avoided in pursuing any particular solution;
❑ they provide the basis for the appraisal of alternative solutions and for monitoring progress in implementation;
❑ they make conflicts, which arise from a solution meeting some of the objectives but undermining others, become transparent; and
❑ they make the transportation planning process more accessible to the public, by helping them to understand what transportation strategies are trying to achieve.

The objectives-based approach cannot remove all conflicts or disagreements. There will always be argument about how best to apply the objectives and it is unlikely that any scheme can satisfy all the objectives. However, objectives provide a structure, within which individual choices can be made allowing priorities to be defined, and a means by which the success or failure of transportation strategies can be assessed. Moreover, the needs of different parts of an urban area, and different groups of users, will often be in competition for scarce resources. It is, therefore, important not just to have a clear understanding of the overall objectives but also their relative importance, so that conflicts can readily be resolved.

Objectives are also required for the development of the transport system. These may be couched in general terms, such as 'to improve accessibility', or they may include quantified targets, such as to 'reduce noise levels to below a specified level'. However they are phrased, the IHT Guidelines (IHT, 1996) argue that the three most important factors to bear in mind are:
❑ the objectives should be comprehensive and cover all aspects relating to the transport system (see Section 3.6 for a possible set of objectives);
❑ they should avoid indications of preferred solutions or strategies, as these may then cause other better solutions to be overlooked in the process of establishing a strategy; and
❑ where quantified targets are stated, care should be taken not to introduce bias by inadvertently setting some targets which are easy to meet while others are very difficult, if not impossible, to achieve.

3.5 Objectives-Led Planning

Two general approaches can be adopted to identifying objectives and related problems. The first is the true objectives-led approach, in which either broad or more detailed objectives, of the kind described in Section 3.6 (below), are specified, typically by elected decision-makers in a local authority and often as a result of wider public consultation. These are then used to identify problems, by assessing the extent to which current or predicted future conditions fail to meet the objectives.

This approach has been adopted in many of the so–called, 'Integrated Transport Studies' (Jones et al, 1990; May et al, 1990; May et al, 1992 and Steer, 1995). Here, predictions of future conditions, assuming a 'do minimum' strategy, have been compared with the objectives, to identify future problems that any transportation strategy would need to address. The main drawback with this approach is that many elected members and members of the public are less familiar with the abstract concept of objectives (such as improving accessibility) than they are with more concrete problems (such as the nearest job centre being 50 minutes away). However, many people are now much clearer about how they want their cities and towns to develop, as thriving and pleasant places in which to live, work and play. Consequently, where objectives are clearly tied to an overall vision, they may be less difficult for politicians and the public to understand. Also, to bridge the gap between objectives and problems, it can be valuable to check the predicted problems that are thrown up with politicians and the public. This can also be a useful way of checking that the list of objectives is complete.

The alternative, problem–orientated, approach starts by defining problems and uses data on current or future predicted conditions to identify when and where these problems are likely to occur. Here, the objectives are implicit in the specified problems and the defining of problems can then be used to develop overall objectives. With either approach. it is essential to be comprehensive in the list of types of problem to be solved.

The problem–orientated approach has been used in a number of recent studies of smaller conurbations (Coombe et al, 1990; Coombe and Turner, 1992; and Ramsden et al, 1992). It is also the approach advocated by the Department of Transport in its guidance on the Package Approach (DOT, 1995b). Although it is more easily understood than true objectives– led planning, it is dependent on developing a full list of potential problems at the outset, to ensure that the strategy tackles 'real' problems and not just symptoms. If particular types of problem (like access to job centres) are not identified, because the underlying objective (accessibility) has not been considered, the resulting strategy will be partial in its impact. As with the objectives–led approach, it is wise to check with politicians and the public that the full set of problems has been identified.

Neither approach is necessarily preferable to the other and both require checks to ensure that the list of identified problems is comprehensive. The key issue, which should determine which approach to adopt, is whether or not the existing policy framework is being questioned at the outset. It is clear that in some urban areas (most notably in larger conurbations, where the issue of the balance of demand for private and public transport is being questioned) a top–down, objectives–led, approach is more appropriate. It is also clear that, in other urban areas (most notably smaller urban areas, where the existing policy framework is an acceptable starting point) a bottom–up, problem–orientated, approach may be more appropriate.

3.6 Setting Transport Policy Objectives

Department of Transport Guidance [NIb]

The Department of Transport encourages local authorities to formulate their own objectives and to decide whether, in doing so, to adopt a visionary approach, identifying long–term goals, or a more detailed short–term problem–solving approach (DOT, 1995b). The DOT has, however, made it clear that it expects those objectives to be compatible with its own. The Secretary of State's overall objectives, revised in December 1996, are: "...to promote safe, efficient and environmentally sustainable transport for people and goods, responding to the individual's need for mobility and supporting economic growth".

The Supplementary Guidance on the Package Approach (DOT, 1995b; and GOL, 1996a) specifies the following objectives as being of importance to the Package Approach [NIc]:

❑ reduction of road congestion;
❑ stimulation of innovation, enhanced quality and choice in the provision of public transport services;
❑ reduction of the number of personal injury accidents on roads;
❑ enhancement of the attractiveness of less environmentally–damaging modes, particularly cycling and walking;
❑ making it easier for people to switch between different forms of transport; and
❑ improvement of provision for the needs of disabled people, pedestrians, pedal cyclists and other vulnerable road–users.

Various UK and European examples of sets of objectives have been defined as part of the formulation of a transportation strategy (Harman, 1995; Jones, 1995; and Roberts, 1995). In particular, there is less concern, in many European cities, for reducing urban traffic congestion than is typically found in the UK. In these European cities, there has been a general shift from maximising mobility to maximising accessibility, with policy documents talking about meeting road traffic needs for 'necessary' rather than 'optimal' mobility (Jones, 1995).

One example of transportation strategy objectives is those for the London Assessment Studies, which were agreed jointly between the Department of Transport, the London Local Authority Associations, the London Planning Advisory Committee, the major transport operators and the police (DOT, 1989b). A more recent formulation of these by the Government Office for London is GOL, 1996b;

❑ to maintain and enhance the quality of London's international transport links;

❑ to enhance the quality of commuter services by train and by Underground;

❑ to promote greater use of less polluting forms of transport, subject to the need to maintain competitiveness and safety;

❑ to facilitate access to the central business districts and ease of movement within them; and

❑ to plug major gaps in the road and rail networks.

A Possible Set of Higher-level Objectives

While the specification of objectives, and of the priorities between them, is the responsibility of the local authority concerned and needs to be tested by a public consultation exercise, the IHT Guidelines (IHT, 1996) argue that it is helpful to define the following nine general higher- level transport objectives.

Accessibility: can be defined as 'ease of reaching' destinations and is concerned about what movement has actually achieved (how much access to goods, services or other people) rather than the objective of more, easier and faster movement (more mobility). Accessibility can be considered from the point of view of the trip–origin, in terms of residents reaching activities like employment, shopping and leisure, or from the point of view of the trip–end, in terms of the catchment area for an employer or retail outlet.

Economic Efficiency: is about the efficient allocation of scarce resources and involves comparing the benefits to travellers with the general disbenefits, such as environmental degradation. In practice, the efficiency objective is primarily concerned with maximising the net benefits, in resource terms, of the provision of transport.

Economic Regeneration: is about supporting or reinforcing the land–use plans for the area, in terms of regeneration and employment. At its simplest, transport can provide new infrastructure and services for areas of new development. However, transport can also contribute to the encouragement of new activity by improving accessibility to an area, by enhancing its environment and, potentially, by improving the image of an area. This objective, therefore, relates directly to those of 'accessibility' and 'environmental protection'.

Environmental Protection: involves reducing the impact of transport facilities and their use on the environment of both users and non–users. The types of impact of concern include noise, atmospheric pollution of differing kinds, vibration, visual intrusion, community severance, fear and intimidation and the loss of intrinsically valuable objects, such as flora and fauna, ancient monuments and historic buildings, as well as damage to buildings through acid attack.

Equity: is concerned with ensuring that the benefits of transportation strategies are reasonably distributed amongst groups in society or focused particularly on those with special needs. One way of considering these equity, or distributional, issues is by reference to an 'impact matrix', which identifies the impact–groups of concern to decision–makers and the objectives and indicators which are of particular concern to them. The IHT Guidelines (IHT, 1996) present such a possible impact matrix, taken from work done for Edinburgh (May *et al*, 1992).

Finance: acts primarily as a constraint on the design of a transportation strategy. In a few cases, the ability to raise revenue may be seen as an objective in its own right. The finance objective can, therefore, be defined as minimising the financial outlay, both capital and revenue, for a strategy or as maximising net revenue.

Practicability: is the other major constraint on strategy design and implementation. The practicability objective can, therefore, be defined as ensuring that policies are technically, legally, financially and politically feasible and adaptable to changing circumstances. Section 5.3.10 of the IHT Guidelines (IHT, 1996) provides a useful checklist of the main factors which govern practicability.

Safety and Security: is concerned with reducing loss of life, injuries and damage to property resulting from transport accidents and with concerns over personal security in terms of anxiety, fear and intimidation, although this latter element can also be covered under environmental protection. Given the common practice of placing monetary–equivalent values on accidents and casualties, the accident–reduction element can also be subsumed within the efficiency objective.

Sustainability: can be considered, in transportation terms, as a higher–level objective which considers the trade–off between efficiency and accessibility, on the one hand, and environment and safety on the other, and includes the growing concern about the health–related effects of transport pollution. Sustainability issues also include consideration of the impact of transport policy on the wider global environment and on the environment of future generations (DOE, 1994).

Figure 3.5 of the IHT Guidelines (IHT, 1996) suggests a set of possible indicators which could be used for each of the above objectives.

Quantified Objectives and Targets

While the higher-level objectives, set out above, indicate the general directions in which a transportation strategy should aim, they say nothing about what would be appropriate to achieve. More quantified objectives can be specified in terms of a series of targets, which can be general or specific. It is important to avoid setting 'solution-specific' objectives, as this constrains the search for solutions. Aspirations for a particular solution, such as more bus-lanes, are not objectives in themselves but are means of achieving them. A transportation strategy which focuses too early on particular types of measure runs the risk of not identifying the most effective package of measures. It is important, therefore, in considering specific targets, that they are measures aimed at the achievement of specific objectives and the overall strategy, rather than being allowed to define the strategy.

3.7 The Role of Targets

Whilst the general visions and objectives explained earlier indicate what a transportation strategy should aim to achieve, more quantified targets may be needed to judge whether a proposed strategy is successful or not. Target-setting is, thus, becoming a growing part of transportation strategy formulation. Thresholds can be useful in providing levels above, or below, which problems can be said to occur.

A target simply indicates a specified degree of attainment of an objective. Lack of achievement of a target does not, however, mean that no progress has been made. For example, the Government has set targets for a one third reduction in casualties between 1985 and 2000 (DOT, 1987) and to reduce emissions of greenhouse gases, including carbon dioxide, to 1990 levels by the year 2000, as part of the Rio Framework Convention on Climate Change (DOE, 1994). It has also endorsed the target to double the use of pedal-cycles by the year 2002 and to double it again by the year 2012, as set out in the National Cycle Strategy (DOT, 1996b). None of these targets is likely to be met but they set policy in the right direction.

The Royal Commission on Environmental Pollution (RCEP, 1994) has gone further, by suggesting 16

targets that, they argue, are necessary "...to achieve a sustainable transport system for the UK". Considerable debate has taken place as to whether these targets are useful, reasonable, mutually consistent and attainable. Some commentators have argued that, beyond having identified traffic-reduction as a goal or objective, a specific target (eg of reducing urban traffic levels to 30% below those of 1990) should be adopted to provide a focus for action (Pharoah, 1992).

The DOT's study of transport policies in European cities found that the most commonly used overall indicator of policy impacts, at the strategic level, for all the cities studied has been the change in modal split over the last ten or 20 years (Jones, 1995). As a result, some European cities are now using modal split as their main target for policy generation and assessment. There are also examples of targets in national transportation planning in Europe (Harman, 1995) and of the link between overall goals or objectives and detailed targets, as in the Dutch National Transport Plan (de Jong, 1995).

However, there does need to be caution in the setting of targets (see Section 17.4; and DOT, 1996a). It is preferable to set targets which are related directly to policy objectives (for example, accident reduction) than to policy inputs (for example, kms of bus-lane) or process indicators (for example, modal split). Care is needed to avoid setting a target for one objective which compromises performance against others or which compromises the achievement of other objectives which do not have explicit targets. It is preferable to specify targets based on a known and proven cost-effective strategy, rather than choose a target for which a strategy then needs to be formulated (May, 1995). More importantly, the cost-implications of achieving planned target-levels should be continuously reviewed, so that resources are not wasted pursuing needlessly high standard of performance.

Why Set Targets?

Six main reasons have been identified why targets are set (Atkins, 1995). These are:
- as signposts – used as policy-signals to indicate intent to have some form of directional change;
- as a means to change bureaucratic culture – focusing activity on 'outcomes' rather than on 'inputs' or processes;
- as measures of achievement – being used as standards against which to measure progress towards the objectives and, hence, useful in the monitoring and review process;
- as a means of management and control – being used to direct and control activities. The Department of Transport are particularly keen on

this use (DOT 1995b);

❑ as standards of public service – being used as standards of public service performance, such as those in the Citizens Charter; and

❑ as a lobbying device – representing an attempt to gain greater interest in, and resources for, a particular issue.

The DOT does not explicitly encourage or discourage the use of targets in packages (DOT, 1995b) but does note that 'targets may have a useful role as a management tool and can play their part in the setting of clear objectives in packages'. The TPP Supplementary Guidance goes on to specify detailed criteria for the definition of targets (DOT, 1995b).

The Attributes of Targets

The Government's TPP Guidance (DOT, 1995b) suggests that targets should be explained in terms of scope, type, level, duration, achievability, relevance and impact of external factors. Four of these characteristics are particularly important (Atkins, 1995). These are:

❑ targets should be measurable – as a means of assessing progress measurement is essential but the implication of target–setting is that a monitoring system, capable of measuring performance in relation to the target, will be established, with consequent resource requirements;

❑ targets should be relevant – they should be pertinent to the objectives being assessed and should flow from the overall strategy;

❑ targets should be time–limited – they should have a definite time–period within which their attainment is sought; and

❑ targets should be realistic – if large changes are being sought then shorter–term or staged targets should be used.

To ensure that target–setting is effective, it is important that achievement and effort are linked. It is also important that the financial implications of setting targets are considered, to ensure that they do not have unrealistic effects.

As the IHT Guidelines (IHT, 1996) argue, the key concern is that the strategy should determine the targets, rather than the targets being allowed to define the strategy. In particular, there will never be any 'correct' set of targets, or thresholds, which can be applied uniformly in all urban areas. Section 3.8 of the IHT Guidelines provides a suggested list of indicators, against which targets and thresholds can be defined. In conclusion, the experience with targets, so far, has shown that sensible and durable targets are often very difficult to define.

3.8 References

ACC (1991)   'Towards a sustainable transport policy', Association of County Councils.

ADC (1991)   'UK Transport Strategy – A Balanced Approach', Association of District Councils.

AMA (1990)   'Changing Gear – Urban transport policy into the next century', Association of Metropolitan Authorities.

Atkins S T (1995)   'Target setting as a project appraisal method', London Transport Planning.

Buchanan C D et al (1963)   'Traffic in Towns', Stationery Office.

Coombe R D (1991)   'Urban transport strategy development', Traffic Engineering + Control 32(1).

Coombe R D, Goodwin R P and Turner D R (1990)   'Planning the transport system for the historic town of Bury St Edmunds', Traffic Engineering + Control 31(1).

Coombe R D and Turner D R (1992)   'Transport strategy in Ipswich', Traffic Engineering + Control 33(3).

de Jong M A (1995)   'National transport policy in the Netherlands', Proceedings of the Institution of Civil Engineers, Transport 111(3).

DOE (1992)   Planning Policy Guidance PPG12 'Development Plans and Regional Planning Guidance', Stationery Office [Sa] [Wf].

DOE (1994)   'Sustainable Development: the UK Strategy', Stationery Office.

DOE (1996a)   Planning Policy Guidance Revised PPG6 – 'Consultation Draft: Town Centres and Retail Developments', Stationery Office [Sd] [Wf].

DOE (1996b)   'United Kingdom National Air Quality Strategy: Consultation Draft', DOE

DOE/DOT (1994)   Planning Policy Guidance PPG13 'Transport', Stationery Office [Sa] [Wf].

DOE/DOT (1996)   Local Authority Circular 'Air Quality and Traffic Management Consultation Paper', DOE/DOT.

DOT (1987)   'Road Safety: The Next Steps' Inter–Departmental Review of Road Safety Policy, DOT.

DOT (1989a)   'National Road Traffic Forecasts (Great Britain)', Stationery Office.

DOT (1989b)   'Statement on Transport in London', Stationery Office.

DOT (1995a)   'Transport Policies and Programme Submissions for 1996–97', Local Authority Circular 2/95, Stationery Office.

DOT (1995b)   'Transport Policies and Programme Submissions for 1996–97 Supplementary Guidance Notes on the Package Approach', DOT.

DOT (1995c)   'Transport: The Way Ahead', DOT.

DOT (1996a)   'Transport: The Way Forward', Stationery Office [Sb].

DOT (1996b)   'National Cycling Strategy', DOT.

Goodwin P, Hallett S, Kenny F and Stokes G (1991)   'Transport: The New Realism', Transport Studies Unit, University of Oxford.

GOL (1996a)   'Transport Policies and Programme Submissions for 1996–97', Supplementary Guidance to London Local Authorities, Government Office for London.

GOL (1996b)   'A Transport Strategy for London', Government Office for London.

Harman R (1995)   'New Directions: a manual of European best practice in transport planning', Transport 2000.

HMG (1995)   'The Environment Act', Stationery Office.

IHT (1992)   'Traffic Congestion in Urban Areas', The Institution of Highways & Transportation.

IHT (1996)   'Guidelines on Developing Urban Transport Strategies', The Institution of Highways & Transportation.

Jones D, May A D and Wenban–Smith A (1990)   'Integrated transport studies: lessons from the Birmingham study', Traffic Engineering + Control 31(11).

Jones P (1995)   'Study of Policies in Overseas Cities for Traffic and Transport' (SPOTT), University of Westminster.

May A D (1991)   'Integrated transport strategies: a new approach to urban transport policy formulation in the UK', Transport Reviews 11(3).

May A D (1995)   Letter entitled 'Transport policy targets: proceed with caution', Local Transport Today (Issue 169).

May A D and Gardner K E (1990)   'Transport policy for London in 2001: the case for an integrated approach', Transportation 16(4).

May A D, Roberts M and Mason P (1992)   'The development of transport strategies for Edinburgh', Proceedings of the Institution of Civil Engineers, Paper P.9865.

Miles J C (1995)   'Recent developments in European transport infrastructure policy', Proceedings of the Institution of Civil Engineers, Transport 3(3).

Pharoah T (1992)   'Less traffic, better towns', Friends of the Earth.

Ramsden J, Coombe R D and Bamford T (1992)   'Transport strategy in Norwich', Highways & Transportation 39(2).

Roberts M (1995)   'Conurbation transport policy: making the most of what we may yet spend', Proceedings of the Institution of Civil Engineers, Transport 111 (3).

RCEP (1994)   'Transport and the Environment: 18th Report', Royal Commission on Environmental Pollution, Stationery Office.

Steer J (1995)   'The formulation of transport policies in British conurbations', Proceedings of the Institution of Civil Engineers, Transport 3(3).

Chapter 4 Roles, Responsibilities and Powers

4.1 Introduction

The legal framework for local transportation planning is set out in Acts of Parliament and modified from time to time in new legislation. Detailed provisions are often set out in statutory 'Regulations', which can be amended and kept up-to-date more readily. EC Directives and Regulations may also be relevant. Major changes to the law are often anticipated in Green Papers (eg DOT, 1996) [Sa] and promulgated in government Circulars and future changes are often reflected in revisions to the current Circulars.

Under the heading of roles, responsibilities and powers, this chapter covers:
- ❑ the respective roles of central and local government (and of the different tiers);
- ❑ responsibilities for public transport services;
- ❑ the development plan system and the place of transportation plans within it;
- ❑ Transport Policies and Programmes (TPPs);
- ❑ methods of funding; and
- ❑ other major statutory provisions needed to give effect to transportation plans.

4.2 Local Government Organisation

In Greater London and the English metropolitan areas (Greater Manchester, Merseyside, Tyne and Wear, the West Midlands and South and West Yorkshire), local government is organised in unitary authorities which have all the local powers and responsibilities for transportation in their area, except for trunk roads. In the metropolitan areas, public transport powers are exercised jointly through the Passenger Transport Authorities (PTAs), which are joint committees of the constituent districts. The Passenger Transport Executives (PTEs) are the executive arms of the PTAs.

Under the Local Government Act 1985 (HMG, 1985a) [Sb], London Boroughs must obtain the Secretary of State's consent to proposals affecting the flow of traffic on 'designated' roads. The Traffic Director for London, appointed by the Secretary of State under the Road Traffic Act 1991 (HMG, 1991a), considers such proposals on his behalf and is responsible for traffic management (including parking and waiting controls) on a network of 'priority routes' designated under the 1991 Act. Public transport planning and provision is the responsibility of London Transport, which is a nationalised industry accountable to the DOT.

Elsewhere in England, there is a mixture of unitary authorities and two-tiers of counties and districts. The unitary authorities were established following a review carried out under the provisions of the Local Government Act 1992 (HMG, 1992a) [Sb]. Where there are two tiers, county councils deal with strategic matters, such as highways (other than trunk roads), transportation planning and the preparation of structure plans, while district councils are responsible for development control and the preparation of local plans (see Section 4.5 below). Off-street parking is generally the responsibility of districts but on-street parking and overall transport policy rests with county councils. Counties and districts have concurrent powers to subsidise local bus services and administer concessionary fares regimes (see Section 4.4 below) but the lead is generally taken by the County Authority.

Unitary authorities are responsible for preparing a local plan for their area and, in most cases, are expected to collaborate with adjacent unitary authorities or county councils on the preparation of a joint structure plan.

There are also 11 English Development Corporations set up by the Secretary of State for the Environment, under the Local Government Planning and Land Act 1980 (HMG, 1980a) [Sb], to help to regenerate run-down urban areas. Development Corporations have defined geographical areas, for which they are the Planning Authority. They are funded by capital receipts from the sale of land, rents, ERDF grants and grant-in-aid from the Department of the Environment (DOE). As they complete their work, they are being wound up – the last one is likely to expire in 1998. Some of them have undertaken major road and public transport projects to open up their areas.

In Scotland and Wales, local government has been re-organised wholly on the basis of unitary authorities, who carry out all the local authority powers and responsibilities for their area [NIa] [Wa].

4.3 Central Government

In England, responsibility for policy on land-use,

including guidance on the preparation of development plans, rests with the DOE, as does responsibility for local government finance. The Department of Transport (DOT) is responsible for transport policy and for supporting certain aspects of local authorities' capital investment in transport, not covered through Revenue Support Grant (RSG) (highways maintenance and other service blocks). The management and improvement of the trunk road network are undertaken by the Highways Agency. The Agency previously delegated responsibility for the management and maintenance of trunk roads to local highway authorities under agency agreements. However, from 1997, the Agency is changing to a system in which there will be competition for the work in new enlarged areas. Both private sector organisations and highway authorities are able to bid. Relationships with individual local authorities are handled mainly through the Government Offices for the Regions (GORs).

DOE also funds the Urban Regeneration Agency (usually referred to as 'English Partnerships'), which was set up under the Leasehold Reform, Housing & Urban Development Act 1993 (HMG, 1993a) [Sc]. The Agency promotes urban regeneration, by helping to bring derelict land and buildings back into productive use, including the provision of improved access, where the lack of it is inhibiting the development of difficult sites. It operates mainly through partnerships and joint ventures with local authorities and the private sector.

In Scotland and Wales, central Government's role is carried out by the Scottish and Welsh Offices [NIb]. Trunk road maintenance in Wales is carried out by 'lead' local authorities, known as 'Trunk Road Agents'. In Scotland, a distinction is drawn between a 'premium' trunk road network (the trunk motorway and dual carriageway network) on which maintenance is carried out by consortia operating companies, and other trunk roads where the works are undertaken by management agents.

4.4 Public Transport Services

While responsibility for adopted, non-trunk, roads rests wholly with local authorities, together with the regulation of traffic on them [NIc], responsibility for local public transport is divided between the public and private sectors [NId].

Bus Services

Following deregulation of bus services by the Transport Act 1985 (HMG, 1985b), the provision of bus services, outside London, is now carried out by companies largely in the private sector. Some former municipal bus undertakings are still owned by their local authorities but they have to be managed at 'arm's length' on a purely commercial basis. The pattern of services and competition on individual routes are no longer regulated by the Traffic Commissioners, with whom services have only to be registered. Most services are now provided on a commercial basis and fare–levels are determined solely by the market. Timetables can be changed frequently with only six weeks notice (see Chapter 24). The role of local authorities has changed significantly as a result: they are no longer the direct providers of bus services and they may not inhibit competition between bus operators. Nevertheless, local authorities retain important responsibilities and powers over the provision of local services. Their role is to ensure that adequate local public transport services are provided, over and above the level provided on a commercial basis by the private sector. As enablers of local transport services, they need to establish and maintain good working relations with local bus operators.

Main Duties and Powers of Local Authorities, Outside Greater London

Under section 63 of the Transport Act 1985 (HMG, 1985b), county councils and unitary authorities have a duty to formulate general policies for the support of public transport services for which, in their view, there is a requirement and which are not being provided on a commercial basis. PTAs have similar duties under the Transport Act 1968 (HMG, 1968), as amended by sections 57–62 of the 1985 Act. The power to enter into a subsidy agreement for a bus service is subject to the requirements of sections 89–92 of the 1985 Act and the Code of Practice on Tendering (HMG, 1985c).

These powers enable local authorities to subsidise services on additional routes (or extensions of existing routes) and at times when there are no commercially–provided services. However, it is difficult to use these powers to increase frequency on a commercially–operated service, because of the legal requirement not to inhibit competition between operators.

Local authorities, including district councils, also have powers under the 1985 Act to take any measures they think appropriate for promoting the availability of public passenger transport (whether commercial or subsidised) and the convenience of the public. These powers are used to provide co–ordinated information, bus stations, etc. PTEs have specific powers, under sections 81–82, to provide and run bus stations and to levy reasonable charges. Powers to provide bus shelters are contained in the Local Government (Miscellaneous Provisions) Act 1953 (HMG, 1953).

Local authorities can use their traffic regulation powers, under Road Traffic Regulation Act 1984 (HMG, 1984), to provide priority to buses and other street–running public transport vehicles, over other forms of road traffic [NIe]. They can ask the Traffic Commissioners to impose traffic regulation conditions on bus services, in order to reduce severe congestion or danger to the public. The PSV (Traffic Regulation Conditions) Regulations (HMG, 1986 and 1994a) spell out the types of restriction available.

Local authorities can use their planning powers to require new developments to be located where they are easily accessible by public transport, to make adequate provision for access and to seek commuted payments from developers (see Chapter 27).

Local authorities have powers, under section 93 of the 1985 Act, to subsidise concessionary fares schemes for the elderly, children and certain categories of disabled people. These schemes may operate on all forms of public transport, not just buses, and, where there is a scheme, all operators have a right to participate. Disputes about the level of compensation to be provided to operators are referred to the Secretary of State.

Under section 106 of the 1985 Act, local authorities can pay capital grants to bus operators to provide or improve vehicles suitable for disabled people.

Train Services

Since April 1994, the responsibility for railway infrastructure has been separated from that for operating railway services [NIf]. Heavy rail passenger services are now provided by numerous different train operating companies. These services are being progressively franchised–out to the private sector, the remainder being the responsibility of the British Railways Board. Rail infrastructure, which was privatised in 1996, is provided by RailTrack.

The franchising of passenger train services is the responsibility of the Office of Passenger Rail Franchising (OPRAF), headed by the franchising director. He administers the subsidy provided by central government to support rail services. Under section 20 of the Transport Act 1968 (HMG, 1968), a Passenger Transport Executive has a special duty to secure the train services which its Authority (PTA) considers necessary for the metropolitan area. These services, formerly operated by the British Railways Board, are usually franchised to private sector operators.

Prior to privatisation, these metropolitan train services received central Government assistance through Revenue Support Grant (RSG) paid to the PTA's constituent metropolitan district councils. The resources were passed on to the PTE through the PTA's levy on the districts. Revised arrangements are being considered for support for these services in 1997/98 and beyond.

Under section 56 of the Transport Act 1968 (HMG, 1968), Ministers may make grants and local authorities may make payments in respect of local rail improvements, such as new and improved stations or other public transport capital schemes.

Section 34 of the Railways Act 1993 (HMG, 1993b) empowers the PTEs to be party to franchise agreements for their services.

The regulation of train services is carried out by the Office of the Rail Regulator (ORR), whose responsibilities include ensuring equity of treatment for operators over charges for access to the rail network and protecting the interests of train passengers.

Light rail systems are mostly owned by the public sector but often operated by the private sector. Recent systems, including Manchester Metrolink, Midland Metro and Croydon Tramlink, have been constructed under design, build, operate and maintain (DBOM) contracts. The operation of two others, South Yorkshire Supertram and the Docklands Light Railway, are being transferred to the private sector. Only the Tyne and Wear Metro and the Glasgow Underground are likely to continue to be operated by local authorities. Future light rail schemes are also likely to be built under DBOM (or similar) contracts.

4.5 The Statutory Planning Framework

The preparation of development plans is the responsibility of local government, having regard to national and regional guidance issued by DOE (in England) and the Scottish and Welsh Offices in their territories [NIg].

The Town and Country Planning Act 1990 (HMG, 1990a) [Sd] (amended by the Planning and Compensation Act 1991) (HMG, 1991b) superseded the previous planning legislation. They established the 'plan–led' approach, under which there is a presumption that development control decisions should accord with the development plan, unless material considerations indicate otherwise.

Previously, the development plan was merely one of the material considerations to be taken into account [NIh]

The Secretary of State for the Environment no longer approves structure plans, which, like local plans, are now adopted by the local planning authorities, but he appoints 'the Panel', which conducts the Examination in Public (EIP) into a structure plan. The Panel consists of a Chairman and one other member, who must be from the Planning Inspectorate Agency. The Secretary of State also issues national and regional planning policy guidance, which local planning authorities are required to take into account in preparing their plans. He has reserve powers to intervene, where there are significant conflicts with national and regional policy which cannot be justified by local circumstances. He can either direct that a plan be modified to meet his concern or, as a last resort, he can call–in all, or part, of a plan for his own decision [NIh].

The Secretary of State's guidance is mainly of two kinds:
❑ Regional Planning Guidance (RPG) Notes (one per region); and
❑ Planning Policy Guidance (PPG) Notes [Se] [Wb].

The PPGs of chief interest to urban transportation planning are [NIi]:
❑ PPG 12 (DOE, 1992) [Sf], which gives guidance on development plans and the procedures for preparing and approving them;
❑ PPG 13 (DOE/DOT, 1994) [Sf], which deals specifically with Transport; and
❑ PPG 6 (DOE, 1996) [Sg] on Town Centres and Retail Developments.

Development plans include:
❑ structure plans prepared by county councils;
❑ local plans prepared by districts and unitary authorities in the shire counties; and
❑ unitary development plans (UDPs) prepared by London Boroughs, Metropolitan Districts and other Unitary Authorities.

Structure plans (and Part I of UDPs) set out key strategic policies as a framework for local planning by district councils. Under the 1990 Act (HMG, 1990a) [Sd] the plan must include land–use policies for the management of traffic (eg through land–use policies which reduce the need to travel) [NIj].

The Act (and regulations made under it) also require a planning authority to have regard to:
❑ regional or strategic guidance by the Secretary of State;
❑ current national policies;
❑ the resources likely to be available;
❑ social, economic and environmental considerations; and
❑ any policies or proposals of a UDA affecting the area.

The Transport content of Plans
Structure plans (or Part 1 of a UDP) should set out policies and proposals relating to the development of the transport network, their timescale and priorities, taking account of national guidance on transport and environmental issues. The Transport Policies and Programme (TPP) for the area, which is updated annually, should be consistent with the Structure Plan.

Local plans (or Part II of a UDP) elaborate these policies in more detail and include local transport proposals of a non–strategic nature, related to development patterns proposed in the plan.

PPG 12 [Sf] sets out, in some detail, the ground that plans should cover. Transport is dealt with specifically in paragraphs 5.22 – 5.36 and 6.12 – 6.14. These issues are spelled out in more detail in PPG 13 [Sf] (see Section 3.3).

4.6 Public Involvement

The involvement of the public, whether through statutory consultation or on a wider basis, is an important aspect of urban policy formulation and implementation and is referred to in several later chapters (see particularly Chapters 10 and 20). There is a role for public involvement, both at the strategic level (eg in deciding the problems to be addressed and the measures to be incorporated in a package bid) and at the local level (eg when implementing a local traffic calming scheme). The public also contribute to traffic and transport studies by taking part in various surveys, which are designed to collect data on traffic/travel patterns and factors affecting local travel–behaviour (see Chapter 7).

Statutory requirements for consultation are laid down in several Acts and Statutory Instruments [NIk]. For trunk road schemes, members of the public are invited to comment on the Draft Order for a scheme and this may be followed by a public inquiry. Local road schemes form part of local development plans and are one aspect covered by the consultation. There are formal requirements to notify the public, once the draft local plan is put on deposit and if there are any modifications following a local inquiry. The public also have to be notified where Traffic Regulation Orders (TROs) are introduced or modified. A recent court ruling, concerning the introduction of a proposed Controlled Parking Zone in Camden, laid a requirement on public authorities to conduct a 'fair and effective' consultation with local residents.

In practice, the public are usually much more widely informed and consulted than the minimum required by statute. The Highways Agency [Wc] normally consults on alternative routes for trunk road schemes, before publishing a Draft Order, and is increasingly using 'round tables' to draw up a short list of options. Local authorities usually consult quite widely before putting a preferred local plan on deposit [Nil]. While there are no formal requirements to consult on urban transportation strategies, or on local traffic management schemes, most highway authorities now place considerable importance on taking account of public views when devising such schemes.

The nature of public involvement, and the degree of influence it has on subsequent decision–making, varies widely from one authority to another. Some mainly inform the public, most consult and an increasing number are encouraging public participation in local decision–making. This can raise important issues concerning the relationship between direct public input and the role of the elected member (see Chapter 10).

4.7 Transport Policies and Programmes (TPPs) [Wd]

These non–statutory documents are the vehicle through which English local authorities put in their bids for capital funding to the Secretary of State for Transport (see Section 4.8) [NIm]. TPP submissions should cover an authority's transportation strategy, across all modes, and may cover bids for major capital road or public transport schemes, minor works, structural maintenance of principal roads, bridge maintainance and strengthening, or for transport 'packages'. Under the 'package' approach, adopted in 1993, authorities, particularly those in urban areas, are encouraged to bid for packages of measures aimed, collectively, at delivering local objectives within a defined area. Originally, TPPs covered both revenue and capital expenditure on all modes of transport. In the 1980s, they were restricted solely to capital expenditure on highways. In 1993, they were widened again to cover capital expenditure on all transport modes. Revenue expenditure, eg on highway maintenance or bus subsidies, is grant–aided by formula through Revenue Support Grant (RSG).

The current policy guidance makes clear that the DOT expects package bids to be prepared for all major urban areas. Where an area is covered by more than one local authority, they are encouraged to collaborate in preparing a single bid and to agree on the priorities. The bid should be based on a clear strategy for the area, covering all forms of transport, including cycling and walking, as well as public transport and road traffic, and the management of traffic and travel demand as well as capital investment. The transportation strategy should be related clearly to the development plan or plans. Guidance on this point is given in paras 5.29 of PPG12 [Sf] and para 5.5 of PPG 13 [Sf].

Together, the 'plan–led' approach to land–use development, the 'package' approach to transport funding and the policies on sustainable development set out in PPG13 underline the need for strategic planning of local transport, including public transport and effective parking policies, and their integration with the development plan process. Some of these policies will involve difficult choices and trade–offs between personal mobility and environmental protection. Local authorities need to ensure that there is effective consultation in the preparation of these plans, both with the general public and also with other organisations whose co–operation and support will be required (see Chapter 10). These include local businesses, developers, public transport operators and the police. The police are responsible for the enforcement of road–traffic law, including traffic management and parking restrictions and should be consulted on the preparation of plans to ensure that they are operationally workable from their point of view (see Chapter 14).

4.8 Funding of Transport Infrastructure in Urban Areas [NIn]

[Wd]

Highway and traffic schemes have normally been financed as public sector projects. Local authorities generally fund them through loans (within the credit limits approved by the DOT) and Transport Supplementary Grant (TSG), although some local authorities fund schemes directly from revenue. Local authorities are normally permitted to use a proportion of capital receipts (eg 50%) to fund capital projects and, in principle, these can be used to fund transport schemes. In practice, however, most receipts go to Districts and most capital expenditure on transport is incurred by Counties. Where a local authority is free of debt, it can use all of its capital receipts to fund capital projects.

Local authorities bid annually through the TPP process for resources for local transport schemes. Funding can take the form of :

❑ Transport Supplementary Grant, for major road and traffic projects (costing more than two million pounds); or

❑ section 56 grant, for major public transport infrastructure projects (costing more than £5m); or

❑ credit approvals, provided as a mix of resources, with either section 56 grant or TSG, or on their own (eg for minor works schemes under £2m). They can be issued in one of the following forms:

❑ Annual Capital Guidelines (ACGs), which are not tied to any particular project or type of expenditure. Most capital expenditure on transport receives credit approval in this form. Once granted, it forms part of the overall borrowing approval for the local authority and can be used for any purpose; or

❑ Supplementary Credit Approvals (SCAs), which are tied to particular schemes or programmes. Some SCAs are given in the form of 'trading' credit approvals. This means that the cost of servicing the loan is not grant–aided through RSG (as is the case with all other credit approvals) but is expected to be met by revenues from trading activities, the sale of capital assets or the receipt of grant from other sources, such as the European Commission.

The annual TPP circular, published by the DOT around Easter, sets out the detailed grant rules, the government's policy priorities and the supporting material required with bids, including economic appraisal for major schemes.

Urban regeneration

Transport expenditure which assists the regeneration of an urban area may be eligible for support as part of a wider proposal funded from the 'Single Regeneration Budget' administered by the GORs. In 1996, the Government ran a pilot competition, the 'Capital Challenge', for awarding SCA support for capital projects on a cross–service basis. Awards, covering the calendar years 1997 to 2000, include a number of transport schemes, bid for, in parallel, under TPPs (DOT, 1996a). It is possible that further cross–service challenge competitions may be held in future.

Other Grants

The DOT occasionally pays 100% grant for road schemes, under section 272 of the Highways Act [Sh], usually where the road is to be 'trunked' when the improvement is complete. Roads to facilitate industrial development may be eligible for grants

under the Industrial Development Act 1982 (HMG, 1982). The allocation of grants is administered by the DOT [We].

Private Funding

Both UK government and EC grants may be conditional on securing co–lateral private sector contributions. These can take a number of different forms.

Under the New Roads and Streetworks Act 1991 (HMG, 1991c), a concession agreement may be reached between a highway authority and the private sector, in which the private firm agrees to finance, build, operate and maintain a road, in return for the right to charge tolls on it [NIo]. After a prescribed period, the road reverts to the highway authority. A road subject to a concession is a special road as defined in Part II of the Highways Act 1980 (HMG, 1980b) [Si]. It does not have to be a motorway but the classes of traffic it can carry may be prescribed by reference to Schedule 4 of the 1980 Act .

The Department of Transport [Wf] has invited private sector bids for roads to be constructed and managed under 'Design, Build, Finance and Operate' (DBFO) procedures. The successful tenderer will be remunerated principally through annual 'shadow' tolls paid by the Department. The level of tolls will be related to the level of traffic using the road and subject to an upper limit. In principle, the same procedures could be used by local highway authorities. The DOE published a consultation document on 20 May 1996 about private funding for local authority projects under the Private Finance Initiative (PFI). To qualify, schemes would have to transfer sufficient risk to the private sector and secure at least 30% private sector funding. Schemes approved (within a quota) would qualify to have their annual payments supported by RSG.

Light Rail

Current government policy is for investment in light rail and similar local public transport systems to be funded in conjunction with significant contributions from the local public sector and the private sector [NIf]. Under the present rules, Government grants (under section 56) towards the capital costs of such systems are available only if justified by external benefits and if the scheme is likely to make an operating surplus. It is normal for a competition to be held to choose a private sector consortium to design, build, operate and maintain the system. Schemes which minimise the level of public sector contribution have a grater chance of receiving government funding.

Heavy Rail

Railtrack can be expected to play a primary role in proposals for investment in railway infrastructure [NIf]. Provision of the trains is the responsibility of separate rolling stock companies (ROSCOs), which moved to the private sector in 1995. Local authorities, wishing to promote improvements on their local rail networks, should normally work through the local train operating company, which is responsible for negotiating access charges and procuring the necessary rolling stock from the ROSCOs. In most cases involving investment in new rail infrastructure (stations, track, signalling, etc), it is advisable for local authorities, at an early stage, to consult Railtrack, who will be able to advise authorities (or alternatively through the train operator) of the infrastructural constraints and requirements which should be considered, according to the type and level of service being proposed. OPRAF may also need to be consulted, for example, if there were a proposal to convert an existing heavy rail line to light rail. Any proposal for heavy and light rail to use the same tracks should also be the subject of separate discussions with the Health and Safety Executive (HSE) Railway Inspectorate (see Chapter 34).

Charges

Some types of expenditure are funded wholly or partly by revenue from user–charges. Parking is an example. Where it is possible to charge for a service, it may be possible to arrange for it to be contracted–out to the private sector. Even where franchising or contracting–out are not possible or not cost–effective, charging reduces the net call on public funds and may release funds for expenditure elsewhere in a local authority's programme. All surplus parking revenues can be retained by a local authority. On–street parking revenue can be applied to the provision of additional parking capacity or to traffic management and other purposes listed in section 55 of the Road Traffic Regulation Act 1984 (HMG, 1984), including measures to help public transport. Off–street parking revenue goes into a local authority's general fund.

Urban Road–Use Pricing

No general powers exist for local authorities to charge directly for the use of roads, other than for the issue of parking permits. In the green paper Transport – The Way Forward (DOT, 1996b) [Sa], the Government announced that it would discus the case for additional restraint measures with the local authorities' associations and other interested parties, with a presumption in favour of introducing enabling legislation to allow local authorities to introduce congestion–charging or area–licensing schemes.

One key issue to be settled in any legislation would be the uses to which revenues from charging might be put. Applying international accounting conventions suggests that it is likely, under existing legislation, that urban road–use charges would have to be treated as a local tax, or levy, and the disbursement of net revenues (after deducting operating costs and repayment of start–up expenditure) would count against the Public Expenditure Control Total. However, the Government has accepted that financial arrangements should ensure that communities that implement congestion–charging should be better off as a result (see Chapter 21).

Developers' Contributions

Where a development benefits directly from improvements to transport infrastructure, it may be possible to negotiate a contribution by that developer towards the capital costs. If the development requires planning permission, the payment of a contribution can be the subject of a planning 'Agreement' under section 106 of the Town and Country Planning Act 1990 (HMG, 1990) [NIp] [Sj]. Contributions can be towards public transport improvements, as well as additional road capacity (see Chapter 27).

Bridge Strengthening

Highway authorities are responsible for the maintenance and strengthening of their own bridges. Where a bridge carrying a local authority road is not owned by the authority the liability for maintaining it falls to the owner. In certain instances, the owner's liability is prescribed in statute. Bridges owned by BR, Railtrack, LRT, and British Waterways are covered by the provisions of section 117 of the Transport Act 1968 (HMG, 1968) and, for BR and Railtrack, the Railway Bridges (Loadbearing standards) Regulations (England and Wales) Order, 1972 (HMG, 1972) [NIq] [Sk]. The effect of these provisions is to limit BR's and Railtrack's liabilities to meeting assessment standard BE4, which equates broadly to a 24–tonne maximum gross vehicle weight (depending on the design and structure of the bridge). Strengthening of bridges to carry vehicle–weights beyond the liability of bridge–owners is the responsibility of the relevant Highway Authority, which may, instead, decide to apply some form of permanent or temporary weight or traffic flow restriction.

EC Grants

European Commission Grants are available under the following programmes:
- ❏ Trans–European Networks (TENs);
- ❏ the European Regional Development Fund (ERDF); and
- ❏ various research programmes sponsored by EC Directorates.

The development and funding of TENs is provided for in Article 129(b)–(d) of the Maastricht Treaty. The networks were adopted by the Council and the European Parliament in 1996. Their purpose is to support schemes of common interest to member states which are provided by a public or quasi–public body and are not economically viable. At present 75% of the funds are allocated to priority projects, which include four of interest to the UK: the Channel Tunnel Rail Link; the upgrading of the West Coast Main Railway Line; the Ireland–UK–Benelux road link; and the Cork–Larne–Stranraer rail link. Bids for the remaining 25% of funds will be sought each year, in the summer preceding the year of spend. Local authority bids should be routed via the GOR to the European Division of DOT. As soon as a bid is approved, 50% of the grant is paid, with the balance later.

ERDF is one of three structural funds. There are three categories:

❑ Objective 1: to promote the development of regions lagging economically behind the rest of the EU;
❑ Objective 2: to assist those regions seriously affected by industrial decline; and
❑ Objective 3: to facilitate the development and 'structural adjustment' of rural areas.

Objective 1 areas, in 1996, include Merseyside, the Highlands and Islands in Scotland and Northern Ireland. Other urban areas may qualify under Objective 2. Measures eligible for grant have to offer substantial benefits appropriate to the needs of the area and not able to be financed without grant. They are promoted mainly by the public sector but private sector schemes can qualify. The Government Offices for the Regions (GORs) co–ordinate applications from their regions.

4.9 Other Important Powers and Responsibilities (see also Chapter 33).

The Highways Act 1980 (HMG, 1980b) [Sl] provides, *inter alia*, for: highway authorities to construct, maintain and improve highways; the stopping up of highways; the making–up of private streets; the acquisition of land; the trunking and detrunking of roads; and the designation of highways as 'Special Roads'[NIr]. New or improved trunk roads are approved under a system of Orders, subject to public inquiry and confirmed by the Secretary of State [NIs]. Local authorities obtain planning permission for their roads under planning law but require 'Side Road' Orders, under the Highways Act, for consequential alterations to side roads that interfere with existing rights of way [NIt].

The Transport and Works Act 1992 (HMG, 1992b) [Sm] provided a new Order–making procedure, similar to that for trunk roads, for the approval of fixed–track transport projects and other public works [NIu]. That took the place of the private Bill procedure in Parliament, which is generally no longer available for this purpose. Hybrid Bills (sponsored by the Government) are still available in exceptional cases. Opposed 'Transport and Works' Orders are considered at a public inquiry, chaired by an independent inspector, who reports to the Secretary of State. Schemes of national significance are subject to a preliminary discussion by Parliament. Orders can confer compulsory purchase powers and the Secretary of State can give 'deemed planning permission' on making an Order. There are arrangements for concurrent consideration of listed building consent and other matters. Detailed advice on the use of the TWA procedures is contained in a Department of Transport publication Transport and Works Act 1992: A Guide to Procedures (HMG, 1992c) [Sm].

The Acquisition of Land Act 1981 (HMG, 1981) sets out procedures which, together with forms prescribed by the Compulsory Purchase of Land Regulations 1994 (HMG, 1994b), provide for most compulsory purchase orders [NIv]. The powers of compulsory purchase will differ according to the purpose or purposes for which the land is required. For example, most powers for compulsory purchase of land for highway purposes are in The Highways Act 1990 (HMG, 1990b) [NIw] [Sn].

The New Roads and Streetworks Act 1991 (HMG, 1991c) replaced the previous legislation on public utilities' works in the highway (PUSWA) [NIx] and provided new procedures for the approval of privately–funded highways and for charging tolls on them [NIy].

The Road Traffic Regulation Act 1984 (HMG, 1984) places a duty on local authorities to secure the expeditious, convenient and safe movement of vehicular and other traffic and the provision of suitable and adequate parking facilities both on and off the highway [NIz]. Section 1 sets out the purposes for which local highway authorities may make traffic regulation orders [NIe]. These are:
❑ avoiding danger to persons or traffic;
❑ preventing damage to the road or any building on or near the road;
❑ facilitating passage of any class of traffic (including pedestrians);
❑ preserving the character of the road, where it is specially suitable for use by persons on horseback or on foot;

preserving the amenities of the area through which the road runs; and

❏ preventing use by traffic unsuitable to the character of the road or adjoining properties.

The Environment Act 1995 (HMG, 1995) [NIaa] added to this list:

❏ regulating traffic for the purposes of air quality assessment or management.

Under sections 1,6 or 9 of the Road Traffic Regulation Act, 1984, local authorities may restrict access to a highway, subject to exemptions for certain classes or descriptions of vehicles [NIe]. These powers can be used to limit access to vehicles with particular characteristics, eg those that meet specified air pollution standards or whose drivers are in possession of a permit, provided that the purposes of the scheme fell within those listed in section 1 of the Act (see above). In practice, any restriction needs to be capable of being adequately signed and enforced if it is to be effective. It is possible to frame restrictions so as to exclude all vehicles, either generally or of a particular type, other than permit holders. It should be noted that current legislation includes no provision that allows local authorities to charge for the issue of such permits.

Part IV of the Act enables local authorities to provide and charge for off street-parking places for vehicles and to authorise use of any part of the road as a parking place. Charging for on-street parking is covered by section 45. Sections 43 and 44 empower local authorities to licence the use of privately-owned, but publicly-available, parking and to regulate the charges and the split between short and long term parking [NIab]. If the operator loses financially by this, he is entitled to compensation.

The Road Traffic Act 1991 (HMG, 1991a) provided new powers to control and enforce on- street parking on 'priority (red) routes' and powers for local authorities in London to take over, from the police, the enforcement of on-street parking in 'special parking areas'. Local authorities in other parts of England and Wales can apply to the Secretary of State for approval to set up special parking areas in their area. Winchester and Oxford were the first authorities to take up these powers, outside London, in December 1996. Fines are replaced by penalty charges and the proceeds can be used by the local authority to fund enforcement. All surplus funds can be used to fund transport expenditure in the categories listed in section 55 of the 1984 Act.

The Traffic Calming Act 1992 (HMG, 1992d) [NIac] and the Regulations made under it (HMG, 1993c)

[NIad] widened the powers of highway authorities to use a range of standard techniques for traffic calming and provided powers for the Secretary of State to approve non-standard designs. The powers to install road humps are contained in sections A–F of the Highways Act 1980 [So]; the wider powers introduced by the 1992 Act constitute sections 90G–I of the Highways Act (see Chapter 13) [NIae].

4.10 References

DOE (1992)   Planning Policy Guidance PPG 12 'Development Plans and Regional Planning Guidance', DOE [Sf] [Wg].

DOE (1996)   Planning Policy Guidance PPG 6 'Town Centres and Retail Developments', DOE [Sg] [Wg].

DOE/DOT (1994)   Planning Policy Guidance PPG 13 'Transport', DOE/DOT [Sf] [Wg].

DOT (1996a)   TPP Circular 2/96 'Supplementary Guidance note on the Package Approach', DOT.

DOT (1996b)   'Transport – The Way Forward', Stationery Office [Sa].

HMG (1953)   'Local Government – Miscellaneous Provisions Act 1953', Stationery Office.

HMG (1968)   'Transport Act 1968', Stationery Office.

HMG (1972)   'Railway Bridges (Load Bearing Standards) Regulations (England and Wales) Order 1972', Stationery Office.

HMG (1980a)   'Local Government Planning and Land Act 1980', Stationery Office[Sj].

HMG (1980b)   'Highways Act 1980', Stationery Office.

HMG (1981)   'Acquisition of Land Act 1981', Stationery Office.

HMG (1982)   'Industrial Development Act 1982', Stationery Office.

HMG (1984)   'Road Traffic Regulation Act 1984', Stationery Office.

HMG (1985a)   'Local Government Act 1985', Stationery Office [Sb].

HMG (1985b)   'Transport Act 1985', Stationery Office.

HMG (1985c)   'The Service Subsidy Agreements (Tendering) Regulations 1985', Stationery Office.

HMG (1986)   'PSV (Traffic Regulations Considerations) Regulation 1986', Stationery Office.

HMG (1990a)   'Town and Country Planning Act 1990', Stationery Office [Sd].

HMG (1990b)   'The Highways Act 1990', Stationery Office [Sl].

HMG (1991a)   'Road Traffic Act 1991', Stationery Office.

HMG (1991b)   'Planning and Compensation Act 1991', Stationery Office.

HMG (1991c)   'New Roads and Streetworks Act 1991', Stationery Office.

HMG (1992a)   'Local Government Act 1992', Stationery Office [Sb].

HMG (1992b)   'Transport and Works Act 1992', Stationery Office [Sm].

HMG (1992c)   'Transport and Works Act 1992 – A Guide to Procedures', Stationery Office [Sm].

HMG (1992d)   'Traffic Calming Act 1992', Stationery Office.

HMG (1993a)   'Leasehold Reform, Housing and Urban Development Act 1993', Stationery Office [Sc].

HMG (1993b)   'Railways Act 1993', Stationery Office.

HMG (1993c)   'Highways (Traffic Calming) Regulations 1993', Stationery Office.

HMG (1994a)   'PSV (Traffic Regulations Conditions) (Amendment) Regulations 1994', Stationery Office.

HMG (1994b)   'Compulsory Purchase of Land– Regulations 1994', Stationery Office.

HMG (1995)   'Environment Act 1995', Stationery Office.

4.11 Further Information

DOE (1988)   Circular 15/88 (WO 23/88) 'Environmental Assessment', DOE.

DOE (1992)   Circular 19/92 (WO 39/92) 'Town and County Planning', DOE [Sf].

DOT (1985a)   Circular 3/85 'Transport Act 1985', DOT.

DOT (1985b)   Circular 5/85 'Transport Act 1985', DOT.

DOT (1992a)   Circular 5/92 'Traffic Management and Parking Guidance (Revised September 1994)', DOT.

DOT (1992b)   Advice Note 'Transport and Works Act 1992 – Guide to Procedures', DOT.

DOT (1995a)   'Local Authority Transport Responsibilities' Local Transport Policy Division, DOT.

DOT (1995b)   Circular 1/95 – 'Guidance on Decriminalised Parking Enforcement outside London', DOT.

HMG (1972)   'Railway Bridges (Load Bearing Standards) Regulations (England and Wales) Order 1972', Stationery Office [Sc].

HMG (1986a)   'PSV (Registration of Local Services) Regulations 1986', Stationery Office.

HMG (1986b)   'The Travel Concession Schemes Regulations 1986', Stationery Office.

HMG (1988)   'PSV (Registration of Local Services) (Amendment) Regulations 1988', Stationery Office.

HMG (1992a)   'Transport and Works
                 (Applications and Objections
                 Procedures) Rules 1992',
                 Stationery Office.

HMG (1992b)   'Transport and Works (Inquiries
                 Procedure) Rules 1992',
                 Stationery Office.

HMG (1992c)   'Road Traffic (Temporary
                 Restrictions) (Procedure)
                 Regulations 1992', Stationery
                 Office.

HMG (1993)   'Local Authorities' Traffic Orders
                 (Procedure) (England and Wales)
                 (Amendments) Regulations
                 1993', Stationery Office.

HMG (1996a)   'Road Hump Regulations 1996',
                 Stationery Office.

HMG (1996b)   'Local Authorities Traffic Orders
                 (Procedure) (England and Wales)
                 Regulations 1996', Stationery
                 Office.

Sweet and   'Encyclopedia of Road Traffic
Maxwell   Law and Practice'.
(annual update)

Chapter 5 Urban Transport into the 21st Century

5.1 Introduction

Travel behaviour takes time to change as travel patterns are set, to a large extent, by the patterns of existing land-use activities as well as by the current norms of society. Infrastructure can take decades to alter and much of the current highway network was laid down on alignments fixed in Roman times, while most railway routes were set out during the last century. Thus, the shape of urban transport in the future will be determined largely by the past and present situations and the majority of the infrastructure that our grandchildren will use exists today. As a result, it is vital that all new initiatives and investments are carefully planned and assessed to ensure lasting value.

However, it is not true to say that, because infrastructure lasts a long time, the nature of transport in the future will simply be a continuation of past and present trends. Technological development will open up new possibilities, economic growth may increase demand, public and political attitudes may shift and life–styles may change. Forecasting how these factors could develop and what impact they would have in the future is fraught with difficulty. The only certainty is that any forecast will almost certainly be wrong in some respect; so policy and planning need to be increasingly flexible and based upon realistic scenarios.

This chapter does not, therefore, attempt to predict or prescribe what will happen to transport into the next century. In particular, it does not espouse any particular policy choices. Instead, it attempts to set out some of the more important influences that may impinge on transport in the early years of the twenty first century and suggests some of the possible effects that may result.

5.2 Transport in the Latter Part of the 20th Century

Since the Second World War, growing prosperity has led to mass motorisation: first for freight movement then for people. High car ownership has given the freedom of personal mobility to the majority of the population. As Buchanan (1963) has said, *a propos* car–ownership: "...what was once the privilege of the minority has become the expectation of the majority". People can now live further from their work and have much greater choice of leisure and shopping activities. Households have chosen where to live, work, shop, educate their children and undertake leisure activities on the assumption of cheap and convenient transport by private car. As a result, life–styles are often dependent on the availability and use of private cars: car–ownership has expanded more than ten–fold and there has been a consequential decline in walking, cycling, bus and train use. Freedom of choice is now a major factor in travel behaviour.

With the end of the 20th Century, it is clear that the demands for personal mobility are most unlikely to decrease. Whether the rise in the demand for travel will continue unabated, however, will depend primarily on three factors: economic prosperity; changes in life–styles (as a result of political and environmental pressures); and the available transport and communications technology.

5.3 Looking Ahead to Transport in the 21st Century

Social Changes

Personal wealth and prosperity seem set to continue to grow in the 21st Century, enabling greater choice of where to live, work, shop and undertake leisure activities. The distribution of prosperity across different sections of society will, however, be a major determinant of both the level and characteristics of travel in urban areas.

Current forecasts suggest that, in Britain, smaller households (comprising one or two people) may be the norm rather than the exception. The average household size is predicted to fall from 2.50 in 1991 to 2.43 after 2001 (a three percent reduction). An ageing population (by 2051 almost a quarter of the population will be of pensionable age and over half of these will be 75 or over), due to improvements in nutrition and medical care, will not only contribute to smaller household sizes but will also change the nature of the demands for travel.

Travel patterns are, nevertheless, determined by norms and fashions of behaviour and these can be influenced, to some extent at least, by policy–makers and professionals in the decisions that they make regarding transportation in urban areas.

Travel Demands

Throughout the early years of the 21st Century, the demand for travel looks likely to continue to rise. If, as expected, car–ownership continues to grow so too will car–use. In 1993, 70% of households had at least one car. Car ownership is forecast to grow to around 550 cars per 1,000 population by 2025, rising towards a possible saturation level of between 600 and 650 cars per 1,000 people. Eventually, 80% of all adults are likely to possess a driving licence, 90% in the case of those below retirement age.

Leisure travel in particular is likely to grow as prosperity increases. Some of this will be longer distance journeys, undertaken by air, but many journeys will continue to require land–based transport. Travel patterns could well become increasingly dispersed both in time and space. However, in the absence of effective demand management, unless additional capacity is provided in the right places to satisfy this growing demand, congestion in urban areas and elsewhere can be expected to worsen. Policy, however, is likely to move away from providing new capacity to meet the forecast demand, to making better use of existing capacity, managing the demand for mobility and seeking to reduce overall travel needs. This could lead to behavioural changes, for example, in commuting patterns, if growing environmental concerns do indeed lead to the introduction of demand management measures designed to curtail the potential growth of road transport. Indeed, finding publicly–acceptable ways of restricting the growth in demand could well be the main challenge for transport policy into the next century.

As the demand for air travel, a key element of leisure movement, continues to grow, surface access to airports is likely to become an increasingly important issue. Despite policies designed to reduce the need to travel by other modes, notably private cars, the demand for air travel continues stubbornly to be linked to economic growth. Policies are therefore likely to be developed that recognise the differing requirements of airline passengers and airport workers. Collective transport is increasingly likely to be used to link urban centres with airports to provide a frequent, reliable and comfortable service which can also accommodate passengers' luggage requirements. Employees' trips and those by 'meeters and greeters' on the other hand, are likely to be subject to demand management measures (eg higher parking charges, park–and–ride and car–sharing) to reduce the environmental effects of additional car traffic.

Changes in Lifestyles

In the future, fewer people may need to travel into town and city centres (or even to the outskirts) to work. Although, as yet, the advent of tele–working and the Internet has not led to significant changes in travel to work patterns, this could well change as more home computers are purchased. Tele–working could enable many people to work (almost) anywhere in the world without the need to commute. Those within the information–based sector of the economy, in particular, would be able to enjoy more flexible lifestyles, including working variable hours from home. As a result, peak–period commuting and the congestion it causes in urban areas may, over time, become less of a problem. Business travel could be the first to be affected. Long–distance journeys for meetings could become less necessary: video conferencing could conceivably replace the need for such travel.

Technological change could affect shopping behaviour also. Tele–shopping enables customers to order and purchase goods direct from their home or workplace. Virtual reality equipment could enable customers to inspect goods in their home and make the sorts of choices that were once only available in shops. Deliveries could be to the door possibly using small, environmentally–friendly, goods distribution vehicles. Nevertheless, tele–shopping is unlikely to account for more than a small proportion of overall shopping needs in the near future.

Although growing prosperity may enable more people to enjoy the benefits of tourism, technological developments may even reduce the need to travel to find them. Although not as good as the 'real thing', virtual reality will enable 'tourists' to enjoy some of the benefits of longer–distance travel without the need to leave their own home.

One of the unknown factors affecting the extent to which travel–substitution may occur is the importance which people attach to cultural diversity and social–interaction in those activities.

Land–use Patterns

Changes in working and shopping practices would, inevitably, lead to changes in land–use patterns, albeit gradually over several decades. New settlements are being encouraged to be concentrated around nodes which can be served by public transport. While locations along radial routes into city centres and city centres themselves are likely to be developed to encourage the use of public transport, in many urban centres all 'non-essential' vehicles could be banned. Indeed, areas where vehicle access is significantly restricted, even if not banned, may well become more common in all types of urban area. Increasingly, those responsible for managing city-centres are concluding that access may be restricted to trams and low-emission buses, with some provision for delivery vehicles, usually at off-peak times and disabled drivers. The problems of diverting through traffic, however, remain unresolved.

Private non-residential (PNR) parking at offices and shopping centres may also become more restricted. This could arise as a result of more stringent planning conditions or from charging for the use of such spaces in some way. PNR parking is a highly land-intensive use and is a major determinant of urban car-use, especially for commuting. If the importance of achieving a modal shift for environmental reasons is recognised, it may become publicly and politically acceptable to take more radical measures to reduce PNR parking. The extension of tele-working may even cause some offices in urban areas to become redundant and be adapted to other uses, such as hotel accommodation or housing.

Planning policies to restrict out-of-town developments and to limit incursions into green fields could lead to higher density developments within town and city centres. These would enable collective transport systems to be more viable than in existing dispersed residential developments. In shopping centres, less floorspace might need to be given over to retailing and the display of goods if tele-shopping were to take off. In contrast, more suburban land may be required for warehouses from which to distribute locally.

Distribution patterns for freight transport could also change. Environmental concerns, particularly in historic centres, may encourage the adoption of schemes to ban large goods vehicles. In order to service shops and businesses in urban centres, out-of-town break-bulk facilities (or urban distribution centres) could be adopted with out-of-hours deliveries into urban areas using small environmentally-friendly vehicles. Some of these facilities are likely to be located near rail depots in order to maximise the use of freight distribution by rail and minimise that by road. Break-bulk facilities could be a common feature in all towns and cities within two to three decades.

Not only is the density of housing likely to change, so, too, could the availability of parking at residential locations. At some locations parking standards are likely to be considerably restricted; many

developments, particularly those close to public transport nodes, may have no residents' parking permitted at all. Indeed such developments are already in existence on the Continent and are planned in, for example, Edinburgh.

The irony is that, even as policy initiatives are increasingly directed towards concentrating developments in centres, which can be efficiently served by public transport, countervailing pressures for dispersal will continue as a result of, for example, greater tele-working.

Fuel Sources

Currently, petrol and diesel are overwhelmingly the most commonly used transport fuels. By the middle of the next century this dominance may be less. If only to reduce reliance on one source of energy supplied from overseas, many countries will aim to encourage the development of alternative fuels. These include ethanol, methanol, seed oil fuels, electric power and hydrogen. Other new fuels include reformulated gasoline, which could lead to reductions of one third in emissions for existing cars with only the need for re-tuning. Another fuel with potential is compressed natural gas (CNG), particularly for buses and urban delivery vehicles. However, for many years, current fuels (diesel and petrol) are likely to continue to predominate, given the huge momentum of the world oil industry and the considerable scope for making conventional internal combustion engines both cleaner and leaner.

Alternative fuels are already being used on an experimental basis by public transport. Their adoption for private vehicles would, however, depend on ensuring that the supplies of such fuels are readily available. Such fuels may not necessarily be the sole source of energy for transport. The adoption of energy-storing hybrid vehicles (using different fuels, perhaps in conjunction with petrol or diesel) could be a foreseeable next step in engine technology. As a result, together with the continuing improvement in the fuel-efficiency of vehicles, there is likely to be significantly less air pollution generated from each vehicle-km but the overall levels of pollution will depend on the volumes of traffic.

Direct Pricing Mechanisms

Payment systems for all modes of transport are changing. Direct 'pay as you go' systems may well become the most common form of payment mechanism, based on smart-card technology. Experiments are already in hand on buses and they are expanding into car-use for paying tolls. Contact-less smart-cards, based on pre-payment, provide for ease of use on public transport and could be adapted to cater for payments for other transport needs, including parking charges where pre-paid cards are already in use.

Payment for road-use by vehicular traffic in Britain would be most likely to be introduced on motorways once the reliability of the technology, including that for enforcement, had been proved, although it is likely to be controversial. The same transponder/smart-card technology could readily be adopted for electronic road-use pricing in urban areas. Once a publicly acceptable package of measures has been devised, in which charging for road-use features, it would be possible for drivers of any vehicle to be charged by time, place or length of journey on both urban and inter-urban roads (see Chapters 18 and 21).

One advantage of the introduction of pay-as-you-go schemes could be the further development of private finance for road construction and maintenance (real tolls replacing shadow tolls) and for the provision of public transport. By providing a predictable and substantial revenue stream to investors, the scope for transferring ownership and responsibility to the private sector becomes much greater. Private finance could eventually fund the construction and maintenance of roads currently operated by both national, as well as local, highway authorities.

Vehicle Technologies

Over the last half century, vehicle technology has changed steadily. New technical developments will continue, probably focusing more on electronic engine-management and control systems, rather than on body design, to reduce the adverse impacts of road transport on the environment. Smaller vehicles for use in towns could, eventually, become mandatory, as a result of pressure to minimise the use of scarce urban road-space by vehicles.

Deteriorating air quality has been a particular cause for public concern. Investment in new technologies will help to ameliorate this; for example, smaller engines can give the same performance while consuming much less fuel and aerodynamic drag can be reduced further. Weight reduction is likely to occur with the replacement of mechanical by electronic units and the extensive use of stronger/more durable materials, such as light alloys and plastics. The use of three-way catalytic converters, which can eliminate up to 90% of noxious gases, will become commonplace and further research should produce solutions to the cold-start problem. Nevertheless, there may still be an opportunity for developing 'lean-burn' engines, which could reduce further the demand for fuel and hence noxious emissions.

Transport Telematics

In-vehicle technology could provide widespread benefits for users. The revolution in communications should ensure that not only will real-time information on congestion be readily available (within the vehicle) but route-guidance may ensure that road-users are encouraged to take the most appropriate route to their destination. Such systems are already available. Messages transmitted to the vehicle from roadside equipment could be displayed, either by dedicated equipment or by head-up displays. The information provided could include messages about current driving conditions, such as whether the road is icy. Vehicles themselves could become information gatherers as well, 'collecting' data from the road and passing it to central computers for distribution to other users.

Intelligent cruise control and other aids for drivers are already being developed. In-vehicle equipment could help improve road safety by vision enhancement (for night driving), collision-avoidance systems and automatic emergency calls (via on-board incident detection equipment) for those vehicles that are involved in an accident. Potentially, acceleration/braking and starting/stopping could eventually all be controlled automatically. The full development of such technology would enable the creation of the 'Automated Highway', whereby control could be passed ultimately from humans to machines. The Automated Highway would need to be limited to motorways, at least initially. Extensive development work is in hand on this concept in the United States but there are many non-technical (eg legal) issues to be addressed before such a concept could be realised.

While engine power is likely to remain high, new technology provides better opportunities for the enforcement of speed-limits. This is already achieved for heavy vehicles by in-vehicle speed-limiters. More interestingly, a communications link with road-side beacons could ensure that no drivers could break the speed-limit in any area. However, before such technology becomes widely available, enforcement is likely to continue to be by the use of camera technology (CCTV) and number plate recognition equipment. Such developments could have significant benefits for road safety.

Automatic sensors, giving on-board diagnostics based within the vehicle, could ensure that a whole range of engine management and other systems is always finely tuned, to optimise the operation of vehicles. These could prevent potentially unsafe vehicles from being used. Furthermore, timely and appropriate maintenance can be undertaken as a result of improved information to vehicle owners. The new technology will be used not only by private cars but public transport vehicles are also likely to include these features.

In urban areas, more sophisticated developments of existing urban traffic management control systems (eg SCOOT, TRANSYT and MOVA) are likely to be implemented. In future, such systems are unlikely to be used solely to maximise the throughput of vehicles. Rather, a more sophisticated approach to traffic management is likely to be adopted to reflect the relevant policies for environmental protection and priorities for pedestrians, cyclists and public transport.

Road-side information, via variable message signs (VMS), is likely to become more commonplace and not just on the motorway network. However, the full benefits of new developments are likely to be reaped only when several different technologies are put together, to provide an all-embracing system of control and information. Perhaps the greatest change is likely to be the development of the Intelligent Transport Environment, integrating roadside technology with vehicle technology. It is possible that future developments could mean that drivers would be able, increasingly, to relinquish control of their vehicle to automatic systems. Such predictions have been made before but, into the 21st Century, the low cost of powerful computing could allow these ideas to be turned into reality. The current constraints on such developments, namely the costs of implementation and concerns about safety, should diminish steadily over time.

Road maintenance should also benefit from technological developments. Not only will it be more efficient to monitor road conditions, via High Speed Road Monitors, but sensors embedded into the road surface could enable highway managers automatically to monitor road characteristics.

Collective Transport

Public transport vehicles are likely to benefit similarly from improvements in technology (eg new non-polluting fuels, lighter bodies and automated control). Fixed-track systems may be used increasingly for radial movements into the larger urban areas as well as for inter-city travel. In addition to the provision of on-street light rail networks, track-sharing, between heavy and light rail, may become commonplace as long as the necessary safety requirements can be met. Widespread use of raised fixed-rail systems, such as monorails or cabin-taxis, running above ground through cities are unlikely to come to fruition due to concern about their visual intrusion.

Better designed public transport may improve inter-changes between different modes of collective transport and with the slower modes (walking and cycling). On-street facilities may also be much improved: bus/tram stops could be more pleasant places providing real-time information about times and travel conditions. On-street or shop-front provision of public transport information centres should become common, starting with major routes into urban areas.

The co-ordination of timetables and the management of public transport operations should be much improved as a result of AVL (Automatic Vehicle Location) systems. Shorter headways may be possible between all types of passenger transport vehicles, as a result of the adoption of AVL for road vehicles and better signalling systems for railways.

Facilities for the mobility-impaired may be enhanced by, for example, low-floor buses/trams with ramp access to platforms becoming ubiquitous. Indeed, the difficulties currently faced by those with impaired mobility should be reduced significantly, as a result of technical developments.

While public transport operators are likely to continue to be from the private sector, the adoption of through-ticketing systems based on pre-paid smart-cards, acceptable for all modes (including the tolling of car-use) could maximise convenience for the customer, while allowing a fair allocation of revenues to operators.

The Urban Streetscape

Many of the changes described above could have a significant visual impact on the urban streetscape. Not only should urban design become better, as the layout of new developments is improved, but traffic, increasingly, may be restricted to specific routes and areas. On local highways, much more space could be allocated to the more environmentally benign modes of travel, notably walking and cycling. Thus, pavements would be wider and many more roads than is currently the case could have dedicated cycle lanes. Furthermore, main roads could have dedicated bus lanes or, in the larger urban areas, there might be on-street tracks for light rail. If all this were to happen, road-space for private cars would be much less than it is today but it would need to be accompanied by either a significant change in choice of travel-mode or a sharp reduction in the demand for travel. Even where ample road space in urban areas is available, priority might be given to higher occupancy vehicles. In residential areas and in locations where there is a mix of different users, speed restrictions may be more rigorously enforced, either by traffic-calming measures or by technological means.

5.4 Conclusions

Given that car–ownership in the UK is only two–thirds of the way to saturation level, continued growth in prosperity is likely to lead to a continuing rise in the demand for travel. However, the nature of demands for movement may change. More flexible working and shopping patterns and, indeed, lifestyles generally may lead to less need for commuting and consequently less travel at peak periods of congestion.

Pre–trip information (via in–house computer terminals) should help travellers to use the most appropriate mode for the particular journey to be undertaken. Public transport is more likely to be part of a multi–modal service, often using slower modes, such as walking and cycling, to reach public transport nodes. Whatever form of transport is used, other than walking and cycling, travel may well be charged for directly, with those who travel furthest and at times of peak congestion paying more. Pre–paid smart–cards are likely to be common for a wide range of transport payments. Driving licences, replaced by smart–cards (validated by in–vehicle units) could restrict an individual's ability to contravene traffic regulations and may even have zonal restrictions for access by car.

Technological developments are likely to lead to more fuel–efficient and safer vehicles and advances in computing and communications are likely to lead to these technologies playing an increasing role in all aspects of urban transport.

The one thing that is clear from the matters discussed in this Chapter is that transportation in urban areas will remain a complex issue, that affects all the population and for which there will often be no easy answers. Perhaps the most difficult, but potentially most important, factors are those affecting policy–choices. Issues of pricing and demand management and the extent to which people are prepared to adapt their lifestyles will all depend on public acceptance. While the debate on such issues will go well beyond just transportation professionals, they will have an important role in both understanding the consequences of different policy–choices and then explaining them to decision–makers and the general public.

Subsequent chapters of this manual should therefore be considered in the light of the possible future transport developments outlined above, taking into account the extent to which it may be desirable to anticipate a range of scenarios, particularly when planning and designing for the longer term future.

5.5 Further Reading

Bannister D and Button K (Eds) (1993) — 'Transport Policy and Management', E&FN Spon.

Bly PH (1996) — 'Researching the Future of Inland Transport', Proceedings of the Institution of Civil Engineers, Transport, November.

Buchanan C D et al (1963) — 'Traffic in Towns – A Study of the Long Term Problems of Traffic in Urban Areas', (Buchanan Report), Stationery Office.

CEST (1993) — Series of 16 Reports. Transport and Communications Centre for Science and Technology, CEST.

DOE (1994) — 'Sustainable Development: The UK Strategy', Department of the Environment CM2426.

The Economist (1996) — 'Taming the Beast: A Survey on Living with the Car', 22 June.

EC (1993) — 'Research and Technology in Advanced Road Transport Telematics', European Commission EC DGXIII Annual Report on Transport Telematics.

Goodwin P (1994) — 'Traffic Growth and the Dynamics of Sustainable Policies', ESRC Transport Studies Unit,University of Oxford.

Moktarian PL (1990) — 'The State of Tele–commuting', ITS Review, 13(4) University of California.

Northcott J (1991) — 'Britain in 2010', Policy Studies Institute, London.

Office of Science and Technology (1995) — Transport Technology Foresight Report 'Progress through Partnership', Stationery Office.

RAC (1992) — 'Cars and the Environment: A View to the Year 2020', RAC Foundation, Royal Automobile Club.

RCEP (1994) — 'Transport and the Environment', Royal Commission on Environmental Pollution, Stationery Office.

The Transportation Planning Process

Chapter 6 Transport Policy Components

6.1 Introduction

The development of any transportation strategy must be set in the context of the role of transport in facilitating the achievement of the wider vision and objectives for an urban area, as described in Chapter 3.

The Green Paper, Transport: the Way Forward (DOT, 1996) published in April 1996, [Sa] was the first for almost twenty years and represented the Government's response to the debate initiated earlier (DOT, 1995c) [NIa]. The debate had shown that people seek a wide range of different ambitions for themselves and for the future development of society. Some of these ambitions are complementary but others require balances to be struck. For instance, it was generally accepted that measures needed to be taken to restrain the growth in road traffic but there was also support for some investment in new roads. The Green Paper suggested that ambitions and demands fall into two main categories. People require access to jobs, to shops, to other people and to all manner of different activities, while businesses want easy access to suppliers and to markets.

More specifically, people want wider choices, including:
❑ public transport which is reliable, clean, safe and reasonably priced;
❑ other modes of personal transport (car, cycle, walk) which can be easily and safely used;
❑ a broad choice of freight transport options, offering quick and reliable deliveries at reasonable cost; and
❑ good choices of location, for homes and businesses, within easy reach of a wide range of other activities.

According to the Green Paper, another strand of ambitions covers wider issues, which can affect many activities other than transport. These include:
❑ the desire for a healthy, sustainable environment;
❑ the aim for a prosperous, competitive, economy; and
❑ careful control over public spending.

People will continue to disagree about the relative weights to be attached to each of these aspirations. The Green Paper argued that the job of Government is to pursue policies which will promote the best outcome for society as a whole. "Where possible, the aim should be to promote solutions which can reconcile different objectives, with a strong economy and a healthy environment complementing each other" (DOT, 1996) [Sa].

A number of key principles were identified as being central to the Government's approach:
❑ increased attention to be paid to the environmental impacts of transport;
❑ transport decisions to be taken at the lowest practical level;
❑ the efficiency of markets to be strengthened, providing the minimum necessary regulation, increasing the role of the private sector and offering the best prospects for meeting users' needs;
❑ strengthening the link between prices and the wider costs of transport, which will tend to increase efficiency and reduce unnecessary environmental impacts; and
❑ the need for improved planning and design of transportation requirements.

Such principles, according to the Green Paper, imply particular emphasis on the following kinds of policies:
❑ better planning of transport infrastructure;
❑ making more efficient use of existing infrastructure;
❑ reducing dependence on car-use, especially in towns;
❑ empowering local decision-making;
❑ switching the emphasis on capital spending from roads to public transport; and
❑ reducing the impacts of road freight vehicles.

The Green Paper also identified three key factors that are fundamental to determining the direction of transport policy:
❑ the relative effectiveness of different measures;
❑ the costs; and
❑ the need for objective analysis.

While the Green Paper is only one in a series of reports from Government and other bodies on transport issues, it is unlikely that the broad principles and recommendations will be changed significantly in the foreseeable future.

6.2 The Components of Transportation Strategies

A wide, and growing, range of different transport policy–instruments or measures can be used to achieve the vision and objectives of a transportation strategy. The following sections group such measures under five basic categories; land–use, infrastructure, management, information provision (including travel awareness campaigns) and pricing measures. This approach follows that set out in Chapter 4 of the IHT Guidelines (IHT, 1996), which provides further details and references on the individual measures. The key question with each measure is its ability to achieve one or more of the objectives identified in Section 3.6. Unfortunately, this is an area of transport policy in which information is often sparse and may not be directly transferable from one urban area to another. Also, some measures (eg congestion pricing) have not yet been tested in real life trials in the UK and, so, evidence on their effects can only be drawn from research studies. Clearly, in this Chapter, it is only possible to provide a brief introduction to each of the 55 measures that are discussed in the IHT Guidelines (IHT, 1996).

In order to provide some indication of the effects of the various measures, each of the five sections has a summary table adapted from the IHT Guidelines. These summary tables are based on work by Professor Tony May at the University of Leeds (May *et al*, 1995). They give a general indication, on a common basis, of the main impacts of each measure, both beneficial and adverse, against the nine main objectives presented in Section 3.6, with the exception of the sustainability and economic regeneration objectives, which are largely influenced by the other objectives. Any such summary tables are only indicative and there will be circumstances and localities in which a particular measure may be more or less beneficial than indicated in the tables. Also, the tables assess measures as if they were operating alone. As indicated in Section 3.8, the key aim in building strategies is to combine measures to produce a result that is better than the sum of their individual impacts.

The measures presented here should not be taken as an exhaustive list of all the measures that are available and new measures are likely to arise, particularly with the developing role of information technology. When developing a transportation strategy, it is important to get both politicians and the public to generate ideas about the nature of the problems and the measures that can best be used to tackle them.

6.3 Land–Use Planning Measures

Land–use planning measures are not generally focused on a particular mode of transport. The main role of such measures is to improve the integration of land–use and transportation planning, particularly by reducing the need to travel both in terms of shorter journeys and less frequent travel. In England, the joint Department of the Environment (DOE)/Department of Transport (DOT) Planning Policy Guidance on 'Transport' (PPG13) (DOE/DOT, 1994), [NIb] [Sb] gives three key aims for local authorities' land–use and transportation planning policies. These are to reduce growth in the length and number of motorised journeys, to encourage alternative means of travel which have less environmental impact (that is walking, cycling and public transport) and, hence, to reduce reliance, in future, on the use of private cars.

PPG13: A Guide To Better Practice (DOE/DOT, 1995b) [Sb] gives further details of both land–use and transportation measures that can assist in achieving the aims of sustainable development. It sets out a framework for the integration of land–use and complementary transportation measures. The DOE and DOT are also carrying out research into the implementation of PPG13 (DOE/DOT, 1995a) [Sb]. Initial results of this research show that PPG13 is having an impact on policy development. However, local authorities are concerned about difficulties in implementing some of the more important policy changes, for example improving public transport, given the limited powers a local authority has over the provision of deregulated bus services. Initial results have also shown resistance by the development sector, who see the PPG13 approach as 'going against the grain' of the market.

Eight land–use measures can be identified:
- ❑ concentrating dense developments within transport corridors – where public transport can provide a viable alternative to the use of cars (see DOE/DOT, 1995b) [Sb];
- ❑ increasing development densities – higher densities may encourage shorter journeys and, thus, the use of walking and cycling, and may help to make public transport more viable (see DOE/DOT, 1993 and 1995b);
- ❑ altering the development mix – a better mix of uses can improve accessibility and, hence, reduce the need to travel (see DOE/DOT, 1995b) [Sb];
- ❑ reducing parking standards – this probably offers the single most direct impact on levels of car–use and can be used in trip–end restraint (DOE/DOT, 1995b) [Sb] (see also Chapters 19 and 21);

Land–use Measure	Accessibility	Efficiency	Environment	Equity	Finance	Practicability	Safety
Corridors	✓✓	✓?	✓?	0	0	X	?
Density	✓	?	?	0	0	X	?
Mix	✓	?	?	0	0	X	?
Parking standards	✓/X	✓	✓	0	0	X	
Developer contributions	0	0	0	✓	✓	X	0
Commuted payments	0	✓	0	✓	✓	X	0
Travel reduction	?	?	?	0	0	XX	?
Flexible hours	✓	✓	0	0	0	X	0

Key: ✓ ✓✓ ✓✓✓   Positive impact of increasing scale   ✓/X   Both positive and negative impacts

  X   XX   XXX   Negative impact of increasing scale   ?   Uncertain impact

  0   No significant impact

Table 6.1: Performance of land–use measures.

❑ increasing developers' contributions for transport infrastructure, including public transport. Alternatively, the provision of public transport services can be required as part of the process of obtaining planning approval for new developments (DOE/DOT, 1995b) [Sb] (see also Chapter 26);

❑ requiring commuted payments – these are a special type of developer contribution in which the normal requirements for private parking provision are waived in return for payment to a local authority of a charge per space, so that the Local Authority can make provision in public car parks or promote park–and–ride schemes (see Chapter 27);

❑ promoting travel–reduction ordinances/company transport plans – travel–reduction ordinances are used in the US and the Netherlands to require developers to produce a plan specifying ways in which they will reduce car-use to their development; this would require legislation in the UK. As an alternative, voluntary company transport plans could be developed (DOE/DOT, 1995b) [Sb]; and

❑ encouraging flexible working hours – a form of land–use measure designed to reduce peak–period demand for travel and the resulting congestion (DOT, 1977).

Table 6.1 summarises the impact of each of these measures against the main objectives set out in Chapter 3.

6.4 Infrastructure Measures

The land–use transportation studies of the 1960s and 1970s were almost invariably concerned with finding the most appropriate infrastructure plan, particularly that for new road construction. The current range of

Integrated Transportation Strategies view infrastructure as a part only of the total package of measures.

Twelve infrastructure measures can be identified, as follows:

❑ pedestrian facilities – these provide a dramatic improvement in the environment for pedestrians and have proved successful in enhancing trade in many town and city centres (see Chapter 22);

❑ cycle routes – these provide dedicated infrastructure for cyclists and thus extend the range of opportunities for safer cycling (see Chapter 23);

❑ guided buses – these provide a lower cost alternative to light rail and can provide greater flexibility in that the buses are able to operate on normal roads in suburban areas (see Chapter 24);

❑ light rail systems – these have been seen as an alternative to conventional railway provision. Although generally similar to conventional rail, light rail systems can also operate on–street, have more frequent stops and achieve better penetration of town centres (see Chapter 34);

❑ conventional (heavy) rail systems – these are now largely limited to the re–opening of closed railway lines and the provision of new stations;

❑ terminals and interchanges – these provide a means of extending the coverage of public transport services, by reducing the time and discomfort involved in achieving interchange between services;

❑ park–and–ride stations – these extend the catchment of fixed track public transport into lower density areas, by enabling car drivers to drive to stations on the main line. Bus park–and–ride can also enhance access to town centres;

❑ new off–street car parks – since lack of parking also acts as a restraint on car-use, any expansion

Infrastructure Measure	Accessibility	Efficiency	Environment	Equity	Finance	Practicability	Safety
Pedestrian facilities	XX	X	✓✓/X	✓	XX	X	✓✓
Cycle routes	✓	0	0	✓	X	0	✓✓
Guided bus	✓✓?	✓?	✓?	✓✓	X	X?	✓
Light rail	✓✓✓	✓✓	✓✓/X	✓	XXX	XX	✓
Heavy rail	✓✓	✓✓	✓✓	✓	XX	XX	✓✓
Terminals/Interchanges	✓/X	✓/X	✓	0	XX	X	✓
Park and ride	✓	✓	✓	0	X	0	✓
Parking supply	✓?	✓?	✓	X	X	X	✓
New roads	✓?	✓?	✓/X	XX	XXX	X	✓✓
Lorry parks	0	0	✓	0	X	0	✓
Trans–shipment	?	?	?	0	X?	XX	0
Other freight modes	?	0	?	0	XX	XXX	?

Key: ✓   ✓✓   ✓✓✓   Positive impact of increasing scale

X   XX   XXX   Negative impact of increasing scale

✓/X   Both positive and negative impacts

?   Uncertain impact

0   No significant impact

Table 6.2: Performance of infrastructure measures. Source: Guidelines on Developing Urban Transport Strategies, IHT (1996).

may encourage additional car–use and lead to increasing congestion. New off–street parking is probably, therefore, best combined with a reduction in on–street parking (see Chapter 19);

❑ new road construction – new roads that bypass sensitive areas can achieve environmental improvements. The DOT's Bypass Demonstration Project (DOT, 1995a) has shown that the effects of bypasses can be enhanced, if they are part of a wider package of measures. Overall, the contributions of new roads in urban areas are limited and, for this reason, the Department of Transport has made it clear that, in general, the problems of an urban area cannot be solved by a programme of road construction alone (DOT, 1995b; and IHT, 1995) (see also Chapter 11);

❑ lorry parks – these provide a means of reducing the environmental impact of on–street overnight parking of lorries (see Chapter 25);

❑ trans–shipment facilities – these aim to provide a means of transferring goods from larger to smaller, less environmentally intrusive, vehicles (ie break–bulk) for distribution in congested town and city centres (see Chapter 5); and

❑ encouragement of other modes for freight – this is likely to focus primarily on rail–borne freight but may also extend to the use of water–borne freight and pipelines (DOE/DOT, 1995b) [Sb].

Table 6.2 summarises the impact of each of these measures against the main objectives set out in Chapter 3.

6.5 Management Measures

A wide variety of management measures can be included in any transportation strategy. These include physical measures (eg traffic calming), legal or regulatory measures (eg bus– only lanes), technical measures (eg intelligent transport systems), financial measures (eg road– use pricing) and social measures (eg car sharing). To a large degree, management measures help to maximise the efficiency of existing transport systems, commensurate with the policy for the area, rather than an approach which relies on new road or rail construction.

Seventeen management measures can be identified, as follows:

❑ conventional traffic management – this includes a wide range of measures normally aimed at increasing the efficiency of the road network. A major practical consideration with all traffic management is that of enforcement. Unless measures are self–enforcing, the costs of enforcement action need to be included in any appraisal (see Chapter 11);

❑ traffic–calming and speed–reduction measures – these are designed to reduce the adverse environmental and safety impacts of vehicles (see Chapter 20);

❑ urban traffic control (UTC) systems – these are a special form of traffic management which aim to co–ordinate traffic signal–control over a wide area (see Chapter 41);

❑ accident remedial measures – these cover a wide

Management Measure	Accessibility	Efficiency	Environment	Equity	Finance	Practicability	Safety
Traffic Management		✓✓?	✓/X	✓/X	0	0	✓✓
Traffic calming	X?	X	✓✓/X	✓/X	XX	0	✓✓
Urban traffic control	✓	✓✓	✓	0	XX	0	✓
Accident remedies	0	0	0	0	0	0	✓✓✓
Parking controls	✓/X	✓	✓	✓/X	0	X	✓✓
ITS	?	?	?	?	?	?	?
Physical restrictions	X?	XX?	X?	?	0	X?	✓?
Regulatory restrictions	✓✓/XX	✓?	✓✓	✓/X	X	XX	✓✓
Car sharing	✓	0	0	0	0	X	0
Bus priorities	✓/X	✓✓	?	✓	0	X	✓
HOV lanes	✓/X	✓?	?	✓	0	X	✓
PT service levels	✓✓	✓	✓	✓	XX	X	✓
Bus service management	✓✓	✓✓	0	✓	X	XX	0
Cycle lanes	✓/X	0	0	✓	0	0	✓✓
Cycle parking	✓?	0	✓?	✓	0	0	✓?
Pedestrian crossings	✓	✓	?	✓	0	0	✓✓
Lorry routes, bans	X	X	✓/X	✓/X	0	X	✓

Key:

✓ ✓✓ ✓✓✓	Positive impact of increasing scale	✓/X	Both positive and negative impacts
X XX XXX	Negative impact of increasing scale	?	Uncertain impact
		0	No significant impact

Table 6.3: Performance of management measures.

range of measures to reduce accidents (see Chapter 16);

❑ parking controls – potentially, these provide an effective way of controlling car–use in terms of trip–end restraint. Controls can include reducing the supply of spaces, restricting the duration or opening hours and regulating use through permits or charging (see Chapter 19);

❑ intelligent transport systems (ITS) – these include a wide range of information and communication technology applied to transport, much of which has been developed under European Commission's research programmes (Keen 1992) (see also Chapter 18);

❑ physical restrictions on car–use – the re–allocation of roadspace to other traffic can reduce car–use. Measures can include extensive pedestrian areas, traffic calming and the use of bus–only lanes (Vincent et al, 1978; and IHT, 1992);

❑ regulatory restrictions on car–use – regulations, including the use of bans and permits, have been used in several European cities, as a way of reducing car–use (GLC, 1979);

❑ car sharing – this offers a means of reducing car traffic, while retaining many of the advantages of private car travel (DOE/DOT, 1995b) [Sb] (see also Section 6.3 on Company Transport Plans, and Chapter 5) ;

❑ bus priorities – these enable buses to bypass congested traffic and thus offer bus–users reduced and more reliable journey times (see Chapter 24);

❑ high occupancy vehicle (HOV) lanes – these can extend the use of bus–only lanes to other vehicles to make more effective use of scarce roadspace (see Chapter 25);

❑ bus and rail service schedules – these can be modified to increase patronage and hence lead to a modal shift away from cars (Webster et al, 1980);

❑ bus service management measures – these can be designed to improve the reliability of bus services, to enhance their quality of service or to reduce operating costs (see Chapter 24);

❑ cycle–lanes and priorities – these serve the same function as cycle–routes and provide safer and more convenient facilities for cyclists (see Section 6.4; and Chapter 23);

❑ secure cycle–parking – this can assist a modal shift to cycling by providing secure and accessible facilities. Some local authorities have also set cycle–parking standards for new developments, in the same way as they set car–parking standards (see Chapter 23);

❑ pedestrian crossing facilities – these not only enhance safety but may also reduce travel–time for pedestrians (see Chapter 22); and

❑ lorry routes and bans – these are primarily designed to reduce the environmental intrusion of heavy lorries (see Chapter 25).

Information Measure	Accessibility	Efficiency	Environment	Equity	Finance	Practicability	Safety
Direction signing	0	✓	0	0	X	0	?
Variable message signs	0	✓?	0	0	X	0	0
Driver information	✓	✓?	0	XX	XX		0
Parking information	✓	✓	?	0	X	0	?
Travel awareness	?	?	?	?	0	0	?
Timetables	✓✓	✓	?	✓✓	0	XX	?
Passenger information	✓	✓	?	✓	XX	XX	?
Operational information	✓	✓✓	0	✓	XX	XX	0
Fleet management	✓	✓	0	0	X	X	0
Telecommunications	0	✓	✓✓	0	X?	X?	✓

Key:
✓ ✓✓ ✓✓✓	Positive impact of increasing scale	✓/X	Both positive and negative impacts
X XX XXX	Negative impact of increasing scale	?	Uncertain impact
		0	No significant impact

Table 6.4: Performance of information measures. Source: Guidelines on Developing Urban Transport Strategies, IHT (1996).

Table 6.3 summarises the impact of each of these measures against the main objectives set out in Chapter 3.

6.6 Information Provision (including Travel–Awareness Campaigns)

Information measures are essential adjuncts to management measures, in that their main aim is to use information to allow the most efficient use of the existing system. They can be categorised as either pre–trip or in–trip. In–trip measures can be further divided into in–vehicle or roadside measures.

Ten information measures can be identified, as follows:

❏ conventional direction–signing – this can provide benefits to all road–users, by reducing journey lengths and travel times (see Chapter 15);

❏ variable message signs (VMS) – these enable drivers to be diverted away from currently known, but unpredictable, areas of congestion. They can also be used to set speed–limits during congested time–periods, thus maintaining overall throughput (see Chapters 15 and 18);

❏ real–time driver information systems and route–guidance – these are rapidly developing forms of Intelligent Transport Systems (ITS) applications. Information is used to provide in–vehicle radio or display messages or to indicate preferred routes to avoid congestion. Dynamic route guidance systems can provide recommended routes allowing for both the vehicle's destination and the prevailing traffic conditions (see Chapters 15 and 18);

❏ parking information systems – these are a further ITS application designed to reduce the high levels of traffic searching for parking spaces in urban areas (see Chapters 15 and 19);

❏ travel–awareness campaigns – these have recently been developed by several local authorities, with the aim of making residents, employees, and particularly car–users more aware of the effects of their travel behaviour on the environment and of the alternatives to car–use that are available (see DOE/DOT, 1995b) [Sb];

❏ timetable and other service information – this is the basic form of information to public transport users (see Chapter 15);

❏ real–time passenger information – this is now being provided at some major terminals, individual stations, bus stops and on–vehicle. The main purpose is to reduce the uncertainty and stress associated with the late–running of services (see Chapter 15);

❏ public transport operational information systems – these use ITS–based fleet–management facilities to identify the locations of buses and to reschedule services to reduce the impact of unreliability (Keen, 1992);

❏ fleet–management systems – these have been widely introduced for freight vehicles, enabling them to respond more rapidly to the changing demands of 'just–in–time' delivery schedules (Keen, 1992); and

❏ telecommunications services – these can, increasingly, provide an alternative to travel through teleworking, on–line shopping and teleconferencing (DOE/DOT, 1995b) [Sb] (see also Chapter 5).

Pricing Measure	Accessibility	Efficiency	Environment	Equity	Finance	Practicability	Safety
Parking charges	/X	✓?	✓?	✓	✓	X	✓?
Congestion charges	✓✓/X	✓✓	✓✓/X	✓✓/X	✓✓✓	XX	✓
Fare levels	✓✓	✓	✓	✓✓	XXX	X	✓
Fare structures	✓✓	✓✓?	✓	✓✓	X	X	✓
Concessionary fares	✓	0	0	✓✓	XX	0	0
Ownership taxes	0	X	X	✓	✓✓✓	0	0
Fuel taxes	0	✓	✓✓	✓	✓✓✓	0	0
Company car tax changes	0	✓✓	✓✓	✓✓	✓✓	0	✓✓

Key:

✓	✓✓	✓✓✓	Positive impact of increasing scale	✓/X	Both positive and negative impacts
X	XX	XXX	Negative impact of increasing scale	?	Uncertain impact
				0	No significant impact

Table 6.5: Performance of pricing measures. Source: Guidelines on Developing Urban Transport Strategies, IHT (1996).

Table 6.4 summarises the impact of each of these measures against the main objectives set out in Chapter 3.

6.7 Pricing Measures

Many professionals have argued that pricing measures should be an integral part of any transportation strategy and that "...if congestion is to be reduced, the choice of method lies between means which increase the cost of private car–use, reduce the cost of public transport use or both" (May et al, 1990b).

Eight pricing measures can be identified, as follows:
❑ parking charges – these provide one of the more widely used forms of parking control and enable demand to be kept at a level below the supply of spaces (see Chapter 19);
❑ congestion charging – this has been proposed in a number of forms, including screen–line or cordon charging, differential licences, toll–rings or fully automated electronic charging. The key barriers to the implementation of congestion charging are concerns about public acceptance and the fact that primary legislation would be needed to permit its introduction (see Chapter 21);
❑ fare levels – these can be adjusted to alter patronage and influence modal shift from car– use. Fares are, to a large extent, outside local authority control but less so where Passenger Transport Authorities exist (Webster et al ,1980);
❑ fares structures – these include the introduction of flat or zonal fares, lower off–peak fares, travelcards and season tickets (Gilbert et al, 1991);
❑ concessionary fares – these provide lower fares, or free travel, for passengers with special needs, such as school children, elderly people and people with disabilities (Goodwin et al ,1988);
❑ vehicle–ownership taxes – these include the annual vehicle excise duty (VED) on road vehicles but are the responsibility of central Government. As a fixed tax, their main impact is on car–ownership, with little or no effect on car use;
❑ fuel taxes – these have a more direct effect on car–use but in a non–selective way. Once again, they are the responsibility of central Government (RCEP, 1994); and
❑ company car taxation – this is also the responsibility of central Government (Lex, 1995) and the general approach is that benefits in kind should be treated in line with payments of income. The system is not intended to subsidise private cars or to provide incentives to drive further but often has these effects.

Table 6.5 summarises the impact of each of these measures against the main objectives set out in Chapter 3.

6.8 Integration of measures

The previous sections highlight the fact that no one measure on its own is likely to provide a solution to urban transportation problems (IHT, 1992). Any strategy will, therefore, need to combine a series of measures together, to achieve the overall vision and objectives for the area. A package of measures can achieve synergy between them, so that the overall benefits are greater than the sum of the parts.

Integrating measures into a package can achieve benefits in three key ways:
❑ first, measures can be brought together which complement one another – for example, the encouragement of new developments in association with rail investment;

Strategy Measures	Development control	PT infra-structure	Park and ride	Parking supply	Highways	Traffic manage-ment	Traffic calming	Parking control	Bus priorities	PT service levels	Information systems
Development control		C			C					C	
PT infrastructure	C							C/P			
Park-and-ride		C						C			
Parking supply	C						C	C			
Highways	C						C		C		
Traffic management	C						C		C	C	C
Traffic calming					C			C/P			
Parking control	C	C	C	C					C	C	C
Bus priorities						C	C			C	
PT service levels	C		C				C	C/P			
Information systems			C			C	C	C		C	
Parking charges		C/F	C/F	C/F		C	C/F	C		C/F	
Road-use pricing		C/F	C/F	F	F		C/F		C	C/F	C
PT fares		C/F	C/F	C			C/P	C	C	C	
Fuel prices		C/F	F	F	F		C/F			C/F	

Table 6.6: A Matrix of interactions between strategy measures. Source: Guidelines on Developing Urban Transport Strategies, IHT (1996) (based on May and Roberts [1995]).

Key: Measures in the left hand column can reinforce the measures in the appropriate column by:
C – complementing it
F – providing finance for it
P – making it more publicly acceptable.

❑ secondly, some measures make other elements of the strategy financially feasible – for example, fares revenue can provide finance for new infrastructure; and

❑ thirdly, some measures are likely to be more publicly acceptable, if combined with others – for example, attitudinal research has indicated that road–use pricing is likely to be more acceptable if some of the revenue generated is used to improve public transport.

Table 6.6 shows the interactions between the main measures discussed in Sections 6.3 to 6.7.

Illustrative Case–Studies of Transportation Strategy Development

There are many examples of transportation strategy development since the late 1980s. This section provides brief references to some illustrative examples, from both the UK and elsewhere in Europe, of typical transportation strategy packages at a variety of levels.

National. Detailed comparisons exist of the practice and administration of national transport policy in Denmark, France, Germany, the Netherlands and Great Britain (Harman, 1995) and of national transport policy in the Netherlands, including the relationship between objectives and targets (de Jong, 1995).

Strategic. Case–studies, at the regional and strategic level in Europe, and 'best practice' principles and techniques for a range of policy measures have been developed (Harman, 1995). In the UK, there have been the major strategic 'scenario–testing' studies for a number of urban areas, generally districts or groups of districts within the major conurbations, such as the Black Country, Coventry, London (May et al, 1990a), Birmingham (Jones et al, 1990) and Edinburgh (May et al, 1992). There is also the study, commissioned by the Government Office for London, of a series of land–use and transportation scenario tests of London and the South East, using the MEPLAN land–use and transportation model (Marcial Echenique and Partners, 1995).

Large Cities. The Transport Research Laboratory (Dasgupta et al, 1994) has investigated the impacts of a range of transport polices in five cities: Leeds, Bristol, Sheffield, Derby and Reading. These looked at six broad policy approaches, namely: halving public transport fares; raising fuel costs by 50%; doubling parking charges; halving the number of parking spaces; applying a cordon charge to the central area; and a 'composite' test, combining halving public transport fares with doubling parking charges. Details also exist of the Leeds strategy (Steer, 1995) and of case studies of traffic and transport in seven European cities (Jones, 1995), which also reviewed six main transport policy–elements, namely, major road investment, public transport provision, traffic management, parking, freight transport and walking and cycling.

Smaller Urban Areas. The report of the DOT's Bypass Demonstration Project (DOT, 1995a) presents case–studies of how the effect of a bypass on a town's environment can be enhanced, if local authorities take advantage of the wide range of opportunities which arise when 'through' traffic is removed or substantially reduced. The six towns studied were Whitchurch, Market Harborough, Petersfield, Wadebridge, Berkhampstead and Dalton–in– Furness. Details also exist of similar studies of free–standing towns, such as Norwich and Ipswich (Coombe, 1995) and Bury St. Edmunds (Coombe *et al*, 1990).

6.9 Funding and Implementation

Attributes of a successful package

Whilst it is not possible to provide a detailed approach to the production of package bids, the IHT Guidelines (IHT, 1996) argue that the following ingredients are likely to add to the chances of success for a Transport Policies and Programme (TPP) Package submission (DOT, 1995b) [NIc] [Wa]:

❑ evidence that transportation planning decisions have been inter–related with land–use planning decisions, ideally encapsulated within an approved development plan;
❑ sufficient insights into the overall economic and environmental consequences of implementing the whole transportation strategy over a credible time–frame;
❑ detailed information concerning the economic and environmental consequences of implementing those individual components which have been included within the Package Bid;
❑ knowledge that likely funding agencies have been identified and broad compatibility demonstrated between their aspirations for transportation and those of the overall strategy;
❑ evidence that opportunities to maximise private sector financial contributions have been fully exploited;
❑ reference to realistic transportation objectives and information on both the mechanics of monitoring and the willingness to make adjustments within the overall transportation strategy, based on the monitoring results of individual components;
❑ confirmation of the significant involvement of members of the public throughout the development of the transportation strategy and evidence that public opinion is broadly in favour of the overall strategy and tolerant of the component schemes which form the current Package Bid;
❑ evidence that the necessary and appropriate level of traffic restraint within the overall strategy has been identified, together with the practical means of achieving this; and
❑ evidence that the Package represents value for money.

Phasing Implementation

A typical transportation strategy will involve a considerable amount of capital expenditure and a large number of concerted actions, spread out over a number of years and possibly involving a range of organisations. Phasing of the implementation of the transportation strategy is, therefore, very important and is often influenced by the availability and sources of funding. The IHT Guidelines (IHT, 1996) point out that, in determining the phasing of a strategy's component parts, it will be necessary:

❑ to decide when each component is required, by analysing when the problems, at which the component is aimed, are likely to emerge;
❑ to understand the relationships between the various components of the strategy, taking account of which elements must come before or after others;
❑ to estimate, realistically, the lead times required to progress each component, taking account of planning procedures and of design and construction times;
❑ to take account also of the capabilities of the transport providers to deliver schemes at the required times; and
❑ to reconcile the ideal sequence of implementation with the likely flow of funding.

Sources of funding

Many potential sources of funding exist for the various elements of a transportation strategy, including income generated from elements of the strategy itself, such as parking or road–use charges and fares. In particular, it is important that the funding sources are used in a way which is consistent with the overall strategy. Details of the funding of transport infrastructure are set out in Section 4.8 and in the IHT Guidelines (IHT, 1996). This has a list of potential funding sources, under the headings of:

❑ Central Government support – covering Annual Capital Guidelines (ACG), Supplementary Credit Approvals (SCA), Transport Supplementary Grant (TSG), Industrial Development Act 1982 grants and grants under Section 56 of the Transport Act 1968;
❑ European sources – covering the European Regional Development Fund (ERDF);
❑ specific funds and Challenges – covering the Single Regeneration Budget (SRB), Capital Challenge, etc; and
❑ private sector contributions, including developers' contributions and the Private Finance Initiative (PFI).

Monitoring, Review and Updating

A transportation strategy should be re–evaluated during the course of its implementation, given that this is likely to take 15 to 20 years. It is sensible, therefore, to include strategy re– evaluation into the overall process from the outset. This creates a dynamic process, with discrete re–evaluation stages phased into the implementation, alongside careful monitoring against the agreed objectives of the strategy. Given the need for a flexible and dynamic approach to strategy development and implementation, it is important to ensure that adequate resources are available to ensure that the monitoring, review and updating process can take place.

6.10 Summary

In conclusion, as the IHT Guidelines indicate, the following issues require attention during the development and implementation of a transportation strategy:

❑ land–use and transportation planning must be seen as complementary components of the overall approach to the management of urban areas;

❑ public opinion and ideas should be sought and used throughout the development of each unique transportation strategy;

❑ there must be a realistic expectation that sources of funding, commensurate with the transport investment required, will be forthcoming over the strategy implementation period and these must be sought over a wide array of possible public, private and mixed funding mechanisms;

❑ realistic transportation targets for each strategy should be set and monitored at regular intervals;

❑ the dynamics of strategy implementation need to be understood and promoted by all those involved in the process;

❑ property blight needs to be identified and, as far as possible, minimised at every stage of strategy development;

❑ the use of limited resources to develop and monitor a transportation strategy should be carefully planned; and

❑ a transportation strategy should not be seen as a 'once and for all activity' but as a continuous and evolving process.

6.11 References

Coombe D (1995) 'Transport policy in free–standing towns', Proceedings of the Institution of Civil Engineers, Transport 111 (3).

Coombe R D, Goodwin R P and Turner D R (1990) 'Planning the transport system for the historic town of Bury St Edmunds', Traffic Engineering + Control 31(1).

Dasgupta M, Oldfield R, Sharman K and Webster V (1994) 'Impact of transport policies in five cities', Project Report 107, TRL.

de Jong M A (1995) 'National transport policy in the Netherlands', Proceedings of the Institution of Civil Engineers, Transport 111 (3).

DOE/DOT (1993) 'Reducing transport emissions through planning', Stationery Office.

DOE/DOT (1994) Planning Policy Guidance PPG13 'Transport', Stationery Office [Sb] [Wb].

DOE/DOT (1995a) 'Implementation of PPG13. Interim Report', DOE/DOT [Wb].

DOE/DOT (1995b) 'PPG13: A Guide to Better Practice', Stationery Office [Wb].

DOT (1977) 'Some effects of flexible working hours on traffic conditions at a large office complex', DOT.

DOT (1995a) 'Better Places Through Bypasses – The Report of the Bypass Demonstration Project', Stationery Office [Sb].

DOT (1995b) 'Transport Policies and Programme Submissions for 1996–97: Supplementary Guidance Notes on the Package Approach', DOT [Sb].

DOT (1995c) 'Transport: The Way Ahead', DOT.

DOT (1996) 'Transport: The Way Forward', Stationery Office [Sa].

Gilbert C L and Jalilian H (1991) 'The demand for travel and travelcards on London Regional Transport', Journal of Transport Economies and Policy 25(1).

Goodwin P B et al (1988) 'Bus Trip Generation from Concessionary Fares Schemes: A

	Study of Six Towns', Report RR 127, TRL.
GLC (1979)	'Area Control', Greater London Council.
Harman R (1995)	'New Directions: a manual of European best practice in transport planning', Transport 2000.
IHT (1992)	'Traffic congestion in urban areas', The Institution of Highways & Transportation.
IHT (1995)	'The role of new and improved roads in future transport policies', The Institution of Highways & Transportation.
IHT(1996)	'Guidelines on Developing Urban Transport Strategies', The Institution of Highways & Transportation.
Jones D, May A D and Wenban–Smith A (1990)	'Integrated transport studies: lessons from the Birmingham study', Traffic Engineering + Control 31(11).
Jones P (1995)	'Study of Policies in Overseas Cities for Traffic and Transport (SPOTT)', University of Westminster.
Keen K (1992)	'European Community research and technology development on advanced road transport informatics', Traffic Engineering + Control 33(4).
Lex (1995)	'Lex Report on Motoring: What Drives the Company Motorist?', Lex.
Marcial Echénique and Partners (1995)	'LASER Scenario Tests for London', Marcial Echénique and Partners.
May AD and Gardner KE (1990a)	'Transport policy for London in 2001: the case for an integrated approach', Transportation 16(4).
May AD, Guest P W and Gardner KE (1990b)	'Can rail–based policies relieve urban traffic congestion?' Traffic Engineering +Control 31 (7).
May AD, Roberts M and Mason P (1992)	'The development of transport strategies for Edinburgh', Proceedings of the Institution of Civil Engineers, Paper 9865.
May AD and Roberts M (1995)	'The design of integrated transport strategies', Transport Policy 2(2).
RCEP (1994)	'Transport and the Environment' Royal Commission on Environmental Pollution 18th Report, Stationery Office.
Steer J (1995)	'The formulation of transport policies in British conurbations', Proceedings of the Institution of Civil Engineers, Transport 111 (3).
Vincent RA andLayfield RE (1978)	'Nottingham Zones and Collar Experiment: The Overall Assessment', LR 805 TRL.
Webster FV et al (1980)	'The Demand for Public Transport', TRL.

Chapter 7 Data Collection

7.1 Introduction

Throughout the design process, from the initial identification of a problem to the final selection of the most appropriate remedy, data are required to assist the judgments being made at every stage. This process requires:

❑ information about the current state of the transport system and how it has changed, and is changing, over time (for problem identification);

❑ specification of alternative design standards and their implications, for application to scheme proposals;

❑ forecasts of the effects that each proposed scheme is likely to have, considered against its objectives, as well as any side-effects that are foreseen; and

❑ the values (and any priorities or weightings) to be used in assessing the overall impact of each scheme on different sections of the community.

Some of these data will be specific to the particular areas in which these problems are occurring and to the alternative schemes being considered. This requires up-to-date local data from scheme-specific surveys and from any regular monitoring surveys which are relevant.

Survey Techniques

A range of survey types is available for different purposes, as set out in Table 7.1. Typically, surveys can cover three broad areas; traffic characteristics, network characteristics and scheme impacts.

Data collection methods differ widely according to the type of survey being undertaken. The needs of traffic-component assessment can usually be met by observational techniques, those of network assessment and travel demand by interview surveys and those requiring personal and qualitative response by interactive methods of attitude measurement, based upon psychometric or market research techniques.

In recent years, the development of 'revealed preference' and 'stated preference' techniques has enabled complex personal trade-offs and their distribution among the population to be more clearly understood, in the context both of the demand for movement by different modes and of the different impacts of transport.

Monitoring Data

The importance of continuous monitoring should not be overlooked. The main objectives of a monitoring programme are to have available the relevant data to allow periodic assessment of transport-related issues likely to be raised within an area and to monitor the performance of existing networks or specific initiatives. It is, therefore, advisable to commit resources to a regular programme of monitoring that provides this information and enables trends to be established.

Ad–hoc sample surveys provide useful data for specific problems but their output may be difficult to integrate within a comprehensive time-series data bank. Thus, regular monitoring surveys can provide the means to relate various ad–hoc surveys to a more substantial base (eg using factors to convert to average annual flows) with an appreciation of the confidence levels of the estimates. Where the data collected are largely qualitative in nature, they can often be interpreted to infer long–term travel–trends or preferences amongst the public. Examples would include attitudinal research surveys or even the results of public consultation. This latter example is a particularly important source of qualitative (and quantitative) data which is dealt with separately in Chapter 10.

The Traffic Appraisal Manual (DOT, 1991) gives details of all the data sources provided from national surveys and of the appropriate methods for converting sample counts into equivalent traffic flow estimates for design purposes. These data are derived mainly from core and rotating censuses and give national factors. Care is required in applying national factors to small areas.

Travel–demand Data

Several other sources of travel–demand data are relevant to urban transportation studies. Examples are: the National Population Census, which gives information on journeys to work; the National Travel Survey (DOT, 1996), which provides information on household travel–patterns; accident databases, compiled from the `STATS 19' record forms (DOT, 1994a) or a locally used variant; and a wide range of planning databases, held at both local and national level and essential for transportation modelling.

Table 7.1: Types and purposes of surveys.

	Types of Survey		Traffic assessment								Network assessment					Community impact assessment		
			Vehicle flow characteristics				Vehicle manoeuvres			Pedestrians' crossing and delay	Pedestrian and cycle movements	Goods movements O-D	Person trips O-D (by mode)	Journey times / operating costs	Accidents	Impacts on environment	Attitudes and choice criteria	Impacts on travel behaviour and activities
			Volume	Classification	Speed	Saturation flows	Turning flows	Parking	Goods access									
Observational	Inventory records		Core and rotating census data	Core and rotating census data	Core and rotating census data							Planning data/household census/NTS, etc.	Planning data/household census/NTS, etc.	Network data NIS, TARA, etc.	Police stats 19 records			
	Continuous monitoring		Automatic sensors-loop-detectors, radar, microwave, infra-red etc.	Automatic sensors	Automatic sensors	Automatic sensors	Automatic sensors	Automatic sensors								Air quality measurements		
	Sample surveys		Manual counts, portable event-recorders, etc. / Video recording image processing							Patrol surveys cordon counts / Video recordings	"Floating" observer	Number-plate matching	Number-plate matching	"Floating" observer methods	Conflict measurement	Noise and pollution measurement		
Interview	Postal questionnaires								Delivery records		Travel diary	Vehicle logs	Travel diary	Vehicle logs			self-completion questionnaires	Travel and activity diary
	Roadside / on-board questionnaires											Specific trip data using portable data capture devices, etc	Specific trip data using portable data capture devices, etc	Specific trip data using portable data capture devices, etc			Subjective response measurements-attribute-scaling semantic differential, etc.	Stated Preference (SP) techniques.
	Home interviews										Travel diary	Delivery records	Travel diary	Travel diary			Repertory grid/Delphi techniques	Household role and decision models
Interactive	Group discussions												HATS gaming-simulation			Public meetings		

70

Many local highway authorities have established road network information systems for monitoring conditions and assessing priorities for the management of their road networks. Developments in information technology are helping to improve the range and accessibility of the information held, as well as offering new opportunities for bespoke systems using proprietary software.

Fundamental to all information systems and to any kind of continuously–updated record of conditions is an accurate basis of referencing. This should include locations and times at which items of data are collected or events of interest occur. Some locational referencing systems are described below.

7.2 Locational Referencing for Road–Based Information Systems

Locational referencing can be achieved either on a geographical area basis or in relation to fixed points on the road network. The location of traffic data can be specified in terms of:
  ❑ the Ordnance Survey Grid Reference (OSGR) system – this covers Britain and is thus capable of giving a unique reference to any location or area;
  ❑ Royal Mail postcodes – these are unique codes allocated to each major property or small group of residential properties throughout the UK;
  ❑ a zoning system based upon local authority boundaries; or
  ❑ by reference to known fixed points on the road network (eg A1 at 100m north of junction with the B100).

Ordnance Survey Grid References (OSGR)

Various procedures have been developed to establish gazetteers (ie lists of addresses that relate to any given zoning system). As an example, proprietary computer software is available which can link addresses and postcodes to their OSGR and local authority ward. This facility can be adapted to allocate addresses to user–specified digitised zoning systems. These zones can be used for referencing information on journeys (eg the origin and destination of a trip).

Road Networks

It is common for traffic–related data to be referenced to the road network, which can be specified as links (ie sections of highway with reasonably homogenous characteristics) and nodes (ie junctions or points where changes in link characteristics occur). In most systems, the nodes are given a reference number, frequently the OSGR location, and links are then specified by the numbers of the nodes at each end. These network codes can then be used to store information relating to highway characteristics, traffic flows, accidents, maintenance records and so on.

Geographic Information Systems (GIS)

Advances in both computer hardware and software technology have led to the increasing use of geographic information systems (GIS). In simple terms, a GIS contains a computerised map on which various database information can be held, displayed, manipulated and reported. Locational referencing is frequently carried out using ordnance survey grid references (OSGR). There are a number of proprietary GIS software packages on the market, which can be tailored to suit the requirements of individual users.

In terms of traffic–related data, a GIS might be used in connection with a traffic–count database, which covers a large number of roads and contains comprehensive historical records. The system could also be used, for example:
  ❑ to identify traffic–count locations and the nature and quantity of data available;
  ❑ to highlight a particular location, or locations, and provide access to the raw data; and
  ❑ to manipulate the raw data to provide, for example, average daily traffic flow levels, heavy goods vehicle content and year–on–year traffic growth trends.

In addition to traffic flows, databases of road standards and accident statistics could be incorporated along with highway inventory information, such as the locations of street furniture and road signs, and maintenance records.

Software is also available that combines the function of GIS and computer–aided design (CAD). As an example, a GIS may contain a database of accidents linked to junction design. By selecting the location of a particular accident, or cluster of accidents, it would be possible to make an inspection of the detailed geometric and engineering design which may have contributed to the poor safety record. The GIS is therefore, potentially, a powerful tool which can provide all the information required to manage a highway network.

7.3 Sources of National Inventory Data

The Department of Transport holds a large amount of data, collected in the course of scheme appraisals [NIa]. More generally, available information held by the Department includes:

❑ the Network Information System (NIS), held by Highways Computing Division; and

❑ the National Traffic Census (NTC) (core and rotating traffic census) data, held by STC Division.

The NIS road network describes the motorway and major road network as a series of digitised links. Certain items of interest are recorded for motorways and trunk roads. These include the prevailing speed–limit, road type (ie whether single or dual carriageway) and road class.

Traffic counts are taken on each of the links making up the NIS network and estimates of annual average flows, classified by vehicle type, are made. The information is held by the Department's Statistics Division 'C' and a subset of the data, annual average daily flows (AADF) of all motor vehicles and the proportions which are heavy goods vehicles, are held in the NIS database. A number of local highway authorities operate similar databases for their own areas.

Publication of Transport Statistics

General information on transport and travel–data can be obtained from a number of publications produced periodically by the Stationery Office (Transport Statistics Reports) and the Department of Transport (Statistics Bulletins) [NIb]. Stationery Office publications include Transport Statistics Great Britain (DOT, annual a); Road Traffic Statistics Great Britain (DOT, annual b) and Road Accidents Great Britain (DOT, annual c), all produced on an annual basis, and the results of the periodic National Travel Survey (DOT, 1996). Of the Department of Transport publications, quarterly bulletins relating to traffic statistics are available which provide a useful source of up–to–date information [Wa].

7.4 Recording of Accidents

(see also Chapter 16)

Details of all injury accidents reported to the police are transcribed onto coding sheets in accordance with the STATS 19 form specified by the Department of Transport (DOT, 1994a) [NIc]. This form is divided into three sections:

❑ attendant circumstances, giving details of the site (eg location, date, prevailing road conditions and weather);

❑ vehicle records, giving details of each vehicle involved; and

❑ casualty records, giving details of each casualty involved.

Full details of the variables and their value in the form are given in the booklet STATS 20. The data are provided by the appropriate Local Processing Authority (LPA), which may be the police force, County Council, Metropolitan District Council or the Scottish or Welsh Office, and are also transferred to the Department of Transport, where they are held for Britain as a whole in a central–data bank. The STATS 19 data are restricted to those reported accidents that involve personal injury. It is estimated that about 4% of accidents involving serious injury are not reported and it is known that, where only slight injury is involved, the records are far less complete (see Chapter 16). In cases where insurance details and proof of vehicle ownership can be exchanged, there is no legal requirement to report an injury accident to the police. Reported accidents are defined and classified as slight, serious or fatal according to the most severe casualty in the accident. Definitions of these categories are, as follows:

❑ 'slight injury' – injuries of a minor nature, such as sprains, bruises or cuts not judged to be severe, or slight shock requiring only roadside attention (medical treatment is not a pre–requisite for an injury to be defined as slight);

❑ 'serious injury' – injuries for which a person is detained in hospital, as an in–patient, or any of the following injuries, whether or not the injured person is detained in hospital; fractures, concussion, internal injuries, crushing, severe cuts and lacerations, severe general shock requiring medical treatment and injuries which result in death more than 30 days after the accident. The 'serious' category, therefore, covers a very broad range of injuries (it is estimated that up to 50% of people with reported serious injuries are not detained in hospital); and

❑ 'fatal injury' – injuries which cause death either immediately or at any time up to 30 days after the accident.

Accident Severity Ratio

Public concern about the occurrence of fatal accidents is understandable and is partly reflected in the high monetary cost attributed to them. However, in any particular locality, fatal accidents may occur in numbers that are too small and variable to give a reliable indication of the accident situation on a localised basis. For this reason, the combined number of serious and fatal accidents is often used as an indicator and should usually be analysed in terms of the involvement of different classes of road–user (eg pedestrians, pedal cyclists). Numbers of slight accidents, though subject to greater uncertainty in reporting, can then provide an indicator of the relative severity of accidents by comparing the ratio of fatal and serious accidents with all accidents;

Severity Ratio (SR)= $\dfrac{\text{Number of Fatal or Serious Accidents per year}}{\text{Total Number of Injury Accidents per year}}$

The ratio will be influenced by the protective attitude of the occupants of vehicles towards vehicles involved in the accident and the aggressive characteristics of the involved vehicles towards vulnerable road–users, if they are also involved in the accident. It will also be influenced by the road environment where the accident occurs and the traffic level on that stretch of the road.

Accident Rates

The frequency of accidents at a particular location (number of injury accidents per year) is not an appropriate indicator of risk, as it takes no account of the degree of exposure to risk. For example, a large number of accidents may simply reflect a large volume of traffic. For this reason, accident reporting is often expressed in terms of accident–rates (ie injury accidents per unit of vehicle movement or total distance travelled). Rates are normally expressed in personal–injury accidents per 100 million vehicle–kilometres. In practice, it is sometimes found that accident–rates can bias investigations towards low traffic flow sites. In some circumstances, it may be better to use the 'Potential for Accident Reduction' (PAR) approach. PAR is designed to estimate, from data at similar sites, the number of accidents expected at the particular site in question, according to its layout and prevailing traffic conditions (McGuigan, 1983). However, accident 'causation factors' are not recorded on the STATS 19 form and are not always related to the limited number of physical features which are recorded (the case for their collection is being assessed). Local data is needed therefore; but, even so, this method must be used with care and with the perception afforded by experienced practitioners.

Processing Accident Data

Accident data records, which are normally stored on computer file for analysis purposes, contain the principal details from the STATS 19 form. Processing and manipulating the raw data usually follows standard computer procedures, often integrated within a GIS [NId]. These perform four basic functions:

❑ assignment of each accident to a node, link, cell or road section on the road network, as defined within an authority's representation of the road system for the purposes of accident location. Within this process, it is important to verify positional accuracy since police STATS 19 accident reports are often imprecise;
❑ extractions of standard tables, showing trends in accidents for the area as a whole, trends for specific categories of accident or accidents at specific site categories;
❑ plotting of the spatial distribution of accidents over the network. This usually reveals clusters of accidents at problem sites and on routes where systematic treatment may be desirable; and
❑ problem site (or cluster) analysis, in which the individual links, nodes, cells or road sections can be examined further.

Lists of links, nodes and road sections are usually compiled in descending order of accident–rate frequency, so that the larger clusters can be identified easily (DOT, 1986). Care must be taken to allow for the fact that sites having the largest numbers of accidents in a given period will usually include a number of sites where the occurrence of accidents has been above average in that period, as a result of random fluctuation. Proper use of the PAR technique, or other methods of identifying sites for application of safety measures, should take this into account.

Accident Analysis

The design of appropriate remedial measures normally involves a detailed analysis of each candidate site, in the form of a 'grid' or 'stick' diagram. These show the characteristics of individual accidents in successive columns, together with a diagram indicating the nature of the conflict. This process can be completed by hand or by computer–based methods, where accident data can be plotted to a relatively fine degree of detail, which is often helpful in planning remedial work. However, there is little point in trying to plot the data to a level of detail which is finer than that attained in the police reporting system. Further, more comprehensive, information on Road Safety is given in Chapter 16.

7.5 Continuous Monitoring

Transport policies, contained within structure plans and unitary development plans, are necessarily expressed in broad terms, mainly because they deal with long–term aims and broad approaches to meeting them. In practice, policies are continually interpreted and translated into specific programmes of short–term action. Both of these activities rely on monitoring the state of the transport system. Monitoring may also reveal that changes, either taking place or in prospect, justify reconsideration of the basic policies themselves.

A system for monitoring the components of traffic (eg, flow, speed and classification) requires a structured sample to be taken from within the study area. In order to achieve this, the road network may

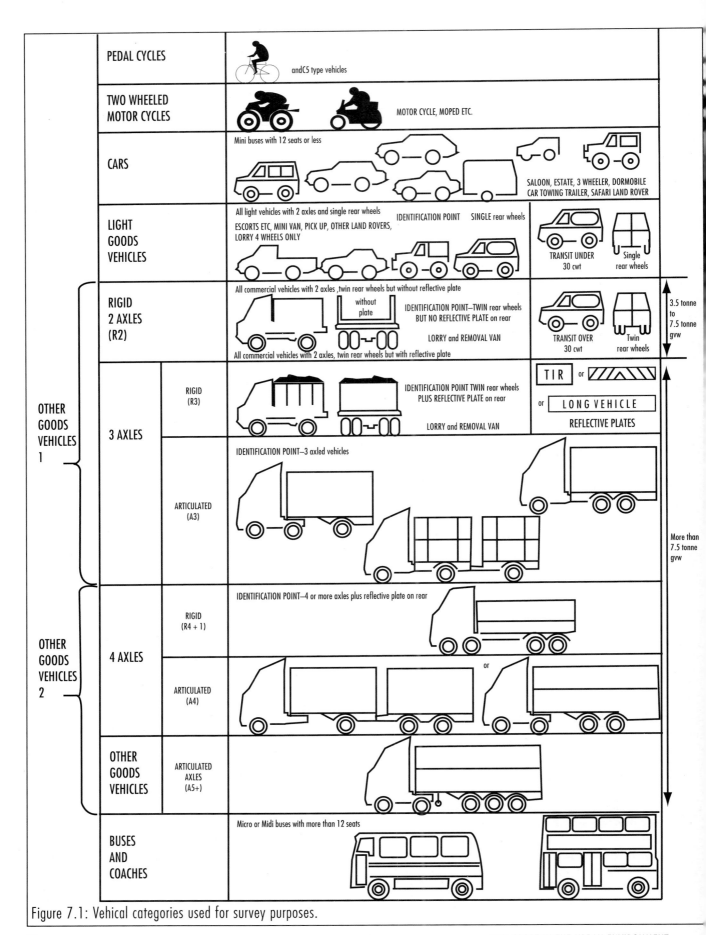

Figure 7.1: Vehical categories used for survey purposes.

be divided into short sections (sometimes only 100 metres long), each section being identified by its type (eg motorway, primary distributor, district or local distributor). The required number of census points for a representative sample is then chosen for each of the road–type sections or for the cordon or screenline. Traffic counting is carried out either continuously at a site or by establishing a rota for a programme of counts. The data from randomly–selected count locations not only provides information on the amount of traffic demand (vehicle–km/day) within the area but may also be used to provide factors to convert sample short–period traffic flows (vehicles/hour), on any link in the network to standard measures. For example, sample counts may be expanded to produce annual average daily traffic (AADT) flows (vehicles/day) for assessment purposes.

National Traffic Data

Most highway authorities have their own systems for monitoring traffic but the Department of Transport performs this task, on a national basis, with its core and rotating (link based) traffic censuses [NIe]. The core census consists of some 130 randomly–selected sites, covering all road classes, which are continuously monitored by automatic counters, which classify by vehicle–type. From these data, expansion factors can be calculated to convert short period counts to their equivalent annual average daily flows (AADF).

The rotating (or link–based) census consists of a comprehensive set of short period counts on every link of the major road network. Each link is visited every three years (in a few cases, at shorter intervals) with about a third of the links (approximately 4000) visited each year. At a randomly–chosen point on each link, counts of each of eleven vehicle types are taken (pedal cycles, two–wheeled motor vehicles, cars and taxis, buses and coaches, light vans, and six separate categories of goods vehicle) for the 12 hours from 0700 hours to 1900 hours.

Figure 7.1 summarises the vehicle–type categories, which are consistent with the Department of Transport's COBA 10 classification (DOT, 1966b). Whilst this has been used informally, as a benchmark, care should be taken, when assembling data from different sources, to ensure consistency in the description of vehicle–types.

These counts are all scheduled to take place on weekdays but not on those at or near to public holidays. In order to minimise the effects of possible seasonal factors, counting is confined to the so-called 'neutral' months (April, May, June, September and October). These counts are converted into AADF, using factors derived from the core census. For those links which are not counted in the year, an estimate of the AADF is made by applying a growth factor, again from the core census, to the previous year's estimate of the AADF. Thus, the Department holds AADF traffic flow estimates for nearly every link in the major road network, although there are always a few links which are not covered, either because they are new links or because a recent change in the road classification of the link invalidates the use of an earlier count.

Traffic flow data can be obtained from the DOT's Directorate of Statistics, which can provide the following for each rotating census count point, on magnetic tape or floppy disc:
- ❑ the original 12–hour count;
- ❑ annual average 24–hour weekday flow; and
- ❑ annual average 24–hour flow.

Subject to the availability of resources, the Department is prepared to produce new analyses from either the rotating or core censuses to meet the specific requirements of clients.

7.6 Inaccuracy and Variability in Traffic Surveys

Automatic Traffic Counts

Little definitive work has been published about the long–term accuracy of traffic counts by automatic traffic counters. Practical experience suggests that errors arise from machine failures, detection problems and poor installation. For longer term counts, the frequency and diligence of survey station monitoring and servicing are crucial to reliability. Permanently installed inductive loops should be more reliable than pneumatic tubes. The best working estimate of the accuracy of measurement of the number of vehicles that pass an automatic traffic counter is that the 95% confidence interval (for a count of longer than 12 hours duration) is of the order of plus or minus five percent of the total count obtained.

All counters should be installed and maintained to the standards laid down in the Manual of Practice on Automatic Traffic Counting (DOT, 1981). When a short–term automatic count is used to predict the average traffic flow for a period longer than the counter was on station, the estimated traffic flow will be subject to sampling errors (see Section 7.7 below).

Manual Traffic Counts

A statistical study of the reliability of manual

classified counts has been carried out by the Department of Transport and is reported in a paper entitled Accuracy of Manual Road Traffic Counts (DOT, 1979a). The conclusion reached was that the true 16–hour flow of all motor vehicles at a given site lies, with 95% confidence, within an interval of about plus or minus 10% of the manual count.

The 95% confidence intervals for some individual vehicle classes were:
- ❏ light goods vehicles   +/– 24%;
- ❏ other goods vehicles   +/– 28%; and
- ❏ all (light plus other) goods vehicles   +/– 18%.

The relatively wider confidence intervals for individual goods vehicle classes results largely from mis–classification between them. The accuracy of the total number of goods vehicles is therefore better than that for an individual class. The confidence intervals for individual hours are likely to be larger but the 16–hour figures shown above can be taken as a guide. The Department of Transport's Highways, Economics and Traffic Appraisal (HETA) Division uses computer–based methods to process count data and to yield the coefficient of variation of any estimate made from any number of 12– or 16–hour counts. The factors are derived from the core census data (see Section 7.5 above).

7.7 Sampling Procedures and Techniques

The problems of scaling–up (or factoring) the number of actual observations, to the total which is representative of the population being measured, is common to all sample surveys. In urban areas, traffic flow through each day, week and year can be very variable and it is, therefore, important to take account of this. Most urban traffic counts can be converted to AADF, in the same way as counts from the link–based census, and Statistics Bulletin 86(7) (DOT, 1979b) describes techniques which rely only on the built–up/non built–up classification of urban roads. When AADF estimates are required from short–period counts, standard expansion factors will give annual average flows and their associated coefficients of variation. The coefficient of variation, expressed as a percentage of the flow, gives a measure of the uncertainty surrounding the estimated AADF.

Every scaling factor has an associated unreliability and the result of factoring is always to worsen the overall confidence interval. Factoring should therefore be kept to a minimum and the factor with the lowest coefficient of variation should always be chosen, where a choice of factors is available. Whilst

it is, in principle, desirable to derive factors locally, a fuller understanding of the accuracy of such factors is necessary, to ensure that local conditions are actually significantly different from national averages. In the absence of this knowledge, national factors should be used. The Department's cost–benefit package, COBA 10 [Sa], gives 'default' values for the scaling factors.

Peak and Seasonal Variations

The difference between the peak–hour flow and the annual average hourly flow (AAHF) is of interest to highway engineers, for various aspects of link and junction design. Table 7.2 gives the factors to convert from AAHF to the peak–hour flow (PHF) and corresponding hourly flows, ranging from 10th highest to 200th highest, for three road types.

7.8 Sample Observational Counts Surveys

The most common requirements of observational surveys are to obtain information on the volumes and types of vehicle passing a particular point on a road link or negotiating a particular junction.

Passage Counts

Different traffic streams can be counted manually by enumerators, using traditional forms and hand–held tally counters or by inputting data directly into portable hand–held electronic data–capture devices. Whatever method is used, sufficient numbers of enumerators should always be employed to ensure adequate cover of the different movements and to allow for regular breaks from what can be a monotonous task. Alternatively, data can be obtained automatically, using detector systems located in or by the carriageway, such as inductive–loop detectors or image–processing.

If manual traffic count methods are undertaken for a sample time–period, it may be necessary to use temporary automatic equipment to collect traffic flows over the whole period, as a `control'. This equipment can operate continuously and unattended. At temporary sites, pneumatic, tribo– or piezo–electric cable sensors may be placed across the carriageway to register the number of axles (and possibly their loads) passing in any particular time–period. It is also possible to use detector loop 'mats' which are laid directly on top of the road surface, as a temporary counter, although these are not often used. Electronic counters are used in association with the detectors, which store the information on solid state RAM. The equipment should be checked at regular intervals, to confirm that

Conversion of Hourly Flow	Types of Road					
	Main Urban		Inter Urban		Recreational Inter–Urban	
	Factor	Coeff* (%)	Factor	Coeff* (%)	Factor	Coeff* (%)
AAHF to PHF	2.630	(11)	2.825	(15)	3.890	(23)
AAHF to 10th highest hour	2.83	(14)	3.231	(20)	4.400	(23)
AAHF to 30th highest hour	2.703	(11)	3.017	(17)	4.974	(21)
AAHF to 50th highest hour	2.649	(10)	2.891	(15)	3.742	(19)
AAHF to 100th highest hour	2.549	(9)	2.711	(12)	3.381	(15)
AAHF to 200th highest hour	2.424	(9)	2.501	(9)	3.042	(13)
AAWF to AADF	0.943	(3)	0.979	(4)	1.015	(4)

Notes: AAHF is assumed to be AADF ÷ 24

AAWF is the Average Annual Weekday Flow (Monday–Friday)

*Coefficient of Variation

Source: Traffic Appraisal Manual (TAM) DOT (1991).

the sensors are still in place and that the counter is working correctly and has sufficient power to last until the next visit. Data can be collected from the counter periodically, by down–loading information onto a data–capture device ,such as a lap–top computer or data module.

At more permanent sites, it may be preferable to install detector loops in the road surface to avoid the heavy maintenance costs associated with temporary sites. At these locations, data may be stored by an electronic counter connected directly to the loops or may be sent to a central computer via a data–transmission line.

All of these automatic detection systems need to be checked at regular intervals by manual counts, to ensure that the recorded counts are compatible with visually observed information. Some sophisticated arrangements of detector systems have been used to count and classify individual vehicles and also to determine axle and gross vehicle–weights with piezo–electric sensors. Manually performed classified counts are still usually carried out to verify the accuracy of the automatic systems.

Junction Counts

The counting of turning movements at junctions may require a large number of field staff but the use of video equipment with subsequent laboratory analysis can also be considered. A video camera might be positioned at a suitable vantage point (eg in a neighbouring building or on a telescopic mast that provides the necessary field of view). If observers are employed, substantial numbers of them may be

Table 7.2 Peak–hour factors by road type classification.

necessary (eg when counting at a four–arm junction, there are twelve possible traffic movements). Saving on staff, by having each numerator observing more than one movement, can lead to a reduction in the quality of the data obtained. Experience suggests that greater accuracy is achieved when vehicles are counted as they leave the junction, because individual traffic streams are identified more easily at this point.

When information on movements within a complex junction, or over a large area, is required a 'number plate' survey may be appropriate. With this type of survey individual vehicles are identified usually by the numerical part of their registration number and a letter, usually the first letter of the alpha code, together with the time when they enter the survey zone. This information can be recorded by an observer, using either an enumeration form, a tape–recorder or an electronic portable hand–held data–capture device. Other observers, placed in a cordon around the junction (or area) note the registration numbers and times as the vehicles leave. Computer programs have been developed to match the registration numbers of vehicles entering and leaving the area in different time–segments, based on the estimated journey times through the zone. Usually, up to 80% of the identified vehicles can be matched in this way. Developments in image–processing techniques allow the recording of number plates to be done automatically, provided that cameras can be adequately positioned to observe the registration plates.

Pedestrian Counts

Pedestrian surveys are usually required to establish the flows along a footway or across a carriageway. The latter will often be required to quantify pedestrian/vehicle conflicts, when assessing the need to install some form of crossing facility. For this purpose, pedestrian counts will usually be carried out over a one hundred metre length of road, ie 50m on either side of the proposed crossing point. Fifty metres is taken as the maximum distance that pedestrians might reasonably be expected to walk to use a formal crossing place, rather than cross where they happen to be. The actual distance that pedestrians are willing to divert will also depend on the intervening traffic flow and on the existence of any physical barriers (eg guard-railing). Origins and destinations of pedestrian trips may only be obtained by personal interviews (see Section 7.9) but surveys of pedestrians' delays at crossing points can be carried out manually or by using video equipment (see Chapter 22).

Cycling Data

Information on existing cycling movements, other demand factors and suppressed demand can be obtained from a number of sources (IHT, 1996). The 1991 Census of Population provides highly accurate and comprehensive transport-to-work data. It is possible to analyse cycle-trips by origin (home) and destination (workplace), zoned as enumeration districts, wards or postcode sectors. The cycling data are a subset of the Transport-to-Work tables, which can be purchased. Short trips by other modes are an indicator of potential cycling trips. Classified traffic counts often provide information on cycle flows. However, where cycle flows are low relative to flows of other classes of vehicle, the results may be inaccurate and should be treated with caution. Cycles should be counted in all manual classified counts and the importance of recording cycles accurately should be explained to the enumerators. Automatic traffic count equipment is available that can detect cycles on segregated cycle-tracks.

Counts of Passengers using Public Transport

Information on the use of public transport can be obtained from manual counts of people boarding or alighting at different stops or from on-board interviews with passengers. Analysis of ticket sales can produce partial, and potentially biased, information, due to the increasing use of railcards and concessionary fares.

Manual counts of public transport modes are undertaken in many urban areas, both of the numbers of vehicles and the numbers of passengers. They can be carried out by either boarding and alighting counts at stops/stations along the whole length of a route (or a group of routes) or they can be conducted by enumerators actually on board each vehicle to be surveyed (see Photograph 7.1).

Photograph 7.1: Travel data for rail-interview off-train.

Other types of survey that are frequently carried out involve recording all buses or trains at a designated cordon and counting how many people are travelling on each vehicle or train. For convenience, the cordon is frequently drawn through a bus stop or station where the vehicles and trains are scheduled to stop in any case.

Another approach to counting public transport passengers (and vehicles) is referred to as 'terminal counts'. These are counts conducted at a terminal point, such as a main railway station, coach terminus or bus station. The survey involves counts of all passengers alighting and boarding vehicles at the terminal point. Terminal count surveys give an indication of the total number of passengers using an urban centre, although they do not take account of any through or cross-centre movements.

All the above surveys can be used for trend analysis or can be used as input to other evaluations, such as corridor analyses or before-and-after surveys.

Speed Measurement

When seeking the average speed of vehicular traffic, it is important to decide how speeds at a point on the road are to be measured. Options include:
❑ use of a radar speed-meter, averaging the individual speeds of vehicles directly; or
❑ timing vehicles over a short distance (L) and calculating the average time taken (t), giving an average speed of L/t.
Of these methods, the former would give the 'time' mean-speed (Vt) and the latter the 'space' mean-speed (Vs). Wardrop has shown that the two definitions of speed are related, thus:

$$V_t = V_s + s^2/V_s$$

where 's' is the standard deviation of the distribution of individual speeds, as measured by the method of timing vehicles over a short distance.

It is possible to calculate the time or space mean–speed, from either set of data, by converting individual speeds into times or vice–versa.

This distinction is important because, in practice, time mean–speed is used for accident analysis at particular sites or the determination of a speed–limit, whilst space mean–speed is used for economic analysis and other applications of speed/flow relationships.

A common method of determining the instantaneous vehicle–speed, measured at a point, is to use a radar speed–meter. The speed meter should be concealed behind street furniture or inside a conveniently parked vehicle, so that drivers are unaware of the observations and do not alter their normal behaviour as a consequence. Speeds can also be measured automatically, using inductive loops spaced a known distance apart and connected to an appropriate electronic counter and by image–processing techniques.

The usual way of measuring link, running or journey speeds is by the so–called 'moving observer' method, in which a car (or light van) travels along the route at the average speed of traffic, while observers record the time taken between different points and the periods during which the vehicle is stopped. A number of runs are necessary to obtain a good estimate for each period of the day being investigated. In this process, the driver attempts to ensure that he passes as many vehicles as pass him, in order to remove bias.

Speeds and Highway Design

The 85th percentile speed (ie the speed up to which 85 per cent of vehicles travel in free–flow conditions) is generally used as a basis for highway design (see Chapter 31). It can be used to determine:

❏ the design speed of minor improvement schemes, by measuring vehicular speeds on the approach to the improvement;
❏ the basis for the design of major/minor junctions;
❏ the basis for the settings of vehicle–actuated traffic signals, at sites with speed–limits of more than 30 miles/h; and
❏ appropriate values for speed–limits.

Surveys to Assess Urban Traffic Conditions

Assessment of urban traffic conditions can be carried out by direct observations, moving–car techniques, aerial photography, time–lapse cinematography and computer analysis of video–tape recordings. It will usually involve measuring one or more of the following:

❏ saturation flows at signal–controlled junctions (see Chapter 40);
❏ cyclic flow profiles (ie the average pattern of traffic flow on a road link during one signal–cycle);
❏ queue–lengths, which can be measured by observers noting at (say) one minute intervals the points at which the queue begins and ends (a distinction must be made between vehicles which are actually stopped and those which are crawling); and
❏ queueing time, as the time between the first stop to the last start but, if the queue is long, an allowance should be made for the time it would have taken to travel, at normal running speed, the length of road covered by the queue. As with queue–length measurements, it is important to distinguish between the time spent delayed (ie the time taken to decelerate to and accelerate from a stop, plus the time spent actually stopped) and the time spent stationary.

Car Parking Surveys

An inventory of the parking spaces available in an area, together with observations of the use made of them, is often required. The number of spaces, including details of where they are and whether they are privately or publicly used, can be recorded on a map of an appropriate scale. Often there is difficulty in establishing the precise number of spaces available. This can be either the number of marked–out parking spaces or the actual number of cars parked (which may be greater than the indicated spaces). The inventory should also include kerbside capacity (estimated where individual bays are not marked), spaces in public car parks and private spaces, including those within the curtilage of individual properties. Distinctions may be made between those spaces for which a charge is made, those with restricted use, such as private to non–residents (PNR), those for permit–holders only and those subject merely to time–limits.

A quick and inexpensive assessment of the demand for parking space in a particular area can be obtained by measuring the accumulation of traffic within the study area by time of day. Using automatic traffic counters, the net accumulation of vehicles entering and leaving the study area can be measured at appropriate time–intervals. The data can then be plotted as a graph, showing the accumulation of traffic for different times of the day, and this provides a good proxy for parking demand. The process may be repeated on different days, to determine the difference in demand for each day of the week or month. In most urban areas, parking demand varies significantly during the week for a variety of reasons. Knowledge of the variation in parking demand assists interpretation of parking occupancy and parking duration surveys, which are normally limited to one day for reasons of economy.

Parking occupancy (ie the number of spaces occupied in relation to the total available) can be obtained by observers patrolling on foot or in vehicles. Video–recording techniques are also feasible. Surveys may be used to compare different days of the week, different times of the day and the effects of different parking policies, when taken over suitable periods. Aerial photographs may also be used to determine parking occupancy but only of open, ground–level, parking areas.

When parking duration (ie the length of stay of individual vehicles) is being surveyed, the parking zone should be divided up into a number of patrols. The frequency of the patrol will depend on the land–use characteristics of the surrounding area. A typical patrol of 60 spaces might take an observer about 30 minutes to complete. Where the land–use generates short–term parking, the patrolling interval should be reduced to perhaps five minutes to achieve an acceptable level of accuracy. Portable data–capture terminals can be used by observers to improve the accuracy and effectiveness of the survey. An alternative is to use video recordings, taken from inside a moving vehicle or from a high vantage point. Parking duration and accumulation can then be determined by comparing consecutive recordings. Information on parking duration in off–street car parks can be obtained from most types of automatic entry/exit ticketing systems. These do not produce the same bias against short–stay parking as do periodic observation methods.

7.9 Origin–Destination Surveys

Given that the objective of an 'origin–destination' survey is actually to gain information on travel–demands by all modes to all activities within the study area, the following is a comprehensive list of data needs (Figure 7.2):
  ❏ AB – intra–zonal trips, within the study area;
  ❏ AC and DB – terminating trips from/to the study area, originated from or destined to a point inside the study area; and
  ❏ XY – extra–zonal through trips, passing through the study area.

While all of these trips can be made both by people who live in the area and by those who do not, in

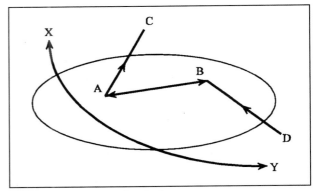

Figure 7.2 Trip–types for O–D surveys

general, travel surveys collect data samples in the following ways;

❏ household surveys	– collect most of the internal trips (AB) – though not all, – collect around half (sometimes less, depending on the study area) of the trips of type AC and DB, – collect none of those passing through the area (XY);
❏ external cordon surveys	– collect no intra–zonal type AB trips, – collect a sample of all type AC, DB and XY trips; and
❏ internal screenline surveys	– collect a mixed sample of all three types – (depending upon the location of the screenline).

Clearly, to get a proper representation of all travel, it is important to gather data from more than one of the above sources. However, a number of factors need to be taken into account:
  ❏ household surveys can collect a much broader range of data and, in general, there is more time to collect more accurate origin–destination data than at screenline or cordon surveys (see Section 7.10);
  ❏ screenline/cordon survey data can duplicate data collected at the household survey, so care needs to be taken when matching the data to household–based data;
  ❏ it is recommended that screenline/cordon data is collected by personal interview since it has been shown (Bonsall et al, 1993) that non–response bias is extreme in the case of self–completion methods;
  ❏ screenline/cordon data needs to be collected for all modes crossing the lines – not just motorised traffic; and

❑ it is essential that screenline/cordon data is collected over the same period as the household survey, since many urban areas experience mis-matches of data, due almost entirely to seasonal, weekly or time-of-day variations between household and other data.

The above recommendations are for comprehensive coverage rather than validation. While it is obviously important that each of the survey-types makes sense in relation to the other, that is not validation. True validation is done within each individual survey type and not between surveys, since each survey method has its own biases.

Each of the surveys mentioned above can be further classified into those which are car-based, those which are public transport based, those which focus on cyclists and pedestrians and those which are household-based.

The key issues in relation to car-based surveys, public transport-based surveys and non-vehicular surveys are:
❑ the sampling must be done rigorously and randomly;
❑ all non-response/refusals need to be recorded and as much information as possible from these retained;
❑ classification counts need to be undertaken over exactly the same period of time; and
❑ data correction/weighting procedures need to be implemented both for non-response and for expansion purposes.

7.10 Household Surveys

This section discusses the choices of survey method available for household-based travel surveys (Richardson *et al*, 1995). The task of selecting the appropriate survey method is crucial to the efficiency of the overall survey effort. The choice of survey method will usually be the result of a compromise between the objectives of the survey and the resources available for it. This compromise, or trade-off, can be neatly illustrated as shown in Figure 7.3.

A trade-off occurs because it is impossible to control all three of the major elements in Figure 7.3; at best, only two of the three can be controlled by the survey designer. Thus, given a fixed budget, as is normally the case, the selection of the survey method, with an associated degree of quality control, will automatically control the quantity of data which can be collected. Alternatively, within a fixed budget, specification of the quantity of data to be collected

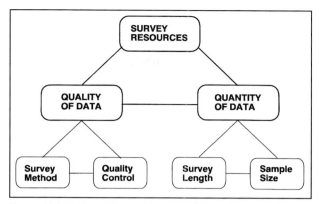

Figure 7.3 Trade-offs in selection of the survey method. Source: Richardson *et al* (1995).

will immediately dictate the quality of data which can be collected. That is, either lots of low quality data can be collected or a limited amount of higher quality data, for a given budget. Generally, the latter course of action is to be preferred.

In determining the total quantity of data to be collected, a further trade-off is present between the sample size from which data is collected and the amount of data collected from each respondent in the sample. Within a limited budget for coding, editing and analysis, it is necessary to trade-off the number of questions against the sample size; which one takes precedent will depend on the purposes of the survey and the length of the survey content list.

Essentially, three different data-collection techniques may be employed:
❑ household self-completion surveys;
❑ household personal interview surveys; and
❑ telephone surveys.

These survey methods vary in complexity, in the types of information which can feasibly be collected and in the level of interaction between the survey designer and the respondents in the survey.

Household Self-Completion Surveys
Self-completion questionnaire surveys are one of the most widely-used forms of survey technique in transportation studies. Self-completion surveys are defined as those in which the respondent completes a questionnaire without the assistance of an interviewer.
Several types of basic survey format can be described, depending on the methods used for collection and distribution of the questionnaire forms. These variations include:
❑ mail-out/mail-back surveys;
❑ delivered to respondent/mailed back; and
❑ delivered to respondent/collected from respondent.

Naturally, the increased response obtained in the two latter methods can only be obtained at considerable extra expense for the personal delivery and collection of the questionnaire forms. However, where a high response rate is essential, as in a National Census, then this method may be the most cost–effective way of obtaining these responses. This method is frequently used when 'long–term' travel diaries (eg 7–day diaries) are distributed (eg in the National Travel Survey in the UK) (Stopher, 1992).

Household Personal Interview Surveys

A personal interview survey is defined as one in which an interviewer is present to record the answers provided by the respondent to a series of questions posed by the interviewer. Personal interview surveys have long been associated with transportation data collection, with home interview surveys providing the major input to the large transportation studies of the 1960s, 1970s and 1980s. Most personal interviews now use computer–assisted personal interviewing (CAPI).

A household personal interview survey may be chosen, in preference to a self–completion survey, for several reasons:
❑ in general, higher response rates may be obtained from personal interview surveys than from self–completion surveys. Response rates of 75% to 80% are not uncommon;
❑ the personal interview survey allows for considerable flexibility in the type of information collected;
❑ the presence of an interviewer means that explanations can be given regarding the meaning of questions or the method in which answers are to be given, thus ensuring consistency of response;
❑ personal interview travel surveys can be carried out over a much shorter time–period than self–completion surveys, which need up to six weeks elapsed time to incorporate sufficient reminder notices into the survey procedure;
❑ since many surveys can be quite long, an interviewer can be effective in maintaining respondents' interest and in ensuring that the full set of questions is completed; and
❑ an interview situation is valuable where it is desired to obtain spontaneous answers from a particular individual.

While being particularly effective in several aspects of transportation data–collection, personal interview surveys are not without their own distinct disadvantages, viz:
❑ they are relatively costly; typically, three to ten times more expensive per returned questionnaire than a self–completion survey (Ampt *et al*, 1994). A consequential problem is the clustering of the

sample in order to keep the costs down;
❑ an interview situation is basically a human interaction between an interviewer and a respondent. Such interactions are rarely, if ever, completely neutral and free of bias; and
❑ they are not suited for situations where questions require a considered response or where factual information is required, which is not immediately available.

In summary, personal interviews are best for surveys where the concepts are complex or where there is a tricky series of sequencing required. They are more costly than their self–completion counterparts and have to be designed thoroughly, to minimise interviewers' bias, but their high response rates and their ability to be carried out within a relatively short time make them ideal in cases where high quality data is required within a medium time–frame. In many cases, a combination of self–completion and personal interview surveys will be the most cost–effective.

Telephone Surveys

The telephone survey have been used for many years outside the area of transportation, mainly in market research.

The growth of telephone interviewing in the 1970s and early 1980s led to the setting up of centralised telephone–interviewing installations for many surveys, a development which revolutionised telephone interviews. Dedicated telephone–interviewing facilities allow for stricter control and closer supervision of interviewers' behaviour than is possible with from–home telephone surveys or with personal interviews (Morton–Williams, 1993).

The telephone survey method has a number of advantages, which include:
❑ it offers the possibility of wide geographical coverage – particularly in a given urban area, where rates for phone calls frequently do not vary with distance;
❑ because telephone interviews are usually performed from a central location, it is possible to have much better supervision of the interviews and, thereby, to maintain a higher level of quality–control on the completed interviews;
❑ by centralising the interview facility, it is possible to use computer–assisted telephone interviewing (CATI). In this method, the interviewer reads the questions from the computer screen and then types the responses directly into the computer,as they are received over the phone; and
❑ telephone surveys are generally cheaper than

personal interviews because of the reduction in labour needed to conduct the survey and the absence of field and travel costs, associated with having interviewers in the field.

The telephone survey method, however, has some potentially serious disadvantages, viz:

❑ there is a limit on the length of survey which can be successfully completed over the telephone (Stopher, 1985);

❑ the number of people in a household with whom it is possible to carry out the interview is almost always limited to one;

❑ with an increasing amount of marketing being done by means of the telephone (some of which are disguised as sample surveys), it is becoming more and more difficult for serious survey researchers to establish their credibility at the beginning of an interview;

❑ because only those households with telephones can be included, the potential for a sample bias is obvious;

❑ because telephone books have usually been used to select the sample, the problem of phone-owning households who are ex-directory being excluded from the sample is added to other problems, such as out-date and non-phone-owning households;

❑ unlike other forms of survey, there is no chance of follow-up for non-respondents in a telephone survey; and

❑ by its very nature, no visual aids can be employed in such a survey.

While telephone surveys are seen by some as having significant potential in the collection of transportation survey data, they should be used with caution, especially for data which is not factual and straightforward.

7.11 Intercept Surveys

Intercept surveys are those which take place at a site which is not in a household – where people are intercepted in the course of carrying out an activity of some kind. They include surveys on-board public transport vehicles and at cordon-points on roads.

The major types of intercept surveys are

❑ on-board vehicle distribution/mail back – in many cases, it is desired to conduct a survey of a particular group of transport system users, eg bus passengers. To attempt to find these people by means of a general household survey would be almost impossible, because they represent such a small percentage of the total population. A more efficient method is to limit the population to include only those people and to use a survey method which will identify only members of that population. On-board vehicle surveys are an effective means of conducting such surveys, where surveyors ride on-board the vehicle and distribute questionnaire forms to all passengers on the vehicle. The passengers may then be required to fill out the questionnaire forms at their convenience and return them by post. A comprehensive description of such surveys is given in Stopher (1985). They have the advantage of being moderately cheap but the disadvantage of generating low response rates, since it is not possible to encourage or remind people to respond in any way;

❑ on-board vehicle distribution/on-board vehicle collection – as an alternative, such as for passengers on train journeys, it may be possible to collect the completed questionnaire forms before the respondents leave the vehicle. For some modes, this poses no particular problems because there will be ample time for passengers to complete the survey before the end of the trip;

❑ on-board distribution/collection plus mail-back – in some studies, hybrid on-board surveys, which combine elements of both the above methods have been used successfully (Sheskin et al, 1982 and Hensher, 1991). The method involves using a two-part questionnaire form. The first part is the more usual postcard style, clearly marked for filling out and return on the bus or train. The second part is a more lengthy form, to be taken away by the traveller, filled out later and mailed back. This method allows for considerably more information to be obtained than can be acquired from the standard on-board/mail-back method;

❑ roadside distribution/mail-back surveys – where the mode of transport under consideration is the private car, then the method of distribution to pin-point those users is often a roadside questionnaire survey. In this survey method, questionnaire forms are distributed to car-users, as they pass by a particular point or set of points on the road. To enable the questionnaire forms to be distributed, it is necessary that the cars are stationary at that point. This can be achieved in one of two ways; either at a natural feature of the roadway (such as a traffic signal or a toll-booth) or deliberately stopped at a census point, with the assistance of local police officers. After the drivers are stopped, they are given a questionnaire form and a brief explanation of the purpose of the survey. Respondents are then asked to fill out the questionnaire form at their convenience and return it by mail (Richardson *et al*, 1981); and

❑ intercept interviews – sometimes intercept surveys involve personal interviews with the drivers of vehicles or travellers as they are stopped at the census point. In these cases, the respondents are stopped by an interviewer who asks them a series of questions – usually about origin, destination, trip purpose and times of travel, with some details on socio–demographic status. The presence of an interviewer generally ensures a much higher response rate than other methods which involve mailing back a postcard.

All mail–back surveys involve an unquantifiable bias in the response rate and should only be used if no other approach is available.

7.12 Stated–Preference (SP) Surveys

SP exercises are a form of attitudinal survey. Two types of multi–dimensional scaling technique are of particular relevance to transport–choice analysis. The first involves the rating of an alternative, overall, by the application of a uni–dimensional scaling technique to a multi–dimensional object. This method is frequently used to ascertain how the uni–dimensional ratings of the individual attributes might be combined into an overall rating of the alternative.

The second method is known by various titles such as 'conjoint measurement' (Luce et al, 1964 and Krantz et al, 1971), 'information integration' (Anderson 1971 and 1974), 'functional measurement' (Meyer et al, 1978) and, in recent years, 'stated preference' or 'stated response' (Pearmain et al, 1991). The principal feature of each of these methods is that they seek the respondent's reaction to a series of hypothetical combinations of attribute levels. The set of questions is determined on the basis of an experimental design, which seeks to present a balanced set of situations to the respondent.

Stated–preference methods are particularly useful in two contexts:
❑ when a substantially new alternative is being introduced and there is little or no historical evidence of how people might respond to this new alternative; and
❑ when the investigator is trying to determine the separate effects of two variables on consumers' choices but where these two variables are highly correlated in practice.

The investigator has control over the combinations of attributes to which the subjects will respond, because of the manner in which the set of questions has been determined by an experimental design. This is particularly important in the second context listed above, because it enables the investigator to isolate the individual effects of the various attributes.

The design of the choice situations to be presented to the respondents is an important component of the overall design of stated preference surveys. Pearmain et al (1991) offer a simple example of an SP design, by considering a situation involving three attributes for a public transport service: fare, travel–time and frequency. If each attribute has only two levels (viz, high–low, slow–fast, frequent–infrequent), then there are eight different combinations of these options, as shown in Table 7.3.

The respondent could then be asked to rank these options in order of preference and, from the combined responses of a sample of respondents, the relative importance attached to fares, travel times and frequency could be determined. Importantly, because of the orthogonal nature of the experimental design (ie where each variable is independent of all other variables in the set of options presented to the respondent), the importance attached to each attribute is a true reflection of the separate effects of each attribute.

Option	Attributes of Public Transport		
	Fare	Travel–Time	Frequency
1	Low	Fast	Infrequent
2	Low	Fast	Frequent
3	Low	Slow	Infrequent
4	Low	Slow	Frequent
5	High	Fast	Infrequent
6	High	Fast	Frequent
7	High	Slow	Infrequent
8	High	Slow	Frequent

Table 7.3 A simple stated–preference experimental design.

As with uni–dimensional scales, the respondent may be asked to perform different tasks with the information presented. For example, they could be asked to:
❑ rank the alternatives in order of their preference;
❑ assign a rating to each alternative, to reflect their degree of preference;
❑ select the single alternative which they prefer the most; and
❑ select choices, in a paired comparison manner, from a series of two–way choice situations.

Each of these methods has its own strengths and weaknesses, both from the point of view of the respondent and the analyst.

One of the problems with stated–preference methods is that, for example, the set of options shown in Table 7.3 is extremely limited. It is likely that more than two levels of each of the attributes would need to be tested and, perhaps, more than three attributes would need to be evaluated. However, as the number of attributes and attribute levels increases, so too does the number of possible combinations. For example, to test three levels of three attributes would result in 27 combinations; three levels of four attributes would require 81 combinations; and so on. Clearly, it is impossible to expect respondents to be able to consider too many different situations. Kroes *et al* (1988) suggest that a maximum of 9 to 16 options is acceptable, with most current designs now adopting the lower end of this range. A maximum of 9 options for the respondent to consider severely limits the number of attributes that can be considered.

To overcome this limitation, and yet be able to consider more attributes and/or more attribute levels, it is necessary to adopt one of the following strategies (Pearmain *et al*, 1991):
❑ use a 'fractional factorial' design, whereby combinations of attributes which do not have significant interactions are omitted from the design. A significant interaction is said to exist when the combined effect of two attributes is significantly different from the combination of the independent individual effects of these two attributes; or
❑ remove those options that will 'dominate', or 'be dominated' by, all other options in the choice–set. For example, in Table 7.3, option 7 is dominated by all other options, while option 2 dominates all others. These options could be removed from the choice–set, on the assumption that all 'rational' respondents would always put option 2 first and option 7 last in any ranking, rating or comparison process; or

❑ separate the options into 'blocks', so that the full choice–set is completed by groups of respondents, but with each group responding to a different sub–set of options. Each group then responds to a full–factorial design within each sub–set of options and it is assumed that the responses from the different sub–groups will be sufficiently homogeneous that they can be combined to provide the full picture; or
❑ present a series of questions to each respondent, offering different sets of attributes but with at least one attribute common to all to enable comparisons to be made. Often the common attribute will be 'time' or 'cost' to enable all other attributes to be measured against easily understood dimensions; or
❑ define attributes in terms of differences between alternatives (eg travel–time difference between car and train). In this way, two attributes are reduced to one attribute in the experimental design. However, they may still be presented as separate attributes to the respondent on the questionnaire.

Adoption of one, or more, of the above strategies will allow more information to be obtained from stated–preference questionnaires, while keeping the task manageable for the respondent.

The major weakness of stated–preference methods, however, is that they seek the reactions of respondents to hypothetical situations and there is no guarantee that respondents would actually behave in this way, in practice. This is particularly the case if the respondents do not fully understand the nature of the alternatives being presented to them. Thus, a high premium is placed on high–quality questionnaire design and testing, to ensure that respondents fully understand the questions being put to them. Unfortunately, this does not always appear to be the case. While a lot of attention has been placed on refining the nature of experimental SP designs and on increasing the sophistication of the analysis techniques to be employed after the data has been collected, relatively little attention has been paid to improving the quality of the questions being put to the respondents. With few exceptions (eg Bradley *et al*, 1994), little research has focused on testing for methodological deficiencies in the survey techniques used to obtain SP data. There are numerous examples of stated preference questionnaires, in which the questions are almost unintelligible. Future work in this area must pay much greater attention to the quality of the survey instrument itself.

7.13 References

Ampt ES and Richardson AJ   'The Validity of Self–Completion Surveys for Collecting Travel

(1994) Behaviour Data', PTRC European Transport Forum, Transportation Planning Methods 2, pp 77–79, Warwick.

Anderson NH (1971) 'Integration Theory and Attitude Change', Psychological Review, (78).

Anderson NH (1974) 'Information Integration Theory: A Brief Survey' in DH Krantz, RC Atkinson, RD Luce and P Suppes (Eds) 'Contemporary Developments in Mathematical Psychology' (Vol 2), Freeman, San Francisco.

Bradley M and Daly A (1994) 'Use of the Logit Scaling Approach to Test for Rank–Order and Fatigue Effects in Stated Preference Data', Transportation 21(2), 167–184.

Bonsall PW and McKimm J (1993) 'Non–Response Bias in Roadside Mailback Surveys', Traffic Engineering + Control 34, pp 582–591.

DOT (annual a) 'Transport Statistics Great Britain', Stationery Office.

DOT (annual b) 'Road Traffic Statistics Great Britain', Stationery Office.

DOT (annual c) 'Road Accidents Great Britain', Stationery Office.

DOT (1979a) 'Accuracy of Manual Road Traffic Counts', DOT.

DOT (1979b) 'Methods of Calculating National, Regional and County Traffic (vehicle basis)', Statistical Bulletin 86 (7), DOT.

DOT (1981) 'Manual of Practice on Automatic Traffic Counting', DOT.

DOT (1986) 'Accident Investigation Manual' (Volumes 1 and 2), DOT.

DOT (1991) DMRB 12.1, 'Traffic Appraisal Manual – Manual of Recommended Practice for traffic forecasting in scheme appraisal on Trunk Roads', Stationery Office.

DOT (1994a) STATS 19 'Road Accident Report Form', DOT.

DOT (1994b) STATS 20 'Instructions for the Completion of Road Accident Reports', DOT.

DOT (1996a) 'National Travel Survey 1993–95', Stationery Office.

DOT (1996b) 'COBA 10 User Manual', DOT [Sa].

Hensher DA (1991) 'Hierarchical Stated Response Designs and Estimation in the Context of Bus User Preferences; A Case Study', in Logistics and Transportation Reviews, 26 (4), 299–323.

IHT (1996) 'Cycle–friendly Infrastructure', The Institution of Highways & Transportation.

Krantz DH and Tversky A (1971) 'Conjoint measurement Analysis of Composition Rules in Psychology', Psychological Review (78), 151–169.

Kroes E and Sheldon R (1988) 'Are there any Limits to the Amount Consumers are Prepared to Pay for Product Improvements?', 15th PTRC Summer Annual Meeting, The University of Bath, England.

Luce RD and Tukey JW (1964) 'Simultaneous Conjoint Measurement: A New Type of Fundamental Measurement', Journal of Mathematical Psychology (1), 1–27.

McGuigan DR (1983) 'Non–junction Accident Rates and their Use in Blackspot Identification', Traffic Engineering + Control 10 (23).

Meyer RJ, Levin IP and Louviere JJ (1978)' 'Functional Analysis of Mode Choice', presented at the 57th Annual Meeting of the Transportation Research Board, Washington, DC.

Morton–Williams J (1993) 'Interviewer Approaches', SCPR, Social and Community Planning Research, Dartmouth Publishing Co, Aldershot.

Pearmain D, Swanson J, Kroes E and Bradley M (1991) 'Stated Preference Techniques: A Guide to Practice', Steer Davies and Gleave and Hague Consulting Group.

Richardson AJ and Young W (1981) 'Spatial Relationships between Carpool Members' Trip Ends', Transportation Research Record (823) 1–7.

Richardson AJ, Ampt ES and Meyburg AJ (1995) 'Survey Methods for Transport Planning', Eucalyptus Press, Melbourne.

Sheskin IM and Stopher PR (1982) 'Surveillance and Monitoring of a Bus System', Transportation Research Record (862), 9–15.

Stopher PR (1985) 'The State-of-the-Art in Cross-Sectional Surveys in Transportation'. In ES Ampt, AJ Richardson and W Brog (Eds) 'New Survey Methods in Transport', VNU Science Press, Utrecht, The Netherlands, 55–76.

Stopher PR (1992) 'Use of an Activity-Based Diary to Collect Household Travel Data', Transportation 19 (2) 159–176.

7.14 Further Information

Ampt ES (1981) 'Some Recent Advances in Large Scale Travel Surveys', PTRC Summer Annual Meeting, The University of Warwick, UK.

Ampt ES (1989) 'Comparison of Self-Administered and Personal Interview Methods for the Collection of 24-Hour Travel Diaries', in Selected Proceedings of the Fifth World Conference on Transport Research, Vol.4. 'Contemporary Developments in Transport Modelling'. Western Periodicals Co: Ventura, Ca, D195–D206.

Brog W and Meyburg AH (1981) 'Consideration of Non-Response Effects in Large-Scale Mobility Surveys' Transportation Research Record (807), 39–46.

Frankel MR (1989) 'Current Research Practices; General Population Sampling Including Geodemographics' Journal of the Market Research Society 31 (4).

Hitlin RA, Spielberg F, Barber E and Andrle SJ (1987) 'A Comparison of Telephone and Door-to-Door Survey Results for Transit Market Research'. Presented at 66th Annual Meeting of the Transportation Research Board, Washington, DC.

LCC (1963) '1961 London Travel Survey Report', London County Council.

Richardson AJ (1986) 'The Correction of Sample Bias in Telephone Interview Travel Surveys'. Presented at 65th Annual Meeting of the Transportation Research Board, Washington DC.

Richardson AJ and Ampt ES (1993) 'South East Queensland Household Travel Final Report: All Study Areas', report to the Queensland Department of Transport. Transport Research Centre Working Paper TWP93/6, Melbourne.

Sammer G and Fallast K (1985) 'Effects of Various Population Groups and of Distribution and Return Methods on the Return of Questionnaires and the Quality of Answers in Large- Scale Travel Surveys', In ES Ampt, AJ Richardson and W Brog (Eds) 'New Survey Methods in Transport', VNU Science Press, Utrecht, The Netherlands 367–377.

Stopher PR and Sheskin IM (1982) 'Towards Improved Collection of 24-Hour Travel Records', Transportation Research Record (891), 10– 17.

Chapter 8 Estimating Travellers' Responses to Changes in the Transport System

8.1 The Role and Relevance of Transportation Modelling

One of the dilemmas facing practitioners is how best to predict the results or estimate the impacts of the various policies and measures, which emerge as part of the process of formulating a transportation strategy for an urban area. Clearly, it is highly advantageous to be able to predict the likely situation a number of years into the future, as this information can provide the basis for economic evaluation of the proposals and can help to establish their value for money, in absolute or comparative terms.

Unfortunately, the history of large–scale land-use/transportation models in the 1960s/70s has had some influence on current attitudes towards their use. Critics would argue that they are extremely time-consuming and expensive to produce, mainly because of the large amounts of survey data required. Also, the outputs, in terms of predicted levels of demand for movement and trip patterns for some specified date in the future, have sometimes proved to be unreliable, particularly with the advantage of hindsight!

Many reasons explain the apparent frailty of the modelling process, which may well extend beyond the suitability of the model itself. Problems may occur with the data–collection process and perhaps with the over–reliance on creating (synthesising) data to fill–in gaps in survey information. The forecasting stage may also prove fallible, since it relies on prediction of such factors as gross national product, related levels of economic activity and expectations of future land–use activity, all of which can be subject to unexpected occurrences. Beyond this lies the difficulty of predicting human behaviour, as evidenced by both trip–making levels and the way these trips are undertaken. As society gradually changes, so do its travel patterns.

All of these factors combine to make long–term forecasting and modelling a hazardous process. For this reason, it is now commonplace to make model–forecasts for a variety of future scenarios, to try to establish a range of feasible situations for which to plan and design. Proponents of modelling theory argue, quite reasonably, that some guide as to what the future might hold in terms of travel patterns is much better than pure conjecture.

Transportation models are, however, constructed for a variety of different purposes. Beyond the traditional 'four–step' land–use models (see Section 8.4), there are other models for assessing the effects of policy–changes (such as changing the volume and/or price of parking and the availability and price of public transport) and models for evaluating the more local traffic effects of particular measures.

The decision on whether or not to undertake a full–scale model analysis needs careful thought, because of the time and expense involved. Factors to be weighed include:

❑ the purpose of the study (eg policy evaluation, all–modes movement study or traffic prediction);
❑ the stage of the study (ie preliminary, feasibility, firm proposals or implementation);
❑ the size of the area being studied (ie major conurbation, city, town or local area);
❑ the overall cost of measures likely to be implemented (as a constraint on the proposals);
❑ the nature of the measures to be implemented (eg physical infrastructure; policy changes);
❑ the type and reliability of data already available; and
❑ the availability and ages of existing models.

For many types of study, some form of model will be essential in order to make an objective evaluation of alternative measures. Once a decision to use a model has been made, then it is necessary to select the type of model most appropriate to the task and to be very clear about what is required of it. The level of accuracy that the model output is likely to produce and the degree of uncertainty associated with the various stages in the modelling process should be fully understood by all those directly involved.

Alternative approaches to the use of mathematical models include framework analysis and trial–and–error evaluation. The process of framework analysis entails setting out the various evaluation parameters which are important to the study and making best estimates of the outcomes for the required timescale, by using either objective or subjective estimation. This process allows a structured comparative assessment to be made for the different scenarios.

For some projects, particularly those involving relatively low investment and which involve little risk if something goes wrong, it may be better simply to implement the scheme on an experimental basis and measure its effects on a 'before–and–after' basis. This is often satisfactory for small–scale traffic management projects, where mathematical modelling may be both time–consuming and likely to produce questionable results, arising from too many unknown factors.

8.2 Types of Transportation Model

Transportation models can be used for a variety of purposes:
- ❑ to forecast the overall demand for travel at some specified date in the future, given predicted changes in factors external to the transport system, such as population, employment and household income;
- ❑ to modify the forecasts of the overall demand for travel, by taking account of the constraints and opportunities provided by the transport system itself;
- ❑ to estimate the way in which land–use patterns could respond to changes in the levels of accessibility resulting from changes in the transport system;
- ❑ to allocate forecast demands for travel to the various modes of transport and, within each mode, to individual roads and public transport services;
- ❑ to calculate the levels of service offered by each mode, when the demands have been allocated to the individual parts of the transport system; and
- ❑ to provide information on vehicle and passenger flows and travel costs necessary for operational, environmental, economic and financial appraisals.

The kinds of transportation model which can be used in the development and appraisal of urban transportation strategies, policies and schemes are, as follows:
- ❑ 'transport demand models' – so called because of their original emphasis on estimating the unconstrained demand for travel;
- ❑ 'strategic' transportation models and 'policy appraisal' models – these also pay considerable attention to estimating the demand for travel; they ensure that the demand for travel matches, or is in equilibrium with, the capacity and level of service offered by the transport system, ie the 'supply';
- ❑ 'land–use/transport interaction' models – these take the notion of equilibrium still further, by forecasting changes in land–uses in response to changes in the accessibility provided by the transport system;
- ❑ 'traffic assignment' models – in contrast to the demand models, these focus on the modelling of

the 'supply' side of the transport system, that is, on the way in which the predicted demand for travel takes place on the transport system; and
- ❑ 'simplified demand' modelling techniques – these provide a compromise between the often complex demand modelling at the macro–level and the more detailed supply modelling.

Transportation models may be used to predict the effects of total transportation strategies, schemes or policies, either individually or in combination. Some kinds of model can be used to appraise schemes and policies, as well as strategies, while other models can deal with only one and not the rest. Section 8.10 summarises the types of urban transportation model and the circumstances under which they are most applicable (see also IHT, 1996).

8.3 An Overview of Transport–Demand Models

Transport–demand models aim to predict the amount of travel which would take place under a given set of assumptions. Broadly speaking, these assumptions can be separated into two components:
- ❑ those related to population, together with other external changes (for example, land–use, household income and car–ownership); and
- ❑ those related to characteristics of the transport system.

Population and land–uses will vary over time, as will the other factors, so transport demand needs to be related to a particular point in time. In addition, in making forecasts there may be different views on how the future population and land–use will develop, so that different assumptions (often in the form of scenarios) may need to be examined for the chosen year.

Changes in the transport system may, themselves, give rise to land–use changes. Although the majority of transportation models do assume independence, a particular characteristic of 'land–use/transport interaction' models (discussed in Section 8.6) is that they attempt to relate the two elements explicitly.

Generalised cost of travel
The majority of models operate on the basis that travel demand is a function of generalised cost (usually, a linear combination of journey time, money cost, walking and waiting time, etc), as perceived by the travellers themselves. Any changes in the transport system can then be represented by average changes in the components of generalised cost between specified geographical zones, at specific times and by specific modes.

Equilibrium between supply and demand

The term 'demand model' implies a procedure for predicting the likely travel–decisions people would wish to make, given the generalised costs to them of all alternatives.These decisions include choices of time of travel, destination, mode, route, frequency of trip, whether to travel in company or alone and even whether to travel at all (trip suppression).

However, if all predicted travel decisions were actually realised, the perceived generalised cost of any particular trip might not stay constant. This is where the 'supply' model comes in. Its function is to reflect how the transport system responds to a change in any given level of demand. The most well–known example is the deterioration in vehicle–speeds, as traffic volumes rise. However, there are a number of other factors, such as the effects of congestion on bus operations, overcrowding on trains and increased parking problems as demands increase.

A model with the built–in requirement that the demand for travel must be balanced against the performance of the network in servicing that level of demand is referred to as an 'equilibrium' model.

Level of detail

The essence of travel demand is the prediction of person–trips between a set of origins and destinations. It is usual to represent this matrix of movement as the flows between pairs of zones. For the majority of cases, these O – D flows can be considered to be the main output of the transport demand model. The question then is the appropriate level of detail to be used on such issues as:

❑ the time period(s) to which the travel matrices relate; ie daily travel or peak hour modelling only – and whether average conditions should be modelled, as opposed to a particular identified day, and, if so, what kind of average (eg average annual weekday, average day within a given month, etc), bearing in mind that average conditions can be exceeded 50% of the time;
❑ the spatial basis (eg the shapes and sizes of zones) at which it is desired to represent the origins and destinations of movements. The zoning system chosen is critical, since the modelling of some effects requires a high level of detail, while others can be much more broadly represented;
❑ the modes of transport by which the movements are made. In urban contexts, it is usual to distinguish between private and public modes (mixed modes like park–and–ride may present problems) but other possibilities, such as walking, cycling and taxis, are not always included. Differences in vehicle type, for example between cars, vans and other goods vehicles, may also be needed;

❑ the types of person making the movements: eg those who have access to cars and those who do not, those who pay different levels of price (for example, season–tickets for commuters, reduced fares for children and pensioners) and those with different levels of income; and
❑ the purposes for which the journeys are made, which relate both to the inherent need to undertake the journey and to constraints on the timing of the journey, such as school hours.

In addition to this detail relating to the matrices of demand, detail of the transport choices, which the model attempts to reflect, must be specified. Typical choices are destination, time of travel, route, mode, frequency and vehicle–occupancy. There is some trade off between the potential for reflecting such choices and the level of detail in the matrices. For example, if the matrices are produced for the whole day only, no scope will exist for considering time–of–travel shifts within the day.

It is not necessarily desirable to set all criteria at the maximum level of detail, as this will require an impractical amount of input data. The complexity and detail of any model must relate to the overall scale and objectives of the exercise. These questions and the resulting compromises are explored further below.

Data requirements

The process by which the structure of a model is 'fitted' to local conditions is known as calibration. Validation is the process whereby a calibrated model is tested, with relevant existing data, to see if it reproduces existing conditions. In practice, however, these become iterative processes, since a model giving poor validation will usually undergo further calibration which, in turn, has to make direct use of the validation data.

The data requirements for models are entirely dependent on the detail specified for the model and it is not possible to produce general rules. At one extreme, it may be argued that every relationship in the model needs to be calibrated against appropriate local data, while, at the other, it is possible (in theory at least) to take all the relationships from another model and merely apply them, for example, to local demographic and network information.

One option is to take a known level of travel–demand, based largely or wholly on observed data, and use the model merely to estimate changes from that base that would result from alternative transportation strategies. This is known as the 'incremental' form (also referred to as 'marginal' or 'pivot point').

Essentially, the base data about person–movements can be derived from three sources (see Chapter 7), as follows:

❑ household surveys – typically involving completion of travel–diaries by each household member;
❑ interviews in course of travel – in the form of either roadside interviews for journeys made by car and commercial vehicles or on–board surveys for public transport journeys – which typically relate only to the journey actually being intercepted; and
❑ counts of people or vehicles crossing particular points on pre–defined 'cordons' or 'screenlines', normally in a specified direction. While relatively cheap and straightforward to collect, these flow data will normally contain no information on trip–purpose, person–type, origin or destination.

Other data will be needed to calibrate the model, including:

❑ zonal populations and workplace numbers; and
❑ network specification – ie details of link–lengths, carriageway widths, average speeds by type of vehicle for roads, and frequency, stage journey times, fares, etc, for public transport.

In some cases, this may be supplemented by attitudinal or 'stated preference' data obtained from interviews with individuals. This is used when it is difficult, on the basis of observed data, to estimate the relevance and contribution of certain variables, for example, the introduction of new parking charges or a new bus service (see Section 7.12).

Finally, other sources of data can be used for validation purposes – for example, journey–time surveys may be used to validate flows on networks and public transport operators' estimates of revenue may be used to validate assumptions about fares and public transport demand.

8.4 Four–Stage Models

The so–called 'four–stage' model has its origins at the beginning of the 1960s and has retained its general form, while benefitting, since then, from considerable improvements in modelling and computing techniques. The reason for the survival of this model form lies in its logical appeal. The four stages relate to trip generation (production and attraction), trip distribution, modal split and traffic assignment. Each stage addresses an apposite question:

❑ how many person–movements will be made (ie trip generation)?
❑ from which origins (production) to which destinations (attractions) will they go (ie trip distribution)?

❑ in which proportions and by what modes will the travel be carried out (ie modal split)? and
❑ what routes or paths through the network will be taken (ie assignment)?

The earliest versions of the model were applied without taking account of variations in trip–purpose, person–type, etc and were often confined to highway networks only. A defining characteristic of a typical four–stage model is its fairly detailed network representation, with consequent implications for the number of zones. This limits the amount of disaggregation that is feasible, depending upon computing capacity or budget, but current versions do include a reasonable amount of variation in input variables.

Trip generation

The aim of the trip generation module is to predict the number of trips likely to enter and leave each zone. While this distinction serves to indicate the direction of the movement, a more useful distinction is between attraction and production of trips, respectively.

From this point of view, all demands for personal travel must be 'produced' at one end of the trip (typically a home) and then 'attracted' to a particular zone which can meet the purpose of the journey (typically a workplace, shopping centre, or school). When working on a production/attraction basis, it is rare that the two quantities are equally well known. With the notable exception of the journey to work, where the attractions are essentially given by the number of workplaces in each zone, it has proved far easier to develop models for productions than for attractions. For this reason, the normal convention is that the productions are taken as well–defined but the attractions are replaced by some measure of the relative attractiveness of different zones to satisfy the given journey–purpose.

The levels of detail (discussed in Section 8.3) are relevant to the modelling of trip productions. Typical assumptions, in the case of urban four–stage models, include all travel by mechanised modes for an annual average weekday, although certain aspects of the model (in particular, traffic assignment) may deal with more restricted time–periods, such as the morning peak–hour.

It is standard to distinguish between journey–to–work trips, in–the–course–of–business trips and other trips, although further distinctions (for example, education and shopping) are often made. As far as person–types are concerned, it is normal only to recognise different levels of household car–ownership but other distinctions could be made.

The standard approach to obtaining the required zonal trip totals is to derive relationships, which estimate household trip generation rates from observed data. The number of trips will be dependent on variables, such as the numbers of persons in the household, the numbers of employed persons and the numbers of children. At the zonal level, these can be related to quantities, such as the numbers of employed residents, the numbers of children of school age, total population and the numbers of households (Ortúzar et al, 1994).

The observed data can derive from household survey information, either specific to the local area or from national sources, such as the National Travel Survey, and demographic statistics from the National Census or local estimates (see Chapter 7) [NIa]. Trip production models have proved reasonably stable over time but the data that they require for application is not always readily available for future years. The result is that trip production procedures often contain additional sub–routines for setting up necessary input data: a particular example of this is the potential need for local predictions of car–ownership or the numbers of employed residents.

In concept, a trip generation model should be able to be applied to zonal planning data for any year/scenario. In practice, however, such models may not always deliver an acceptable fit for the base year at the zonal level and, therefore, an incremental approach may be used to forecast the zonal productions for a future scenario. This involves using the observed numbers of productions in each zone for the base year (assuming these are available) and then using the model, in conjunction with planning data for base and future scenarios, to derive growth–rates which are then applied to the observed productions.

Trip distribution

Whilst trip generation is at the start of the model process, the trip distribution and modal split components will follow but need not occur in a fixed sequence and, to achieve equilibrium, may need to be invoked several times within an iterative process.

The process for distribution modelling may be viewed as that of building up (or reproducing) a matrix of person–movements and, in general terms, the number of trips (T) in the (i–j) matrix cell is likely to be related to:

❑ characteristics of the origin/production zone (i);
❑ characteristics of the destination/attraction zone (j); and
❑ characteristics of the separation, ie the perceived 'generalised cost' of travel, between zones i and j.

This suggests a model of t
$$T_{ij} = a_i b_j f_{ij}$$

Separate models are us specified journey purpose.

Some early forms of this function:
$$f_{ij} = d_{ij}^{-n}$$

where d_{ij} is the distance between i and j and n is the 'distance deterrent' exponent.

The general form of the model is widely known as the 'gravity' model because of similarity to Newton's law of gravitational attraction, even though the strictly Newtonian formula for f_{ij} (where n = 2) is no longer used. Several alternative functional forms have been suggested for f_{ij}, of which the simplest and most well–known is:
$$f_{ij} = \exp(-\lambda c_{ij})$$

where c_{ij} is the perceived generalised cost between i and j and λ is a positive valued parameter, variously referred to as the 'concentration' or 'sale' or 'spread' parameter. This is known as the 'deterrence function'. A common problem with deterrence functions is that they are attempting to explain a wide range of variation (effectively, the distribution pattern among N^2 cells, where N is the number of zones), using a very small number of parameters. Hence, the distribution component of the four–stage model may not deliver an estimated trip matrix which is sufficiently robust to carry forward to the remaining stages of the model, particularly where calibration of the deterrence function is poor, due to lack of data.

For this reason, it is sometimes necessary to make more use of data from observed origin/destination matrices. This can be done either on an incremental basis, so that the model merely predicts changes relative to an observed base, or by introducing a number of specific constants (sometimes referred to as K–factors), to ensure a satisfactory fit of the estimated trips to observed data in different areas of the matrix.

An improvement upon this is the form of distribution model associated with techniques referred to as 'matrix estimation from counts' (van Zuylen et al, 1980). This consists of a distribution model form, for the matrix T_{ij}, whose parameters are estimated under a selection of constraints. In addition to the row and column constraints, which are implicit in the a_i and b_j parameters referred to earlier, constraints may relate to the totals within different sub–matrices or total traffic flows across defined screenlines.

possible to improve the fit of the model to [...] data by introducing additional (sometimes [...]rary) parameters: the aim is, however, to find a [...]gical, principled, way of doing this, which satisfies statistical rigour. For further details, reference should be made to Chapter 5 in Ortúzar and Willumsen (1994). The limitations of distribution models are explained in Bates and Dasgupta (1990).

Modal split

The majority of four–stage models do not distinguish beyond 'private' and 'public' modes and, moreover, certain person–types are assumed captive to public transport, usually meaning those in households without a car, although in some cases a more refined definition of 'car availability' is used. Thus, choice of mode tends to be confined to predicting the proportion of persons, assumed to have access to a car, who actually use public transport.

In the cars versus public transport case, there is a range of possibilities for a suitable function, based on a comparison of the perceived generalised costs of the private and public modes. Indeed, it is possible to produce entirely empirical functions relating the proportion choosing public transport to the difference, or ratio, of the generalised costs of the two modes. In most cases, the generalised cost is pre–specified in terms of average 'behavioural' weights attached to each component.

The desire to broaden the modal split model to more than two modes led to the development of the 'logit' model. Because the structure of the model is not always suitable when some of the modes are inherently similar (for example, buses and trains may be considered more similar to each other than either of them is to cars), the so–called 'nested logit' may also be used to reflect these similarities. Within a nested structure, the first choice may be between a car and public transport, while, for those choosing public transport, there may be a further choice between bus and train. Further details of nested logit models are given in Chapter 6 of Ortúzar and Willumsen (1994).

Traffic Assignment

An assignment model takes a matrix of trips (on an origin–destination basis), which is the output of the trip–distribution process, and assigns (or 'loads') it as traffic onto an appropriate network. While the underlying principles are not mode–specific, the different characteristics of highway networks and public transport networks lead, in practice, to a rather different set of practical problems.

Although assignment is treated as a single component

of the model, it actually entails a number of separate processes; which are:

❏ choice of route (or path) for each i–j combination;
❏ conversion of person–trips into vehicle–trips by applying average vehicle–occupancies;
❏ aggregation of i–j vehicle–flows on the links of the chosen paths;
❏ introduction of supply–side constraints (eg capacity restraint), as a result of the volume of link flows nearing or exceeding capacity; and
❏ estimation of the resulting generalised cost for each i–j combination.

While the travel matrices will, typically, be produced on an 'annual average day' basis, the matrix will normally be factored before assignment (eg to a peak hour), since it is necessary to relate flows on links to operational definitions of capacity (ie capacity restraint).

The operation of capacity restraint, or similar supply interactions, will change the costs of travel both actual and perceived. Therefore, it will generally be the case that the costs output from the traffic assignment process are inconsistent with those used to drive the distribution and modal split models. Earlier versions of the four–stage model usually ignored this. However, it is now common to allow a certain amount of iteration, although the scale of a model may make iterating to convergence a computationally costly process. Further details of assignment procedures can be found in Ortúzar and Willumsen (1994).

Concluding remarks about four–stage models

The main criticisms of the four–stage model relate to practical difficulties and points of detail rather than to their basic structure. The following points may be noted:

❏ the level of zoning will, typically, be sufficient to estimate traffic flows on all classified roads but the local distributor and access road network will be less fully represented;
❏ the detail will be sufficient to test the general impact of specific road schemes but not sufficient for detailed design and, in particular, the detailed working of junctions and the analysis of turning movements will not normally be possible;
❏ on the public transport side, the detail will generally be sufficient to represent different public transport routes but not individual bus or tram stops, although most individual railway stations may be included;
❏ account is not usually taken of changes in the

time–of–day trip–profiles, for example to represent 'peak spreading';

❑ personal factors affecting modal choice are not generally taken into account, as average behavioural values of generalised costs are used;

❑ usually no treatment of walk or cycle modes is incorporated, apart from the walking involved in accessing public transport routes; and

❑ in most cases, only limited iteration through the model stages, in search of equilibrium, is undertaken because of the heavy computational burden.

The four–stage model was primarily designed for the analysis of large–scale urban investment projects. It is the level of detail provided by the networks which allows the four–stage model to investigate reasonably precise locations, in terms of the impact of changes in accessibility between specific zones. By contrast, the four–stage model is much less suitable for the investigation of global, highly flexible policies (such as changes in public transport fares) or policies likely to involve substantial changes in travellers' responses, such as times of travel and road–use pricing.

8.5 Strategic Transportation Models

The main features of a model, which might be classified as 'strategic', are:

❑ the model must take into account a wide range of transport choices, including changes in the frequency of travel; and

❑ the model does not aim at great spatial detail and the zones are chosen to reflect only broad variations in land–use and traffic density.

In many cases, the zones will be so large as to make it either difficult or inappropriate to make use of a transport network, although this is not necessarily a feature of a strategic model.

The recent interest (in the 1990s) in using strategic models for urban studies can be ascribed to three factors:

❑ dissatisfaction with many of the assumptions of the traditional transportation planning models, many of which are, essentially, variants on the four–stage model;

❑ the cumbersome nature of inputting general policy–options in four–stage models and the, often, long turn–round time for computing the outcomes; and

❑ the substantial investment in data–collection and preparation that is necessary for such models

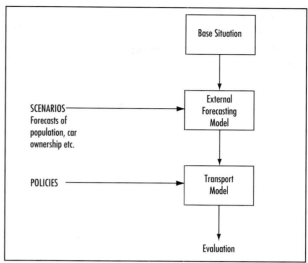

Figure 8.1: Main components of a strategic model.

before even quite broad issues can be quantitatively assessed.

An example of a strategic model is the TRL's strategic transport model (STM) originally developed as the London Area Model (LAM) (see Oldfield, 1993).

The essential components of a strategic model are illustrated in Figure 8.1.

In contrast with most conventional transportation models, a strategic model is not calibrated to reproduce the base situation: rather, it relies on externally–derived matrices built to reflect the current pattern of travel, making use of existing survey data. The model is essentially an extrapolatory device, applied to a reliable base of current travel patterns and backed–up by as much understanding of travel behaviour as possible. Clearly, the quality and coverage of existing data is of crucial importance in this respect.

The forecasting component is, in essence, the same as a trip generation stage of the standard four–stage model. Its aim is to take account of changed travel–patterns, due to zonal changes in population, employment and, in particular, car–ownership. This results in calculations of origin and destination growth–factors for each journey–purpose matrix that is distinguished.

In this case, forecasting is assumed to operate independently of any transport changes. However, it is not implied that such an estimated future demand is, in fact, realistic and it is a function of the transportation model to modify it in the light of changing transport conditions. It is, of course, possible also that transport conditions could themselves affect the growth of demand by, for example, influencing land–use decisions. Models which deal directly with such interaction are discussed in Section 8.7.

The core of the strategic model is the 'transport demand/supply interaction model', which, as shown in Figure 8.2, contains two essential components – the demand component, in which a given change in transport conditions is reflected in generalised cost and modifies the current travel–demand estimate accordingly, and the supply component, which, given the current estimate of demand, modifies the generalised cost by consideration of that level of

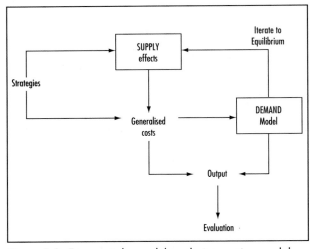

Figure 8.2: Transport demand/supply interaction model.

demand relative to the available capacity. These two components are iterated until the level of demand is compatible with the generalised cost matrix, that is, until equilibrium is reached.

Policies for testing are input to the process in one of two ways: either by intervening directly in the generalised cost matrix (as might be the case, for example, with pricing policies) or by modifying the supply process, to represent changes in capacity. In either case, these result in changes to both demand and generalised cost matrices, which can then be evaluated against other strategies or against a suitable base case.

With the strategic model, certain crucial simplifications are made, as compared with the traditional four–stage approach (described in Section 8.4), as follows:
- ❏ the spatial detail is greatly reduced and, correspondingly, the supply side is modelled by zonal relationships (though the possibility of using a coarse transport network is not precluded); and
- ❏ the underlying problems of the distribution model are mitigated by putting the model on an incremental basis and, while this relies on a

suitable set of base matrices, the fact that the zonal detail is reduced makes this a less exacting requirement.

These key features allow the model to be set up and operated in considerably shorter times than the traditional four–stage model. However, there are limitations on the kinds of policies to which strategic models should be applied, largely connected with the zone size and specification. Examples of policies difficult to assess with these models are:
- ❏ road–use pricing, where a cordon bisects a zone in the strategic model, since behaviour may well be critically affected by which side of the cordon the travel begins; and
- ❏ a new light rapid transit (LRT) line, in a case where the catchment area is small relative to the size of the zones through which it passes.

Hence, strategic models are best used to investigate flexible policies of relatively global impact (especially land–use and pricing options). They can be used to analyse a large number of different variants and, because of their fast turnround, give broad impressions of changes in key quantities like average journey speeds, total vehicle–kms and a limited amount of spatial variation. They are not suitable for the assessment of detailed infrastructure schemes or for aspects of network management.

8.6 Policy Appraisal Models

The increasing need for demand management is leading authorities to consider the use of direct pricing policies. The types of model referred to already have general facilities for modelling response to price changes but they are less well suited to cases where the price levels vary during the course of a day and/or widely across different categories of travel.

Strategies involving significant increases in transport prices and significant variation in those prices over the day require a detailed account of time–of–day choice, since such strategies are designed to bring about material shifts in the times at which people travel. In addition, because real price changes are being modelled, explicit treatment of income–group variation is desirable. Taken together, these two requirements greatly increase the complexity of the model.

Where the policies being considered are likely to lead to major shifts in time–of–day choice, then units of travel should be based upon 'tours', which analyse outbound and inbound movements at the same time. In practical terms, however, models based on tours have only been implemented in disaggregate approaches, which are a particular variant of the four–stage model (IHT, 1996). While the notion of tours has a certain theoretical superiority, its implementation in most cases (apart from pricing policies) is unlikely to compensate for its additional complexity and the break with convention.

Three major issues in assessing pricing policies are likely to require the use of tours, since it is essential to deal with all the trips made in the course of a round tour as one, to ensure temporal logic and consistency in aspects such as the duration of stay at certain destinations. These are:
❑ road–use pricing, where there is a significant variation in the charge by time–of–day and/or direction of travel;
❑ peak period fare–surcharges (or off–peak fare–concessions) on public transport; and
❑ parking charges, which vary not only by duration but also by time of arrival.

It is convenient also to treat cases of restricted access, such as when car parks are closed or traffic is banned from entering defined areas at certain times, as a variant of pricing policy.

The role of the time–of–day choice model thus becomes to predict the simultaneous choice of the time of the outward and return portions of the tour, as a function of a comparison of the differences in the perceived generalised cost of the round tour. A recent example of a model which has attempted to implement these ideas is APRIL (Williams *et al*, 1993), which was developed for the Department of Transport's Congestion Charging in London study.

These models are highly specialised tools. While they are derivations of the strategic models, described in Section 8.5, they are generally much more complex and involve substantially longer computing run–times for each scenario tested.

8.7 Land–Use/Transport Interaction Models

All the transportation models described in the previous sections assume that land–uses (primarily, the locations of population and employment) can be predicted independently of the transport system and that they can be assumed fixed for the purposes of investigating alternative transport policies. Although there is the opportunity of undertaking additional tests with different land–use assumptions, this is not the same as a fully interactive model.

A number of models have been developed in which the 'dynamic' interaction between transport provision and land–use activities is synthesised. By their general nature, they are more complex than the 'pure transport' models described so far and some of the concepts may be unfamiliar to most transportation planners. A key investigation into the properties of land–use/transport interaction (LUTI) models was carried out in the latter part of the 1980s, under the auspices of the Transport Research Laboratory. The report of the investigation, known as ISGLUTI (International Study Group on Land–Use/Transport Interaction), has been published (Webster *et al*, 1990). Nine candidate models were investigated in detail and, for seven of the models, their reactions to a series of policies was tested extensively against a wide range of appropriate indicators. The report summarises these and makes recommendations as to the way forward.

8.8 Simplified Demand Modelling

Recent changes in national and local transport policy have resulted in:
❑ a revision of the assumption that travel demand in urban areas can be forecast without recognising the constraining influences of supply; and
❑ a desire to develop more 'sustainable' transport systems, which, in practice, means that the effects of public transport improvements and demand management mechanisms need to be reflected in the way models are structured.

Nevertheless, some form of model to estimate likely potential demand is essential where demand is properly determined by price and limited by supply constraints. Fully–specified demand models are costly to create and the circumstances do not always justify the expense. Therefore, the simpler, but generally more approximate, techniques described in this section can be applied, where full demand models are not feasible or are seen as too costly.

Forecasting traffic growth
Whatever processes are used to estimate changes in demand, in relation to changes in travel costs and availability of the supply of transport, some means of forecasting the unrestrained or underlying demand for travel is required. Thus, car–ownership and trip–end models will normally be required to derive growth factors to apply to base year trip–matrices. A description of the Department of Transport's procedures, which are generally applicable (as a first step) in small and medium–sized towns with low public transport usage, can be found in Chapter 6 of the 1994 SACTRA Report (SACTRA, 1994). Having derived an initial forecast of demand in this way, the

simplified techniques described below can be used to estimate the likely responses of drivers to the perceived changes in travel costs.

Induced and suppressed traffic

The first reason for wanting to consider potential changes in demand arises from the phenomenon of generated or (more strictly) induced traffic. The term 'induced traffic' is used to embrace all the possible responses to new roads and road improvements. SACTRA, in their 1994 Report, concluded that induced traffic was a real phenomenon of some considerable importance in certain cases, notably in congested and complex urban areas. The strict definitions of induced traffic and induced trips are complicated but need to be understood (see Hills, 1996). For simplicity, it can be said that induced traffic includes the following reactions of travellers, both actual and potential, to a given improvement in accessibility:

❑ rescheduling of existing trips to take advantage of improved conditions at peak periods (time of day choice);

❑ decreases in vehicle occupancy, with former passengers using their own private vehicle for some trips currently made with other people (vehicle–choice);

❑ switching from public transport, cycling and walking to the use of private vehicles for existing trips (mode choice);

❑ travelling to new destinations further afield for the same purpose as existing trips (destination choice);

❑ travelling from new origins further away to the same destinations for the same purpose as existing trips (origin or location choice); and

❑ increasing the frequency of existing vehicle–trips between any given origin and destination for any given trip purpose, including entirely new trips (trip frequency).

As SACTRA reported, little empirical evidence exists, as yet, which enables the magnitudes of these effects to be identified.

In cases where road or parking supply is insufficient to cater for the current demand, the mechanisms listed above work in the opposite direction; this is known as 'suppressed' traffic. In this case, by increasing the supply, the suppressed demand will be released, eventually to the point where there is no suppression at all. Increases in supply beyond that point will then begin to 'induce' further traffic. This phenomenon applies to public transport capacity–improvements or price–reductions, as well as increases in road–space and parking supply. Reducing traffic–demand, through some means

external to the road system, such as improvements in public transport, may result in decreased congestion on the road network but the phenomenon of extra road traffic then being induced can be expected, unless complementary traffic–restraint measures are applied.

Establishing a realistic base–case

The first problem faced in undertaking an appraisal of either a road scheme or a public transport scheme designed to relieve traffic congestion or, for that matter, a scheme to manage road traffic demand, is how to establish a reasonable base–case or 'do–minimum' situation. Often the initial demand forecasts, when assigned to the do–minimum network, will result in what are judged to be unrealistically high levels of congestion.

Ways in which the initial demand forecasts might be tempered include:

❑ adjustment of the demand for parking in central areas, so that the parking supply would not be exceeded;

❑ detailed examination of the network, to identify the scope and opportunity for low–cost improvements, of the kind which a highway authority might undertake as a matter of course, aimed at reducing local congestion hot–spots; and

❑ estimation of the extent to which the peak demands could spread into the inter–peak and off–peak periods, without creating even longer periods of congestion.

For further advice, see the Advice Note on 'Traffic Appraisal in Urban Areas' (DOT, 1996a).

Peak–spreading techniques

After re–routeing, the primary response to changes in road congestion seems to be change in time of travel; ie peak–spreading in the face of rising congestion and peak–contraction again as congestion decreases. This is the first demand–response which should be considered when an overloaded do–minimum network is encountered or when any major change to the transport system is being appraised. Unfortunately, no satisfactory simple method is yet available. Pending further research, models based on counts taken at local bottlenecks should be developed and applied. Failing this, the general relationship between the 'peakiness' of the peak and average speed, reported by Goodwin et al (1991), may be used but with care, and bearing in mind the very limited data on which the relationship was based. Further advice is given in 'Traffic Appraisal in Urban Areas' (DOT, 1996a).

Whatever model is applied, it is important that it is used to estimate both the spreading of the peak–period in the do–minimum case and the corresponding contractions of the peak in the do–something cases.

Assignment–based techniques for modelling demand–responses

In the event that these procedures fail to produce what are regarded as reasonable or credible levels of congestion on the forecast year do–minimum network, some form of demand–restraint (traffic suppression) is required. However, if it is logical to suppress demand on the do–minimum network, then it must also be logical to let the demand expand to some extent on a do–something network, in response to the extra capacity provided by the scheme under appraisal.

The constraining effect of insufficient highway capacity has been reflected in the past through a variety of assignment–based techniques. The main ones are, as follows:

❑ 'matrix–capping' methods – these include:
    ❑ growth cut–off methods in which growth is assumed to stop at some point in the future, across the whole trip–matrix; and
    ❑ matrix estimation methods – in which growth is restricted to the capacity of the network by means of techniques where capacities are used as constraints instead of traffic counts, as in its more usual application;
❑ 'shadow network' methods – in which duplicate networks are created and connected to the 'real' network by zone–centroid connectors only. As traffic assigned to the real network begins to approach capacity, traffic is diverted to a 'shadow' network and subsequently ignored;
❑ 'speed limitation' methods – in which traffic is loaded incrementally, until a pre–specified threshold average journey speed is reached, at which point loading of the affected movements ceases; and
❑ 'elastic user–equilibrium assignment' methods – which include 'disaggregate elasticity models'. In essence, these methods adjust the trip–matrix on a cell–by–cell basis, on the basis of the changes in travel costs for each zone–to–zone movement and in accordance with a specified elasticity of demand with respect to travel cost.

These methods are discussed in detail in section 6.9.5 of the IHT Guidelines (1996) and some data on travel–cost elasticities is given in the Department's advice about induced traffic (DOT, 1997).

Stated preference models

The move towards more sustainable transport systems means that the effects on demand of traffic and parking management and improvements to public transport and cycling all need to be estimated. A common approach is to undertake stated preference surveys (see Section 7.12) to gauge people's reactions to specific changes proposed, due to a policy or scheme. Providing that the results can be scaled, by reference to some revealed behaviour, stated preference surveys can provide a useful means of estimating the effects of specific proposals.

8.9 Road Traffic Assignment Models

A common response of vehicle–users to changes in conditions on a road network is to change route. For this reason, assignment models, especially of the road system, are popular with transportation planners. The broad principles of assignment have been discussed in section 8.4 but, in this and the next section, road traffic and public transport person–trip assignment models are discussed in more detail.

The Department of Transport's Traffic Appraisal Manual (DOT, 1991) and Advice Note on Traffic Appraisal in Urban Areas (DOT, 1996a) contain much useful guidance on road traffic assignment modelling in urban areas. A more detailed overview can be found in Section 6.10 of the IHT's Guidelines (IHT, 1996).

The uses of road traffic assignment models in urban transportation appraisal can be summarised, as follows:

❑ they provide generalised cost information for the demand models and, for this purpose, the coarseness of the representation can vary from the notional (with some kind of area speed/flow relationships) to the detailed, with small zones and sometimes even including junction modelling;
❑ detailed models, often congestion assignment models, can be used for the analyses of road–based proposals, which form the basis for certain kinds of transportation study; and
❑ whatever their scale, traffic assignment models can provide travel–cost information for economic and financial evaluation and flow, delay and speed information for operational, safety and environmental appraisals.

Traffic assignment models may be:
❑ either part of a more complex modelling system involving demand components; or
❑ the main modelling stage, as is likely to be the case in smaller urban areas, where complex demand modelling is rarely justified and where

some relatively simple method of estimating changes in demand may be acceptable (as described in Section 8.8).

The main types of road traffic assignment model are:
- ❏ strategic transportation and policy appraisal models;
- ❏ conventional four–stage models; and
- ❏ congestion assignment models (for detailed analyses of congestion problems and solutions).

The main features of these differ according to the importance of the purpose for which they are used. The principal purposes are set out below.

Capacity restraint

Capacity restraint occurs because, as traffic volumes rise, average vehicle–speeds fall. Variation of modelled speeds in response to increases in flow is the basis on which capacity restraint effects are represented in traffic assignment models. This kind of relationship can be effected through area speed/flow relationships, as found in certain types of strategic and policy appraisal model, link–based speed/flow relationships, as used with networks in four–stage models, and junction flow/delay relationships, as employed in congestion assignment models (see IHT, 1996).

Multi–Routeing

Changing route is the simplest option available to drivers faced with congested conditions. Multi–route modelling procedures allow trips between the same origin–destination pair to be distributed across a number of alternative routes. Several techniques are available for introducing multi–route procedures into a road traffic assignment model (see Ortúzar et al, 1994).

Equilibrium

For urban areas, the more commonly–used assignment processes seek to fulfil the conditions of Wardrop's First Principle, which states that no driver can reduce his or her generalised cost of travel by changing route – that is, all routes used by drivers from any given origin to any given destination will have equal travel costs and all routes not used will have greater travel costs. Processes which are based on this principle are known as 'equilibrium' assignments. A number of different algorithms are available for the assignment process, as described in DOT 1996a and Ortúzar et al (1994),

Convergence

It is important that the iteration between traffic assignment and changes in average speed converges to a satisfactory degree, bearing in mind the use which is to be made of the model output. Guidance on convergence, based on the latest research, is given in van Vuren et al (1995). However, it is important to recognise that perfect convergence is usually elusive in a congested network and that statistical tests should be undertaken, to ascertain whether or not the differences between the do–minimum and do–something are discernible against any residual non–convergence in the two cases (see DOT, 1996a).

Time–periods

Traffic conditions on urban roads vary markedly over any given day. Techniques for modelling the effects of congestion are most effective if traffic flows are relatively stable over the period being modelled. Capacity restraint, therefore, implies the need for short time–periods for modelling. In strategic transportation and policy appraisal models, and some more aggregate four–stage models, peak periods of two to three hours' duration may be appropriate. In more detailed four–stage models and congested assignment models, time–slices within time–periods are recommended (see DOT, 1997). For operational analyses, only models of the morning and evening peak–periods are required. However, inter–peak models are necessary in circumstances where the potential for transfer from cars to new or enhanced public transport is being considered.

Derivation of a trip–matrix

Demand data for trip matrices are normally obtained from travel surveys. Individual trip–purposes and vehicle–types may need to be assigned separately, so statistically reliable matrices could be required by vehicle–type and trip–purpose. These requirements mean, generally, that roadside interviews are required. Missing movements can be infilled using the partial matrix approach (Kirby, 1979) and refined using matrix estimation (van Zuylen et al, 1980).

Network coding and specification

Networks for road traffic assignment models should be of sufficient extent to include all realistic choices of route available to drivers. This often involves using 'skeletal' or 'buffer' networks around the area of primary interest, within which networks are coded in some detail. Models in which the network is 'cordoned' close to the boundary of the area of primary interest may artificially constrain the modelled route–choices. They may also hamper the application of methods to estimate induced and/or suppressed traffic, leading to misleading results. The primary sources of data for the development of network definitions are usually maps and on– site surveys.

Components of generalised cost

The usual elements of generalised cost for car travel are time- and distance-related costs. Distance travelled is used to represent vehicle operating costs, such as fuel, tyres and depreciation. Traffic assignment models have the facility to vary the relative importance of the time and distance cost-components. Usual practice is to start the calibration process using values of time and vehicle operating costs obtained from the Department of Transport's Highways Economics Note No 2 (DOT, 1996b) and to adjust these, possibly by vehicle-type and (more unusually) by trip-purpose separately, to obtain a satisfactory calibration.

Road traffic assignment in strategic transportation and policy appraisal models

Strategic transportation and policy appraisal models are characterised by having a small number of relatively large zones. The primary purpose of traffic assignment within these models is to estimate average inter-zonal travel costs. Multi-route assignments are usually required within these models but the route-choice set can be limited in number. In some models, the inter-zonal paths available are pre-defined, thus reducing model run-times by eliminating the path-building stage. Capacity restraint is represented by area speed/flow relationships on a notional network or by link based speed/flow relationships on a strategic road network, with a complementary (and usually notional) network to represent the remainder of the highway capacity.

Convergence tests for the traffic assignment element of these models can be limited to consideration of average inter-zonal travel costs, as the aggregate nature of networks makes flow-based estimation unreliable. In addition to average inter-zonal travel costs, such models are generally capable of outputting, on a zonal basis, aggregate data such as average vehicle-speeds, vehicle-kilometres and vehicle-hours.

Road traffic assignment in four-stage models

Traffic assignment models in the context of the four-stage modelling process are typified by relatively small zones and a highway network that represents all main roads. Capacity restraint is normally represented by highway link-based speed/flow relationships. Simple representation of junction effects, usually in terms of turn penalties and banned turns, are also a common feature of such models. In smaller models, or for critical areas of larger models such as city centres, more comprehensive junction-based modelling can be included.

Four-stage model applications require multi-route modelling and equilibrium assignment procedures. This is because of the importance of such models in the evaluation of strategies and schemes that affect the capacity of specific links in the highway network. Model convergence is measured using stability and proximity criteria (van Vuren et al, 1995). Aggregate outputs are available, as for the more strategic models, but spatially more detailed outputs, such as corridor flows and journey times, are also a requirement.

Congested road traffic assignment models

A detailed zoning system and a network that includes all roads that carry significant volumes of traffic characterise congested road traffic assignment models. Multi-routeing and equilibrium assignment are essential features. Capacity restraint is effected through the explicit modelling of junctions, taking account of turning capacities, signal-timings and the interaction of conflicting traffic movements. Average link speeds are generally fixed, that is, all delays are assumed to be as a result of conflicts at junctions. Use of link-based speed/flow procedures is sometimes made in the peripheral parts of the network, to provide realistic routeing into and out of the area of junction-modelling.

In congested assignment models, the junction-modelling procedure is used to represent the interaction between adjacent junctions. For example, where modelled queue-lengths exceed the available queueing capacity of a link, these models represent the effects that this 'blocking back' will have on the workings of the upstream junction. Similarly, the effects of bottlenecks in the network in 'metering' the flow of traffic to downstream junctions are also represented.

Convergence is measured using stability and proximity criteria, as defined by van Vuren et al (1995). While aggregate outputs are readily available from congestion assignment models, it is the ability of these models to produce a wide variety of detailed junction performance information that distinguishes them from other types. Possible outputs include: expected average flows on links, by direction; main turning movements at main junctions; total delays at junctions; average delays for main turning movements; and queue-lengths at junctions.

However, even with a well-converged model, the queue and delay information can display considerable instability from iteration to iteration and great care is required to avoid over-interpretation of the model output.

8.10 Public Transport Passenger Assignment Models

The two primary uses of public transport passenger assignment models in urban strategy development are:
- ❏ to provide travel–cost information as input to demand models; and
- ❏ to provide information for operational, economic and financial appraisals.

The average inter–zonal travel cost information required by demand models can often be provided by relatively coarse public transport network models. However, operational, economic and financial appraisals of public transport strategies require spatially–detailed assignment models, as the passenger–loadings can be crucially dependent on such detailed matters as stop locations and route alignments.

As with road traffic assignment (Section 8.9 above), the main types of public transport passenger assignment model are:
- ❏ strategic transportation and policy appraisal models;
- ❏ conventional four–stage models; and
- ❏ models for detailed public transport appraisals.

The main features of these differ according to the importance of the purpose for which they are being used. The principal purposes are set out below.

Sub–modal choice

Allocation of passengers between public transport sub–modes, such as buses, trams and trains, can be carried out either at the demand modelling stage or via the passenger assignment model. Mode–specific average travel costs can be generated by assignment models for use in the demand–modelling process. Use of the demand model for this task is most common in strategic transportation and policy appraisal models (Sections 8.6 and 8.7). Use of the assignment model is the more common technique for the more detailed models.

In public transport assignment models, there are two methods for apportioning trips to the alternative sub–modes:
- ❏ on the basis of the relative frequency of service provided by each mode; or
- ❏ by application of a probability model to apportion trips between sub–modes, followed by a further division within the sub–modes based on the relative frequency of the alternative services available.

Multi–routeing of Origin–Destination trips

Multi–routeing is a common feature of urban public transport networks, particularly those in large urban areas. It is the result of phenomena such as alternative bus corridors serving the same passenger movements, 'competition' between orbital and radial services for cross–town trips and competition between bus and train sub–modes. For the appraisal of public transport proposals, an assignment model needs to reflect these multi–routeing opportunities, although the need for multi–routeing can be reduced by the employment of a fine–zoning system.

Over–crowding on vehicles

If passenger demand is such that elements of the public transport network become overloaded, the resulting delay and discomfort may cause passengers to change their route, in order to minimise their perceived generalised travel costs. This effect can be represented within a public transport passenger assignment model and is analogous to the capacity restraint process within a road traffic assignment model. If vehicles are operating at, or close to, capacity, passengers are affected in two ways:
- ❏ the probability of being able to board the first vehicle to arrive is reduced; and
- ❏ the comfort of passengers on the vehicle is reduced through the lower probability of obtaining a seat or, in more severe cases, 'crushing' of standing passengers.

The first case can be represented in an assignment model by increasing the passengers' waiting time. In the second case, where the assignment process has allocated more trips per unit time to the service than there are seats available, the in–vehicle time can be increased to reflect the reduced comfort levels. The introduction of 'crowding' into the modelling process introduces an iterative loop between assignment and the calculation of revised crowding penalties. This adds realism but also adds to overall model run–times. The incidence of 'crowding' in UK local public transport systems outside London is currently minimal. Over–supply, resulting from bus deregulation and continuing patronage decline, means that passenger–loads on buses and trains rarely reach their seating capacity, even at peak times.

Time–periods

Public transport passenger assignment should cover at least one of the peaks and an inter–peak time–period. This provides a basis for deriving daily and annual patronage and revenue forecasts. The inter–peak period, though usually less significant in terms of hourly flows than peak periods, is important in terms of total patronage.

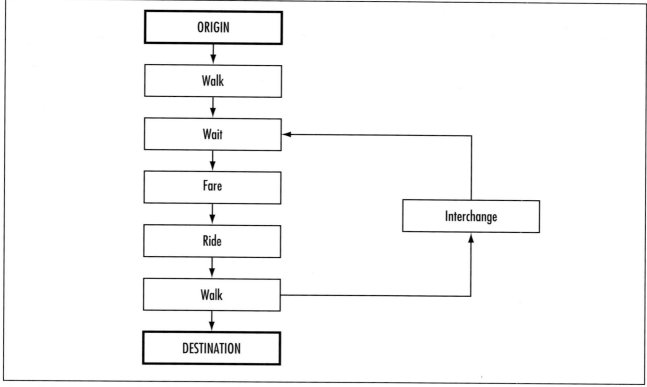

Figure 8.3: Components of a public transport journey.

Derivation of a trip–matrix

Two generic types of observed trip–data are used for the construction of matrices for public transport assignment models: surveyed observations of passenger movements; and data–outputs from electronic ticket machines. Passenger surveys are most widely used and can be divided into three types: on–vehicle surveys; surveys at stops; and household interview surveys (see Section 7.11). Missing movements can be infilled using the partial matrix approach (Kirby, 1979) and refined using matrix estimation (van Zuylen *et al*, 1980).

Network, route and service derivation

The main sources of data for the development of public transport network representations are published timetables, service maps and computerised public transport databases developed by local authorities.

Components of generalised cost

Figure 8.3 shows the most significant quantifiable components of a public transport journey and the way they interact. Trip–makers do not value the various components of travel time in an identical manner. However, a common base is required in order that the total 'behavioural' generalised cost of travel can be calculated for competing paths through the public transport network. The most common practice is to convert all travel cost components to either in–vehicle time or money–equivalents. An indication of the relative valuations of perceived travel–time that have been derived from transport modelling studies and SP surveys is shown below:

- ❑ walk time –   1.5 to 2.0 times weighting relative to in–vehicle time;
- ❑ wait time –   1.5 to 2.5 times weighting relative to in–vehicle time; and
- ❑ interchange – penalty – 5.0 to 10.0 minutes equivalent of in–vehicle time.

The high valuations of walk and wait time with respect to in–vehicle time reflect passengers' dislike of these elements of a public transport journey. The interchange penalty represents the perceived inconvenience of having to change from one vehicle to another, over and above any extra walking and waiting time incurred when changing between services. Fares also need to be converted to the same units as the travel–time components and, for this, standard practice involves the derivation of a monetary–equivalent value for in–vehicle time.

Public transport passenger assignment in strategic transportation and policy appraisal models

The primary purpose of public transport passenger assignment, within strategic transportation and policy appraisal models, is the forecasting of average inter–zonal travel costs; routeing issues tend to be secondary. Network appraisal is restricted by the small number of large zones and coarse network representations, which characterise such models, and detailed assignment facilities, such as multi–routeing, are not usually included. In some strategic models, such as the START model (Bates *et al*, 1991), public transport passenger paths are an input to the process and cannot be varied during the course of a model run. Sub–mode choice is generally carried out at the demand modelling rather than the assignment stage. Depending on the nature of the study area, modelling of crowding may be appropriate. Also, it is sometimes appropriate to allow the frequency of bus–based public transport to vary in line with forecast demand, with consequent effects on the perceived generalised cost of travel. The primary outputs of assignment models, used within strategic transportation and policy appraisal models, are average inter–zonal (and sometimes intra–zonal) travel costs. Other outputs are generally at an aggregate level, such as vehicle–kilometres, patronage and revenue by mode.

Public transport passenger assignment in four–stage models and for public transport appraisals

Public transport passenger assignment models for these applications are required both to supply representative average inter–zonal costs, for demand–modelling purposes, and to route passenger flows accurately, such that the patronage and revenue impacts of infrastructure and service investments can be forecast. There is little difference between the models developed for four–stage models and those required for public transport project appraisal; indeed, they may often amount to the same model. The need to appraise infrastructure investment means that the multi–routeing and sub–mode choice elements of the assignment process are usually required. Model capabilities in this respect are enhanced by fine–zoning systems, in which zone sizes are ideally small and relate to the distances trip–makers have to walk to gain access to public transport. The facility to model overcrowding is most likely to be of use in London, where physical and financial constraints, combined with the absence of bus deregulation, tends to limit the supply of public transport. However, local train services elsewhere could develop crowding problems, where transportation strategies encourage rail travel in the absence of investment to increase network capacity.

Integration with road traffic models

Whilst the road traffic and public transport passenger assignments are separate processes, there are a number of circumstances in which they can interact. At a matrix level, compatible zone systems are required to allow the demand–modelling to be carried out. This usually implies common zone–sets, although the requirement for small zones in public transport assignment modelling means that a hierarchical zone system is worth considering.

Interaction at a network level is desirable, so that:
❑ forecast changes in traffic speeds can be reflected in the modelled journey speeds of buses; and
❑ bus flows, as coded within the public transport model, can be used as pre–loaded fixed flows within the road traffic model.

If network integration is to be achieved, a common network specification is required and a compromise between the network requirements of the two assignment processes may have to be struck. For example, road junctions may need to be added to the public transport model and stops and bus–only roads to the road traffic model.

8.11 Summary of Model Applications

Models of the transport system are required in order to:
❑ forecast travel–demands and consequent levels of service on the do–minimum transport system; and
❑ forecast the responses of travellers to changes in the transport system and the consequent levels of service on the do–something transport system.

Forecasting future travel–demands requires sub–models of car–ownership or availability and of trip–ends. The trip–end sub–model may relate to a single mode or all modes, depending on the structure of the rest of the transport model. These components are required in all transportation models that are used for transport system appraisal.

The main responses of travellers to changes in the transport system and levels of service are:
❑ change of route (re–assignment);
❑ change of time of travel (time of day choice or peak–spreading and contraction);
❑ change of destination (redistribution);
❑ change of mode (modal split);
❑ change of vehicle–occupancy (car sharing);
❑ change in trip frequency (trip suppression and generation); and
❑ change in the location and/or intensity of activities.

With the exception of the last, which takes place over a longer time, some means of representing all the other responses is usually required for the appraisal of changes to an urban transport system. Estimation of the relocation of activities, while of increasing importance, requires very specialised models which represent fully all the interactions between land–use and transportation.

The precise way in which travellers' responses may be estimated will depend on a number of factors, including the following:

❏ the nature of the area, the size and the variety of available transport modes and their market shares;
❏ the type of study; and
❏ the availability of data and resources.

For small and medium–sized urban areas, where the vast majority of motorised personal travel is made by private car, the most appropriate model structure is likely to be, as follows:

❏ car–ownership and trip–end sub–models for forecasting future demand. The Department of Transport's models may be used here or, alternatively, the Department's trip–end forecasts may be used directly (DOT, 1991);
❏ a spatially–detailed congested road traffic assignment model for assigning the future demand to the do–minimum and do–something road systems (Section 8.9);
❏ a simplified means of estimating drivers' responses to changes in travel costs, including congestion, such as a peak–spreading/contraction relationship and a simple elasticity model (Section 8.8); and
❏ modal choice models derived from stated preference data, to estimate the effects on drivers of new initiatives (Section 8.8).

For the inner areas of large cities and conurbations, where a significant proportion of personal travel is made by public transport, multi–modal models will be required for a fuller understanding of the consequences of strategic options. Two kinds of multi–modal structure are in use:

❏ the spatially–detailed traditional four–stage model (trip generation, distribution, modal split and traffic assignment), in which only a limited number of traveller–types are treated individually (Section 8.4); and
❏ the spatially–aggregate strategic transportation model, in which all the main responses (changes of route, vehicle–occupancy, time of travel, destination, mode and trip frequency) are represented and in which a more extensive number of traveller–types are treated individually (Section 8.5).

The conventional four–stage approach requires a spatially–detailed road traffic assignment model, in which capacity–restraint effects are represented (Section 8.9), and a detailed public transport model, in which the effects of overcrowding may need to be included (Section 8.10). The approach suffers from three main drawbacks:

❏ the costs of data–collection and model–development are substantial;
❏ computer run–times for these models are relatively long; and
❏ time–of–day choice is difficult to incorporate satisfactorily, data on vehicle–occupancy is often poor and trip–frequency choice is rarely included in practice.

The strategic transportation model approach employs a coarse zoning system and cannot, therefore, represent the detailed effects of transport infrastructure changes. With its disaggregate treatment of types of traveller and full range of main demand responses, this kind of model is best suited to analyses of policies, especially those involving pricing. A strategic model will usually be much cheaper to set up and quicker to use than a typical four–stage model. However, for the appraisal of infrastructure changes and the further exploration of policy effects, the demand changes derived from a strategic transport model need to be fed down to spatially–detailed road traffic and public transport assignment models. The development of these more detailed assignment models requires considerable resources.

Policy appraisal models are spatially aggregate and pay even more attention than strategic transportation models to the responses of individual traveller–types, with particular emphasis being given to modelling time–of–day choice (Section 8.6). They are required for the detailed appraisal of policies, such as congestion charging and parking, where one significant effect is that travellers would change their times of travel willingly.

Lastly, the development of land–use/transport interaction models may be justified, where the changes to the transport system being appraised are likely to cause significant shifts in the locations and/or intensities of activities (Section 9.6).

8.12 References

Bates JJ and Dasgupta M (1990)   'Review of Techniques of Travel Demand Analysis: Interim Report', Contractor Report 186 TRL.

Bates JJ, Brewer M, Hansen P, McDonald D and Cements D C (1991)
'Building a strategic model for Edinburgh' Proc PTRC Summer Annual Meeting, Transportation Planning Methods Seminar, PTRC.

DOT (1991)
'Traffic Appraisal Manual' DMRB 12.1, Stationery Office.

DOT (1996a)
'Traffic Appraisal in Urban Areas', DMRB 12.2.1 Stationery Office.

DOT (1996b)
'Highways Economics Note No 2: Values of Time and Vehicle Operating Costs for Use in Economic Appraisal', DMRB 13.2 Stationery Office.

DOT (1997)
'Induced Traffic Appraisal' DMRB 12.2.2, Stationery Office.

Goodwin RP and Coombe RD (1991)
'Dealing with the National Road Traffic Forecasts in Urban Areas', Proc PTRC Summer Annual Meeting. PRTC.

Hills PJ (1996)
'What is induced traffic?' in Special Issue (ed D Coombe) of Transportation 23(1).

IHT (1996)
'Guidelines on Developing Urban Transport Strategies', The Institution of Highways & Transportation.

Kirby HR (1979)
'Partial matrix techniques', Traffic Engineering + Control, 20(8/9).

Oldfield RH (1993)
'A Strategic Transport Model for the London Area', Research Report 376. TRL.

Ortúzar J de D and Willumsen LG (1994)
'Modelling Transport', (Second Edition) Wiley.

SACTRA (1994)
'Trunk Roads and the Generation of Traffic', Standing Advisory Committee on Trunk Road Assessment, Stationery Office.

van Vuren T, Harris R and Emmerson P
'Convergence monitoring for congested assignment in road scheme appraisal', Proc PTRC

(1995)
European Transport Forum, Seminar E. PTRC.

van Zuylen H and Willumsen LG (1980)
'The most likely trip matrix estimated from traffic counts', Transportation Research 14B(3).

Webster FV, Bly PH and Paulley NJ (eds) (1990)
'Urban Land–Use and Transport Interaction', Report of the International Study Group on Land–Use Transport Interaction (ISGLUTI) Avebury.

Williams IN and Bates JJ (1993)
'APRIL – a strategic model for road–use pricing', Proc PTRC Summer Annual Meeting, PTRC.

Chapter 9 Economic and Environmental Appraisal

9.1 Introduction

Proposals for improving traffic and transport systems in urban areas should be developed within the context of urban development plans (see Chapter 4), so that they are brought forward as part of the planning of the area as a whole and their objectives are compatible with such plans. This is true whether the promoter of the transport scheme is a local authority, a government department (for trunk roads) or a private developer. The planning process may identify parts of the transport network where new investment in roads or public transport infrastructure or major redesign of existing facilities or traffic management would be beneficial on economic, traffic, environmental or safety grounds. However, as there are limits on available resources, it is necessary to select those schemes and measures which will achieve the greatest net benefit to the community as a whole, as part of a co-ordinated policy. Increasingly, these priorities are being established within the context of the 'Package Approach' (see Chapter 6). Priorities need to be determined at many different levels eg.local authority, government regional office, national and EU level, so the appraisal process needs to be capable of meeting the requirements of these different perspectives.

The central purpose of appraisal is to compare the advantages and disadvantages of any proposed course of action against relevant alternatives, so as to demonstrate fitness for purpose and value for money of the chosen option. Appraisal is relevant to:
 ❑ internal decision-making (by transportation professionals and politicians);
 ❑ external decision-making (by government and funding agencies); and
 ❑ the public (within the consultation and public inquiry processes).

Appraisal is, therefore, a point of interface between the often complex technical material required by planners and scheme-designers and the needs of the public and political decision-makers to understand the broad picture. The ability to refine technical material and present and explain the balance of advantage clearly to a wider audience, as well as to feed-back to technical reassessment, is therefore critical. Successful appraisal must not only inform and aid decision–making but should also be an integral part of the planning process. It must inform, yet should never replace, political decision–making.

9.2 The Appraisal Process

In its guide for Government Departments, HM Treasury sets out the requirements for the economic appraisal of expenditure decisions, as follows:

"Systematic appraisal entails being clear about objectives, thinking about alternative ways of meeting them and estimating and presenting the costs and benefits of each potentially worthwhile option. Used properly, appraisal leads to better decisions by policy–makers and managers. It encourages both groups to question and justify what they do. It provides a framework for rational thought about the use of limited resources." (HMG, 1991).

The appraisal process can be represented as a series of linked steps. Although presented in a sequential manner, there should be strong feedback between the steps.

Step 1: Definition of objectives and constraints
It is desirable to begin by setting down the objectives because, from this, will follow the criteria against which policy measures are to be judged. Possible objectives for an urban transportation strategy are:
 ❑ economic efficiency;
 ❑ environmental conservation and improvement;
 ❑ safety improvements;
 ❑ accessibility;
 ❑ sustainability;
 ❑ economic regeneration; and
 ❑ equity considerations.

Principal constraints are the lack of finance, impracticability and inherent conflicts between objectives. See also Chapter 6 and IHT (1996)

Step 2: Problem identification
The need for action arises either because of perceived inadequacies in the performance of the existing system or because of the need to cope with new problems and opportunities, resulting from the development plan, which require prior investment.

Step 3: Policy/scheme identification and design
The development of policies or schemes, designed to

meet the objectives, follows in response to the problems identified. At the early stages, a range of options will usually need to be considered, which may include various mixes of management and investment (see Chapter 3). Progressively, the poorer options can be discarded and the better options developed.

Step 4: Measurement and forecasting
Identification of the relevant factors (inputs and outputs) will lead to the measurement of those factors in the base case, model calibration, validation and forecasting of the outcome for alternative scenarios over the life of the plan, using indicators relevant to the objectives. The appropriate forecasting methods will vary according to the size and character of the proposal (see Chapter 8).

Step 5: Evaluation
The measurement and forecasting process necessarily concerns physical performance, such as journey times, accidents, environmental and other attributes. Evaluation is the process of applying weights to the many dimensions of performance, so as to add up the relevant benefits and costs. The process may be either rule–based, using standard values, or judgement–based, relying on implicit rather than explicit weights. If explicit weights are used, they may be monetary in nature (as in cost–benefit analysis) or based on rating scales, points or scores (as in multi–criteria analysis). There is a strong case for consistency in the approach to evaluation across the relevant range of projects and policies. Accepted methods are discussed later in this Chapter.

Step 6: Decision–support
For the chosen evaluation method, it is necessary to have procedures for summing the benefits and costs over the project life and for assessing risk. The precise requirements depend upon the financing arrangements, especially if private as well as public finance is involved. Finally, the outcome of the technical appraisal process must be presented to decision–makers in a comprehensible manner, with the main assumptions and qualifications spelled out in plain English and free of jargon.

Economic and environmental appraisal is required for various purposes. These include the consideration of:
❏ 'strategies' – alternative strategic plans for an urban area;
❏ 'packages' – alternative means of dealing with particular identified problems; and
❏ 'schemes' – alternative scheme designs (capacity, layout, alignment etc).

In principle, the appraisal methods used for these purposes should be completely consistent. In practice, this poses formidable problems, because of the huge range of options potentially available at the strategic level and the impossibility of assessing all of them in detail. Therefore, a practical approach is to proceed in a hierarchical manner, by developing a transportation strategy for appraisal, first, and then assessing particular measures and schemes within the context of the strategy. This hierarchy of appraisal, whilst eminently sensible, has implications for appraisal style, which needs to be rather different at strategic and scheme levels.

9.3 Strategic Appraisal

At this level, the emphasis is on developing broad options for consideration (sometimes referred to as 'scenarios' or 'cartoon strategies'). For example, these might include options involving significant capital expenditure on infrastructure, demand management through physical and/or pricing measures, priorities for public transport, environmental measures, such as pedestrianisation and traffic calming, and an appraisal of key environmental constraints and opportunities, such as conservation areas, historic sites, townscape and urban design opportunities. These need to be assessed, both separately and in combination, against a 'do–nothing' baseline option, which itself requires careful definition. The breadth of the appraisal means that detail has to be sacrificed, in terms of the spatial representation of the urban area and of the environmental characteristics of particular locations within it. Correspondingly, a broad–brush style of appraisal, against the objectives set out in Section 9.2, is appropriate. Although at this stage the appraisal may be primarily directed to satisfying objectives and solving known problems, analysts should nevertheless be aware that some options, though attractive in strategic terms, may nevertheless involve negative local effects, for example, on land–take, amenity or access.

Performance measurements of strategies, in relation to their objectives, poses the central problem of appraisal. Given that some elements, such as capital and operating costs, are readily measured in monetary terms, while others, such as accessibility and environmental performance, may be represented by quantitative indicators and yet others, such as equity, may involve elements of judgement, how is the overall balance sheet to be added up? Various methods are suggested, in the multi–criteria analysis literature, along the following lines:
❏ identify suitable performance indicators, by which each objective is to be measured;
❏ measure the performance of each strategy using

the chosen set of indicators;
❑ score each strategy–option against each indicator within each objective;
❑ define appropriate weights to be applied to each indicator;
❑ sum the weighted scores across all objectives to give an overall assessment; and
❑ test the robustness/sensitivity of the strategy appraisal to different priority weights for the various objectives.

A particular choice within this process is between using monetary values or other numerical weights to bring the results to a single unit of measurement, on the one hand, or presenting the outcome as a series of bullet–points or ticks, representing the achievement level of alternative strategies against objectives, on the other. Whichever method is chosen, close attention must be given to the pattern of impacts achieved by alternative strategies in coming to an overall assessment.

A more detailed description of the strategic appraisal process is to be found in the IHT Guidelines (IHT, 1996).

9.4 Appraisal within the Package Approach

The development of the package approach to local transport funding (see Chapter 6) has created a new context for the appraisal of urban transportation plans [NIa]. The package approach requires analysts:
❑ to think clearly about the nature of problems and objectives for an entire area;
❑ to relate proposed policy measures to the defined problems and objectives; and
❑ to consider a range of policy and scheme options, in combination, in terms of the impact on the area as a whole.

The most innovative aspect is the opportunity to design options which range across policy and spatial boundaries, including mixes of both management and investment measures within the same package.

From an appraisal point of view, a package strategy is best seen as a set of policy measures designed to secure defined objectives or to overcome known problems. The assessment is, therefore, of a set of measures acting together. Important elements within the strategy may also need to be appraised individually to demonstrate their value for money. Therefore, the appraisal, at the package level, is designed primarily to enable assessment of overall coherence, as well as the roles played by the various

elements. Guidance on the appraisal requirements is provided by the DOT; this is essential reading for local authority staff engaged in package bid preparation (DOT, 1995a). The guidance requires applicants to submit an overview of how the measures making up the strategy interrelate and contribute to the objective, together with an assessment of the role of specific schemes and policies within the package, with the aid of three 'framework tables' specified in DOT (1995). These are:
❑ a Physical/Environmental Assessment Table (all schemes);
❑ an Economic and Financial Assessment Summary (for schemes of two million pounds to £20m); and
❑ a Comparative Economic Assessment (only for schemes exceeding £20 m).

9.5 Scheme Appraisal

Scheme appraisal should take place within the logical framework outlined above (Section 9.2). Facilities for pedestrians and cyclists, particularly segregated footpaths and cycle paths, should be incorporated in the 'physical infrastructure' and appraised within the same logical framework, although techniques for the economic appraisal of targeted investment in those modes is in a rudimentary state and requires further development. Schemes should be conceived within a context of defined objectives and identified problems and the design, forecasting and evaluation stages of the appraisal should be undertaken in an internally consistent manner. The word 'scheme' refers not only to physical infrastructure but also to policy and management measures which affect the performance of parts of the network.

Consideration of priorities between schemes, and comparisons of mutually–exclusive options within schemes, should be based on the costs and benefits resulting from traffic, economic, environmental and safety effects on the community, both direct and indirect, as well as in the short and long term. The essential point in any comparison is that the basis should be fair and consistent. This means, in particular, that scrupulous care must be given:
❑ to the definition and forecasting of the do–minimum case against which scheme options are to be tested. For example, the traffic flows and speeds forecast within the do–minimum scenario must be realistic;
❑ to the consideration of intermediate low–cost options, such as a series of junction improvements as against a full by–pass;
❑ to the comparison of all alternatives on the same set of appraisal rules and values; and

❑ to the consideration of network effects and other consequences, outside the immediate study area.

Where a series of related schemes may have a larger impact than the sum of their individual effects, they should be assessed as a group (DOT, 1996a).

The forecasting and design aspects of appraisal are discussed elsewhere (Chapter 8 and Part V). Scheme appraisal depends for its quality and accuracy on the forecasting techniques used, so the forecasting procedures must be chosen with the overall appraisal in mind. This affects the following:
  ❑ the choice of traffic appraisal methodology;
  ❑ the definition of the study area;
  ❑ the treatment of different time–periods within the forecasting procedure;
  ❑ the sophistication of junction modelling;
  ❑ the choice of economic appraisal methodology (ie fixed or variable matrix); and
  ❑ the choice of environmental indicators and prediction method.

In making these choices for important schemes, reference should be made to the DOT Advice Note on Traffic Appraisal in Urban Areas (DOT, 1996b). The remainder of this chapter considers the analysis and decision stages of the appraisal process.

9.6 Cost–Benefit Analysis

One approach to the economic appraisal of transport schemes and measures is to use a form of social cost–benefit analysis. This involves applying monetary values to the relevant physical forecasts, so as to derive the monetary benefits and costs over the assumed life of the scheme, and then aggregating the costs and benefits over time in an appropriate manner. The DOT's COBA and URECA programmes are examples of such a procedure (DOT, 1996a) [Sa]. COBA is primarily aimed at inter–urban road schemes and URECA at urban schemes.

To assess the relative cost–benefit performance, the profiles of costs and benefits, over the life of a scheme, must be built up, as follows:
  ❑ estimates of the benefits, at least for the opening year and the design year, must be found by running the traffic model and the economic evaluation for the traffic and economic conditions predicted for those years; and then
  ❑ interpolate between the modelled years and extrapolate beyond the design year to the final year of the scheme life (30 years for road schemes). This gives a complete series of costs and benefits for that scheme in relation to another (or the do–minimum case) for each year of its assumed useful life. The Department of Transport recommends that neither extrapolation or interpolation should cover periods greater than about 10 years (DOT, 1996b).

Discounting Costs and Benefits

The typical profile of costs and benefits involves costs of design and construction at the start, together with any subsequent capital renewal costs, with annual benefits to travellers and savings in maintenance and operating costs spread over the life of the scheme. A formula is required to weight and sum the benefits and costs occurring at different points in time. It is worth stressing that the entire appraisal is conducted in real terms, that is, at constant prices ignoring inflation. Only if the relative price of a resource (such as fuel) is forecast to change over time do actual price changes need to be incorporated in the model.

The formula used to weight the value of future costs and benefits, in relation to their present value, derives from the compound interest formula $S = P(1 + i)^t$ where 'i' (% per annum) is the interest rate and 't' is the interest period in years. When re–arranged, $P = S/(1+i)^t$ gives the present value (P) of a corresponding future sum (S), where 'i', in this context, is known as the discount rate.

Thus, the Net Present Value (NPV) of the profile of costs and benefits

$$NPV = (B_0 - C_0) + \frac{(B_1 - C_1)}{(1 + i)} + \frac{(B_2 - C_2)}{(1 + i)^2} + \ldots + \frac{(B_n - C_n)}{(1 + i)^n}$$

$$= \sum_{t=0}^{t=n} \frac{(B_t - C_t)}{(1 + i)^t}$$

where B_t = benefit in year t; C_t = cost in year t; and n is the assumed useful life (in years) of the scheme.

This NPV is the net surplus of discounted benefits over discounted costs, over the whole life of the scheme. In terms of the effects expressed in monetary terms, the bigger the NPV, the better the scheme. The choice of discount rate is a matter for national economic judgement. The Treasury provides guidance on this for public sector projects and, currently (1997), a discount rate of eight percent per annum is mandatory for schemes for which DOT funding is sought. For a privately–financed scheme, the discount rate is likely to be higher, to reflect the profit–margin required and the attendant scheme–specific risks.

A number of situations commonly arise in capital investment appraisal. These are summarised briefly.

Accept/Reject Criterion

The simplest case – is a project acceptable? If the NPV is positive, after discounting at the appropriate rate, then the scheme is acceptable in terms of the monetary benefits and costs.

Mutually–Exclusive Alternatives

Very often several different options exist for a project, only one of which can be selected. In the absence of capital rationing (see below), the option with the highest NPV is preferred in monetary terms.

Capital Rationing

Frequently, insufficient capital is available to enable all economically worthwhile projects to be undertaken. In this situation, some form of priority–ranking is required. Although practical considerations, such as land acquisition and readiness to start construction, may be relevant, projects can be ranked, from an economic point of view, in order of their Net Present Value to Capital Cost ratios (NPV/C_0) (where C_0 is the cost at year zero) and selected in that rank–order, until the budget is exhausted. A similar approach, used by DOT, ranks by the ratio of Present Value of Benefits to Present Value of Costs. At that point, the marginal (or cut–off) benefit/cost ratio is defined, where B/C ratio is given by $(NPV/C_0 + 1.0)$.

Mutually–Exclusive Alternatives under Capital Rationing

To choose between such alternatives, it is necessary to check that:
❏ the chosen option has a high enough NPV/C_0 ratio for acceptance; and
❏ that the incremental NPV/C_0 ratio of the chosen option, over the best cheaper alternative, is higher than the marginal cut–off ratio.

Timing of Investment

A simple test of whether a project should be implemented as soon as possible or deferred to a future date is provided by the first year rate of return (FYRR) ie $(B_1-C_1)/C_0 \times 100\%$, where year 1 is the first year of operation.

If this value is less than twice the discount rate, this suggests that benefits in the early years of the scheme's life are modest and deferment should be considered. However, the FYRR test is not a reliable substitute for NPV or benefit/cost ratio when assessing overall project worth, as it ignores all the future costs and benefits beyond the end of year 1.

Presentation to decision–takers

The concepts of NPV and benefit/cost ratio are often unfamiliar to political decision–takers, who may require a more intuitive indicator of the worth of a project. For this purpose, the Internal Rate of Return (IRR) can be useful. The IRR is that rate of interest which just equalises the discounted costs and the discounted benefits, so that:

$$\sum_{t=0}^{t=n} \frac{B_t}{(1+i)^t} = \sum_{t=0}^{t=n} \frac{C_t}{(1+i)^t}$$

Hence, NPV $= \displaystyle\sum_{t=0}^{t=n} \frac{(B_t - C_t)}{(1+i)^t} = 0$

Then, if the internal rate of return on the project is greater than the discount rate, the project is economically worthwhile. For decision–takers to know that the predicted rate of return on the project is, say, 15% pa while the cost of capital is eight percent per annum provides a ready indicator of the project's worth and is therefore a useful presentational device.

To summarise, cost–benefit analysis indicators, such as the Net Present Value and other measures of worth, are important but should not be presented in isolation, because:
❏ they give no indication of the pattern or distribution of costs and benefits over different groups within the community; and
❏ they omit entirely those equally real costs and benefits, such as environmental impacts, which are not valued in monetary terms.

For these reasons, the DOT has developed the so–called 'framework' approach.

9.7 The Framework Approach

A useful model for reporting major road and traffic management schemes is provided by the DOT's 'Assessment Summary Report' procedure for trunk roads. The Report brings together the engineering, environmental, traffic and economic assessment, so as to identify clearly the advantages and disadvantages of the preferred route or scheme option. Within the Report, the Framework is used to organise and present relevant data. It consists of a matrix setting out the expected effects on different groups in the community, for each route or scheme option.

The Framework data are a mixture of descriptive, quantified and costed entries. The use of the Framework is intended to enable these various quantified and non–quantifiable costs and benefits to be compared in a comprehensive and consistent way.

The Framework method follows the recommendations of the Leitch Committee (ACTRA, 1978). The methods recommended by DOT are set out in DOT (1993a). An example of an earlier version of the Framework approach, applied in an urban context, is to be found in SACTRA (1986).

Under the Department's methodology, the impacts of schemes are presented in four Appraisal Groups:

Group 1: Local People and their Communities;
Group 2: Travellers;
Group 3: The Cultural and Natural Environment; and
Group 4: Policies and Plans.

The 'environmental impact table' is followed by an 'economic performance table', which reports the results of the cost–benefit analysis, for those impacts which are evaluated in monetary terms. No attempt is made, within the Framework, to apply weights to the non–monetary entries. It is, therefore, a balance sheet, in which the balance is assessed by judgement. An illustrative example of the Assessment Summary Tables is given in the Design Manual for Roads and Bridges Vol. 5 (DOT, 1993a). In the remaining sections of this chapter, assessment of the key traffic, environmental and economic impacts is reviewed.

9.8 Traffic Effects

The principal methods for assessing the traffic effects of schemes are set out below. The choice of technique is a matter for professional judgement and will depend on individual circumstances. Expenditure on appraisal should always be small in relation to the capital expenditures involved and the complexities of the analysis should be in proportion to those of the scheme. In some cases, especially for small–scale traffic management schemes, experimental implementation may be the best test.

Simple tabular methods (eg Roads Note 502) [Sa]

These are appropriate for small schemes, which do not warrant any modelling of traffic and involve limited measurement and assessment based on simple indicators having a high judgmental content.

COBA (Cost Benefit Analysis)

COBA is the longest established DOT method of economic appraisal of inter–urban road schemes. The latest version is COBA 10 (DOT, 1996a) [Sa]. When used in an urban context, COBA has a number of limitations which can be serious in certain circumstances. These are:

❑ COBA takes the assigned traffic flows on links and calculates average speeds, using COBA speed/flow relationships, derived from link–specific geometric data. Although different speeds are calculated for peak and off–peak conditions, these may be inconsistent with the speeds in the traffic assignment model, from which the assigned flows were derived;

❑ COBA was designed for inter–urban scheme appraisal, where junctions are usually widely–spaced. The program is not capable of modelling interaction between junctions, such as blocking back and similar features of heavy urban congestion; and

❑ COBA contains 'cut–offs' for both minimum speed and maximum delay, which may be inappropriate in urban conditions. However, this can be overcome by the use of local values validated for the model area.

URECA (Urban Economic Appraisal) [Sa]

URECA is an alternative to COBA, designed to overcome these shortcomings (DOT, 1996e) [Sa]. It is used in association with congested assignment models, such as CONTRAM and SATURN. URECA performs the economic calculations directly on the speed and flows output from the traffic assignment models and, therefore, overcomes the limitations of COBA when applied to urban conditions. However, the realism of the estimated flows and delays, critical to the credibility of the economic appraisal, must be examined carefully. Congestion assignment models may also be used, together with URECA, to estimate the cost of delays to traffic during construction, which can be particularly important for urban schemes. The use of the Department of Transport's QUADRO program for estimating the effects of roadworks is not recommended in urban conditions (DOT, 1996f). QUADRO uses a single diversion route when assessing traffic delay/costs at roadworks and this is not always appropriate where numerous potential diversion routes exist. Although in use since the early 1990s, experience is still being gained in the practical application of URECA in scheme appraisal. In most cases, expert advice is likely to be required.

Earlier versions of COBA and URECA were restricted by a fixed demand or 'fixed trip–matrix' assumption. This assumes that the volume and pattern of trip–making is given and fixed with network conditions. This is a convenient simplifying assumption, because the benefits of road improvements are then the travel–time and cost–savings of catering for a given, assumed fixed, amount of traffic. Figure 9.1 shows the cost–saving, for a representative origin – destination pair, as travel

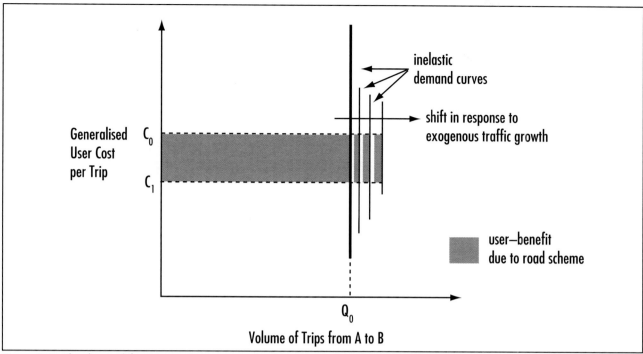

Figure 9.1: The demand for trip–making assumed independent of its cost.

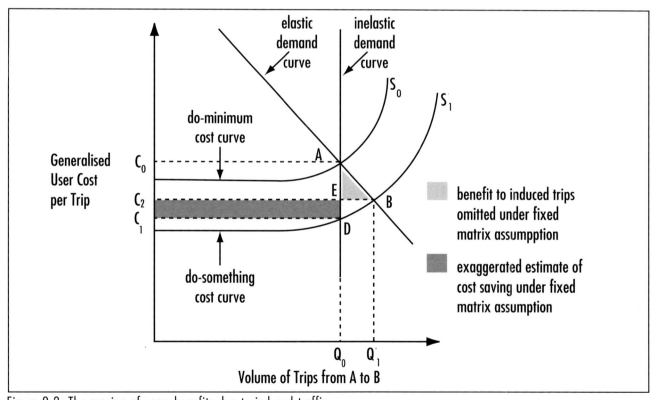

Figure 9.2: The erosion of user–benefits due to induced traffic.

cost for the Q_0 trips between A and B falls from C_0, in the do–minimum, to C_1, in the do–something case. The overall scheme benefit is the sum of the cost–savings, over all the O–D pairs whose travel costs are affected by the scheme.

Since the volume of travel is assumed to be fixed, the benefits may also be calculated by aggregating the cost–savings over the links in the network. This is how earlier versions of COBA and URECA used to operate. They ignored the effects that new road

Average Resource Values of Travel–time Savings	
By Vehicle–occupant	Value (pence per person–hour)
(a) Working Time	
Car driver	1289.8
Car Passenger	1070.6
LGV Occupant	1003.1
OGV Occupant	945.0
PSV Driver	983.1
PSV Passenger	1064.1
Train Passenger	1619.3
Underground Passenger	1593.9
Average All Workers	1277.5
(b) Non–Working Time	
Average all vehicle–occupants	
(standard appraisal value)	315.0
By vehicle–type*	Value (pence per vehicle–hour)
(a) Cars	
Working time	1407.6
Non–working	549.1
Average	673.6
(b) LGVs	
Working time	1424.4
Non–working time	504.0
Average	1166.7
(c) OGVs Average	945.0
(d) PSVs	4900.8
(e) All Road Vehicles Average	784.4
*assuming average vehicle–occupancy rates	

Table 9.1: Average resource values of savings in travel–time (1994 prices).

capacity might have on travel demand.

However, in recent years, it has been recognised that the calculations of user–benefits can, in some circumstances, lead to over–estimates where it is assumed that new capacity does not induce additional vehicle–kms on the network (the fixed demand assumption). SACTRA (1994) concluded that induced traffic does occur and is potentially of greatest importance in any of three circumstances:

❑ where the network is operating, or is expected to operate, close to capacity; or

❑ where the elasticity of demand with respect to travel costs is high; or

❑ where the implementation of a scheme causes large changes in travel costs.

The first of these criteria will apply to those urban schemes which enhance system capacity. The second will apply particularly where there is lively competition between public and private transport modes. The third is typical of a new river–crossing, where no link existed previously. In such cases, variable demand methods are required, in which traffic volume is allowed to respond to improved network quality (ie a variable trip matrix).

In Figure 9.2, user–costs fall initially from C_0 to C_1 but the volume of traffic responds, partially recongesting the network and giving a final benefit of C_0ABC_2. This benefit accrues partly to the existing users (Q_0) and partly to the induced traffic ($Q_0 Q_1$). Relative to Figure 9.1, the aggregate benefit may be higher or lower but, in congested conditions, it is likely to be lower (SACTRA, 1994). The traffic modelling, in such cases, will need to be undertaken using a conventional four–stage model or by using 'elastic' assignment methods (see Chapter 8). In response to the SACTRA Report, the Department of

Years	Economic Forecast	
	Low Growth (% pa)	High Growth (% pa)
1994–1995 (Actual)	1.900	1.900
1995 – 2001	1.625	2.875
2002 onwards	1.805	3.055

Table 9.2: Annual compound rates of growth (% pa) assumed for the real values of time–savings.

Transport issued guidance on 'Induced Traffic' (DOT, 1997) and investigated the most appropriate methodology for undertaking variable trip–matrix economic appraisals. Versions of the URECA program, later than 3.1, can perform both fixed and variable trip–matrix appraisals. The COBA Manual (DOT, 1996a) [Sa] provides guidance on performing variable trip–matrix appraisals with the COBA program. For further reading, see Section 9.14.

All these methods produce estimates of the travel–times and distances for each scenario. These are converted into benefits using standard values of time, operating costs and accidents. Full details of these values are given in DOT (1996c); a brief summary of each appears below.

Travel–time savings

Travel–time benefits are by far the most important component of the costed benefits of road schemes, generally accounting for over 80% of these benefits. Travel time is split into two categories, working (or employers' business) time and non–working time. The value of savings in working time is based on the assumption that employees' time can, on average, be valued at a rate equal to their gross employment costs per hour, including relevant overheads. The valuation of savings in non–working time spent travelling is based on a wide variety of studies of travellers' responses to road improvements, tolled crossings and public transport improvements. Current appraisal values and their projected rates of growth are shown in Tables 9.1 and 9.2. The values of time given here are based on research work reported in 1987 (MVA/ITS/TSU, 1987). These values may change as a result of recent research for the Department of Transport (Gunn et al, 1996).

Savings in vehicle–operating costs

Vehicle operating costs are normally a relatively small component of the traffic effects and may either rise or fall, depending on the distance, speed and queueing effects of the scheme. Any net savings in fuel, wear and tear and depreciation should be seen as a tangible benefit of the new road. The standard formula for valuing operating cost–savings considers all the distance, time and speed–related components of cost.

Accident–reduction benefits

Accidents–savings are another important component of overall benefit. Scheme assessment must include consideration of the likely net effects on road safety, that is, the change in the expected level of accidents of all kinds (see Chapter 16). Accident reductions can be estimated and valued. The Department of Transport (DOT, 1996d) gives values (regularly updated) for the

Type of Link	COBA 10 Link Accident type	Average Accident Rate (PIA per million vehicle–kms)	
		Link Only	Combined Links and junctions
New Links (Urban) Single Carriageway	4 to 7	0.231	0.822
Dual (AP) Carriageway	10 to 12	0.266	0.901
Dual (3) Motorway	13 to 15	0.333	1.310
Existing Links (Urban) Urban A	8	0.231	0.822
Urban B/Other	9	0.303	0.864

Table 9.3: Average accident–rates used in COBA 10.

Type of Casualty	Value (£ per casualty)
Fatal Casualty	784,090
Serious Casualty	89,380
Slight Casualty	6,920
Cost per damage–only accident (urban areas)	987

Table 9.4(a): Values of casualty–avoidance (1994 prices).

values of accident–reduction, which are based on three components:

❑ the direct financial costs to individuals and organisations involved (eg damage to vehicles and allocated police and medical costs);

❑ the opportunity cost of lost net output to the economy for those killed or injured; and

❑ a community–average value based on individuals' willingness to pay for marginal reductions in the risk of death or injury.

Although the Department presents figures of accident–rates for each road type (see Tables 9.3 and 9.4), these may be too general for use in specific urban schemes and local data should be used, wherever possible. However, it is important to ensure that the local data is statistically reliable and represents normal operating conditions. Thus, the Department recommend that local data should normally cover the five years prior to the assessment and must cover at least three years. It should also relate to a period when conditions on the road have been broadly unchanged (DOT, 1996d) [Sa].

9.9 Costs

The capital, maintenance and other costs of providing the road need to be set against the estimated travel–time, operational and safety benefits of schemes.

The main capital costs arise from:

❑ surveys, design and administration, including site–supervision, and materials testing;

❑ the costs of acquiring the land to construct the scheme and of any necessary accommodation works;

❑ the costs of construction, including the main engineering and ancillary contracts, such as the costs of altering or accommodating public utilities, and including contingency allowances;

❑ compensation costs, for example for severance or noise nuisance, and also costs of mitigation, such as noise barriers and visual screening; and

❑ capital renewal (as opposed to routine maintenance) costs incurred at various times during the life of the scheme.

The routine maintenance costs will include provision for

minor repairs, cleaning, lighting and snow–clearance but exclude structural repair and resurfacing, which should be regarded as a capital renewal item.

9.10 Economic Development Effects

An objective of many transport schemes is to promote economic regeneration. By improving accessibility, new roads may help to create local employment, especially in 'footloose' industries such as distribution. Where effects on local employment are expected, these should be included in the Framework. However, appraisal should be cautious and careful on this point, because:

❑ it is necessary to avoid double–counting, since the transport cost savings to existing traffic are already taken into account through the user–benefit calculations;

❑ it is possible that some of the predicted increase in activity may be diverted from elsewhere within the district rather than be entirely new; and

❑ it is important to recognise that the local and national economic perspectives may be different.

Thus, although a new road may increase one district's share of regional economic activity, this often comes at the expense of other districts. The net national increase in economic activity as a result of road investment is a subtle and sometimes elusive concept. Nevertheless, where new roads enable land to be reclaimed or regained for industrial and commercial use or where roads can be used to rejuvenate the image and confidence of an area, it is legitimate to consider those planning benefits within the Framework. If the anticipated effects on the pattern of land–use are significant, this must be considered within the transportation modelling for the scheme.

9.11 Environmental Effects (see

Chapter 17)

With rare exceptions, where actual financial costs are incurred, such as the costs of mitigation, monetary values are not attributed to the environmental impact

Type of road	Casualties per injury–accident			Cost per accident (£)	
	Fatal	Serious	Slight	Link–only	Combined link and junction
New Links Single carriageway (30 or 40 miles/h)	0.021	0.223	1.030	63,130	54,300
Dual(AP) carriageway (30 or 40 miles/h)	0.025	0.211	1.029	65,180	55,060
Motorways (70 miles/h)	0.036	0.221	1.382	75,950	75,950
Existing Links Urban A (30 or 40 miles/h)	0.021	0.223	1.030	63,130	54,300
Urban B and other (30 or 40 miles/h)	0.014	0.215	0.972	56,390	50,920

Table 9.4(b): Average casualties per injury–accident and costs per accident (1994 values and prices).

of road schemes. Yet these environmental impacts represent real effects on society, which must, somehow, be incorporated within the assessment. SACTRA (1992) stated that there is no fundamental impediment to the monetary valuation of most environmental impacts and recommended that progress should be made towards valuing some of the more homogeneous impacts, such as noise and some pollutants. The DOT has initiated research to help to determine appropriate monetary values. However, some environmental impacts (eg loss of natural assets and landscape effects) are site-specific, so that their monetary values are not transferable to or from other locations. Comprehensive monetary valuation of the environmental impacts of transport schemes is,

therefore, unlikely to be achievable for many years, if ever.

The Environmental Impact Assessment (EIA) will comprise, therefore, a number of qualitative and quantitative, but non–monetary, measures. The EIA needs to be tailored to the specific characteristics of the type of scheme or strategy being tested. Hence, the precise format of the EIA will depend on its purpose, such as to support a TPP submission or to provide a statutory Environmental Statement (ES) or just to give general guidance to decision–makers, and on its objectives, such as to justify construction of a new road or the introduction of bus–lanes or to test options at the strategic level. This will also determine

Air Quality.
Cultural Heritage – including effects on archaeology, conservation areas, and built heritage.
Disruption due to Construction.
Ecology and Nature Conservation.
Landscape Effects – including townscape effects and visual impact assessment.
Land–Use including effects on agricultural land and production, and effects on development land and designations for future development.
Traffic Noise and Vibration.
Pedestrians, Cyclists, Equestrians and Community Effects – including pedestrian delay, pedestrian amenity, drivers' delay, pedestrian fear and intimidation, community severance and accidents.
Vehicle Travellers – including driver stress and the view from the road.
Water Quality and Drainage.
Geology and Soils.
Policies and Plans.

Table 9.5 Checklist of environmental issues. (Based on the issues included in the Design Manual for Roads and Bridges)

the level of detail of information needed. It is important that a rigorous and consistent approach is adopted to all of these assessments and the Framework approach is a useful tool to achieve this aim (SACTRA, 1992). Within this approach, specific attention may need to be given to air quality in the context of obligations placed on local authorities, under the Environment Act (HMG, 1995), to monitor air quality and, if necessary, to take remedial action.

The EIA process can be summarised, as follows:
- ❏ define the purpose and objectives of the EIA and determine the format required to support the decision–making process and the level of detail of information needed;
- ❏ in light of the objectives, identify the environmental issues which are likely to demonstrate a significant effect, due to the scheme or option, using a checklist of issues as a guide (various checklists exists, an example is given in Table 9.5) and taking into account the environmental sensitivity of the area;
- ❏ identify units of measurement or indicators for each environmental issue (for example, specified changes in traffic volume may be used as indicators of traffic and air quality change) and then define the methodology to predict and assess each issue, in some cases this may simply be a qualitative statement;
- ❏ calculate (where it is possible to quantify, such as with noise levels) or estimate the change in effect due to the scheme or option, compared with the do–nothing situation, which can be either beneficial or adverse;
- ❏ define thresholds, expressed in terms of absolute values and percentage change, based on UK, EU or international standards, where they exist. For example, in the case of air pollution, standards exist for the main primary traffic–related pollutants (see Chapter 17);
- ❏ compare the predicted values against the thresholds, to determine the significance of the effect (eg slight, moderate or severe) and then indicate the likely number of people affected and the geographical extent, to illustrate the magnitude of the effects; and
- ❏ draw the results of the issue–based assessment into a summary Framework, grouped in terms of the effects on different user–groups, in such a manner that it allows the different effects to be balanced one against the other, together with a written statement. Effects during the construction phase, as well as when the scheme is operational, should be included, if relevant. Where adverse environmental effects are predicted, proposals to mitigate these effects should be drawn up and any 'residual' effects, which cannot be mitigated, should be identified.

The Design Manual for Roads and Bridges Volume 11 (DOT, 1993b) is designed to provide guidance, primarily, on the assessment of all new trunk road and motorway schemes and can be used for appropriate urban road improvement schemes. It provides a useful starting point for any road–based EIA, as it gives a checklist of environmental issues to be covered, sets out assessment methods and thresholds for individual issues and provides the template for an assessment Framework. The DMRB approach comprises three assessment stages, reflecting the development of the scheme and the level of detail required: Stage 1 is at the initial option-selection stage, when a large number of options are tested; Stage 2 is when the number has been narrowed down to two or three options for public consultation; and Stage 3 is the assessment of the 'preferred route', for which an Environmental Statement may be required.

Other guidance is available for the assessment of road traffic impacts from the Institute of Environmental Assessment (IEA, 1993) and for public transport projects from the Passenger Transport Executive Group (PTEG) (Lee *et al*, 1991).

The EIA is also the first stage in producing an Environmental Statement (ES). An ES should be an unbiased and objective document, produced in accordance with the European Directive (EC, 1994). This is enacted in England and Wales through the Town and Country Planning (Assessment of Environmental Effects) Regulations 1988 and Amendments and Highways (Assessment of Environmental Effects) Regulations 1988 and 1994, which amend the previous section 105A of the Highways Act 1980 [NIb]. The requirement for an ES can be either mandatory (for Schedule 1 projects) or voluntary (for Schedule 2 projects), depending on the scheme, as follows:
- ❏ Schedule 1 projects – construction of motorways, express roads and long distance railway lines and airports, with a runway length of 2100m or more; and
- ❏ Schedule 2 projects – tramways, elevated and underground railways, suspended cable–cars or similar, used exclusively or mainly for passenger transport.

Also, within urban areas, any scheme where more than 1,500 dwellings lie within 100m of the centre line of the proposed road (or an existing road, in the case of major improvements) may require an ES. Outside urban areas, new roads or major improvements over 10km in length, or over one kilometre in length if through a National Park, or within 100m of a site of special scientific interest, a national nature reserve or conservation area, may require an ES.

Proposals to amend the EC Directive are with the Council of Ministers of the European Union, which could add construction of a new road or widening of 10 km or more of dual carriageway to the list of Schedule 1 projects.

Guidance on the information to be included in an ES is given in a number of DOE and DOT documents (DOE, 1988; 1989; 1991; and 1995; and DOT, 1993b and 1994).

The DMRB (DOT, 1993b and 1994) states that the ES should consist of three parts:
❑ Volume 1 – a comprehensive and concise document, drawing together all the relevant information about the scheme, including baseline information, predicted environmental effects and mitigation. An Environmental Impacts Table (developed as part of the Stage 2 and Stage 3 assessments) is included, which itself has three parts:
  ❑ 'appraisal groups', which is essentially the same as the appraisal Framework showing the impacts on four groups (see Section 9.7);
  ❑ land–use table, stating land–take and land–uses affected; and
  ❑ mitigation table, setting out agreed mitigation measures, their costs and predicted benefits;

❑ Volume 2 – a detailed assessment of the significant environmental effects under each issue and a non–technical summary giving a balanced and accurate statement of the key information and conclusions in Volume 1 of the ES, expressed in non–technical language for wider circulation to members of the public. It should include a 'constraints map' of the scheme and the surrounding area.

Weighing up the advantages and disadvantages of schemes is undoubtedly a difficult process for decision–makers. For that reason, there need to be checks and balances, internally within the project team and externally through the processes of public consultation and Public Inquiry, to minimise the risk of bias in the assessment.

9.12 Public Transport Impacts

Some projects will have significant impacts on public transport operators and users. These will range from projects, such as town–centre pedestrianisation, which affect public transport, among other interests, to road projects, which create bus–priorities or improve public transport running, or comprise major public transport schemes, such as local rail, LRT or guided–bus schemes.

In principle, public transport appraisal should be no different from highway appraisal. The Framework Approach should be followed, with public transport users and operators identified as specific interest–groups within the appraisal. However, for the appraisal, this will have particular implications, as follows:
❑ modelling of modal split, so as to estimate increases in demand for public transport arising from faster or more reliable services;
❑ assessment of implications for service levels, fares, patronage, revenues and bus–operating costs; and
❑ consideration of the effects of public transport improvements on traffic congestion levels for remaining road–users.

Where projects are in the hands of a local authority, as will be the case with traffic management and bus–priority schemes, the Framework Approach (incorporating a cost–benefit analysis, which considers public as well as private transport effects) is the appropriate analytical tool. New railway stations on local lines may also fall into this category. However, in practice, not all public transport projects are of this nature and different considerations then arise.

It should be remembered that local bus and train operations in the UK are commercially organised [NIc]. This means that public transport operators will base their decisions on the expected commercial return to them, through the revenues and costs, when considering investment decisions relating to vehicle replacement, depot provision and service levels. Local authorities may choose to supplement commercial provision, by adding 'socially desirable but unprofitable' services and by offering concessionary travel to eligible groups, such as senior citizens and people with disabilities. However, the commercial nature of the deregulated service places severe constraints on a local authority's freedom to use public transport fares and/or service levels as a regulatory or planning tool. In practice, within the current (1997) regulatory environment, local authorities must try to focus on measures which are both socially advantageous and acceptable to commercial operators.

Also, with major public transport investment, for example in new LRT services, special appraisal rules must be followed, if the scheme is to be considered for grant–aid from the government under section 56 of the Transport Act 1968 [NId]. This applies to substantial improvements 'of at least regional significance'. In such cases, the Department of Transport requires a restricted form of cost–benefit analysis to be carried out. The operator is required to set service levels and fares in as commercial a manner as possible, raising prices (for example) to convert a premium quality service into enhanced revenue.

The distinction between this form of appraisal, which excludes user–benefits, and the appraisal of other transport schemes, where they are included, needs to be understood when the results of such appraisals are compared.

Any residual public transport user–benefits are excluded from the appraisal. In addition, private sector funding through developers' contributions must be sought. However, it is recognised that there are some wider social benefits, which the operator cannot capture and which may therefore be added into the revenues and costs. These non–user benefits include road 'decongestion' benefits, environmental benefits and employment–generating benefits. Thus, the appraisal takes the form of comparing the present value of net revenues plus the non–user benefits less the capital, maintenance and operating costs of the project against the baseline case. For relevant references on the appraisal of such projects, see Section 9.14.

9.13 References

ACTRA (1978)	'Report of the Advisory Committee on Trunk Road Assessment', Stationery Office.
DOE (1988)	Circular 15/88 (WO 23/88) 'Environmental Assessment', Stationery Office.
DOE (1989)	'Environmental Assessment – A Guide to Procedures', Stationery Office.
DOE (1991)	'Monitoring Environmental Assessment and Planning', Stationery Office.
DOE (1995)	'Preparation of Environmental Statements for Planning Projects that require Environmental Assessment – A Good Practice Guide', Stationery Office.
DOT (1993a)	TD37/93 (DMRB 5.1.2) 'Scheme Assessment Reporting', Stationery Office.
DOT (1993b)	'Environmental Assessment', (DMRB 11) Stationery Office.
DOT (1994)	'Environmental Assessment Amendment' (DMRB 11), Stationery Office.
DOT (1995)	TPP Submissions for 1996/97 'Supplementary Guidance Notes on the Package Approach', DOT.
DOT (1996a)	'COBA 10 Manual' (DMRB 13.1), Stationery Office, [Sa].
DOT (1996b)	Advice Note (DMRB 12.2.1) 'Traffic Appraisal in Urban Areas', Stationery Office.
DOT (1996c)	Highways Economics Note 2 (DMRB 13.2) 'Values of Time and Vehicle Operating Costs', Stationery Office.
DOT (1996d)	COBA 10 Manual (DMRB 13.1.2) 'The Valuation of Accidents', Stationery Office [Sa].
DOT (1996e)	'The URECA Manual', DOT [Sa].
DOT (1996f)	'The QUADRO Manual' (DMRB 14.1), Stationery Office.
DOT (1997)	'Induced Traffic Appraisal' (DMRB 12.2.2), Stationery Office.
EC (1994)	European Commission Proposal for a Council Directive amending Directive 85/337/EEC on 'The Assessment of the Effects of Certain Public and Private Projects on the Environment', Official Journal of the European Communities Volume 37 (C130).
Gunn H, Bradley M and Rohr C (1996)	'The 1994 National Value of Time Study of Road Traffic in England', PTRC 'Value of Time' Seminar, October 1996
HMG (1991)	'Economic Appraisal in Central Government. A Technical Guide for Government Departments', Stationery Office.
HMG (1995)	'The Environment Act', Stationery Office.
IEA (1993)	'Guidelines for the Environmental Assessment of

Road Traffic' Guide Note No. 1, Institute of Environmental Assessment.

IHT (1996) 'Guidelines on Developing Urban Transport Strategies', The Institution of Highways & Transportation.

Lee N and Lewis M (1991) 'Environmental Assessment Guide for Passenger Transport Schemes', Passenger Transport Group, Manchester.

MVA/ITS/TSU (1987) 'The Value of Travel Time Savings', Policy Journals, Newbury.

SACTRA (1986) 'Urban Road Appraisal' Report of the Standing Advisory Committee on Trunk Road Assessment', Stationery Office.

SACTRA (1992) 'Assessing the Environmental Impact of Road Schemes'. Report of the Standing Advisory Committee on Trunk Road Assessment, Stationery Office.

SACTRA (1994) 'Trunk Roads and the Generation of Traffic'. Report of the Standing Advisory Committee on Trunk Road Assessment, Stationery Office.

9.14 Further Information

On the Common Appraisal Framework for the package approach, see:

Farrell P and Pearman AD (1994) 'Cross–modal evaluation within Integrated Transport Plans', Transportation Planning Systems 2 (1).

May AD (1994) 'The Development of the UK Package Approach and the Common Appraisal Framework', Transportation Planning Systems 2 (3).

MVA, Oscar Faber TPA and ITS, Leeds University (1994) 'A Common Appraisal Framework for Urban Transport Projects', Birmingham City Council.

May AD and 'The Design of Integrated

Roberts M (1995) Transport Strategies', Transport Policy 2 (2).

On induced traffic, see:

Hall MD and Merrall AC (1996) 'Variable Trip Matrix Scheme Assessment', Traffic Engineering + Control 37 (4).

Coombe D et al (1996) Special issue of 'Transportation' 25(1).

On environmental assessment, see:

Bateman I et al (1993) 'External cost–benefit analysis of UK highway proposals: environmental evaluation and equity', Project Appraisal 8 (4).

Institute of Environmental Assessment (1995) 'Guidance on Baseline Ecological Assessment', IEA.

Rendel Planning and Environmental Appraisal Group (UEA) (1992) 'Environmental appraisal: a review of monetary evaluation and other techniques', TRL Contractors Report 290.

On evaluation of public transport projects, see:

DOT (1989) 'Section 56 Grant for Public Transport', Circular 3/89, Stationery Office.

Halcrow Fox and Associates (1991) 'The Application of Section 56 Criteria', PTE Group with the Department of Transport.

Nash CA and Preston JM (1991) 'Appraisal of rail investment projects – recent British experience', Transport Reviews 11(4).

Chapter 10 Involving the Public

10.1 Introduction

This chapter gives an overview of the requirements for, and processes of, public involvement in the planning of transport policies and schemes. It begins (Section 10.1) with fundamental questions: about who constitute the 'public'; why should they be involved in the planning and implementation of transport schemes; and to what extent? Sections 10.2 and 10.3 outline the statutory consultation procedures that highway authorities and others need to observe in relation to specific transport schemes and in the broader Development Plan context, respectively. In Section 10.4, various ways in which the minimum statutory requirements might be extended into non-statutory public involvement are discussed. Techniques of public involvement are outlined in Section 10.5. Section 10.6 concludes by considering transport consultation in a wider context and, in particular, considers the role of elected members in relation to direct public involvement.

Why Involve the Public Beyond the Minimum Requirement?

A minimum level of public involvement is assured through the various statutory requirements for the direct involvement of particular groups of the public in defined circumstances. These include the preparation of structure plans, which inevitably feature road schemes and sometimes railways and other fixed track systems, and of local plans and development applications (see Sections 10.2 and 10.3). In addition, there are requirements to notify the public of proposed Traffic Regulation Orders. However, there are no statutory requirements for public involvement when developing local transport strategies or TPP package bids. Nevertheless, the Department of Transport's Package Bid Guidelines recommend that 'public support be demonstrated' and that the overall policy framework be subject to consultation in the development plan process.

There are good reasons, however, for involving the public more fully than the minimum statutory requirement and to extend this involvement into areas where there is no statutory requirement to consult at all (Gyford, 1991). These reasons range from the philosophical to the pragmatic, including:

❑ traffic engineers or transportation planners working in a local authority are unlikely to know

any particular area as well as those who use it regularly. Often, local people are not only well aware of the problems but also have positive ideas about what kinds of measure might be appropriate;
❑ increasingly, transportation strategies require changes in public attitudes and behaviour, if they are to succeed in their objectives. Indeed, the more a scheme is supported by the groups affected, the more likely will the implementation of the scheme be successful;
❑ where there are financial constraints, such that a highway authority has to choose between conflicting claims on resources, schemes are more likely to be adopted if there is evidence of strong local support; and
❑ transportation strategies and investments will have a major impact on the areas where people live, work and carry out their daily activities. Increasing numbers of people feel that they have a democratic right to be consulted, especially if the Local Authority is going to take any action which would materially affect their way of life.

Although the general public does not formally have to be consulted on most traffic management proposals, they are increasingly being approached on such matters and a legal ruling has increased pressure to do so. A High Court challenge by a residents' association in the London Borough of Camden over a Controlled Parking Zone produced a ruling from the judge, in 1996, that the Council had not conducted 'fair and effective' consultation with residents' groups and that the scheme should be considered afresh.

For these reasons, coupled with the more general interest in community involvement among local authorities brought about, for example, by the adoption of Agenda 21 on sustainability (see Section 17.3), highway authorities increasingly undertake more than the statutory minimum level of consultation for transport proposals [NIa].

Who Are 'The Public'?

The 'public', in the broadest sense, includes all those who have an interest in the development of an urban transportation strategy and the implementation of its component parts but excludes the team of local authority professionals and any consultants charged with the task of preparing the strategy. They comprise three broad categories of people:

❑ those who are users of transport services and networks, by any mode of transport, in relation to both passenger and freight movement;

❑ those who are affected by transport provision and its use; principally those who live, work, shop, run their businesses etc. in the area concerned; and

❑ those who provide transport infrastructure or operate it on behalf of others.

The 'man in the street' represents a sub–group of the first two categories and usually requires different means of contact and methods of communication than the various professionals involved in, or affected by, the transport proposals. However, some people may find themselves in two or three of those groups, by virtue of the different roles they play in society. While this may mean that different processes of involvement are, in part, duplicated themselves, by reaching the same people, it does suggest that some of the difficult trade–offs that have to be made are conflicts within, rather than between, groups of people, which may make it easier to identify and resolve certain problems.

Groups of people to be considered

The following groups of people would normally be included, at some stage and to varying degrees, in a local public involvement exercise:

❑ local elected members (MPs, MEPs, County, District, Town and Parish Councillors);

❑ council officers (County, District, Town, Parish);

❑ Residents' Associations and 'action groups', which are often only formed when change is threatened;

❑ Conservation Area committees, amenity societies etc;

❑ local residents, within the area affected by the proposals;

❑ motoring organisations and relevant trade associations;

❑ other special interest–groups (eg cyclists, horse riders, walkers);

❑ minority groups (eg the elderly and the mobility–impaired);

❑ owners/managers of firms and enterprises affected, including local employers, places of worship, places of education, etc;

❑ shoppers and other visitors to the town (including workers who commute into the area);

❑ emergency services, including the police;

❑ public transport providers;

❑ local business people and shopkeepers; and

❑ Chambers of Commerce.

An important spatial issue concerns the extent of the area in which to consult. This is important, not just in terms of the resources needed but because the balance of views may vary according to the size of the area covered.

When preparing transportation packages for urban or rural areas, public involvement should encompass the whole of the package area but should also provide opportunities for visitors from outside the area and those travelling through the area to make their views known. Similarly, when dealing with specific transport schemes, an attempt should be made to engage those who will actually use, or be directly or indirectly affected by, the scheme.

To What Additional Extent Should the Public be Involved?

The notion of 'involving the public' can encompass a wide range of processes and relationships, from seemingly ratifying decisions that have already been made, through to active participation in shaping decisions or to the devolution of power itself.

Broadly speaking, three levels of public involvement can be identified (IHT, 1996):

❑ **information**: where the public is notified about the proposed implementation of a transportation strategy, perhaps using public relations techniques to `sell' the proposals to the public. This is essentially a one–way process, in which information is disseminated from an authority to the public via press releases, etc;

❑ **consultation**: where the views of the public are sought, at various stages of the strategy formulation and implementation process, and are input to a process which remains under the full control of the relevant professionals. This can typically be characterised as a 'one–way process with feedback' and usually involves the use of leaflets, exhibitions or questionnaires; and

❑ **participation**: where the public are brought into a two–way dialogue with the professionals and have a direct influence on the outcome of the process; as a result, changes in attitudes and perceptions are likely to occur on both sides. Here, at least part of the process requires direct discussions between the various parties.

Note that the distinction between consultation and participation is that the latter mechanism tries to ensure a stronger public influence over the final outcome whereas, with consultation, those in charge of the study have much greater discretion as to whether, and to what degree, to take account of the results of the exercise.

In general, the higher the level of public involvement in the process, the greater are the time and other resources required during strategy development but this is offset by the improved quality of the strategy/scheme and the likely success of implementation. Success can be measured in terms both of a smoother administrative implementation (with a reduced number of formal objections) and of higher levels of public compliance with traffic regulations after a scheme is introduced.

One example is in Sheffield (Taylor, 1996), where there had been considerable local reaction against a residential traffic–calming scheme that had been introduced with little public consultation. The City Council then decided to undertake an extensive consultation exercise over three years, based upon an 'awareness campaign', involving market research, open days, exhibitions, meetings with local groups, general public meetings and questionnaires. Not surprisingly, conflicts were revealed and proposals put forward to overcome them. This process received general support and resulted in a set of measures that could then be implemented smoothly.

More innovative local authorities now use the selective application of information, consultation and participation, within the same strategic study, targeted at different groups. Information provision extends to the widest possible target audience but this can be something of a 'scatter gun' approach, with information reaching many people who have no local interest in the process. Consultation should be more targeted at particular groups (businesses, local residents, visitors, etc), although it may involve large numbers of people in the structured process of information provision and feedback. Participatory techniques generally work best by involving far fewer people than is potentially the case with consultation or information procedures. Thus, there is a role for all three forms of involvement, particularly where an authority is keen to maximise input from the public.

Examples are provided by Hertfordshire County Council (HCC, 1994) and Surrey County Council, who have used techniques ranging from information leaflets and consultative household and other questionnaires, through to participatory weekend 'brainstorms', in the drawing up of their transportation strategies (see Figure 10.1)

10.2 Statutory Consultation Procedures: Transport Schemes

This section describes the statutory requirements for involving the public and the legal rights of redress in planning processes for highway and fixed track schemes in England. The associated development plan processes are outlined in Section 10.3.

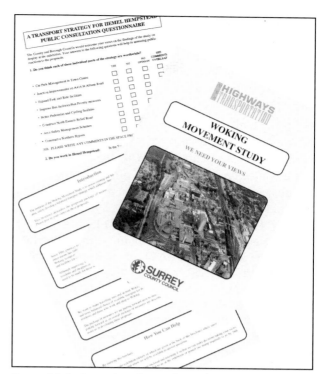

Figure 10.1: Examples of consultation leaflets and questionnaires.

Trunk Road Proposals

The Department of Transport, under the Secretary of State for Transport [Sa], is responsible for overall transport policy and, through the Highways Agency, for motorway and trunk road construction and maintenance in England. Trunk road proposals (including most motorways) are therefore made by the Secretary of State for Transport [Wa]. Other road proposals are usually pursued by the Local Highway authority [NIb].

At the national strategic level, the Secretary of State's intentions are signalled through Roads Policy White Papers (which are published every few years) and through Ministerial statements. The indicative line of a trunk road improvement is shown in strategic Development Plans for each area and in each highway authority's Transport Policies and Programmes (TPP) document [Wb]. Statutory consultation is covered by the Highways Act 1980 (HMG, 1980) [Sb].

Transport – the Way Forward (DOT, 1996) [Se] signalled a move to integrate trunk road planning more closely to the development of regional planning guidance.

Statutory Publication of Orders and Schemes

Once the Secretary of State has decided to implement a specific scheme, it is necessary to prepare a draft of the Order and to publish a notice of the Order in at least one local newspaper circulating in the relevant area. The Order notice must also be published in the London Gazette [Sd]. The notice must explain what the Order is about, where the public can view it and how long members of the public have to lodge an objection with the Secretary of State [NIc]. Parties who will be directly affected by the building of the road have to be notified individually, in particular, those whose properties may be affected or who will be subject to compulsory purchase procedures.

Public Inquiries (see Chapter 27) [Se]

Circumstances in which an inquiry must be held follow, in particular, receipt of written objections from 'statutory objectors', which include local authorities, public utilities and affected property owners. However, even in the absence of a legal need to have a public inquiry, the Secretary of State may nevertheless decide to order one, to give the general public the chance to voice their views [NId]. In practice, public inquiries are the norm rather than the exception.

The decision to hold a public inquiry must usually be made, and announced, within four weeks of the closing date for objections to the draft Order(s)[Se]. The Inquiry would usually begin within 22 weeks of that announcement [Sm]. During that time, members of the public who have objected should receive a 'Statement of Case' from the Highways Agency, explaining the proposals. The Public Inquiry is run by an Inspector appointed by the Secretaries of State for the Environment and Transport, from an independent panel drawn up by the Lord Chancellor's Department [NIe] [Sf] [We].

The purpose of the Inquiry is for the Inspector [Sf] to ascertain the relevant facts and arguments and to decide what should be recommended to the Secretary of State. The Rules of Inquiries entitle statutory objectors to be heard at the inquiry and allow them to cross-examine others. The Inspector can refuse to hear repetitious or irrelevant evidence but will usually let anyone speak who has a relevant contribution to make. Objectors may also submit evidence in writing, without presenting it in person. Everybody who so wishes is normally given time to put their case. Inspectors make a great effort to ensure that members of the public are not overwhelmed, either by the procedure or by clever lawyers, although some people may still not feel comfortable with the formal quasi-judicial style. Consequently, groups of objectors sometimes employ lawyers and/or professional advisers to present their case.

After the Inquiry, the Inspector [Sf] writes a Report, including recommendations on whether or not the Order(s) should be made or confirmed or how it should be modified, and submits it to the Secretary of State.

The Secretaries of State consider the Inspector's report and the recommendations made in the light of objections and representations made to the Inquiry and reach a decision on whether or not the Orders should be made or contfirmed. Their decision is set out in a letter to all interested parties and the Inspector's report is published simultaneously. Their decision may not necessarily reflect all the detailed conclusions arrived at by the Inspector but, if they are significantly different, they will justify these differences in their decision letter. If the decision is to make or confirm the Orders, the Secretary of State for Transport [Wd] will then make or confirm the statutory Order(s) authorising the scheme.

Appeals and Legal Challenges

If the Secretaries of State are disposed to disagree with an Inspector's recommendation, as a result of taking into consideration some new evidence or new matter of fact not considered at the Inquiry, then statutory objectors must have a chance to make representations before a final decision is made. If, within 21 days, a request is made, under this procedure, by a statutory objector to reopen the Inquiry, it must be reopened [Sn].

If new information comes to light after the Inquiry, that objectors feel would have led to a different outcome, they can make representations on the matter to the Secretaries of State. Such representations must be considered by the Secretaries of State alongside the Inspector's report. It is possible that such representations could lead to the re-opening of the Inquiry. However, objectors have no statutory right to ask for the Inquiry to be re-opened. Objectors can only obtain a re-opening if they can demonstrate that an issue was misrepresented to the Inquiry or that figures were falsified or that the Inspector wrongly excluded some important evidence. In such cases, the Inspector may re-open the Inquiry only for the purposes of considering these issues.

When Orders are made or confirmed or other authorisations are granted, objectors who still are aggrieved have six weeks in which to apply to the High Court [So] for the Orders of consents to be quashed. However, the grounds for challenge are limited and the procedure is expensive, so it is not commonly used.

Railways and Other Fixed-Track Schemes

Railways and other fixed track systems are authorised by means of Orders made by the Secretary of State under the Transport and Works Act 1992 (TWA) (HMG, 1992a) [NIf] [Sh]. The TWA procedures apply to a number of different types of project, such as light and heavy rail schemes, preserved railways, guided bus systems, canals, tidal barrages and works in the sea which interfere with rights of navigation. Prior to January 1993, such schemes were authorised either by Parliament, under the Private Bill procedure, or (in the case of light railways) by Orders made by the Secretary of State under the Light Railways Act 1896 (HMG, 1896) [Sh]. The TWA procedures are open to both public-sector and private-sector promoters.

The essential purpose of the TWA procedures is to enable the Secretary of State to make properly informed decisions about those projects which need statutory authority. A key feature of the procedures is their accessibility to those affected by proposals, such as local authorities and other statutory bodies. Public involvement is encouraged by means of non-statutory guidance on consultation before an application is made and by the statutory requirements of the Transport and Works (Applications and Objections Procedure) Rules 1992 (HMG, 1992b) [Sh].

Promoters of all schemes are strongly advised, in the TWA Guide to Procedures (HMG, 1992b) [Sh], to consult thoroughly and widely before they proceed formally to apply to the Secretary of State for an Order to authorise their scheme. This is not a statutory requirement but failure to do so is likely to lead to stronger opposition, once an application has been made, and result in a less well-formed scheme. It is for promoters to decide how and when to consult but it should usually involve close liaison with local authorities and relevant statutory bodies and public discussion, for example, in the form of exhibitions and surveys. If a proposed scheme is included in a local authority's draft Development Plan, it will also be open to public examination in the procedures which apply to the adoption of such Plans, including consideration at a local inquiry.

The 1992 Rules set out the detailed requirements for making a formal application to the Secretary of State for a TWA Order. These include a number of publicity requirements (in rules 8, 9 and 10), such as publishing a notice about the application in a local newspaper, notifying those whose property is required to be purchased for the scheme and serving a copy of the application on local authorities. Other statutory bodies have to receive details of a scheme depending on its effects, for example, the Environment Agency where deep tunnelling is involved. In most cases, the promoters of a scheme have also to submit, with their application, an assessment of the environmental effects of the works that are involved.

Anyone may make representations to the Secretary of State about an application during the 42 day objection period [Sh] and, if an inquiry is held, may present oral evidence and cross-examine the promoters. Where a promoter has not carried out adequate consultation prior to application, this is likely to be reflected in the number of objections received and the consequential length of the Inquiry.

The responsibility for public roads other than trunk roads (including some motorways) lies with the appropriate local authority as the designated Highway Authority (County Council, Unitary Authority, Metropolitan District Council or London Borough Council).

Local Roads Policies

Local authorities set out their long term land-use policies and transportation proposals in their development plans [NIg]. There is a statutory requirement to consult about the policy behind local road schemes (including 'need' and not just over implementation) during the process of preparing development plans. At this initial stage, many authorities take advantage of the discretion afforded them to exceed the minimum statutory requirements for public consultation.

The Structure Plan (and UDP Part 1) should specify the network of major roads 'of more than local importance' [NIh]. It should set out any major improvements to the national primary route network proposed by the local Highway Authority and its broad policy on priorities for minor improvements. At this stage of plan formulation, the public has its first (and most strategic) opportunity to influence events. However, aside from particular interest groups, the public are not as used to thinking in terms of broad strategies as they are of local measures and may not be forthcoming with their views so early in the process.

Local Plans (and UDP Part IIs) should elaborate the proposals for improving the primary route network and must also indicate other proposed new roads and improvements of a non-strategic nature, as they relate to the development patterns proposed in the Plan. The timescale for implementing policies set out in Structure Plans is 10 to 15 years, whilst for Local Plans it is normally 10 years. Plans are usually reviewed at least every five years.

A highway authority's Transport Policies and Programmes (TPP) [We] document is prepared annually as a bidding document for implementing transport policies and, although this is not subject to statutory public consultation, most authorities make them available for comment [NIi]. The TPP is used to support bids for various forms of capital funding for local transport infrastructure. Government planning policy guidance advises that they should also be consistent with transport policies and proposals in the relevant Structure and Local Plans (DOE/DOT, 1994) [Si].

Local Road Schemes

Proposals for local road schemes can take several forms. For example, a highway or planning authority may seek planning approval for a new road or may apply to have a road, or an area, pedestrianised. A developer could also apply for a road to be closed, or apply for permission to build one, in connection with a new site development (eg a housing estate). The Secretary of State for Transport [Wf] or a highway authority may publish a statutory Order or scheme to define the line or route of a new road (the 'Line Order'), to build connecting roads or to close or alter associated side–roads.

Planning permission is required for local road proposals, unless the changes occur within the existing highway boundaries. For local authority roads, the Authority makes an application for planning consent under Regulation 3 of the Town and Country Planning General Regulations 1992 (HMG, 1992c) [NIj] [Sj]. The procedure for processing such applications is identical to that for other types of planning proposal (see Section 10.3).

Local authorities may prepare Classified Side–Road Orders and Compulsory Purchase Orders. These must be advertised for a six–week period and are submitted to the Secretary of state for Transport [Sg] for confirmation. The Secretary of State has the power to order an inquiry into the confirmation of such Orders.

Development plans must include land–use policies for the management of traffic, which will be subject to consultation as part of the plan. However, where proposals involve only traffic calming or traffic management measures, there are no requirements for statutory consultation with the general public. Nevertheless, as with other such proposals, it is mandatory to consult statutory undertakers and emergency services (eg water suppliers and the police). Furthermore, for schemes which include road humps, it is a requirement to consult with the police, the fire and ambulance services and organisations or groups representing people who use the road. For other traffic calming features, installed under the powers contained within the Traffic Calming Regulations, consultation is required with the police and with such other people who use the road affected as the Highway Authority think fit (see Chapter 13) [NIk].

10.3 Statutory Consultation Procedures: Development Plans

Opportunities for public involvement in the preparation of strategic and local development plans occur at several stages of the process (COI, 1992) [NIl]:

❏ at the consultation–draft stage, before the plan is placed on deposit;
❏ at the deposit stage;
❏ at the Public Local Inquiry (PLI) or Examination in Public (EIP);
❏ at the stage when modifications are proposed or where the Authority propose to reject the recommendation of an Inspector or a panel; and
❏ finally, there is a right to challenge the plan in the courts.

Statutory Requirements Prior to Deposit

Government planning policy guidance (DOE, 1992) [Si] emphasises that it is important for authorities to give their proposals adequate publicity and to ensure that people have an opportunity to comment on them, at the earliest stages of the process.

The Regulations require authorities to consult a number of bodies, such as Government departments, local authorities and providers of infrastructure. Beyond this, however, there are no statutory requirements for the extent and length of pre–deposit publicity and consultation, although a statement of pre–deposit publicity and consultation is required. In addition, the Secretary of State has reserve powers to carry out (further) public involvement exercises. However, authorities often undertake extensive publicity and consultation before placing their plans on deposit, for reasons similar to those mentioned in Section 10.1. For example, the London Borough of Haringay circulated a draft of its Unitary Development Plan to interested parties for more than a year before finalising its Deposit Draft in 1993.

Statutory Requirements at the Deposit Stage

The Regulations require an authority to put its preferred plan 'in deposit' for six weeks in publicly accessible buildings, such as libraries and other local authority buildings, and to give notice of this to the general public. They will already have given individual notice to prescribed consultees and to anyone else they consider should be given notice. In addition, authorities have to publish a notice once in the London Gazette [Sd] and for two successive weeks in a local newspaper. This notice specifies where copies of the plan may be viewed and the timescale and procedure for making comments.

Objectors who make their objections to a Local Plan or Unitary Development Plan (UDP) within the six–week period have the right to be heard at a Public Local Inquiry (PLI). Objections must be put in writing and must specify which matter is being objected to and why. However, those submitting representations in support of a Local Plan or UDP have no automatic right to appear at the PLI. They will normally be invited to attend by the Inspector only if their representatives raise relevant issues, which would not otherwise be presented in evidence by the authority. Similarly, there is no legal right to appear at Examinations in Public (EIPs), which are held to review draft Structure Plans; participation in the discussion of issues is by invitation. In both cases, consideration of objections received outside the six–week period is a matter for the Authority's discretion.

Some objectors may not wish to appear in person at an Inquiry or EIP; their written representations, the Authority's response to them and any further comments they make before the end of the inquiry or EIP will be taken into account by the Inspector or EIP panel. Written representations carry the same weight as those made orally at an inquiry or EIP – but they lose the opportunity to explain or elaborate on them and do not allow for cross–examination.

The number of objections to plans varies widely but has increased dramatically since the Planning and Compensation Act 1991 (HMG, 1991). Many of these objections come from developers and land–owners but there has been an increase also in the number of objections from the general public, local community groups, national amenity and environment groups and special interest groups.

The Inquiry or EIP Stage

For Local Plans or UDPs, the Local Authority must hold a Public Local Inquiry (PLI), unless none of the objectors wants one and the Authority is informed of this in writing. In such cases, the Local Authority must consider each objection and prepare, and make publicly available, a statement of their decision on each objection with the reason for reaching it. They will then proceed either to modify the plan or to adopt it without modification.

For Structure Plans, an Examination in Public (EIP) must be held, unless an authority obtains consent from the Secretary of State to dispense with one. This dispensation is only available where an authority can demonstrate to the Secretary of State that the proposals can be decided fairly and justly, without the need for additional information and discussion.

Public Local Inquiries and Examinations In Public differ considerably in format. At a PLI, an independent Inspector from the Planning Inspectorate hears 'objections' and the procedure is quasi–judicial, with the Authority and objectors having the opportunity to cross–examine witnesses. The EIP is a more flexible 'probing discussion' of selected matters chosen by the Authority, which are considered by a panel appointed by the Secretary of State.

In recent years, there has been a substantial increase in the average length of Inquiries, as a result of the increase in public involvement, as registered by the number of objections. Certain changes of procedure have been introduced to try to speed–up the process. Prior to an Inquiry or EIP, an authority may suggest changes to the plan to meet objections or as a result of new information becoming available. The Authority should advertise the changes, giving reasons for suggesting them, and allow a 'reasonable time' (this is not defined by statute) for further objections and representations. If objectors submit a statement saying that they are satisfied by the changes proposed, they will not need to appear at the PLI or EIP.

Following the PLI (or EIP), the Inspector (or panel) prepares a report which sets out reasoned conclusions and makes recommendations. The report is submitted to, and considered by, the Local Authority, who must then prepare a statement explaining, and giving reasons for, the action it intends to take on each recommendation. This stage provides a further opportunity for public involvement.

If the Inspector's [Sf] (or the Panel's) report recommends no modifications and the Authority proposes no modifications, then the Authority may adopt the plan after 28 days. If the report recommends modifications but the Authority does not propose to accept them, then anyone may object to the absence of modifications within six weeks of the Notice of Intention to Adopt issued by the Authority.

Requirements if Modifications are Made

If the Authority proposes modifications, it must publish a notice announcing its intention to modify the plan and make its modifications available for inspection. The public then has six weeks to respond to the list of proposed modifications. The Authority is bound to consider the representations and may decide to hold a further PLI or reopen the EIP. This is not normally necessary where the matters raised have already been considered at a PLI or EIP.

The Secretary of State has powers to direct the modification of a plan at any time prior to its adoption. He has additional powers to call-in a plan or part of a plan at any time before adoption. Any modifications he proposes are dealt with in the same way as modifications proposed by the Authority. However, the Authority cannot proceed to final adoption of the plan until they receive notification from the Secretary of State that he is satisfied with the modifications. If he proposed modifications, the Authority follows similar procedures to those applied to its own modifications. Final approval of the plan is the responsibility of the Local Authority.

Upon adoption of the plan, the Authority must publish a notice in the London Gazette [Sf] and in a local newspaper for two successive weeks. The notice must be sent also to anyone who asked to be notified of the adoption. This gives the public the opportunity to challenge the plan in the courts.

Legal Challenges
Section 287 of the Town and Country Planning Act 1990 (HMG, 1990) [Sk] gives an 'aggrieved person', who questions the validity of the plan, the right to apply to the High Court to have the plan quashed. If the validity of the plan is questioned on procedural grounds, it is necessary to show 'substantial prejudice'. The application must be made within six weeks of the notice that the plan has been adopted.

10.4 Non-Statutory Public Involvement

This section discusses ways in which the public can be involved in giving views on different kinds of proposals, beyond the minimum statutory requirements set out in Sections 10.2 and 10.3 above.

Trunk Roads
Before publishing draft Orders, the Highways Agency [Wg] may present as many as three or four alternative possible alignments for the road for consideration by the public. However, the consultation is based on the premise that the need for the road is accepted in principle, so that the public will not normally be expected, or invited, to express an opinion on that matter, although this is now under review.

The non-statutory consultation procedures normally consist of:
- ❑ an exhibition showing the alternative route plans;
- ❑ brochures showing the alternative routes and their various impacts. These may be distributed house-to-house or provided at the exhibition or at a local public site, such as a library; and
- ❑ a self-completion questionnaire, so that the public can make its views known at this early stage.

Public meetings may also be held but these are not a statutory requirement and are not commonly organised by the Highways Agency [Wg]. A period is specified (again, usually six weeks) during which comments on the scheme can be sent to the Agency [Wg].

A decision on the preferred route is made internally in the Agency [Wg] and is recommended to the Secretary of State for Transport [Wh] for approval. This route is then announced through various media, such as the local press. All those who filled in questionnaires or sent other submissions to the Agency [Wg] are notified by means of a fairly detailed circular, which indicates the general route chosen (this may not be precisely the same as the final route).

Figure 10.2: Example of publicity produced for information about a scheme already in progress.

TRANSPORT IN THE URBAN ENVIRONMENT

The process of involving the public does not, however, finish at this stage. Once the final route has been chosen and the scheme is in progress, the public needs to be informed about precisely what will happen and when, so that they can avoid possible inconvenience and so that misunderstandings are avoided. By its nature, this is generally a one–way information–giving process, which often can be simply carried out through the distribution of informative material such as leaflets (see Figure 10.2).

Round Tables

There have been many complaints that the Public Inquiry procedures, commonly used in the past (see Section 10.2), have been long, cumbersome and expensive and that it might be useful if other ways of reaching decisions could be found. The 'Round Table' is one such attempt and is being tried in a few cases to help with the determination of routes for trunk roads, by dealing with strategic issues before proceeding to a detailed Inquiry.

The inclusion of a Round Table in the process has several implications for the involvement of the public. Participants consist of individuals or representatives of groups who are chosen at the discretion of the Chairperson, who is selected by the Department of Transport. This may mean that a large number of the public, who might wish to have their views heard, are excluded. On the other hand, since it is conducted by means of discussion, rather than the quasi–judicial method of a Public Inquiry, it may be more accessible (and intelligible) to those who do participate.

As of spring 1996, only three Round Tables had been held; in each case, it was after the 'need for a road' had been established. However, a number of observers have suggested that the correct use for a Round Table would be at the stage when the problem was being identified, rather than after a specific type of solution, such as a new road, had been identified. The role and possible function of Round Tables is therefore not yet clear. However, if the Round Table discussions can resolve some of the possible conflicts, in advance, by negotiation, then any subsequent Public Inquiry might be considerably shortened.

Local Schemes in Development Plans

Authorities may undertake extensive consultation exercises as part of the process of drawing up a draft plan. Commonly used methods for eliciting views include public meetings, travelling exhibitions, leaflets, notices in public buildings and special editions of Council newspapers (see Section 10.5) (DOE, 1994).

Some authorities have attempted to gain the views of those sections of the population which do not usually respond to consultation exercises, by targeting these groups with appropriate literature (eg translated into other languages) or by setting up special meetings. Videos have been used to encourage responses to the plan and a few authorities have included representatives of local groups on steering groups involved in the production of successive stages of the plan (Wilcox, 1994).

Local Transportation Strategies and Packages

There are five possible stages at which the public might become involved in the transportation strategy development process (IHT, 1996):
- ❑ in setting objectives, at the outset of the process;
- ❑ in identifying current and likely future problems in the area;
- ❑ in developing ideas for measures to resolve problems and meet objectives;
- ❑ in indicating levels of support for different policy measures; and
- ❑ in deciding on the preferred strategy for the area.

Following the adoption of a strategy, there will be a more local–level process of public involvement, as part of implementing certain types of scheme and measures contained in the package (eg traffic calming and bus priorities), in the course of which techniques of public involvement may have a role to play other than the ones described here (eg street participation exercises) (see Chapter 20).

Several local authorities are now involving the public in the development of 'Vision Statements' for urban areas. Often associated with Local Agenda 21 initiatives and the Local Plan development process, these are concerned with issues wider than just transport. However, in practice, transport–related issues figure largely in these debates and such a process might provide an important input to the setting of objectives for an urban transportation planning process, if the timing of these various exercises coincide.

A useful way of getting public input to setting objectives is either via the 'visioning' process noted above or by starting off the definition of objectives, in terms of problems that need to be tackled in the area.

People are generally able to identify transport–related problems, through their own local experiences in using the transport system and in hearing of the problems experienced by friends, employees, customers, etc. It is important to obtain views 'across the board' and not to allow a few people to dominate. Additional effort may sometimes be needed to obtain views, for example, from non–car owning households, who mainly travel by public transport and on foot and often feel powerless to voice opinions.

Encouraging people to give ideas on how to resolve these problems or meet other objectives may lead to suggestions that are either highly strategic (eg, subsidies for public transport) or very local in nature (eg, relocating a bus stop). However, if well handled, the process invariably produces some useful ideas, which can add to the options that have been thought of by the professionals.

Inviting the public to indicate support for different measures can help to narrow down the range of options and forestall later problems, especially during implementation. Where certain measures do not appear to have much public support, but are central to a strategy, there is time to try and convince people that including the measure is crucial to the overall success of the strategy.

Finally, public involvement in the selection of the final strategy (or endorsement of the recommendation, if there is only one option) is important in getting both local political support and central Government funding. It is most crucial,

however, at the implementation stage, where individual measures may disadvantage particular individuals or groups. Here, evidence of strong local support may be needed to carry through an element of the scheme that might receive vociferous local reaction among a minority and hence strong local media coverage.

10.5 Techniques for Encouraging Public Involvement

Techniques under this heading may be intended simply to inform, with no further action on the part of the recipient, or they may invite people to become more actively involved in the consultation or participation aspects of the process. Information may be targeted at particular groups, such as commuters or families with school–age children, in relation to particular aspects of a strategy, such as journeys to work or to school.

Media Coverage
A wide audience can be reached through articles in local newspapers and radio and television coverage, although it is difficult to get across much factual information in this way and the Authority has little control over what is presented. The media may be looking for an `angle' that can give an unbalanced picture.

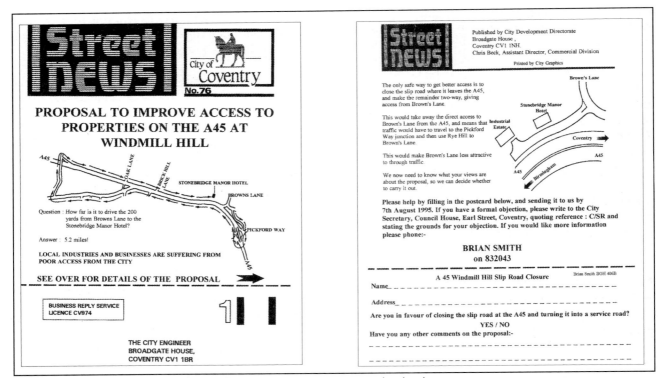

Figure 10.3: An example of a leaflet used to gauge reactions to a very localised measure.

A well–prepared press release, launched at a press conference in the case of a large scheme or strategy, and clearly written back–up material can be of help to journalists and presenters and ensure that facts and figures are correctly quoted. Such material can also form the basis for leaflets and other publicity materials.

Despite their limitations, the media can convey simple messages to a wide range of people, can help to raise the importance of a study in the public mind and can provide a useful, free, source of information–dissemination. In some circumstances, it may be appropriate to pay for an advertisement, to ensure full control over content and style of presentation.

Posters, Leaflets and Information Sheets

These provide ways in which the public can be informed of the study and its progress. In the context of urban transportation strategy development, posters are often used to draw the attention of the public to meetings, the availability of literature, etc. Leaflets can provide more information and can be taken away and referred to later. Information sheets can be produced as a series and can report on particular aspects of the study. Where appropriate, authorities can translate these for local people for whom English is not their first language. Decisions have to be taken as to where to display or distribute them. Normally, they are made available at Council offices, in local libraries, health centres, post offices, etc. Longer opening hours can maximise the opportunities of the public to view proposals. Some authorities provide details of planning applications on the Internet.

The distribution of leaflets to households, in connection with strategic level transportation issues, is normally part of a consultation exercise (see below) rather than just an information strategy. However, some councils regularly circulate households with news–sheets, giving information about what the Authority is doing, and it may be possible to include an insert there or in a commercial local 'free' newspaper (see Figure 10.3).

Public Meetings

Public meetings are the traditional means of public consultation, though they are often unwieldy in terms both of the numbers of people who attend and of the dynamics of the event. People with strongly–held, but unrepresentative, views can dominate the proceedings and it may be difficult to gauge the balance of opinion. Firm chairmanship is required. Even where the meeting is well–ordered, there are questions as to whether those who attend public meetings are representative of the local community. Nevertheless, public meetings are an expected form of consultation and, if well organised, can provide valuable information.

One successful method of getting feedback from those attending and of encouraging a consultative rather than a confrontational attitude is to divide the meeting into several groups. After a brief introduction to the event and an overview of the study, those attending are asked to join one of a number of groups dealing with particular problems or issues (eg tackling congestion or the needs of cyclists), each of which appoints its own rapporteur. Each group then reports back to the whole meeting on its conclusions, which may lead on to a constructive debate about some of the points that have been raised. Not only is this a much more productive use of time but, because feedback is coming directly from members of the public, reactions from others attending tend to be less confrontational.

Presentations

Presentations are usually made to groups smaller than public meetings, perhaps representing local businesses, residents associations, etc. This is usually done on an invited basis (although the study team may be proactive in encouraging invitations). These meetings may use material similar to that shown in the larger public meetings but there is also an opportunity to go into greater depth on more specialist aspects of particular interest to that group. Because each group has more of a common view, it is easier to gauge opinion and the smaller group–size gives more opportunity for in–depth discussion.

Focus Groups

Focus groups are a widely used form for qualitative social research, used to gauge attitudes and perceptions and to obtain input to the planning process. They are used by central and local government and are widely used throughout the market research industry.

This technique also involves meetings with small groups of people (typically 8 to 10) but they differ from presentations, in that:
  ❏ participants do not know each other and, often, the mix of people cuts across normal social groupings; and
  ❏ moderating of the group is normally carried out by an independent person and not someone directly associated with the study team.

The discussions can be informative, providing insights into prevailing public concerns and

Photograph 10.1: An example of material for a public exhibition.

perceptions and helping those involved in the study to understand the public viewpoint(s). They can uncover a depth and quality of information that is difficult to obtain using questionnaire surveys or other written forms of feedback. If the groups are carefully selected to be representative of different population sub–groups, experience has shown that the broad findings can replicate closely those of a large scale random sample survey in the area.

Exhibitions

Exhibitions provide an opportunity for a local authority to set out, in simple terms, what is being done and to invite feedback on particular issues. Both the presentational format and the siting of the exhibition are important factors in determining who sees it and how effectively messages are got across.

Many people find it difficult to absorb large quantities of figures or to visualise what a transport scheme marked on a plan would look like when constructed; some also find it difficult to read maps. At the same time, others are interested in the details of particular proposals. In general, therefore, the purposes of an exhibition are:
- ❑ to direct information at several different target audiences, highlight key facts and provide back–up figures for those who are interested;
- ❑ to present quantitative information, in diagrams rather than tables where possible;
- ❑ to supplement maps with aerial photographs, which people find interesting and informative; and
- ❑ to help people to visualise the consequences of introducing particular measures, through the use of photographs, sketches and computer–generated images.

Feedback from those attending an exhibition is best obtained through a combination of report– back from the professionals, who are in attendance to hear views and answer queries, and short self–completion questionnaires.

Exhibitions need to be well–publicised to attract a wide audience. The location(s) of the event(s) can have a major impact on the number and profile of attendees. Shopping centres are good examples of places where a cross section of the public can be found. However, care is needed in their selection, since different centres attract different socio–economic mixes of customers. There is also a need to cover weekdays, evenings and weekends (see Photographs 10.1 and 10.2).

Public Participation

Active citizen participation involves a two–way process, in which the public and the professionals

Photograph 10.2: A public exhibition.

learn from each other and modify their perceptions and attitudes as a result. Techniques that are essentially 'participatory' are interactive and involve face–to–face meetings, although participants may be given exercises to complete outside the meetings.

Transport Forum

A number of local authorities have set up 'transport forums', as part of the process of developing a local transportation plan. Their exact role and responsibilities vary but, essentially, they provide a mechanism for a regular exchange of views between the professionals, the general public and local businesses. Although they have neither a right of veto nor the formal status of a Steering Committee, in practice a forum generally has a strong influence on the outcome of a study. In this setting, the professionals working on the study interact on an equal basis with people representing different facets of the local community, exchanging views and attempting to reach an outcome, or a set of alternative outcomes, that satisfy opinion within the forum. At the end of the day, a strong contrary message from representative public consultation or from local politicians may override the forum but, in most cases, this is unlikely to happen. By working closely together, over extended periods of time, strong factional interests can often be overcome and a degree of consensus achieved.

Transport Community Groups

In larger urban areas, or smaller towns with large catchment areas, a transport forum of manageable

size is unlikely to encompass the broad range of views to be found in the local community. Within the study area, there may be sub–areas with very different characteristics and requirements, which need to be understood and included with their own voice in the participation process if local `ownership' of resulting plans is to be achieved.

One solution to this problem is to convene a number of Community Groups, in different parts of the Study Area, which interact directly with the study team and also have one representative each on the main

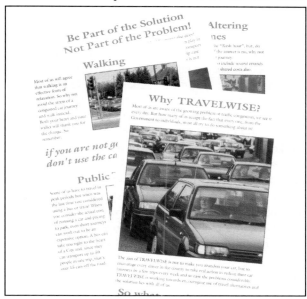

Figure 10.4: A leaflet explaining the "Travelwise" campaign.

Transport Forum. This provides a two–level participatory process.

Community 'Workshops'

This may involve up to 80 or more community representatives, together with local professionals and facilitators, usually in a two–day (eg weekend) meeting, which has a particular goal in mind. This may be to agree on a strategic vision for a local community or a set of objectives to guide local planning (Weisbord *et al*, 1995). The approach could be used specifically to shape the overall direction of a strategic plan for local transport.

10.6 Public involvement in the Wider Context

Links with Other Public Involvement Exercises

As this chapter has indicated, there are a variety of ways in which the public might become involved in local planning–related decisions, from new road schemes to development plans and development application submissions. It is important that an authority ensures consistency between the various `messages' sent out to the public and takes account of any scope for linking some of these processes, where appropriate, both to achieve financial savings and to avoid public information/co–operation overload.

An increasing number of local authorities are developing 'Transport Awareness Campaigns', such as 'Don't Choke the City', 'Headstart' and 'Travelwise' (Ciaburro *et al*, 1994). They aim to raise public understanding of the existing and future transport–generated problems in their local area and to make people aware of the actions that they can take to reduce these problems by modifying their own behaviour (eg through peak–hour staggering or changing mode). These campaigns also provide information on the existing modal alternatives to car–use and on new initiatives that are being taken by the Local Authority to improve these alternatives (eg by extending a cycle network). In this way, they provide a useful educative background, against which the development of transportation strategies can be debated with the public. Awareness campaigns can also complement TPP packages and other means of implementing these strategies (see Figure 10.4).

Anecdotal evidence suggests that a group that has experienced a local awareness initiative may be able to contribute more readily and fully to the traditional public involvement process, as they are better 'educated' about the issues involved and have had time to think them through. In future, it is possible that awareness campaigns may become directly linked with public involvement processes concerning transport issues at the strategic and local levels.

The Role of Elected Members

As with the formulation and implementation of other local authority policies, elected Members have a key role to play in the process of developing and implementing transport schemes and strategies [NIm]. They represent one element of a three–way partnership between the Members, professionals (local authority officers and their consultants) and the general public. The balance of influence of this partnership may vary from one authority to another, depending, for example, on the level of Members' interest in a particular scheme or strategy and on the level of public involvement that has been adopted (eg whether the emphasis is on consultation only or full participation).

Whatever this balance, however, elected Members are the final political arbiters, in weighting the relative importance of problems and issues, in setting and prioritising study objectives, in agreeing the overall emphasis and composition of transportation strategies and in approving the necessary expenditures from local sources for scheme implementation. They also play an important role in the scrutiny of proposals and in assessing the effectiveness of schemes, once they are implemented. If local residents or business people are unhappy about the outcomes, local Members have to bear the brunt of their concerns.

It is a matter of judgement for individual Members as to the extent to which they become involved in the consultation/participation process itself (eg by chairing meetings or observing the processes), although if they are directly acquainted with the issues and the ranges of views expressed in the course of this process, this may help the relevant committees in making final judgements.

The Association of Metropolitan Authorities' Discussion Document on Transport: The Elected Member Role (AMA, 1994) stresses the importance of decision–taking founded on local accountability: "Most transport decisions result in gainers and losers. They require a sensitive balancing act. A range of different factors are often involved. The needs of long–distance travellers have to be weighed against the interests of local communities. Travelling times and development potential have to be balanced against environmental considerations. The people who take these decisions must be democratically accountable to local people."

Any process of public involvement is likely to identify a range of views, concerns and preferences among the different interest groups, some of which may conflict or compete for available resources (eg roadspace). Ultimately, it is a political judgement as to what weight should be given to each of these views and to factors that may have been of less immediate concern to some people, such as the need for long–term solutions compatible with a sustainable transport policy.

Representativeness of Public Views

It is sometimes suggested, particularly by those who have reservations about public involvement, that the only people who take the trouble to express their views are those who are opposed to a proposal. Public consultation, it is argued, cannot claim to be representative of the 'silent majority'. For example, the low rate of response to self–completion questionnaires, commonly experienced, suggests that the sample response is likely to be biased towards those with a strong view on the subject.

There is some support for this view. Lowe (1989) reports on a comparison (in connection with a proposed rapid transit scheme in Southampton) between the views expressed by residents who returned a questionnaire distributed with the City Council newspaper and the views obtained from a random household survey of residents. Those who returned the questionnaire did not match the profile of the population and they were more likely to be opposed to the scheme than the population as a whole.

Whether the aim of consultation is to arrive at a 'democratic' solution, to obtain feed–back and ideas or to pre–empt later objections, the comments above underline the need to make sure that, when a public involvement exercise is undertaken, a broad range of techniques is used. The widest possible number of interested parties needs to be reached and encouraged to contribute in order to obtain a 'community' view. Even then, the results need to be interpreted with caution.

10.7 References

AMA (1994)   'Transport: The Elected Member Role', Association of Metropolitan Authorities, London.

Ciaburro T, Jones P and Haigh D (1994)   'Raising public awareness as a means of influencing travel choices', Transportation Planning Systems 2 (2), pp 5 –21.

COI (1992)   'Planning' Central Office of Information, Stationery Office.

DOE (1992)   PPG12 Planning Policy Guidance 'Development Plans and Regional Guidance', DOE [Si] [Wi].

DOE (1994)   'Community Involvement in Planning and Development Processes', DOE.

DOE/DOT (1994)   PPG13 Planning Policy Guidance 'Transport', DOE/DOT [Si] [Wi].

DOT (1996)   'Transport – the Way Forward', DOT [Sc].

Gyford J (1991)   'Citizens, Consumers and Councils', Macmillan, London.

HCC (1994)   'Public Consultation – Notes for Guidance', Transportation Department, Hertfordshire County Council.

HMG (1896)   'Light Railways Act 1896', Stationery Office.

HMG (1980)   'The Highways Act 1980', Stationery Office [Sb].

HMG (1990)   'Town and Country Planning Act 1990', Stationery Office [Sk].

HMG (1991)   'Planning & Compensation Act 1991', Stationery Office.

HMG (1992a)   'Transport and Works Act 1992', Stationery Office [Sh].

HMG (1992b)   'Transport and Works (Applications and Objections Procedures) Rules 1992', Stationery Office [Sh].

HMG (1992c)   'Town and County Planning General Regulations 1992', Stationery Office [Sj].

IHT (1996)   'Guidelines for Developing Urban Transport Strategies', The Institution of Highways & Transportation.

Lowe CE (1989)   'Effective Public Consultation in Transport Planning', Proceedings

of the PTRC Summer Annual
Meeting Seminar B, PTRC.

Taylor D (1996)   'Small Steps, Giant Leaps',
                        Transport 2000, London.

Weisbord MR and   'Future Search' Berrett–Kochler,
Janoff S (1995)   San Francisco.

Wilcox D (1994)   'The Guide to Effective
                        Participation', Brighton
                        Partnership.

10.8 Further Information

DOT (Annual)   Annual Circulars for TPP
                        submission, including Package
                        guidelines, DOT.

Fagence (1977)   'Citizen Participation in
                        Planning', Pergamon Press,
                        Oxford.

HMG (1994a)   Statutory Instrument 1994 No.
                        3263 'Tribunals and Inquiries: the
                        Highways (Inquiries Procedure)
                        Rules 1994', Stationery Office
                        [SI].

HMG (1994b)   Statutory Instruments 1994 No.
                        3264 'Tribunals and Inquiries;
                        Compulsory Purchase by
                        Minister (Inquiries Procedure)
                        1994', Stationery Office [SI].

HMG (1995a)   'Town & Country Planning
                        (General Permitted
                        Development) Order 1995',
                        Stationery Office.

HMG (1995b)   'Town & Country Planning
                        (General Development
                        Procedure) Order 1995',
                        Stationery Office.

Part
III

Traffic, Safety and Environmental Management

Chapter 11 Managing Use of the Road System

11.1 Introduction

This Chapter provides an overview of the issues affecting the management of the road system, covering:
- ❑ all uses, including by those not actually travelling;
- ❑ institutional issues and policies;
- ❑ design and management techniques; and
- ❑ objectives and consultation.

In particular, it addresses the importance of developing a functional hierarchy for road systems in urban areas.

11.2 Use of the Road System for Transport

The historic purpose of the public highway is for the passage of people and goods. The public at large enjoys long–established rights to pass and repass and to gain access from the public highway to adjacent property. In urban areas, however, the exercise of these rights has to recognise that users of vehicles and adjacent property often need to load and unload passengers and goods on the highway and would like to park their vehicles there. Moreover, highways often incorporate paved public open space and they are used for many non– transport purposes, as discussed in Section 11.3. The highway is also one of the principal physical elements of the urban environment, influencing its layout, shape, form and appearance.

This mixture of uses, the pressures of traffic growth, the importance of providing for walking and cycling as well as motor traffic and public concern for the urban environment have led to a range of powers and techniques for managing the use of the road system for transport. Rights of passage on foot are largely unrestricted but vehicular use can be regulated in many ways, as well as being influenced by layout and detailed design. For these powers and techniques to be used to good effect, it is important to bear in mind some basic characteristics of movement in urban areas.

All premises and areas of land require access on foot and sufficiently close access by particular categories of motor vehicle to enable people with impaired mobility, the emergency services and those delivering heavy or awkward items of goods to reach them. Providing such access will usually also enable access

by bicycle. The need and justification for more general access by motor vehicles of various kinds depends upon the nature of the activities carried out at the premises or on the land, the space available for vehicles within and near the site and the competing demands of other uses for the relevant parts of the highway. Where the scope for vehicular access is limited, it may be appropriate to provide closer access for public transport and service vehicles than for private cars. Service vehicles and private cars used for access by their drivers usually need to be parked near their destinations, which often results in parking being a major use of highway space.

A vehicle which needs to gain access via particular streets near to the origin and destination of its journey is, nevertheless, part of the 'through' traffic on each other road or street that it uses on the way. The essential difference between 'through' and 'access' traffic is that, whereas traffic requiring access to premises served by a given street has (by definition) to use that street, through traffic can usually avoid that particular street by taking a different route, if there is sufficient reason to do so.

In general, the longer the journey a vehicle is making, within or through an urban area, the wider is the range of alternative routes and the greater the scope for managing the driver's choice of route, in the interests of the local environment and traffic safety. This can involve encouraging through traffic to use longer routes than the most direct that was previously available. The extra distance involved is usually only a small proportion of the whole journey for long–distance traffic but the proportionate inconvenience and extra cost can be larger for more local traffic, Even so, the inconvenience of diversion is less for users of motor vehicles than for pedestrians and cyclists, whose routes should be kept as direct as is practicable. Design, to provide appropriate routes for traffic of all kinds, is helped by the concept of a functional hierarchy of roads and routes, as discussed in Section 11.7.

11.3 Non–Transport Uses of Street–space

Streets can provide valuable communal space, where people can congregate, sit or take a stroll (see Photograph 11.1), and where trading, entertainment and ceremonies can take place, children can play outside and other activities can spill over from adjacent premises. Indeed, relaxed informal use of streets and squares is one of the hallmarks of a mature society and, in more densely developed areas, the street has to be seen as part of the living space for people who live and work in the area, as well as a thoroughfare, and local access streets should be adapted to give priority to these non–transport uses. Deliveries, servicing and access by car can take place from parking places suitably located in the street or nearby.

Some main thoroughfares are also likely to be centres of attraction for non–transport uses. Shopping streets

Photograph 11.1: Communal street space in the centre of a small town.

Photograph 11.2: Space for window–shopping, meeting and talking – all vital functions of a footway.

need sufficient footway space for window shopping, as well as movement (see Photograph 11.2). These are also the places where entertainment attracts crowds and there may sometimes be parades or large crowds of, say, Christmas shoppers. Good design and management can help to reduce conflict between these activities and the use of the street for movement of people, public transport, essential services and other vehicular traffic.

Public utility companies (ie gas, water, electricity, telephone etc) enjoy rights to occupy space beneath the surface of the highway and to gain access to that space from time to time. This underground use of the street can sometimes be just as important as the traffic passing over it and the presence of underground services can severely constrain tree planting and lead to the physical restriction of access by vehicles. The statutory undertakers' occupation of the highway and access to apparatus is regulated through the New Roads and Streetworks Act (HMG, 1991) and they are required to reinstate the street as they find it.

Recreational use of vehicles on the highway can be tolerated, regulated or prohibited depending on the type of road and vehicle–use. Bicycle time trials and some racing is conditionally permitted. Some motor sport is allowed but road–racing and the closure of roads to traffic to facilitate such activities is not usually permitted.

Roadside advertising is under the control of the Local Planning Authority and is permitted on some parts of some highways in consultation with the Highway Authority.

11.4 Context for Managing Use of the Road System

The broad policy framework for each area is set out in the Authority's approved Plan (see Chapter 4). This will generally cover key areas of transport policy and, in particular, the relationship between new developments and transport [NIb].

Each local highway and traffic authority must submit, to its regional Government Office, its annual Transport Policies and Programme (TPP), in which its current policies are summarised [NIc] [Sa] [Wa]. District Councils are encouraged to address the need for transport policies, in support of sustainable development, including provision for alternative modes of transport to the car, in their Local Plans, through the Structure Plans for that area and policy guidance by the Departments of Transport and the Environment (see Chapter 4) [NId] [Sb] [Wb].

Towns and cities have developed as centres of economic and other activities which depend on transport, largely by road. But road traffic is only a means to an end. Unmanaged, it can choke the very urban life that it serves. The dilemma is how to manage demand for access and movement and yet maintain prosperity and economic growth. Traffic management is one of the principal means of managing the demand for use of the roads and the context for its use for this purpose is illustrated in Figure 11.1. Traffic management techniques have been developed to manage congestion, to give more space on the streets to people on foot, to improve road safety, to reduce the adverse impact of road traffic on the local environment and, in all these ways, to improve the quality of urban life and help to maintain the viability of town centres.

Concern for the environment and the effect of traffic pollution on health have combined with concerns about road safety to create public demand for more radical and innovative approaches to tackling the perceived threat from increasing traffic. To convert these concerns into public support for practical and broadly acceptable traffic management policies and schemes in the local circumstances of each town or city requires effective partnerships between highway and planning authorities, developers, local businesses, public transport providers, community groups, road–user groups, the emergency services and enforcement agencies.

11.5 Objectives, Priorities and Resources

Objectives
Traffic management can be used to achieve some or all of the following objectives:
- ❑ reduction in casualties and physical damage through traffic accidents;
- ❑ assistance to those walking, cycling and using public transport;
- ❑ improvement of the local environment;
- ❑ improved access for people and goods; and
- ❑ improved traffic flow on main roads.

These objectives often conflict and balances have to be struck, which are appropriate for a particular road or area, but almost all traffic management has a bearing on road safety and the environment. The secondary effects and consequences of schemes designed to achieve specific objectives should always be considered.

Traffic management may also form an integral part of schemes to achieve wider objectives, such as urban regeneration, environmental improvement or development projects, or to complement the construction of new road or public transport infrastructure.

Priorities
The general policies and needs of a particular area are likely to have been identified, in broad terms, through the structure and local planning process and

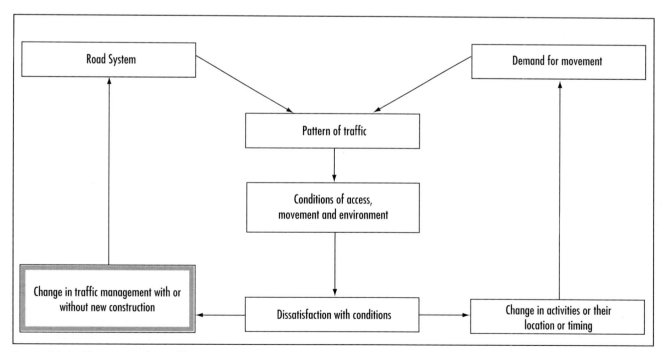

Figure 11.1: The context for traffic management.

transportation planning policy statements. However, it is essential to identify priorities in some detail and to test public acceptance of these at an early stage in the design of specific traffic management schemes.

It may help to identify the relative priority to be afforded to different groups of affected people, such as residents, shoppers, employees, service traffic, cyclists or people with impaired mobility. Without clearly expressed objectives for a particular scheme or set of regulations, it can be difficult to resolve objections and to resist pressure for exemptions.

With respect to some objectives, such as road accident reduction, effectiveness can be quantified and measured with confidence. Other objectives, such as environmental improvement or encouragement of cycling, are more nebulous or may not show results in the short term. Since prioritisation of these objectives is necessarily subjective, it is essential that decisions as to the most appropriate allocation of resources is guided by clear analysis of likely cause and effect.

There will often be pressing calls for traffic management work to meet short–term needs and this can frustrate the achievement of broader or longer term objectives. Clear establishment of objectives and priorities is essential if progress is to be achieved.

Resources

Traffic management is essentially low cost compared with the provision of new infrastructure. The most significant input is often skilled human resources. The resource cost of staff time required to research, survey, specify and consult on regulations may well exceed the actual cost of implementation of the scheme concerned. However, even when these costs are taken into account, traffic management schemes can offer high rates of return. Moreover, modest resources allocated to, say, closure of a shopping street to through traffic may provide the key to releasing other resources for environmental improvement.

11.6 Legislation, Regulatory Powers and Procedures and Funding Arrangements

Legislation

Traffic management is undertaken under powers given by, and subject to, the provisions of the Highways Act 1980 (HMG, 1980) [Sc], the Road Traffic Regulation Act 1984 (HMG, 1984), the Road Traffic Act 1988 (HMG, 1988) and the New Roads and Streetworks Act 1991 (HMG, 1991). Related powers

are also contained in the Town and Country Planning Act 1990 (HMG, 1990) [NIe] [Sd]. These Acts are amended and supplemented from time to time by more specific Acts and are implemented in detail through secondary legislation, especially Commencement Orders, that bring particular provisions into effect on specified dates, and Statutory Instruments (SIs), which often include detailed regulations, specify procedures and keep matters of detail up to date. Much of the legislation is interpreted by the relevant Government Departments, in circulars and other forms of advice to local authorities.

Because of the complexity and changing nature of the law relating to highways and traffic, practitioners are strongly advised to consult one of the regularly updated Law Encyclopedias (eg Sweet and Maxwell). Not only do the encyclopedias give convenient cross–references to commencement dates, amendments and SIs but they also include relevant case law, which illustrates how statutes are to be interpreted. In this field, simplified text books on highway and traffic law should be used with care, as they cannot present an up–to–date statement of the law (Orlik, 1993).

Regulatory Powers and Procedures

Local authorities, in their role as traffic and highway authorities, may only exercise powers and carry out duties specifically laid down in legislation. Government Ministers and their Departments are similarly constrained. In general, central Government avoids direct involvement in purely local issues but it is sometimes possible for local authorities to act outside their normal powers, with specific approval of Ministers. Secretaries of State generally have reserved powers, which are much wider than those given to local authorities [NIc]. Details of powers and procedures are given in Chapters 4 and 13.

Regulations which cannot be enforced effectively or which seriously overload the capacity of the police and the court system, force these agencies to determine priorities. Close co–operation with the enforcement agencies is essential, if the objectives and priorities of highway authorities are to be achieved in practice. The police generally prefer the use of physical controls and the use of simplified procedures to consolidate traffic regulations.

Funding

Central Government influences local transport expenditure, after considering the merits of proposals submitted [Sa] in each authority's annual (TPP) submission and the available national resources [Wc].

Local transport packages and local safety schemes are selected for funding approval through Supplementary Credit Approval (SCA), which authorises borrowing. Central funding, called Transport Supplementary Grant (TSG) [Se], may be available for substantial traffic management schemes (over two million pounds) and currently runs at 50%. It therefore needs to be supplemented from local taxation and other sources to enable an authority to carry through a balanced programme of transport management work on the roads for which it is responsible [Nlf].

Authorities have also come to depend on other sources of revenue income, private sector funding and assistance for research and demonstration projects to supplement grants and local authorities' own funds (see Chapter 4). These sources may include:

❑ income from car parking charges, but this can only be used for specific purposes (see Chapter 19);

❑ developers' contributions, generally to relieve existing constraints or to accelerate projects so as to enable traffic–generating developments to proceed;

❑ central Government and EU–sponsored innovative, experimental or demonstration projects;

❑ European Regional Development Fund (ERDF) grants, in eligible regions (and transport infrastructure is likely to be eligible from 1997); and

❑ transport investment that is part of a wider plan, such as for regeneration or housing improvement, where authorities may be able to draw in funds from other Government sources.

11.7 Hierarchy of Roads for Motor Vehicles and Routes for Pedestrians and Cyclists

The design or adaptation of the road system in a town or city, to provide for the mixture of transport and non–transport uses mentioned in Sections 11.2 and 11.3, can be realised by applying the concept of a functional hierarchy of roads for motor vehicles, integrated with a system of routes for pedestrians and cyclists (see Photograph 11.3). The functional hierarchy of roads stems from the need to reconcile, in a single design for a road and pathway network, the functions of providing for efficient movement of motor vehicles with those for other transport and non–transport uses. It is strongly influenced by the idea that any motor vehicle travelling between a particular origin and destination should intrude as

Photograph 11.3: Segregation of pedestrians, cyclists and road vehicles.

little as is practicable into the neighbourhoods and living areas that it has to pass through on its journey. However, this idea may be fully achievable only in new development areas (see Part IV).

Principles and Advantages of a Functional Hierarchy

The provision of access to sites and buildings and their immediate surroundings should, if possible, be separated from the provision for the through movement of motor vehicles. More generally, the conflicts between different functions of the road system can be reduced by designing or adapting the various roads and paths in an urban network, so that mutually–incompatible functions are, as far as is practicable, separated onto different roads and the network as a whole provides safely and conveniently for the desired mixture of uses.

With this in mind, the mixture and balance of functions to be performed by each road needs to be identified, so that it can be designed or adapted accordingly. The resulting design should reinforce the intended balance of functions, by encouraging the appropriate uses and behaviour and by discouraging incompatible uses and behaviour. Features that can contribute to the design include: alterations to the width and alignment of carriageways; footways; cycle–tracks; the layout of the carriageway, including any parking places; signing; marking; lighting and street furniture; speed–limits and other traffic regulations; and the choice of surface materials and landscaping.

Among the advantages of this approach are that:

❑ more space can be given to activities most closely related to frontage buildings, where the functions of communal space and access predominate;

Functions	Routes for non-motor traffic			Roads for motor vehicles:			
	Cycle Routes	Pedestrian Routes	Pedestrian Streets	Access roads	Local Distributors	District Distributors	Primary Distributors
Predominant Activities	Cycling	Walking Some cycling in shared space	Walking Meeting Trading	Walking Use of highway by frontagers. Vehicle access. Delivery of goods and servicing of premises. Slow moving vehicles	Vehicle movements near beginning or end of all journeys	Medium distance traffic to Primary Network Public Transport services. All through traffic between different parts of the urban area	Fast moving long distance through traffic No pedestrians or frontage access
Pedestrian Movement	Considerable freedom to cross and priority over cyclists on shared pathways	Complete freedom - priority over cyclists and motor vehicles in shared space	Complete freedom	Considerable freedom with crossing at random and some areas shared with vehicles	Careful consideration of opportunities for crossing	Minimum pedestrian activity with positive measures for their safety	None - vertical segregation between vehicles and pedestrians
Cycling	Priority over pedestrians and motor vehicles in shared space	Permitted in shared space but priority there should be given to pedestrians	Generally acceptable but may be restricted if pedestrian usage is high	Shared use of area available to vehicles	Assisted by provision of cycle lanes if possible	Cycle lanes, and help in turning right should be provided, if alternative cycle routes are not available	None - alternative routes for cyclists should be provided
Stationary Vehicles	Only in parts of carriageway not designated specifically for cyclists	Only in shared space and not so as to obstruct pedestrians	None except for servicing, emergency and access by people with mobility handicaps	Considerable, if space not provided off the highway	Appreciable, if space not provided off the highway	Some, depending on traffic flow conditions	None
Heavy Goods Vehicle Activity	Only in parts of carriageway not designated specially for cyclists	None	Essential servicing and frontage deliveries only	Only for delivery of goods and services	Only to serve frontage premises and reach nearby access roads	Movement between different parts of the urban area	Suitable for all through movements
Vehicle Access to Individual Properties	Only with consideration for cyclists	Only via shared space or across footways	None, except for emergency vehicles and necessary access for servicing and by people with mobility handicaps	Predominant	Some to frontage premises and sites	None, apart from sites generating vehicle flows at levels similar to those on local distributors	None apart from sites of national traffic importance, with purpose-designed access.
Local Vehicular Movement	Some on lightly trafficked routes with consideration for cyclists	None	None except where public transport vehicles are admitted	None	Predominant	Some, according to spacing of junctions	Very little - junction spacing should preclude local movements
Through Vehicular Movement	None	None	None	None	None	Predominant role for medium distance traffic	Predominant role for long distance traffic
Vehicle Operating Speeds and Speed Limits	Less than 20 miles/h with speed-reducing measures	Less than 5 miles/h except for careful cycling on shared pathways	Less than 5 miles/h (vehicles enter on sufferance)	Less than 20 miles/h with speed-reducing measures	Subject to 30 miles/h limit but layout should discourage speed	Subject to 30 or 40 miles/h limit within the urban area	More than 40 miles/h depending on geometric constraints

Table 11.1: Functions of an urban hierarchy.

Source: Adapted from USM Guidelines.

Photograph 11.4: A pedestrianised street – part of the civilising of towns.

❏ pedestrians, cyclists and public transport can be provided with more convenient and attractive routes, in reduced conflict with other traffic;
❏ activities incompatible with traffic flow increasingly can be accommodated elsewhere than on designated traffic routes;
❏ acceptance can be gained for segregation of traffic and limitation of access, to increase the capacity of designated traffic routes;
❏ accidents can be reduced by minimising the number of intersections and vehicular conflicts on designated traffic routes and making safe crossing–places for pedestrians where they find it natural to cross;
❏ environmental impacts of motor traffic can be limited by concentrating flows onto fewer routes designed for smoother flow; and
❏ where it is decided to invest in extra infrastructure for moving traffic, the rate of return on such investment can be increased by concentrating traffic onto a few selected corridors.

A Form of Hierarchy

The form of hierarchy which has evolved from the one originally set out in the Buchanan Report (Buchanan *et al* 1963) is summarised in Table 11.1. Among the four main categories of road for motor vehicles, the two middle categories are less clearly distinguished than the two 'extremes' of access roads and primary distributors and their roles have to be interpreted flexibly in relation to local circumstances. It is more important to achieve a good match between design and intended mix of functions than to strive to comply with

the description of a preconceived category of road. Many small and medium–sized towns, for example, will have no need for primary distributors within the urban area, although they will require good access to the local primary network. The number of categories of road in the hierarchy will depend on the size and population density of the urban area concerned.

Routes for pedestrians usually comprise a mixture of footpaths apart from roads, pedestrian–only streets (see Photograph 11.4) and footways along–side roads, with different degrees of separation from the carriageway. Crossing of carriageways can be assisted in ways ranging from dropped kerbs to grade separation. Similarly, routes for cyclists comprise a mixture of cycle–paths apart from roads, special facilities along roads and at road junctions and, ideally on quieter roads only, cycling in mixed traffic (see also Chapters 22 and 23).

Developing a Hierarchy in Practice

A new settlement or a substantial area of redevelopment in an existing town or city can be provided from the outset with a network of roads and paths, in which design and resulting functions are closely matched, especially if there are few constraints and if densities are low.

In a more dense town or city, the existing pattern of land–use has to be serviced by an affordable adaptation of the existing network, much of which may have developed in haphazard ways over the history of the settlement. The initial definition of a hierarchy will necessarily have to be a practical compromise but identification of the current mixture of functions of each road and the scope for modifying it, over a period of time, is an important starting point for subsequent decisions about traffic management and development control, especially as it affects frontage access (see Chapters 13 and 27). Coherent policies and decision–making in these two respects over a period of time can do much to reduce the conflict between incompatible functions and inconsistency between physical form and the mix of functions on the various roads in the network and to enhance the provision of safe and convenient routes for pedestrians and cyclists.

As part of this process, incompatible functions, that still have to be performed by the same road, can sometimes be separated in 'time. For example, on a stretch of main radial road, which also serves as an inner suburban shopping high–street, a peak–hour bus–lane may, at other times, provide kerbside space for delivery vehicles servicing the shops and short–stay parking for shoppers.

The hierarchy as currently defined, and the identified

scope for changing it, together provide one of the starting points for the design of traffic management schemes, including traffic calming and local–area safety schemes (see Chapters 20 and 16). As with all such schemes, however, the proposed hierarchy within the area affected by the scheme should be open to modification in the light of consultation with local people and affected road–users.

Statutory Definitions and Administrative Categories of Road

It is important not to confuse the functional hierarchy of a road network with the various statutory and administrative categorisations. In the functional hierarchy, roads and paths are categorised in terms of actual or intended uses within the network as a whole and as an aid to design, adaptation and management, irrespective of which authority is responsible for any particular road. Statutory and administrative categorisations are made for quite different purposes, some of them according to which authority is responsible for maintaining particular roads within the network. A number of relevant definitions are brought together for reference in Tables 11.2 [NII] and 11.3 [NIm]. Any association between these and the functional categories in Table 11.1 is circumstantial and subject to many exceptions.

Definitions	Additional Information
A <u>Highway</u> is a way over which the public have the right to pass and repass.	On some highways, this right of passage may only be exercised on foot, on horseback or by specific classes of vehicles (as described by any Order which may be applicable). Public highway refers to those highways which are maintained at public expense (ie by the highway authority). Highways not maintainable at public expense are none the less highways. A highway may also be a waterway, a navigable river or road ferry. The term highway maintainable at public expense roughly corresponds to roads maintained by the road authority in Scotland.
A <u>Carriageway</u> is a highway or part of a highway over which the public have right of way for vehicles.	These rights may be restricted by the implementation of a traffic regulation, speed–limit or other Orders, such as one giving special road status (see Table 11.3). The right of way for vehicles does not detract from the established right of pedestrians to cross the carriageway.
A <u>Footway</u> is that part of a highway which also comprises a carriageway, over which the public have right of way on foot only.	———
A <u>Footpath</u> is a highway over which the public have right of way on foot only, not being a footway.	The essential difference between a footway and a footpath is that the former is adjacent to a carriageway. A way which is exclusively for passage on foot is a footpath.
A <u>Bridleway</u> is a highway over which the public have right of way on foot and on horseback.	The right of way may also apply to leading horses or driving animals, in which case it may be known as a driftway.
A <u>Cycle Track</u> is a way which is part of a highway over which the public have right of way on pedal cycles, other than pedal cycles which are motor vehicles, with or without the right of way on foot.	———
A <u>Road</u> in England and Wales is any length of highway or any other road to which the public have access and includes bridges over which a road passes.	However a highway may be designated as a 'Special Road', a 'Trunk Road' or a 'Principal Road' (see Table 11.3).

Table 11.2: Statutory definitions for terminology (England and Wales) [NII] [Si]. Source: RTUA, (IHT, 1987) page 35.

11.8 Elements of Urban Road Layout Contributing to Management of Use

The principal components of the urban road network are lengths of road (Chapter 36), road junctions (Chapters 37 to 43), footpaths (Chapter 22) and cycle–paths (Chapter 23). A length of road comprises some or all of carriageway, footway, cycle–track, surface shared by different users, verges and accesses to frontage or other nearby premises. There may also be a central reservation and a reserved area for light rail vehicles. All of these elements of a length of road have their counterparts at road junctions, although accesses to premises at a junction are exceptional. Footpaths and cycle–paths provide rights of way for pedestrians and cyclists, along alignments where there is no right of way for motor vehicles. A footpath and cycle–path can share the same surface, provided that the relevant powers are invoked (see Chapter 23).

The design and operation of each of these elements is discussed in the relevant Chapters. The principles

Definitions	Additional Information
A Trunk Road is a highway which constitutes part of the national system of routes for through traffic and for which the Secretaries of State for Transport, Wales and Scotland respectively are the highway authority.	———
A Principal Road is a non-trunk road, which is classified as such by the appropriate Secretary of State as being sufficiently important in the national highway system to justify principal status.	Principal road status is currently less important than it once was. It is still used as one element in the determination of the level of grant to local highway authorities and is important in some legal procedures, such as the making of or alteration to, speed–limit 0rder and is used by some highway authorities to specify and delegate functions to agency authorities.
A Classified Road is a highway, which is agreed by the Secretary of State and where appropriate the local highway authority as being of importance to the movement of traffic.	Classified Roads may be either I, II, or III. All Class I and II roads and some Class III roads are given numbers. Class I, II and III generally coincide with the prefix A, B or C respectively although there are some exceptions to this. Numbers are allocated according to the relationship between the origin of the road and the sectors of the country created by the main routes from London (A1-A6) and Edinburgh (A7-A9).
A Primary Route is a route that is designated by the appropriate Secretary of State as the most satisfactory all-purpose route for through traffic between two or more places of traffic importance.	Primary routes have green background direction signs for non-local destinations. A list of primary destinations is given in the Traffic Signs Manual Chapter 2. Motoways are not primary routes because they are 'special roads' and therefore not available for use by all traffic. A trunk road need not necessarily be a primary route.
A Special Road is a road designated by an Order made by the appropriate Secretary of State, which restricts its use to certain classes of traffic.	The most common type of special road is a motorway whereby pedestrians, animals, pedal–cycles etc. are prohibited. Roads other than motoways may be designated a 'special road'. Another important feature of special roads is that public utilities do not have a statutory right to place their apparatus within a special road.
A Motorway is a particular type of special road where motorway regulations apply.	All motorways are special roads but not necessarily trunk roads.
An All Purpose Road is any road other than a special road i.e. a road which is not restricted by a Special Road Order.	———
A Designated Road (within London) is a non-trunk road designated by the Secretary of State because of its importance	All highway and traffic proposals on or affecting these roads have to be cleared by the Secretary of State.

Table 11.3: Classification of highways according to their status as traffic routes [NIm]. Source: RTUA (IHT, 1987) page 36.

and standards applying to their design allow each element of a particular network to contribute appropriately to the management of use of the network as a whole, in accordance with the functional hierarchy.

Junctions are critical in determining the capacity of urban networks and the ways in which queues of vehicles form at the various junctions can have a strong influence on the routes that drivers choose to take. The layout and, where there is signal control, the signal–timings at each junction can be used to favour some movements of vehicles and pedestrians and to discourage others. This can include giving priority to particular types of vehicle, such as buses or bicycles. At congested times and places, they can also be used to cause the resulting queues to form at places where they are less disruptive and intrusive than they might otherwise be. The design of junctions can also be used to alert drivers to the need to adopt a different style of driving, notably in respect of speed. A roundabout, for example, can usefully mark the change from a relatively spacious suburban radial road to more constricted conditions prevailing nearer to the centre of a town. The junctions between a main radial road and local distributors leading from it into residential areas can be designed to ensure that vehicles entering the residential area do so at a low speed, appropriate to their new surroundings. In appropriate cases, wide vehicles can be excluded physically, provided that alternative means are available for necessary access, especially by the fire service (see also Chapters 37 to 43).

Most lengths of road in urban networks can carry more vehicular traffic than can enter the next junction downstream, so there is scope for reallocating some of the road–space between junctions to balance the functions that the length of road is intended to perform. Kerbs, surface colour and texture, street furniture and landscaping can all play a part in this, as well as any necessary signs and markings. Even on roads where the vehicle–flow function predominates, there is often room, except on the approaches to major junctions, to allocate space to extensions of the footway, bus–boarders, sheltered parking and loading bays, landscaping to shield pedestrians from motor vehicles and cycle– lanes or separate cycle–tracks, as well as to accommodate those frontage accesses for which no alternative can yet be arranged. Where local traffic distribution or access functions predominate, non–transport functions, such as providing communal space, may be correspondingly more important. Space can accordingly be allocated to footways, to cycling (see Photograph 11.3), to parking spaces that contribute positively to the layout as a whole (see Photograph 11.5), to appropriate speed–reducing features (see Photograph 11.6) and to

Photograph 11.5: Road–space allocated to parking.

landscaping (see Photograph 11.7). On access roads, the use of functionally–designed shared surfaces is often appropriate.

On roads of all kinds, except motorways or sections of near–motorway standard, the layout of footways, cycle–tracks and other facilities for cyclists, pedestrian and cycle–crossings should all contribute to natural, direct and attractive routes for the journeys that pedestrians and cyclists wish to make, minimising their conflicts with heavy flows of motor traffic and including good access by pedestrians to public transport stops. These elements of pedestrian and cycle routes should be complemented by

Photograph 11.6: Features designed to reduce vehicular speeds.

Photograph 11.7: Landscaping incorporated with traffic calming.

footpaths and cycle–paths away from roads. Such separate paths should be well–surfaced and well–lit, and should be designed to achieve a correct perception of personal safety that will encourage their use at all times of the day and night. Where appropriate, the needs of equestrians also require to be considered and accommodated.

11.9 Quality of Design and Construction

Opportunities to change, substantially, the visual character of an existing street or urban space do not occur often. Major traffic management schemes may provide this opportunity but almost always affect some users adversely. If the public are to accept and support change, the deciding factor for many may well be the way in which the opportunity to change the character of the environment is presented. Less convenient vehicular access and movement for some users may be considered a price worth paying if the environment can be improved markedly. However, this can be difficult to steer through the political decision–making process.

Experimental schemes, to determine the acceptability of measures, tend to reveal the importance of using quality materials and street furniture. Unfortunately, delays in funding or in agreement on a permanent scheme can result in temporary structures or materials being retained for too long and requiring labour–intensive maintenance. The poor appearance of a temporary or experimental scheme can often militate against its retention in permanent form. Clearly, there is a need continuously to monitor experimental schemes and to deal with problems as they arise. Traffic–calming schemes, often readily accepted in principle, can also founder if their detailing is poor. The first test of any material used in the harsh public highway conditions must be its durability and ease of maintenance, including reinstatement by utility companies.

The changing nature of city centre streets, as places to linger and as places for street entertainment and shopping, calls for a variety of street furniture.

Photograph 11.8: Co–ordinated design in an area–wide traffic–calming scheme.

Photograph 11.9: Prioritised allocation of road space.

Co-ordination of the design and appearance of street furniture can do much to create a sense of place. Throughout urban areas, the design of street lighting, bus shelters, guard-rails, bollards, seats, cycle-tracks and so on merits attention because co-ordinated design can do much to raise the quality of the street scene (see Photograph 11.8). Well-designed robust planting boxes and guarded semi-mature trees can help to define urban spaces. All this costs money but, through careful choice of materials and good co-ordinated design, a new sense of place can be created to revitalise pride in a town, neighbourhood or street, to encourage people to enjoy their urban surroundings and to linger (DOE, 1992 and MOT Denmark, 1993) (see also Chapters 12 and 17).

11.10 Management of Access and Movement

Access to and from the public highway network is a basic requirement for virtually all activities. Urban areas consist of a complex mixture of interacting land-uses, which usually include well-defined residential, shopping and industrial areas but also areas of mixed and, sometimes, conflicting land-uses. Often schemes are needed which will achieve the best compromise between access requirements and an acceptable environment (see Photograph 11.9).

Priorities for Access
Where there is particular conflict between access and quality of environment, it is often helpful explicitly to identify and to agree priorities for access. These may differ by time of day and reflect factors such as availability of public transport, alternative car

parking provision and the need for vehicles to continue to use streets for goods deliveries, for example. Chapter 21 discusses various techniques for the management of demand. When determining priorities for access, consideration needs to be given to emergency services, people with impaired mobility, servicing, goods deliveries and to parking, whose management is discussed in Chapter 19.

In residential areas, access needs are likely to predominate over those of through traffic. Comprehensive traffic calming, considered in Chapter 20, can create conditions in which some road-space can be reallocated for parking by residents and their visitors, as well as for cyclists and pedestrians and for planting.

Priorities for Movement
The degree to which freedom of movement by private car can be accommodated will be determined by the capacity of the street network and local environmental objectives. In most cases, encouragement of public transport, cycling and walking will be seen as priorities. Restraint through parking policies, reductions of road-space for general traffic and, possibly, road-use pricing may be necessary to discourage excessive car traffic. Selective management of goods vehicles may also be required, as discussed in Chapter 25, to reduce obstruction of movement and to improve the environment. In some areas, where acceptable alternative routes exist or can be provided, through traffic may need to be removed from sensitive areas and the opportunity taken to improve the area. Where public transport is positively encouraged, priority can be provided, as discussed in Chapter 24.

Once priorities have been clearly established and agreed, traffic management schemes can be developed to restrain as much of the low priority

Photograph 11.10: Car/bus transport interchange — sparsely equipped and exposed to the elements.

movement as is necessary to inhibit congestion. Personal movement on foot or bicycle should be as direct as possible, to encourage the use of these modes, and convenient routes for people with impaired mobility should be provided.

Good location and design of transport interchanges can encourage the transfer of journeys from private cars to forms of transport which make less demands on urban road space (see Photograph 11.10). Access to town centres by bus can also be given preference, by providing stopping places closer to the main attractions than the central area car parks.

Conflict between Movement, Access and Parking

The degree to which increased frontage access and parking can be accommodated will generally be a reflection of the balance of functions of a particular road. Uncontrolled access from the kerbside can impede traffic flow and reduce safety (see Photograph 11.11). Frequent turnover of kerbside parking space can also disrupt through movement. Parking on verges can be regulated but uncontrolled use of land adjacent to the highway can be more difficult to deal with. A clear designation of the road–hierarchy (see Table 11.1) should underpin policies on access and parking control.

Emergency Services

The emergency services must be able to respond to incidents quickly. Local authority fire services are required to arrive at fires within response times recommended by the Home Office. Similarly, NHS

ambulance services are required to respond to emergency calls within the response times in the Patients' Charter, issued by the Secretary of State for Health.

After the conceptual and objective–setting stages have been completed, fundamental conflicts may still exist between the needs and wishes of different groups, which can lead to difficult decisions at the political level. Once again, a clear designation of the road–hierarchy (see Section 11.7) can be of considerable help in this situation.

For some traffic management schemes, there is a statutory requirement to consult with the police and for others, for example, under the Highways (Road Humps) Regulations 1996 (HMG, 1996) [NIg] [Sf], the fire and ambulance services as well (see Chapter 13). However, even where no statutory requirement to consult exists, the emergency services should be kept fully informed of traffic management proposals and given the opportunity to comment.

The Department of Transport has published a Traffic Advisory Leaflet (DOT, 1994) setting out a code of practice for taking account of the needs of the emergency services in traffic–calming schemes. Although specifically concerned with traffic calming, the code is also relevant to traffic management schemes generally. The code recommends that highway authorities should establish a dialogue with the emergency services on the broad principles for introducing traffic management measures. Emergency service strategic routes should be integrated into the functional hierarchy of main

Photograph 11.11: Uncontrolled access to frontages and violation of waiting restrictions.

roads, local roads and access roads. Proposed traffic management measures should then take account of the strategic function of the route and the possible impact on emergency service response times. If appropriate, journey times should be monitored and the impact of the scheme reviewed in the light of experience.

11.11 Safety Management

About three quarters of road accident casualties occur in built–up areas. Whilst fatalities and serious injuries to vehicle occupants are less than on roads where speeds are higher, the presence of pedestrians and cyclists and the complexity of activity on the typical urban street makes it even more important to adopt a comprehensive approach to the achievement of safe conditions in urban areas for all kinds of road–user. Urban safety management provides such an approach, in which a safety management strategy is developed for each urban area (IHT, 1990). National, regional and local targets can thus be translated into specific local initiatives for each urban area and related to wider policy objectives, as discussed in Chapter 16. In particular, the establishment of a functional hierarchy of roads is combined with analysis of the accident record and perceived risks for all parts of the network, in the development of local area safety schemes. These are traffic management schemes, primarily with safety objectives, which can often, with advantage, be augmented to contribute also to improving the local environment.

Funding of local safety schemes, through the TPP/TSG/SCA system [Sa], requires a structured approach to road accident problems in an area [NIh] [Wd]. A policy of treating high–risk sites ahead of others shows progressively lower rates of return, once the worst sites have been treated, and the relevance of local area–wide safety schemes increases accordingly.

Safety management strategies also need to reflect the wider objectives of transport, health and environmental policies. For example, the need to reduce both the actual and perceived risks involved in walking and cycling may influence attitudes towards the use of cars for short journeys.

Safety management requires a highway authority to involve many other agencies, particularly those responsible for education, health and enforcement. Programmes to influence the attitudes and behaviour of road–users are at least as important as physical measures and regulation. Urban safety management strategies need to be integral parts of each authority's Road Safety Plan (LAA, 1996) [NIi] (see Chapter 16).

11.12 Environmental Management

All highway improvement and traffic management schemes should seek to make a positive contribution to the local environment. In many cases, the major aim of a traffic management scheme will be to achieve significant environmental benefits, for a specific street or area, in terms of visual appeal and the reduction of intrusion, noise and exhaust emissions (see Photograph 11.8). In others, traffic throughput, road safety or improved access may be the major objective. But, in all schemes, the aims of limiting traffic flows, on local distributor and access roads, to levels compatible with acceptable environmental standards and providing safe and attractive routes for pedestrians and cyclists, should be kept firmly in mind (DOT, 1987) [Sg].

Creation of a high quality environment is not necessarily incompatible with provision for movement. Opportunities for redevelopment exist along many major urban traffic corridors, where tree–lined streets and fine buildings can be combined, by good design, with access and movement by all modes of transport. Traffic management needs to be compatible with retention of the best from the past and the creation of new opportunities for high quality urban design. Environmental quality should be an objective in all design for traffic management and this requirement usually reinforces that of safety, especially the safety of vulnerable road–users. This subject is considered in Chapter 17.

11.13 Use of Transport Telematics in Managing the System and Informing its Users

Transport telematics (also called Intelligent Transport Systems or ITS) entails the collection, transmission and dissemination of information about traffic and its application for control and information systems and is made possible by modern technology. New aspects of traffic control, signalling, guidance and driver–information are increasingly available for general application. Sophisticated presence– and movement–detection systems provide information of increasing quality about traffic and a variety of data–transmission media are available at decreasing cost. Locational monitoring using satellites has created opportunities that would have been prohibitively expensive using terrestrial telecommunications. As discussed in Chapters 15 and 18, these developments make better roadside and in–vehicle information systems possible, together with the high levels of reliability required for road–use pricing in urban networks. Detection of traffic flows and flexible, responsive, control of traffic movements through junctions have traditionally been favoured over predetermined fixed–time signal plans for the irregular street networks in most British cities. Advances in technology have enabled interactive control of extensive networks to be developed and priority to be assigned to chosen routes and vehicles. Locational control of congestion and selective management of traffic movements can now be achieved.

The police have radar and inductive loop devices for detection of speed–limit and red–light contraventions, used in conjunction with 35 mm and video cameras. This equipment can be linked to computer systems to process data for the issue of fixed penalties or court summonses (see Chapter 14).

For drivers, up–to–the minute roadside information can be displayed using variable message signs. The display of route and incident information and car parking space availability can make a valuable contribution to urban traffic management. The technology for in–car information systems is available and prices are falling. For bus passengers, electronic information displays, triggered by approaching vehicles, can help to encourage bus travel by reducing the uncertainty associated with waiting. Information systems are discussed in Chapter 15.

11.14 Charging for Road–Use

Regulation of traffic by price presupposes that demand for road–use is 'elastic'; ie that demand will fall as the price rises. Much economic evaluation is based on the assumption that travel–time is a commodity that can be 'traded'. This may be true only to some extent, as travel is only a means to an end not an end in itself. Car commuters, for example, may tolerate additional imposed costs, if the value to them of moving between home and work is much greater than the overall cost of travel. But, although use of a car for certain journeys may be regarded as 'essential', the times at which they are undertaken may be varied, for example, to avoid peak–hour congestion charges. This suggests that car–use, even for journeys to and from work, is not inelastic. Demand for travel for many other purposes is clearly more elastic, for instance car–use for leisure, shopping and recreational travel.

One key objective of road–use pricing is to reduce the total amounts of traffic using urban road networks at congested places and times. This will usually result in some transfer from car–use to public transport and other modes. An adequate public transport system with capacity to respond to resulting increases in demand is thus a prerequisite but, in large cities, the likely increase may only be a modest percentage of existing patronage. Moreover, the supply– capacity (seat–km per hour) of an existing bus fleet will go up automatically if congestion in the road network is reduced.

Car parking can also be regulated by price. Differential parking charges are used to encourage long– stay parking only in the more remote car parks. An integrated charging regime for on– and off–street parking can enable off–street public car parks to be managed coherently with the demand for on–street parking. When combined with real–time information systems, such a regime can reduce the hunt for spaces and the interruption of traffic movement by the high turnover of on–street parking.

Local authorities must take care not to see on–street car parking charges, or charges for road–use, as general income. Charging under existing powers for use of on–street car parking and enforcement of regulations may produce incidental surpluses. An authority must then determine whether further provision of off–street car parking anywhere in its area is 'necessary or desirable' before surpluses are used for other transport purposes, such as public transport or road improvement (see Chapter 19).

Demand–management policies and techniques of all kinds are considered further in Chapter 21.

11.15 Public Involvement and Consultation

Regulation of use of the road system is one of the more frequent and direct contacts that people have with the local democratic system in action and most users feel competent to express an opinion, based on personal experience. Not surprisingly, public involvement has developed and been protected in law. Locally–elected representatives, who are given responsibility for making decisions affecting their electors' lives in this respect, and the officers advising them, need to take care to ensure that consultation is both thorough and structured. This helps them to appreciate and to gauge public opinion on specific issues. Hopefully, it results in better decision–taking with more consensus as to what the problems are and how best they can be tackled (see Chapter 10).

Consultation Arrangements
Direct public consultation and related correspondence both make heavy demands on staff

Photograph 11.12: A traffic management scheme chosen by local people.

resources devoted to traffic management. Public involvement needs to be organised systematically and appropriate procedures need to be identified for each project. In small towns and city districts with their own identity, regular advisory or consultative meetings with a cross–section of elected and community representatives can provide a valuable sounding–board for informed local opinion. For major studies and schemes, formal steering groups and consultative committees may be appropriate. Examples can be found in the report on Bypass Demonstration Projects (DOT, 1995).

Public meetings and exhibitions, structured to draw out clear responses, including from single–issue interest groups, are particularly helpful in exposing local concerns before detailed proposals are formulated. Later consultation can help to identify adverse effects on individual interests, before formal Orders are drawn up or land acquisition procedures are begun.

Where major transportation policy objectives are being pursued, for example through integrated transport package schemes, full public consultation at the formative stages is particularly advisable. Comprehensive traffic management can affect many people and measures need to command a broad measure of public support, if they are to be carried through successfully to implementation. Involvement of the local media is invaluable in reaching a wider public than those who attend public meetings.

Attitude surveys and Stated Preference (SP) exercises provide useful indications of the acceptability and effectiveness of proposals intended to bring about changes in travel habits and can supplement impressions gained at public meetings and exhibitions (see also Chapter 7).

For measures affecting the residential environment, such as local area safety schemes and traffic calming, local involvement in the development, and even the final design, of schemes often makes helpful and significant contributions (see Photograph 11.12). Even though demand for such local schemes remains high, acceptance of particular features in different neighbourhoods can be quite hard to assess in advance of consultation.

Statutory Requirements
The Traffic Regulation Order (TRO) procedures require formal consultation [NIj], ranging from 'posting' informative notices, ie displayed in the street or area concerned, through to full public inquiries , as discussed in Chapters 10 and 13.

Informal consultation with representatives of affected road–users should expose major objections and should provide an opportunity for constructive modifications, before embarking on the statutory procedures. Public representations to councillors, before they approve, modify or reject an Order in committee, may provide a valuable democratic safeguard.

11.16 Appraisal of Schemes to Assist Decision–Making

Implementation of agreed policies depends on the allocation of financial and staff resources. Elected members, advised by their officers, need to take decisions on the priorities between a range of competing programmes. In the case of major projects, complex appraisal techniques are required, for example, to secure financial support against competition from elsewhere (see Chapter 9). In the field of traffic and environment management, however, local decisions tend to predominate.

Local authority transport programmes typically group together measures which fulfil similar objectives. For example, accident remedial and small works programmes may be targeted almost exclusively at accident reduction and schemes can be appraised and prioritised primarily by their contribution to achieving a single policy–objective. On the other hand, a programme for works associated with private development will, of its nature, need to be reactive to external opportunities. Major benefits could be missed, if it was not possible to `top up' developers' contributions at the appropriate time.

Quantification of the likely cost of meeting policy–objectives, such as reducing traffic accidents, developing comprehensive cycle–networks or periodically carrying out systematic revisions of parking controls, helps to put these budgeting decisions into perspective. One of the most telling ways of assisting the process is to make clear which projects will be foregone, as financial cut–off levels are applied. This allows comparison of the value of different types of project to be made by reference to concrete examples.

Within specific programmes, councillors are more likely to rely on advice derived from technical assessment of the relative priority of similar measures. Rigorous technical appraisal should be applied, quantifying and evaluating as many factors as is practicable, using cost–benefit analysis where appropriate, including a simple first year rate of return for small schemes with similar profiles of cost and benefit over a common lifetime. Other factors need to be assessed, often subjectively, and public opinion tested. Comprehensive check–lists of the factors to be evaluated can speed the appraisal process and assist comparison (see Chapter 9).

Appraisal of the benefits and disbenefits of Traffic Regulation Orders (TROs) requires particular care, as inaccurate summaries of objections can leave the Authority's Order–making procedure open to legal challenge.

Thorough appraisal of the financial, technical and environmental effectiveness of schemes builds confidence that resources are being directed appropriately. Effects on policy objectives, other than transport ones, may need to be incorporated into the overall appraisal process, for example, the effect on disabled people, on people with impaired mobility, on those living in poverty or on the wider environment. It is necessary to demonstrate sensitivity to the interaction of transport policies with the wider needs of the community.

11.17 Implementation of Schemes

Formal processes have to be completed to implement TROs (see Chapter 13). Where physical works are involved, experimental Orders may be used. However, formal contact arrangements between the contractor or site engineer and affected occupiers is advisable. Press and radio coverage of new works and regulations is particularly helpful in retaining the public's tolerance of temporary disruptions. Close liaison with police and the emergency services is also essential.

The need to programme alterations to statutory undertakers' plant, and to lay in new electricity supply points or telephone lines, is often the most critical element of small works. Delivery times for equipment, such as traffic signals and lighting apparatus, often exceed the tender and works period and these may need to be ordered in advance. No matter how small the project, an implementation programme should always be prepared.

The sequence of traffic flow arrangements, whilst urban traffic alterations are being introduced, can be crucial. Weekend implementation is often preferred, with extensive roadside information signs displayed on the days before a change is brought into effect. Even so, some drivers will find themselves in the wrong place at the wrong time and temporary turning areas or escape routes may need to be provided.

Works planned by local authorities and their contractors need to be recorded in advance on the Streetworks Register and, in some cases, there may be requirements to notify the Local Land Charges Register [NIk] [Sh], particularly where access rights are affected.

Finally, the effectiveness of all traffic management projects should be monitored and reviewed, to determine whether anticipated benefits have been realised in practice and whether unforeseen effects have arisen. Readiness to modify schemes (promptly, if necessary) is important in promoting public confidence in subsequent proposals (see Chapter 20).

11.18 References

Buchanan C *et al* (1963)	'Traffic in Towns', Stationery Office
DOE (1992)	Design Bulletin 32 'Residential Roads and Footpaths – Design Considerations', DOE.
DOT (1987)	Local Transport Note 1/87 'Getting the Balance Right', Stationery Office [Sg].
DOT (1994)	Traffic Advisory Leaflet 3/94 'Fire and Ambulance Services – Traffic Calming: A Code of Practice', DOT.

DOT (1995)	'Bypass Demonstration Projects', DOT.
HMG (1980)	'The Highways Act 1980', Stationery Office [Sc].
HMG (1981)	'The Highways Act 1981', Stationery Office [Sc].
HMG (1984)	'The Road Traffic Regulations Act 1984', Stationery Office.
HMG (1988)	'The Road Traffic Act 1988', Stationery Office.
HMG (1990)	'Town and Country Planning Act 1990', Stationery Office[Sd].
HMG (1991)	'The New Roads and Streetworks Act 1991', Stationery Office.
HMG (1996)	'The Highways (Road Hump) Regulations 1996', Stationery Office [Sf].
IHT (1987)	'Road and Traffic in Urban Areas RTUA', The Institution of Highways & Transportation.
IHT (1990)	'Guidelines for Urban Safety Management', The Institution of Highways & Transportation.
LAA (1996)	'Road Safety Code of Good Practice', Local Authorities Association.
MOT Denmark (1993)	'An Improved Traffic Environment – A Catalogue of Ideas', Road Data Laboratory Report 106, Ministry of Transport, Denmark.
Orlick M (1993)	'An Introduction to Highway Law', Shaw and Sons.
Sweet & Maxwell	Local Government Library Encyclopedias on 'Road Traffic Law and Practice' and 'Highway Law'.

11.19 Further Information

ACC (1991)	'Towards a Sustainable Transport Policy', Association of County Councils.
Borg W, and Erl E (1993)	'Short Distance Passenger Travel', European Conference of Ministers of Transport, Paris.
CSS (1994)	'Traffic Calming in Practice', County Surveyors Society.
CT (1990)	'Lorries in the Community', Civic Trust, County Surveyors Society, Department of Transport, Stationery Office.
EU (1988)	'European Charter of Pedestrian Rights', Report of Committee on the Environment etc. Adopted by the EU Parliament 1988.
HMG (1994)	Royal Commission on Environmental Pollution, 'Chapter 18, Transport', Stationery Office.
ICE (1994)	'Managing the Highway Network', Institution of Civil Engineers.
LAA (1989)	'Highway Maintenance : A Code of Good Practice', Local Authorities Association.

Chapter 12 Town Centres

12.1 Introduction

This chapter alludes to the wider aspects of town centre management but concentrates on transport aspects. Nevertheless, decisions on traffic and transport systems in town centres should be made in the context of an agreed overall management strategy [NIa]. The control of traffic and its undesirable effects, such as noise and fumes, should be one specific aim. At the same time, traffic management and the promotion of alternatives to the private car contribute to other aims – including encouraging investment, environmental improvement, generating a coherent image for the town centre and allowing more imaginative use of public space. The underlying assumption is that provision should only be made in town centres for traffic that is essential to the economic survival and development of that centre, so that the maximum space is retained for core activities and the highest possible level of amenity provided for pedestrians. A co-ordinated approach to town centre initiatives, including traffic restraint measures, is needed to put these principles into effect (ICE, 1993).

12.2 Transportation Planning Objectives

Planning Policy Guidance (PPG) 6 Town Centres and Retail Developments (DOE, 1996a) [NIb] [[Sa] sets out objectives for transportation planning in town centres, under the following three headings.

Access to Town Centres:
❑ to manage access by car and parking, as part of an overall strategy for the town centre;
❑ to promote improvement in quality and convenience of alternative modes to the car; and
❑ to meet the access and mobility needs of disabled people.

Traffic Management:
❑ to take an integrated approach to transport in town centres, which complements the strategy for development;
❑ to provide good access by car, public transport, bicycle and on foot;
❑ to review the allocation of space for pedestrians, cyclists and public transport in shopping areas and the scope for implementing priority measures;

❑ to protect and enhance the pedestrian environment;
❑ to reduce the impact of through traffic and address the need for deliveries; and
❑ to be consistent with advice in PPG13 Transport (DOE/DOT, 1994) [Sb].

Car Parking:
❑ to produce a comprehensive parking strategy for the town centre;
❑ to set appropriate maximum car parking guidelines for new developments;
❑ to ensure that car parking in the town centre serves the town centre as a whole;
❑ to promote the provision and use of car parks for shoppers and other short–term parking; and
❑ to set policies and parking charges, which give priority to short–term parking for visitors and shoppers, rather than commuters.

12.3 Vital and Viable Town Centres

The design and management of town centres requires the skilled application of many of the techniques described elsewhere in this book. The fundamental challenge is to create a high quality environment for pedestrians, whilst retaining access for essential private traffic, public transport and vehicles used for servicing and maintenance (see Photograph 12.1). Hard choices have to be made about the extent to which access is to be allowed and what exemptions from restrictions are to be made, for example for cyclists or people with disabilities. Where vehicle access is permitted, at all times or only at certain times of the day, the detailed design of the areas shared between vehicles and pedestrians is critical to the way in which space will be used and to the success of the town centre. This need for balance is clearly established in Government policy (HMG, 1987). The advice in PPG6 (DOE, 1996a) [Sa] emphasises a significant shift in the nature of that balance.

In addition to PPG6, a range of Planning Policy Guidance from central government emphasises the importance of recognising the links between, and the co-ordination of, land–use and transport decisions. Some of the key PPGs cover General Policy and Principles (DOE, 1996b), Transport (DOE/DOT, 1994),

Photograph 12.1: An example of a high quality street environment.

Planning and the Historic Environment (DOE, 1994a) and Tourism (DOE, 1992) [Sc]. In the case of transport, further guidance is provided in PPG13 – A Guide to Better Practice (DOE/DOT, 1995). The importance of a strategic planning framework is examined in more detail in Chapters 1, 3 and 6.

Town Centre Strategies are devised to facilitate the continuing evolution of the centres, keeping them competitive with out–of–town complexes and neighbouring town centres and meeting the needs of the local population. The character of a town centre will be determined by its history, the type of community it serves and the nature of competing or complementary centres. There is no magic formula for a successful town centre but the Urban and Economic Development Group (URBED) publication Vital and Viable Town Centres (HMG, 1994) identifies three broad indicators of health. These are:
- ❑ attractions – the core facilities in the town centre;
- ❑ amenity – the supporting facilities and general environmental standards; and
- ❑ accessibility – how easy it is to reach and to move around within the centre.

Focal points for activity, including town centres, inevitably attract traffic if they are successful. They will not thrive if users cannot access them with comparative ease. The medium– to long–term viability of a town centre may, therefore, depend as much on the development control decisions made, within government guidance, by a planning authority, as on traffic management or environmental

initiatives designed to bring about immediate improvements (BDP, 1992). Statutory Local Plans, supported by PPGs [Sc] and other central government policies, provide the basis both for resisting unsuitable out–of–town proposals and for supporting appropriate development in town centres. Approvals for town centre sites should ensure that the applicant has addressed not only land–use issues, such as purpose, density, scale, appearance and orientation of the development, but also its likely traffic generation, level of parking provision, accessibility by public transport, cyclists and pedestrians and when and how it is to be serviced. The Institution of Highways & Transportation (IHT) has published useful guidelines for assessing such effects of development proposals (IHT, 1994) and Part 4 of this book examines the topic in more detail.

A particular characteristic of most town centres, and the key to their long–term economic survival, is the diversity of activity they provide. Retailing, employment and services, such as banking and refreshment, will be found in all centres. Most also provide leisure facilities and some have residential accommodation. Historic towns can have the added advantage that the fabric of the town represents an attraction in itself. With such a variety of activity, town centres can be in use by the public for 16 or more hours each day. Add to this the out–of–hours servicing and maintenance and 24–hour activity is becoming increasingly common in town centres (Comedia, 1991). Control and management of access may be needed throughout the day and night in such circumstances, to protect the amenity of residents or to minimise the risk of accidental or criminal damage to public and private property. Conversely, provision must be made for all legitimate servicing and maintenance activity and for users of facilities to gain access, so that the diversity essential to healthy town centres can be maintained.

There must also be at least a minimum level of activity in a town centre for it to be successful. People in a town centre expect a choice of shops or to be able to perform a range of tasks in one location or both. In other words, there needs to be a 'critical mass' of facilities, without which a town centre is likely to enter into a spiral of decline. The scale of this will vary from centre to centre and will depend, crucially, on the attractiveness of alternative centres (BDP, 1992).

The need to provide both diversity and critical mass in a town centre will influence the area defined as the centre. Apparent fringe areas may, in fact, be an intrinsic part of the town centre. A street of estate agents or specialist shops may appear to be detached from the main shopping area but, nevertheless, make an important contribution to the centre as a whole. This must be considered when planning changes to access.

Some town centres benefit from particular features or characteristics, which add to the attractiveness for visitors. River–fronts and wharf areas often act as a focus for leisure activity and historic areas or cultural buildings may draw users to a centre. Indeed, every effort should be made to promote such features as a component of the town's quality and heritage. Doing so can, however, produce its own traffic management problems, including the need to accommodate peaks of activity on a seasonal basis, at weekends or at certain times of the day, and to provide more than usually comprehensive information (such as pedestrians' direction–signing) for occasional or one–off visitors. Many nationally and internationally recognised historic centres, such as York and parts of London, have responded by developing tourist management strategies, including provision for coaches and marketing public transport as a means of accessing major attractions. The English Historic Towns Forum has published guidance on the management of visitors in large numbers (EHTF, 1994a).

A key pointer to success in a town centre is the level of pedestrian activity and the environment in which it takes place. Whatever means of travel a visitor uses to access the centre, he or she will almost certainly spend some time as a pedestrian, a definition which, here, includes wheelchair and pushchair users. It is essential that pedestrians' needs and aspirations are given the highest priority. Doing so is also likely to increase the overall performance of a town centre (TEST, 1988). Public expectations of town centres have heightened over recent years. Out–of–town shopping centres, with their contained and managed environments, have set standards that users want to see replicated, as far as practicable, in town centres. The IHT has published guidelines on moving towards meeting these aspirations, through more wide–spread and effective pedestrianisation (IHT, 1989), and the needs of pedestrians are explored in more detail in Chapter 22. Providing a high quality environment is the essential complementary step to meeting basic access needs for vehicles and ensuring good accessibility for pedestrians.

12.4 Creating a High Quality Environment

Two components of a high quality environment are the appearance of the area, both public and private buildings and spaces, and the way it is used. Both are crucial in a town centre. Users will feel more at ease and will respect an area better if it has a coherent overall image, has been well designed at the detailed level and is well maintained and enforced. This applies everywhere but is particularly important in the case of Conservation Areas. All well–kept areas will be more likely to encourage users to spend time there and will contribute further to the positive feel of the town (see Photograph 12.2). The importance of this basic maintenance of standards is expanded upon by Mitchell (1986) and Hillman (1988).

As well as the core facilities that a town offers, such as shops, employment services and commercial and leisure facilities, users in high quality town centres typically look for:
❑ places to rest and relax (see Photograph 12.3);
❑ informal entertainment;

Photograph 12.2: An attractive and well maintained city centre space.

Photograph 12.3: A place to rest and relax.

Photograph 12.4: An example of a managed market.

❑ managed markets/street trading (see Photograph 12.4);
❑ facilities to amuse children;
❑ information on facilities or events (see Photograph 12.5);
❑ feelings of safety and security;
❑ cleanliness and tidiness, including the prompt removal of graffiti and flyposting;
❑ attractive planting, landscaping, features and use of materials;
❑ spaces free from clutter;
❑ freedom from heavy traffic; and
❑ freedom from noise and fumes.

Safety and security aspects are also significant to users and all projects should address these at the design stage (LBM, 1994).

Measures to meet these requirements will be more effective if they are introduced, within an overall urban design framework, which provides:
❑ a coherent image and sense of place;

❑ emphasis on focal points and points of interest, such as buildings and spaces;
❑ positive/systematic treatment of the links between spaces; and
❑ integration of the public and private domains.

The framework documents should cover how the area is to be used and how it is to look when complete. A good example of this is the Glasgow City Centre Public Realm Study (Gillespies, 1995a). The process of achieving the vision is more appropriately covered by the type of action plan, advocated in Section 12.5, on Town Centre Management.

The focus of this Chapter on traffic and transport issues means that specific reference is warranted to three aspects of physical design. These are:
❑ space–sharing between pedestrians and vehicles;
❑ the geometric design of shared spaces; and
❑ the choice and maintenance of materials for areas of street–space used by people and vehicles.

Definitive rules for sharing space between pedestrians and vehicles in town centres are difficult to provide. There will always be an element of professional judgement as to acceptable arrangements and appropriate designs for shared spaces. Furthermore, public consultation may show a particular combination to be acceptable to users of one town centre but not of another. At one extreme, half a dozen service or maintenance vehicles per hour, driven carefully in an otherwise fully–pedestrianised street, will not cause undue danger or disturbance to pedestrians and are unlikely to affect the way pedestrians use the street. In these circumstances, people crossing the street are likely to walk in straight lines towards their next destination, ignoring vehicle paths. On the other hand, pedestrians in a street carrying 30 or more buses an hour are likely to treat the space used by the buses as an exclusive

Photograph 12.5: Town centre information display.

TRANSPORT IN THE URBAN ENVIRONMENT

vehicle–path, irrespective of the way it is delineated. They will tend to cross this path at right angles and walk near the edges of the street. Most decisions about space–sharing lie somewhere between these examples.

One guideline is to provide clear delineation of vehicle–paths, where vehicles are predominantly passing through, rather than gaining access to, a street. This would apply, for example, in the case of:
- ❑ bus routes;
- ❑ through cycle–routes; and
- ❑ access roads to car parks.

More extensive guidelines on Buses in Shopping Areas are included in the report of a study carried out for London Transport (TEST, 1987).

Even where vehicles are being allowed access for local servicing only, separate paths may be justified where the flow or speed of vehicles would otherwise cause anxiety to pedestrians. Nevertheless, emphasising by design that a street is intended to be primarily for

Photograph 12.6: Shared space where pedestrians are not threatened by the presence of occasional vehicles.

Photograph 12.7: Good town centre traffic management – Town Hall Square Reading. Courtesy Babtie Group.

pedestrians can influence the way vehicles use the street (see Photograph 12.6). Reducing carriageway widths, raising pedestrian crossing points, introducing chicanes and tightening turning radii can all help to calm intruding traffic. Some of the most effective street designs manage to control drivers' behaviour, by combining speed–reducing features with more subtle measures, such as the visual treatment of entry points to the street, the placing of street furniture within the street, the choice and detailed layout of surfacing materials and even the presence of large numbers of pedestrians. Most of these elements combine in the 20 miles/h zone in Newbury's main shopping street, for example, where flat–topped road humps have become informal pedestrian crossings. Traffic Calming Guidelines (DCC, 1991) and Traffic Calming in Practice (CSS, 1994) provide illustrated examples of many of the measures available and the DOT has published both a brief catalogue of road safety measures (DOT, 1994) and a bibliography of relevant Government guidance on traffic calming (DOT, 1996f) [Sd].

Traffic calming is an evolving technique with innovative schemes being trialled by imaginative local authorities. Such innovation is to be encouraged but it is essential that the measures are introduced within an overall strategy for calming and safety, that a full audit is carried out of the likely safety performance of individual schemes (see Section 16.11) and that the necessary statutory approvals are obtained. The IHT has published guidelines on some of these aspects (IHT, 1990 and 1996). Road Safety as a whole is considered in more detail in Chapter 16.

A location which has combined many of the principles of good town centre traffic management is Town Hall Square in Reading (see Photograph 12.7). A road which separated the Old Town Hall, now a museum, conference and entertainment venue, from

Photograph 12.9: An example of the purposeful use of bollards.

the town centre was closed to through traffic and incorporated into the Town Hall forecourt to form a public square, whilst retaining vehicular access for servicing. Although the short link that was closed offered a saving of hundreds of metres over the alternative route, subtle design has resulted in few contraventions of the restrictions and has led to the square being a focus for leisure activity in the town centre.

The longer–term integrity of shared or pedestrianised areas is heavily dependent on the performance of surfacing materials. It is critical that these are capable of carrying the loads and withstanding the uses to which they will be subjected in practice. Full account must be taken of the type of maintenance, delivery and service vehicles that will use each part of the street and of the maintenance methods that will be used on the surfaces themselves (see Photograph 12.8). For example, refuse vehicles may be the heaviest vehicles using a particular surface and high–pressure water, used for cleaning, can quickly wash out bedding sand for artificial stone–paving or blockwork, producing subsidence or failure. The designer can provide protection and prevent access to parts of a street by the strategic placing of street furniture or the purposeful use of bollards (see Photograph 12.9). This, may however, add

Photograph 12.8: an example of an appropriate maintenance method.

unnecessary clutter and a better environment is provided by designing and constructing surfaces to withstand the types of vehicles that are likely to use them, for whatever reason.

The choice of materials, and their detailed deployment, should take full account of the overall townscape. Footway and carriageway surfaces, which complement and blend in with the surrounding buildings, help to create a coherence to the area as a whole. Users are more aware of being in a functioning town centre and that they should give priority to the activities and pedestrian movement taking place there. Again, particular attention is warranted to the design of street works in historic areas, a need emphasised by English Heritage (EH, 1993).

Although the choice of appropriate materials and construction, to withstand expected loading and to maintain or enhance the visual coherence of an area, are essential steps, more is needed to guarantee a high quality environment. Careful attention must be paid to the detailing of construction and to the overall layout of surfaces, signs and landscaping, if the maximum value is to be obtained. Attention to detail is important in any scheme but its absence is most likely to be noticed in historic areas, a point recognised in a series of publications by the Civic Trust (CT *et al*, 1993 and 1994), the English Historic Towns Forum (EHTF, 1994b) and the DOT (DOT, 1996a) [Se].

However well–designed and constructed a scheme is, regular care and maintenance are essential to the quality of the town centre. Town Centre Management Strategies need to provide for regular and systematic inspection of the fabric of the area, to encourage and facilitate public reporting of damage, vandalism (including graffiti) or failures, using 'Defects Cards' or 'Hotlines' for example, and to act promptly when such damage is reported. At the same time, litter–picking, sweeping, gully–emptying, landscape maintenance and other periodic maintenance activities need to be programmed at appropriate intervals. In times of highly constrained budgets, the period between programmed maintenance visits, or the degree of deterioration which is deemed to warrant reactive maintenance, tends to increase but excessive relaxation of standards can prove to be a false economy. If left unattended, relatively minor damage to surfacing can lead to major deterioration of the surface, requiring significant investment to restore its integrity at a later date. Similarly, failure to repair broken street furniture, or to remove graffiti, can lead to 'copy cat' damage, either through carelessness or as deliberate acts of vandalism. The

need for good maintenance strategies for town centres has been acknowledged for many years (Hillman, 1988). Comprehensive guidance on the maintenance of physical assets is contained in the Highways Maintenance Handbook (Atkinson, 1990).

The foregoing analysis of designing effectively for vehicles using and sharing pedestrian space illustrates that traffic management, urban design and engineering design should be carried out as a co–ordinated exercise, if the end result is to contribute fully to a high quality town centre environment. Much valuable work in recognition of these principles has been carried out in Scotland, on the initiative of the former Strathclyde Regional Council (SRC, 1994 and 1995) and more recently by Scottish Enterprise.

12.5 Town Centre Management

Town Centre Management, which derives from the need to view town centres as complex and integrated entities, manifests itself in a wide range of forms, from the purely janitorial, keeping streets clean and tidy, to the long term strategic, facilitating and promoting major change, with many combinations in between (Gillespies, 1994 and 1995b). Each is an equally valid approach, if based on an analysis of the Strengths, Weaknesses, Opportunities and Threats

Photograph 12.10: Enforcement of parking regulations.

that apply to the centre (a SWOT analysis). Town Centre Management may be the responsibility of a nominated Town Centre Manager, with a permanent staff or an ad–hoc team of people. The Association of Town Centre Management advocates tailoring the approach to the needs of the particular location and suggests the following definition for Town Centre Management (Donaldsons *et al*, 1994):

'Town Centre Management is the effective co–ordination of the private and public sectors, including local authority professionals, to create, in partnership, a successful town centre – building upon full consultation.'

Whatever form of Town Centre Management is adopted, some common elements should be:
- ❑ a vision or overall strategy;
- ❑ an Action Plan for achieving the vision;
- ❑ partnership with users, businesses, service providers and investors, with clear agreements over responsibilities;
- ❑ active leadership and promotion of the overall plan, including fund–raising;
- ❑ day-to-day management and control; and
- ❑ enforcement power and duties (see Photograph 12.10).

The overall action plan, supplementing the vision or strategic plan, should cover changes to traffic management, servicing and access. It should detail the desired end–scheme, as outlined in the strategic plan, and set out how its implementation is to be phased. Restrictions on access are often initially opposed by businesses and may have to be introduced in a less stringent form, or over a smaller area, than originally proposed. Once successful, pressure is then frequently brought to bear to extend the restrictions. It is, therefore, worth adopting a stage–by–stage approach, helping to retain public support for the strategy at all stages.

Any long–term plan should also allow for updating in the light of developing circumstances. For example, land–use changes may occur which dictate alterations to access arrangements.

Once changes have been effected to access arrangements, as part of an overall plan, a key town centre management function is to ensure the operation and enforcement of those access restrictions and priorities. This involves establishing working relationships with groups, such as traders and market–stall operators, to ensure compliance with voluntary agreements or permit systems. Similar relationships are needed with bus and taxi operators, over issues such as temporary parking and the

Photograph 12.11: The pedestrianised town centre of Reading. Courtesy: Babtie Group.

layover of vehicles in the centre. Liaison is needed with the police to enforce access restrictions and on–street parking regulations, unless these have been taken over by the local authority. Town centre managers are usually directly responsible for controlling issues such as street trading, markets and off–street council car parking and for liaising with private car park operators. An effective means of establishing and maintaining contact is the creation of town centre working groups, with membership drawn from the spectrum of organisations involved in service provision. The groups require careful management to avoid degeneration into 'talking shops'. Making them task–focused can prevent this happening.

The pedestrianisation of Reading Town Centre is one example, out of many well–planned and implemented town centre improvements (RBC, 1989) (see Photograph 12.11). Promoted by both the County and Borough Councils, its radical impact on traffic circulation, including bus–routeing, was endorsed, following extensive consultation and promotion. It was the culmination of strategic transportation and land–use planning over many years, yet its eventual acceptance required extensive on–the–ground negotiation, with individual businesses and service providers, and consequent changes to the detail of the initial proposals. Much of this negotiation was carried out by the Town Centre Manager. The initial scheme, in its modified form, was so well received that it has subsequently been extended, with further access restrictions coming into effect.

A key lesson from the Reading project is that a successful and thriving town centre is dependent on sensible implementation and enforcement of controls

166

Photograph 12.12: Restrictions on vehicular access at particular times.

and restrictions and management arrangements must be capable of adapting quickly to change. However thorough the pre–planning of access restrictions might be, unforeseen circumstances can arise which call for a relaxation of the formal restrictions. The system must be capable of accommodating these on a flexible and equitable basis.

12.6 Managing Town Centre Access

Previous sections have stressed that high quality town centres require restrictions on vehicle access. At the same time, their success is dependent upon significant numbers of people being able to gain access to the facilities offered and on the servicing and maintenance of these facilities. Demand for private vehicle access will usually exceed the levels compatible with a high quality environment and some form of demand–management will need to be employed. It is essential, however, that restraint measures are considered as part of a co–ordinated package of measures. The elements of demand–management will include 'sticks' and 'carrots', such as:

❑ restricting access to particular areas and/or spaces;

❑ restricting access at particular times (see Photograph 12.12);

❑ regulating demand by parking cost and supply;

❑ prioritising access for specific types of user, such as permit holders and Orange Badge holders;

❑ prioritising access for specific vehicle–types, such as buses, light rail, taxis, and cycles;

❑ making alternatives to car–use more attractive (eg park and ride) (see Photograph 12.13); and

Photograph 12.13: Park and Ride as an alternative means of access for car–users.

Photograph 12.14: Specific provision for cyclists.

❑ regulating demand by road–use pricing.

Measures for deterring and restraining traffic are explained more fully in Chapters 13, 19, 20 and 21 and their enforcement is addressed in Chapter 14.

Implicit in this approach is the concept of need. Most people would accept that disabled people should be given privileged access. The extent to which other users should be assisted or discouraged may vary according to local priorities. A typical list of users to be included in a priority hierarchy might be, as follows:
   ❑ emergency services;
   ❑ pedestrians;
   ❑ disabled people;
   ❑ cyclists;
   ❑ public transport users;
   ❑ drivers of delivery vehicles;
   ❑ public utility operators;
   ❑ taxis; and
   ❑ private cars.

Chapters 22, 23 and 24 detail how walking, cycling and public transport uses can positively be encouraged and appropriate provision made for people with impaired mobility.

The need for positive measures for pedestrians has been stressed throughout this chapter and specific reference has been made to the IHT Guidelines on pedestrianisation (IHT, 1989). The particular needs of people with impaired mobility are also covered by IHT Guidelines (IHT, 1991) and the DOT has published more specific guidance on Parking for Disabled People (DOT, 1995a) [Se].

The National Cycling Strategy (DOT, 1996b) [Sf] sets out how the specific needs of cyclists can be met and how cycling can play a significant role in urban transport policy, including town centre strategies (see Photograph 12.14). The DOT has also published a Cycling Bibliography (DOT, 1995b) [Sd] and guidance on specific issues such as Bike–and–Ride (DOT, 1996c) [Se].

The role of public transport in serving town centres is addressed in the Chartered Institute of Transport's 'Better Public Transport for Cities (CIT, 1996) and useful practical guidance on designing for buses is given in Better Buses – Good Practice in Greater Manchester (AGMA, 1992).

Crucial to the vitality of town centres is the need to provide for effective servicing. Businesses have become increasingly flexible in their acceptance of loading restrictions, especially where these have been introduced as a component of an environmental improvement. Nevertheless, there is a limit to the times when businesses can accommodate deliveries, especially in the case of small businesses where the proprietor has to be present to receive the goods. There will always be a need to provide for emergency service access. In the case of markets, stall–holders may also need to top–up stock during the day, again requiring special arrangements. Acceptance of a town centre proposal often depends crucially on careful negotiation with individual businesses or their representatives, such as Chambers of Commerce, over restrictions on access and on patient explanation as to why they are needed. Guidance on managing lorry movement is offered by the Freight Transport Association (FTA, 1983) and the Civic Trust (CT *et al*, 1990) and the topic is addressed in more detail in Chapter 25.

In addition to influencing the balance between modes, prioritisation can occur between users of the same mode, for example in the allocation of, and charging for, parking space. Typically, this would be:
   ❑ disabled people;

Photograph 12.15: Vehicular access to rear service courts only.

- residents;
- short–term shoppers;
- users of leisure facilities;
- people on personal business;
- local employees on unsocial hours;
- other local employees; and
- onward commuters (eg parking–and–riding).

Chapter 19 explores the possibilities for parking management in more detail.

This approach has, for example, been used in the prioritisation of both on– and off–street parking space in Wimbledon Town Centre in the London Borough of Merton (Nicholson, 1995). More general advice on parking provision is available in Parking Policy, Design and Data (Young, 1991), through the British Parking Association (BPA, 1988) and in PPG13 – A Guide to Better Practice (DOE/DOT, 1995) [Sg].

In allocating space, by whatever method and for whatever mode of travel, it is important to look individually at the component areas of a town centre. A town centre square is always a focus for pedestrian activity and should be kept as free of traffic as possible. All traffic could be banned throughout the working day, for example, with servicing and maintenance traffic only being allowed outside these hours. Elsewhere, a street on the edge of the main shopping area might be designated for use as the principal setting down and picking up location for

buses serving the centre. This could result in an increase in traffic flow, through the concentration of activity in one street. This may actually be seen as an advantage by businesses in the street, who experience increased passing trade.

The creation of different categories of roads and spaces, with clear rules about the types of vehicle allowed access and at what times of day, is a fundamental element of co–ordinated town centre design. Some typical categories would include:
- a pedestrian zone with no vehicular access;
- servicing traffic only, but time restricted;
- a bus–only street;
- cycles and Orange Badge holders only;
- special permit holders only, such as market stall–holders; and
- access to specific premises only, such as car parks and service yards (see Photographs 12.15).

Generally, through traffic should be excluded. To achieve this, in many city centres, alternative routes have to be built, or designated, to by–pass the town centre. The location of these all–purpose routes must be considered carefully, as they can form barriers to movement into, and out of, the town centre, particularly for pedestrians and cyclists. Part 5 of this book examines the techniques for major infrastructure construction and Chapters 15 and 18 respectively look at information for transport users and at the management of the transport network for maximum efficiency.

Under certain circumstances, it may be appropriate to provide for through movement in a town centre for modes other than private cars. A light rapid transit route may run on–street, serving cross–town movement as well as access to the town centre. The same may apply for bus or cycle routes. Chapter 34 looks at the provision of fixed–track routes.

Photograph 12.16: Vehicles powered by CNG minimise pollution.

Photograph 12.17: Signing of town centre car parks with real–time information.

Public concern about the effects of noise and pollution on health and quality of life is reflected in Government guidance (DOE 1994b [Sh] and 1994c [Si]). The Government has also suggested ways in which positive action, through planning decisions (DOE/DOT, 1993) and traffic management (DOT, 1996d), can control pollution and has also recognised the need to consider the possible adverse effects of traffic–calming measures (DOT, 1996e).

With heightened public awareness of traffic pollutants, consideration could be given to requiring, or favouring, vehicles which minimise pollution, when considering who should be allowed access to restricted areas. Light rail vehicles are quiet and non–polluting at their point of use and therefore suitable for town centres (as in Manchester City Centre, for example). Similarly, hybrid electric buses offer an environmentally–friendly alternative, in certain circumstances; electric minibuses are used on the 'Downtown Shuttle' in Charlotte, North Carolina, USA, for example. Vehicles powered by compressed natural gas (CNG) might be of more general and immediate applicability in the UK (see Photograph 12.16). Similar in many ways to petrol or diesel vehicles, they are quieter and emit virtually no noxious gases or particulates. Many types of vehicle are available, with CNG engines as an 'off–the–peg' option.

An analysis of the benefits of different types of alternative fuel and of other methods of reducing vehicular emissions is contained in the Royal Commission on Environmental Pollution's report on Transport and the Environment (RCEP, 1994)

Any vehicle–user, public or private, has to park or get off their vehicle at some point in a town centre. They will need either secure and safe facilities, in which to park their own vehicle, or safe and comfortable areas in which to wait for public transport. In all cases, the route from their parking or alighting point to the main areas of town centre activity must be well signposted, safe, easy to use and well lit. Parking facilities and pedestrian links to them must be seen by users to be part of the town centre, so that a positive image is conveyed from the moment of arrival. Chapters 15, 19, 22 and 24 address requirements for users of public and private vehicles.

This concept of a town centre being welcoming to all arriving users starts even before the alighting point. For private vehicle–users, routes to car parks or cycle–parking facilities should be clearly signed, with real–time information about available spaces where possible (see Photograph 12.17). Pedestrians and cyclists should have safe and comfortable routes on

Photograph 12.18: Information and consultation.

TRANSPORT IN THE URBAN ENVIRONMENT

the approaches to the centre, especially where their routes cross major vehicle routes. The ways of achieving this are detailed in Chapters 15, 22 and 23.

12.7 The Importance of Consultation

Throughout this chapter, references have been made to the need to generate acceptance among users of proposals, especially those involving restrictions. This will only be achieved if effective consultation is carried out (see Photograph 12.18), at all stages of the process, from identifying existing shortcomings and creating the vision or overall plan, through promoting specific proposals within the plan and implementing agreed proposals, to enforcing regulations designed to make the implemented scheme work well. Advice on consultation is presented in Chapters 10, 11 and 13.

12.8 References

AGMA (1992)   'Better Buses – Good Practice in Greater Manchester', Association of Greater Manchester Authorities.

Atkinson, K (Ed) (1990)   'Highway Maintenance Handbook', Thomas Telford.

BDP (1992)   'The Effects of Major Out of Town Retail Development', Building Design Partnership, Stationery Office.

BPA (1988)   'BPA Seminar Papers', British Parking Association 1988 onwards.

CIT (1996)   'Better Public Transport for Cities', Chartered Institute of Transport.

Comedia (1991)   'Out of Hours – A Study of Economic, Social and Cultural Life in twelve Town Centres in the UK', Comedia.

CSS (1994)   'Traffic Calming in Practice', County Surveyor's Society.

CT/CSS/DOT (1990)   'Lorries and Traffic Management – A Manual of Guidance', Civic Trust/County Surveyors' Society/DOT.

CT/EHTF (1993)   'Traffic Measures in Historic Towns', Civic Trust/English Historic Towns Forum.

CT/EHTF (1994)   'Traffic in Townscape – Ideas from Europe', Civic Trust/English Historic Towns Forum.

DCC (1991)   'Traffic Calming Guidelines', Devon County Council.

DOE (1992)   Planning Policy Guidance Note 21 'Tourism', DOE [Wa].

DOE (1994a)   Planning Policy Guidance Note 15 'Planning and the Historic Environment', DOE [Sj] [Wb].

DOE (1994b)   Planning Policy Guidance Note 23 'Planning Pollution Control', DOE [Sh] [Wa].

DOE (1994c)   Planning Policy Guidance Note 24 'Planning and Noise' DOE [Si] [Wa].

DOE (1996a)   Planning Policy Guidance Note 6 'Town Centres and Retail Developments', DOE [Sa].

DOE (1996b)   Planning Policy Guidance Note 1 'General Policy and Principles', DOE [Sk].

DOE/DOT (1993)   'Reducing Transport Emissions Through Planning', DOE/DOT

DOE/DOT (1994)   Planning Policy Guidance Note 13 'Transport', DOE [Sb] [Wa].

DOE/DOT (1995)   'PPG13 – A Guide to Better Practice' DOE/DOT [Sg] Wa].

Donaldsons/ Healey and Baker (1994)   'The Effectiveness of Town Centre Management', The Association of Town Centre Management.

DOT (1994)   'Safer by Design – a Guide to Road Safety Engineering ', DOT.

DOT (1995a)   Traffic Advisory Leaflet 5/95 'Parking for Disabled People', DOT [Se].

DOT (1995b)   Traffic Advisory Leaflet 9/95

'Cycling Bibliography', DOT [Sd].

DOT (1996a) Traffic Advisory Leaflet 1/96 'Traffic Management in Historic Areas', DOT [Se].

DOT (1996b) 'The National Cycling Strategy', DOT [Sf].

DOT (1996c) Traffic Advisory Leaflet 3/96 'Bike and Ride', DOT [Se].

DOT (1996d) Traffic Advisory Leaflet 4/96 'Traffic Management and Emissions', DOT [Se].

DOT (1996e) Traffic Advisory Leaflet 6/96 'Traffic Calming: Traffic and Vehicle Noise', DOT [Se].

DOT (1996f) Traffic Advisory Leaflet 10/96 'Traffic Calming Bibliography', DOT [Sd].

EH (1993) 'Street Improvements in Historic Towns', English Heritage.

EHTF (1994a) 'Getting it Right – a Guide to Visitor Management in Historic Towns', English Historic Towns Forum.

EHTF (1994b) 'Traffic in Historic Town Centres', English Historic Towns Forum.

FTA (1983) 'Designing for Deliveries', Freight Transport Association.

Gillespies (1994) 'Streetscape Best Practice Guide', Scottish Enterprise.

Gillespies (1995a) 'Glasgow City Centre – Public Realm', Strathclyde Regional Council.

Gillespies (1995b) 'Comparative Study of Streetscape Works', Scottish Enterprise.

Hillman J (1988) 'A New Look for London', Stationery Office.

HMG (1987) Local Transport Note 1/87 'Getting the Right Balance – Guidance on Vehicle Restriction in Pedestrian Zones', Stationery Office.

HMG (1994) Urban and Economic Development Group (URBED) 'Vital and Viable Town Centres – Meeting the Challenge', Stationery Office.

ICE (1993) 'Tomorrow's Towns – An Urban Environment Initiative', Institution of Civil Engineers.

IHT (1989) 'Pedestrianisation Guidelines', The Institution of Highways & Transportation.

IHT (1990) 'Guidelines for Urban Safety Management', The Institution of Highways & Transportation.

IHT (1991) 'Reducing Mobility Handicaps: Towards a Barrier–Free Environment', The Institution of Highways & Transportation.

IHT (1994) 'Guidelines for Traffic Impact Assessment', The Institution of Highways & Transportation.

IHT (1996) 'Guidelines for Road Safety Audit', The Institution of Highways & Transportation.

LBM (1994) 'Designing Out Crime', London Borough of Merton/Metropolitan Police.

Mitchell G (1986) 'Design in the High Street', The Architectural Press.

Nicholson J (1995) 'Parking in Town Centres – the Key to Sustainability', Highways & Transportation (November)

RBC (1989) 'Centre Plan', Reading Borough Council.

RCEP (1994) Royal Commission on Environmental Pollution Eighteenth Report – 'Transport and the Environment', Stationery Office.

SRC (1994) 'Streetscape – a Design Guide', Strathclyde Regional Council.

SRC (1995) 'Footways Development Guide', Strathclyde Regional Council.

TEST (1987) 'Buses in Shopping Areas', TEST.

TEST (1988) 'Quality Streets – How Traditional Urban Centres Benefit From Traffic–Calming', TEST.

Young, W. (1991) 'Parking Policy, Design and Data', Monash University, Australia.

12.9 Further Information

Barton H (1995) 'Sustainable Settlements – A Guide for Planners, Designers and Developers', University of the West of England/The Local Government Management Board.

Civic Trust Regeneration Unit (1994) 'Caring for Our Towns and Cities', Boots the Chemists.

DOE (1996) 'Quality in Town and Country – Urban Design Exhibition', DOE (September).

Evans R (1993) 'Liveable Towns and Cities', Civic Trust.

Hass–Klau C (Ed) (1988) 'New Life for City Centres', Anglo–German Federation.

HMG (1994) 'Sustainable Development – The UK Strategy', Stationery Office.

Chapter 13 Procedures for Implementing Traffic Management Measures

13.1 Basic Principles

The management of traffic can be achieved through the implementation of a variety of procedures and measures. Some of these will require the making of one or more Orders or other statutory procedures, to permit regulation of traffic, whilst others will not (see Photograph 13.1).

Photograph 13.1: Traffic management Shepherd's Bush Green, London.

Highway authorities and some other authorities in Britain [NIa] are empowered under the Road Traffic Regulation Act 1984 (RTRA) (HMG, 1984a) to make Traffic Regulation Orders (TROs), to regulate the speed, movement and parking of vehicles and to regulate pedestrian movement. The Highways Act 1980 (HMG, 1980) [Sa], including the amendments made by the Traffic Calming Act 1992 (HMG, 1992a), give highway authorities powers to introduce measures which regulate the movement of vehicles, reduce accidents or improve the local environment and which do not require Orders to be made.

As indicated in Chapter 14, enforcement will be easier to achieve if the purpose underpinning an Order is clear to road–users and is supported by clear, unambiguous traffic signs. There will, however, be locations where compliance with the requirements of the Order is improved by the introduction of physical measures or restrictions.

When drivers do not comply with Orders and are convicted of an offence, the courts may impose penalties, such as fines, the endorsement of a licence or even disqualification from driving (HMG, 1972). The procedures for making Orders are laid down by the Secretary of State and must be observed strictly by the Order–making authority, to avoid legal difficulties [NIb]. The procedures generally involve consultation on, and publishing of, proposals and the consideration of objections. In some circumstances, public inquiries must (or may) be held. Local authorities also have powers to make Orders, similar to Traffic Regulation Orders, which require the same or a similar procedure to be followed (see Sections 13.6 *et seq*).

Even where an Order is not required, there may still be a statutory requirement to advertise and consult on proposals and to consider objections. Where there is no statutory requirement to do so, it is still advisable to consult widely in the development of a traffic scheme [NIc]. Consultation with the public is discussed in detail in Chapter 10.

All traffic signs must be in accordance with the current Traffic Signs Regulations and General Directions (TSRGD) (HMG, 1994b) or else be specially authorised by the Secretary of State [NId]. Advice on the design and use of traffic signs is given in the Traffic Signs Manual (DOT, 1977), supplemented by other publications issued by the Department of Transport (DOT, 1994) [Sb] and Working Drawings for Traffic Signs Design and Manufacture (HMG, 1994a).

13.2 Procedures for Making Traffic and Parking Orders

Several different regulations specify the procedures for making traffic and parking Orders, under the RTRA 1984 [NIb]. These are:

❑ The Local Authorities' Traffic Orders (Procedure) (England and Wales) Regulations, 1996 (HMG, 1996b) [Sc];

❑ The Secretary of State's Traffic Orders (Procedure) (England and Wales) Regulations 1990 (HMG, 1990b), which prescribe the procedures that the Secretary of State must use [Sd]; and

❏ The Road Traffic (Temporary Restrictions) Procedure Regulations 1992 (HMG, 1992c).

All of these regulations specify, in rather precise terms, the procedures which are to be used. There are variations between them for different Order–making authorities and between the procedures required for Orders of different types or purposes, such as permanent, temporary or experimental Orders. There are, however, certain general similarities.

The 1996 Regulations [Sc]are intended to be quicker and cheaper to use and easier to understand for Order–making authorities than previous versions while, at the same time, continuing to safeguard the interests of road–users, frontagers and the public at large.

Item	Case	Consultee
1.	Where the Order relates to, or appears to the Order-making authority to be likely to affect traffic on, a road for which another authority is the highway authority or the traffic authority.	The other authority.
2.	Where the Order relates to, or appears to the Order-making authority to be likely to affect the traffic on, a Crown road.	The appropriate Crown authority.
3.	Where the Order relates to, or appears to the Order-making authority to be likely to affect traffic on, a road subject to a concession.	The concessionaire.
4.	Where the Order relates to, or appears to the Order-making authority to be likely to affect traffic on, a road on which a tramcar or trolley vehicle service is provided.	The operator(s) of the service(s)
5.	Where the Order relates to, or appears to the Order-making authority to be likely to affect traffic on,– (a) a road outside Greater London, which is included in the route of a local bus service; or (b) a road in Greater London, which is included in the route of a London bus service.	In case (a), the operator(s) of the service(s) In case (b), the operator(s) of the service(s) and London Regional Transport.
6.	Where it appears to the authority that the Order is likely to affect the passage on any road of- (a) ambulances; or (b) fire-fighting vehicles.	In case (a), the chief officer of the appropriate NHS trust. In case (b), the chief officer of the fire brigade of the fire authority.
7.	All cases	(a) The Freight Transport Association. (b) The road haulage Association. (c) Such other organisations (if any) representing persons likely to be affected by any provision in the order as the Order-making authority thinks it appropriate to consult.

Table 13.1: Consultation Requirements [Se], Source HMG (1996b).

Before an Order is made, the chief officer of police must be consulted. In addition, the Authority must follow the consultation requirements set out in tabular form in Regulation 6 of The Local Authorities' Traffic Orders (Procedure) (England and Wales) Regulations 1996 (see Table 13.1) [NIe].

Under these arrangements, an Order–making authority must, before making an Order in a case specified in Table 13.1, consult the appropriate persons specified.

In addition to the consultation requirements set out in Table 13.1, it is usual to consult with local residents and representatives of commerce and industry, but it is for the Order–making Authority to decide on the extent of additional consultation. The benefits of wider consultation are set out in Chapter 10.

Notices which set out proposals and invite objections must be placed in the local press. In addition, London authorities must advertise Orders, under section 6 of the Road Traffic Regulation Act 1984, in the London Gazette [NIf]. It remains open to local authorities outside London to advertise in the London Gazette, if they so wish. At the same time, local authorities may, if they think fit, also display street notices and deliver notices or letters to premises likely to be affected by the provisions of the Order.

The display of street notices is not an absolute requirement, as there are places where it would serve no useful purpose, such as in an underpass where there is no right–of–way for pedestrians, but the Department of Transport [Sf] has made it clear to authorities that they should display street notices wherever the public might reasonably be expected to see them. During this period, and until six weeks after the Order is made, a copy of the draft Order (including, where the road is a restricted one, the length of road affected, a statement of the Authority's reasons for making the Order and a map of the roads affected) must be placed on deposit for inspection by the public at all reasonable times. Any objections must be considered by the Order–making Authority.

Under certain circumstances, a Public Inquiry must be held [NIg] [Sg]. These are:
  ❑ where an objection has been made by a bus operator who provides a local service on a road to which the Order relates; or
  ❑ where an objection has been made to an Order which would prohibit, or have the effect of prohibiting, loading or unloading in a road, unless:
    ❑ the prohibited period is between 07.00 hours and 10.00 hours or 16.00 hours and 19.00 hours, or
    ❑ the authority is satisfied that the objection was

frivolous or irrelevant.

There is no requirement to hold a public inquiry if the Order would have the effect of:
  ❑ prohibiting loading or unloading by the creation of a disabled persons bay; or
  ❑ prohibiting loading or unloading up to 15m either side of a junction unless the effect of the prohibition, taken with prohibitions already imposed, is to prohibit loading or unloading of vehicles for a distance of more than 30 metres out of 50 metres on one side of any length of road. This is to ensure that objectors may require a public inquiry to be held, where the effect of prohibiting loading around junctions would be to prohibit loading over longer distances. This situation could occur, for example, where the spacing between junctions was 50 metres or less, which would provide insufficient space for loading by heavy goods vehicles.

In other circumstances, it is open to authorities to hold a Public Inquiry if they so wish. The procedures for holding a Public Inquiry are set down in the procedure regulations (HMG, 1996b; and DOT 1996c).

Any objections to an Order must be considered by the elected members of an Order–making authority. If, after considering objections, an authority decides to proceed with the making of the Order, then the objectors must be informed of the reasons, in writing, and the appropriate notices placed in the press. The chief officer of police must also be informed.

If an Order is modified to take account of objections, it may not be necessary to re–advertise, provided that the revisions do not substantially change the proposals. When the Order is made, the Order–making Authority must advertise a 'notice–of–making' in a local newspaper, and in the London Gazette for Orders by London authorities, under sections 6 and 9 of the Road Traffic Regulations Act 1984 (HMG, 1984a) [NIf]. Steps must be taken to provide the appropriate traffic signs to bring it into effect and these must be in accordance with the current Traffic Signs Regulations and General Directions, although other signs may be specially authorised by the Secretary of State [NIh].

The validity of an Order may be questioned during the six weeks after it is made, on the grounds either that it is outside the Authority's powers or that the interests of the applicant were prejudiced by the Authority's failure to follow the specified procedures. It may not be questioned after this period. An Order must be made within two years from the date of publication of the 'notice–of–making'.

Traffic Orders may be permanent, experimental (up to 18 months) or temporary (up to 18 months). However, the 18 month time–limit for temporary Orders does not apply if the Order–making Authority is satisfied, and it is stated in the Order that they are satisfied, that the execution of the works will take longer. The Secretary of State may, at the request of an Order–making authority, extend the duration of a temporary order by up to, but not more than, 6 months.

The Road Traffic (Temporary Restrictions) (Procedure) regulations (HMG, 1992c) describe the procedures to be followed for the making of temporary Orders and Notices [NIj].

A highway authority, or a concessionaire of a toll road, may take immediate action to prohibit or restrict the use of a road by the placing of an appropriate notice in circumstances where there is a likelihood of a danger to the public or of serious damage to the highway. Such a notice can continue in force for between five and 42 days, depending on the reason for making of the Notice, and the Notice may be followed by the making of a temporary Order. There is no right of objection to a temporary Order or Notice. See sections 14 and 15 of the road Traffic Regulation Act 1984, (HMG, 1984a), as substituted by the Road Traffic (Temporary Restrictions) Act 1991 (HMG, 1991b) [NIj].

Experimental Orders may be made for up to 18 months, to test a scheme before deciding whether to make it permanent. They are subject to different and shorter procedural requirements. These are that:
- ❑ Orders must be advertised at least seven days before the Order comes into effect;
- ❑ consultation requirements must be followed before the Order is made [NIi];
- ❑ there is no requirement to advertise an Order inviting objections before the Order is made;
- ❑ where an Order is intended to become permanent, objections may be made within six months of the Order being made, or subsequently varied, by the authority;
- ❑ an authority must not vary or modify an Order more than 12 months after the Order is made (if it does so, it must follow the normal, and not the shortened, procedures for making the Order permanent); and
- ❑ an authority must consider all objections made during the objection period, before deciding whether or not to make an Order permanent and they may, in certain circumstances, be required to hold a public inquiry.

To avoid the possibility of a successful challenge in the High Court, an authority must be able to demonstrate where the element of experiment or uncertainly lies, as an experimental Order can be made only for the purpose of carrying out an experimental scheme of traffic control.

It is common practice to include exemptions, for certain classes of vehicles or for particular purposes, within a traffic regulation or parking Order. As examples, these exemptions often include:
- ❑ emergency services (ie the police, fire and ambulance services);
- ❑ statutory undertakers and other public bodies involved in the construction and maintenance of the highway and the services located within it;
- ❑ vehicles needing access for weddings, funerals or removals; and
- ❑ post office and security vehicles making collections and deliveries of mail, cash or valuables.

It is also common to exclude, from the effect of waiting restrictions, vehicles which are stopping for the purpose of picking up or setting down passengers and their luggage and also vehicles being loaded or unloaded. Each Order will require careful consideration on its own merits, to take account of local circumstances. Other exemptions may also be necessary to meet local requirements but do not need to be shown on the signs introduced to effect the regulations.

The police have powers to control traffic and to regulate the use of the highway in emergencies and on other special occasions.

The Road Traffic Act 1991 (HMG, 1991a) provides for the decriminalisation of most non–endorsable on–street parking offences in London (where the system has already been introduced) and it permits similar arrangements to be introduced elsewhere [Sh]. Local authorities may apply to the Secretary of State for Orders decriminalising the offences within particular geographical areas. Section 43 of, and Schedule 3 to, the Road Traffic Act 1991 enables eligible local authorities outside London to apply to the Secretary of State for Orders creating Permitted Parking Areas (PPAs) and Special Parking Areas (SPAs). Within a PPA, contravention of Orders designating permitted on–street parking places, such as meter–bays, residents' and disabled persons' bays and free parking bays, will no longer be criminal offences and will become subject to new enforcement arrangements. Within an SPA, most other non–endorsable parking offences will be decriminalised and enforced by the Local Authority [NIk]. More detail on this topic is given in Chapter 19.

13.3 The Purposes of Traffic Regulation Orders

Traffic Regulation Orders (TROs) constitute a major category of Traffic Order. The powers provided by the RTRA 1984 (sections 1 and 2 for local authorities outside London) (HMG, 1984a) allow a highway authority to make a TRO to control the movement and waiting of vehicles, for the following reasons:

❑ to avoid danger to persons or other traffic using the road or any other road or to prevent the likelihood of any such danger arising;

❑ to prevent damage to the road or to any building on or near the road;

❑ to facilitate the passage on the road or any other road of any class of traffic (including pedestrians);

❑ to prevent the use of the road by vehicular traffic of a kind which (or its use by vehicular traffic in a manner which) is unsuitable, having regard to the existing character of the road or adjoining property;

❑ to preserve the character of the road, in a case where it is specially suitable for use by persons on horseback or on foot;

❑ to preserve or to improve the amenities of the area through which the road runs; and

❑ in the interests of conserving air quality (HMG, 1995).

The criteria appropriate for making Orders, similar to the TROs in London under section 6 of the RTRA 1984 (HMG, 1994a), are more specific and are set out in Schedule I of the 1984 Act and by cross–referencing to section 1.

Outside London, if a TRO prevents vehicular access to premises for more than 8 hours in any 24 hours and there is an unwithdrawn objection to such an Order by the owners or occupiers of those premises, the Secretary of State's approval is required before the Order can be made (see Part II of Schedule 9 to the RTRA, 1984).

13.4 Common Types of Traffic Regulation Order

These TROs are commonly introduced to manage traffic flow, or as part of a broader traffic management scheme, and may form part of a package of measures, including prohibitions on turning movements at junctions. Orders may be made to prohibit all vehicles, or certain classes of vehicle only, from using a road for all or part of the time. Typical examples are:

❑ all vehicles prohibited – this can be used to prevent all vehicles from using shopping or other streets for certain periods of the day, where and when pedestrian activity is high. Alternative, and sometimes more appropriate, powers are contained in section 249 of the Town and Country Planning Act 1990 (HMG, 1990a) [NII] [Si];

❑ all motor vehicles prohibited – this allows pedal cycles and horse–drawn vehicles to continue to use the road. Again the use of section 249 can be more appropriate (HMG, 1990a) [Si];

❑ prohibitions of specified classes of vehicle – which may be by weight, width, length or by description, like 'buses', 'cycles' or 'horsedrawn vehicles'; and

❑ prohibitions with exemptions for specified classes of vehicles (see Photograph 13.2)– commonly used for providing priority for buses or cyclists, such as 'no right turn except for buses'.

Vehicular turning movements can be controlled in two ways, using Traffic Regulation Orders, supported by:

❑ restrictive signs – which prevent certain manoeuvres being carried out and are indicated by a sign within a red roundel; or

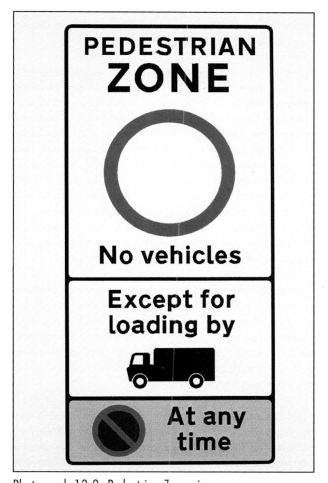

Photograph 13.2: Pedestrian Zone sign.

❏ positive signs – which make certain manoeuvres mandatory and are indicated by a sign with a blue background.

Orders of this kind may be introduced at junctions to control turning movements or to create one–way streets. Where a number of turns are to be banned at a particular junction, then, for clarity, consideration should be given to making the permitted manoeuvres mandatory rather than describing the restricted turns. Exemptions to these Orders may be provided, as with other Orders (eg 'except for buses').

Certain types of regulatory signs do not require the support of a traffic regulation Order. These include 'Stop', 'Give–Way', 'Keep Left or Right', yellow–box markings and some others (HMG, 1994b) [NId]. In addition to the standard exemptions for emergencies, other exemptions may be applied to suit local conditions, although the extent to which this is possible may be constrained by the regulations relating to the use of the prescribed signs. It can also lead to complex signing, which is difficult to read and understand quickly. The most common exemptions are for buses (to give priority in some locations), for cyclists (to improve the safety and convenience of a cycle trip) and for access or loading (to maintain reasonable access to adjacent premises).

Another alternative is to specify exemptions by permits issued to particular persons, for example service–vehicle drivers, whose work takes them into areas where there are widespread parking restrictions. However, the use of permits should be considered carefully, as they can be administratively cumbersome and can result in a wide selection of different coloured permits being displayed. It should also be noted that excessive use of exemptions can result in a general perception that an Order is not being enforced. This may bring the Order into disrepute and create enforcement difficulties as, for example, with some 'access only' exemptions.

Traffic Regulation Orders, which specifically prohibit or restrict heavy commercial vehicles, are sometimes introduced to protect particularly sensitive roads or structures or larger areas for environmental purposes, or to specify through routes. Orders relating to general traffic may also be introduced for environmental reasons.

The Environment Act 1995 (HMG, 1995) explicitly provides for Orders to be made in the pursuit of national or local air quality management strategies. The Courts have ruled, however, that section 13 of the Road Traffic Regulation Act 1984, which provides for immediate closure of roads to prevent danger to the public, does not allow for such closures to be made on general air quality grounds. Section 87 of the 1995 Act provides for new regulations to be made on air quality management grounds. Among other things, these might prohibit or restrict the access of prescribed vehicles to prescribed areas, either generally or in prescribed circumstances. To date, no such regulations have been made.

13.5 Control of Waiting and Loading (see Section 13.14 on Priority Routes)

In theory, any vehicle parked on the highway, other than in a designated parking place, could be considered to be causing an obstruction, although in practice evidence of the obstruction being caused would normally be necessary if prosecution were envisaged. Custom and practice in Britain has been that vehicles are permitted to park at the kerbside, where it is safe to do so, except where parking is specifically prohibited. There are several prohibitions on waiting, including:

❏ in the vicinity of pedestrian crossings (ie within the controlled areas of a Zebra, Pelican or Puffin [NIk] crossing, as indicated by the extent of the zig–zag markings, or on the approach to a Pelican or Puffin crossing, as indicated by studs across the approach lanes);
❏ where double white lines are provided in the centre of the carriageway, to reduce the risks associated with overtaking where visibility is poor, even although loading is still permitted;
❏ in parking places reserved for a specific type of vehicle, such as motorcycles, or class of vehicle–user, such as disabled drivers or residents;
❏ where a TRO preventing waiting has been made and is indicated by the appropriate yellow lines and supplementary signs; and
❏ on a bus, tram, or mandatory cycle–lane during its period of operation.

In addition, drivers are advised not to wait at places indicated by advisory markings, which may be used at entrances to schools (local authorities can use mandatory signs outside schools), hospitals and ambulance or fire stations or in the vicinity of junctions and other hazards identified by the Highway Code (HMG, 1987) [NIm]. Although failure to observe the Highway Code is not an offence in itself, it may have a bearing on any subsequent criminal or civil proceedings.

Traffic Regulation Orders may be introduced to prohibit waiting at any time or to restrict waiting at certain times of the day or on certain days of the week or to limit the length of stay. There are no statutory exemptions for passenger-carrying vehicles, except those provided specifically by Order-making authorities.

It is common practice for a TRO to include an exemption from waiting restrictions for picking up and setting down passengers and their luggage and for the purposes of loading and unloading. However, there will be circumstances in which these exemptions will not be appropriate and they should not be included automatically. Each situation must be considered on its merits. If loading and unloading are prohibited, this must be indicated by appropriate signs and road-markings.

If a proposed no-waiting Order (outside London) prevents loading/unloading and there are unwithdrawn objections, a public inquiry must be held unless the prohibited period is between 07.00 and 10.00 hours or 16.00 and 19.00 hours.

Bus stop clearway Orders are a special type of waiting restriction, under which no vehicles other than buses can stop. They are employed to keep bus stopping-places free from other traffic and are indicated by special bus stop clearway prohibition signs and bus stop road-markings incorporating a wide, single yellow line.

Urban clearways may be appropriate, when no waiting is to be permitted during morning and/or evening peak periods. They include the exemption for picking up and setting down and are indicated by the appropriate signs but road markings are no longer used for urban clearways. If necessary, they can be used to cover more extensive periods of the day.

Clearway restrictions, which do not include the exemption for picking up and setting down passengers, can be used in urban areas and apply at all times of the day and on all days of the week. They are appropriate to primary distributors and dual carriageway roads, where stopping of any kind should be prevented. Rural clearways are not indicated by road-markings and rely solely on the provision of traffic signs, which should be erected at regular intervals.

Extensive use of standard yellow line road-markings may be visually intrusive in environmentally-sensitive areas. This effect can be reduced by the use of the narrower yellow lines (50mm wide), permitted in TSRGD (HMG, 1994b) [NId], and/or a paler shade of yellow (as set out in Chapter 5 of the Traffic Signs Manual (DOT, 1977) [Sb]. However, care needs to be taken to lay the lines neatly and different shades of yellow should not be used on adjacent lengths of line.

TSRGD 1994 allows yellow lines and kerb blips to be dispensed with in pedestrian zones, where the prescribed 'Pedestrian Zone' entry signs and repeater plates are erected, the footways and carriageways are paved level, the entry of vehicles into the zone is restricted, at the times during which waiting and loading are prohibited, and the waiting and loading prohibition applies uniformly throughout the zone [NIk].

In other environmentally-sensitive areas, the Department of Transport is prepared to consider authorising special 'Restricted Zone' signing and allowing yellow lines to be dispensed with. Roads in such restricted zones should not be major through routes and should be neither so narrow as to make parking impracticable nor so wide that it would be unreasonable to prohibit waiting on both sides. It is important to secure the agreement of the police to this type of arrangement. Early consultation with the Department of Transport is recommended where it is proposed to introduce a designated 'Restricted Zone'.

13.6 Other Types of Order

Local authorities are empowered to make a range of Orders under The Road Traffic Regulation Act 1984 provisions but they also have powers to make Orders, similar to Traffic Regulation Orders, which require the same or a similar procedure to be followed [NIa]. Some of the more commonly used Orders are described in the remainder of this chapter.

13.7 Parking Orders

Many different types of on-street parking schemes can be created under the powers provided in Part IV of the RTRA 1984 (HMG, 1984a) as amended by the Road Traffic Regulation (Parking) Act 1986 (HMG, 1986c) [NIn]. The procedures to be followed when making Orders are prescribed in the procedure regulations listed in Section 13.2. More detail on the types of schemes which can be introduced is given in Chapter 19.

13.8 The 'Orange Badge' Scheme for Disabled Persons

Section 21 of the Chronically Sick and Disabled Persons Act 1970 (HMG, 1970) provides for a prescribed form of badge, the 'Orange Badge', to be issued by local authorities for motor vehicles used by disabled persons [NIo], (see Photograph 13.3).

Photograph 13.3: Parking sign for disabled badge holders.

Local authorities are required, under the Local Authorities' Traffic Orders (Exemptions for Disabled Persons) (England and Wales) Regulations 1986 (HMG, 1986b), as amended by the Local Authorities' Traffic Orders (Exemptions for Disabled Persons) (England and Wales) (Amendment) Regulations 1991 (HMG, 1991c) [Sj], to include exemptions in Traffic Orders, for orange badge holders, which enable them:

❑ to park for as long as they wish, where others may wait for a limited time only;
❑ to park free of charge and without time–limit at parking meters on–street and pay–and–display on–street parking; and
❑ to park for up to three hours on single or double yellow lines, when no–waiting restrictions are in force.

The last of these does not apply where loading and unloading is also prohibited.

The criteria for issuing badges is covered by the Disabled Persons (Badges for Motor Vehicles) Regulations 1982 (HMG, 1982), as amended by the Disabled Persons (Badges for Motor Vehicles) (Amendment) Regulations 1991 (HMG, 1991d) and the Disabled Persons (Badges for Motor Vehicles) (Amendment) Regulations 1992 (HMG, 1992d) [NIp]. Circular 3/91 (DOT, 1991) [Sk] also describes the issuing criteria. The scheme does not apply in areas of central London and certain town centres, where local authorities have applied their own scheme to suit local circumstances.

13.9 Stopping–Up and Diversion of Highways

A highway may be stopped up in several ways. Two of the more common are:

❑ using Highways Act powers – section 13 of the Highways Act 1980 (HMG, 1980) [NIq] [Sl] (section 18 for special roads) provides for the making of

Orders authorising a highway authority, *inter alia*, to stop up or divert a highway in connection with the construction or improvement of a trunk or classified road. Alternatively, in England and Wales, a highway authority can apply to a magistrates court, under section 116 of the Act, on the grounds that the highway is either unnecessary or can be diverted, so as to make it nearer or more commodious to the public [Sl]. Footpaths and bridleways may be stopped up or diverted under powers provided in sections 118 and 119; and
❑ using Planning Act powers – section 247 of the Town and Country Planning Act 1990 (HMG, 1990a) [Sm] gives the Secretary of State for Transport [Wa] power to stop up any kind of highway, if he is satisfied this is necessary to enable development to be carried out in accordance with planning permission. By virtue of section 253 [So], the Secretary of State may, in certain circumstances, anticipate the granting of such planning permission [NIr]. The use of this power is extremely rare. Conditions may be attached, such as the provision or improvement of another highway. An Order under this section removes all public rights of way.

Section 248 [Sp] gives the Secretary of State for Transport [Wb] power to stop up a highway crossing the route of the main highway, where it is expedient to do so in the interests of the safety of users of the main highway or to facilitate the movement of traffic on the main highway.

Section 257 [Sq] allows local authorities to stop up footpaths and bridleways, if they are satisfied that this is required to enable development to be carried out. Again, a valid planning permission is required and conditions can be attached regarding the provision or improvement of another footpath or bridleway. Orders under this section are confirmed by the Local Authority, if there are no objections. Where objections are received, the Orders are determined by the Secretary of State for the Environment [Wc].

13.10 Extinguishment of Vehicular Rights

Where a local planning authority adopts, by resolution, a proposal for improving the amenity of part of their area which involves a highway, an Order can be made extinguishing the existing vehicular rights (under section 249 of the Town and Country Planning Act 1990) [NIl] [Sr]. Such an Order can apply to any non–trunk road and can provide for specific vehicles or classes of vehicles to be exempted.

This method is commonly used when streets are to be pedestrianised. The Order would be made by the appropriate Secretary of State. Where vehicles are still physically able to gain access to such an area, it may be necessary to institute a Traffic Regulation Order creating relevant offences, to back the intention of the Section 249 Order (see DOT, 1987) [NIs] [Si].

13.11 Facilities for Cyclists

Cyclists are entitled to cycle on the highway, including carriageways (except motorways and other roads from which cyclists have been excluded), bridleways (so long as they give way to pedestrians and horseriders using the `way') and designated cycle–tracks.

It is an offence, under Section 72 of the Highways Act 1835 (HMG, 1835), to cycle on any footway [NIt] [Ss]. Cyclists do not have the right to cycle on a footpath and an offence may be created in respect of specific footpaths where Traffic Regulation Orders or local bylaws exist.

The wide variety of ways in which facilities can be provided to assist cyclists are discussed in detail in Chapter 23. Statutory procedures to convert all or part of a footpath to a cycle–track are contained in the Cycle Tracks Act 1984 (HMG, 1984b) and the Cycle Tracks Regulations (HMG, 1984c). Statutory powers to convert footways are contained in the Highways Act 1980 [St] (detailed guidelines on conversions are given in DOT, 1986a and DOT, 1986b) [NIu].

The Toucan crossing is a signal–controlled crossing, designed for unsegregated use by both pedestrians and cyclists (see Chapters 22 and 23), (see Photograph 13.4). Implementation of a Toucan crossing requires

Photograph 13.4: Toucan crossing.

the use of a modified push–button plate and a green cycle symbol aspect, both of which require special authorisation from the Secretary of State, as these are not prescribed in TSRGD 1994 (HMG, 1994b) [NIk]. Trials are being carried out into the use of nearside signals, as used for Puffin crossings.

13.12 Pedestrian Crossings

The provision of a formal pedestrian crossing imposes restrictions and duties on road–users, such as giving precedence to pedestrians under certain circumstances and prohibiting waiting on the approaches to the crossings. Before a formal crossing is introduced, the Highway Authority must carry out certain procedures.

They must:
❑ consult the chief officer of police;
❑ give public notice of the proposal; and
❑ inform the Secretary of State in writing.

The layout of formal pedestrian crossings must conform to the current regulations. Local authorities may also make arrangements for the provision of school crossing patrols, to assist children in crossing the road between the hours of 08.oo and 17.30. More detailed information on the introduction of formal pedestrian crossing facilities is provided in Chapter 22.

13.13 Public Transport Facilities

Section 122 of the Road Traffic Regulation Act (HMG, 1984a) requires local authorities to have regard to 'the importance of facilitating the passage of public service vehicles and of securing the safety and convenience of persons using or desiring to use such vehicles', when exercising any of the functions which the Act confers upon them. The location of bus stopping–places, the provision of shelters (possibly with real–time passenger information) and the maintenance of unimpeded access to bus/tram stops for buses/trams at all times are among the issues which should be considered. Off–street parking at public transport interchange sites, including those used for park–and–ride, can be provided by local authorities, using powers under sections 32–35 of RTRA 1984 (HMG, 1994a).

Sections 1,2 and 4 of RTRA 1984 provide powers to implement bus–lanes, bus–only roads, exemption from prohibited turns and selective priority at signal–controlled junctions [NIs]. In some cases, other vehicles, such as taxis and cycles, may be included. These powers may also be used to designate certain sections of the highway for guided buses or light rail vehicles, which may be of particular value in very congested central urban areas, where specifically–designed priorities aid operational efficiency. Where complete segregation of such systems is possible, implementation under the Transport & Works Act 1992 (HMG, 1992b) [Su] may be advantageous. Section 122 of the RTRA 1984 is also of relevance here.

Measures which can be implemented to assist public transport are discussed in more detail in Chapter 24.

13.14 Priority Routes

The Road Traffic Act 1991 (RTA) (HMG, 1991a) provides for the introduction of priority routes in London and for the appointment of a Traffic Director for London. The Secretary of State has designated a network of 315 miles of the more important local and trunk roads in London as priority routes, with a view to reducing congestion and improving the movement of traffic in the city. Priority routes are more generally known as 'Red Routes' because of the red kerbside 'no–stopping' controls and special red–bordered signs (see Photographs 13.5 and 13.6). The Traffic Director for London has a general duty to co–ordinate the introduction and maintenance of the priority–route measures and to monitor their operation.

Traffic Management and Parking Guidance (DOT, 1992b) has been issued by the Secretary of State, as required by section 51 of RTA 1991. This document sets the framework for the London Local Authorities to develop traffic management and parking policies and, in particular, their role in the introduction of priority 'red' routes. The Traffic Director's network plan has been published, as required by section 53 of RTA 1991. This document sets the strategy for priority routes and provides the basis for the preparation of a series of priority route local plans, setting out the detailed proposals for individual parts of the network. These local plans have been prepared by local authorities and the Traffic Director, under sections 54 and 55 of the RTA 1991.

The traffic management measures in the local plans are introduced, generally, through the provisions of the Highways Act 1980 (HMG, 1980) and the Road Traffic Regulation Act 1984 (HMG, 1984a). Traffic Orders for the new red line controls are introduced, using experimental powers under section 9 of the Road Traffic Regulation Act 1984, to allow for on–street adjustment before Orders are made permanent. Priority routes use a marking and signing

Photograph 13.5: Priority (Red) Routes, designated kerbside parking spaces.

Photograph 13.6: Priority (Red) Routes; with bus–only lane.

system, which requires special authorisation from the Secretary of State, although, in future, these may be prescribed in regulations. Local authorities elsewhere can introduce similar schemes, using their existing powers and any necessary signs authorisations.

13.15 Measures Affecting Goods Vehicles

The Road Traffic Regulation Act 1984 (HMG, 1984a) enables highway authorities to control all types of traffic, including lorries, by means of Traffic Regulation Orders [NIs]. A TRO could be made where, for example, the Authority believes that local roads are too narrow or not appropriate for use by large goods vehicles (see Photograph 13.7). TROs can be used in very flexible ways, including specifying through routes for heavy commercial vehicles and

TRANSPORT IN THE URBAN ENVIRONMENT

prohibiting or restricting their use (either with or without exemptions)in particular zones or on particular roads (see Section 13.4).

Sometimes, different measures and Orders can be introduced together so as to create environmental areas that eliminate or physically discourage through traffic. These can include road closures, width restrictions and traffic–calming features. Lorries in the Community (DOT, 1990a) [Sv] gives advice on ways of alleviating the pressure caused by lorries on local communities. The minimum size of lorry affected by such bans is standardised to a choice of those vehicles over 7.5 tonnes gross weight or over 17 tonnes gross weight. The over–7.5 tonne ban affects all lorries displaying a red and yellow rear reflective plate. The over–17 tonne ban does not affect two–axle lorries; thus every lorry affected by it will either be an articulated lorry, a lorry with a trailer or a 3– or 4–axle rigid lorry (see Chapter 25).

13.16 Speed–Limits

Orders may be made to establish either a maximum or a minimum speed–limit (see Chapter 20). The procedures followed are prescribed in the regulations listed in Section 13.2. All roads in Britain are subject to a maximum limit, whereas very few have a minimum limit. In addition, various classes of vehicle are subject to a vehicle maximum speed–limit (see Schedule 6 of RTRA, 1984) [NIv]. Signs may also be erected to indicate a local advisory maximum limit, such as those used on motorways during adverse weather conditions or roadworks. Failure to comply with an advisory sign is not itself an offence but it

Photograph 13.7: Lorry restriction advance direction sign.

may provide strong evidence of criminal liability, such as under sections 2 or 3 of RTA 1972, or civil liability in a subsequent prosecution.

The Secretary of State is responsible for speed–limits on trunk roads but retains reserve powers for all roads. Local authorities are responsible for speed–limits for all roads in their area, other than trunk roads.

A national maximum speed–limit applies to all roads, unless a lower local limit is in force. The national limits are, currently:

❑ motorways and dual carriageways – 70 miles/h; and

❑ single carriageways – 60 miles/h.

A road will be a 'restricted road' if, on it, there is a system of street–lighting by lamps not more than 183m apart or if, by an Order under sections 82 and 83 of the RTRA 1984, the status of a 'restricted road' is imposed [NIw]. The speed–limit on a restricted road is 30 miles/h. In Scotland, the distance between street lamps must not exceed 185m and the road must be either Class C or unclassified. An Order under section 84 of this Act may impose a speed–limit of more than 30 miles/h on a road which would otherwise be a restricted road. Whilst that Order is in force, the road will not have the status of a restricted road (see Section 84(3)) but that status will automatically revive if the Section 84 Order is revoked and there is a system of street–lighting, as mentioned above. All roads without such a system of lighting, and not covered by an appropriate Order, are not restricted. A maximum limit of less than 30 miles/h requires the consent of the Secretary of State and is normally only granted in the case of a 20 miles/h zone [NIx]. Procedures for the implementation of 20 miles/h zones are described in Circular Roads 4/90 (DOT, 1990c) [Sw].

Speed–limit signs must be provided at the terminal points of an Order. On trunk and principal roads, they are required to be lit, whereas, on other roads, they should be lit or made of reflective material [NIk]. In addition, reflectorised repeater signs are required:

❑ on roads other than restricted roads; and

❑ on restricted roads, where lights are more than 183m apart, or where there is no street lighting.

Special care should be taken to ensure that the appropriate speed–limit signs are erected at the junctions between roads where different limits apply and, if necessary, when new roads are adopted. At roundabouts, between roads where different limits apply, the junction should be restricted to the limit applying on the majority of approaches. If there is an equal number, then the lower limit should apply.

13.17 Traffic Calming

Traffic calming is a means of controlling vehicle speeds, reducing accident–risks and improving the environment, using self–enforcing traffic engineering measures such as road humps, chicanes and carriageway narrowings (see Photograph 13.8). Further guidance on design matters is given in Chapter 20.

The powers to introduce road humps are contained within the Highways Act 1980 (HMG, 1980) [NIy] and their design is governed by the Highways (Road Humps) Regulations 1996 (HMG, 1996a) [NIz] [Sx]. The Road Traffic Act 1991 (HMG, 1991a) amended the Highways Act 1980 [St] to enable the Secretary of State to authorise the use of road humps which do not conform to the requirements of the Regulations. Useful information about road humps and the procedural requirements is contained in Department of Transport Circular Roads 2/92 (DOT, 1992a) [Sz] and 4/96 (DOT 1996a) [Sab] and in Traffic Advisory Leaflet 7/96 (DOT, 1996b) [Sab].

Photograph 13.8: Speed cushions on Foxwood Lane, York.

The Road Humps Regulations require consultation with the District Council [Sac] [Sy] (where the Highway Authority is not a unitary authority) fire and ambulance services and one or more organisations representing people who use the affected road. Section 90C of the Highways Act 1980 [St] requires highway authorities to consult with the police and to place notices at appropriate points on the highway and in one or more local newspapers, to give notice of the proposals and to allow objections to be raised and considered [NIk]. Road hump regulations now allow greater flexibility in their use (see Photograph 13.9). It is important that local authorities exercise care in the design and location of

Photograph 13.9: Road humped Zebra crossing, London Borough of Hounslow.

any road humps, to ensure that schemes do not prejudice the safety of road–users.

The Traffic Calming Act 1992 (HMG, 1992a) [Sad] amends the Highways Act 1980 by the addition of sections 90G, 90H and 90I in England and Wales, to provide for regulations to be made to clarify the power of highway authorities to install traffic–calming measures other than road humps. The Highways (Traffic Calming) Regulations 1993 (HMG, 1993 and, 1994c) [Sae] provide certainty in the use of build–outs, chicanes, gateways, islands, overrun areas, pinch–points and rumble devices [NIaa]. Where a particular traffic–calming feature is not covered by the Regulations, special authorisation may be sought from the Secretary of State.

The Traffic Calming Regulations require consultation with the local police force and with such other people, who use the road affected, as the highway authority think fit. It is advised that the views of bus service operators and local fire and ambulance services should be sought.

13.18 References

DOT (1977)   'Traffic Signs Manual', Stationery Office [Sb].

DOT (1986a)   Circular 1/86 (WO 3/86) 'Cycle Tracks Act 1984 The Cycle Tracks Regulations 1984', DOT.

DOT (1986b)   Local Transport Note 2/86, 'Shared Use by Cyclists and Pedestrians', DOT.

DOT (1987)   Local Transport Note 1/87,

'Pedestrian Zones – Getting the Right Balance', DOT.

DOT (1990a)   'Lorries in the Community', DOT.

DOT (1990b)   Circular – Roads 3/90 (WO 54/90)'Road Humps', DOT.

DOT (1990c)   Circular – Roads 4/90 (WO 2/91) – '20 mph Speed Limit Zones', DOT.

DOT (1991)   Local Authority Circular 3/91,'Orange Badge Scheme of Parking Concessions for Disabled and Blind People', DOT.

DOT (1992a)   Circular – Roads 2/92 (WO 46/92)'Road Humps and Variable Speed Limits', DOT.

DOT (1992b)   Local Authority Circular 5/92, 'Traffic Management and Parking Guidance', DOT.

DOT(1994)   Traffic Advisory Leaflet 8/94, 'Traffic Signs, Signals and Road Markings Bibliography', DOT.

DOT (1996a)   Circular – Roads 4/96 (WO 32/96), 'Road Humps', DOT.

DOT (1996b)   Traffic Advisory Leaflet 7/96. 'The Highways (Road Hump) Regulations', DOT.

DOT (1996c)   Local Authority Circular 5/96 'New Procedures for Traffic Orders', DOT.

HMG (1835)   'Highways Act 1835' Stationery Office, [Ss].

HMG (1970)   'Chronically Sick and Disabled Persons Act 1970', Stationery Office.

HMG (1972)   'Road Traffic Act 1972', Stationery Office.

HMG (1980)   'Highways Act 1980', Stationery Office.

HMG (1982)   Statutory Instrument 1982 No. 1740 'Disabled Persons (Badges for Motor Vehicles) Regulations 1982', Stationery Office.

HMG (1984a)   'Road Traffic Regulation Act 1984', Stationery Office.

HMG (1984b)   'Cycle Tracks Act 1984', Stationery Office.

HMG (1984c)   Statutory Instrument 1984 No. 1431, 'Cycle Tracks Regulations 1984', Stationery Office.

HMG (1986a)   Statutory Instrument 1986 No. 1078, 'Road Vehicles (Construction and Use) Regulations 1986', Stationery Office.

HMG (1986b)   Statutory Instrument 1986 No. 178. 'Local Authorities' Traffic Orders (Exemptions for Disabled Persons) (England and Wales) Regulations 1986', Stationery Office [Sj].

HMG (1986c)   'Road Traffic Regulation (Parking) Act 1986', Stationery Office.

HMG (1987)   'The Highway Code', Stationery Office.

HMG (1990a)   'Town and Country Planning Act 1990', Stationery Office [Si].

HMG (1990b)   Statutory Instrument 1990 No. 1656, 'Secretary of State's Traffic Orders (Procedure) (England and Wales) Regulations 1990', Stationery Office [Sd].

HMG (1991a)   'Road Traffic Act 1991' Stationery Office [Sh].

HMG (1991b)   The Road Traffic (Temporary Restrictions) Act 1991, Stationery Office.

HMG (1991c)   Statutory Instrument 1991 No 2709 'Local Authorities' Traffic Orders (Exemptions for Disabled Persons) (England & Wales) (Amendment) Regulations 1991', Stationery Office [Sj].

HMG (1991d)   Statutory Instrument 1991 No 2709 'Disabled Persons (Badges for Motor Vehicles) (Amendment) Regulations 1991', Stationery Office.

HMG (1992a)   'Traffic Calming Act 1992' Stationery Office [Sad].

HMG (1992b)   'Transport and Works Act 1992', Stationery Office [Su].

HMG (1992c)   'The Road Traffic (Temporary Restrictions)(Procedure) Regulations 1992', Stationery Office.

HMG (1992d)   Statutory Instrument 1992 No 200 'Disabled Persons (Badges for Motor Vehicles) (Amendment) Regulations 1992', Stationery Office.

HMG (1993)   'The Highways (Traffic Calming) Regulations 1993', Stationery Office [Sae].

HMG (1994a)   'Working Drawings for Traffic Signs Design and Manufacture' – Volume I, Stationery Office.

HMG (1994b)   Statutory Instrument 1994 No. 1519, 'Traffic Signs Regulations and General Directions 1994' (TSRGD), Stationery Office.

HMG (1994c)   Statutory Instrument II 1994 No. 2488. 'The Roads (Traffic Calming) Scotland Regulations 1994', Stationery Office [Sae].

HMG (1995)   'Environment Act 1995', Stationery Office.

HMG (1996a)   'The Highways (Road Humps) Regulations' Stationery Office, [Sx].

HMG (1996b)   Statutory Instrument 1996 No 2489 'The Local Authorities Traffic Order (Procedure) (England and Wales) Regulations 1996' Stationery Office [Sc].

13.19 Further Information

ALBES/DOT(1986)   'Highway and Traffic Management in London,

A Code of Practice', Stationery Office.

DOT (1974a)   Circular – Roads 31/74 (WO 214/74), 'Resident Parking Schemes' DOT [Sal].

DOT (1974b)   Circular – Roads 23/75, 'Car Parking for the Medical Profession', DOT [Sam].

DOT (1980)   Circular – Roads 1/80 'Local Speed Limits', DOT [San].

DOT (1981)   TA 19/81 (DMRB 8.2), 'Reflectorisation of Traffic Signs', Stationery Office.

DOT (1982)   'Door to Door; Guide to Transport for Disabled People', DOT.

DOT (1983)   Circular – Roads 4/83 (WO 47/83), 'Speed Limits', DOT [Sao].

DOT (1984a)   Circular – Roads 2/84 (WO 4/84), 'Orange Badge of Parking Concessions for Disabled and Blind People' DOT.

DOT (1984b)   Circular – Roads 6/84, i) 'Road Traffic Regulations Act 1984' and ii) 'Parking for Disabled People', DOT.

DOT (1991)   Traffic Advisory Leaflet 3/91 – 'Speed Control Humps', DOT [Sab].

DOT (1993a)   Circular – Roads 2/93 (WO 46/93) – 'The Highways (Traffic Calming Regulations) 1995', DOT [Sap].

DOT (1993b)   Traffic Advisory Leaflet 7/93 – 'Traffic Calming Regulations', DOT [Sab].

DOT (1994a)   Local Transport Note 1/94 – 'The Design and Use of Directional Informatory Signs', Stationery Office [Saq].

DOT (1994b)   Local Transport Note 2/94 – 'Directional Informatory

	Signs Interim Design Notes', Stationery Office [Saq].
DOT(1994c)	Traffic Advisory Leaflet 4/94 – 'Speed Cushions', DOT [Sab].
DOT (1994d)	Traffic Advisory Leaflet 11/94 – 'Traffic Calming Regulations (Scotland)', DOT.
DOT (1994e)	Traffic Advisory Leaflet 3/94, 'Emergency Services and Traffic Calming: A Code of Practice', DOT [Sab].
HMG (1971a)	'The Zebra Pedestrian Crossings Regulations', Stationery Office.
HMG (1971b)	'Vehicles Excise Act 1971', Stationery Office.
HMG (1987)	Statutory Instrument 1987 No. 16, 'The 'Pelican' Pedestrian Crossing Regulations and General Directions 1987', Stationery Office.

Chapter 14 Enforcement

14.1 Basic Principles

Compliance with traffic regulations by road-users is essential for road safety and efficiency of movement. Most road-users comply with most traffic regulations most of the time. However, enforcement is necessary because road–users sometimes perceive sufficient immediate advantage in breaking traffic regulations as outweighing the potential disadvantages, including the risk of accidents both to themselves and to others and the risk of incurring penalties.

Little enforcement is required where the disadvantages of breaking traffic regulations are significant and obvious, such as driving against the prescribed direction of flow in a busy one–way street or driving through a substantial road–closure barrier. Such regulations are generally referred to as self–enforcing. Substantial enforcement and deterrence are required where the benefits of contravention are clear and the disbenefits less so, such as speeding on a clear road on a fine day and most parking offences.

Compliance with traffic regulations, where not self–enforcing, depends largely on drivers' perception of the risks and implications of being subjected to enforcement action. For example, regular enforcement of speed–limits at particular locations usually results in better compliance, at least at those locations where the risk of being caught is perceived to be high. The deterrence effect can be enhanced by appropriate publicity; for example, the effectiveness of speed enforcement cameras has been increased by publicising and signing their presence.

Deterrence can also be enhanced by making the penalties for contravention severe. For example, wheelclamping illegally parked cars is an effective enforcement method because of the severity of the penalty, both financially and in terms of inconvenience. However, the severity of penalties which can be imposed for traffic offences is limited by the acceptability of such penalties to society as a whole.

Many drivers do not generally regard ordinary traffic offences as serious because, unlike most other crimes and misdemeanours, a high proportion of them will have committed a motoring offence at some time in their lives. However, this perception is often not shared by, for example, pedestrians who tend to have a greater awareness of the anti–social consequences of inappropriate speed. As a result, enforcement of traffic regulations, especially parking regulations, does not enjoy the same level of public support as other enforcement activities, such as prosecuting burglars. Nevertheless, public opinion can be influenced by sustained and targeted media campaigns. For example, the publicity given to the risks associated with drinking and driving has had an effect on the public acceptability of such behaviour. Public support is likely to be enhanced if authorities ensure that any traffic or parking regulations which are introduced are appropriate and justified by the prevailing traffic conditions. Even so, objections are relatively common and decisions may need to be made in respect of the general public good rather than individual inconvenience.

14.2 Policy Issues

The increase in traffic volumes, with its associated adverse effects on traffic flow, road safety and the quality of the environment, has led to a reassessment of transport policy (eg DOE/DOT, 1994) [NIa] [Sa]. In particular, increased emphasis is being given to making the best use of existing road infrastructure and to reducing the number of casualties arising from road accidents. In London, a network of Priority (Red) Routes (DOT, 1992) has been identified, on which the movement of all classes of traffic, including buses, is given priority, so that congestion is reduced and people and goods can reach their destinations more easily, reliably and safely. The Government has also set a national target for reducing road–accident casualties (see Chapter 16) [NIb]. The effective enforcement of traffic regulations has a major part to play in achieving these objectives.

In most areas of the country, no single agency is responsible for traffic management [NIc]. The responsibility for introducing traffic regulations lies with local authorities, while the police are responsible for the enforcement of such regulations. Concentration of limited police resources on core activities, such as crime prevention and detection, has resulted in lower priority being given to the enforcement of traffic regulations. Consequently, legislation now enables local authorities to enforce parking regulations (see

Chapters 13 and 19) [NId]. In London, where these arrangements have been implemented, the Boroughs are responsible both for the introduction of parking controls and for their subsequent enforcement. Significantly, the Local Authorities receive the income from the imposed fines to pay for the enforcement activity. This arrangement focuses overall responsibility for parking matters in each area largely within one organisation, resulting in better designed controls and improved compliance (PCfL, 1995).

The contribution that effective enforcement of traffic regulations can make to the achievement of traffic management objectives can be enhanced significantly, if enforcement is seen as an integral part of overall traffic policy. For example, the problem of drinking and driving has been tackled by a combination of police enforcement, hard–hitting media campaigns funded by the Department of Transport [Sb], local authority inputs through the education efforts of road safety officers and by encouraging public houses to provide information on bus/taxi transport facilities. Likewise, excess speed can be addressed by different agencies working on a partnership basis. Some highway authorities have decided to fund the installation of speed–detection cameras, to enable the police to achieve enforcement objectives more efficiently and effectively.

Local authorities can contribute to traffic enforcement in other ways, such as by altering the design of roads, so as to discourage speeding and thereby reduce the need for active enforcement, and by ensuring that the problems and cost of enforcement are minimised when introducing new traffic or parking regulations.

14.3 Legislative Responsibilities

The police and traffic wardens (who are part of the police service) have the primary responsibility for enforcing traffic regulations, including waiting and loading restrictions. For example, the police have power to issue Fixed Penalty Notices (FPNs) for parking and other traffic offences. They also deal with offences which involve driving–licence endorsement. Traffic wardens have more limited powers to enforce parking offences by issuing FPNs. They can also deal with some endorsable parking offences and with Vehicle Excise Act offences. They can also assist with fixed–point traffic duties (AC, 1992). Further details of powers and responsibilities are given in Section 14.5 and in Chapter 19, including reference to the particular parking control arrangements in London. In some areas, local authorities employ their own officers to enforce on–street and off–street parking controls but this does not extend to 'yellow–line' offences.

14.4 Enforcement of Speed–Limits
(see Chapter 20)

Inappropriate speed for the prevailing conditions is a major cause of road accidents and better compliance with speed–limits has been shown to reduce accidents significantly. The police, who are responsible for enforcing speed–limits, use a variety of enforcement methods. A number of technical aids are also employed and generic systems have been approved by the Home Office as providing reliable evidence for prosecution purposes. Home Office 'type approval' is a requirement before any particular device can be used for enforcement. Examples are:

❑ hand–held radar – which is a self–contained radar device, which directs a radar beam at approaching vehicles and calculates their speed from the reflected signal. More accurate laser–beam devices are also in use;

❑ 35mm cameras – used at mobile or permanent sites on the roadside, which measure vehicle speeds using radar or piezometric tubes and automatically photograph vehicles exceeding the speed–limit. Offending vehicles are identified so that owners, and thence drivers, are traced through DVLA records, although this involves a significant amount of administrative work; and

❑ video cameras – which photograph a traffic stream continuously and which are linked to a speed–detector. The speed–detector identifies speeding vehicles and the camera 'reads' the relevant registration numbers so that, after analysis, the offending owner/driver can be prosecuted. Trials have also been held to display the registration numbers of offending vehicles on downstream variable message signs. Video cameras can also be mounted in police vehicles and on motorcycles.

14.5 Enforcement of Parking Regulations

Parking regulations are a widely–flouted category of traffic regulations and effective enforcement is essential to secure the compliance of drivers.

The responsibility for enforcement of parking regulations is divided between the police and local traffic authorities. Where a decriminalised parking enforcement regime is in place, the Local Authority is responsible for most parking enforcement (see Chapter 19) [NId].

In those areas where decriminalised powers have not been taken up, waiting restrictions are enforced by the police, with or without the assistance of traffic wardens. Enforcement at parking places, such as meter bays, is carried out by traffic wardens, in which case the Local Authority reimburses the police for part of their costs, or by local authority–employed parking attendants. The powers of the latter are more limited than those of traffic wardens. They can issue Excess Charge Notices (ECNs) or Notices of Intent to Prosecute (NIPs) for contraventions at designated parking places but they cannot issue Fixed Penalty Notices (FPNs). Further details are given in Chapter 19.

The Road Traffic Act 1991 gives local authorities in London wheelclamping and removal powers within areas where decriminalised parking applies. Corresponding powers can be given to authorities outside London to operate a decriminalised regime, subject to approval by the Secretary of State [NId]. This is in addition to authorities' previously–held powers, under the Road Traffic Regulation Act 1984, to remove vehicles from designated parking places. The Road Traffic Act 1991 also empowers local authorities to impose charges on vehicle–owners before unclamping their vehicles or releasing them from the pound [NIe]. The police also have powers to remove vehicles which are either parked in contravention of a parking or waiting Order or which are parked in a dangerous or obstructive manner.

Experience in London has shown that wheelclamping has a powerful deterrent effect, leading to a substantial improvement in compliance with parking regulations. A vehicle–removals operation can deal with fewer vehicles than a clamping operation, with similar resources, so there is less probability of an offending driver being caught. The deterrent effect is also lessened because a removed vehicle, unlike a clamped one, is not visible to other drivers. However, wheelclamping is not a suitable method of enforcement where illegally–parked vehicles are causing an obstruction or are parked so as to be a hazard to road–users. In these situations, the vehicles in question need to be removed to a vehicle pound or to a more suitable parking place in the vicinity.

The enforcement of traffic regulations can be effective only if fines or penalties, which are not paid, are followed up immediately. These usually involve fixed sums and it is important to ensure that payment arrangements are made as convenient as possible, to encourage prompt payment of fixed penalty notices.

Normally, large numbers of penalties have to be processed and sophisticated systems are required to ensure that processing is timely and efficient. Processing usually involves obtaining the name and address of the vehicle–keeper from the DVLA in Swansea and establishing, from him or her, the identity of the driver involved in the offence [NIf]. If the penalty is not cancelled and the driver does not pay, he or she can be prosecuted through the magistrates' court (where criminal traffic offences are involved) or pursued for payment of a civil debt (where decriminalised offences are involved).

Specialist software packages are available to streamline the processing of notices, especially for parking offences, and specialist firms also undertake this activity on a commercial basis. A number of local authorities have successfully used commercial debt–collecting agencies to improve the effectiveness of recovery of penalty payments. However, enforcement authorities should maintain strict control of the process, to ensure that any commercial firms involved comply with the highest standards in dealing with the public on their behalf.

The Road Traffic Act 1991 requires traffic authorities outside London to set the level of penalty charges (Penalty Charge Notices) at one of three specific bands. The corresponding levels adopted in London are 33% to 50% higher. A standard 50% discount is specified for payments received within 14 days (DOT, 1995) [NIg] [Sc].

14.6 Enforcement of Other Traffic Regulations

A range of traffic regulations, other than those related to parking and speeding, require high compliance levels for safety (eg banned turns) or to promote policy objectives (eg bus–only lanes). The introduction of more sophisticated control measures, such as lorry permits and other entry–permit systems aimed at imposing charges and/or restricting the types or levels of traffic in an area, require careful consideration of the related enforcement issues (see Chapter 21). This will include methods to be used to identify individual vehicles, in order to establish whether or not they fall into the category covered by any restrictions in force. For example, difficulties have been experienced in enforcing bus and taxi–only lanes in towns where taxis look like ordinary saloon cars. Examples of 'other' control measures are:

❑ video–cameras, mounted on buses, to monitor and enforce the use of bus–only lanes;
❑ the use of dynamic weighbridge/video cameras to enforce weight restrictions;
❑ the use of infra–red detectors to prevent bridge strikes by high–sided vehicles; and
❑ the use of video cameras to ensure proper use of level crossings on railway lines.

14.7 Design Issues

Whenever possible, traffic schemes should be designed to minimise the need for enforcement. The enforcement of traffic regulations takes up resources, which could be used to combat other more serious crimes, and represents a continuing financial cost to society. Measures which are self–enforcing are therefore more likely to be both operationally efficient and cost–effective. Traffic schemes should aim to 'design out' both the ability and the inclination of drivers to commit traffic offences. Examples include:

❏ traffic calming schemes such as speed humps, speed tables, chicanes and rumble strips (see Chapter 20), which make it difficult and uncomfortable to drive at excessive speed;

❏ measures, such as traffic islands and kerb realignment, designed to prevent or deter prohibited movements;

❏ carefully placed bollards and other street furniture used to enforce road closures, lorry bans and parking restrictions; and

❏ guardrailing used to discourage illegal kerb–side parking.

It is important to consult the police at the design stage, to seek their views on any proposals. This applies equally to traffic measures considered to be largely self–enforcing, as well as to those where police enforcement will be necessary to ensure compliance.

Clear and correct signs and road markings are necessary for drivers and other road–users to understand the traffic regulations and to abide by them. Particular attention should be paid to the design and maintenance of regulatory, prohibitory and warning signs. The absence of clear signing can make traffic regulations technically unenforceable. Signing which is over– complicated, and therefore not easily understood by drivers, contributes to the degree of non–compliance and adds unnecessarily to the burden of enforcement agencies. Drivers are also more likely to resent enforcement action attempted in such circumstances.

14.8 Financial Considerations

The enforcement of traffic regulations is intended to achieve better compliance, the benefits of which can be quantified in terms of a reduction in accidents and other benefits. Enforcement activity should be targeted so as to maximise these benefits at an acceptable cost. Traffic authorities should consider monitoring compliance levels, as part of their regular monitoring of traffic conditions in their areas.

Revenue considerations should not determine policy but need to be considered, together with the benefits of achieving compliance, in deciding on the appropriate level of resources to be allocated to traffic and parking enforcement. In general, the cost of enforcement rises in direct proportion to the amount of resources deployed, for example hours of patrolling by traffic wardens. The total revenue from fines, which may or may not accrue to the enforcing authority, also increases as more enforcement resources are deployed. However, the amount of additional revenue generated as enforcement levels are enhanced is likely to decrease as compliance improves. Ultimately, there must be a point at which enforcement is so effective that drivers are deterred from offending and revenue from fines becomes minimal. Other revenue considerations might include, for example, an increase in revenue from legitimate paid–for parking as unlawful parking is deterred by increased enforcement levels and, in the case of increased enforcement of speed–limits, a reduction in police and health service costs as the number of accidents decreases. Effective enforcement of traffic regulations, by whatever agency, requires clear funding mechanisms, if it is to achieve its true potential in road safety and traffic management.

14.9 Possible Future Changes

The Road Traffic Act 1991 empowers local authorities to take over, from the police, most of the responsibility for the enforcement of parking regulations [NId]. There is also scope for change in the role of the police in other traffic–related duties, such as escorting abnormal loads, dealing with broken–down and abandoned vehicles, policing roadworks and obstructions, accident investigation, tachograph examination, excise licence checks and some routine traffic patrols (HMG, 1995). However, there are wider issues involved in granting enforcement powers to agencies other than the police, especially if this requires stopping vehicles on the road. Careful consideration has to be given to such issues in evaluating the feasibility of any changes and this is a matter for central government.

Significant scope exists for new technology to be used in traffic enforcement. Cameras are already used to detect speeding offences and red–light violations at traffic signals. Video cameras have also been used to enforce bus–lanes and weight–limits and closed–circuit television (CCTV) could be used to enhance enforcement at signal–controlled junctions. Developments in intelligent image–analysis may enable the widespread use of cameras to detect a number of different traffic offences, using automatic vehicle detection/number plate recognition systems perhaps linked to coded tags or electronic number plates. Furthermore, if electronic charging for road–use is introduced, it is likely that the identification and enforcement systems for such a regime would make the enforcement of other offences easier and more effective.

14.10 References

AC (1992) The Audit Commission 'Fine Lines: Improving The Traffic Warden Service', Stationery Office.

DOE/DOT (1994) Planning Policy Guidance Note 13: 'Transport', DOE/DOT [Sa] [Wa].

DOT (1992) Traffic Management and Parking Guidance, Local Authority Circular 5/92, 'Traffic In London', DOT.

DOT (1995) Circular 1/95 (WO 26/95), 'Guidance On Decriminalised Parking Enforcement Outside London', DOT [Sc].

HMG (1995) 'The Review of Police Core and Ancillary Tasks', Stationery Office.

PCfL (1995) 'Annual Report and Accounts 1994/95', Parking Committee for London.

14.11 Further Information

Chick C (1996) 'On–Street Parking – A Guide to Practice', Landor Publishing.

DOT (1995) DOT Circular 1/95 (WO 46/95) 'Guidance on Decriminalised Parking Enforcement outside London', DOT.

PCfL (1993) 'Code of Practice on Parking Enforcement', Parking Committee for London.

SPA (1995) 'Policing Plan for Surrey 1995–1996', Surrey Police Authority.

Chapter 15 Information for Transport Users

15.1 Strategic Role of Information for Users

Traffic and travel information can be used in a variety of ways to influence the behaviour of the travelling public. Accurate, relevant and timely traffic and travel information can affect behaviour by encouraging travellers to make informed journey choices. The choices may include:

❑ the best time at which to make a journey;
❑ the best mode(s) of transport to use;
❑ the best route to take;
❑ whether or not to share a car with someone else;
❑ whether or not to go to an alternative destination; and
❑ whether or not to abandon the journey;

At its most basic, traffic and travel information provides essential data to warn of hazards or to advise of speed–limits, directions or restrictions and other alternatives. More advanced information can be provided before and during a journey to assist travellers and could include:

❑ current and predicted transport network conditions;
❑ expected disruption, if any, to the journey;
❑ alternative modes of transport with relevant journey details;
❑ alternative routes for any given mode; and
❑ alternative times for public transport journeys

There are essentially three ways in which information can be presented to travellers. One is by providing information en–route via signs at the roadside with fixed or variable messages. The others are by using radio or other advanced communication technologies to provide information in the vehicle itself or to provide advance pre–trip information, to enable users to make travel–choices prior to starting their journey.

15.2 Role of Signing in Traffic Management and Control

Signs and road markings are a visual means of conveying information to a driver relating to the highway on which he or she is travelling, or wishes to travel, and should promote the safe and efficient use of the highway. The information has to be seen, read, understood and acted upon in a short period of time

and so must be presented in as clear and concise a manner as possible. To achieve this, a range of standard sign–types has been developed for rapid assimilation by drivers (see Section 15.4).

Under the Vienna Convention (1968), a similar style of signing has been adopted throughout most of Europe and drivers should be able to understand the basic meaning of a sign in any European country. Recent developments in microprocessor control and matrix displays have provided greater flexibility in the way information can be disseminated. Known by their generic name of Variable Message Signs (VMS), these signs are able to display a variety of symbols and textual messages, including colour representations of mandatory signs.

15.3 Principles of Signing Road Traffic

Careful provision of prescribed signs and markings can make a considerable contribution to the safe and efficient operation of the highway network.

Traffic signing is effected by the use of road markings, road studs, traffic signals, lamps, cones, cylinders and beacons, as well as by various types of upright signs with textual or graphical images (see sections 64 to 80 of the Road Traffic Regulation Act 1984) (HMG, 1984) [NIa].

The circumstances in which each of these devices is permitted, or required to be used, and illustrations of most of them are contained in the current Traffic Signs Regulations and General Directions (TSRGD) (HMG, 1994) and subsequent amendments (HMG, 1995) [NIb]. Detailed specifications of most prescribed signs are given on working drawings published by the Stationery Office (HMG, 1994/95) [Wa]. Advice on the use and design of signs is given in the Traffic Signs Manual (DOT/TSM), supplemented by Department of Transport Circulars (DOT, 1995) [Sa] Advice Notes and Standards and Local Transport Notes (DOT, 1987; 1994a and 1994b) [Sa].

Generally speaking, all traffic signs used on the highway must conform to the Regulations, which are periodically updated, but special signs can be individually authorised by the Secretary of State in appropriate circumstances.

Signs should be sited so that the information is given to road–users precisely when they need it; neither too soon, lest it be forgotten before it is needed, nor too late for the safe performance of any consequent manoeuvre. Appropriate signing, therefore, needs to take account of how fast traffic is travelling. Subject to the siting constraints for different types of sign, they should not be placed so as to be environmentally intrusive and cause sign clutter. Further information on the size, design and mounting of traffic signs is given in the Traffic Signs Manual (DOT/TSM).

Signs which do not conform to the appropriate regulations, or are unauthorised signs or advertisements, may distract the attention of road–users to the detriment of road safety (DOE, 1984) [Sb]. They might also be held by a court to be unlawful obstructions of the highway.

15.4 Categories of Traffic Signs

Apart from traffic signals, which are discussed in Chapter 18 and Part 5, traffic signs may be divided into three broad categories. These are:
❑ upright signs – which are themselves divisible into warning, regulatory and informatory signs (Schedules 1 to 5, 7, 10 and 12 of TSRGD) (HMG, 1994) [NIc];
❑ road markings (Schedule 6 of TSRGD) (HMG, 1994) [NId]; and
❑ miscellaneous signs, including traffic–light signals (Schedules 8, 9 and 11 of TSRGD) (HMG, 1994) (see also DOT/TSM) [NIe].

Warning signs, which give warning of hazards ahead, are usually either triangular (black symbol within a red border) or rectangular (white legend on a red background). The latter are used mainly for temporary warnings. A variety of supplementary plates, giving further information, is available for use with certain prescribed warning or regulatory signs.

Regulatory signs give notice of restrictions or prohibitions on the speed, movement and waiting times of vehicles. They are mostly circular with a red border, indicating a negative instruction, or a blue background, indicating a positive instruction. Exceptions are the 'Stop' sign (octagonal) and the 'Give Way' sign (inverted triangular). Waiting restrictions and zonal restrictions signs are rectangular. Most regulatory signs may be used only if an appropriate Traffic Regulation Order (TRO), such as a one–way street or no–waiting restriction, has previously been made by the local traffic Authority (see also Chapter 13) but some can be used without an Order, for example 'Give Way', 'Keep

Left' or 'Keep Right'. 'Stop' signs do not require an Order but do require the consent of the Secretary of State. Supplementary plates may also be used and must accord with any associated Traffic Regulation Order, such as 'except for access' or 'except for loading'.

Information signs give information about routes, places and facilities of particular value and interest to road–users. Directional signs come into this category; they are either rectangular or 'flag' type (ie pointed at one end). The colours used depend on the road on which they are used and the information they give (see Section 15.5).

Road markings are provided to convey a warning, a requirement or basic information. They are usually white, except where associated with waiting or loading restrictions, in which case they are yellow, except for the London Priority (Red) Routes, which use red lines. In certain circumstances, regulations require that reflecting material should be used (HMG, 1994). Studs incorporating reflectors may be provided to supplement road–markings. Reflectors are usually white in colour but red, green or amber studs may also be used, depending on their location on the carriageway.

Apart from traffic signals, other miscellaneous signs include:
❑ temporary signs placed on or near a road to warn, inform or regulate traffic for special events, roadworks or other temporary situations;
❑ flashing beacons (usually amber or, in the case of police, blue), warning drivers to take special care (eg beacons at fire stations and level crossings);
❑ cones used to define routes around obstructions or road–works;
❑ cylinders indicating the temporary division of a carriageway;
❑ indicator lamps at refuges, to warn drivers of their presence;
❑ school crossing–patrol signs and warning lights; and
❑ flashing amber danger lamps, to define the extent of a temporary obstruction.

15.5 Direction Signing

Hierarchy
Direction signing is used to guide drivers to their destinations by the most appropriate routes. The broad approach adopted in Britain is first to guide traffic towards a general destination then, at the appropriate point, to direct it to more specific areas and finally to local destinations. These may be simply

street name–plates (though street name–plates and house numbers are not classed as traffic signs) or could, in the case of important traffic attractors, be individual buildings or car parks. Signing, therefore, becomes increasingly specific, moving down a hierarchy of destinations, as decision points are reached.

The hierarchy of destinations used is:
❑ 'regional' destinations – major geographical areas such as 'The North' and 'The South West' (DOT, 1994a) [Sa];
❑ 'primary' destinations – such as important locations on the motorway and primary route network, and towns and cities which are important destinations for longer–distance traffic;
❑ 'non–primary' destinations – towns and smaller cities of less importance to traffic;
❑ 'local destinations' – such as small settlements, city suburbs, environmental areas, industrial estates and destinations such as transport interchanges or other public buildings, which may be significant attractors of traffic; and
❑ 'tourist' attractions and facilities – attractions with over 150,000 visitors a year, and which meet certain other criteria, may be signed from the nearest motorway junction(s) within 20 miles [NIf]. On all–purpose roads, tourist signing should be in accordance with criteria set by the Local Authority and may include signing to all tourist destinations which attract visitors to an area (DOT, 1995) [Wb].

Colours
The colours used for direction signs depend on the status of the traffic route on which they are placed and the type of information given. A detailed explanation of the colour coding of direction signs can be found in LTN 1/94 (DOT, 1994a) [Sa]. A multi–tier system is used in Britain, as follows:
❑ on motorways – blue background with white border, symbol, legend and route number – main destinations;
❑ on the primary route network – green background, with white border, symbol and legend, and yellow route number. Note that a primary route is not a road classification as such but rather a route designated, by the Department of Transport, as the most appropriate between places of major traffic importance. These signs help drivers to identify, and follow, the primary route network;
❑ on non–primary routes – white background with a black border, symbol, legend and route number. These signs are used on all roads which are not part of the motorway or primary route network;
❑ tourist attractions and facilities – brown background with a white border, symbol, legend

and mileage. Approved symbols may be used to denote the type of attraction or facility. These signs can be used on all classes of road including motorways (DOT, 1995). Traffic signs on trunk roads and motorways, to tourist attractions and facilities in England, must follow the criteria of the Highways Agency [Wc]. It is common practice for the applicant to pay the cost of designing, erecting and maintaining tourist signs; and
❑ traffic diversions – signs using a yellow background with a black border, symbol and legend are for temporary use, most commonly in connection with a diversion caused by road–works or when used by a motoring organisation, such as the Automobile Association (AA), for directions to a specific event, subject to approval by the local traffic Authority. Temporary signs can also bear a blue legend on a white background or a white legend on a blue background, as normally used by the Royal Automobile Club (RAC) (see Regulation 41 of TSRGD) (HMG, 1994).

In addition to the above signs for general traffic, other colours can be used for specific themes. These include lorry routes (black background with a white lorry symbol), pedestrians (blue background with a white symbol of a walking man) and cyclists (blue background with a white cycle symbol). See the Traffic Signs Regulations and General Directions 1994 (HMG, 1994) for further information.

Location
The choice between three types of direction sign depends on where they are to be sited in relation to junctions. An 'Advance Direction Sign' (ADS) is placed before a junction is reached, to give drivers advance information about their possible route–choices. On high speed roads, a 'Forward ADS' may be provided some distance ahead of the ADS and, on motorways, signs are provided one mile and half a mile ahead of, as well as at, the exit itself. Tourist signs on motorways may be provided, if appropriate, ¾ mile and ¼ mile in advance of the exit. Advance direction signs may be mounted overhead on bridges or gantries or by the side of the carriageway.

The layout of individual advance direction signs may be of the map–type, where the layout of the junction is represented diagrammatically. These are especially appropriate at roundabouts and may display other indications to traffic, by incorporating certain warning or regulatory signs. They may also be of the stack–type, where the individual destinations are stacked above each other on the sign face, or the lane–destination type, where the carriageway is clearly marked into traffic lanes, which are appropriate for different destinations.

A 'Direction Sign' (DS) repeats the information on the ADS but is placed at the junction where the turning manoeuvre is actually to be made. Direction signs are normally of the flag–type but may incorporate an inclined arrow, depending on the type of junction.

A 'Route Confirmatory Sign' is placed after the junction, to give confirmation of the route number and the distances to major destinations ahead.

A 'Boundary' or 'Place–Name Sign' complements the direction signing system, by informing drivers that the destination previously signed has been reached. Additional information signs, such as for rivers and county boundaries, assist travellers in locating their position and correspond with information commonly found on road maps.

The precise layout, design and siting requirements for the different types of direction signs are given both in the Department of Transport's Traffic Signs Manual (DOT/TSM) and in Local Transport Notes 1/94 (DOT, 1994a) and 2/94 (DOT, 1994b). General points (DOT, 1994a) [Sa], which should be followed as far as possible, are:
- ❑ once a destination has been signed, all subsequent signs along that route must indicate that destination until it is reached;
- ❑ the number of destinations shown on any one sign should not exceed six, preferably not more than five on motorways;
- ❑ not more than two destinations should be signed in each direction on any one sign. Exceptionally, a third destination may be used but only if absolutely necessary;
- ❑ destinations should not be signed beyond a destination of a higher category in the hierarchy;
- ❑ mileage to destinations should not normally be shown on ADSs but may be shown on DSs within junctions, unless indicated on a following route confirmatory sign;
- ❑ route numbers should appear on both ADSs and DSs, unless they are of purely local significance;
- ❑ signs should be sited where there is good visibility for approaching drivers and away from vegetation, which may grow to obscure them; and
- ❑ the size of signs and lettering should relate to 85 percentile traffic speeds, to ensure legibility from an adequate distance and to allow drivers time to react and to take appropriate action safely.

15.6 Variable Message Signs (VMS)

Variable message traffic signs have been in use for many years but the introduction of microprocessor control and matrix displays has provided increasing opportunities for their use in the management of traffic.

Advice on suitable applications for VMS, criteria for selecting the most suitable type of sign, operating methods and standards can be obtained from the Design Manual for Roads and Bridges, Volume 8 (DOT, 1990).

Regulation 46 of the TSRGD (HMG, 1994) provides that variable message signs may display either signs shown in Schedules 1 to 5, 7, 11 or 12 or the range of messages prescribed in Schedule 15 [NIb]. Any signs not included in these Schedules must be authorised by, or on behalf of, the Secretary of State. Applications for authorisation should, in the first instance, be addressed to the relevant Government Regional Office or the Highways Agency for trunk roads in England [NIg] [Sc] [Wd]. As more experience of VMS is gained the range of prescribed messages is likely to be increased in Amendment Regulations.

Authorisation is required for:
- ❑ the location of the sign(s);
- ❑ the size, colour and character of the sign(s);
- ❑ the display characters (if not the standard character in Schedule 13 of TSRGD); and
- ❑ the legends proposed for display using the above characters.

The Traffic Signs Regulations and General Directions (HMG, 1994) also require that the equipment used for all variable message signs be of a type authorised in writing by, or on behalf of, the Secretary of State before being placed on, or near, any road.

This authorisation is commonly referred to as 'Type Approval' and is usually held by a manufacturer for each type of sign produced. If an engineer specifies signs of a novel design, the manufacturer may need to obtain a specific Type Approval from the Department of Transport [NIg] [Sc] [We], which can take several months. Allowance for this should be made in the contract programme.

The main types of variable message sign technology are described in Section 6 of Chapter 18. VMS can be beneficial in a variety of situations, either where a message is not required to be displayed permanently or where several alternative messages, which are not interdependent, are required at the same site under different circumstances. The following are examples.

Regulatory signs (time–dependent, tidal and peak–hour schemes) such as:
- ❑ banned turns;

- no entry;
- restricted access (eg pedestrian zones);
- lane–control (eg tidal flow);
- warning signs;
- over–height vehicles (in conjunction with diversion signs);
- gate(s) closed; and
- weather related (eg flood, fog, ice/snow, wind).

Informatory signs, such as:
- route–status, providing information about prevailing route conditions, such as incidents and roadworks (see Photograph 15.1);
- car parks, where matrix signs can display the actual number of vacant spaces in real–time, when linked to a monitoring system, such as UTC (see Photographs 15.2 and 15.3);
- direction signs, which show major destinations;
- diversions (see Photograph 15.4);
- lane–controls;
- facilities/services (eg open 24 hours or closed); and
- motorway services, petrol–price signs.

Warning signs, such as:
- detection of over–height vehicles approaching a bridge; and
- detection of long vehicles crossing a carriageway at a junction ahead.

15.7 Illumination of Traffic Signs

The requirements for the illumination of signs, whether internal, direct external or by reflectorisation, are contained in Regulations 18–21

Photograph 15.1: An example of a route–status information system.

Photograph 15.2: An example of car parking information with real–time data on vacant spaces.

and Schedule 17 of the current Traffic Signs Regulations and General Directions 1994 (HMG, 1994) [NIh]. TA 19/81 (DOT, 1981) gives guidance on the use of different types of reflective material. The illumination requirements for traffic signs are related to whether or not a system of street–lighting, as defined in Schedule 17 of TSRGD 1994, is present within 50m of the sign. In general, all motorway directional signs should be directly lit, even in areas with street–lighting. The requirement also to light primary route signs is left to the discretion of the engineer, after considering the position of the sign and the materials to be used in its construction. The precise lighting requirements for all prescribed signs

Photograph 15.3: An example of car parking information with real–time data on vacant spaces.

Photograph 15.4: An example of a temporary diversion information sign.

are found in Chapters 1 to 8 of the Traffic Signs Manual (DOT/TSM). The current Traffic Signs Regulations and General Directions (TSRGD) should be consulted, since requirements change from time to time.

All other direction signs must be 'reflectorised', if not directly lit, except signs for pedestrian and cycle routes and tourist attraction signs, where reflectorisation is not required by the regulations but is nevertheless generally recommended.

15.8 Environmental Impact of Direction Signs

The number and size of direction signs should be kept to a minimum, commensurate with the need to provide adequate information for drivers. The location of direction signs should take account of other prohibitory signs, traffic signals, street name–plates and other street furniture. Sign supports should also be designed with care, to avoid ugly and overlarge structures. In dense urban areas, it is sometimes difficult to find suitable locations for signs, especially where footways are narrow, and it may be necessary to compromise on the number of signs provided. In these cases, directional signs should take precedence over other informatory signs but not over regulatory signs. Special treatment may

be necessary in conservation and other environmentally–sensitive areas to consider, for example, cast fingerpost signs rather than standard direction signs.

15.9 Traffic Signs at Roadworks

Proper temporary signing is mandatory at roadworks, to give adequate warning to drivers and to enable them to take appropriate action. Consistent application of good temporary signing practice will help achieve high standards of safety at roadworks. To this end, the Department of Transport has set out detailed requirements for the layout and siting of temporary signs for different types of roadworks in Chapter 8 of The Traffic Signs Manual (DOT/TSM) [NIi]. These requirements cover all aspects of temporary traffic arrangements, including detailed layouts and recommendations for the use of cones, cylinders and temporary lighting. Working drawings for all signs required for road–works are available as a package (HMG, 1994/5 – Vol 3) [Wf].

Under the New Roads and Streetworks Act 1991 [Nij] certain undertakings must, by law, conform to the signing provision in the Code of Practice on Roads and Streetworks (DOT, 1992).

15.10 Design of Signs

The requirements for the design of direction signs are given in the Traffic Signs Regulations and General Directions 1994 (HMG, 1994) [NIh] and Local Transport Notes (DOT, 1994a and 1994b) [Sc]. The Department of Transport has also published working drawings for traffic signs and these can be used with computer–aided design (CAD) systems to produce individual designs (HMG, 1994/5) [Wg].

15.11 In–Vehicle Information Systems

Information can be presented to the driver in a vehicle in a number of ways. The most popular form of in–vehicle travel news is provided by standard radio broadcasts. All local, regional and national radio stations carry regular travel bulletins. Broadcasters obtain information through regular contact with local highway authorities,with local and regional police and with public transport control centres, or by electronic exchange of travel information between these organisations.

All methods of information provision should be relevant, timely and accurate, in order to maintain credibility with users. Effective use of the media is particularly important when planning road–works that may disrupt traffic.

One dedicated in–vehicle travel information system, widely used in the UK is 'Traffic*master*'. This is a commercial system that uses a large network of infra–red detectors placed at strategic locations on the UK motorway and trunk road network to monitor traffic speed. These detectors relay information on slow–moving traffic, via a control centre, to small in–vehicle units, where they are displayed on a digitised map of the motorway and trunk road network. Drivers are thus forewarned of potential problems on their journey and can then decide on their best course of action (see Section 18.12).

The Radio Data System (RDS) is a development of the standard radio broadcast system, based on technology to carry additional digital data in tandem with normal broadcasts. This additional data can take the form of text–based information like teletext, for example, to show the name of the broadcast station on the radio display. It can also take the form of control codes, which can provide instruction to the radio receiver. During travel news bulletins, for example, a radio station can simultaneously transmit control codes to indicate that travel news is being transmitted. These control data are picked up by the radio receiver, which then automatically interrupts whatever is currently being played (radio, cassette, or CD) to broadcast the bulletin. Similarly, control data can be transmitted to indicate when the broadcast is over and thus return the receiver to whatever was originally being played.

The Traffic Message Channel (TMC) is an additional feature of RDS. It extends the use of the data transmission capability to provide encoded travel messages, which are decoded by the in–car radio receiver, and either displayed as text on the receiver's digital display or, when used alongside a voice synthesised database, played back as an audio message. The messages are encoded using a European standard protocol known as ALERT C+. Any organisation with the relevant software can construct and transmit a travel message, using the ALERT C+ protocol. Cooperation with radio broadcasters will provide the necessary transmission technology. Encoded travel messages can then be forwarded by the organisation to the transmission point and broadcast to travellers. The messages can be filtered to provide information for the chosen route only and can be in the driver's native language.

Digital Audio Broadcasting (DAB) offers the prospect of improved transmission capacity and scope for dynamic (ie. interactive) information provision, through a Europe–wide DAB–TMC network.

15.12 In–Vehicle Navigation Systems

In–vehicle information can be provided through the use of route–planning and navigation systems. There are basically two kinds of route–guidance system – autonomous (ARG) and dynamic (DRG). Both make use of digitised road maps within the receivers. Some autonomous systems allow the selection of journey parameters, such as origin, destination and arrivsal time; a preferred route is then calculated, usually based upon time or distance criteria. Others combine the use of digitised mapping with terrestrial or satellite navigation services. These pinpoint the location of the car in relation to the chosen route and guide the driver, using simple visual and/or spoken instructions at relevant moments (eg 'take the next left turn' or 'stay in the right–hand lane'). Additionally, dynamic systems incorporate real–time information. The calculated route depends on actual traffic conditions and may be recalculated in response to these and other incidents during the journey.

Fully dynamic route–guidance (DRG) systems rely on loops, infra–red or microwave beacons or digital cellular telephone (GSM) to provide a two–way communication link between vehicles and the roadside. The basis of these systems is that equipped vehicles act as 'floating cars' and report their journey times on roads back to a control centre. The information from all equipped vehicles can then be collated at the control centre, to provide a continuously–updated picture of traffic conditions over the entire network. When a driver requests a route, this is effectively computed at the control centre, taking into account overall network conditions prevailing at that time.

15.13 Electronic Pre–Trip Information Systems

The circumstances under which pre–trip information is used vary considerably, depending on the type and destination of information to be distributed and the type of recipient. Pre–trip information is needed in three broad locations:
  ❑ in the home;
  ❑ in the office and similar workplace; and
  ❑ on–street or in other public places.

Many of the systems provide information on a range of travel modes, which gives travellers the opportunity to seek alternative arrangements for any given journey, particularly for public transport trips. They also help travellers to make a particular journey using more than one mode of transport.

No statutory guidelines exist, as at 1997, on the introduction of pre–trip information services, other than the requirement for planning permission for on–street information provision. However, foremost in the provision of pre–trip information is the requirement for source information of the highest quality.

A range of media exists by which electronic pre–trip information can be disseminated. The list below provides an indication of the methods available. These are:

❏ television and teletext (cable/satellite/terrestrial);
❏ radio broadcasts;
❏ telephone and fax services;
❏ on–street displays and terminals; and
❏ computer network systems.

It may be possible to make use of existing communications infrastructure, such as via television/teletext and radio broadcasts, in order to disseminate information. This provides the most widespread coverage to the population but is likely to require at least some form of communications link to the main broadcaster, as well as an information dissemination system, such as dedicated PC software. Portable electronic travel–guides are also under development, based on 'palm–top' micro–computers.

Telephone services offer easy accessibility to travel information, as the majority of the population own, or have access to, a telephone. A wide range of telephone services is available, including pre–recorded message services and fax services. Consideration has to be given to the level of service offered, the expected demand, publicity and whether the service is expected to generate revenue (eg using premium–rate lines).

Trip–planning information systems have been developed which enable pre–trip planning of journeys, made by public or private transport, to utilise the network more efficiently. If the aim is to promote the use of public transport, then a multi–modal trip–planning facility should be adopted. This allows enquiries involving combinations of bus, train, aeroplane or ferry to be processed, thus avoiding the need to search through a series of timetables or make a number of telephone calls to obtain the required information.

Before offering such a trip–planning facility, permission must first be obtained from all public

transport operators whose information is required, to produce the database for the catchment area under consideration. Any subsequent alterations to schedules must be notified as soon as these become available, to prevent out–of–date and inaccurate information being provided to travellers.

Commercial software packages provide nationwide trip–planning facilities by road and rail. Less commonly, packages are available for bus journeys within geographical limits. The software can be used as–purchased or may be modified and combined with other software to provide multi–modal information. However, public access to these packages usually requires the payment of an annual licence fee to software copyright–holders.

Terminals holding the information can be touch–screen or keyboard operated and should be located in places of high public usage, such as major transport interchanges (rail or bus stations, ferry terminals, airports), shopping centres, hospitals, libraries and tourist information bureaux. Permission must be sought from site–owners prior to the installation of such a system. Other useful facilities which could be incorporated include a printer to enable a record of the journey to be obtained and a 'through the glass' terminal which is accessible 24 hours a day. An example of a trip–planning system is shown in Photograph 15.5.

Consideration must also be given to the type of

Photograph 15.5: An example of a trip–planning system.

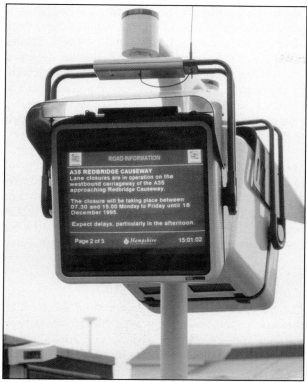

Photograph 15.6: A typical display screen associated with a trip–planning system.

information displayed. This depends largely upon the information–display medium used. Particular attention should be paid to the volume of information which the user is expected to digest and the process by which the user gains access to the information. Circumstances should be avoided where the user is required to search through large volumes of unrelated information in order to get to a relevant section. An example of a typical display screen is shown in Photograph 15.6.

15.14 Bus Passenger Information

The use of real–time public transport passenger information systems for bus services has become an important tool, adopted by local authorities and passenger transport organisations across the UK, to promote the use of public transport.

Such a real–time system can be considered in two distinct parts :

❑ vehicle location and communication of data to a central control computer; and

❑ communication from control computer and the displaying of real–time information on expected arrivals on electronic message signs. This could be at bus stops or, if intended to influence potential users, at other places such as shopping malls, offices, factories and even in the home.

Vehicle Location (see Section 18.5)

Continuous updating on a vehicle's location can be achieved by a number of means, such as radio–polling, dead–reckoning (beacons and bus–oedometer) or the global–positioning system (GPS). Whichever method is adopted, equipment will be a necessity on board the vehicles to enable location data to be processed and transmitted to a central control computer. Usually, this will involve an on–board computer and radio, although some systems operate without the radio link. If a beacon–based system is used, then planning permission from the Local Planning Authority is required to erect the beacons on existing street furniture, such as lamp columns, and a radio–transmission licence must be obtained, if transmitting power exceeds 500mV.

Transmission of data to the control computer can be via Band III data radio or X25 packet–switched radio. For Band III radio, a licence must first be obtained from the Confederation of Passenger Transport to use a pre–defined frequency/channel. This process normally has to be undertaken by a bus operator participating in the system, as it is difficult for a local authority to be granted such a licence.

Real–Time Information Displays

Various technologies can be deployed to provide real–time information to passengers at bus stops but the value of displaying a conventional, up–to–date printed timetable should not be under–estimated. The alternative sign–types currently available are liquid crystal displays (LCD), light emitting diode (LED) or transflective LCD. The former two are restricted, in terms of their ability to accommodate fluctuations in ambient light conditions, and are, therefore, not very suitable in situations where direct sunlight falls on the screen. However, the transflective LCD reacts to such fluctuations and will adjust its light–output to allow the information on the screen to be legible in all conditions.

'Talking' signs have been developed for visually–impaired bus–users and provide audible, as well as visual, bus information. The signs are activated either by a push–button or a hand–held infra–red key–fob trigger. When active, a digitally–recorded voice announces the approaching bus, the route number, final destination and the estimated time to arrival at the stop.

All displays should be designed to an environmental standard, such as BS EN 60529 (BSI, 1992), which specifies the environmental conditions under which equipment should operate satisfactorily. Signs must also be designed to withstand vandalism and screens should be manufactured from either clear polycarbonate or toughened glass.

While electronic variable message signs could be

Photograph 15.7: An example of real–time 'countdown' information in a bus shelter.

Photograph 15.8: Real–time information on the estimated time to arrival of buses at a particular stop.

installed at a variety of locations, within the bus infrastructure and outside it, the majority have been designed to be installed inside bus– shelters. Shelter manufacturers are tending towards shelters designed with this in mind and, with little or no modification to the unit, a sign can be accommodated safely. However, for older shelters, this is often not possible and, in these situations, replacement with a modified shelter is required (see Photographs 15.7 and 15.8).

The communications network for the signs can be, for example, via radio–paging, low power (up to 500mW) radio or telecommunication land–lines. For radio–paging, a licence will be required and this is issued by the Department of Trade and Industry. No licence is required for low–power radio transmission.

15.15 Rail Travel Information

The majority of rail travel information is provided through paper timetables, available at all railway stations and usually on display at various public places, such as Tourist Information offices in town centres. Information is also provided to travellers through enquiry bureaux at stations and through telephone enquiry services. These can offer person–to–person enquiries or pre–recorded voice announcements. A 'National Rail Passenger Information Service' is in place, following the introduction of a single national telephone number for rail enquiries. Callers whose local enquiry bureau is busy are automatically redirected to unoccupied lines elsewhere in the country.

Other electronic systems are dedicated to the provision of rail information. The main system is that which is used at all main railway stations in the UK and provides on–screen details, on platforms, of current rail network conditions, such as next arrivals and late–running services. Journey–planning software is also commercially available to travellers with access to a computer. Once installed on the computer, it allows each user to enter the details of the proposed journey, such as origin, destination and time of departure, for any train service in the UK. It then calculates the most appropriate journey, based on the shortest overall travel time.

15.16 References

BSI (1992)   British Standards Institution – BS EN 60529 'Degrees of Protection Provided by Enclosures', BSI.

DOE (1984)   Circular 11/84 ' (WO 18/84) Town and Country Planning (Control of Advertisements) Regulations 1984', DOE [Sb].

DOT (1981)	TA 19/81 (DMRB 8.2) 'The Reflectorisation of Traffic Signs', DOT.
DOT (1987)	Local Transport Note 2/87 'Signs for Cycle facilities', DOT.
DOT (1990)	TD33/90 (DMRB 8.2) 'The Use of Variable Message Signs on All–Purpose and Motorway Trunk Roads', DOT.
DOT (1992)	'Code of Practice on Roadworks and Streetworks', Stationery Office.
DOT (1994a)	Local Transport Note 1/94 'The Design and Use of Directional Informatory Signs', DOT [Sa].
DOT (1994b)	Local Transport Note 2/94 'Directional Informatory Signs: Interim Design Notes' (to be superseded in 1997 by Chapter 7 of the Traffic Signs Manual), DOT [Sb].
DOT (1995)	DOT Circular – Roads 3/95 (WO 3/96) 'Traffic Signs to Tourist Attractions and Facilities in England', DOT.
DOT/TSM	Traffic Signs Manual: Ch 1 (1982) 'Introduction' Ch.2 (1982) 'Directional Informatory Signs' Ch.3 (1985) 'Regulatory Signs' Ch.4 (1986) 'Warning Signs' Ch.5 (1985) 'Road Markings' *Ch.6 'Illumination and Reflectorisation' *Ch.7 'The Design of Traffic Signs' Ch.8 (1991) 'Traffic Safety Measures and Signs for Road Works'. (*not yet published as at 31 March 1997).
HMG (1984)	'Road Traffic Regulation Act 1984', Stationery Office
HMG (1994)	Statutory Instrument No.1519 'Traffic Signs Regulations and General Directions 1994', Stationery Office.
HMG (1994/5)	Vols.1, 2 and 3 'Working Drawings for Traffic Sign Design and Manufacture', Stationery Office.
HMG (1995)	'Traffic Signs (Amendment) Regulations, General Directions', Stationery Office.

15.17 Further Information

Colby FJM (1984)	'Variable message traffic signs' Highways and Transportation, 31(7).
DOT (1975)	Circular – Roads 3/93 (WO 15/94)'Street Name Plates and Numbering of Premises', DOT.
DOT (1978)	H1/78 'Design Criteria for Sign/Signal Gantries', DOT.
DOT (1994a)	Circular – Roads 4/94 (WO 50/94) 'Revision of Traffic Signs Regulations and General Directions', DOT.
DOT (1994b)	'Traffic Signs Bibliography', DOT.
DOT (1995)	'Know your Traffic Signs', Stationery Office.
ERTICO	Information Service, Rue de la Requence 61, 1000 Brussels, Belgium.
HCC (1993)	'ROMANSE Project – Final Report' Hampshire County Council
ITS Focus	Information service and 'Current Issues in Telematics' – Conference Papers ITS Focus (June 1996), TRL, Crowthorne.

Note: All current TAs and TDs on traffic signing are now incorporated into the Design Manual for Roads and Bridges (DMRB) Vol. 8 'Traffic Signs and Lighting'.

Also, in DMRB, TA 61/94 'Currency of the Traffic Signs Manual' describes the history and current status of the constituent chapters of the Manual and draws attention to sections which have been superseded by later guidance.

Chapter 16 Road Safety

16.1 Nature and Scale of Traffic Accident–Risks

About three quarters of traffic injury accidents and nearly half of fatal accidents occur in built–up areas, ie on roads subject to speed limits up to and including 40 miles/h (DOT, 1995). The complexity of the urban environment leads to many conflicts between road–users, especially at junctions, which account for more than two–thirds of all urban accidents.

Outside town centres, up to one half of all accidents occur on primary and district distributor roads and about one quarter each occur on local distributor and residential access roads. Accidents may form clusters, at particular places of conflict or be concentrated on some busy lengths of road but in many parts of the road network, especially on residential roads, they are more scattered.

The numbers of deaths, seriously injured and all casualties reported for different classes of road–users in 1995 are given in Table 16.1. These figures relate to injury–accidents reported to the police but many accidents go unreported. Comparisons of police data

with reports from Accident and Emergency departments of hospitals (James, 1991) indicate that only 76% of seriously injured and 62% of all casualties are reported in the national statistics. Thus, the true figures for casualties in urban areas are likely to be of the order of 37,000 seriously injured and 350000 in total.

Pedestrians are the most vulnerable road–users, making up over half of the deaths in urban areas. A further quarter of deaths are amongst car–occupants. The position is reversed for injuries, with car–occupants making up over one–half and pedestrians one–fifth of casualties. The vehicles involved in the largest numbers (over three–quarters) are cars.

The risk of becoming a casualty, per kilometre travelled in urban areas, is highest for pedal cyclists and riders of two–wheeled motor vehicles (of the order of 645 and 750 per 100 million vehicle–km respectively) and next highest for pedestrians (410 per 100 million vehicle–km walked), compared with 73 per 100 million vehicle–km for car occupants (Ward et al, 1994).

Categories of reported accidents and casualties in the UK (1995)	Built–up roads			Non–built– up roads	All roads
	Fatal	Serious	Total (all categories)	Total	Total
Accidents	1,438	25,785	169,716	53,254	222,970
Vehicles involved	2,046	40,666	298,127	100,072	398,199
Casualties	1,497	28,430	214,970	83,667	298,637
Sub-division of casualties					
Pedestrians	797	10,537	44,972	1,939	46,911
Pedal cyclists	115	3,109	22,599	2,299	24,898
Two wheel motor vehicles	162	3,793	17,050	6,050	23,100
Car occupants	390	9,692	116,831	67,252	184,083
Bus or coach occupants	8	639	8,170	909	9,079
Light goods occupants	12	399	3,654	2,894	6,548
Heavy goods occupants	9	153	999	1,762	2,761
Others	4	108	693	562	1,257
Totals	1,497	28,430	214,970	83,667	298,637

Table 16.1: Numbers of Reported Accidents and Casualties in 1995. Source: DOT (1996a)

16.2 Factors Contributing to Accidents

Most accidents are due to a combination of factors relating to human failings, road deficiencies and vehicle defects. In depth multi–disciplinary studies have shown that human factors contribute in 95% of accidents in urban areas, road factors in about 20% and vehicle factors in one percent. The interaction between human failings and road features has important implications for applications of remedial measures to aid and influence road–users.

A comprehensive study in Leeds (Carsten *et al*, 1989) identified the main failings amongst drivers, which precipitated the accidents, as failure to give way, lack of anticipation, loss of control, wrong positioning and improper overtaking. The underlying reasons are, mainly, going too fast or following too close, perceptual errors (failing to look or see) and errors of judgement. These driver–errors are compounded by adverse features of road design, such as unsuitable layout or poor visibility, slippery roads and obstructions due to parked or stationary vehicles.

In accidents involving pedestrians, failures amongst both pedestrians and drivers are evident. A high proportion of pedestrians fail to give way, in circumstances where it is not possible for drivers to anticipate their actions. In other accidents, both drivers and pedestrians misjudge the situation or fail to see the other. Parked or stationary vehicles play an important part through obscuring inter–visibility between drivers and pedestrians.

16.3 Legislation and Obligations

The two most important statutory duties that are relevant are the Highways Act 1980 (HMG, 1980) [Sa] and the Road Traffic Act 1988 (HMG, 1988).

The Highways Act 1980, section 41(1) [NIa] [Sb], places a statutory duty–of–care on the Highway Authority towards road–users of all kinds:

❏ 'The authority who are for the time being the Highway Authority for a highway maintainable at public expense is under a duty, subject to subsections 2 and 4 below, to maintain the highway.'

The Road Traffic Act 1988, section 39(3) [NIb], makes provision for the Highway Authority to carry out studies into accidents and to take steps to both reduce and prevent accidents:

❏ '...each local authority must carry out studies into accidents arising out of the use of vehicles on roads or parts of roads, other than trunk roads, within their area';

❏ '...must, in the light of those studies, take such measures as appear to the authority to be appropriate to prevent such accidents...'; and

❏ '...constructing new roads, must take such measures as appear to the authority to be appropriate to reduce the possibilities of such accidents when the roads come into use.'

Recommended good practice is provided by three IHT Guidelines; namely, those on Accident Reduction and Prevention (IHT, 1990a), Urban Safety Management (IHT, 1990b) and Road Safety Audit (IHT, 1996), which are discussed in Sections 16.7, 16.8, and 16.11, respectively

The Highways Act 1980, section 58 [NIc] [Sc], provides the defence to a claim that a Highway Authority has taken such care as is reasonably required:

'In an action against a highway authority, in respect of damage resulting from their failure to maintain a highway maintainable at the public expense, it is a defence (without prejudice to any other defence or the application of the law relating to contributory negligence) to prove that the Authority had taken such care as in all the circumstances was reasonably required to secure that the part of the highway to which the action relates was not dangerous to traffic.'

It is likely, therefore, in a case against a highway authority, that the argument will be based on whether reasonable care has been taken by the Authority to protect road–users. In this respect, authorities could be judged on the basis of respected published advice.

16.4 Government Policy and Local Authorities' Code of Practice

Government policy
The main thrust of government policy in the 1990s has been the determination to achieve the target, set in 1987, of reducing accident casualties by a third by the year 2000, compared with the annual average for 1981–85 (DOT, 1987) [NId]. The first monitoring report set out the implications of the targets for particular groups, ie 40% reduction in deaths and serious injuries for pedestrians and child cyclists; 60% in deaths and 65% in serious injuries for motorcyclists; and 50% in drink–related deaths and serious injuries (DOT, 1989). These groups are particularly relevant in urban areas.

The strategy for achieving the casualty–reduction targets focuses on four key areas for improvements in

Situation	Remedial options	Potential savings in accidents
Wet Road		
❑ skidding or loss of control	anti–skid treatment to restore micro–macro texture	30% – 60% (80% on wet road)
❑ darkness	improving surface texture	no figures available
❑ spray obscuring visibility	restore macro texture	no figures available
❑ poor delineation	texture of markings to contrast with road surface	no figures available
*For details of road surface treatment benefits see County Surveyors' Society, Report 1/5 (CSS, 1988)		
Darkness		
❑ lit road poor luminance	improve quality; match surface texture with lighting	30% – 50% dark
❑ unlit road	instal lighting	30% – 50% dark
	or delineate with reflectorised lane and edge markings	50% – 70% dark
*For details of road surface treatment benefits see County Surveyors' Society, Report 1/9 (CSS, 1990)		
Running off road		
❑ cross carriageway		
or down slope	safety barriers	15% fatal or serious
❑ hitting rigid object	safety barriers or crash cushions	15% – 65% in severity
Junctions		
❑ excess conflicts	instal roundabout	30%–40% at small or mini
		40%–60% fatal or serious
	instal traffic signals: new turn facility, phasing	10%–50%
❑ turning traffic	traffic signals	30% – 50%
	waiting lanes or 'ghost islands'	20% – 30%
❑ overshoot from minor road	traffic islands in minor road	50% – 80%
❑ overshoot at roundabout	yellow bar markings	50%
❑ running red lights	camera detection	20% – 60% red light running
❑ restriction of sight	realign or relocate junction or remove obstruction	no figures available
*For details of roundabout or traffic signal benefits see County Surveyors' Society, Reports 1/4 and 1/6 (CSS, 1987 and 1989)		
Excessive Speed		
❑ too fast for conditions	camera detection	30% – 40%
	traffic calming (specific roads)	up to 70%
Vulnerable road–users		
❑ cyclists	cycle lanes or tracks	20% (30% pedestrians)
❑ pedestrians		
in road	controlled crossings	20%
stepping out	guard–rails	10%
obscured by		
parked vehicles	controls on parking	no figures available

Table 16.2: Accident situations and remedial options.

safety: vulnerable road–users; influencing drivers' behaviour; vehicular safety; and roads. The Road Safety Report 1995 (DOT, 1995) [NIe] identifies the range of measures being taken, or planned, to make roads safer and acknowledges the major contribution of organisations outside central government; namely, local authorities, the police, motoring organisations, the insurance industry, schools and colleges, the motor manufacturing industry, private sector firms and voluntary organisations.

By 1995, it became clear that it was time to review and to consider developing new targets, far beyond the year 2000, building on the limited success of the original target. An essential element of setting targets is the identification and quantification of measures which will have impact on reducing casualties to different groups of road–user in different road environments. Options for developing further targets were set out in a consultative conference Targets 2001: Where do we go from here? led by the Parliamentary Advisory Council for Transport Safety (PACTS, 1995).

Local Authority Associations' Road Safety Code of Good Practice

The role of local authorities in reducing road casualties is crucial. Those having highway powers are responsible for almost 96% of the total mileage in Great Britain. These roads carry nearly 70% of all traffic and account for 86% of all accidents. Local highway authorities have a prime responsibility for reducing death and injury on the roads. Those local authorities without such powers can also assist road safety, both directly through their various services and indirectly by influencing public opinion and attitudes.

The Code of Practice was developed by the Local Authority Associations in 1989 (LAA, 1989) to underline the considerable importance which they attach to road casualty reduction. This was against the background of support for the Government's casualty reduction target. The strategy recommends action by local authorities in seven key areas: road safety planning; information; engineering; road–user education and training; enforcement; encouragement; and co–ordination of resources. The essence of the Code is the provision of an integrated road safety service.

Revision of the Code (LAA, 1996) was made to allow for the changing level and nature of responsibilities, consequent on the reorganisation of local government, and the wide application of compulsory competitive tendering (CCT). It reinforces the Association's determination to continue to achieve a substantial improvement in road safety in the foreseeable future.

16.5 Collaboration between Agencies

The multi–disciplinary nature of road safety requires that there is strong collaboration across many boundaries of responsibility, both within local authorities and with outside organisations.

Each of the professional disciplines involved in engineering, education and enforcement should work together. The prime responsibility for promoting co–operation rests with the Local Highway Authority, to establish organisational arrangements and to ensure co–ordination between the various services involved in the road safety [NIf] strategy. Internally, this involves accident investigation, road safety engineering and road safety education staff. Externally, it involves liaison with local education authorities, teachers and the police. Voluntary workers also have an important part to play.

A new dimension to co–operation in road safety was added, in 1992, when the Department of Health announced its strategy for the Health of the Nation (DOH, 1992), with accidents being one of five key issues (DOH, 1993) [Wa]. Targets have been set for reducing deaths by the year 2005 for three age–groups: children (under 15 years); young people (15–24 years); and the elderly (65 and over). A major proportion of these deaths arise from motor vehicle accidents: 50% for children; 75% for young people; and 20% for the elderly. While acknowledging the need for 'healthy alliances', health organisations do not necessarily recognise the experience, expertise and application of successful measures within the highways and transportation fields. Thus, it is important that local highways authorities also take prime responsibility for involving health workers in their strategic plans.

16.6 Public Perception and Attitudes

Perception of risk

While risks of accidents and injuries can be quantified, to identify priority needs and remedial actions, it is the perception of risk as seen by individuals, rather than the actual risk, which frequently determines the acceptability or success of countermeasures. False perceptions of risk may hinder progress towards improving road safety.

Perceived risks are biased by the drama of the accident or by aggressive driver behaviour. The relatively rare occurrence of a multiple pile–up on a motorway or a multiple–fatality coach crash attracts more press publicity than the thousands of road–users killed in urban areas each year. Likewise, 'road rage' and 'joy–riding' make headlines, yet in–depth studies have shown that aggressiveness contributes to only about two per cent of accidents. In consequence, there is public demand for resources to be devoted to such problems at the expense of treatments in areas where returns are likely to be far greater.

It is important to keep the public informed of objective assessments of accident risk, the potential of different remedial measures and how priorities for action are determined. At the same time, a better understanding of road–users' perception of risk, risk–assessment and risk– taking behaviour can lead to more effective remedial measures, especially through education.

Attitudes to risk

Attitudes to risk and the expectation that road–users have of the safety of the road system are increasingly influential in determining the acceptability of proposed countermeasures. While the balancing of resource costs and potential benefits are important considerations in determining needs and priorities, it is attitude to risk, which is subjective and unpredictable, that can sway the balance in decision–taking one way or the other. Public participation in local road safety schemes is an essential part in promoting road safety, to explain the risks and the relevance of the proposed action, to identify any incidental consequences, to listen to their concerns and to allay their fears.

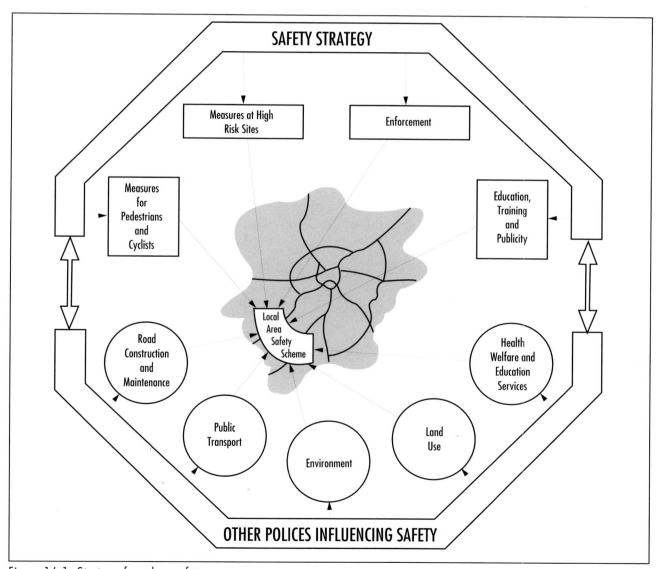

Figure 16.1: Strategy for urban safety management.

16.7 Urban Safety Management and Road Safety Plans

Urban safety management (USM) is the term used to describe a structured approach to accident prevention and casualty reduction. It can help authorities to develop road safety plans and to implement the LAA Road Safety Code of Good Practice (LAA, 1989 and 1996).

Urban Safety Management Strategy

A strategy for each urban area as a whole allows national, regional and local targets for accident reduction to be reflected in specific safety initiatives. These initiatives need to be related to wider policies for the area, balancing safety, traffic, environmental and land-use objectives. The strategy allows for consistent local safety objectives to be developed for each part of the area.

In terms of local area-wide safety schemes, the strategy approach is based on the Urban Safety Project carried out by the Transport Research Laboratory and two Universities, in collaboration with five local highway authorities (Mackie *et al*, 1990). This showed that, if a new approach to reducing accidents were adopted nationally in urban areas, about 15,000 injury-accidents could be avoided each year.

The IHT Guidelines on Urban Safety Management (IHT, 1990b) proposed that local highway authorities adopt a safety management strategy for each of their urban areas and they describe the use of modest highway and traffic engineering measures, akin to traffic calming (see Chapter 20), to improve road safety on an area-wide basis through local area safety schemes.

Safety Management within each Urban Area

Within each urban area, the aim should be integrate all activities affecting safety. Both the direct effects of safety programmes and the indirect effects on safety of other policies should be taken into account. Having developed an area strategy, more detailed studies are needed to develop safety objectives for each local area (see Section 16.10). The strategy is represented in Figure 16.1.

Principles of good safety management

The potential for reducing deaths, injuries and damage in accidents on urban roads and helping people to feel safer in traffic can be realised by applying ten principles. These are:

❑ to consider all kinds of road-user, especially the most vulnerable;
❑ to consider the functions and use of different kinds of road;
❑ to formulate a safety strategy for each urban area as a whole;
❑ to integrate existing accident-reduction efforts into the safety strategy;
❑ to relate safety objectives to other objectives for the urban area;
❑ to encourage all professional groups to help to achieve safety objectives;
❑ to guard against adverse effects of other programmes upon safety;
❑ to use the scarce expertise of road safety specialists effectively;
❑ to translate strategy and objectives into local area safety schemes; and
❑ to monitor progress towards safety objectives.

Implementation of good safety management

The focus for the implementation of good safety management is the adaptation of the way in which the road network is perceived and used. The process involves four steps:

❑ to identify the current and possible future hierarchy of primary and district distributors, local distributor and access roads, together with associated pedestrian and cycle- routes;
❑ to appraise the extent and characteristics of all recent accidents and the public perception of safety on all parts of the network;
❑ to assess traffic flows and performance on each route, in relation to the functions expected from its role in the hierarchy; and
❑ to set safety objectives for each part of the road network.

Road safety plans

Development of a road safety plan, setting out a strategy for casualty-reduction, is a key recommendation of the LAA Code of Good Practice. Furthermore, such a plan is now expected to accompany any application for funding for local safety schemes by the Department of Transport.

Road safety plans provide a means of stating, clearly, objectives, targets and actions. They almost invariably cover the role of road safety officers and of safety engineers but may extend to other departments. They should also monitor progress. A guide to good practice has been developed on the basis of the experience of local authorities throughout England, Scotland and Wales (Oscar Faber TPA, 1993).

Category of road	(Pedestrian casualty–rate per 100 million–km walked)			
	All ages	Children		Elderly People
		5–9 years	10–15 years	65+ years
Primary and district distributor roads	1026	1981	3634	2067
Local distributor roads	487	2027	883	310
Residential access roads	164	475	375	146

Table 16.3: Pedestrian casualty–rates on different types of road.   Source: Ward *et al* (1994).

16.8 Accident Analysis and Investigation

A systematic approach to accident analysis and investigation makes best use of resources for improving road safety. The basis for the management of such a system is outlined in the IHT Guidelines Accident Reduction and Prevention (IHT, 1990a). The objectives are:

❑ the application of cost–effective measures on existing roads, as a basis for accident reduction; and

❑ the application of safety principles in the provision, improvement and maintenance of roads, as a means of accident prevention.

The former is achieved through detailed accident investigation and the latter through the operation of a safety audit (see Section 16.11).

Accident investigation essentially comprises a phased operation of:

❑ identification of problems from accident analysis;

❑ diagnosis of sites and situations;

❑ selection of treatment;

❑ design and implementation of measures; and

❑ evaluation of net benefits;

The basis of the structured system is the use of four main approaches, set out below.

❑ **Single sites**: treatment of specific sites or short lengths of road at which accidents cluster.

❑ **Mass action**: application of a particular type of remedy to locations having common accident factors.

❑ **Route action**: application of remedies to a length of road having above average accident–rates for that type or class of road.

❑ **Area action**: aggregation of remedial measures over an area with an accident–rate above a pre–determined level.

All these actions are appropriate to urban situations. Their application in Local Safety Schemes and relative returns are indicated in Section 16.10.

Detailed procedures, including techniques for investigation, effectiveness of available remedial measures and needs for monitoring and evaluation are laid down in RoSPA's Road Safety Engineering Manual (RoSPA, 1992). A summary of accident remedial measures appropriate to different situations and an indication of potential savings in accidents is given in Table 16.2. Other sources of information on the safety benefits of a range of measures are summarised in a series of reports compiled by the County Surveyors' Society (CSS, various).

16.9 Involvement of Different Kinds of Road–User

While the accident investigation approaches, outlined in Section 16.8, are generally aimed at seeking remedial measures of an engineering nature, the involvement of different kinds of road–users must not be overlooked. Particularly vulnerable in urban areas are young and elderly pedestrians, young cyclists and motorcyclists. The mass action approach is useful to identify locations where additional facilities for pedestrians and cyclists are desirable but it is also applicable to publicity and road–user training schemes, to identify the needs for guidance on crossing strategies and advice on how to use roads more safely.

Particular attention needs to be paid to the risks for pedestrians on different types of road. An in–depth study in Northampton (Ward *et al*, 1994) sheds light on casualty–rates per distance walked. Table 16.3 indicates overall rates for primary and district distributor roads, local distributor and residential

Local Safety scheme approach	Accident–reduction (before – after)/before	First year economic rate of return (FYRR)
Single sites	33%	50%
Mass action	15%	40%
Route action	15%	40%
Area action	10%	20%

Table 16.4: Typical expected benefits from the four approaches (based on results for 1985–95).

access roads and comparable rates for the more vulnerable, young and elderly, pedestrians.

Relative to the overall rate on all types of road, accident–rates are two and a half times as high on primary and district distributor roads but less than half on residential access roads. Particularly high rates are observed for five to nine year old children on primary and district distributor roads and local distributors; and for 10–15 year old children and those aged 65 and over on primary and district distributor roads.

Specific aspects of pedestrian and cyclist safety are dealt with in Chapters 22 and 23.

16.10 Local Safety Schemes

The potential for accident–reduction in an urban area encompasses all four approaches set out in Section 16.8. The initial step in developing a framework for setting priorities for action is to make an overall assessment of the accident problem and to classify locations falling within these categories. The next step is to prioritise locations for follow–up study and diagnosis, leading to a phased plan for implementation over a period of (say) five years. With the pattern of allocation of resources to road safety engineering that was prevalent in the decade 1985 to 1995, the expected benefits from the four approaches were, at least, as shown in Table 16.4.

The first year economic rate of return (FYRR) has been adopted for small local safety schemes but for large, area–wide, schemes benefits are discounted over longer periods to calculate the Net Present Value (NPV) of each scheme (see Chapter 9).

In economic terms, extra resources should be allocated to the types of scheme for which the highest rates of return are achievable. This should continue until the highest achievable rates, from schemes not yet undertaken, are close to those obtainable in other areas of local government spending.

When first setting out a programme of local safety schemes, it is appropriate to start with single site treatments, which are likely to give the best returns at least cost, but even at the early stages it is important

to plan a programme which recognises the potential for action at the other levels. A review of local safety schemes undertaken in the 1992/93 Transport Supplementary Grant (TSG) allocation (Tootill *et al*, 1995) showed that, at that time, area–action schemes were rarely undertaken and accounted for only three per cent of the total number of schemes.

Example of area–action scheme – Cornerhall, Hemel Hempstead

Figure 16.2 illustrates how area–action treatments were implemented in the Cornerhall area of Hertfordshire, where 208 road traffic accidents involving injury were reported over a five–year period between 1986 and 1990: casualties included 48 pedestrians, 32 of whom were children. Speeds of 40 miles/h were common on a majority of the local roads in this area, with a significant amount of rat–running during the peak hours, to avoid queues on the main roads.

Applying the urban safety management approach, a clearly–defined hierarchy of roads and their functions was identified and measures were selected appropriate to those roads. Consultation to gain the views of local people resulted in few objections and widespread support for the scheme.

Implementation of the measures was phased over three years, to spread the cost of £441,000. Since completion of the scheme in 1994, accidents have been reduced by 35% and rat–running has reduced substantially.

16.11 Road Safety Audit

Road safety audit is a formal procedure for assessing the accident–potential and safety performance in the provision of new road schemes and the improvement of existing roads. The basis for audit is the application of safety principles to new scheme design, to prevent accidents occurring or to reduce their severity. It requires an objective approach to the assessment of accident–risk, through the independent checking of schemes by people unconnected with the original design but with experience and expertise in road safety engineering and accident investigation.

The practice of road safety auditing has only developed

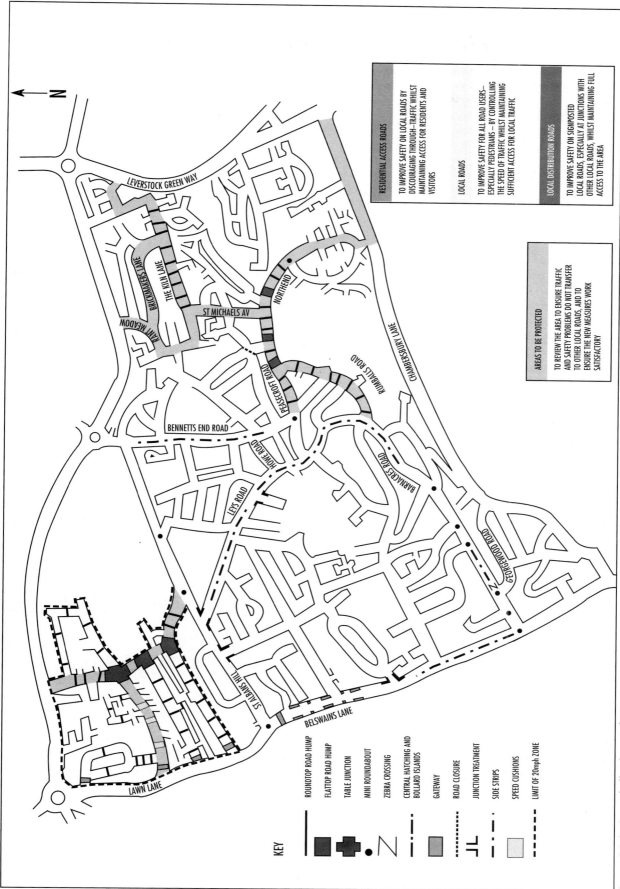

Figure 16.2: Cornerhall Urban Safety Managmement Project, Hemel Hempstead.

KEY

- ROUNDTOP ROAD HUMP
- FLATTOP ROAD HUMP
- TABLE JUNCTION
- MINI ROUNDABOUT
- ZEBRA CROSSING
- CENTRAL HATCHING AND BOLLARD ISLANDS
- GATEWAY
- ROAD CLOSURE
- JUNCTION TREATMENT
- SIDE STRIPS
- SPEED CUSHIONS
- LIMIT OF 20mph ZONE

RESIDENTIAL ACCESS ROADS

TO IMPROVE SAFETY ON LOCAL ROADS BY DISCOURAGING THROUGH–TRAFFIC WHILST MAINTAINING ACCESS FOR RESIDENTS AND VISITORS

LOCAL ROADS

TO IMPROVE SAFETY FOR ALL ROAD USERS– ESPECIALLY PEDESTRIANS – BY CONTROLLING THE SPEED OF TRAFFIC WHILST MAINTAINING SUFFICIENT ACCESS FOR LOCAL TRAFFIC

LOCAL DISTRIBUTION ROADS

TO IMPROVE SAFETY ON SIGNPOSTED LOCAL ROADS, ESPECIALLY AT JUNCTIONS WITH OTHER LOCAL ROADS, WHILST MAINTAINING FULL ACCESS TO THE AREA

AREAS TO BE PROTECTED

TO REVIEW THE AREA TO ENSURE TRAFFIC AND SAFETY PROBLEMS DO NOT TRANSFER TO OTHER LOCAL ROADS, AND TO ENSURE THE NEW MEASURES WORK SATISFACTORY since 1990, when the first IHT Guidelines and DOT Standard were published. In 1991, safety audit of trunk roads became mandatory and, at the same time, many local authorities started undertaking audits on local roads in both urban and rural areas [NIg]. Subsequent practice is reviewed in Review of Road Safety Audit Procedures (Crafer, 1995), which provided a basis for revised IHT Guidelines for the Safety Audit of Highways (IHT, 1996). The DOT standard has also been revised and incorporated into the Design Manual for Roads and Bridges (Volume 5, Section 2).

Road safety audits should be an integral part of highway planning, design, construction and maintenance. To ensure that all highway schemes operate as safely as practicable, safety should be considered throughout the whole preparation and construction of any project. The IHT Guidelines outline the principles, procedures and practice of a road safety audit and make recommendations on good practice. Safety audits for grade–separated junctions are discussed in Chapter 43 and safety considerations in the design and evaluation of performance at junctions are discussed in Chapters 38, 39 and 40.

16.12 Education, Training and Publicity

Education, training and publicity programmes use educational methods, skills, training schemes and publicity activities, often in conjunction with other professionals and volunteers, to seek to influence attitudes and bring about safer behaviour. A range of programmes is needed to address different groups and interests. The main responsibility for implementing these programmes lies with Road Safety Officers. The RoSPA publication Road Safety: A Managers Guide (RoSPA, 1996) provides a practical guide on methodology and sources of relevant information.

Road safety education for children

Road safety education in schools can play an important part in developing children's attitudes and behaviour in a way which should make a positive impact on safer behaviour for life. A comprehensive programme covering all aspects of the safe use of roads should be developed jointly between road safety officers, teachers and other agencies, including the police and health promotion officers. An essential requirement is that it must be amenable to integration into the national curriculum.

Good practice guidelines for education of 5–16 year olds have been issued (DOT, 1994) and major

demonstration trials have shown that road safety education can provide a real and very relevant context for the national curriculum (Sykes *et al*, 1995). To be effective, pupils should receive small, but frequent, regular and purposeful inputs. An imaginative lead is required but one which also reflects realism, sensitivity and awareness to current pressures within the primary and secondary schools. Photograph 16.1 illustrates an appropriate resource for 8–11 years olds, developed on these principles (Clayton *et al*, 1995).

Important elements in promoting road safety education in schools are:
❑ school–based in–service training (INSET), which is by far the most effective way of promoting good practice in road safety education;
❑ road safety officers liaising with the major organisations, to provide advisory services to schools;
❑ road safety officers taking the lead in developing and maintaining liaison with other support agencies, such as the police and health promotion officers;
❑ evaluating and keeping records, which should be a priority for both education co–ordinators and the road safety department;
❑ road safety 'theatre' in schools, such as 'Drivetime' for the pre–driver age–group; and
❑ teachers and road safety officers being aware that needs and opportunities are arising

Photograph 16.1: On the Move... teaching resources for eight to 11 year olds.

continually for the further development and updating of road safety education.

Most importantly, the success of any programme for road safety education in schools depends on the commitment and co-operation of all concerned. In preparation for such school education, pre–school children can benefit from programmes such as Traffic Clubs, developed to help parents to guide their children's first footsteps (Bryan–Brown, 1994) [Wb].

Young drivers' training
New initiatives (DOT, 1995) have been directed at young drivers, who are disproportionately involved in accidents. These include:
❏ re–testing of all relatively–new drivers convicted of a serious driving offence, within a specified period of passing the driving test;
❏ post–test training of young drivers;
❏ an enlarged element of theory in the driving test; and
❏ education of young people of below driving age.

Influencing road–users' behaviour
Education programmes, mounted at both local and national level, should aim to raise the levels of understanding and awareness of road–users to road safety issues and to influence their behaviour. Publicity campaigns led by central Government are supported by local authorities but local campaigns in themselves have an important role. Local publicity programmes should complement traffic engineering measures aimed at accident–reduction. They should be designed to achieve specific objectives, in terms of improved behaviour, knowledge and attitudes for specific target groups of road–users. Priority objectives and target groups should be determined by rigorous accident–analysis and the impact and effectiveness of campaigns evaluated.

Other important applications of education programmes relate directly to: engineering and enforcement programmes; the consultative element in safety engineering schemes, especially in urban safety management (see Section 16.7); the informative and persuasive element in enforcing legislation; and the use of new technology (see Section 16.13).

16.13 Enforcement

The most successful road safety legislation in recent years has been that on drinking and driving. First, the 1967 legislation, which introduced the 'breathalyser', brought about a substantial reduction in deaths and injuries almost overnight (more than 10% of the national casualty toll) but the benefit was short-lived.

Secondly, the 1983 legislation, which simplified procedures and was accompanied by new technology for breath–testing, resulted in a steady decline in drink–related casualties and to a halving of drink–related deaths over the next decade. The essential difference between the two situations was that enforcement of the 1983 legislation was accompanied by continued high–profile publicity and education campaigns, while the 1967 legislation relied on enforcement only. Drinking and driving has now become socially unacceptable.

One of the main concerns in road safety is how to control speed. Not only do many drivers exceed speed–limits but many accidents arise from drivers going too fast for the situation even below the limit. Initiatives to combat this problem include: the use of speed–detection cameras, in both urban and rural areas, to help to enforce speed limits; the use of cameras to detect red light running offenders; and the fitting of speed–limiters to buses, coaches and heavy goods vehicles. However, the level of non–compliance is often so great that the available resources are quite inadequate to process the large number of offenders. Enforcement cannot be fully effective unless major changes in the attitude of road–users to speeding can be effected, through extensive publicity and education.

In urban areas, where 20 miles/h limit zones have been introduced, the requirement to have appropriate physical measures to calm traffic acknowledges the impossibility of enforcement by police presence alone. Traffic calming and the management of speed are dealt with in Chapter 20.

16.14 Relevance of Developments in Vehicle Safety

Good vehicle design and construction are essential for road safety. Vehicles are increasingly being designed to help prevent accidents, for example through better lighting and braking. They are also being designed to prevent or minimise injury, through such devices as air–bags and improved restraints.

Of particular relevance to urban safety are developments to improve the protection of pedestrians through better design of the fronts of cars, for example through the elimination of sharp edges, optimum shape and crush characteristics of front–end design. It is at the lower range of speeds (20 miles/h and less) that these improvements have greatest effect, adding urgency to the need to use traffic–calming measures to lower speeds of traffic in urban areas.

Interactive control of speed through the use of speed–limiters, activated remotely by roadside control systems using short–range communications links, has potential for reducing speeds and casualties in urban areas.

16.15 Valuation of Accident–Prevention

The valuation of traffic accidents and casualties is regularly updated and reported (DOT, 1996b). The methodology has been developed to give more meaningful valuations on human costs, resource costs and medical costs. The elements costed (in decreasing order of magnitude) cover human costs, lost net output, property damage, medical and ambulance services, police services and insurance administration.

Separate average valuations are made for built–up and non–built–up roads and motorways and for different levels of severity. Average values per accident in built–up areas (at June 1995 prices) are:

Fatal	£87,6440
Serious injury	£10,5470
Slight injury	£10,380
All injury	£32,160
Damage–only	£1,020

The overall average value of preventing an injury–accident on built–up roads, together with an allowance for the cost of unreported vehicle–only crashes, is estimated to have been £50,250 at June 1995 prices.

16.16 Appraisal of Safety Measures and Schemes

Appraisal of the effectiveness of different kinds of safety measure and of individual remedial schemes is important to provide feedback, to improve future procedures and practice.

For engineering schemes, the main criterion for assessment is change in frequency of accidents or casualties. The most widely–used technique is by before–and–after analysis, an essential element of which is comparison of changes which introduce an appropriate control, matched for type of site or area.

Factors which can affect the validity of any comparison are:
❑ changes in other factors, not associated with the scheme treatment, which may also cause changes in accident frequency. For example, changes in

legislation or national publicity campaigns or changes in adjacent infrastructure, which might produce marked changes in traffic flows; and
❑ 'regression to mean' effects. Within the statistical fluctuation expected in average values over a number of years, locations with high levels of accidents in one year have a tendency to exhibit lower levels in the following year.

Thus, choice of before–and–after periods (at least three years of each) and choice of control data are critical to the soundness of the evaluation. Suitable statistical tests for different types of scheme are well documented – see the RoSPA Road Safety Engineering Manual (RoSPA, 1992).

For area–wide schemes in particular, it is also desirable to aid interpretation of changes by making before–and–after comparisons of a series of factors, not simply accident or casualty numbers. Of particular relevance are measurements of speed distribution, traffic flow and composition, travel times and public perception of the safety of the scheme. It is also important to examine whether the scheme has led to an increase in accidents, traffic speeds and volumes in adjacent areas and to allow for these consequential changes in the evaluation.

For education, training and publicity activities, the effect on accidents or casualties is likely to be long–term and appraisal of schemes in these terms is difficult. Nevertheless, it is important to undertake before–and–after evaluation in terms of other appropriate measures. These may include questionnaire surveys of awareness of problems and changes in attitudes towards safety issues, as well as observations of behaviour on the road.

16.17 Monitoring and Evaluation

Monitoring enables overall accident–trends in an area to be compared with national trends and targets. It also gives early warning if something goes wrong with an individual scheme. Monitoring the effect of particular measures at a number of sites makes it possible to build up local control data that can improve safety engineering work in the future.

Monitoring should be carried out at two levels: for the whole region covered by the local highway authority and for individual safety schemes or groups of schemes.

Regional Monitoring
In regional monitoring, it is useful to disaggregate the accident data into groups of factors and locations that

may help to highlight particular problems. The main groups should cover:

- ❑ where? – accidents per district, class of road, speed–limit, junction or non–junction;
- ❑ who? – type and class of road–user and severity of injury;
- ❑ what? – type of vehicle involved and type of collision;
- ❑ when? – time of day, day of week, month; and
- ❑ why? – state of the road (dry, wet, icy) and light (daylight, darkness).

Trends should be monitored and reported annually, to provide a basis against which all road safety workers can judge performance.

Scheme monitoring

Monitoring of individual schemes is essential for three main reasons. These are:

- ❑ to ensure that, if an accident situation worsens following implementation of measures, further steps are taken to reverse the situation;
- ❑ to quantify the change in accidents and to discover if there is any change over time in the effectiveness of the treatment; and
- ❑ to evaluate the benefits of the scheme, in relation to the original objectives.

Monitoring of schemes should add to the overall knowledge and experience of accident investigation and should help to build up control–data for future appraisals. A summary list of individual schemes, grouped by type, should be produced on a regular basis.

16.18 References

Bryan–Brown K (1994)   TRL Project Report 99 'The effectiveness of the General Accident Eastern Region Children's Traffic Club', TRL.

Carsten O M J, Tight M R, Southwell M T, and Plows B (1989)   'Urban accidents: why do they happen?', AA Foundation for Road Safety Research.

Clayton A B, Platt V, Colgan M, and Butler G (1995)   'A child–based approach to road safety education for 8–11 year olds', AA Foundation for Road Safety Research.

CSS (various)   'Reports from the Accident Reduction Working Group' (formerly SAGAR), County Surveyors' Society.

CSS (1987)   'Small and Mini–Roundabouts', County Surveyors' Society.

CSS (1988)   'Road Surface Treatments', County Surveyors' Society.

CSS (1989)   'Automatic Traffic Signal Installations', County Surveyors' Society.

CSS (1990)   'Street Lighting Installations', County Surveyors' Society.

Crafer A (1995)   'Review of road safety audit procedures' Occasional Paper No 1, The Institution of Highways & Transportation.

DOH (1992)   'Health of the nation: A strategy for health in England', Stationery Office.

DOH (1993)   'Health of the nation: Key area handbook Accidents', Department of Health.

DOT (1987)   'Road safety: the next steps', DOT.

DOT (1989)   'Road casualty reduction annual progress report 1988/89', DOT.

DOT (1994)   'Road safety education 5–16: Good practice guidelines' DOT.

DOT (1995)   'Road safety report 1995', DOT.

DOT (1996a)   'Road accidents Great Britain: The casualty report', Stationery Office.

DOT (1996b)   '1995 Valuation of Road Accidents and Casualties', DOT.

HMG (1980)   'The Highways Act 1980', Stationery Office [Sa].

HMG (1988)   'The Road Traffic Act 1988', Stationery Office.

IHT (1990a)   'Highway safety guidelines: Accident reduction and prevention', The Institution of Highways & Transportation.

IHT (1990b)   'Guidelines for Urban safety management', The Institution of Highways & Transportation.

IHT (1996)   'Guidelines for the Safety Audit of Highways', The Institution of Highways & Transportation.

James H F (1991)   'Under–reporting of accidents', Traffic Engineering + Control 32(12).

LAA (1989)   'Road safety code of good practice', Association of County Councils.

LAA (1996)   'Road safety code of good practice', Association of County Councils.

Mackie AM, Ward HA, and Walker RT (1990)   TRL Research Report 263 – 'Urban safety project 3: Overall evaluation of area wide schemes', TRL.

Oscar Faber TPA (1993)   'Road safety plans: A guide to good practice', DOT.

PACTS (1995)   'Targets 2001: Where do we go from here?' Conference Proceedings, Parliamentary Advisory Council for Transport Safety.

RoSPA (1992)   'Road safety engineering manual', Royal Society for the Prevention of Accidents.

RoSPA (1996)   'Road Safety: A Managers Guide', Royal Society for the Prevention of Accidents.

Sykes J, Broome W, O'Leary K, and Harland G (1995)   TRL Report 148 'Road safety education: Good practice in Hertfordshire', TRL.

Tootill W J and Mackie A M (1995)   TRL Report 127 'Transport supplementary grant for safety schemes: Local Authorities' schemes from 1992/93 allocations', TRL.

Ward H, Cave J, Morrison A, Allsop R, Evans A, Kuiper C and Willumsen L (1994)   'Pedestrian activity and accident risk', AA Foundation for Road Safety Research.

Chapter 17 Environmental Management

17.1 Introduction

It is now recognised that many towns and cities cannot provide the amounts of road–space necessary to accommodate unrestricted traffic growth. Consequently, the need to manage the demand for travel is more widely accepted (see Chapter 21). Moreover, there is widespread public concern with the global, as well as local, damage caused to the environment by vehicular traffic. The transport policy implications of this are discussed in Chapters 1 and 6.

Traffic management has an important role to play in managing the environmental impact of vehicular traffic. Environmental management schemes are employed in town centres and residential areas to limit the more apparent and potentially damaging effects of vehicular traffic on vulnerable road–users and residents. The schemes are usually very localised and generally discourage the use of vehicles, either by banning them or by making their use more costly or difficult. However, these measures may have other impacts, which are not immediately apparent and which may even be damaging to the environment in other ways. Accordingly, the total environmental impact needs consideration and proposed schemes should not be assessed in isolation or in terms of just the local area.

The effects on the environment must also be considered in the development of all traffic management or transport improvement schemes. Any scheme which changes traffic patterns will have an effect on the environment, even if not specifically designed to do so. Practitioners need to be aware of the total impact of their actions, particularly where options are being evaluated, to ensure that the outcome contributes to the global objectives of achieving sustainability, as well as addressing the local dimension (see Chapter 3).

17.2 Objectives of Environmental Management

Many of the techniques examined in this chapter have commonly been labelled 'environmental management' and concentrate on alleviating local problems. Many local environmental management schemes promote the key recommendations in the 17th Report of the Royal Commission on Environmental Pollution (RCEP, 1994) and, in particular, the recommendation to '...improve the quality of life, particularly in towns and cities, by reducing the dominance of cars and lorries and providing alternative means of access'.

However, in its wider context, environmental management should examine all aspects of sustainability. Environmental management ultimately deals with wider issues, including strategic transportation policy, by, for example, the promotion of walking, cycling and public transport, together with improved air quality. Consequently, environmental management should be a key mechanism for achieving a sustainable environment, whilst providing acceptable levels of accessibility for all users, in the context of global, national and local guidelines.

17.3 Framework for Environmental Management Schemes

Global and National Guidelines
Agenda 21 defines in detail the global strategy for moving towards a sustainable future in the 21st Century, as set down by the Rio Earth Summit in 1992 (UNCED, 1992b).

The European Commission's strategy is set out in the EC Fifth Action Programme on the Environment (EC, 1992), which proposed a strategy to reduce transport needs, improve network coordination, support environmentally–friendly modes, integrate public transport modes and reduce the use of cars. The EC policy on sustainable mobility was detailed further in an EC White Paper published in 1993 (EC, 1993), which supplements the programmes in the Fifth Action Programme.

UK national policy planning guidance is set out in the White Paper 'This Common Inheritance' (DOE, 1990 to 1996), which introduced the concept of stewardship of the environment and which applies directly to transportation planning and traffic management. This is reviewed on an annual basis. Sustainable Development – the UK Strategy (DOE, 1994) set out the British Government's approach and recognised the importance of developing indicators by which progress towards sustainability can be judged. The Government has produced a preliminary set of indicators of sustainable development (HMG, 1996a).

PPG13 (DOE/DOT, 1994) [NIa] [Sa] is the principal source of guidance to local authorities on the content of, and the integration between, transportation plans and land–use policies. The guidance emphasises three main aims:

❑ to reduce growth in both the length and number of motorised journeys;

❑ to encourage alternative means of travel which have less environmental impact; and

❑ to reduce reliance on the private car.

PPG6 (HMG, 1996b) [Sb] on Town Centres and Retail Developments sees fostering development in town centres as having an important role in reducing the need to travel and the reliance on cars. PPG6 [Sb], like PPG13 [Sa], advises against development which is likely to add significantly to the overall number and lengths of car trips.

In 1995, the Government published its strategic policies for air quality management, Air Quality – Meeting the Challenge (DOE et al, 1995), which sets out conclusions on the policies needed to prevent the UK's air quality from deteriorating. The Environment Act 1995 (Part IV) (HMG, 1995) sets terms for a National Air Quality Strategy to be developed by the Government (DOE, 1996) and places a duty on local authorities to address air quality problems and to produce Action Plans for 'Environmental Management Areas' where needed.

Local Guidelines

One of the key elements of the Development Plan process, involving Unitary Development Plans (UDP) [NIb], Structure Plans or Local Plans, as indicated in PPG13 [Sa], is to ensure integration between transport considerations and land–use policies. Taken together, the plans should cover various aspects, from strategic transportation issues through to more detailed policies on traffic and transport, which reflect the underlying philosophy both of national advice on sustainable development and local priorities as reflected in Transport Plans. They should also contain a wide range of environmental policies, on topics such as air quality, visual impact, impact on heritage features, landscape considerations and noise (see also Chapter 3).

Transport Plans identify the priorities for action on the measures required to encourage alternatives to private cars and to restrain their use. The creation of targets, against which progress can be monitored, can

be an important part of the process and can include environmental management targets (see also Chapter 6).

The concept of Local Agenda 21 (LA21) (UNCED, 1992a and 1992b) is a community–based approach to setting the agenda for sustainable development. More than half of the actions identified in Agenda 21 are the responsibility of, or are significantly influenced by, local government. In the UK, the Local Government Management Board (LGMB) is co–ordinating work on LA21 and had aimed at establishing, by the end of 1996, local agendas to promote sustainable development within communities. One aspect of their work is the development of 'sustainability indicators', which includes the theme that '...access to facilities, services, goods and other people is not achieved at the expense of the environment or limited to those with cars'. A number of indicators reflecting this theme are identified in the Framework for Local Sustainability (LGMB, 1993).

17.4 The Role and Purposes of Targets

Establishing targets can be a useful tool in policy development and evaluation and is stressed by the Royal Commission on Environmental Pollution's (RCEP) report on Transport and the Environment (RCEP, 1994). Target–setting can have considerable value as an objectives–based approach towards environmental management, so long as the full economic implications of achieving each target level are understood. Moreover, they must be practical and achievable.

National and Local Target–Setting

A number of individual national targets have been adopted following international agreement at the Earth Summit (1992) or via European Commission Directives (for example, those concerning limits on 'greenhouse' gas emissions). The Royal Commission on Environmental Pollution (RCEP, 1994) suggested a series of targets concerning transport policy and local planning authorities are encouraged to develop their own targets, covering areas such as the overall need to travel, location of development, modal shift, accessibility and safety (SCC, 1995).

Types of Target

A distinction can be made between two different types of targets and how they are developed. On the macro scale, targets can be set at the international or national level, such as carbon dioxide emission standards set at the Rio Earth Summit (UNCED,

1992b). 'Top–down' planning systems can then translate these into local authority targets, for example in Transport Plans and local Agenda 21s. Although individual schemes may make only a marginal contribution to the overall international target, the cumulative effect can have a marked impact. At the same time, local conditions can be improved and good practice established.

The reverse process can also be followed, with targets set locally, possibly through public participation. These targets can be scheme–specific, varying within the limits of absolute targets defined in the Transport Plans or local Agenda 21s. This process can loosely be seen as 'bottom–up' planning. Environmental management schemes have traditionally followed this route.

The Value Of Targets

The Local Government Management Board (LGMB, 1993) identified the benefits from setting targets, as follows:

❑ targets can articulate a vision and can begin to move perceptions towards it. For example, if targets were expressed to reduce overall levels of car mileage or car dependence, this could help to move perceptions towards more sustainable transport habits;

❑ targets can initiate and focus a debate. For example, the Government's target for recycling 25% of domestic waste generated a wide debate among local authorities and waste management companies. Thus, the actual process of setting targets helps to focus on the likely outcome or policies; and

❑ targets can encourage technological change and give an incentive to private industry to research and develop the means of achievement. For example, the Government's declaration of the target of increasing energy efficiency by 20% caused an upsurge in industrial investment in energy–efficiency technologies in the early 1980s.

The use of targets means that performance can be monitored by both professionals and the general public. Monitoring can have substantial time and resource implications, especially if targets are complex and/or numerous. Nevertheless, targets can provide a basis for driving projects and actions, for evaluating the environmental impact of schemes and for highlighting how individual items relate to the overall plan and to each other.

Limitations on the Effectiveness of Targets

It is vital that targets are realistic and achievable. Targets that are unlikely to be met run the risk of losing credibility and commitment. It may be argued that target–setting is inappropriate, on the basis that achieving a number of small changes is more important than reaching a specific level on one particular dimension. Moreover, devoting resources to the pursuit of one target is likely to be at the expense of pursuing others. Without estimating the marginal economic return from achieving each target separately, it is almost impossible to allocate limited resources to several different targets in an efficient way. For this reason, targets need to be re–evaluated regularly and, where appropriate, progressively increased from low initial levels towards an optimum rather than an ideal.

Examples of national and local targets are set out below:

National Targets:

❑ a recommendation by RCEP to reduce emissions of carbon dioxide from surface transport by the year 2000 to the 1990 level (RCEP, 1994);

❑ a recommendation by RCEP to reduce daytime exposure to road and rail noise to not more than 65dBA (18h L_{10}) at one metre from the facades of houses (RCEP, 1994); and

❑ approved Government policy to reduce total road casualties by the year 2000 by 30% from the annual average between 1981 and 1985.

Local Targets:

❑ to halve the number of child casualties by the year 2000 (A Transport Strategy For Lothian);

❑ to reduce the number of residents subjected to daytime road traffic noise, over a level defined in 1996, by 1% by 2006 (A New Transport Plan for Surrey); and

❑ a doubling, to five percent, by the year 2011, of the 1991 proportion of residents travelling to work by bicycle (Buckinghamshire County Council – Draft Integrated Transport Strategy).

17.5 Environmental Management Measures

Schemes such as pedestrianisation, park–and–ride and area–wide traffic management have been the basis of many environmental management schemes. The assessment has generally focused on the original objectives and, therefore, concentrated on the immediate benefits in terms of both location and time. However, studies which have examined the wider and longer term implications of such schemes have shown that the short–term local benefits have, in certain circumstances, been undermined by other unforeseen disbenefits. Therefore, when assessing the environmental impact of a specific scheme, consideration should not be confined to just the immediate area and the short term but the effects should also be evaluated over a wider area and for the longer term. The significance of individual impacts will vary according to the location, the time and for other reasons; for example, air quality in areas of high pedestrian activity, visual intrusion in conservation areas and night–time noise in residential areas. Such variations in sensitivity should be considered in the evaluation of measures.

When evaluating schemes on this broader basis, the process and the issues can become complex. Traffic noise is an example. It is perceived as more of an intrusion where a low ambient noise level is interrupted by high peaks. Whilst diverted traffic will create the same amount of noise at similar speeds, it will not be so apparent or considered such an intrusion if it is displaced onto a road with higher ambient noise levels.

With the requirement for local authorities to examine the air quality in their area (HMG, 1995), environmental management measures can be air–quality led, thus defining new objectives. However, to achieve an overall improvement in air quality, everything else being equal, a reduction in the total volume of pollutants being discharged into the air is required. A scheme which diverts traffic to an already congested road may improve conditions in the immediate 'calmed' area but, if it adds to the congestion on the alternative route, overall, more pollutants may be discharged into the atmosphere. Therefore, overall improvement in air quality can only be achieved in the short–term by a reduction in vehicle mileage or by reducing delays. Other factors must also be considered. If, by reducing delays, additional traffic is induced, the net effect on vehicular delays could be negative. Conversely, an increase in delays to vehicles, caused by the provision of facilities for pedestrians, cyclists or public transport, may increase pollution in the short–term but, in the long–term, if it were to achieve a modal shift, could have a positive net effect.

The above examples illustrate the complexity of evaluating the consequences of applying environmental management measures and that the task of setting targets and monitoring their effects can be complicated. It also shows that remote effects are important and should be assessed as part of the value of a local scheme.

Environmental Factors
Environmental management should be concerned with the whole picture of sustainability. When consulting on quality of life issues in a neighbourhood, some recurrent themes are found. Some of these relate to

specific local impacts arising from any proposed scheme and are briefly described below. Further information can be found in The Design Manual For Roads And Bridges – Volume 11 (DOT, 1993).

Noise
Traffic noise comes from the interaction of tyres with the road surface; from engines, exhausts, brakes and unsecured loads; and, in respect of goods vehicles, from vibration and body–rattle. Noise is accentuated by braking and acceleration, by high speed and by travelling in low gears. Situations where there are noticeable peaks against a low ambient background noise are perceived as a particular problem. Vehicle maintenance, driver behaviour and road surface conditions all have an additional influence.

Vibration
There are two forms of vibration; air–borne vibration caused by low frequency sound waves and ground vibration caused by the contact of the vehicle with the ground. Vehicle design, engine speed, road surface condition and the underlying ground strata all have an influence.

Air Quality
Vehicular traffic emits a mixture of chemicals that can damage the environment and may be harmful to health. They include oxides of carbon, oxides of nitrogen, sulphur dioxide, hydrocarbons (including benzene), 1,3–butadiene and fine particles. Under the Environment Act 1995 (HMG, 1995), local authorities are responsible for local air quality management within the framework of the National Air Quality Strategy (DOE, 1996). Traffic management measures may change vehicle flows and operating conditions in such a way that emissions are increased or decreased (DOT, 1996) [Sc].

Visual intrusion
Visual intrusion is a subjective factor and difficult to measure. It relates to traffic (composition, volume, whether parked or moving) and to the quality of the built environment (street furniture including traffic signs, materials, lighting source) both in terms of quality and volume.

Accident–Risk
Actual accident–risk is an important aspect of a good quality environment. Personal injury accidents are recorded but consideration should also be given to damage–only crashes, to the high percentage of unreported accidents and to vulnerable road–users (see also Chapter 16).

Severance
Heavily–trafficked roads through a community can

divide that community and change the social behaviour of the people affected. In particular, it can affect the way in which people choose to make trips, particularly where the road is seen to be a barrier to safe walking and cycling and a cause of detours and delays. The introduction of safe facilities for pedestrians, such as signal–controlled crossings or traffic–calming measures, which slow the speed of traffic, can reduce community severance.

Parked Vehicles

Inconsiderate parking causes danger and nuisance and reduces pedestrians' amenity. When parked on footways, vehicles inconvenience and endanger pedestrians and damage the fabric of the environment. Heavy goods vehicles' activity can result in visual intrusion and excessive noise and pollution. This is most disturbing during unsocial hours (see Chapter 19).

Intimidation.

The perception of accident–risk and the fear and intimidation caused by large fast–moving vehicles are detrimental to the quality of life. The situation is made worse where footways are narrow and heavily used.

Practical Application

The importance and weighting given to particular factors will be for individual authorities to determine,given their local objectives and targets, together with the more global objectives and targets that may be set out in their Transport Plan. The success or otherwise of a scheme is determined by assessing the performance of the scheme against global, national and local targets relating to the factors listed.

The following examples in Table 17.1 indicate some of the environmental benefits (+) and disbenefits (–) resulting from introducing typical environmental management techniques. They are an indication of the complex issues involved in assessing a scheme and are not intended to be exhaustive.

17.6 Monitoring and Review of Environmental Measures

As with traffic management, environmental measures need to be reviewed to assess their effectiveness. The review process should identify where further studies or modifications may be needed following the initial implementation of environmental measures. The approach could be based on the application of evaluation criteria, the use of guidelines, consultation, public participation and environmental monitoring systems. This information may be published as part of a review report, stating clearly the objectives or targets against which the effectiveness was measured and whether the implemented measures are still effective and appropriate. This process may be used as a tool for the measurement of a scheme's sustainability by:

❑ assessing the effectiveness of environmental measures against objectives or targets, by comparing original estimates with the performance once the measure becomes operational;

❑ detecting adverse effects of measures at an early stage, so that remedial action can be taken before changes becomes a serious problem requiring time–consuming and expensive mitigating action; and

❑ using information obtained to add to knowledge and to encourage best environmental practice.

The review and monitoring process itself should be evaluated, from time to time, to ensure that it continues to meet appropriate objectives. Resource constraints may require monitoring of environmental measures to be selective, concentrating on those aspects that give rise to the most concern and for which tried and tested techniques are available. The use of readily–available data as a proxy may be considered in appropriate cases, for example traffic flow and composition by engine–type as an indicator of local air quality. Results from the review and monitoring process may help to improve methodologies for appraising schemes prior to their introduction.

Table 17.1: Positive and negative outcomes of environmental measures.

Factors	Positive Outcomes (+)	Negative Outcomes (−)
(a) Traffic Calming (see also Chapter 20 and photographs 17.1 and 17.2)		
Noise	❑ Where traffic volumes are reduced without an increase in speed. ❑ Where traffic is displaced to roads with already high noise levels (unless noise is increased to an unacceptable level). ❑ Where traffic speeds are reduced	❑ Where vehicles are caused to accelerate and decelerate ❑ Where vehicles are caused to use a low gear ❑ Where insecure vehicle loads are displaced.
Vibration	❑ Where the weight and volume of vehicular traffic are reduced. ❑ Where the location of road humps avoid proximity to residential properties.	Ground borne: ❑ Where ramp gradients are too steep ❑ Where road humps are located close to properties on soft soils Air borne: ❑ Where vehicles are caused to travel in low gear ❑ Where vehicles are caused to accelerate and decelerate
Air Quality	❑ Where a scheme achieves a modal shift ❑ Where traffic volumes are reduced without detrimental change in speed	❑ Where vehicles are caused to travel in low gear ❑ Where vehicles are caused to accelerate and decelerate ❑ Where vehicles are diverted onto an already congested road ❑ Where total vehicle mileage is increased
Visual intrusion	❑ Where traffic volumes are reduced	❑ Design of schemes can be intrusive
Accident Risk	❑ Has been shown to reduce accidents by up to 70%	
Severance	❑ Where safe crossing points are provided ❑ Where traffic volumes are reduced	
Parked Vehicles	❑ Where parking is regulated	
Intimidation	❑ Where speed is reduced ❑ Where traffic volumes are reduced	
(b) Pedestrianisation (see also Chapter 22 and photographs 17.3 and 17.4)		
Noise	❑ Within pedestrianised area ❑ Where traffic is diverted to major traffic routes	❑ Where displaced traffic creates additional noise on alternative routes
Air Quality	❑ In the immediate area.	❑ Where alternative routes are longer and/or more congested ❑ Adds to already near congested route
Visual intrusion	❑ Where vehicular traffic is substantially reduced or eliminated	❑ Where more people are affected along the diverted route ❑ Where buses and heavy good vehicles are concentrated on particular routes
Accident Risk	❑ Where the accident rate on the diverted route is lower, especially for pedestrians and cyclists	❑ Where cyclists are forced onto heavily trafficked peripheral roads
Severance	❑ Where vehicular traffic is substantially reduced or eliminated	❑ Where the diversion route is longer and more people are affected ❑ Where traffic is diverted past schools or other sensitive land uses ❑ Where cyclists have a more circuitous route ❑ Where cyclists access to shops and services is reduced

Factors	Positive Outcomes (+)	Negative Outcomes (−)
Parked Vehicles	❏ Easier access for delivery vehicles and disabled vehicles	❏ Where parking problems are created in surrounding roads
Intimidation	❏ Where vehicular traffic is substantially reduced or eliminated	❏ Where diverted traffic uses residential roads or other sensitive routes

Note: The positive outcomes of pedestrianisation may be offset if access by buses and service vehicles is increased.

(c) Park & Ride (see also Chapter 24 and photograph 17.5)

Factors	Positive Outcomes (+)	Negative Outcomes (−)
Noise	❏ Where traffic volumes in town centres are reduced	❏ Where bus traffic is increased
Vibration	❏ Where traffic volumes in town centres are reduced	❏ Where bus traffic is increased
Air Quality	❏ Where overall vehicle mileage is reduced and congestion in the central area is reduced	❏ Where additional car trips are generated as a result of easier parking ❏ Where bus traffic is increased ❏ Where traffic queues occur
Visual intrusion	❏ Where traffic volumes in town centres are reduced ❏ Where associated priority measures allow buses a smoother journey	❏ Poor design of park and ride car park
Accident Risk	❏ Where vehicle mileage is reduced ❏ For cyclists and pedestrians where traffic volumes are reduced	❏ Where new accesses to car parks create danger particularly for pedestrians and cyclists
Severance	❏ Where traffic volumes in town centres are reduced	

(d) HGV Controls (see also Chapter 25 and Photographs 17.6 and 17.7)

Factors	Positive Outcomes (+)	Negative Outcomes (−)
Noise	❏ Where HGVs are diverted from sensitive areas	❏ Where displaced traffic creates additional problems on alternative routes
Vibration	❏ Where HGVs are diverted from sensitive areas	❏ Where vehicles are forced to accelerate and decelerate ❏ Where vehicles are forced to use low gear
Air Quality	❏ Where volumes of HGVs are removed from sensitive areas	❏ Where vehicles are forced to accelerate and decelerate ❏ Where vehicles are forced to use low gear
Visual intrusion	❏ Where volumes of HGVs are removed from sensitive areas	❏ Where overnight parking is banned
Severance	❏ Where overnight parking is banned	
Intimidation	❏ Where HGVs are removed from sensitive areas	

(e) Landscaping (see Photograph 17.8)

Factors	Positive Outcomes (+)	Negative Outcomes (−)
Noise	❏ Where it provides a barrier to noise (real or perceived)	
Air quality	❏ Appropriate planting may trap or absorb pollutants	
Visual intrusion	❏ Where good design is used and visual appearance is enhanced	
Accident Risk	❏ Where visibility is improved and drivers adopt lower speeds	❏ Where visibility is obstructed

Table 17.1 continued.

Factors	Positive Outcomes (+)	Negative Outcomes (−)
Severance	❑ Where used to segregate pedestrians and cyclists from other traffic and give them priority	
Parked Vehicles	❑ Where parking is removed or controlled ❑ Where used to provide modest barriers to noise pollution and shade for parked vehicles	
Intimation	❑ Acts as barrier to vehicles and protects pedestrians	
(f) Direction Signing (see Chapter 15 and Photograph 17.9)		
Noise	❑ Where journey lengths are reduced	❑ Where traffic along some routes is increased
Air Quality	❑ Where journey lengths are reduced	❑ Where traffic along some routes is increased
Visual Intrusion		❑ Where size and number of signs are increased and views of the townscape obscured
Accident Risk	❑ Where they provide positive advice and guidance for drivers and reduce uncertainty	
Severance	❑ Where traffic is diverted from sensitive areas	❑ Where diverted traffic aggravates existing severance
(g) Parking Controls (see Chapter 19 and Photographs 17.10 and 17.11)		
Noise	❑ Where on-street parking is reduced	❑ Inconsiderate drivers and passengers (door slamming, radio, excited voices)
Air Quality	❑ Where parking information reduces journey length or encourages modal shift	❑ Where journeys are lengthened by drivers searching for parking places
Visual intrusion	❑ Where on–street parking is removed or limited	❑ Where the number of signs and carriageway markings are increased
Accident Risk	❑ Where parking is restricted to safe locations	❑ Where an increase in speed is encouraged
Severance	❑ Where areas free of parked vehicles are provided for pedestrians and cyclists	
Parked Vehicles	❑ Where footway parking is controlled	
(h) Traffic Signals (see Chapter 40 and Photograph 17.12)		
Noise	❑ Where co-ordinated signals promote free flow	❑ Where there is an increase in acceleration, deceleration and idling, (reving engines)
Vibration		❑ Where low frequency vibration is generated by large vehicles accelerating and idling
Air Quality	❑ Where co–ordinated signals promote free flow	❑ Where there is an increase in acceleration and deceleration
	❑ Reduced emissions (compared with priority junctions)	❑ Where traffic is required to queue ❑ Where additional traffic is induced by increasing capacity
Visual Intrusion		❑ Where additional street furniture is provided
Accident risk	❑ Should be reduced	
Severance	❑ Where signal timings are used to reduce traffic volume at priority junctions ❑ Where pedestrian crossings and cycle facilities are provided	
Parked vehicles	❑ Where parking is banned in the vicinity of junctions	

Table 17.1 continued.

Photograph 17.1: Traffic calming in a village street.

Photograph 17.2: Traffic calming: entry treatment.

Photograph 17.3: Pedestrianisation of a high street.

Photograph 17.4: Pedestrianisation of a side street.

Photograph 17.5: Park & Ride.

Photograph 17.6: HGV controls.

Photograph 17.7: HGV control using width restriction.

Photograph 17.8: Visual and noise barriers under construction.

Photograph 17.9: Direction signing – size can be intrusive.

Photograph 17.10: Parking control in an historic centre.

Photograph 17.11: Parking control in an outer suburb.

Photograph 17.12: Traffic signals in a rural location.

17.7 References

DOE (1990 to 1996)   UK Annual Report 'This Common Inheritance', Stationery Office.

DOE (1994)   CM2426 'Sustainable Development: the UK Strategy', Stationery Office [Sa] [Wa].

DOE et al (1995)   Department of the Environment: Welsh Office, The Scottish Office, Department of Environment (NI) and Department of Transport 'Air Quality: Meeting the Challenge – The Government's Strategic Policies for Air Quality Management', Stationery Office.

DOE (1996)   'UK National Air Quality Strategy: Consultation Draft' Stationery Office.

DOE/DOT (1994)   Planning Policy Guidance Note 13 'Transport', Stationery Office.

DOT (1993)   Design Manual for Roads and Bridges Volume 11 – 'Environmental Assessment', Stationery Office.

DOT (1996)   TAU Traffic Advisory Leaflet 4/96 'Traffic Management and Emissions', DOT [Sc].

EC (1992)   Commission of the EC 'Towards Sustainability: A European Community Programme of Policies and Action in relation to the Environment and Sustainable Development' The EC Fifth Environmental Action Programme, Official Publication of the EC (COM (92) 23/11 Final).

EC (1993)   Commission of European Communities (1993) 'The Future Development of the Common Transport Policy: a global approach to the construction of a Community framework for sustainable development', EC White Paper.

HMG (1995)   'The Environment Act', Stationery Office.

HMG (1996a)   'Indicators of Sustainable Development for the UK', Stationery Office.

HMG (1996b)   Planning Policy Guidance Note 6 'Town Centres and Retail Developments', Stationery Office [Sb] [Wa].

LGMB (1993)   'A Framework for Local Sustainability', Local Government Management Board.

RCEP (1994)   Royal Commission on Environmental Pollution 'Transport and the Environment', Stationery Office.

SCC (1995)   'A New Transport Plan for Surrey', Surrey County Council.

UNCED (1992a)   'Agenda 21 – Action plan for the next century', UNCED.

UNCED (1992b)   'Agenda 21 – Rio Earth Summit', endorsed at UNCED.

17.8 Further Information

Buchanan, C et al (1963)   'Traffic in Towns', Stationery Office.

Civic Trust   'Caring for our Towns and Cities', CT.

CSS (1995)   'Sustainable Transport Indicators and Transport', County Surveyors' Society (February).

DOE (1988)   Circular 15/88 (WO23/88) 'Environmental Assessment', DOE.

DOE (1994a)   Planning Policy Guidance Note 23 'Planning and Pollution Control', Stationery Office [Sd] [Wa].

DOE (1994b)   Planning Policy Guidance Note 24 'Planning and Noise', Stationery Office.

DOE/DOT (1992)   Departments of the Environment and Transport to the Royal Commission on Environmental

Pollution 'Transport and the Environment', DOE/DOT.

DOT (1993)   Department of Transport: Design Manual for Roads and Bridges Volume 10 'The Good Roads Guide', Stationery Office.

DOT (1996a)   Traffic Advisory Leaflet 6/96 'Traffic Calming: Traffic and Vehicle Noise', DOT [Sc].

DOT (1996b)   Traffic Advisory Leaflet 12/96 'Road Humps and Ground–borne Vibrations', DOT [Sc].

Dutch MOT (1990)   Netherlands Ministry of Transport. Public Works & Water Management, 'Second Structure Plan', Den Haag,

EC (1985)   Council Directive 85/337/EEC of 27 June 1985 on 'The assessment of the effects of certain public and private projects on the environment', EC.

EC (1990)   Commission of the European Communities (1990) Green Paper on 'The Urban Environment', EC.

EC (1992a)   Commission of European Communities (1992) Green Paper on the 'Impact of Transport on the Environment. A Community Strategy for 'Sustainable mobility' COM (92) 494.

ECMT (1990)   European Conference of Ministers of Transport: 'Transport Policy and Environment' ECMT Ministerial Session, OECD.

Harman R (1993)   'Transport Planning Practice in Other European Regions, Manual of Information', Surrey County Council.

Harman R (1995)   'New Directions: a manual of European best practice in transport planning', Transport 2000.

HMG (1988a)   'Town and County Planning (Assessment of Environmental Effects) Regulations 1988',

Stationery Office [Sf].

HMG (1988b)   SI No. 1241 'The Highways (Assessment of Environmental Effects) Regulations 1988', Stationery Office [Sf].

HMG (1991)   Royal Commission on Environmental Pollution '15th Report on Emissions from Heavy Duty Diesel Vehicles', Stationery Office.

SACTRA (1986)   'Urban Road Appraisal', Stationery Office.

SACTRA (1992)   'Assessing the Environmental Impact of Road Schemes', Stationery Office.

Stringer B (1994)   'Setting targets for transport: policy making by numbers?', Local Transport Today (17 August).

TRL (1995a)   Project Report 103 'Vehicle and Traffic Noise Surveys Alongside Speed Control Cushions in York', TRL.

TRL (1995b)   Report 174 'The Environmental Assessment of Traffic Management Schemes: A Literature Review', TRL.

TRL (1996a)   Report 180 'Traffic Calming: Vehicle Noise Emissions Alongside Speed Control Cushions and Road Humps', TRL.

TRL (1996b)   Report 235 'Traffic Calming: Vehicle Generated Ground–borne vibrations alongside Speed Cushions and Road Humps', TRL.

UNCED (1987)   World Commissions on Environment and Development 'Our Common Future', Oxford University Press.

Chapter 18 Technology for Network Management

18.1 Introduction

It is now widely acknowledged that it is necessary to make better use of the existing road network and Intelligent Transport Systems (ITS) can help in this respect. Transport telematics (ie the combination of computers and telecommunications technologies) can provide improved methods of network management information systems and services for travellers. The purpose of these intelligent telematics–based systems and services is to manage the network better and to give travellers advice, which is accurate, reliable and timely. The technology now exists to make information available as it actually occurs – known as real–time.

18.2 Operational Objectives

Intelligent Transport Systems can improve traffic operations in the areas of:
- ❑ traffic management and control;
- ❑ public transport management and operations;
- ❑ travel and traffic information systems;
- ❑ automatic toll–collection and congestion–pricing systems;
- ❑ freight operations and fleet control; and
- ❑ integrated urban and inter–urban network management, which combines all these activities and more.

18.3 Scope of Co-ordinated Signal Systems

(Refer also to Chapter 40 on 'Traffic Signal Control' and Chapter 41 on 'Co–Ordinated Signal Systems').

Co–ordinated traffic signal systems, in the form of Urban Traffic Control (UTC), were one of the earliest and most successful applications of telematics for network management. At the network level, they control traffic by adjusting the signal parameters which determine the red and green signal–time allocations to control traffic flows at junctions and pedestrian crossings. The aim is often to produce the minimum total queue–length on the network or the minimum total vehicle–hours for a given amount of travel. Such strategies may be modified to give preference on certain route–corridors (green waves) and/or priority on pre–determined routes for emergency service vehicles or buses.

There are two main types of co–ordinated traffic signal system: fixed time plans, usually developed using the TRANSYT software package, or dynamic plans, usually based on SCOOT (Split Cycle Offset Optimisation Technique) software. Fixed time plans must be updated over time, as traffic patterns change. Dynamic plans do this automatically.

Co–ordinated signal systems can produce large savings in vehicular–delays (typically about 20% – see Chapter 41). They are also able to respond to changing traffic conditions. Priority can be given to selected vehicles (buses, trams, emergency vehicles), although with some delays to other traffic. They are also able to provide 'gating' of vehicles (ie deliberately limiting the volume of traffic passing through a junction) at the periphery of particularly congested areas. However, this requires attention to queue–management to avoid knock–on–effects on other parts of the network..

SCOOT can respond to changing traffic conditions in real time but only relatively slowly. The speed of response can be improved by using 'expert systems', such as CLAIRE (see Chapter 41).

Co–ordinated signal systems are merely demand–responsive and, therefore, cannot control traffic in space, in the sense of actively re–distributing traffic over alternative routes, to cope with changes caused by incidents. Also, the benefits of demand–responsive systems are reduced when dealing with very congested traffic, as the scope to allocate green times to utilise 'spare' capacity is limited. On their own, they are incapable of explaining to drivers what strategies are in operation, and why, or of advising them on alternatives that would contribute to relieving particular congestion conditions. However, these systems do generate an enormous amount of information, collected by the detection loops it uses to obtain information on vehicles and queues. Because of this, other telematics systems are frequently added, as described below, leading ultimately to a fully–integrated Intelligent Transport System (see Section 18.15).

18.4 Selective Detection and Control

Selective Vehicle Detection (SVD) can be employed at

a signal–controlled junction, to provide priority at that junction for certain types of vehicle. Usually it will be a component of a UTC system and is generally used for the following purposes:

❏ to assist buses and trams in keeping to schedule without undue delay, usually aided by priority lanes; and

❏ to give priority to emergency vehicles requiring response on demand – often provided directly from a control centre (eg 'green waves' for fire services).

Vehicles need to be identified and classified as one of the types of vehicle warranting priority and 'located' in relation to the signal (Chandler *et al*, 1985). Identification and classification can be made in several ways, such as:

❏ using the vehicle's encoded number or 'signature', which can be recognised by existing detection systems, such as induction loops or video–image detectors (see Section 18.10); or, more usually,

❏ using roadside equipment (so–called 'beacons') that may employ infra–red, microwave or ultrasonic technologies to communicate with tags or transponders on the vehicles.

Communication to operate the signal controller will either be directly from vehicles or via the control centre and the UTC computer. Examples of applications of Selective Vehicle Detection for public transport vehicles can be found in London (Meekums *et al*, 1995) and Southampton (CEC, 1994a) (see also Chapter 24). These systems bring benefits for buses but the disbenefits to other traffic may need to be considered. For example, problems can be caused if too many buses demand priority at the same time. One strategy is to detect and respond to those vehicles that are running late and to ignore those that are on schedule.

The positional accuracy required for detection needs to be better than about plus or minus two metres. This tends to favour the use of above–ground detectors or beacons for detection, in preference to buried induction loops. The benefits of SVD for bus operations are that:

❏ it provides improved reliability in running public transport vehicles and gives a better service for passengers, who have more confidence in schedules;

❏ priority improves the performance of public transport vehicles, in comparison with private transport, and thereby encourages modal shift;

❏ improved utilisation of the vehicle fleet leads to reduced bunching and fewer 'lost' seat–kms for operators; and

❏ it offers the opportunity to provide on–board information for passengers, such as the name of the next stop and estimated time to reach the terminus.

18.5 Vehicle Location and Polling

The next step up from selective detection is continuous Automatic Vehicle Location (AVL), combined with two–way communication between fleet drivers and their control centre ((see Chapters 24 and 25). The aim of AVL is to locate, at frequent intervals, individual vehicles in a fleet, such as buses, trams, police, ambulance, fire, freight delivery vehicles and AA/RAC patrols, and to pass emergency and status messages between these vehicles and their control centres. This enables better management and deployment of vehicles and resources by fleet operators. A major application of AVL has been with public transport fleets, where it has special use for public transport information systems. Examples are:

❏ information (arrival/departure) displays at termini;

❏ information displays at bus stops; and

❏ enquiry services, including 'public access' terminals (see Chapter 15).

AVL systems use roadside loops, beacons or proprietary navigation systems to locate the vehicles. Vehicles are generally polled in rotation, from the control centre, over a mobile radio or cellular telephone system, for location and status information, including emergency alarms.

Vehicle locations and status can be displayed in the control centre, using a Geographic Information System (GIS), which enables the vehicles' positions, movements and messages to be displayed on a map background. Software is required to monitor performance against schedules, to recognise incidents and to raise alarms requiring an operator's intervention. For public transport information systems, travel–time to the next stop(s) must be estimated and communicated to remote information displays. Examples of the use of AVL in public transport operations can be found in London (see Photograph 18.1 and Balogh *et al*, (1993)) and Southampton (Mansfield *et al*, (1994)) (see also Chapter 24).

Issues which are relevant include the following:

❏ some interpolation between the periodic 'fixes' taken by the AVL system is generally necessary, to give the required confidence in location accuracy;

❏ problems arise if not all buses are equipped, because unequipped buses will arrive at stops and termini unannounced;

❏ as with all information systems, high accuracy

Photograph 18.1: Bus stop information display in London Transport's Countdown System. Courtesy: London Transport Buses.

and reliability of information is needed to sustain credibility with the public;

❑ information at stops is not so necessary if buses always run to time or at very frequent intervals;

❑ information accuracy of one to two minutes is generally adequate;

❑ the system must be programmed with the correct and latest information on scheduling of services; and

❑ the cost of continuously–polling vehicles can be considerable unless a privately–owned communications system is used.

The benefits which can be expected from an AVL system are that:

❑ it offers the facility for emergency alarms from vehicles, for example, when the driver or a passenger is in trouble;

❑ improved reliability of running to schedule gives a better service for passengers;

❑ it offers improved information for passengers who are reassured about operations and can make better use of any spare time (see Figure 18.1);

❑ there are clear indications that passengers appreciate and value information (HCC, 1996);

❑ it offers indications that patronage can be increased (HCC, 1996); and

❑ improved utilisation of fleet–resources gives higher returns to operators.

18.6 Variable Message Signs (VMS) (see Chapter 15).

VMS can be used where greater flexibility is required than can be offered by fixed direction or advisory signs. Where their normal state is 'off' or 'blank', they are sometimes referred to as 'secret signs'. Use of VMS is increasing in response to more complex traffic management requirements and the need for more information to be provided to drivers. As with other traffic signs, VMS is governed by the Traffic Signs Regulations and General Directions (TSRGD) (HMG, 1994) [NIa].

The aim of using VMS is to provide drivers with mandatory and/or advisory information, at the roadside, relating to situations ahead or in the immediate vicinity. Applications, some of which might require special authorisation, could include:

❑ hazard warning information, for example 'accident', 'congestion', 'delays ahead', 'speed restriction', and 'lane(s) closed';

❑ diversion route advice;

❑ tidal flow lane–allocations;

❑ car parking guidance/occupancy (see Chapter 19);

❑ lane/road/bridge closed to selected or all vehicles;

❑ low bridge/over–height/over–weight vehicle warnings;

❑ weather or other environmental warnings of bad road conditions, for example, fog, ice, wind or flood; and

❑ estimated travel–times as, for example, on the Boulevarde Peripherique in Paris and the South–Eastern Freeway in Melbourne.

Three general types of technology are employed for VMS: electro–mechanical, reflective flip–disk and light–emitting. It is feasible to combine technologies within the same sign. When used as warning signs, it is usual for them to be fitted with amber–flashing lanterns. The three types are described below.

Electro–Mechanical signs involve rotating planks with two faces or prisms with three faces which are usually used to give versatility to a standard fixed–face traffic sign.

Reflective flip–disk signs are made up of a matrix of disks, one side black, the other fluorescent. The momentary application of an electrical current will magnetically 'flip' a disk between the 'on' and 'off' states. These signs are well suited to showing combinations of letters or symbols as a message.

Light emitting signs normally use fibre–optic or light–emitting diode (LED) technologies. The major advantage of these signs is that a greater range of messages can be displayed than for reflective technology signs. LEDs, being solid–state devices, can also produce very good reliability with minimal maintenance. Representations of standard traffic signs can be made using coloured light sources to generate pictograms, red rings or triangles. Some pictograms, however, can be difficult to show effectively and, as they are not covered by TSRGD, need to be specially authorised. To achieve sufficient brightness, the viewing angles of both LED and fibre–optic signs are comparatively narrow. This can make light–emitting signs difficult to align satisfactorily at some sites but, where aligned correctly, their presence (though not necessarily the detail) can be seen at long range. Drivers must be able to read the signs easily from any approach lane at the required distance.

Examples of VMS and their applications are numerous and it is only possible to give a flavour here. Photograph 18.2 shows an application for parking guidance and information. Photograph 18.3 shows a sign set used on the M25. This is not strictly an urban application but is included to indicate the capabilities of VMS for displaying message–sets involving characters and pictograms. An extensive system (CITRAC/FEDICS) has been introduced in Central Scotland.

In addition to requirements for siting and mounting (see Chapter 15), the sign designer needs to consider a number of factors, including sign–size, character height, legibility, contrast and viewing angle, for a wide range of ambient illumination levels andexpected approach speeds. Messages must be comprehensible to the vast majority of drivers. There is also a need to ensure consistent messages in a series of signs, ie a 'fail–safe' operational strategy is needed. There is usually a requirement for VMS to be able to communicate back to a control centre, to confirm that the correct message is set, although this can also be done using a CCTV surveillance system.

Signs, including the messages, must conform with TSRGD or be specially authorised by the Secretary of State [NIb]. As more experience is gained in the use of VMS, it is likely that Amendment Regulations will include more features and messages within TSRGD, thus reducing the need for special authorisations.
VMS messages can expect to reach a high proportion of drivers. However, their effect is localised as the signs can only affect those who pass by them while the message is displayed. They can be used to slow drivers down and to divert them around problem areas or, for example, to Park–and–Ride (P&R) interchanges. Drivers can also benefit from some, limited, indication of the reason for, and extent of, likely queues and delays. Parking management

Photograph 18.2: Parking guidance VMS in Southampton.

Photograph 18.3: VMS used for warning and speed control on the M25.

systems can improve traffic flow in town centres, by directing drivers to free spaces, and thus reduce the amount of circulating traffic searching for parking spaces They can also reduce the length of car park queues (see Chapter 19).

More generally or, for example in the case of a VMS used for tidal flow systems or to indicate a weather hazard, the benefits are relevant to the particular installation. VMS are fairly costly to instal and are generally used only where particular problems occur and/or where the accident risk is high and the costs can be justified. For instance, messages relating to a low bridge, height or weight restrictions or a road closed due to wind or snow have specific application to accident 'black' spots.

18.7 Ramp-Metering and Access Control

Sometimes also referred to as 'motorway access control', ramp–metering is used to assist the merge of two streams of traffic or to control flow through a bottleneck downstream of the merge. The usual form of operation is to meter traffic from motorway on–ramps, as it seeks to merge with traffic on the main carriageway, with the following aims:

❑ to reduce the likelihood of merging vehicles causing flow breakdown in the traffic on the main carriageway;

❑ to help to restore free flow after flow breakdown has occurred; and

❑ to deter traffic from using that particular on–ramp when flow on the main carriageway is close to capacity (CEC, 1992).

Ramp-metering is occasionally used dynamically to detect incipient gaps in traffic and then to allow vehicles to join the main carriageway by fitting them into the available gaps. The control systems for this need to be quite sophisticated. Vehicle–detection systems, usually pairs of detector–loops, are provided on the on–ramps and also on the main carriageway to monitor the relative flows of traffic in real–time, both up and downstream of the merge. A computer control algorithm is then used to relate the data from the detectors to the speed–flow curve for the main carriageway. The algorithm operates the ramp meter signals and aims to keep the main carriageway operating at flow/capacity (v/c) levels which are less than the critical 'knee' (v/c = 0.95) value, at which flow–breakdown is likely to occur (Owens *et al*, 1990).

Ramp-meter traffic–signals in the UK are of the conventional three–aspect type, following the standard sequence. In other countries, notably the US and the Netherlands, two–aspect signals (red/green) are used to 'gate' the inflow with rapid alternations.

A system has been installed on the M6 in the UK, using conventional signals (Owens *et al*, 1990) (Photograph 18.4). In France, on the Boulevarde Peripherique around Paris, conventional signals are used but, perversely, the system is arranged to give priority to vehicles on the ramp rather than on the main carriageway (CEC, 1992). In the Netherlands, traffic is deterred from using the on–ramp at peak times using VMS and red/green signals are used to 'gate' traffic, releasing as few as one vehicle at a time (CEC, 1992).

Ramp-metering may not be applicable where there is insufficient ramp–length to store queueing vehicles,

where sight–lines are poor or where gradients are too steep. In these cases, ramp–queues, caused by the metering signals, may back–up onto surface streets. So, the designer, in conjunction with the relevant authorities, must decide the balance of priorities between the traffic on the main carriageway and on the ramp. To minimise congestion overall, priority is usually given to keeping free–flow on the main carriageway.

Ramp-metering can give priority to a main carriageway, or a ramp, leading to a better balance of flows within a network. It can reduce conflicts, prevent flow–breakdown on the main carriageway and, probably, reduce accident–risks. Above all, ramp–metering can reduce congestion and delays. Results from the M6 sites show benefits of over 20% in vehicle–hours saved, due to delaying or preventing the on–set of congestion on the main carriageway, plus an increase in maximum throughput of about three per cent. On the M6 junctions, where ramp metering is used, the ramps are too far apart for co–ordination of flows but, on urban motorways and ring roads, linking may be warranted and beneficial.

Photograph 18.4: Ramp metering system on the M6.

18.8 Automatic Incident Detection (AID) and Management

The aim of AID is to detect incidents automatically and quickly, in order that the problem can be dealt with and the roads returned to normal operation as soon as possible. Almost all AID systems work by detecting an abnormality in the expected traffic conditions, such as a queue, a stopped vehicle or an empty lane, and comparing it against 'normal' conditions using a computer algorithm and then raising an alarm and/or putting into operation a strategy to deal with the problem.

Methods for detecting queues are essentially the same as those used to detect vehicles (see Chapter 40 Traffic Signal Control; and Chapter 41 Co-ordinated Signal Systems). Methods include:

❑ detector-loops in the carriageway;
❑ above ground detection (AGD), using short-range infra-red, or microwave communication links; and
❑ CCTV, combined with image-processing.

Examples include the MIDAS incident-detection/speed-control system, which is widely deployed on UK motorways (HA, 1994); Traffic*master*, which originally used AGD for motorways and trunk roads, but is extending to include urban applications (Photograph 18.5 and Traffic*master*, 1996); and SCOOT, with extensions, which utilise the SCOOT detector-loops to collect historic data as a basis for detecting incidents (Palmer *et al*, 1995). Image-processing systems (see Section 18.11) also have an application for automatic incident- detection in urban areas.

The management strategies adopted by highway authorities, in response to an incident-alarm, rely mainly on using VMS to warn, slow down or divert drivers (see Section 18.6). A traffic-responsive UTC system, such as SCOOT, can itself respond either, gradually, to help to alleviate the situation or more quickly if, for example, an 'expert system' (see Section 18.3) has been developed in advance. Integrated systems which combine this approach with VMS to warn and divert drivers are also possible (McDonald *et al*, 1995). At the urban/inter-urban interface, variable speed restrictions may be used, as on the M25 (see Photograph 18.3) to calm traffic approaching congested urban areas.

Photograph 18.5: Traffic*master* in-vehicle unit.

Some of the issues which arise are that:

❑ unexpected queues can give rise to increased accident-risk and secondary accidents;
❑ queues can build up extremely quickly, especially in busy periods, to spread out and affect the rest of the road network;
❑ these queues may persist, and extend, long after an incident is cleared and the front of the queue starts to dissipate, because traffic continues to join at the back;
❑ VMS only give information locally and to those drivers who see them;
❑ other driver information systems can give information over a wider area, for example traffic message broadcasting (see Chapter 15); and
❑ CCTV may be a necessary adjunct to AID, for sites with a high incident rate (see Section 18.10).

AID systems alert operators automatically to a problem and to the need to initiate action. The quicker the incident is detected, the quicker it can be cleared and the less are the secondary accident risks and accumulated delays. VMS can be used to warn drivers at known trouble- spots, that they are approaching the back of stationary or slow-moving queues. This leads to reduced accident-risk.

The UTC can be switched rapidly to a new plan to cope with the new traffic situation and thus help to alleviate congestion and delays. Drivers further afield can be informed, for example using traffic broadcasts, and advised to avoid the area and so reduce the build-up of traffic around the incident. Integration of the AID, UTC and VMS systems is possible, using expert systems that will, for example, automatically deduce and implement the best traffic management strategy in response to an alarm.

18.9 Weather Monitoring and Response

The aim is to detect and predict adverse weather and potentially poor road conditions, in order to warn drivers or to initiate treatment of the affected roads (eg to salt or snow plough) or, in extreme cases, to close a road or bridge.

The likelihood of road-icing is generally deduced from measurements of air and road surface temperatures, which are taken using suitable thermometers or thermistors. Measurement of electrical capacitance can be used to test for the presence, and concentration, of salt and to show if further salting is needed.

Wind speed can be measured using an anemometer.

Strong or gusty winds can pose a particular danger for high-sided vehicles, especially when unexpected, such as when crossing a bridge over a valley or estuary which acts as a Venturi funnel.

Fog is usually detected using instruments that measure either the amount of light transmitted or the amount scattered by the air. These measurements can then be related to visibility distance (Jeffery *et al*, 1981). Precipitation (eg rain and snow) is not actually measured, although warnings can be given locally if it causes reduced visibility and poor road conditions. Even relatively shallow floods produce danger of aquaplaning and extreme ones can make roads impassable.

Most weather detection systems are operated in conjunction with Meteorological Office forecasts and aim to provide local refinements, through selecting pre-cursor sites or black spots where, for example, fog or ice is likely to form first. Techniques of thermal infra-red mapping may be used to help identify these precursor sites (Stansfield, 1995). Examples are on the Britannia Bridge (van der Heijden *et al*, 1994) in Wales and, in Devon, where the Icealert system is used (Stansfield, 1995).

One key response of road authorities to advance notice of adverse weather and poor road conditions is to warn drivers of the hazard and to trigger remedial action, such as salting or gritting if ice or snow is forecast. Warnings and advice can be given locally using VMS for fog, ice, wind, snow or flood conditions, together with advisory speed limits or diversion information, if appropriate. Traffic message broadcasting (see Chapter 15) is used to reach drivers further afield and can often be supported by additional TV, radio and even newspaper reports, when sufficient warning can be provided by the Meteorological Office. Adverse weather and poor road conditions are particularly dangerous when they are unexpected; for example, black ice or patchy fog (Jeffery *et al*, 1981). The main benefits from weather monitoring are reduced accidents, delays and congestion and these can be obtained either by reacting quickly to alleviate the problem, for example by salting and gritting roads, or by warning drivers when and where a specific problem exists. However, roadside equipment for weather monitoring tends to be expensive, involving both a number of detectors and VMS. Applications are, therefore, usually confined to black spots, where the costs or risks of accidents can justify implementation.

The Meteorological Office can provide early warnings of bad weather conditions only at a general level. Monitoring-sites can often be identified that will give more local detail and can be used to activate VMS automatically.

18.10 CCTV Surveillance

The aim of CCVT surveillance is to provide network managers with a pictorial view of the operation of key parts of their network. Operators use the pictures as a basis for altering traffic control strategies, for confirmation of incidents reported by the public, and to record conditions or events over a period of time.

Cameras are usually of the solid state Charge Coupled Device (CCD) type and mounted on high masts or buildings. Costs depend on the performance and facilities required, which may include:
- ❏ monochrome or colour picture displays;
- ❏ low-light performance capability;
- ❏ broadcast quality or high definitions;
- ❏ zoom capability; and
- ❏ fixed or with pan-and-tilt manoeuvrability.

The communications link between the camera and the control centre generally uses co-axial or fibre-optic cable or microwave radio transmission.

Control centre monitors are normally set up to rotate through a sequence of cameras, with only a small number dedicated to looking at particular fixed sites. The ratio of control centre monitors to roadside cameras is usually about one-to-five but image-processing techniques, which give an automatic alert for possible incidents, can allow much higher ratios to be used. Photograph 18.6 shows the set-up in the urban traffic control centre for the city of Leicester.

CCTV systems are expensive to deploy, so they are generally only installed where a case can be made in terms of the likelihood of severe congestion, traffic incidents or accidents. Also, cameras must be positioned so that the privacy of people in nearby homes or business premises is not compromised.

Pictures are generally recorded so that visual records of incidents can be obtained and saved. There are usually more cameras than monitors so not all views are available at any one time and preferred views are generally set by the operators from experience.

CCTV is also thought to assist public relations because it provides a highly visible and easily understood feature in a high technology control room. The pictures can also be readily shared between a UTC centre and a police control room and, sometimes, relayed to local TV stations. The systems

Photograph 18.6: TV monitors in Leicester's urban traffic control centre.

thus provide a highly public indication of investment in traffic management tools

CCTV provides the opportunity for an operator to see an incident occur and to get early warning of problems. If the operator does not actually witness the incident, he or she will nevertheless be able quickly to get visual confirmation of the situation and of the consequences of the incident. This avoids the need to dispatch a police patrol and speeds up response–time, because getting a patrol vehicle close to an incident surrounded by congestion can often be difficult and slow.

The operator can also assess the situation and determine the need for emergency services, the nature and extent of action required and, subsequently, can continue to monitor the progress of the incident and any further requirements for action or withdrawal. CCTV, therefore, allows a control centre to make the best use and deployment of all available resources. Recordings also provide visual records of incidents and how they were treated.

18.11 Image-Processing Systems

Image–processing equipment can be used to analyse CCTV camera pictures, to determine a wide range of traffic parameters, including:

❑ to detect traffic incidents, differentiating between various types of situation and whether traffic is moving or stationary;
❑ to monitor conditions in bus–lanes for infringements by banned vehicles, whether moving or parked;
❑ to emulate detector–loops to count vehicles and measure flow, speed, headway, etc., and to classify vehicles;

❑ to substitute for loops in traffic control detection systems;
❑ to monitor the paths of vehicles and pedestrians through junctions or on crossings; and
❑ to identify 'probe' vehicles for transportation planning (O/D) surveys, for journey–time measurements or for the enforcement of traffic and parking legislation, using number– plate recognition techniques.

Image–processing systems work by analysing the picture elements (pixels) of successive frames taken by a CCTV camera and looking for differences in grey–levels due to movements, usually of the 'edges' of vehicles (Hoose, 1991). Examples of systems are given in Photograph 18.7; and in Hoose (1994); Sowell *et al* (1995); and HCC (1996).

Image–processing systems generally need to be calibrated for each camera–location and view. Fixed cameras are, therefore, preferred or default settings are needed for pan–tilt–zoom cameras, so that they return to a fixed view. Where the system is to be used for Automatic Incident Detection (AID), a threshold must be set for an alarm to be raised. Local conditions can make image–processing difficult, including:

❑ dull weather, in fog and at dawn and dusk, when contrasts are low;
❑ in strong sunlight, when shadows might easily be confused for moving vehicles; and
❑ headlamp glare from wet road surfaces.

Image–processing techniques can also be used for reading vehicle number–plates automatically (Hill *et al*, 1994). Most automatic debiting systems rely on cameras for recording and identifying violators and useful benefits can be obtained from automating the process (see Section 18.13 on 'Automatic Debiting').

Image–processing provides substantial added value to existing CCTV installations, because it has the ability to look at a continuous stretch of road (typically 200m to 300 m), compared with other incident–detection systems using loops, for example, which effectively detect at a single point.

Systems can inspect all of the pictures transmitted to a control centre, not just those displayed on the monitors, and can select automatically the most interesting or relevant views to display. Fast incident–detection, typically less than 30 seconds, is achievable and can be used automatically to warn the operator when an incident occurs, thus relieving him or her of the responsibility of watching the monitors all the time. Consequently, there is a better use and deployment of resources. Also, systems can readily be deployed in temporary situations, such as at a roadworks.

Photograph 18.7: Visual display of the IMPACTS image processing system.

18.12 Route–Guidance and Journey–Planning

The aim is to help drivers to plan and to follow the best routes, using reliable trip information which can be obtained before setting out (pre–trip) or during a journey (in–trip).

Pre–trip planning (see also Section 15.13) is now feasible using proprietary software packages, which enable routes to be planned, using a personal computer in the home or office, before setting out on a journey. Terminals (see Chapter 15), usually in public places, can also be used to plan journeys, although the emphasis here is more usually on travel by public transport modes.

In–trip route–guidance, on the other hand, requires the appropriate equipment to be installed in the users' vehicles. The technology for in–vehicle information systems is described in Section 15.11. VMS direction signs (see Section 18.6) are also sometimes referred to as route–guidance systems where they are used to give route and diversion advice, although the term is more commonly used for in–vehicle systems, where the range of information and the scope for its delivery are vastly greater.

Trip–planning software is commercially available, for example, the 'Milemaster' package from the Automobile Association or 'Autoroute' from Microsoft. The first commercial system to provide dynamic traffic information was Traffic*master* (Traffic*master*, 1996) (see Section 15. 11). It now covers most of the UK motorway network.

Autonomous navigation systems are widely available in Japan. Take–up and development in Europe and the US has been slower, although work supported by the CEC Framework research programmes has produced autonomous navigation systems with RDS–TMC updating, as described in Section 15.11 (see also Photograph 18.8 and CEC, 1996a). Further development of these systems, using two–way GSM cellular telephone links, has produced the SOCRATES dynamic route–guidance (DRG) system (Catling, 1994).

A DRG system, operating in conjunction with a traffic responsive UTC, offers the potential to control traffic, both in time and in space, and so produce a new generation of traffic control systems.

Most other trip–planning and guidance systems, even if dynamic, essentially give all drivers on the same O–D trip the same response to the same question.

Care is needed, therefore, in the way these systems are used. If all traffic is advised to divert, this could result in a worse problem somewhere else. Provided the proportion of vehicles equipped with route–guidance systems is relatively low, both equipped and non–equipped drivers can benefit. This is because the few equipped vehicles will be diverted around an incident and so reduce pressure on the affected route. As the proportion of equipped vehicles rise, so do the control problems.

Systems should be able to handle a range of definitions for the 'best' route including, for example, minimum time, distance or generalised cost, as well as the most scenic, suitable for towing a caravan or suitable for an HGV. Equipment manufacturers need also to exercise care in the way information is presented to drivers. A complicated visual display may cause distractions or take so long to assimilate that it gives rise to increased risks of accidents.

Most autonomous navigation systems require access to a detailed digital road map. These can be expensive to produce and to maintain. Commercial versions are available but they need to be augmented with information about temporary traffic restrictions. Standards are required for systems, such as RDS–TMC, so that they are inter–operable. This will ensure that a vehicle unit produced to receive messages in one region or member–state of the EU will also be able to receive and decode messages in other regions or member–states of the EU and even in foreign languages.

Significant benefits of DRG are estimated to accrue from systems that can help drivers to plan and follow optimum routes (Jeffrey, 1994). From the driver's point of view, these systems are more attractive if they can provide real–time traffic information and

Photograph 18.8: CARIN in–vehicle navigation unit.

they are of most use on unfamiliar journeys. Potential benefits, worth around 10% of journey time, have been estimated for a fully dynamic route–guidance system operating in London. Route–guidance systems, combined with traffic responsive UTC, provide a new generation of traffic control tools that enable traffic to be controlled both in time and in space.

18.13 Automatic Debiting Systems (ADS) and Decremented Pre-Payment

The aim of automatic debiting systems (ADS) is to charge a toll automatically from a vehicle for use of a facility, such as a toll road, tunnel or bridge without requiring the driver to stop. Similar technologies are used for road–use and congestion pricing systems (see Chapter 21), for automatic charging in car parks (see Chapter 19) and can be interfaced to in–vehicle information systems.

A large number of technical possibilities exist, ranging from a 'vignette' (the equivalent of a paper tax disc) in the vehicle that can be inspected visually or using a bar–code type reader from a toll booth, to a transponder with an 'electronic purse' (smartcard) that can be credited through pre–payment with 'electronic cash' units and debited automatically by short–range communication from a roadside unit at each toll point.

The basics of a system require:
❑ an effective means of detecting the passage of vehicles and of classifying them;
❑ a two–way communication between a device in the vehicle and a roadside unit;
❑ a roadside unit that will either interrogate the in–vehicle device for its identity (AVI tags) or instruct a transponder to deduct the toll charge from the smartcard;

❑ that all successful transactions using transponders will be anonymous but the identity of the vehicle needs to be recorded (eg by video analysis of the registration plate), whenever the transaction fails, to ensure enforcement after the event; and
❑ that the in–vehicle device must be unique to the vehicle, driver or company, depending on whether the system applies different tolls to different classes of vehicles and on how payment is made.

For access control purposes, passage may then be granted or denied using a barrier. For tolling and road–use pricing purposes, all vehicles are allowed to pass unhindered but with violations being recorded and followed up later, or not, depending on the enforcement policy.
Payment can be made:
❑ in advance (pre–payment) using the equivalent of a season ticket (vignette) or by stored credit units (smartcard); or

❑ at the time, using the equivalent of a debit card with automatic deductions from a subscription account; or
❑ in retard (post–payment), where the bill accumulates and the driver (or company) is invoiced later.

Most systems favour pre–payment smartcards linked with a transponder in the vehicle. This is known as ETC (Electronic Toll Collection). If smartcards are used for payment of many different services from many different providers, a sophisticated 'clearing–house' system will be needed to ensure a fair allocation of revenues.

Cards are bought by the driver in advance and can be re–charged with credit units at suitable terminals. Simpler payment systems include:
❑ a season ticket, recognised by the system as valid until an expiry date has passed;
❑ credits for a fixed number of passages – the credit being incrementally debited by the roadside unit for the cost of each passage; and
❑ in the case of a toll road, debits which may be made dependent on the distance travelled.

Communication between the roadside and vehicle units can be achieved using induction loops, radio or infra–red beacons for access control but, increasingly, will use a microwave link. CEPT, the European frequency allocation authority, has defined a microwave band at 5.8 GHz for automatic debiting systems for high–speed tolling (Hills et al, 1994).

Examples of ADS are to be found on the Dartford

Crossing, the Mersey Tunnels and increasingly throughout the world (Hills, 1996).

Tag–based systems of ADS with a central account are in use in a number of countries, notably France, Portugal and the US, which have a long history of toll roads. Similar systems are used in the UK but only at major esturial crossings, such as the Dartford Crossing (see Photograph 18.9), the Severn Bridges and the Mersey Tunnels.

Historically, most tolling sites involved extensive toll–plazas, with many lanes and toll–booths, where tolls were taken manually or via a coin–collection machine. Interest in auto–tolling systems is increasing because of the potential for increased vehicle throughput, with less land–take, shorter queues and less pollution (Morton *et al*, 1994).

With most AVI tag systems, the communications range is fairly short (2 m to 5 m) so the vehicles must pass through a controlled gap (ie a single lane).

Likewise, the transaction time needed in most AVI tag–implementations precludes the driver approaching at speeds greater than about 20 miles/h (30 km/h) and this limits throughput.

More advanced developments, such as ADEPT (Photograph 18.10)) and systems developed for the German and UK tolling trials, involve systems that are fully automatic, using transponders with up to 20 m range, and enable drivers to pass unhindered at high running speeds (on motorways in Germany up to, perhaps, 160 km/h) and without lane control.

Adequate reliability is required for public acceptance. Avoiding charging in error is more important than ensuring that everyone is charged. Hence, stringent requirements for reliability and enforcement are needed, which have to rely on a system such as image–video processing for automatic number–plate recognition and reading (see Section 18.10).

Concerns about civil liberties can arise with AVI tag systems because, the identities of vehicles have to be recorded at the roadside and could be traced unless there are data–protection safeguards built into the system. Transponder/smartcard transactions, being anonymous, avoid these problems.
Automatic debiting systems provide a highly convenient method of paying for transport services. They avoid the need for elaborate and expensive toll–plazas and high manpower requirements. Improved speed of transaction and off–line enforcement mean less delay for traffic and greater convenience for drivers.

Photograph 18.9: Toll plaza on QEII bridge at Dartford where the M25 crosses the Thames.

Opportunities for fraud, especially by employees of the toll operator, are virtually eliminated. Automatic systems provide the ability to vary tolls with time of day but the prevailing tariff needs to be displayed prominently (eg using VMS).

The electronic 'purses' on smartcards may also be used, for example, to pay public transport fares and parking charges as well as for in–vehicle information services. In principle, this could herald the onset of a cashless economy.

18.14 Environmental Monitoring

The aim of environmental monitoring is to detect, or to predict, adverse environmental conditions and to warn drivers or, in extreme cases, to initiate a ban on traffic entering sensitive areas.

Photograph 18.10: The ADEPT trial site near Thessaloniki in Greece (1993).

Environmental pollution includes exhaust emissions (see Section 17.5). The rate of build–up and concentration varies considerably with the prevailing weather and atmospheric conditions. With current traffic volumes in some cities noxious gases can, on occasions, approach and even exceed 'acceptable' limits defined by the World Health Organisation (WHO) or the CEC, especially in the summer months (Simms, 1994). Emissions can be monitored and overall conditions estimated from the roadside (Sonnabend, 1994; and Bell *et al*, 1996) but only at a highly aggregated level. The variables affecting the 'micro–climate' in a particular street are too complex to model accurately.

Proprietary sensors are available for measuring and analysing emission chemicals and particulate concentrations in the atmosphere. Regular monitoring is undertaken by the DOE (Hayman, 1994). The results are augmented by additional stations set up by local authorities, designed to monitor background levels.

Leicester, as the 'Instrumented City', is being used to investigate an overall UTC and traffic management strategy, designed to minimise the number of stops/starts, smooth out acceleration profiles and reduce trip–times and distances in order to minimise vehicular emissions (Bell *et al*, 1996). However, the relationship between traffic control and local pollution levels is both complex and variable.

Other management strategies that can be considered when pollution levels become too high are:
❑ alternative routeing strategies (using VMS), to divert traffic around the more congested areas of towns; and
❑ encouraging drivers to park–and–ride and, thus, to use a less polluting collective mode of transport.

Whatever strategy is employed, it is crucial to make the changes known in advance, using additional signing, radio, TV broadcasts and advertisements in newspapers, to avoid serious congestion occurring at the periphery of the affected area and to minimise the likelihood of a political reaction or even a legal challenge against the strategy.

Demand management, generally, and urban road–use pricing or congestion charging particularly (see Chapter 21) may offer opportunities for reducing the overall impact of exhaust emissions.

Gaseous emissions are the main problem from petrol engines and particulates from diesel engines. For cars, the combination of lead–free petrol and catalytic converters reduces emissions markedly but is not a complete solution, especially for car–journeys that are short and/or involve cold starts. Gases are normally dispersed by winds and dissolved by rain. In fine, still weather, particularly during summer months, they tend to accumulate in street 'canyons'. Local weather conditions will strongly influence the rate at which exhaust emissions accumulate and will, therefore, influence the choice of a control strategy.

Improved traffic control can, in principle, make a useful, though probably only marginal, contribution compared with other methods, such as catalytic converters, reduced emission (lean–burn) engines, alternative fuels, including Compressed Natural Gas (CNG), and electric/hybrid vehicles. However, vehicle emissions can be reduced by avoiding stop/start conditions and smoothing traffic flows. Banning vehicles altogether is a desperate measure and would require public acceptance, which may be achieved in particular locations through public awareness campaigns (see Chapter 10).

The benefits from environmental monitoring and control strategies are mainly in improved air quality in urban areas, with possible reductions in health risks for residents. By–products can also be expected, including reduced noise, vibration, accidents and visual intrusion, all of which contribute to an improved quality of life.

18.15 Integration of Systems and Links to External Systems

The aim of systems–integration is to operate traffic and transportation as one overall system, rather than a collection of separate components, and thereby to achieve a system–optimum. In principle, the synergy that can be obtained from this will result in overall benefits that exceed those from the sum of the parts.

The technical components include many of those already discussed in this chapter and these can be combined in a number of ways, including:
❑ UTC, coupled with motorway control systems, to control traffic entering built–up areas and to restrain traffic within towns – perhaps in combination with demand management and road–use pricing systems (see Chapter 21);
❑ linking together traffic, AID, weather and environmental monitoring systems to provide short–run forecast information about road and traffic conditions;
❑ improved trip planning and information systems that will inform travellers of routes and conditions and enable them to plan journeys by car or public transport before they set out (see Chapter 15);

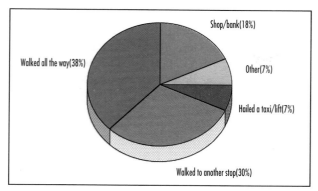

Figure 18.1: Results from ROMANSE survey of use of bus waiting time.

❑ improved information and route guidance *en–route*, to tell drivers when part of an urban area is congested and to guide them to alternative routes or to a P&R site so that they can leave their cars in a safe place and transfer conveniently to another mode for the remainder of the journey;
❑ improved travel–information systems, to tell travellers what collective transport options exist, and where and how to use them; and
❑ integrated (smartcard) ticketing, to enable travellers to use single pre–payment smartcards at toll facilities, parking (P&R) and any combination of public transport modes, using the same card.

Opportunities also exist for using a common infrastructure as the basis for integrating automatic–tolling, road–use pricing, driver information and dynamic route–guidance systems, all using the same communication link and a single smartcard as a means of payment (Hills *et al*, 1994).

The concept of an integrated Intelligent Transport System (ITS) was promoted from 1988, by the EU in the Second and Third Framework (DRIVE) programmes (CEC, 1989). Two major UK projects began in 1992 to develop and implement urban pilot systems (CEC, 1996b). Hampshire County Council's ROMANSE project in Southampton (HCC, 1996) involved:
❑ an integrated Travel and Traffic Information Centre (Photograph 18.11);
❑ all services and information brought back to the centre for processing; and
❑ decisions made and implemented at the centre regarding strategies, messages to disseminate, and so on.

Birmingham City Council's BLUEPRINT project (CEC, 1995) involves a distributed system and various stand–alone sub–systems, such as UTC, motorway control centres, motoring clubs, and emergency services, all connected over a network that enables knowledge and information to be shared.

The ROMANSE and BLUEPRINT pilot projects were completed in 1995. They went some way to realising the concept of an integrated ITS. Both use Geographic Information Systems (GIS) systems to co–ordinate and display the information provided by the sub–systems (see Figure 18.1). Current traffic and travel information can be compared with historic data and/or related to other geographic, ie spatial, information. This can then be combined with accident records, land–use statistics and population distribution giving added value for planning purposes (see Chapter 6 'Transport Policy Components' and Chapter 8 'Estimating Travellers Responses').

Supported by both the EC and the Department of Transport, the projects involved public/private partnerships of local authorities, electronics, telecommunications and traffic systems manufacturing industries, consultants, motoring clubs as information service providers, universities and research institutes. All those involved learned to work together and to deal with some of the many problems related to institutional, legal, and Intellectual Property Rights (IPR) issues.

The ITS concept requires either a distributed system, as in BLUEPRINT, or a centralised system, as in ROMANSE. Either way, the systems only know and disseminate information about conditions in their local area. Longer distance travellers need to know more about conditions on the roads beyond the local area and *en–route* to their final destination. For a national system, therefore, links between local area control centres are needed.

Work on this is also promoted by the EC, in co–operation with the member–states (CEC, 1989 and

Photograph 18.11: The ROMANSE Travel and Traffic Information Centre in Southampton.

1994b). For the various partners in the European projects, such as ROMANSE and BLUEPRINT, and including central and local governments, the EC has supported the development of common standards, aimed at achieving inter–connectivity and inter–operability of systems. These include:

- ❏ systems architecture;
- ❏ message set and protocols;
- ❏ databases; and
- ❏ communications and data–exchange.

This work will enable information–services and control centres to be designed and constructed to common functional specifications and to agreed interface designs.

Through the Fourth Framework Programme (1995–1998), CEC has co–funded demonstration projects including taking forward the work already done in, for example, ROMANSE and BLUEPRINT. The resulting systems will facilitate common levels of service to be achieved in towns and cities across Europe, common information to be presented on, for example VMS, and for inter–operability of driver information (RDS–TMC), trip–planning (TRIPlanner) and route–guidance (SOCRATES) systems.

18.16 References

Balogh S and Smith R (1993) 'London Transport's COUNTDOWN System – A Leader in the Bus Transport Revolution', Digest of IEE Colloquium on Public Transport Information and Management Systems, IEE Digest No 1993/123, Institution of Electrical Engineers.

Bell MC, Reynolds SA, Gillam WJ, Berry R and Bywaters I (1996) 'Integration of Traffic and Environmental Monitoring and Management Systems', Proc 8th International Conference on Road Traffic Monitoring and Control, IEE, London April.

Catling I (1994) 'SOCRATES' in Advanced Technology for Road Transport: IVHS and ATT (Ed. Catling I), Artech House.

CEC (1989) 'DRIVE '89': the DRIVE Programme in 1989, Commission of the European Communities, DG XIII, Document DRI 200, Brussels, CEC.

CEC (1992) 'Motorway Traffic Flow Monitoring and Control': Final Report of the DRIVE Project V1035 (CHRISTIANE), CEC R&D Programme – Telematics Systems in the Area of Transport, CEC.

CEC (1994a) 'Report on AVL/UTC Trials in Southampton', Deliverable No 34 of the DRIVE II Project V2049 (PROMPT), CEC R&D Programme – Telematics Systems in the Area of Transport, CEC.

CEC (1994b) European Commission, 'Telematics Applications Programme (1994–1998)', Call for Proposals, Official Journal of the European Communities, CEC.

CEC (1995) BLUEPRINT IRTE Pilot, Deliverable No 28 of DRIVE Project V2018 (QUARTET), CEC R&D Programme – 'Telematics Systems in the Area of Transport' CEC, Brussels.

CEC (1996a) 'PLEIADES Final Report', DRIVE II Project V2047 (PLEIADES), CEC R&D Programme – 'Telematics Systems in the Area of Transport', CEC, Brussels.

CEC (1996b) 'Transport Telematics in Cities: Experiences Gained From Urban Pilot Projects of the Transport Telematics Sub–Programme Under the 3rd Framework Programme (1992–1994) of the CEC', published by DG XIII, CEC.

Chandler MJH and Cook DJ (1985) 'Traffic Control Studies in London – SCOOT and Bus Detection', Proc PTRC Summer Annual Meeting University of Sussex.

HA (1994) NMCS2 'MIDAS System Overview', DOT/HA Instruction MCH 1696B, Highways Agency.

Hayman GD (1994) 'An Overview from the UK Government', Proc BICS Conference on Monitoring and Reducing Air Pollution from Traffic, Birmingham, UK, BICS.

HCC (1996)   'The ROMANSE Project Final Report', Hampshire County Council.

Hill G and Adaway W (1994)   'Machine Vision Applied to a Speed Violation Detection–Deterrent System', Traffic Technology International '94, UK & International Press.

Hills PJ (1996)   'Tolling Targets and Technologies', ITS Intelligent Transport Systems International Issue No 4, pps 47 – 50 et seq. Surrey.

Hills PJ and BlytheP (1994)   'Automatic Toll Collection for Pricing the Use of Road Space', Advanced Technology for Road Transport: IVHS and ATT (Ed. Catling I), Artech House.

Hoose N (1991)   'Computer Image Processing in Traffic Engineering', Research Studies Press.

Hoose N (1994)   'Incident Detection and Traffic Surveillance Using the IMPACTS Video Analysis System', Traffic Technology International '94, UK & International Press.

HMG (1994)   'Traffic Signs Regulations and General Directions', Stationery Office.

Jeffery DJ (1994)   'Mobile Information Systems', Artech House.

Jeffery DJ and White ME (1981)   'Fog Detection and Some Effects of Fog on Motorway Traffic', Traffic Engineering+ Control 22(4).

Mansfield RS and Bannister D (1994)   'The Use of ATT Applications for Public Transport in SCOPE Cities', Proc. First World Congress on Applications of Transport Telematics and Intelligent Vehicle–Highway Systems, Vol 6, pp 2920–2927.

McDonald M, Richards MA and Shinakis EG (1995)   'Managing an Urban Network Through Control and Information', Proc. Vehicle Navigation and Information Systems Conference, pps 516 – 522, Seattle.

Meekums R, di Tarranto C, Davidsson F, and Bretherton D (1995)   'The PROMPT Project', Digest of IEE Colloquium on 'Urban Congestion Management', IEE Digest No 95/207, Institution of Electrical Engineers.

Morton, TW and Lam, WK (1994)   'The Effect of Automatic Vehicle Identification on Toll Capacity at the Dartford River Crossing' Traffic Engineering + Control 35(5).

Owens, D and Schofield MJ (1990)   'Motorway Access Control: Implementation and Assessment of Britain's First Ramp Metering Scheme', TRL Research Report RR 252, TRL.

Palmer JP, Powers JP and Wall GT (1995)   'Automatic Incident Detection and Improved Traffic Control in Urban Areas', IEE Colloquium on 'Urban Congestion Management', IEE Digest No 95/207, Institution of Electrical Engineers.

Simms KL (1994)   'Modelling the Impact of Traffic on Air Quality', Proc BICS Conference on 'Monitoring and Reducing Air Pollution from Traffic', Birmingham, BICS.

Sodeikat H (1994)   'Co–Operative Transport Management with EUROSCOUT', in 'Advanced Technology for Road Transport': IVHS and ATT (Ed. Catling I), Artech House, 1994.

Sonnabend P (1994)   'KITE Kernel Project on Impacts of Transport Telematics on the Environment', Proc. First World Congress on Applications of Transport Telematics and Intelligent Vehicle–Highway Systems, p 665, Paris.

Sowell WH and LaBatt JS (1995)   'Video Detection Takes a New Track', Traffic Technology International, UK & International Press, Surrey.

Stansfield K (1995)   'Weather Prediction – The Way Ahead', Highways & Transportation 42(9).

Traffic*master* (1996)	Company Literature from Traffic*master* PLC, Milton Keynes MK14 6DX, 1996.
Van der Heijden WFM and Haandrikman R (1994)	'ROSES Central Monitoring System',Proc. First World Congress on Applications of Transport Telematics and Intelligent Vehicle–Highway Systems, p 1097, Paris.

18.17 Further Information

Catling I (Ed) (1994)	'Advanced Technology for Road Transport': IVHS and ATT, Artech House.
CEC (1994)	First World Congress on 'Applications of Transport Telematics and Intelligent Vehicle–Highway Systems', Proc Vols 1 to 6, Paris, Nov/Dec 1994.
BICS (1994)	Proc BICS Conference on 'Monitoring and Reducing Air Pollution from Traffic', Birmingham, UK.
Walker J (Ed) (1990)	'Mobile Information Systems', Artech House.

Chapter 19 Parking

19.1 Introduction

Drivers usually assume that they will be able to park their vehicles within a reasonable distance of their final destination, accepting that sometimes, in congested areas, this might involve some time searching for a space.

Drivers' personal judgements of what constitutes an acceptable place to park vary considerably in terms of location, size of space and whether or not parking fees are charged. Judgements are influenced by the purpose and urgency of the trip, the ownership of the car and personal affluence, as well as by individual attitudes and behaviour patterns. Drivers will also consider the security of their cars and, possibly, their own personal security when choosing where to park, for example, late at night.

Parking availability and characteristics can strongly influence a driver's choice of destination. The success of out-of-town shopping centres is often attributed to the provision of large areas of free car parking near the shops. Competition for retail business between town centres and out-of-town shopping malls, and also between different towns in the same region, is such that adequate car parking may be an important factor in securing the future economic viability of a town centre, provided that the availability of parking does not attract more traffic into the centre than the road network can accommodate. Traffic congestion, of course, will detract from the attractiveness of the centre (DOE, 1996) [NIa] [Sa].

The growth of vehicular traffic, with its associated problems of congestion, accidents, noise and pollution, has led to a significant reappraisal of transport policy. It is no longer considered either feasible or desirable to cater for unlimited growth of traffic in many town centres. Management of the amount and type of parking is, therefore, an important means of influencing overall levels of traffic demand, as well as the balance between different purposes of car trips which are generated, such as shopping journeys being given preference over journeys to work (DOE/DOT, 1994 [Sb] and 1995 [NIb] [Sc]). Parking regulations, as a means of influencing traffic demand, have been in use for many years and are understood and accepted by most vehicle-users. Regulation of on-street parking may also be used for traffic management reasons, to reduce the risk of accidents and to safeguard the traffic-carrying capacity of the carriageway. For example, the introduction of Priority (Red) Routes in London is intended to reduce traffic congestion caused by inappropriately and illegally parked vehicles (LAA, 1992). Effective enforcement is essential if the objectives of introducing parking controls are to be achieved (see also Chapter 14).

19.2 Types of Parking

On-street kerbside parking space is usually regarded as the most convenient place to park, particularly for physically disabled people who are unable to walk long distances. Where there is no provision for servicing off-street, space for loading and unloading on-street is also required for delivery and service vehicles.

Publicly available off-street parking spaces are normally provided and operated either by local authorities or by private operators, to cater for demand in town centres and at other major destinations. Whereas private parking spaces are available only to authorised users and may include private residential garages and forecourts, as well as private non-residential parking (PNR) attached to, or incorporated within, commercial and public authority buildings.

For each type of parking, spaces may be provided for pre-defined classes of vehicle, including bicycles and motor-cycles, as well as cars or coaches, and/or other classes of users.

19.3 Parking Policy

Parking policy should be determined as an integral part of a local authority's transport policy and within the planning framework provided by structure plans, local plans and unitary development plans [NIc]. The objectives of local parking policies should be clearly identified and may include those which contribute to wider transport policies, such as traffic restraint or accident prevention. Parking policies can be particularly effective in helping to achieve overall traffic restraint whilst, at the same time, providing adequate parking spaces for, say, residents and customers of local shops. Parking for motor cycles and bicycles should be considered as part of an overall policy, especially if these modes are to be encouraged.

The total amount and balance of parking 'stock' in an area should be considered in devising appropriate parking policies. On–street parking and off–street parking, including private parking, should be considered together, as complementary parts of the total parking stock available.

In dense urban areas, where the demand for spaces at peak parking periods is likely to exceed the supply, policy decisions are needed on the allocation of the available space amongst the various categories of potential users. Priority is often given to the demands of local residents and short–stay shoppers first, with long–stay parking for commuters and local workers being regarded as less essential. The need for new off–street car parks should also be considered within overall transport policy and the adequacy of the local highway network.

The establishment of clear parking policies should lead to plans for the effective management of both on–street and off–street parking. Time limits may be imposed on different categories of parkers and differential charges levied to optimise the use of the available space. When demand exceeds supply and where a local authority's policy is to give priority to short–stay parking, it may be necessary to impose maximum parking stays at on–street spaces and to adopt steeply graduated charges at off–street car parks to discourage long–stay parking. Where it is necessary to provide some long–stay parking, a system of car park season tickets can be introduced, to enable the allocation of parking spaces to be better targeted, perhaps by limiting the issuing of season tickets to the occupants of commercial premises with little or no off–street parking provision. However, the extent of privately controlled and operated 'public' car parks in an area may limit an authority's ability to implement its parking policies. The extreme case is where local authorities have no control over existing private non–residential (PNR) parking which often accounts for a large proportion of the parking stock (ADC, 1993).

Supply and Demand

Regular assessment should be made of the existing and projected future demand for parking spaces in an area, both on–street and off–street. A comparison of supply and demand figures, preferably broken down into short–stay and long–stay parking, enables the likely overall balance, surplus or deficit, to be established at any time. The parking requirements of local residents should be assessed separately and appropriate provision identified. The use of a simple

computer spreadsheet model should enable alternative scenarios to be evaluated quickly and easily.

The intensity of demand for parking and the duration of stay will vary with the time of day, day of week and season as well as by the type of vehicle and the purpose of the journey. For example, parking demand by short–stay shoppers is likely to be at its peak on weekdays at 11 am and 3.30 pm, Saturday mornings and during the Christmas season, whereas long–stay parking demand by residents tends to be greatest early in the morning and in the evening. Future parking demand should be quantified in the light of planning policies and anticipated developments in the area. The effects of future development on the existing parking stock, for example by the possible loss of temporary off–street parking sites due to rebuilding or the need to restrict on–street parking on a busy road, should also be assessed. Account should also be taken of any existing and projected additional private non–residential (PNR) parking in the area.

Public Consultation

Parking policies have direct, and often significant, effects on people's lives. In particular, the availability of parking spaces, both on–street and off–street, is of great concern to local residents, retailers and other local businesses. Residents may be concerned about the accident–potential and environmental implications of indiscriminate on–street parking in their streets, as well as the availability of parking for their own and their visitors' use. It is important, therefore, to consult widely before embarking on any significant changes in parking arrangements, not least to avoid the risk of having to alter costly measures after their introduction because they prove to be unpopular. Moreover, legal requirements for some types of measures, such as on–street parking controls and some off–street parking charges, have to be publicised before they can be implemented. Local authorities may find it helpful to carry out more extensive public consultation exercises when major changes are proposed. Some examples of consultation leaflets are shown on Montage 19.1 (see also Chapter 10).

Parking Standards

Under planning legislation, local planning authorities have extensive powers to control development, including the provision of parking [NId]. Redevelopment presents opportunities to bring parking provision into line with land–use and with parking policies for the area. For example, the minimum adequate parking can be stipulated for a new residential development or, conversely, the

Details of the Controlled Parking Zone (CPZ)

The purpose of the CPZ would be to reduce and control the problems of all day parking by commuters and office workers between 8.30 a.m. and 6.30 p.m, Monday to Saturday. The CPZ would make it easier for residents, short-stay visitors and shoppers to find parking spaces. CPZ schemes are in operation in Richmond and Hampton Wick.

The main features of the CPZ are:

1. All kerbside space where it is safe and convenient to park, would be marked out with special parking bays for residents, shoppers, visitors and business parkers. Everywhere else would be covered by yellow lines where parking would be banned. Council contractors would enforce the bays, and the Police and Traffic Wardens would continue to enforce yellow lines.

2. Residents who might want to park in 'residents only' bay, or in a 'dual-use' visi[tor] bay would need to buy a permit. The propo[sed] charge for the residents' permit is £25 a yea[r] pence per day) to cover printing, administr[ation] and monitoring the permit scheme. The cost of enforcing and administe[ring] [the] bay including, if nec[essary] offenders

parking would be provided in the off-street car parks, and in some business spaces on-street, with the exact balance still to be decided. Season tickets would be available for long term parking at a proposed charge of £90 per quarter. There should be enough spaces for all those working in the area (including Council staff) who use their cars at work or can't use public transport. However, about 6 out of every 10 motorists who park long term now, including [ma]in commuters and many Council staff, would [pu]blic parking space in Twickenham to make alternative

THE ROYAL BOROUGH OF

North Kensington Controlled Parking Scheme

K'

WHAT ARE YOUR VIEWS

We would like to find out your views about the proposed controlled parking scheme in North Kensington. You can make your views known by filling in this questionnaire and returning it to us using this prepaid return slip, by the 20th October.

Do you think there are parking related problems in your area?

Yes ☐ No ☐ Don't know/No opinion ☐

Would you like to see a controlled parking scheme introduced to help alleviate these problems?

Yes ☐ No ☐ Don't know/No opinion ☐

Comments

If controlled parking is introduced in North Kensington, between which hours do you think it should operate?

9.00am-5.00pm Monday to Friday ☐
8.30am-6.30pm Monday to Friday ☐
8.30am-1.30pm Saturday ☐
8.30am-8.30pm Monday to Friday ☐
8.30am-1.30pm Saturday ☐
Other (please specify)

It would hel[p]

next page.

Montage 19.1: Some examples of consultation leaflets on parking proposals.

maximum allowable parking provision for a new office block in a congested town centre with good public transport access can be restricted to that required for operational and service needs only (see also Chapter 30).

Parking standards prescribe the amount and type of parking provision required by the Local Planning Authority for different categories of development. The requirement can be expressed as an absolute figure, per unit of floor area of development or similar parameter, or as a permissible range. Some local planning authorities have refined their parking guidelines further, by relating them to the characteristics of the area. For example, a larger number of spaces might be required per unit of floor area of new office space to be located at some distance from public transport services than would be required (or permitted) near a major public transport interchange (see also Chapter 28). The proximity and quality of public transport services at a particular location can be quantified and summarised in the form of public transport accessibility indices. Careful monitoring is required to ensure that such standards do not provide an incentive for developers to seek out sites that are less well served by public transport, so that they will be allowed more parking provision.

Where it is not desirable or feasible for parking spaces to be provided actually on the site, it is possible for the developer to pay an agreed sum of money to the Local Planning Authority, to be used by the Authority to fund the provision of the required parking off–site. Such payments, known as 'commuted' payments, are provided for by planning legislation and are negotiated between a developer and a local planning authority. Further details are included in Chapter 27 (see also DOE/DOT, 1995) [Sc].

In order to reduce dependence on private cars for a large proportion of trips, local planning authorities are encouraged to adopt reduced requirements for parking for locations which have good access to modes of transport other than the private car (DOE/DOT, 1994) [Sb]. Strategic policies on parking should be included in Regional Planning Guidance and Structure Plans to avoid the potential for competitive provision of parking by neighbouring authorities [NIe].

19.4 Legislative Background and Responsibilities

The Road Traffic Regulation Act 1984 (RTRA)
The powers to control waiting and loading and to provide and charge for on–street parking are provided by the Road Traffic Regulation Act 1984 (HMG, 1984) amended by the Road Traffic Regulation (Parking) Act 1986, the Road Traffic Act 1988, the Parking Act 1989, the Road Traffic Act 1991 and the Road Traffic Regulation (Special Events) Act 1994 (HMG, 1986; 1988; 1989; 1991 and 1994a) together with their associated Regulations and Orders [NIf].

Highway authorities may prohibit waiting, on–street,

for all or part of the day and may limit the duration of any waiting permitted. Restrictions may also be applied to prevent loading/unloading. Usually, loading restrictions are only applied during peak traffic hours but they can be used more extensively if necessary. A traffic Order must be made in accordance with the requirements of the current procedure regulations (see Chapter 13).

The RTRA 1984 (HMG, 1984) also contains powers to enable the enforcement of parking controls at off–street parking bays. These powers are used by many local authorities to enforce the traffic Orders associated with 'pay and display' control (see Section 19.9) in off–street car parks. Further details are set out in Chapter 13.

The Road Traffic Act 1991
Under the Road Traffic Regulation Act 1984 (HMG, 1984), all parking offences were regarded as criminal offences and subject to criminal law. The Road Traffic Act 1991 (HMG, 1991) provides for the decriminalisation of most non–endorsable parking offences in London and, subject to the approval of the appropriate Secretary of State, elsewhere in the United Kingdom [NIg]. The Department of Transport's Circular 1/95 (DOT, 1995a) [Se] gives guidance to local authorities outside London seeking to apply for decriminalised enforcement powers.

The essence of the Road Traffic Act 1991 is that, in those areas where the new arrangements apply, parking offences are no longer criminal [NIg]. Other provisions are that:
❏ enforcement of non–endorsable offences ceases to be the responsibility of the police and becomes the responsibility of the Local Traffic Authority [NIg];
❏ parking attendants, also known as Parking Control Officers or PCOs, are empowered to place parking tickets (Penalty Charge Notices or PCNs) on vehicles contravening parking regulations and can, in appropriate cases, authorise the towing away or wheel–clamping of vehicles [NIg];
❏ the penalty charges associated with PCNs are civil debts, due to the Local Authority and enforceable through a streamlined version of the normal civil debt recovery process [NIg];
❏ vehicle owners wishing to contest liability for a penalty charge may make representations to the Local Authority and, if these are rejected, they may have grounds to appeal to independent adjudicators whose decision is final. Groups of local authorities, such as the Parking Committee for London, are responsible for setting up and operating the adjudication arrangements [NIg]; and

❑ the Local Traffic Authority retains the proceeds from the penalty charges, which are used to finance the adjudication and enforcement systems. Any surpluses must also be used for traffic management purposes, under the provisions of section 55 of the Road Traffic Regulation Act 1984 [NIg].

However, criminal law and police enforcement remain applicable to endorsable parking offences (broadly those involving dangerous or obstructive parking), to parking in areas outside those where decriminalised parking applies and to some other specific parking offences.

The Road Traffic Act (RTA) 1991 makes a distinction between 'Permitted Parking Areas' (PPAs), where contravention related to parking places, such as meter bays, are decriminalised, and Special Parking Areas' (SPAs), where other parking offences, such as parking on yellow lines or on cycle tracks, are decriminalised. The latter offences also include contravention of off–street parking Orders, which means that off–street parking enforcement is brought into a decriminalised parking regime. In practice, local authorities usually find it necessary to apply for Orders creating both SPAs and PPAs, with the same boundaries, in an area to be controlled, in order to achieve efficient and effective enforcement [NIg].

Further details relating to the RTA 1991 are set out in Chapter 13. The Act also deals with local authorities' setting parking charges and parking penalty charges. These issues are covered under Parking Finance in Section 19.10 below and in Chapter 14.

19.5 Management of On–Street Parking

Waiting Restrictions
Waiting restrictions and parking–control regulations should complement each other. Restrictions can govern where and when drivers are prohibited from waiting, whereas parking controls can establish places where drivers may park, subject to stipulated conditions. These measures can influence, directly, the volume and nature of traffic in an area, by giving more roadspace to moving vehicles or by providing sufficient parking spaces to avoid cruising and reversing manoeuvres by drivers searching for parking spaces. The function and character of a road can be greatly affected by determining where and when different categories of vehicles are permitted to park (see Chapter 20).

Waiting and loading restrictions are used widely in urban areas but care should be taken to ensure that they are used only where they are really needed and at times when they are justified. Measures of this kind are most beneficial in shopping streets and near junctions. They can also protect bus stops, allow vehicles access to the kerb to pick up and set down passengers and allow loading and unloading to take place at the kerb during defined periods. Restrictions, introduced incrementally over many years in response to isolated circumstances, should not be inconsistent with each other. For the same reason, it is beneficial to standardise on the hours of operation. Waiting restrictions should be signed and marked in accordance with the current Traffic Signs Regulations and General Directions (HMG, 1994b) [NIh].

Time Limits and Charges
Time limits may be imposed and charges levied so as to maximise the use of the space available. This also ensures that those who use the facilities contribute to the costs of their provision and maintenance. Where demand for parking exceeds supply, demand can be regulated and reduced by raising the level of parking charges. However, care should be taken not to raise charges to levels which might excessively discourage vehicle–users from visiting the area.

Particular kerbside areas may also be designated for use by specified classes of users, such as doctors or diplomats, or groups of people, such as residents or disabled persons. Appropriate charges can generally be made within schemes of this kind.

Other areas may be designated for bicycle or motor–cycle parking. Cycle parking needs to be considered carefully in relation to the desired destinations. Further information on cycle parking is given in Chapter 23 (see also IHT, 1996).

Shared Spaces
Spaces can be designated for more than one use at different times with or without charges. This provides a flexible form of management in which, for example, residents exhibiting a permit might park free of charge, or with a charge, and visitors might have a time–limit and/or have to pay (HMG, 1986 and DOT, 1986). It should be noted that the more complex the arrangements are the more difficult they are to sign and hence to be understood by drivers. This can lead to problems of enforcement.

Control and Collection of Charges
Any device for control and collection of charges for the use of parking spaces should:
   ❑ be reasonably cheap to install and to maintain, in relation to the estimated revenue;
   ❑ be simple to use and easily understood by drivers;

Method	Advantages	Disadvantages
Parking meters.	❏ Enforcement is straightforward. ❏ Help to impose physical parking discipline. ❏ Generate revenue. ❏ Useful for short–stay. ❏ Help match demand to supply. ❏ Potential of electronic versions.	❏ Relatively expensive to install, operate or adjust to new charges. ❏ Environmentally intrusive. ❏ Cannot be used to favour specific user–groups.
Ticket dispensing machines (Pay and Display meters).	❏ Enforcement is relatively easy. ❏ Cheaper and less intrusive than meters. ❏ Suitable for short– and long–stay. ❏ Potential for separate residents' tariffs.	❏ Drivers have to walk to meter. ❏ Extra signing is required.
Parking discs.	❏ Relatively cheap to operate. ❏ Environmentally unobtrusive.	❏ Enforcement is difficult. ❏ Generate no revenue. ❏ Can discriminate against visitors.
Parking permits/ Season tickets.	❏ Enforcement is easy. ❏ Availability can be restricted to specific types of user. ❏ Generate revenue. ❏ Can be issued for varying time–periods.	❏ No control over duration. ❏ Fraud is possible as holders can allow others to use them. ❏ Fraudulent requests. ❏ Administration efforts is required.
Pre–purchase cards cancelled and displayed by user.	❏ Enforcement is relatively easy. ❏ Cheap to implement and operate. ❏ Environmentally unobtrusive. ❏ Generate revenue. ❏ Price can be changed easily.	❏ Risk of fraud. ❏ Need for outlets to sell cards reduces income.
Limited waiting.	❏ Cheap to install and modify.	❏ Enforcement is very difficult. ❏ Markings and signs can be environmentally intrusive. ❏ Generates no revenue. ❏ Need substantial patrolling.
Specific permitted–vehicles (eg vehicles for disabled, motorcycles, car pools).	❏ Spaces can be marked out.	❏ Enforcement can be difficult ❏ Permits need to be displayed when vehicles are used in a specific way.

Table 19.1: Control devices and systems for on–street parking.

- ❏ be secure and reliable;
- ❏ be capable of providing information for management and auditing purposes;
- ❏ deter fraud and assist enforcement;
- ❏ comply with relevant regulations; and
- ❏ be flexible enough to allow the charges and time periods to be readily adjusted and possibly to allow payment by debit/credit cards and/or stored value smart–cards.

A number of generic control devices and systems for levying parking charges are listed in Table 19.1, together with some of their advantages and disadvantages.

Controlled Parking Zones (CPZ)

The purpose of a Controlled Parking Zone (CPZ) is to provide a uniform set of waiting restrictions over a given length of road or a street network and to reduce sign clutter by removing the need for 'time–plates' within the zone, except on lengths of road where the restrictions apply at different times from the rest of the zone. The zone–entry signs give details of the times and the days when the restrictions operate and yellow line–markings indicate where waiting restrictions are in force. Designated parking bays within the zone can provide for a variety of different types of waiting or loading facilities.

It is important that CPZs should not cover too large an area, as this can lead to problems for drivers who find it difficult to remember the restrictions listed on the zone–entry signs. CPZs should be limited to, for example, shopping areas or similarly well–defined areas. Conventional time–plate signing, without zone–entry signs, should be used with the yellow line markings where the waiting restrictions are complex or the areas subject to control are too large.

Restricted Zones

In environmentally sensitive areas, such as special heritage or conservation areas, the Department of Transport [Se] may authorise a restricted zone–signing scheme which dispenses with the use of yellow line markings to denote waiting restrictions. This type of dispensation tends to be confined to old and picturesque town or village high streets, where there are uniform restrictions and adequate provision for adjacent off–street parking. In environmentally sensitive areas, where this dispensation is not allowed, conventional yellow line waiting restrictions can nevertheless use a narrower paler line (see Chapter 13).

Loading/Unloading

Consideration should be given to providing other designated spaces to meet particular needs. For example, loading bays may be necessary along streets with commercial and/or industrial premises. Special provision may also be required at places where cash, mail or other valuables are delivered or collected.

Doctors and Diplomats

Medical practitioners with residences or surgeries in densely built–up areas may also need designated spaces (DOT, 1975) [Sf] and some parking spaces, in central London and elsewhere, are reserved for the use of diplomats. Schemes also exist, in some areas, whereby parking attendants have discretion in relation to vehicles parked illegally by doctors, nurses and midwives when visiting patients.

Public Transport

Vehicles used to provide public transport services, for example for tourists, often need to wait for extended periods, apart from the usual requirement for stopping places for picking up and setting down passengers. In some cases, it may be appropriate to designate particular places on–street for use as bus stands, to serve as crew–change, terminus or schedule adjustment points. Special facilities, preferably off–street, should be provided near places attracting many tourists or visitors and on–street coach meters may be appropriate in some circumstances. Each case should be treated on its merits and useful guidance is provided in the ALBES Code of Practice (ALBES/DOT, 1986).

Lorries

Lorries, and especially large goods vehicles (HGVs) away from their base and particularly overnight, require properly designated parking accommodation. They can create access and environmental problems if they are parked indiscriminately on–street, especially in residential areas. Where local authorities identify lorry–parking as a problem in an area, then customised off–street lorry parks should be provided and signed and/or appropriate on–street lorry–parking areas designated, for example in industrial estates (see Chapter 25).

Layout of Parking Spaces

Careful consideration must be given to the siting of any on–street parking places:
- ❏ to avoid creating a road safety hazard, by obstructing visibility near bends, junctions or places where significant numbers of pedestrians cross the road;
- ❏ to create suitable crossing points for pedestrians, to avoid the inconvenience and danger caused by long unbroken rows of parked vehicles;
- ❏ to avoid danger to cyclists from vehicular traffic

passing close to the designated parking bays, especially where a narrowing of the carriageway results;

❑ to avoid impeding the free flow of traffic at places where this is important to the role of the street in question;

❑ to maintain reasonable and adequate access to premises, including access for loading and unloading, particularly where there are security considerations, for example for mail or bank deliveries; and

❑ to avoid obstructing access to fire hydrants and interfering with detection loops or other traffic monitoring equipment.

Individual bays should be large enough to permit drivers to park reasonably quickly, thereby reducing the risk of significant interruption to traffic flow. If individual bays are not marked, the number of cars that can park in a length of road may be greater than if it were marked out. If there is a charge for use of a particular bay then the bay must be marked. Parking spaces should be signed and marked in accordance with the current Traffic Signs Regulations and General Directions (HMG, 1994b) [NIh].

Size and Position of Bays

Parking bays can be parallel to the kerb or angled to it. Typical layouts for waiting and parking controls applied to different categories of road are indicated in Table 19.2. The minimum size for a bay, parallel to the kerb, should be 1.7 m in width and 4.5 m in length but variations up to 2.5 m and 6.0 m respectively are common, to allow for different site conditions and sizes of vehicles. Wider bays should be provided where the space is for the use of those with a physical disability (IHT, 1991). In addition, there may be some local need for bicycle and motorcycle parking (HMG, 1994b and DOT, 1980) [NIh].

Wide streets give scope for both moving and stationary vehicles to be accommodated and making streets one-way can often allow additional parking spaces to be provided. Layouts should minimise environmental intrusion, for example, by arranging parking on one side of a street only and using appropriate landscaping.

Clear Road-Widths for Traffic

The extent to which it is necessary to preserve a clear width of carriageway for traffic flow depends on the type of road in question (see Chapter 11 on Road Hierarchy). Suggested widths for locations remote from road junctions, where some disruption to traffic movement may be more safely accommodated, are:

❑ on district distributor roads, with 24-hour flows

in excess of 5000 vehicles, and roads carrying HGVs with three or more axles and/or frequent two-way bus flows, the minimum clear running width should preferably be seven metres, with 6 metres as the absolute minimum;

❑ on local distributor roads, with 24-hour flows between 2000 and 5000 vehicles, the preferred minimum clear running width is six metres, with five metres as the absolute minimum;

❑ on access roads, with 24-hour flows between 500 and 2000 vehicles, the preferred minimum clear running width is six metres, with 4.5m as the absolute minimum; and

❑ on minor residential access roads, including short culs-de-sac, with 24-hour flows less than 500 vehicles, the clear running width should be at least 3.5m. This does not permit the free flow of two-way traffic but it is sufficient to allow unhindered access for emergency and service vehicles.

19.6 Residents' Parking Schemes

Residential streets on the fringes of town centres, near suburban railway stations and other significant destinations, often attract commuters, shoppers and other visitors to park for long periods. This results in local residents having difficulty in parking near their homes, if they do not have sufficient space in private driveways or garages.

Although there is no inherent legal right for any vehicle owner to park on the public highway, residents' parking schemes are often introduced to assist those living in the area and to make town centres and fringe areas more attractive places in which to live. They are particularly applicable to areas with older terraced housing, where there is seldom any off-street parking available within the curtilage of the dwellings. These schemes impose constraints on both residents and non-residents and considerable care must be taken to ensure that they are justified (DOT, 1974) [Sg].

Investigations into such schemes should consider:

❑ the size of the area which would need to be treated, bearing in mind the alternative locations which might be used by displaced parkers and the effects on streets just outside the area, whose residents may not have hitherto experienced any parking problems and are likely to resent what they may see as unnecessary controls;

❑ the types of measure to apply, whether they should be applied to whole streets or only to short lengths of street and whether restrictions by time of day are appropriate;

ROAD HIERARCHY	TYPICAL CONTROLS WHICH MAY BE APPLIED	TYPICAL LAYOUTS
PRIMARY DISTRIBUTOR	Waiting and loading prohibited to ensure traffic flow. Clearway arrangements may also be considered	Waiting and loading prohibited
DISTRICT DISTRIBUTOR	Waiting prohibited to ensure traffic flow. Loading ban may also be applied especially during peak hours. Urban Clearway arrangements may be considered.	Waiting prohibited—loading restricted during peak hours
LOCAL DISTRIBUTOR	Waiting prohibited or restricted on at least one side to ensure traffic flow where parking demand is sufficient. Peak hour controls may be sufficient where parking is less. Bus stops may be protected by waiting restrictions or Bus Stop Clearway.	Peak hour waiting restriction / Waiting restricted / Single wide yellow line indicates Bus stop clearway / STOP / Bus stop clearway
SHOPPING OR COMMERCIAL STREET	Waiting may be prohibited or restricted on at least one side to ensure traffic flow. Waiting may be limited and/or charges applied to ensure adequate turnover of space. Charging by parking meter or pay & display meter. Spaces reserved for Orange Badge holders should be provided at points most convenient for the source of parking demand. Loading bays may be provided at intervals or where a particular problem has been identified. Bus stops may be protected by waiting restrictions or Bus Stop clearway. Taxi ranks may be provided where a local demand has been identified. Similarly, parking places for special users e.g. motor cycles, could be provided.	Ticket machine / Parking for specific vehicles / Loading bay / DISABLED / Angled parking where width permits / Metered parking places / Motor cycles / TAXI / LOADING ONLY / Waiting prohibited to ensure traffic flow / Parking for disabled / Taxi rank
RESIDENTIAL ACCESS ROAD	No restrictions where parking demand is low. Elsewhere junction radii and other sensitive sites may be protected by localised restrictions. Where parking demand is high resident permit holder spaces may be provided. Adjacent to, for example town centres or commuter railway stations, peak hour or limited waiting may be applied to deter long stay parking.	Bus stop boarder / Resident permit holder parking / Junction radii protected if problem identified / Limited waiting

Not to scale.

Table 19.2: Typical waiting and loading controls applied to a hierarchy of roads.

❏ the enforcement implications resulting from the types of measure to be introduced, including the cost;

❏ the advantages and disadvantages to residents and any inconvenience to non–residents;

❏ criteria for allocating permits for residents, for example allocating permits only to the occupants of dwellings with no off–street parking available, and whether or not to charge for the parking permits; and

❏ how to cater for visitors' cars, service deliveries and emergency vehicles.

Four methods of control are described in Table 19.3, together with their advantages and disadvantages.

19.7 Parking for Disabled People

The most convenient spaces, on–street and in public car parks, should be allocated for the exclusive use of people who are physically disabled. Lowered kerbs at adjacent footways should be introduced to assist wheelchair users and advice should be sought from this group in designing and installing equipment, such as parking meters and ticket machines (DOT, 1984b [Sh] and IHT, 1991).

Provision of on–street parking for disabled people is also made under the Orange Badge Scheme, whereby vehicles being used by a badge–holder may park for up to 3 hours on yellow lines, without time–limit, where others may park only for limited periods and without charge or time–limit at parking meters. However, parts of Central London are excluded from

Photograph 19.1: Unauthorised parking on footways.

this scheme (DOT, 1982 [Si], 1984a [Sj] and 1995b [Sk]).

In cases of special hardship, holders of Orange Badges may apply to the local Highway Authority for a designated disabled person's space outside their home or business premises. Even if such a space is provided, however, it does not give an exclusive right to park and must be available for use by other badge holders. A possible alternative to overcome this problem (though not yet tested in the courts) might be to designate the space for permit holders only and then to issue only one permit.

19.8 Parking on Footways and Cycleways

Unauthorised parking on footways and cycleways causes problems (see Photograph 19.1). Parked vehicles obstruct pedestrians and are a hazard to cyclists and to disabled, blind and elderly people. Heavy vehicles can damage pavements and underground services. Moreover, it is illegal to park on the footways where yellow line waiting restrictions operate, if the vehicle is a heavy commercial or if a vehicle of any kind is left in a dangerous or obstructive position. It is also an offence under section 72 of the Highways Act 1835 to drive along footways or cycleways [NIi].

London, Worcester and Hereford have taken private Act powers to ban pavement parking. Only the London ban has so far been introduced (1966), although some other authorities have used traffic regulation Orders under the Road Traffic Regulations Act (HMG, 1984) to achieve the same end [NIj].

Under the 1991 Road Traffic Act (HMG, 1991), the parking of heavy goods and other kinds of vehicles on the footway becomes a decriminalised offence, in those areas where a local Act of Parliament is in force. In addition, local authorities are able to make traffic regulation Orders banning footway–parking, which can be enforced by them under a decriminalised system [NIg].

When contemplating the introduction of footway or cycleway parking bans, it is important for local authorities to consider the wider effects, including the availability of alternative legal parking facilities, especially in older areas where off–street parking space may be limited.

The installation of bollards, guardrail or planting of trees can provide an effective physical means of preventing parking on footways or cycleways.

Method	Description	Advantages	Disadvantages
Parking for residents or permit holders only.	❑ Permits to park are issued with or without charge to residents.	❑ Usually ensures that residents can park in the streets at any time, subject to total residential demand.	❑ Inflexible ❑ Can be over–restrictive and affect normal activities, such as servicing and visitors' parking. ❑ Does not necessarily guarantee residents a space.
Limited waiting exemptions (not recommended, see DOT Circular Roads 31/74).	❑ A Traffic Regulation Order is made, imposing waiting restrictions in the area, with specific exemptions for residents.	❑ Permits reasonable level of access for other vehicles. ❑ Sometimes preferred by the police, as it reduces the number of complaints from residents about offenders.	❑ Does not necessarily guarantee residents a space ❑ Residents still subject to the laws of obstruction and under certain circumstances, they may be prosecuted. ❑ Parked vehicles on a restricted street may encourage other drivers to disregard restrictions. ❑ May be difficult to enforce.
Parking places with exemption from charges (or separate charges) for residents.	❑ Parking places are designated by a Regulation Traffic Order, with exemption from charges, or separate permit charges, for residents.	❑ Does not remove all the kerbside parking from the on–street parking stock. ❑ Offers considerable flexibility.	❑ Does not necessarily guarantee residents a space ❑ Equipment costs are incurred
Restriction by time of day.	❑ Waiting is limited for specific periods (for example, during a morning peak period to discourage commuter parking).	❑ Relatively easy to enforce. ❑ Does not impinge too heavily on normal activities.	❑ Only effective where problems are caused by all–day parking. ❑ Residents are also unable to park during the restricted periods.

Table 19.3: Measures to implement resident's parking schemes.

19.9 Coach Parking outside London

Many towns, but particularly those with historic attractions, suffer from coach–parking problems. Several requirements should be considered, apart from the mere provision of adequate space and where they can set down and pick up passengers. These include:

❑ information for drivers;
❑ facilities for parking;
❑ facilities for passengers; and
❑ waste disposal.

The provision of adequate facilities for coaches, drivers and passengers are an important factor in attracting visitors by coach to a particular town or city. It is worth remembering that coach drivers frequently have considerable discretion as to where they stop to let their passengers spend money.

Drivers' Information

Coaches are large vehicles which require considerable room to manoeuvre. Annoyance can be caused to residents when coaches use unsuitable roads for access or parking. Drivers should therefore be provided with adequate, clear, information on routes into and around the urban area, including where they can park, for how long, what facilities are available and any charges that will be made.

Clear signing should be provided to coach parks with, if possible, current information on the availability of spaces. It is important that, if one park is full, the route to the next available one is clearly signed and easily accessible by a coach. Attention must be given to routeing, particularly in respect of any height restrictions. Many single–deck coaches are 3.5m high and double–deckers, at 4.0m or 4.2m, are quite prevalent.

Special information leaflets should be made available to interested organisations and operators. A limited on–street survey will provide information on the more common operators entering the urban area. The Confederation of Passenger Transport (CPT) (see Section 19.14) is an organisation which can give advice as to who else should receive information leaflets. Liaison with the organisers and operators of major tourist attractions in the locality may be a useful source of information, in identifying the likely demand for coach parking.

Parking Charges

If designated parking spaces are provided, either on–street or off–street, authorities may wish to levy a charge. It is important that drivers do not perceive any charge as being excessive for the facilities provided, otherwise they will attempt to park free, usually at an unsuitable location. Since coach parking charges are usually higher than equivalent car parking charges, it is important that payment meters give change or that there is a change facility close by.

Passenger Facilities

A high proportion of coach passengers have impaired mobility. Facilities at coach parking places should therefore be carefully considered or, alternatively, the parking place relocated to a more suitable site. Toilet facilities, shelter and seating are needed close to the set–down point. Refreshment kiosks or cafes should be encouraged, especially if passengers are numerous or likely to be faced with a long wait.

Clear pedestrian signs to the town centre or other attractions are important, with distances clearly stated. Information on facilities available, together with simple maps, is useful but it should be remembered that many people find maps difficult to understand.

Waste disposal

Waste from coaches can include toilet waste, as well as general litter. Drivers also often like to use stop–overs to wash the outside of their vehicle. If large numbers of coaches are attracted to a particular location, it is worth providing waste disposal facilities, since this will obviate unauthorised disposal and assist in keeping the area clean and attractive.

19.10 Coach Parking in London

In central London, there is a high demand for coach facilities, particularly at the major tourist attractions. However, these locations are generally not suitable for providing facilities for coaches and their passengers. In particular, the provision of overnight parking is inappropriate in central London and, as a result, authorities have introduced a number of coach bans in residential areas. Authorities do have the ability to wheel–clamp coaches that park illegally.

The London Coach–Parking Map is produced on a regular basis by the Metropolitan Police, including translations into the major European languages, to assist coach drivers in the area. Parking facilities should be provided, where possible, and the needs of coaches should be considered when new developments, adjacent to areas with high coach usage, are being considered. There are, however, particular difficulties in collecting parking charges, due to the number of foreign coach operators.

19.11 Off–Street Parking

Off–street parking is normally provided, either in the form of open surface car parks, in purpose–built multi–storey car parks or within buildings on one, or more, parking floors of a mixed–use building, such as an office building.

Surface car parks are relatively inexpensive to construct and are generally preferred by car–users, although some prefer the protection of a roofed building in both hot and cold weather. However, they do not make efficient use of land, in terms of the area that has to be devoted exclusively to parked vehicles.

Multi–storey car parks permit a more intensive use of space, which is important in areas where land is in short supply and land values are high, but their construction, operating and maintenance costs are

Photograph 19.2: Example of a well–designed multi–storey car park building. Courtesy: David Nicholls.

significantly greater than those of surface car parks. However, it may not be necessary to justify the cost of construction independently if a multi–storey car park is being provided as an integral part of some other development, such as high density housing, a retail centre or a prestige public building.

In the design of car parks, close attention must be given to the external appearance and to the operational characteristics and efficiency. There are many examples of well–designed, bright and efficient multi–storey car parks (see Photograph 19.2).

Size and Layout

The minimum practical size of site suitable for a multi–storey car park is determined by the need to provide ramps between floors and is generally accepted to be 35 m square. The size of a multi–storey car park will be determined by such factors as:

❑ the amount of land available;
❑ the number of spaces required, bearing in mind the need to justify, at least in part, the capital costs involved in terms of the expected net revenues; and
❑ the impact of the traffic generated by the car park on the external road network.

Short–stay, usually higher–priced, parking in the more central locations will have a greater turnover for a given level of occupancy and will therefore attract more traffic throughout the day. Long–stay parking,

especially when directly associated with a large office or factory building, will produce high traffic flows only in the morning and evening peak periods.

The number of spaces provided in any single facility will be constrained by the following:

❑ the generally accepted maximum capacity for an integrated car park, with several aisles accessed directly by ramps, should be 1600 spaces and a single search path should not exceed 500 spaces;
❑ local planning authorities often impose a limit on building–heights and, hence, the number of parking levels for a particular site, on planning and environmental grounds; and
❑ the costs of construction underground are likely to be significantly greater than those above ground.

The total number of spaces available in a car park is termed the storage, or static, capacity, as distinct from the dynamic capacity, which is the maximum in–flow or out–flow of vehicles from the whole car park (DOE/DOT/TRL, 1969). The most important determinant of dynamic capacity is usually the type of control employed at entry and exit, including the method of collecting any charges. With minimal formalities on entry or exit, the dynamic capacity is determined by the capacity of the circulatory aisles, 800 to 900 vehicles per hour for most layouts. As a general rule, the dynamic capacity should be sufficient to permit up to 25% of the static capacity to

Photograph 19.3: VMS signing for car parks.

enter or leave the car park within 15 minutes (ie up to 100% turnover in an hour).

The maximum practical occupancy is likely to be lower than the theoretical static capacity, particularly where there are no marked–out bays or car park staff to ensure disciplined parking. In addition, since cars are arriving and departing simultaneously, newly–vacated spaces may be missed by those already in the car park searching for a space. Where entry is controlled, deliberate under–capacity margins of about five per cent, depending on size and turnover, are sometimes introduced to overcome this problem. Where parking discipline is particularly poor and spaces between columns are badly designed, actual occupancy can be as much as 50 % below the theoretical storage capacity. Conversely, in some small private parking areas, where drivers are known to one another and parking is even tolerated in circulation areas actual occupancy may be up to 125% of static capacity.

Entry and Exit Controls and Payment Systems

The type of control, if any, to be used on entry and/or exit is most important and usually determines, or will be determined by, the method of collecting any charges. In general, entry to a car park should not be permitted unless an appropriate space is available. Entry may be controlled by a lifting–arm or a rising–step barrier. Rising–step barriers should be supplemented by traffic signals, which show red when the barrier is raised to reduce the chance of damage either by equipment malfunction or driver error. Exits may be controlled in a similar way or by using collapsible plates, hinged on their leading edge, to ensure that vehicles can only pass over them in one direction. Where parking is free, or where payment is

made on entry or using a pay–and–display system, exits need not be controlled.

A variety of payment systems is in common use, including:

❏ 'Fixed Charge' – where payment of a fixed charge is made to a cashier or using an automatic machine on entry to, or exit from, the car park;

❏ 'Pay–On–Exit' – where a ticket is issued on entry and payment is made to a cashier or automatic machine on exit, according to the scale of charges and the time spent in the car park. When automatic machines are in use the failure of a driver to have the correct change, or mechanical breakdown in the system, can result in serious congestion. Equipment is available which allows payment by an electronic device, such as a stored–value smart card;

❏ 'Pay–and–Display' – where, after a space has been found, a ticket is purchased from a machine within the car park and displayed on the vehicle. This system eliminates delays at the entrances and exits but, where parking is permitted for more than one fixed period, the driver must decide how much time to purchase before leaving the vehicle. A Traffic Order is required to enable those drivers who do not pay the correct charge to be fined and, if necessary, prosecuted. This system is often

Photograph 19.4: VMS signing for town centre car parks.

criticised because it penalises drivers who may genuinely have misjudged their length of stay; and
❏ 'Pay–On–Foot' – where a ticket is issued on entry and payment is made to a cashier or using an automatic machine on departure before the driver returns to his vehicle.

Credit cards and decrementing cards are increasingly being used for payment of car–park charges and new devices, such as in–car transponders with smart cards, are likely to become available in future (see also Chapter 18).

Parking charges should be clearly displayed at the entrances to car parks along with other information about the terms and conditions of use, such as maximum length of stay, excess charge offences, and whether Orange Badge holders may park free. An 'escape route' should also be provided for drivers who choose, at the last moment, not to enter and pay.

Access and Circulation
To prevent queueing at the point of entry, the entry capacity should be equal to, or greater than, the maximum anticipated arrival rate. An access road should provide a queueing reservoir, for those occasions when the entry to the car park is operating at or near its dynamic capacity, and it should be designed to assist the transition from the higher speed travel on the external road network to walking speed within the parking area. Access roads should be used exclusively for entry into the car park so that traffic on the adjacent roads is not unnecessarily delayed.

The rate of out–flow at the exit from the car park should not exceed the reserve capacity of the road onto which it discharges and priority must be retained on the external road system, so that any queueing takes place within the car park (IHT/IStructE, 1984).

Floor Levels and Ramp Arrangements
The circulation system within a multi–storey car park depends upon the type of structure. There are four main types: flat deck, split level, ramped floor and warped slab (IHT/IStructE, 1984).

Ramps may be used solely to distribute traffic between levels (so–called 'clearway ramps') or they may also act as parking aisles, giving direct access to parking bays. Ramps may be one–way or two–way, although the latter generally require higher design standards for visibility and clearance to structures.

Aisles give direct access to individual parking–bays.

The minimum recommended width for a one–way aisle is six metres, and 6.95m for two–way operation, although this may be reduced if parking bays are angled. Parking–bay widths will depend on the use made of the parking facility; 2.3m is the minimum, 2.5m is desirable for shoppers and between 3.2m and 3.6m for use by disabled people.

Increased bay–width permits easier and quicker manoeuvring into and out of the bays, does not impair aisle–capacity and makes getting into and out of vehicles more convenient. Any columns between bays should be positioned so as not to obstruct the opening of car doors. The additional width for disabled parkers may be shared between two adjacent bays.

Signing
It is important that public car parks are adequately signed to assist and direct drivers who are unfamiliar with the area. This helps to avoid congestion and reduces the amount of time and fuel wasted while searching for places to park. Where a choice of car parks is available, signs should direct drivers to the one most appropriate for their purpose, such as long–stay or short–stay or parking provided in conjunction with a particular event. Consideration could be given to introducing computer–controlled variable message signing (VMS), linked to entry and exit ,to direct drivers to car parks where spaces are still available. Two types of variable message signs are illustrated in Photographs 19.3 and 19.4. It is essential that the information given by variable message signs is reliable if drivers' confidence and compliance is to be maintained (see Chapter 15). Direction signs to car parks should not be used as a means of advertising for the benefit of the operator, whether public or private.

A comprehensive system of signing and road marking should be provided on routes within the car park to assist circulation, to achieve the most appropriate search path and to find the quickest exit. Where several search paths are available, it may be helpful to indicate which levels have spaces available.

Automated Car Parks
A number of different mechanical devices have been developed for parking and storing cars, although they have not been widely adopted in Britain. These range from simple devices for placing one car above another to complex computer–controlled systems, which usually require the cars to be placed on pallets or plates, which are then closely stacked using a combination of lifts and rollers, thereby reducing, or even dispensing with, the need for circulation ramps

and aisles. The disadvantages of these devices are the increased costs of maintenance and, particularly, the delay in parking and recovering vehicles, especially at peak times.

Pedestrian Facilities in Multi–Storey Car Parks

A new car park may affect existing pedestrian routes and there may be a need for replacement or additional footpaths, pedestrian crossings and signing for pedestrian routes.

Within the car park, ticket machines and entrances to lifts and stairways should be demarcated from parking areas. Signs should direct pedestrians to the appropriate exit and each level should be given a unique identity to help drivers to find their cars on their return. Letters or numbers are often used but colour schemes or pictorial signs such as animals or flowers may be easier to remember.

The holders of Orange Badges for disabled people (see Chapter 13) should have the most convenient spaces in a car park reserved for their use and ticket machines must be easily accessible to them, unless charges for them are waived. Care must be taken to ensure that disabled people can leave the car park easily, preferably without having to rely on lifts as these may occasionally be out of order.

Management of Car Parks

Car parks must be carefully managed if they are to provide a high standard of service to users. Long–term maintenance plans, covering the fabric of the building, running surfaces and equipment, must be drawn–up so that appropriate budgetary provision can be made. Day–to–day attention to cleansing, removal of graffiti, repair of defective lights, signs, lifts and ticket machines is essential. Carefully drawn–up maintenance contracts can ensure that service levels are maintained at relatively low cost. Staff training is also important and specific training courses for parking attendants are available.

19.12 Parking Finance

Parking policy should be regarded as an integral part of traffic management and not simply as a revenue–raising activity. Nevertheless, substantial costs and revenues are often involved and these require careful financial management.

Monitoring of income, occupancy levels and ticket sales is essential, as car parks are valuable assets which should be intensively used. However, a car park which is frequently full may indicate a need for parking charges to be increased to bring demand more closely into line with available supply. In principle, it is desirable that, at the very least, the costs of providing and maintaining parking facilities should be met by the users of those facilities. Revenues in excess of this 'break–even' should be set to fulfil transport policy objectives.

Off–street car parks are normally provided and funded either by local authorities, under powers contained in the Road Traffic Regulations Act 1984 (RTRA) (HMG, 1984) [NIf], or by the private sector, as commercial ventures or in conjunction with other developments which the car park serves, or jointly by the private and public sectors in partnership. Local authority funding sources include accumulated funds provided by developers as commuted payments (see Section 19.2), as well as funds arising from surpluses from on–street or off–street parking operations.

Under the RTRA 1984, local authorities are entitled to retain the net revenues resulting from on–street parking charges, after allowing for maintenance and, where appropriate, local authority enforcement costs. These revenues derive from normal parking charges and excess charges. Revenues from fines imposed by the courts on vehicle–owners who contravene the regulations accrue to the Exchequer, although prosecution costs may be awarded to the Local Authority in certain circumstances. The proceeds of Fixed Penalty Notices (FPNs), issued by the police, also accrue to the Exchequer. Some local authorities, who carry out their own on–street enforcement, have adopted the practice of issuing 'Notices of Intent to Prosecute' (NIPs), which offer the offenders the opportunity to pay a fixed amount to the Local Authority, so as to avoid being prosecuted for the offence.

Under the RTA 1991 (HMG, 1991), local authorities are able to retain the revenue they receive from decriminalised parking penalty charges and to use it to fund their enforcement activities [NIg]. Any surpluses can be used to improve off–street facilities in their area or, where this is unnecessary or undesirable, for certain other transport–related purposes. Circular 1/95 (DOT, 1995a) [Sd] requires that each local authority operating a decriminalised parking regime '...should aim to make the new system overall at least self–financing as soon as practicable'. In assessing its performance against this objective, a local authority may take into account costs and revenues from its off–street parking operations. However, the attainment of this objective should not be '...at the expense of the safety and traffic management objectives of decriminalised parking enforcement or be achieved by setting unreasonable levels of penalty and other charges'.

Authorities are thereby encouraged to treat their off–street and on–street parking operations as a single financial account, which they should seek, at least, to balance. '...Local authorities should avoid using parking charges as a means of raising additional revenue or as a means of local taxation' (Circular 1/95) (DOT, 1995a) [Se].

19.13 References

ADC (1993) 'Parking Policy': Association of District Councils Seminar, December.

ALBES/DOT (1986) 'Highways and Traffic Management in London: A Code of Practice', Stationery Office.

DOE (1996) Revised Planning Policy Guidance Note 6: 'Town Centres and Retail Developments', DOE [Sa] [Wa].

DOE/DOT (1994) Planning Policy Guidance Note 13: 'Transport', DOE/DOT [Sb] [Wa].

DOE/DOT (1995) 'PPG 13 – A Guide to Better Practice', DOE/DOT [Sc] [Wa].

DOE/DOT/TRL (1969) Report–LR221, 'Parking: Dynamic Capacities of Car Parks', DOT.

DOT (1974) Circular – Roads 31/74 (WO 214/74), 'Resident Parking Schemes', DOT [Sg].

DOT (1975) Circular – Roads 22/75 'Car Parking For the Medical Profession', DOT [Sf].

DOT (1980) 'Traffic Signs Manual', Chapter 5, Stationery Office.

DOT (1982) Circular – 4/82 (WO 49/82), 'Orange Badge Scheme of Parking Concessions for Disabled and Blind People', DOT [Si[.

DOT (1984a) Circular – Roads 2/84 (WO4/84). 'Orange Badge Scheme Parking Concessions for Disabled and Blind People', DOT [Sh].

DOT (1984b) Circular – Roads 6/84 (WO 57/84), 'Parking for Disabled People', DOT [Sj].

DOT (1986) Circular – Roads 6/86 (WO39/86), 'Road Traffic Regulation (Parking) Act', DOT.

DOT (1995a) Circular 1/95, WO 20/95. 'Guidance on Decriminalised Parking Enforcement Outside London', Stationery Office [Sf].

DOT (1995b) Traffic Advisory Leaflet 5/95, 'Parking for Disabled People', DOT [Sk].

HMG (1984) 'Road Traffic Regulation Act 1984', Stationery Office.

HMG (1986) 'Road Traffic Regulation (Parking) Act 1986', Stationery Office.

HMG (1988) 'Road Traffic Act 1988', Stationery Office.

HMG (1989) 'Parking Act 1989', Stationery Office.

HMG (1991) 'Road Traffic Act 1991' Stationery Office.

HMG (1994a) 'Road Traffic Regulation (Special Events) Act 1994' Stationery Office.

HMG (1994b) 'Traffic Signs Regulations and General Directions 1994', Stationery Office.

IHT (1991) 'Reducing Mobility Handicaps: Towards a Barrier–free Environment', The Institution of Highways & Transportation.

IHT (1996) 'Cycling–friendly Infrastructure', The Institution of Highways & Transportation.

IHT/IStructE (1984) 'Design Recommendations for Multi–storey and Underground Car Parks' (2nd Edition), IHT/IStructE.

LAA (1992) Local Authority Circular 5/92: 'Traffic In London: Traffic

Management and Parking Guidance', Local Authorities Association.

19.14 Further Information

AA (1993)   'Women and Cars: Emancipation, Enrichment and Efficiency', AA Public Policy Department.

ACLA (1992)   'Enforcement of Parking regulations'; The Audit Commission for Local Authorities and the NHS in England and Wales. No. 11.

ACTO (1986)   'A Guide to Car Parking Management', Association of Chief Technical Officers.

CPT   Confederation of Passenger Transport, 52 Lincoln's Inn Fields, London, WC2A 3LZ.

CSS (1995)   'Parking Decriminalisation', Desk Top Study, County Surveyors' Society.

DOT (1992)   LA Circular 5/92 'Traffic in London: Traffic Management and Parking Guidance', DOT.

DOT/LAA (1987)   'Traffic and Parking: Tapworth Report', DOT/Local Authorities Association.

GOFL (1994)   'Strategic Guidance for London for Planning Authorities' (Consultative Draft); Government Office for London, Stationery Office.

Hill JD, Shenton CC and Arnold AJ (1987)   'Multi–storey Car Parks', British Steel Corporation.

Kerrigan M (1992)   'Measuring Accessibility – a Public Transport Accessibility Index', paper to PTRC Summer Annual Meeting.

LPAC (1993)   'Advice on a Parking Strategy for London (Supplementary Parking Advice)', London Planning Advisory Committee.

NEDO (1991)   'Company Car Parking', National Economic Development, Office Traffic Management Systems Working Party (April), NEDO.

Nicholson, J (1995)   'Parking in Town Centres – The Key to Sustainability', Highways & Transportation, 42(11).

PCfL (1993)   Parking Committee for London, 'Code of Practice on Parking Enforcement;' Shaw & Sons Limited.

Trani C et al (1991)   'Parking Policies', PIARC.

Chapter 20 Traffic Calming and the Control of Speed

20.1 Introduction

Speed is a significant factor in about one third of accidents in the United Kingdom. This is particularly so in urban areas, where speeding vehicles also adversely affect the quality of life of many communities. This does not necessarily imply that vehicles simply exceed the speed–limit applying in the area in question, although many often do, but rather that the speed is inappropriate for a particular section of road or set of conditions. For example, all roads within a 30 miles/h limit are not suitable for speeds of 30 miles/h, regardless of the time of day, state of the weather, level of pedestrian activity or prevailing traffic conditions. However, speed–limits are often interpreted this way by drivers. The speed perceived as appropriate at any time depends also on the viewpoint of the individual ie as a driver, as a local resident or as a vulnerable pedestrian or cyclist.

The quality of life in towns and villages can be adversely affected by growing volumes of traffic and traffic travelling at inappropriate speeds. The demand for on–street parking often competes with the need for delivery facilities and space for pedestrians. Many urban roads were not built to accommodate the level and speed of traffic to which they are now subjected. Pedestrians, especially children, the elderly and the disabled, and cyclists are particularly intimidated by traffic travelling too fast in urban areas. Not only does traffic speed increase the potential for accidents but it also creates a barrier that divides communities.

Road layout can give drivers the impression that they are travelling at a safe speed, even when exceeding the speed–limit, but alterations to layout, which emphasise the overall environment of a street, can change and correct such mistaken perceptions. Blanket speed–limits of 30 miles/h or 40 miles/h are often applied but are not widely respected and create heavy demands for enforcement on police resources. Roads with generous dimensions encourage, rather than inhibit, speed. The link between speed and accidents is discussed in Chapter 16.

The largest single contributory cause of accidents (95%) is human error. The attitude, culture or behaviour which leads to such errors can be influenced not only by physical changes to the road

Photograph 20.1: Calmed traffic and enhanced enviroment – Teignmouth, Devon.

layout (see Photograph 20.1) but also by enforcement, education, training and publicity .

To be effective, the management of speed has to include the different elements of:
- ❑ traffic calming;
- ❑ speed–limits;
- ❑ enforcement; and
- ❑ education, to change attitudes to speeding.

20.2 Traffic Calming

The term 'traffic calming' covers a range of traditional and new techniques, designed to reduce the adverse effect of traffic in urban streets. The Buchanan Report (Buchanan *et al*, 1963), in defining 'environmental areas', advanced the concept of applying traffic management techniques, on an area–wide basis, to achieve safety and environmental targets. Traffic calming has given that concept a renewed emphasis.

Powers to construct traffic–calming measures are detailed in The Traffic Calming Act 1992 (HMG, 1992), which amended the Highways Act 1980 (HMG, 1980) [Sa], by the addition of sections 90G, 90H and 90I, which allow works to be carried out "...for the purposes of promoting safety or preserving or improving the environment through which the highway runs" [NIa]. Other powers contained in Part V of the Highways Act 1980 [Sa] include those under sections 64 (roundabouts), 68 (pedestrian refuges), 75 (variations in relative width of carriageways and footways) and 77 (alteration of the level of a highway). The Highways (Traffic Calming) Regulations 1993 (HMG, 1993) [NIb] [Sb] and The Highways (Road Humps) Regulations 1996 (HMG, 1996) also apply in England and Wales (see Chapter 13) [NIc] [Sc].

Where a traffic-calming feature is not covered by the Traffic Calming or Road Hump Regulations, or where a highway authority is uncertain whether existing powers would allow the installation of certain traffic calming measures, applications for 'Special Authorisation' can be obtained from the Department of Transport [NId] [Wa]. However, this should seldom be necessary, as the Regulations cover a wide range of measures.

Traffic calming in the UK owes much to earlier continental European practice and experience, especially in Denmark, the Netherlands and Germany. The shared-space 'Woonerf' design was pioneered in the Netherlands and uses unkerbed level paving shared by both vehicles and pedestrians and is designated with a special sign indicating that the surface is shared. However, the high cost associated with Woonerfs has constrained their widespread use. Many continental schemes in major cities involve the reallocation of highway space and these have been particularly successful where wide roads or avenues existed previously. There are now numerous excellent examples of shared surfaces in the UK (CSS, 1994) (see Photograph 20.2).

Photograph 20.2: Shared surface – Sidmouth, Devon.

20.3 Speed–Control and the Environmental Objectives Of Traffic Calming

The advent of traffic calming, including the extensive use of speed-reduction and speed-control techniques, marked a significant change in the approach to traffic engineering. The traditional approach was to design road layouts to accommodate increasing volumes of traffic and this led to an open-road impression which, outside congested periods, allowed excessive speeds and has contributed to the domination of urban roads by motor vehicles. Traffic calming has allowed this approach to be re-examined, leading to schemes which benefit all road-users. In this context, 'road-users' includes pedestrians, disabled people, cyclists, buses, commercial vehicles, delivery vehicles, local residents, shop-keepers, those who work in the area and those using the area for rest, play or entertainment, as well as private cars.

The objectives of traffic calming include:
❑ improvement of the environmental quality of streets;
❑ general improvement of safety and reduction of accidents
❑ improvement in safety and convenience specifically for vulnerable road-users (including people with impaired mobility, elderly people, pedestrians, children and cyclists);
❑ reduction in noise, disturbance and anxiety;
❑ restoration of communities divided by speeding traffic;
❑ enhancement of the appearance of streets;
❑ discouragement of the use of unsuitable routes by heavy vehicles and 'through' traffic;
❑ changes to the attitude of many drivers towards speed; and
❑ tangible demonstration that streets are for people as well as for traffic.

20.4 Policy Framework

Traffic-calming schemes should not be designed in isolation but need to be part of an overall policy framework for an urban area and should fit within the overall transportation strategy for that area.

The Policy Framework should be reflected in a planned road hierarchy (see Chapter 11), taking account of the incidence of accidents, safety and environmental targets and the management of speed. The guidelines for Urban Safety Management produced by The Institution of Highways & Transportation (IHT, 1990) give advice on a structured approach to the development of such a policy framework. Further advice can be found in Traffic Calming Guidelines, published by Devon County Council (DCC, 1991), and Traffic Calming in Practice (CSS, 1994).

When developing a road hierarchy for an area, account should also be taken of current and planned development in the area. The function of each road should be examined in the context of traffic needs, environmental sensitivity, type and layout of road, pedestrian usage, the presence of schools and shops and its place in the overall approach to road safety measures for the area. It is essential that all types of road–users are considered, not just drivers of vehicles. The sensitive development of a road hierarchy can promote many of the traffic calming objectives and can play a major role in ensuring that the dominance of vehicles is reduced, in those roads where such a reduction would provide maximum benefit. As land–use is taken into account in its development, the hierarchy then forms the basis of future land–use, transport and development control policy and should influence the design of new roads and estates roads (see also Chapter 11).

Once a carefully considered policy–framework has been established, a strategic approach can be adopted in the investigation, not only of traffic–calming schemes but also of local safety schemes, street–lighting design and layout, maintenance standards, management of the environment and the allocation of resources. Such a framework is a prerequisite for the justification of 'package bids'.

20.5 Specific Policy Issues

Traffic calming can be a way of resolving potential conflicts and competition for road–space but it has to be developed in an integrated way, taking account of the needs of all users of the roads in question. Specific policy issues which arise when traffic calming schemes are being designed are addressed below.

Buses
Specific attention needs to be paid to the design of traffic–calming measures on bus routes and early consultations need to be held with the bus operators (see Chapter 24). The buses need to be able to negotiate the routes safely at a reasonable operating speed. The following guidelines should be adopted when introducing road humps on bus routes, although legislation may allow different design criteria:
- ❏ the hump height should not exceed 75 mm (in the case of raised junctions, 100 mm may be

acceptable but, with flatter ramp gradients, 10% may be desirable);
- ❏ the ramp gradient generally should not exceed 10% for flat–topped humps;
- ❏ the plateau length should not be less than the largest wheelbase of any public service vehicle likely to use the route, with a six–metre minimum length;
- ❏ speed cushions can be a preferable alternative to full width humps on bus routes; and
- ❏ ramp gradients for speed cushions should not exceed 12% and side ramps should not be steeper than 25%.

The design may also need to incorporate those requirements that give priority to public transport. Information is given in Current Issues in Planning Vol. 2 (Trench, 1995).

Pedestrians and Cyclists
One of the main objectives of traffic calming is to improve conditions for pedestrians and cyclists and to encourage these modes as alternatives to travel by car. Care is needed in the design of measures, if these objectives are to be achieved (see Chapters 22 and 23).

Wherever possible, provision should include:
- ❏ adequate widths and special routes, often giving priority to pedestrians and cyclists;
- ❏ central islands and clearly designated crossing arrangements;
- ❏ footways that maintain the same level at formal pedestrian road–crossings, such as flat topped road humps, together with tactile surfaces;
- ❏ designated cycle facilities throughout routes and as bypasses at pinch points; and
- ❏ arrangements to make access for mobility impaired people as easy as possible.

Vulnerable Road–Users
The introduction of traffic–calming measures should provide facilities to assist and protect vulnerable road–users, such as children, the elderly and mobility–impaired people. Traffic–calming measures, which distract the attention of these vulnerable road–users from their purpose of either crossing or walking along the road, may become hazards which cause accidents rather than reduce them. Care in the detail of layout and design is essential.

Emergency Services
It is important to maintain good access and a rapid response time for emergency services. Consultations are required with the emergency services for road humps (HMG, 1996) [Sc]. For other measures, under the Highways Act 1980 (HMG, 1980) [Sa] and the Highways (Traffic Calming) Regulations 1993 (HMG,

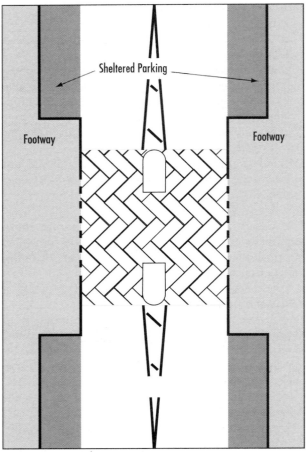

Figure 20.1 Footway build–outs reduce the distance for pedestrians to cross, improve visibility and can provide sheltered parking.

1993) [Sb], consultation is required with the police, although other services would also normally be consulted [NIe]. The improvement of road safety is a common goal and the emergency services and the local Highway Authorities should work together in setting and achieving accident–reduction targets. Advice on a code of practice and on consultations with the fire and ambulance services is given in Department of Transport Traffic Advisory Leaflet 3/94 (DOT, 1994) [Sd], which emphasises the need to identify strategic routes.

Routeing of Heavy Goods Vehicles (HGVs)

Heavy Goods Vehicles should be encouraged, by road signs, to remain on the highest available category of route for as much of their journey as possible. Traffic calming can be used to control speeds but the largest size of vehicles involved needs to be taken into account. In every case, the environmental impact of heavy goods vehicles should be reduced to a minimum (see Chapter 25).

On–Street Parking

Provision for the required levels of on–street parking should form an integral part of the design of all traffic

calming and parked vehicles themselves can sometimes assist in reducing traffic speeds, if they are located in appropriate locations. Footway build–outs can be used to define parking areas and at road crossing points, to improve visibility for pedestrians. Care must be taken, perhaps by paving demarcation, to discourage pedestrians from standing too close to the carriageway where they might be at risk of being struck by projections from passing vehicles (see Figure 20.1). Planting can also be used to reduce the visual intrusiveness of parked vehicles. However, care must be taken to avoid planting schemes which, in themselves, may obscure pedestrians.

Traffic Restraint

The promotion of, and provision for, alternative forms of transport to private cars are unlikely to provide a solution to urban transportation problems, in isolation. Complementary measures to manage demand for use of private cars, and to maximise the use of alternative forms of transport, are necessary as part of an overall strategy (see Chapter 21) [NIf]. However, traffic calming, applied over wide areas of a town, can have a considerable effect on the overall demand for car–use, as part of a traffic restraint policy, and can contribute to changing public attitudes to car–use, safety and the environment.

The Environment

Traffic–calming schemes are likely to be more successful and popular where the overall appearance of the street scene is improved. The objective must always be to achieve an environmentally sympathetic scheme, which complements highway safety requirements and does not give rise to maintenance problems. This requires working across disciplines and organisations and with the local people (DOT, 1995b) [Sd]. The appearance and design of streets should make drivers aware that slow speeds and tolerant behaviour are necessary. An improved environment allows pedestrians, cyclists and residents to enjoy the freedom that traffic calming offers and encourages more social activities in residential and shopping areas (see also Chapter 11). Particular attention to design is needed in environmentally sensitive areas, such as historic centres and conservation areas (EHTF, 1994). Investigations are being carried out, by the DOT and TRL, into the effects of traffic calming on vehicle emissions. No conclusive evidence has yet been obtained but it seems clear that schemes need to be designed to encourage smooth driving behaviour, without excessive acceleration and deceleration.

Economic Development

Traffic–calmed areas have potential for economic growth and development and this can be a positive encouragement to shopping and other commercial activity, as the progressive pedestrianisation of town centres has demonstrated.

When development takes place, it often places an extra burden on the existing highway infrastructure, which can be mitigated by measures to improve the capacity of the network. However, new accesses and junctions, with their attendant visibility splays, require more land to be taken into the highway, sometimes to the detriment of the environment. In some instances, it may be possible to reduce visibility splays and sight lines by introducing traffic calming on the existing highway, so that lower speeds prevail and tighter geometric design standards can be adopted, but care is required to ensure that the measures will actually result in reduced speeds and that appropriate visibility splays are adopted.

Within new development, whether commercial or residential, opportunities exist for the introduction and financing of traffic–calming measures that are integral with the design process, and not merely extras added at a later stage. As a result, the road layout may look significantly different from traditional designs. For instance, frequent changes of direction and a lateral shift of road alignments can be incorporated, allowing for the inclusion of well–designed areas of landscaping at an early stage, rather than added later to infill spare spaces (see Part IV).

Walking and Cycling

The main aim of PPG 13 (DOE/DOT, 1994) is to plan for less travel, especially by car [NIg] [Se] [We]. The provision of alternative modes of transport, such as bus, train, cycling or walking, are seen as being necessary for new development to be acceptable. However, cycling and walking will only be attractive in a pleasant environment, where cars are not allowed to dominate. Communities severed by wide, fast and busy roads can be 're-united' by imaginative and well designed traffic–calming schemes that involve the reduction of traffic speeds. This serves the aim of PPG13 [Se], by making walking and cycling more attractive (see Figure 20.2).

Traffic Calming on Main Roads

Successful traffic calming requires a road–hierarchy framework. It has to be recognised that the function of major routes and main distributor roads is to carry the majority of the traffic. However, despite this main function as a traffic route, traffic–calming techniques may still be appropriate where drivers need to be encouraged to proceed at a pre–defined speed in a calm and safe manner. Pedestrians, cyclists and vulnerable road–users need protection with the provision of specific facilities, such as road crossings, and where a main road passes through a small village centre there may be a case for comprehensive traffic calming. The Village Speed Control Study (Wheeler,

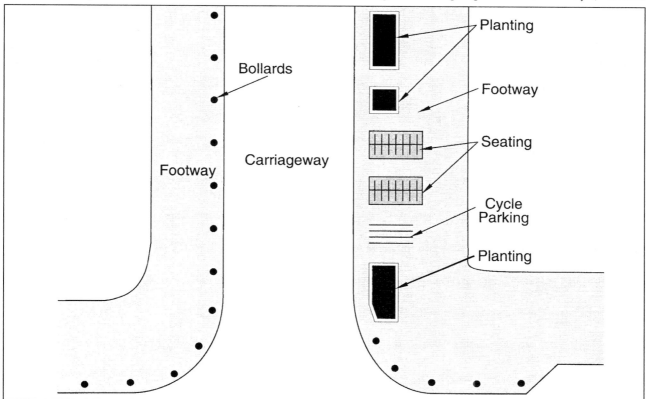

Figure 20.2: Facilities added to the street scene and reduced space for vehicles.

1994) concluded that modest traffic calming had a limited effect. The TRL Report on Craven Arms, Shropshire (Wheeler *et al*, 1996) gives examples of the more extensive type of traffic calming that can be adopted but, clearly, the particular character of each village must be respected.

Motorcyclists

Balance and control are essential for the safety of motorcyclists and obstructions or low–speed deviations, posed by traffic–calming measures, can be difficult to negotiate. Conversely, some measures, such as humps and chicanes, may tempt some motorcyclists to negotiate them with high acceleration and disregard for other road–users.

Safe Routes to School

Community representatives and school authorities should be consulted to ensure that traffic–calming proposals can assist children on their journey to school. 'Safe–Routes–to–School' programmes can be developed that include the introduction of traffic–calming measures.

20.6 Public Participation and Consultation

Public participation, consultation and dialogue are all particularly important in the context of traffic calming, not least because of the public interest in the measures. To be successful, traffic–calming schemes need to fit into their local environment, be locally distinctive and be locally 'owned', in the sense that they are regarded by local occupiers and users as an asset and not a liability.

The most successful approaches to public involvement are those where local people and users feel that they are genuinely involved in the conception and design of the scheme. Care must be taken at the start by defining the problem with the help of local people, to avoid presenting them with the solution as a *fait accompli*. However, the approach to consultation should be appropriate to the size and significance of the scheme, the type of community in which it is sited and the numbers and types of affected users. It should be appreciated from the outset that the costs of consultation can be high, even for relatively small schemes (see Chapter 10).

Depending on the nature and extent of the scheme involved, appropriate approaches could include:
- ❑ starting with a clean sheet of paper, as far as proposals or options are concerned, and involving the public in discussion, comment, suggestions and the evolution of the scheme, by providing

alternative plans, accident data, route information and traffic flows to inform the debate;
- ❑ spending time consulting on the problems of the wider area;
- ❑ encouraging discussion through a consultation steering group, which should include representatives of all affected sectors of the community; and
- ❑ adopting a standard consultation approach, including publicity and a display of draft proposals or options.

A consultative steering group might involve appropriate representation by some of the following:
- ❑ the County Council or Unitary Authority;
- ❑ the District Council;
- ❑ the Town or Parish Council;
- ❑ local residents;
- ❑ traders or business people;
- ❑ school staff and pupils;
- ❑ the ambulance service;
- ❑ the fire brigade;
- ❑ the police;
- ❑ bus companies;
- ❑ taxi operators;
- ❑ cycling organisations;
- ❑ road haulage associations;
- ❑ civic societies;
- ❑ the chamber of trade and commerce; and
- ❑ an official with multi–disciplinary experience.

However, it must be stressed that consultation with appropriate interest–groups should be considered even where the nature and/or scale of the scheme does not warrant a full 'steering group' approach.

In each situation, the key to success is to keep all the 'players' informed about what is happening and also to ensure that all those who live or work in the area have an opportunity to express a view. Objectives need to be agreed, targets set and measurements taken and reported both before and afterwards (see also Chapter 17).

Although bus companies and emergency services should be engaged in the debate, this should be to ensure only that their concerns are properly considered, as important inputs to the overall balance of the scheme.

20.7 Scheme Design

To be successful, the 'design team' should be multi–disciplinary and, for larger schemes, may need to call upon the skills of some or all of the following professionals:

- a design engineer;
- a maintenance engineer;
- a town planner;
- a road safety practitioner;
- a landscape architect;
- a listed building expert; and
- an artist.

The team should include representation from the District Council, as well as from the Highway Authority, where the scheme is not in a Metropolitan District, London Borough or Unitary Authority. Before work begins on a specific scheme, the Highway Authority should be clear about the function of the affected street in the proposed local road hierarchy. The objectives of the proposal should be defined and a broad approach developed which should then be the subject of consultation and iteration, until the optimum design is achieved. Input to the process includes traffic characteristics, safety issues, environmental factors, physical characteristics of the site and an overall evaluation of the scheme and any alternatives.

The success or otherwise of a scheme will be judged not only by the resultant accident and speed reductions but also by the overall environmental feel and quality of the scheme and its acceptability by the local population. Attention to detail is, therefore, essential, in terms of the appropriateness of the materials, their technical performance and the practicalities of future maintenance.

Some specific policy issues that need to be taken into account during the design process are dealt with in Section 20.5. Other issues to be addressed are:
- 'noise', which can be a problem near residential properties;
- 'signing', which should not be too intrusive, whilst still retaining its function;
- 'narrowings', if the gap is kept to a minimum length the incidence of drivers attempting to beat each other through the gap may reduce;
- 'gateways', which often result in a decrease in speed at the gateway but this is not always sustained through the rest of the scheme, so that additional features may be needed;
- 'visual narrowing', using natural materials such as hedges and planting, can sometimes be more acceptable than hard kerb–extensions but may have a limited effect on speed;
- 'cycle routes', which require careful design; and
- 'attention to detail', which is essential, in order to gain public acceptance.

Changes in the character of noise can be annoying, even if the overall level is reduced. Hence, it is

Photograph 20.3: Build out and sheltered parking.

important to encourage smoothly flowing traffic at low speeds. Speed–cushions should be located so that they are generally straddled by vehicles rather than partly over–run. Rumble–strips are not recommended in urban areas.

Responsibility for the maintenance of the various elements of the scheme should be agreed in advance by the relevant authorities. Schemes should be designed for ease of maintenance, with particular attention being given to the drainage of the carriageway, following the introduction of horizontal and vertical variations, to accessibility for sweepers, gully–emptiers and paving machines and to the possibility of damage to signing and other street furniture.

Attention to detail in design is important as schemes often include a number of different materials. The siting of street furniture, and the surface laid around and under it, needs careful consideration to ensure that maintenance problems do not arise.

20.8 Traffic–Calming Measures

Effective traffic–calming schemes are made up of a combination of measures. Care must be taken to ensure that the measures used are appropriate to the site and to the defined objectives. Details on the types of measures may be found in CSS (1994), in DCC (1991) and other references (see Sections 20.15 and 20.16). An alphabetical listing of most of these is set out below.

Bar Markings, although mainly used to draw attention to an approaching junction or roundabout

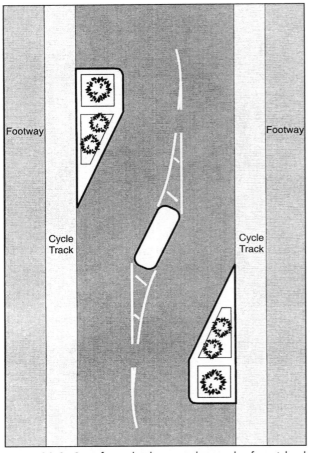

Figure 20.3: Specific cycle–lanes and central refuge island included in the design of a chicane.

on high speed roads, in the form of yellow transverse bar markings, are sometimes used prior to a change in speed–limit, possibly combined with, or part of, a 'gateway' feature. Yellow transverse bar–markings require special authorisation [NIh] but other bar–markings generally fall within the powers provided under the Highways (Traffic Calming) Regulations (HMG, 1993) [Sb].

Build–Outs are a narrowing of the carriageway, constructed on one side of the road as an extension to the verge or footway, and are often combined with sheltered parking or flat–topped crossing facilities (see Photograph 20.3).

Chicanes consist of two or more build–outs on alternate sides of the road, but not opposite one another, and create horizontal deflections. Speed cushions may be used in conjunction with chicanes to make the chicane more effective, by precluding a 'racing' line.

Cycle Measures are not specific traffic–calming measures but are features that should be provided to ensure the safety of cyclists, when negotiating particular traffic–calming measures, such as cycle–lanes or cycle tracks by–passing chicanes or pinch points (see Figure 20.3).

Entry Treatment consists of a change of surface, a ramp, a narrowing or some other features at a junction or change of road characteristic (see also gateways).

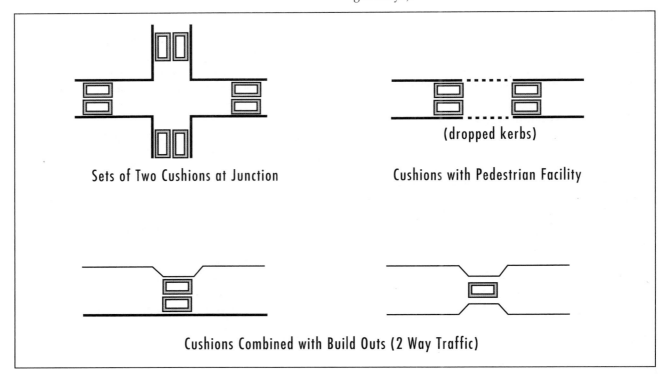

Figure 20.4: Speed cushions.

TRANSPORT IN THE URBAN ENVIRONMENT

Photograph 20.4: Gateway with over-run areas.

Environmental Road Closures can be used as part of an area-wide scheme to reassign traffic.

False Roundabouts involve the creation of a small roundabout where there is no actual road junction and its purpose is to modify traffic speed. Legislation does not permit mini roundabouts, formed by road markings, to be used for this purpose.

Footway Crossovers are the continuation of an existing footway across the mouth of a side-road, with vehicles allowed to cross the footway but giving way to pedestrians.

Footway Widening, often as part of the redefinition of road space, is used to give more space to pedestrians or for planting. It may be part of a build-out and can be particularly effective at formal or informal pedestrian crossing points.

Gateways are combinations of natural or man-made features at the entry to, or exit from, areas where the rules or drivers' expectations change, such as at the introduction of speed-limits.

Photograph 20.5: Pedestrian refuge at a school.

Photograph 20.6: Pinch point with flat-topped road humps and priority signing.

Horizontal Deflections occur at build-outs, chicanes and pinch points, often with priority signing.

Islands usually take the form of a longitudinal island, built in the carriageway, with or without facilities for pedestrians, to improve lane-discipline, restrict overtaking or lower vehicle speeds by reducing lane-width and separating cyclists from other vehicles.

Junction Priority Changes are used as part of an area scheme to interrupt long stretches of 'through' road. Care is needed in signing when introducing a change such as this.

Junction Treatments can incorporate a variety of measures as part of an overall scheme, including flat-topped road humps, narrowing and removal of excess areas of carriageway and the introduction of ramps, chicanes, horizontal deflections and tight curves.

Mini-Roundabouts are used at junctions on long straight roads to break up the road into shorter sections which slows traffic; also used at T or Y junctions to reduce the dominance of one particular flow.

Narrowings consist of short or long pinch points, often combined with priority signing.

One-Way Streets may be used as part of an area-wide scheme to break up a road into short sections and indirect routes. By creating detours, they can discourage 'rat running' but may encourage higher speeds because of the absence of opposing traffic. Contra-flow bus lanes or cycle routes may be incorporated.

Over-run Area is a part of the road which is textured

or coloured, so that it appears to narrow the carriageway but can be used by large vehicles to complete turning manoeuvres; but attention needs to be paid to the needs of cyclists and pedestrians should be discouraged from waiting on these areas before crossing (see Photograph 20.4).

Parking involves redefining the road space to provide defined parking and to reduce the area of carriageway.

Pedestrian Refuges are used to aid pedestrians' crossing movements, by allowing a carriageway to be crossed in two stages. They are also used to control overtaking and to improve lane–discipline (see Photograph 20.5).

Pinch Point consists of a pair of build–outs on opposite sides of a road to create a narrowing, thus helping to modify vehicle speeds and to reduce the risk to pedestrians when they cross the road (often combined with speed cushions and priority signing) (see Photograph 20.6).

Planting can be used to change the perceived width of a road, to define a gateway and to improve the overall environment.

Raised Junctions consist of a plateau or flat–topped road hump, built across the whole area of a junction.

Red Light Cameras are automatic cameras, which record red traffic light violations – 'red–running'.

Reallocation of Road Space entails the definition of road space to accommodate all users of the space in question, so as to reduce the dominance by motor vehicles.

Road Humps are used to reduce vehicle speed and, in the case of flat–topped humps, may provide a level surface for pedestrians to cross.

Road Markings are used to hatch out areas of carriageway, to define traffic lanes and to create the visual effect of narrowing of the carriageway.

Rumble Devices are part of a carriageway made of materials which create a noise or vibration as vehicles pass over. They are useful as an alerting device before a hazard but may not reduce speed. They may attract objections when sited close to houses.

Shared–Use Roads are short lengths of road, mostly in new estates, which can provide an attractive appearance in living areas. When used in shopping streets, it may be preferable to have a 25mm kerb upstand, except at crossing points, to assist pedestrians who are visually impaired.

Sheltered Parking consists of parking spaces protected by a build–out and can be in–line, angled or in echelon form (see Photograph 20.4).

Speed Cameras are automatic cameras, which record speed violations in excess of a pre–set threshold value.

Speed Cushions are a form of road hump, occupying only part of a traffic lane, which, generally, can be spanned by buses and HGVs but not by cars and can be used in conjunction with chicanes (see Figure 20.4).

Street Furniture, properly used, can help to redefine road space, create the visual effect of narrowing and contribute to gateways and other features.

Surface Treatments consist of change in the colour or texture of a carriageway, to denote where the character or use of the area changes (see also gateways).

'Thumps' are thermoplastic road humps not less than 900mm wide and 30mm to 40mm high.

20.9 Speed–Limits

Introduction and Legislative Framework

The speed of vehicles is an emotive issue, which often generates intense local concern and debate, partly because the perception of appropriate speed often differs greatly between, for example, drivers, pedestrians and cyclists. Measures for influencing the speed of vehicles generally fall into two categories, legislative and physical. However, there is an emerging third category, namely, influencing driver behaviour through education and social peer–group pressure, although this must be viewed in a longer timescale.

Powers to restrict vehicle speeds are provided in Part VI of the Road Traffic Regulation Act 1984 (HMG, 1984) [NIi]and Schedule 6 of that Act defines maximum speeds for motor vehicles of certain classes [NIj]. The procedure for a highway authority in making a speed–limit Order is laid down in The Local Authorities Traffic Orders (Procedure) (England and Wales) Regulations 1989, as amended (HMG, 1989) [NIk] [Sf]. Advice on speed–limits is provided in Department of Transport Circular Roads 1/93 (DOT, 1993) [Sg] and Traffic Advisory Leaflet 1/96 (DOT, 1996) [Sd].

Before deciding to change an existing speed–limit, the appropriate authority must consider all the relevant factors. The police view is important and account should also be taken of the characteristics of the road, such as its alignment, the level and type of frontage activity, the accident record and the degree of severance caused to a community by the speed of vehicles. In urban areas, speed–limits should fit into a rational and easily understood hierarchy if they are to be observed by drivers (see Chapter 11).

Before introducing a speed–limit for the first time, at a particular location, highway authorities need to consider the benefits and the disbenefits. Factors which should be included in such an assessment are:

❑ expected accident savings;
❑ improvement to the environment;
❑ improvement in amenities;
❑ reduction in public anxiety;
❑ improved facilities for vulnerable road–users;
❑ delays to traffic;
❑ costs of implementation;
❑ costs of engineering measures and their maintenance; and
❑ costs of enforcement, especially where the speed–limit is regarded as unreasonable by the drivers.

General Principles

Local speed–limits are normally unnecessary where the character of the road, in practice, limits the speed of most vehicles to a level appropriate to the conditions. Also, mandatory speed–limits should not be used to solve the problem of isolated hazards, such as a single road junction or a bend.

Safety should be a major factor in determining the need to impose a speed–limit and can often help to justify the cost of engineering measures. While speed–limits may be introduced because of a poor accident record, care should be taken to assess the underlying causes, as this may well indicate other more effective measures. For example, the provision of a footway may be a more effective means of ensuring pedestrian safety than a speed–limit. Also, speed–limits should be lowered only when a consequent reduction in vehicle speeds can reasonably be expected and when the police consider that the lower limit is enforceable.

Speed–limits, on their own, cannot be expected to reduce vehicle speeds, if they are set at a level substantially below that at which the majority of drivers would otherwise choose to drive. It is not necessarily the case that a lower speed–limit will result in lower speeds. However, current research suggests that, when speed–limits are backed–up by extensive publicity and education campaigns, more disciplined driving results.

The most important factor when setting a speed–limit

is how the road looks to drivers. Any feature which provides drivers with a perception of increased risk is likely to result in a reduction in their speed. The road geometry, ie width, sightlines, bends, frequency of crossings and the environment through which it passes, such as rural, residential, shop frontages and schools, all influence a driver's choice of speed.

If lower speeds are to be achieved, the character of the road will need to be altered so that drivers will perceive more potential risks and reduce their speed accordingly. Additional measures, other than signing, are required if speeds are to be reduced to a level where drivers perceive and accept the need for the limit. Police enforcement can then target the irresponsible minority of drivers.

There is little point in establishing a speed–limit, however desirable from an environmental or safety point of view, if it is not going to have a significant effect on actual vehicle speeds.

The Department of Transport recommend a staged approach to determine the appropriate speed limit for any road (DOT, 1993), as follows:

❑ consider the desirable speed–limit, taking into account the general environment of the road, its place in the proposed road hierarchy and its recent accident history;
❑ if the measured 85 percentile speed is within either 7 miles/h or 20% of the desired speed–limit (whichever is the smaller), then it would be appropriate; but
❑ if the measured 85 percentile speed is outside this acceptable range, then either:
  ❑ impose a higher speed–limit; or
  ❑ alter the road layout and environment so as to reduce speeds to within the acceptable range of the desired limit.

Measures to control vehicle speeds are broadly of two types; those that involve changes to the road geometry, such as narrowings, build–outs and chicanes, and those that involve variations to the carriageway surface, such as road humps and speed cushions. It should be noted that the use of some of these measures has legal implications and reference should be made to the appropriate Department of Transport Circulars and Advice Notes (see Section 20.16).

20.10 20 miles/h Speed–Limit Zones

Full–time 20 miles/h speed–limit zones are most appropriately used in areas where an urban safety strategy has been applied, including traffic–calming measures. These are often residential in character although, increasingly, shopping streets are being treated in a similar manner. Zone signs have to be erected at every entrance into the area and should normally be used in conjunction with the creation of gateways. Zones should not be too large nor should they impede access to hospitals or fire–stations. There should also be alternative routes available for 'through' traffic. Proposed schemes have to be submitted to the Department of Transport for approval [Wb]. Speed–control in the zones should be self–enforcing and it is necessary to demonstrate that the average vehicle speed, following the introduction of speed reducing features, will be 20 miles/h or less (DOT, 1990).

Specific signs, which differ considerably from normal terminal signs associated with speed–limits, have to be erected at each entrance and exit to the zone. A degree of flexibility of design and colour is permitted, to reflect the local distinctiveness of the area. Emergency services and bus operators have to be consulted and their responses have to be included in the submission to the Department of Transport [Wc] for the Speed–Limit Order (see Chapter 13) [NIl].

20.11 Variable Speed–Limits

Section 45 of the Road Traffic Act 1991 (HMG, 1991) [Sg] amends section 84 of the Road Traffic Regulation Act 1984 (HMG, 1984) (speed–limits on roads other than restricted roads) to remove the requirement that a single speed–limit must apply to each section of road [NIm]. Thus, it is possible for speed–limits to vary by time of day or according to traffic flow conditions. Typical uses would be outside schools, when children are going to school or returning home, or as part of a speed/flow management scheme on heavily trafficked roads. Proposed schemes have to be submitted to the Department of Transport for approval [NIl] [Wd].

20.12 Enforcement of Speed–Limits using Cameras.

The Road Traffic Act 1991 (HMG, 1991) extended the powers available to the police for the detection of moving offenders, by allowing a photograph of an offence to be used as evidence in court uncorroborated by a police officer at the scene [NIn]. The provisions of the Act enable enforcement cameras to be used as a means of influencing driver behaviour and improving road safety. Advice regarding implementation of the Act has been produced in the Home Office Circular 38/1992 (HO, 1992) [Sh] and the Department of Transport Circular – Roads 1/92 (DOT, 1992) [Si].

Speed Cameras

Local authorities are generally responsible for selecting the location of speed camera sites and for installing and maintaining sites. Such sites are based on Department of Transport Guidelines given in Circular Roads 1/92 (DOT, 1992). The police are responsible for the provision of speed camera and radar equipment installed at the fixed sites and for the collection and processing of the speeding offence evidence.

Site selection considerations include:
- ❑ a history of speed–related accidents;
- ❑ police agreement on site location and operation;
- ❑ large numbers of vehicles travelling in excess of the speed–limit; and
- ❑ site conditions which would affect radar operation, such as gradients, the presence of parked cars, high voltage power–lines and metal fencing.

Special carriageway markings must be laid near the camera cabinet to enable the police to provide a secondary means of checking the accuracy of the radar equipment.

The police install a radar unit, a flash–camera control unit and a camera at each site. Two photographs are taken of each offence at 0.5 to 1.0 second intervals. The built–in flash unit allows the camera to be used at night or in poor visibility conditions. The radar uses the 'doppler' principle to measure closing/opening speed and calculates the true speed of the target vehicle at a 20° offset. The film is periodically removed by the police and analysed to identify offending vehicles. The registered keeper of the vehicle is then issued with a 'Notice of Intended Prosecution'. Offending drivers receive a 'Conditional Offer of Fixed Penalty' or a 'Summons' depending on the severity of the offence. Trigger speeds for the camera are set by the police.

Due to the cost of camera equipment, it is normal practice for the police to 'rotate' the camera equipment between a number of fixed camera sites. Experience has shown that a ratio of 1 camera to 6 fixed sites is preferred, although 1 to 8 is acceptable. Dummy flash equipment can also be installed at the fixed sites to enhance the deterrent effect. Dummy units can be used to record the number of infringements and thus help to monitor driver behaviour.

Results have shown that speed cameras have a significant effect on vehicle speeds and accident rates. Cameras have been found to influence driver behaviour not only at the site but both upstream and downstream as well. In addition, the use of both fixed and portable speed cameras has been shown to influence overall driver behaviour. More importantly, results have shown a significant reduction in the number of vehicles travelling well in excess of the speed–limit, ie 15 miles/h or more above the limit. The reduction in the number of vehicles travelling at these higher speeds, so called 'fliers', is known to reduce accident–rates significantly.

Red–Light Cameras

Automatic enforcement cameras are very effective in reducing accidents attributed to drivers failing to comply with a red signal. Although there may be a slight increase in the number of rear–end collisions, this does not outweigh the overall benefit. The camera takes two photographs at one second intervals after a predetermined period of red signal. The second photograph ensures that the driver proceeds through the junction and is not turning right.

Top priority sites for these cameras are those sites with a high 'failure–to–comply' accident record. However, these accidents also occur at other signal–controlled junctions and a highway authority should also consider the installation of cameras at 'high profile' sites, perhaps accompanied by warning signs, in order to achieve a change in drivers' behaviour.

20.13 Signing for Speed and Red–Light Cameras

The purpose of the informatory sign is to warn drivers that speed and/or traffic–signal cameras are present and operating in the areas through which they are driving.

Experience indicates that the effectiveness of cameras in reducing speeds and traffic signal violations is enhanced by the introduction of informatory signs. However, the erection of signs is optional and the absence of signs does not invalidate enforcement founded on speed–detection and signal–violation cameras. Highway authorities have flexibility in the use of the signs but should not use them indiscriminately. Information on this is given in DOT Circular – Roads 1/95 (DOT, 1995a) [Sj].

While it is important that signs are used to supplement cameras, some of the effectiveness is lost if signs are erected only at or near camera sites. The signs are best used on an area–wide or route basis. It is not possible to define precisely the extent of the area, but it should normally relate to the geographical area covered by the cameras. Generally, an area will be the whole, or part, of an urban area, which has a number of camera sites within it. In such cases, signs would be sited at the boundary of the area, with repeater signs within the area, where appropriate. If a number of cameras are used along a route, then the route should be regarded as the 'area' and signs should be erected accordingly.

When static or portable speed–detection cameras are used at road works on a temporary basis, the road works site should be considered as the 'area' and signs provided accordingly, using the standard black on white signs.

20.14 Framework for Influencing Drivers' Behaviour

Potentially, the most productive way to overcome the problems of excess speed is to influence drivers' behaviour by making speeding socially unacceptable. Ways of achieving this range from national campaigns to local initiatives. Topics suitable for national campaigns might include the level of injury at different speeds, persuading car manufacturers to direct advertising away from top speed and acceleration, Insurance Companies taking more note of speeding convictions in setting premiums, restricting new drivers to lower speed levels and ensuring that fleet operators do not set schedules or offer incentives which encourage speeding. Local initiatives could include campaigns which encourage local communities to become involved in the process of identifying the nature and extent of speeding, by setting up their own 'community response team' and involving employers in sponsoring community–inspired schemes. The aim should be to create a culture in which any violation of speed–limits is regarded as socially unacceptable.

20.15 References

Buchanan CD et al (1963)	'Traffic in Towns', Stationery Office.
CSS (1994)	'Traffic Calming in Practice', County Surveyors' Society.
DCC (1991)	'Traffic Calming Guidelines,' Devon County Council.
DOE/DOT (1994)	PPG13 'Transport', Stationery Office [Se] [We].
DOT (1990)	Circular – Roads 4/90 (WO 2/91)'20 mph Speed Limit Zones', DOT [Sd].

DOT (1992) | Circular – Roads 1/92 (WO 22/92)'Use of Technology for Traffic Enforcement: Guidance on Deployment', DOT [Si].

DOT (1993) | Circular – Roads 1/93 (WO1/93) 'Road Traffic Regulation Act 1984 Sections 81 – 89 Local Speed Limits', DOT [Sg].

DOT (1994) | Traffic Advisory Leaflet 3/94 'Fire and Ambulance Services – Traffic Calming: a Code of Practice', DOT [Sd].

DOT (1995a) | Circular – Roads 1/95 'Traffic Signal and Speed Camera Signing', DOT [Sj].

DOT (1995b) | 'Better places through By–passes', Stationery Office [Sd].

DOT (1996) | Traffic Advisory Leaflet 1/96 'Traffic Management in Historic Areas', DOT [Sd].

EHTF (1994) | 'Traffic in Historic Town Centres', English Historic Trust Forum/Civic Trust.

HMG (1980) | 'Highway Act 1980' (and amendments), Stationery Office [Sa].

HMG (1984) | 'Road Traffic Regulation Act 1984', Stationery Office.

HMG (1989) | SI 1989 No. 1120 'The Local Authorities Traffic Orders (Procedure) (England & Wales) Regulations 1989', Stationery Office [Sf].

HMG (1991) | 'Road Traffic Act 1991', Stationery Office.

HMG (1992) | 'Traffic calming Act 1992', Stationery Office.

HMG (1993) | SI 1993 No. 1849 'Highways (Traffic Calming) Regulations 1993', Stationery Office [Sb].

HMG (1996) | SI 1996 No. 1483 'The Highway (Road Humps) Regulations 1996', Stationery Office [Sc].

HO (1992) | Circular 38/1992 'The Use of Automatic Detection Devices for Road Traffic Law Enforcement', Home Office [Sg].

IHT (1990) | 'Guidelines for Urban Safety Management', The Institution of Highways & Transportation.

Trench S (Ed) (1995) | 'Current issues in Planning' Vol. 2, Gower Press.

Wheeler AH (1994) | 'Speed Reduction in 24 Villages: Detail from the VSIP Study', PR85 – TRL.

Wheeler AH, Abbott PM, Godfrey NS, Lawrence DS and Phillips SM (1996) | Traffic Calming on Major Roads: 'The A49 Trunk Road at Craven Arms, Shropshire', TRL.

20.16 Further Information

Abbott PG, Phillips SM and Layfield RE (1995) | Project Report 103 'Vehicle and Traffic Noise Surveys Alongside Speed Control Cushions in York', TRL

Abbott PG, et al (1995) | TRL Report 174'The Environmental Assessment of traffic management schemes: A Literature Review', TRL.

Abbott PG, Tyler J and Layfield RE (1995) | TRL Report 180'Traffic Calming: Vehicle Noise Emissions Alongside Speed Control Cushion and Road Humps', TRL.

CT (1990) | 'Lorries & Traffic Management', Civic Trust.

DOT Circulars – | Roads 2/92(WO 46/92) 'Road Traffic Act 1991: 'Road Humps and variable Speed Limits' – Roads 4/96(WO46/92) 'Road Humps' DOT 1996 [Sd].

DOT Traffic Advisory Unit – Traffic Advisory Leaflets [Sd]:

1/87 'Measures to Control Traffic for the Benefit of Residents Pedestrians and Cyclists', DOT 1987.
2/90 'Speed Control Humps', DOT 1990.

3/90 'Urban Safety Management – Guidelines from IHT', DOT 1990.

3/91 'Speed Control Humps', (Scottish Version) DOT 1991.

7/91 '20 miles/h Speed Limit Zones', DOT 1991.

2/92 'Carfax Horsham – 20 miles/h Zone', DOT 1992

3/93 '20 miles/h Speed Limit Zones', DOT 1993.

7/93 'Traffic Calming Regulations',DOT 1993.

8/93 'Advanced Cycle Stop Line for Cyclists', DOT 1993.

9/93 'Cycling in Pedestrian Areas', DOT 1993.

11/93 'Rumble Device', DOT 1993.

12/93 'Overrun Areas'. DOT 1993.

13/93'Gateways', DOT 1993.

1/94 'VISP A Summary', DOT 1994.

2/94 'Entry Treatments', DOT 1994.

4/94 'Speed Cushions', DOT 1994.

7/94 'Thumps – Thermoplastic Road Humps', DOT 1994.

8/94 'Traffic Signs Signals and Road Marking Bibliography', DOT 1994.

9/94 'Horizontal Deflections', DOT 1994.

1/95 'Speed Limit Signs A Guide to Good Practice', DOT 1995.

2/96 '75mm High Road Humps', DOT.

4/96 'Traffic Management and Emissions 1996', DOT 1996.

6/96 'Traffic Calming: Traffic and Vehicle Noise', DOT 1996.

7/96 'Highways (Road Humps) Regulations 1996', DOT 1996.

12/96 'Road Humps and Ground–Borne Vibrations', DOT 1996.

DOT (1986) LTN 2/86'Shared use by cyclists and pedestrians', Stationery Office.

DOT (1989) LTN 1/89 'Making way for cyclists' LTN 1/89, Stationery Office.

DOT (1992) 'Killing speed and saving lives', DOT.

DOT (1993) 'The Highway (Traffic Calming) Regulations 1993', DOT [Sb].

DOT (1995a) 'Safer by design' 2nd Edition, Stationery Office [Sd].

DOT (1995b) 'Better Places through Bypasses', Report of the Bypass Demonstration Project ISBN 0–11–551749–9, DOT [Sd].

HMG (1994) SI 1994 No. 1519 'Traffic Signs Regulations and General Directions 1994', Stationery Office.

Sayer IA and Parry DI (1994) TRL Project Report 102, 'Speed Control using Chicanes – A Trial at TRL', TRL.

TRL (1994) (1990–1992), 'Traffic Calming No. 1' 'Current topics in transport: a selection of abstracts', TRL.

TRL (1994) (1992–1994) 'Traffic calming update no. 1.1 'Current topics in transport: a selection of abstracts', TRL.

Watt GR and Harns GJ (1996) TRL Report 246 'Traffic Induced Vibrations in Buildings', TRL.

Webster DC (1993a) 'Road humps for controlling vehicle speeds', PR18, TRL.

Webster DC 1993b) on 'The Grounding of Vehicles Road Humps' Traffic Engineering + Control, 34(7/8).

Webster DC (1994) TRL Report 101 'Speeds at 'Thumps' and Low Height Road Humps', TRL.

Webster DC (1995a) TRL Report 177 'Traffic Calming – Vehicle Activated Speed Limit Reminder Signs', TRL.

Webster DC (1995b) TRL Report 182 'Traffic Calming – Four Schemes on Distributor Roads', TRL.

Webster DC and Layfield RE (1993) 'An assessment of rumble strips and rumble areas', PR33 TRL.

Webster DC and Layfield RE (1996)	'TRL Report 186 'Traffic Calming – Road Hump Schemes using 75mm High Humps', TRL.
Webster DC and Mackie AM (1996)	TRL Report 215 'Review of Traffic Calming Schemes in 20 mph Zones', TRL.
Wheeler AH (1992)	'Resume of traffic calming on main roads through villages', TRL.
Wheeler AH, Taylor M and Payne A (1994)	'The Effectiveness of Village 'Gateways in Devon and Gloucestershire', TRL.
Wheeler AH and Taylor M (1995)	'Reducing Speeds in Villages', Traffic Engineering + Control, 36(4).
Wheeler AH, Taylor M and Barker J (1995)	Annex to Project Report 85 'Speed Reduction in 24 Villages: Colour Photographs from the VISP Study', TRL.
Wheeler AH, Abbott PM, Godfrey NS, Phillips SM and Stait R (1996)	TRL Report 238 'Traffic Calming on Major Roads: The A47 Trunk Road at Thorney, Cambridgeshire', TRL.
Windle R and Mackie A (1992)	'Survey of public attitude acceptability of traffic calming schemes', TRL.

Chapter 21 Demand Management

21.1 Introduction

It is now widely accepted that unrestrained demand for travel by car within large urban areas, in the UK and elsewhere, cannot be accommodated. This is due to a combination of financial constraints and concerns about the impacts of traffic on local communities and their environment.

Permitting traffic to grow to levels at which there is extensive and regular severe congestion is economically inefficient. Congested conditions also aggravate the social and environmental impacts of traffic, by exacerbating both noise and polluting emissions, impeding road-based public transport and service vehicles and making conditions unpleasant for walking and cycling. Congestion in the inner parts of towns and cities encourages the relocation of activities, jobs and people out to the urban fringe, resulting in greater dependence on motor vehicles and compounding the environmental impacts. Indeed, increasing awareness and concerns about the potential environmental and health consequences is creating pressure for demand management, independently of that based on congestion.

This chapter sets out the principal measures for the management of traffic demand and outlines the institutional factors which can affect their application.

21.2 Why Manage Demand?

Historically, the main reason for seeking to manage demand for transport has focused on economic efficiency. The most logical tool to improve economic efficiency is a direct form of road-use pricing, as argued by Vickrey in evidence he gave to the US Congress in 1959 (Vickrey, 1959). The economic rationale is that unless the price directly incurred by someone in making a journey covers the full costs of the journey their travel will impose a net cost on the community. The full costs include both the personal costs which the traveller incurs (vehicle running costs, fuel, parking etc) and the social costs which the traveller imposes on the community, through adding to congestion and increasing the potential for accidents, as well as the adverse impacts on the environment, through creating noise, atmospheric pollution and contributing to severance. As the

marginal costs imposed on others vary by location and traffic conditions so, it is argued, the charges incurred by vehicle–users should also vary (Downs, 1992). Gomez–Ibanez and Small (1994) provide further insights into the rationale of using road–use pricing to manage urban traffic–demand, while Jones and Hervik (1992) set it in a wider context, with a review of alternative measures; Downs (1992) also includes an extended discussion of other policy measures to manage congestion.

The argument for managing travel–demand has, however, broadened beyond that based on economic efficiency. This reflects increasing concern about the impacts of congestion, in particular on urban communities, but also of traffic more generally and the consequences for communities of increasing highway capacity. Limitation on funds available for investment in urban transport has also contributed to the debate about the extent to which demand should be restrained to match the supply which can be provided.

Evidence from a range of surveys in the UK shows that traffic congestion is widely recognised as a serious problem (University of Westminster, 1996). Fifty percent of respondents in one survey in London cited 'traffic jams and congestion' as one of the main problems of life (NEDO, 1991) and it is perceived to be worsening. Deteriorating air quality and traffic accidents are also viewed as serious problems (Gallup, 1989). Jones (1992) contains a review of a number of sources of such material. Similar views are held throughout Europe, with 59% of respondents to a survey conducted in all EU member–states describing traffic, in their local community, as being either 'unbearable' or 'hardly bearable' (UITP, 1991). The costs of congestion to the nation are very large; the Institution of Civil Engineers estimated the cost to exceed £10bn in 1989 (ICE, 1989), while the Confederation of British Industry calculated that congestion cost the economy about £15bn a year (CBI, 1989).

Awareness of the impacts of traffic on the environment has been growing. The most immediate public concerns are with local effects, in terms of noise and air pollution and their effects on the quality of life, including possible threats to health. The Royal Commission on Environmental Pollution has argued that there is a need to restrain the future use of motor vehicles, through a combined policy of pricing and measures to promote the use of alternatives to private motor vehicles (RCEP, 1994). The National Air Quality Strategy is intended to provide a framework within which improvements can be achieved especially through local action (DOE, 1996).

There is also a growing recognition of the need to achieve 'sustainability', ie to manage development and transport in such a way "... that meets the needs of today without compromising the ability of future generations to meet their needs" (UN, 1987). The World Bank has distinguished between 'economic and financial' sustainability, which requires the efficient use of resources and the proper maintenance of assets, 'environmental and ecological sustainability', which requires that the external effects of transport are taken into account fully in determining future development, and 'social sustainability', which requires all sections of the community to benefit from improved transport. Each of these is relevant to demand management (World Bank, 1996).

Although the construction of additional or alternative highway capacity can alleviate some of the effects of congestion and of other traffic problems, the benefits may be offset, unless growth in traffic volumes is restrained. However, it is evident that the impacts of such schemes on local communities limit the extent to which they can be implemented, even if funds were available. Constraints on public expenditure and the demands of other sectors of the economy have severely restricted the availability of public funds and private finance is not likely to be a complete alternative general source of funds for urban areas.

Thus, the objectives of demand management policies are:
❑ to reduce congestion and thereby improve economic efficiency;
❑ to improve the quality of life through improvement of the local environment;
❑ to provide a stimulus for the local economy; and
❑ to reduce the local and global impacts of atmospheric pollution.

Not all of these objectives are necessarily complementary. Measures designed to satisfy one objective can, in practice, be counter–productive with respect to others. There can also be significant differences between the short– and long–run effects of some measures. Thus, a carefully designed package of measures, which addresses a balance of objectives relating to the particular needs of a local area, will usually be needed.

While conscious of environmental concerns, government, both local and national, tends to place a strong emphasis on maintaining or strengthening the economy. In particular, local government seeks to ensure that their town or city is an attractive place in which to work, to do business and to live. Towns are rarely so isolated that they are not in competition for economic activity with others. For some, the competition is regional, but the development of the Single European Market and global economies means that the competition is also with urban areas elsewhere in Europe and the rest of the world. While fears of competitive disadvantage may inhibit the adoption of radical measures to control traffic, it is also possible that radical measures will stimulate the local economy or, at least, prevent it from deteriorating in the face of competition from other towns and cities which are seen as being more attractive.

Despite concerns about the impacts of congestion, and traffic, there is little evidence that there is a general public willingness to accept the significant reductions in personal mobility implied by many of the measures necessary to the achievement of policy objectives to reduce traffic congestion and to improve both the local and global environment (University of Westminster, 1996).

Any single demand management measure is unlikely, by itself, to be either adequate or acceptable. A successful policy would require a combination of measures (CIT, 1996). Some measures would act as 'sticks', others as 'carrots'. Although there is some evidence that the cumulative effects of a series of small changes could be important, it is probable that any set of measures intended to achieve a substantial change in the use of private motor transport will have to be severe, particularly if they are to be sufficient to meet the targets proposed by The Royal Commission on Environmental Pollution (RCEP, 1994). Considerations of what politicians are prepared to put forward, and what the public will accept, are likely to result in a gradually increasing restraint, allowing transport users and other parties time to adapt.

One of the major challenges is to devise measures which do not unduly restrict personal mobility and which do not put local economies at risk. Ideally, these should also enhance the local quality of life and economy. Possible measures include:
❑ congestion charging and road–tolling;
❑ road–user charges levied on fuel and vehicle ownership;
❑ controls on vehicle–use;
❑ controls on vehicle ownership;
❑ parking controls and pricing;

- ❏ physical measures of traffic restraint;
- ❏ controls on land–use development;
- ❏ public transport improvement;
- ❏ encouraging more travel by foot and cycle; and
- ❏ encouraging greater use of telecommuting and of transport telematics.

21.3 Congestion Charging and Road–Tolling

Although road–use pricing is the term which has long been used to describe direct charging for the use of roads in urban areas, that term can also apply to other circumstances, such as motorway tolling. The term 'congestion charging' is therefore used here to represent the particular application of road–use pricing to manage demand in congested conditions.

Since, arguably, congestion is the result of imperfect pricing, congestion charging would appear to be the most rational means by which demand can be balanced with supply. Pricing can also be used to reduce the adverse effects of motor vehicles on the environment, both through encouraging the use of more fuel–efficient vehicles and reducing their use overall (RCEP, 1994). The principles of congestion charging have a long history, strongly rooted in economic theory, dating back at least to the advice of Vickrey to the US Congress in 1959 (Vickrey, 1959).

In 1962, the Minister of Transport appointed a committee, under the chairmanship of Reuben Smeed, to investigate the technical feasibility of introducing congestion charging. The Committee concluded that the methods of vehicle taxation then in use (which have changed little since) had deficiencies, "...notably their inability to restrain people from making journeys which impose high costs on other people", and suggested that "...road charges could usefully take more account than they do of the large differences that exist in congestion costs between one journey and another". The Committee examined various forms of taxes and charges, including differential fuel taxation, an employee tax, a parking tax, daily licences and direct pricing. They concluded that "...considerably superior results are potentially obtainable from direct pricing systems" (DOT, 1964). A particular strength of the Smeed Report was the establishment of a set of principles with which any congestion charging system should comply; these have stood the test of time, although extensions have been suggested (May, 1994).

In 1967, the Department of Transport reported on a study of Better Towns with Less Traffic. The report concluded that "...the most promising form of restraint, at least for the shorter term, would be to intensify control over the location, amount and use of parking" and that "...direct road–use pricing, of the type described in the Smeed Report, is potentially the most efficient means of restraint. Although it is by no means certain that a workable system could be devised, its advantages would be so great as to justify further research and development" (DOT, 1967).

In the early 1970s, the Greater London Council examined a wide range of options for restraining traffic demand. The conclusion of this work was that the most effective would be a pricing system for Central London, operating during the working day and based on the use of supplementary licences (May, 1975).

The first urban congestion charging system to be implemented was the Singapore Area Licensing Scheme (ALS), which came into operation in 1975. A paper supplementary licence was required to enter the Central Business District during the morning peak period (Holland et al, 1978). This scheme has subsequently been extended, so that it includes the whole working day, with differential peak and inter–peak charges (Turner et al, 1993). The paper–based ALS system is being replaced by Electronic Road Pricing (ERP). Initially, this will cover the same area as the ALS but the single, daily, charge will be replaced by a charge each time a vehicle enters the charged area (the Central Business District) (Menon et al, 1993). It is intended that ERP will subsequently be extended throughout the Island, transferring some of the current ownership taxes (see Section 21.6) to direct charges for road–use (LTA, 1996).

The Singapore ALS is the simplest form of congestion charging, with users incurring a fixed charge for entering an area during the charged period, regardless of the number of entries they make. Any vehicle within the charged area at the commencement of the charged period incurs no charge, unless it leaves and re–enters the area before the end of the period. This system is classified as an 'entry licence'. A supplementary licence, which is required to be displayed on the vehicle's windscreen within the charged area and time–period, is classified as an 'area licence' (Menon et al, 1993). Thus, strictly defined, the Singapore ALS is an 'entry licence', not an 'area licence', system.

Although the system in Singapore has proved satisfactory for over twenty years, concerns about potential fraud and difficulties of effective enforcement led the London Congestion Charging Research Programme (the London Programme) to conclude that a paper–based entry licence would not be appropriate for London. Indeed, the London Programme concluded that the only paper–based system which might be used in London would be an area licence required for any vehicle on the public highway, whether parked or moving, during the charged period (DOT, 1995a).

The ERP system which has been selected for Singapore is based on charging vehicles when they pass a charge point and so is classed as a 'point–based charge'. The charge points can be linked to form cordons, cells or screenlines. Alternative charging systems include:

- ❑ 'time–based charging', in which charges are a direct function of the time spent travelling within the charged area and period;
- ❑ 'distance–based charging', in which charges are a direct function of the distance travelled within the charged area and period; and
- ❑ 'congestion metering', in which the charge is based on the amount and degree of congestion encountered during a journey, determined by the moving average speed calculated over a preceding defined distance.

Each of these systems requires the use of automatic fee–collection (AFC) technology, which can also be used for the implementation of both entry and area licensing systems, thus reducing the fraud and enforcement difficulties associated with paper–based systems.

Automatic fee–collection systems need to be integrated with highly–automated enforcement systems, which require that:

- ❑ high levels of both accuracy and reliability are provided, to ensure credibility and thus encourage compliance. In general, failures should favour the user;
- ❑ the system should operate without interrupting the flow of traffic or affecting its speed. It must also operate accurately at speeds well in excess of the highest prevailing speed–limit, as well as under stop/go conditions;
- ❑ the system must operate with vehicles located at random across the full width of the carriageway and at very close spacing both across and along the carriageway;
- ❑ it must be possible to detect and identify all potential violators;
- ❑ privacy must be assured for all individual transaction records;
- ❑ the system must be secure against fraud, tampering, evasion and vandalism;
- ❑ the system should be easy to understand and use, thereby facilitating high levels of compliance;

- ❑ information must be available to users on the charges being incurred, as well as current credit or debit balances on their accounts;
- ❑ the system must easily accommodate occasional users, including visitors, both foreign and domestic;
- ❑ the system should accommodate 'charge privileges', which could take the form of exemptions, credits or upper limits on the charges over a specific period; and
- ❑ if motorway tolling is introduced in the UK, mutual compatibility with urban congestion charging systems is highly desirable, as is compatibility with comparable systems elsewhere in the EU.

With automatic fee–collection for point–based systems, charges can be varied by time of day, location and direction of travel. Charges could also vary by vehicle type provided that the technology for enforcement is adequate.

Automatic fee–collection can be based on the use of either off–vehicle, central, accounting systems or on–vehicle accounting. The former operates with relatively simple, in–vehicle, tag technology. The London Programme concluded that it would be necessary to have a 'read–write' tag, which can convey information to the driver, rather than the simpler 'read–only' tag, which is widely used where manual toll–collection systems have been automated (DOT, 1995a). On–vehicle accounting systems make use of smart cards; these may contain pre–paid travel credits or cash in electronic form. The in–vehicle unit (IVU), described as a transponder, has to have an interface with the smart card and must have the intelligence to manage the charge transaction. The IVU is, therefore, considerably more sophisticated and expensive than a read–write tag but it can also provide for a wide range of in–vehicle information services, as well as automatic road–use pricing.

The Singapore ERP system is based on transponders with electronic cash, using smart cards. The London Programme concluded that to meet the various needs of different users, it would probably be necessary to have a system which offers both on– and off–vehicle accounting, ie, offering both read–write tags and transponders with smart cards.

The London Programme identified that the management of occasional users and visitors would present particular difficulties, unless (or until) there is a common nationwide system. The provision of charge privileges (discounts, exemptions, credits or maximum capped charges) for some users would also complicate the design and management of a congestion charging system (DOT, 1995a).

The enforcement of congestion charging would rely largely on automatic video systems to capture the information required. These systems would retain a video image of the licence plate of all vehicles which might have violated the system, through either not having a valid IVU or for not having sufficient credit for the transaction. The vehicle's keeper would then be traced through licensing records. This system has three key weaknesses. First, it depends on the image being legible; some are not, due to damage, dirt or being obscured by other vehicles. Secondly, it depends on the central licensing records being accurate and up–to–date, which requires a sustained high level of performance by DVLA. Thirdly, it depends on the licence plate being valid; some may be deliberately false. The credibility of the system would depend on the enforcement process being able to identify and pursue a high proportion of violators successfully. It would therefore need to be:

❏ accepted by road–users as accurate;
❏ regarded by road–users as being administered fairly; and
❏ designed to ensure that the probability of being identified as a violator and having to pay the associated penalty, taken together, provides an effective deterrent to non–compliance.

It is likely that violations of a congestion charging system would be treated in much the same manner as parking offences, with most violations being handled under civil law, and only serious offences being subject to criminal law.

The implementation and operation of a congestion charging system with automatic charge collection would be relatively costly. For London, it was estimated that the implementation costs of an inbound cordon for Central London with read–write tags would be of the order of £85m. This increased to some £140m for a system of transponders with smart cards. These latter costs would be lower with a system using an established electronic cash card, since some of the costs would be borne by the electronic cash card operator. The annual operating costs were estimated at some £55m per annum, for both the tag and smart card based systems. These costs are largely independent of the level of charge. However, even with a point–based charge of two pounds, the financial payback period would be less than four years (DOT, 1995a).

Although not designed as congestion charging systems, the experience of implementing 'toll rings' in Bergen, Oslo and Trondheim would usefully inform preparation for any planned urban congestion charging system (Gomez–Ibanez et al, 1994).

In addition to the reports on the London Programme, documents which provide further insights include CIT (1990 and 1992), Downs (1992), Gomez–Ibanez et al (1994), Hewitt et al, (1995) and TRB (1994). Blair (1994) provides an extensive annotated bibliography.

Although congestion charging and road–tolling are well supported by economic theory, their implementation has not been easy or popular; nor is direct road–use pricing necessarily the optimum measure for every circumstance. Other principal demand management measures are described in the subsequent sections.

21.4 Demand Management through Fuel Prices

The economic rationale for congestion charging is that the users of motor vehicles do not perceive the true costs of using their vehicle, either at the point of use or at the time they make individual travel decisions. Given the impacts of vehicle–use on the community, even when traffic is flowing reasonably freely, this rationale applies more generally. There is, therefore, a case for increasing the generalised cost of car–use closer to the point of use by increasing the price of fuel through taxation. Indeed, Government policy in the 1990s has been to increase the real level of motor fuel duty by five percent per annum, for environmental reasons. However, it is thought that the effects of this increase on vehicle operating costs are being offset by increased fuel efficiency or by trading down to smaller engines and that there may be little lasting effect on the use of motor vehicles, as long as the scope for downsizing and increased fuel efficiency remains. Indeed, the Royal Commission on Environmental Pollution concluded that, to be effective, much greater price increases would be required and recommended that motor fuel duty be increased year by year, so as to double the price of motor fuel, relative to the price of other goods, by the year 2005 (RCEP, 1994).

Unfortunately, policies based on increasing the price of fuel are indiscriminate in their effect. Thus, the remote rural dweller, who has no real alternative to the use of a car for essential travel, is affected much more than the city resident, who may well have reasonable alternatives. Differential pricing is not feasible, since fuel bought in lower price zones can be used in those where prices are higher. Furthermore, it is likely that the primary impact of increased fuel prices would be a reduction in 'optional' travel, such as leisure journeys in the evenings and at weekends, and that the impact on congestion in urban centres may be quite limited (Dasgupta *et al*, 1994).

Thus, although fuel prices could well be an element of a national demand management policy, they are not an appropriate measure for local demand management, when the objective is to reduce motor vehicle use in particular localities and at peak times.

It can be argued that, without fiscal and other measures which encourage or require reductions in the more harmful vehicular emissions, there will be neither the incentive for industry to design less polluting vehicles nor for the public to buy them, particularly if there is any price or performance penalty. Yet, there is technical scope for reducing the main pollutants by at least 50% (Bly *et al*, 1995) and also reducing the specific fuel-consumption of both petrol- and diesel-engined vehicles.

21.5 Regulating the Use of Vehicles

Demand can be managed by using regulations to control the use of vehicles. In Bologna, and some other Italian cities, permits to use motor vehicles within the city centre are issued to essential users only; others have to park outside the restricted area or travel by some other mode. The Bologna scheme forms part of a comprehensive policy to reduce traffic in the city centre, the old town. The policy includes public transport improvements, parking controls and pedestrianisation. The traffic-limited zone was introduced in 1989 and extends from 0700 hours to 2000 hours. Although there is a large number of exemption permits (for residents, delivery vehicles and those with private off-street parking), the overall policy reduced the number of vehicles entering and leaving the city centre by about 50%. However, this has since been eroded by increases in the number of exemption permits issued (Topp *et al*, 1994).

An alternative to permits is the 'odds and evens' scheme, in which use of vehicles is permitted on alternate days for registrations ending in odd and even numbers. Athens is one city which has applied this scheme. Sao Paulo has implemented a variant, whereby vehicles with registrations ending in certain digits are prohibited on certain days. One risk of the 'odds and evens' policy is that it encourages an increase in the number of vehicles owned, to provide households with both odd- and even-numbered vehicles.

Evidence suggests that, although permits and related schemes can, in principle, achieve a reduction in car-use, effective and tight control is required to avoid diminution of the effects through fraud and evasion. Indeed, a permit scheme is a variant on the entry or area licence schemes with user privileges (see Section 21.3).

In California, measures have been introduced which require larger employers to introduce ride-sharing and other schemes, to reduce the volume of 'drive alone' commuting journeys to their employment sites. However, it is not clear that these measures have any significant effect across a region (Wachs, 1990). Nottinghamshire County Council is promoting a similar concept, through a travel-reduction programme, but without the legislative support applying in California. Other UK authorities are sponsoring 'travel awareness' campaigns, to increase appreciation of the unfavourable impacts of traffic, and travel more generally, and of the actions which can help to mitigate those impacts.

The US Federal Clean Air Act, as amended in 1990, requires communities in the USA with serious air pollution to take measures to reduce vehicular emissions and not to pursue infrastructure and development policies which could exacerbate vehicular emissions (Shrouds, 1995). In practice, this requires measures which reduce car-use or, at least, discourage increases in vehicle-kilometres in the region.

21.6 Restraint on Vehicle Ownership

It is possible to contain growth in car-use by restraining growth in vehicle ownership. For many years, this has been a key feature of Singapore's transport policy, which has also included increasing the capacity and the quality of the Island's public transport and highway systems, as well as the management of demand through road-use pricing and parking controls. A similar policy on restraining car-ownership has also been pursued in Hong Kong, where first registration costs and annual taxes are high.

Initially, Singapore exercised control on car ownership through high import duties and high annual charges, with a charge-structure designed to encourage the scrapping of older cars and to discriminate against company car-ownership. However, continued rapid growth of vehicle ownership led to a decision, in 1990, to introduce absolute limits on the number of vehicles which could be registered. Buses for use on scheduled services and emergency vehicles were the only classes excluded. The target was to reduce the annual increase in vehicles owned to three percent. Prospective buyers of new vehicles have to bid for a 'Certificate of Entitlement' (COE) in a monthly tendering process. Depending on the category of vehicle, the premium has been between some £4000 and £9250 per vehicle, adding between 12% and 25% to the total price paid (Olszewski *et al*, 1993; and Koh *et al*, 1994).

While fiscal measures, such as those adopted by Hong Kong and Singapore, may be effective in controlling car–ownership, there is a view that, as disposable income increases, pressure will arise for a change in the balance between charges for vehicle–ownership and charges for vehicle–use. In Singapore, it is anticipated that the introduction of Electronic Road Pricing will facilitate this (LTA, 1996). Significantly, there are no local vehicle manufacturing interests in either Hong Kong or Singapore. They are also city states, in which the economy of rural areas is not an issue. The impacts of such fiscal policies in countries with strong manufacturing interests and/or extensive rural economies would be markedly different and their pursuit more difficult.

However, other demand management measures could be used to deter car–ownership in congested urban areas. In parts of Japan, for example, it is necessary to prove that a parking space is available before a car can be purchased. Car–clubs, through which access to a car is provided when required rather than direct ownership, can serve to reduce demand, by curtailing immediate access to a car. For example, Auto Teilet Genossenschaft (ATG), based in Lucerne, Switzerland, has over 5,000 members, including a number of businesses, and Stadt–Auto, a similar organisation based in Bremen, has over 3,000 members (see Section 29.10). Research suggests that ATG members have reduced their annual mileage from, typically, 15,000km to about 5,000km and also made significant savings in total transport expenditure.

To restrain car–ownership and use, the City of Edinburgh Council has sponsored an inner city housing development, in which the purchasers of houses covenant not to own a car and arrangements are being made for ready access to rented vehicles on favourable terms, based on the 'Stadt–Auto' car–club concept. Indeed, lack of convenient, or secure, parking space (particularly in inner urban ares) can be a deterrent to car–ownership and space problems can also influence the sizes of cars owned.

21.7 Parking Controls and Pricing

The control of parking is principally addressed in Chapter 19; it is relevant here to consider parking in the context of managing traffic demand.

Parking controls, including pricing, can be used to influence vehicle ownership but their primary use as a demand management measure is to regulate parking capacity and to allocate the available space between different groups of user. However, the control of parking affects only trips with a destination within the area subject to the controls. Used by itself, therefore, parking control can reduce congestion for those vehicles passing through the controlled area, with the result that 'through' traffic flows can increase. This can be because of diversion from longer but previously quicker routes, which avoided the congested area, or from other times of day or because trips which were previously unattractive become feasible.

The use of parking controls as a demand management measure is weakened by the fact that, in almost all urban areas, a high proportion of parking spaces are in private non–residential (PNR) car parks not under public (local authority) control. Recognising that parking controls could only be effective if PNR could also be managed in accordance with an area–wide parking and transport policy, studies were undertaken in the 1970s (DOE, 1976) to determine whether it would be feasible to extend public control to such spaces. The problems were found to be considerable, not only in the definition of a PNR space but also in enforcing regulations on their use. The conclusion was that such policies would be difficult both to introduce and to operate. However, there is scope for the use of parking controls as an element of urban transport policy and re–examining the case for public control of PNR car parking.

Much of the PNR stock of parking spaces is used for employees' parking, normally free of charge. It can be argued that anyone with a free, and guaranteed, parking space provided at work does not bear the full costs of their journey to work. In consequence, those travelling to work in major urban areas may well make modal choices and, possibly, residential location choices which, in a community context, are economically inefficient (Shoup *et al*, 1992). If it is not possible to control such parking directly, through PNR controls, a case can be made for treating it as a 'benefit in kind' on which income tax is payable, as with cars provided by employers.

At some employment locations in the United States, the use of parking space is controlled to give preference to those commuting by car–pool or van–pool. These vehicle–pools are sometimes organised by the employers themselves, to save the cost of providing parking accommodation.

21.8 Physical Restraint Measures

Traffic can be managed through the use of physical measures, designed to make the use of motor vehicles less attractive. These may reduce speeds or extend travel distances, as described in Chapter 20.

While the creation of traffic–free areas in urban centres may remove traffic from some streets, they do not necessarily reduce demand overall, unless coupled with other measures. Some cities have implemented more comprehensive measures (Topp *et al*, 1994). For example, York has a policy of reducing the use of motor vehicles across a wide part of the City. This is achieved through a combination of physical and regulatory measures and by positively encouraging non–motorised modes. Gothenburg has sought to limit traffic within the city centre, by creating a system of cells between which there is no direct access for cars. To move between cells, drivers have to return to a ring road which encircles the controlled area.

Roadspace can be reallocated, either to disadvantage car–users explicitly, for example, by allocating space to public transport and multiple–occupancy vehicles, or to deter short distance travel by car. An experiment in Nottingham, in 1975, is an early example of the former that proved unsuccessful and was withdrawn (Vincent *et al*, 1978). However, Zurich has been particularly successful in using traffic management and control measures to secure reliable and quick travel times for public transport, with positive discrimination against other motor vehicles.

While the primary objective of bus–priority schemes in many cities is to enhance public transport services, there is frequently a secondary objective of seeking to discourage car–use. Positive discrimination against 'inefficient' use of roadspace by vehicles with only a driver, or only a driver and one passenger, is extensive in the US through the use of 'HOV' (high occupancy vehicle) facilities, both on the highway and at employer's parking lots. Critical to the success of HOV facilities is that users can travel more quickly and easily door–to–door than those driving solo. The system is reinforced if they can be seen by–passing long queues of those not permitted to use the facility. In the UK, extensive bus priorities can achieve some of the benefits of HOV lanes, through giving priority to those in buses, but they do nothing to encourage reduction in demand through 'ride–sharing' in cars or through the formation of car– and van–pools.

21.9 Land–Use and Development Controls

Since travel is a derived demand, it should be possible to reduce demand, overall, through changes in land–use location policies. Indeed, it can be argued that much of the increase in the use of cars is a direct result of policies which have permitted, even encouraged, the dispersion of major activity centres to the fringes of urban areas and beyond. Many of these locations are not readily accessible by public transport and, with concentration into larger units for retail, education, healthcare and recreation, few people live near enough to access them by foot or bicycle.

The need, at least, to curtail, if not reverse, some of these trends has now been recognised through the publication of revisions to the Department of the Environment's Planning Policy Guidance documents, one of the most important of which is PPG13 (DOE/DOT, 1994) [NIa] [Sa]. A key principle of these policies is to locate new developments so as to facilitate access by public transport, bicycle and on foot, with a preference for locating developments in existing town and city centres. There is also a presumption in favour of more mixed development.

Development policies can be used to control the extent of parking in new developments. One possibility is to increase the proportion of parking capacity, under public rather than private control, through the provision by the developer of commuted payments to the Local Authority in lieu of providing spaces within the development. The Local Authority is required to use such revenues to provide parking spaces in a public facility. By this means, it is possible to set limits on the number of spaces provided within developments (Sanderson, 1994) (see also Chapter 28).

Important though land–use and development policies are, the general pace of development and redevelopment is such that significant benefits across a large urban area are only likely to be achieved over the medium to long term. Furthermore, there is little evidence that mixed development can have a significant effect in reducing travel demand. With increasing affluence, choice has become important to consumers and greater job mobility means that, although an initial housing location decision may be based on a convenient journey to work, few people are willing to change their house every time they change their job. The effectiveness of land–use policies in reducing travel demand will depend, to a significant extent, on the pursuit and effectiveness of other complementary policies (Barrett, 1995).

As with many other demand management measures, competition between adjacent localities for economic strength can seriously reduce the effectiveness of well–intended policies. Given the choice between accommodating the requirements of a major project, which will enhance the local economy, by relaxing their more stringent policies or maintaining those policies and seeing the project go elsewhere, many authorities will opt for the former. While a firm national, or regional, policy framework might help to avoid such 'bidding' situations, it would be at some cost to local autonomy on key decisions.

21.10 The Roles of Public Transport, Taxis, Cycling and Walking as Alternatives to Travel by Car

For many of those with access to a car, public transport is now seen as the choice of last resort. The reasons for this include perceptions (whether true or not) of poor information, uncertainty and unreliability, unacceptable travel times, discomfort and inconvenience, concerns about threats to personal security and price. For many people, their car is an extension of their private space. Yet, in some urban areas, public transport still carries a significant proportion of commuters, as well as travel for other purposes, notably into and out of central London.

Demand management can cause some trips to be made at different times of day or to different locations, some not to be made at all and others to be combined. Some people will switch to public transport, if there is a sufficiently convenient service, and others may switch to taxis, cycling or walking. Evidence from various studies (Dasgupta et al, 1994; MVA, 1991 and DOT, 1995a) suggests that the extent to which demand for travel by car can be reduced, by feasible restraint policies, is limited. However, these and other studies suggest that restraint measures in combination with improved public transport can increase the shift away from car–use.

The combination of public transport improvement with traffic restraint, acting as 'carrot and stick', is particularly important if restraint policies are not to affect local economies adversely, particularly those of town and city centres which development policies are intended to strengthen. Seeking to make an entire journey 'seamless', with easy transfer between different modes and operators, is of particular importance. This should include arrangements for common (smart–card) ticketing between all the local public transport operators, as well as for park–and–ride charges (CIT, 1996).

In addition to ensuring that the public transport system is of sufficient quality to complement any restraint on the use of private vehicles, it is important to recognise that, because some car–users are willing to switch to taxis, adequate and convenient facilities for taxis will be required at major activity centres.

Many urban journeys are short, well suited to walking or cycling, which can be quicker than motorised alternatives. To encourage greater use of these modes, better facilities are necessary, designed to meet their specific needs, providing safe, direct and easy routes to activity centres. The provision of improved facilities for cyclists and pedestrians is likely to be a particularly important element of any policy designed to restrain the use of cars in urban areas (see Chapters 22 and 23).

21.11 Telecommuting

The information revolution will undoubtedly lead to fundamental changes in both the ways in which people work and where they work, transforming the relationship between work and home locations and, thus, the need for travel. By working from home, or from local telecommuting centres, travel to and from work will be reduced. It has been estimated that by the year 2010 up to 10 million workers in the UK will be teleworking (Gray et al, 1993).

Research in California (where telecommuting is probably most highly developed) and the Netherlands indicates that telecommuting can lead to a net reduction in travel (Pendyala et al, 1991 and Hamer et al, 1991). Research also suggests that the information revolution will reduce the need to travel for some purposes but that this could be complemented by increases in travel for other purposes, although net reductions were found by Koenig et al (1996).

Although not intended as a demand management measure, it would seem that traffic reduction benefits could be obtained by encouraging the development of telecommuting and through exploiting the potential of information technology to reduce the need to travel for a variety of purposes.

21.12 The Potential for Transport Telematics

As noted in Chapters 15 and 18, the development of advanced transport telematics (ATT) to promote intelligent transport systems (ITS) is expected to contribute to increasing efficiency in travel and transport, not least through the provision of real–time information. For those using roads, this should enable them to plan their journeys to avoid congested times and places. For those travelling by public transport, transport telematics will provide access to up–to–date routeing, timetable and fare information, coupled with real–time information on service operations to allow for late running. Ideally, the system should provide optimal advice for the traveller, regardless of the mode of travel or the operator of the system, but the achievement of this will depend on arrangements for funding of the development and implementation of ATT systems.

The basic systems to permit travel–planning and real–time journey information have been implemented through demonstration projects, such as London Transport's 'Count–Down 'for real–time information at bus stops, and the more extensive, EC–supported projects in Birmingham and Southampton (see Chapter 24). Traffic*master* provides network–wide real–time information on motorway traffic conditions and variable message signs serve this purpose more locally (see Chapter 15).

The introduction of smart–card based payment systems could contribute towards the creation of the seamless journey, in which a traveller may use different systems, provided by different operators, with all fares and charges paid by use of a common smart–card. This could be a pre–payment card or an electronic cash card. For maximum convenience on public transport, it should be a contactless card. However, for automatic road–use pricing systems (tolling and congestion charging), a contact card is likely to prove more suitable, to provide the speed of data–transfer required (Blythe *et al*, 1995).

21.13 Legislation

Many elements of a demand management plan can be implemented within existing legislation but others would require new legislation. These include the introduction of any form of road–user charging (including congestion charging) and local authority control of PNR parking. Existing regulations can be used to control entry, although there are doubts about their suitability for area–wide control schemes, such as the permits used in Bologna.

21.14 Financial Considerations

As explained in Chapter 4, most finance for local road schemes and transport measures depends ultimately upon Treasury funding. Although the Package Bid approach in England and, in some respects, Challenge

Funding in Scotland, offer local authorities greater freedom in determining local priorities, their scope and options are constrained by the total funding provided. Major investment in public transport infrastructure is still very largely dependent upon central government. The actual arrangements vary between England, Scotland and Wales (IHT, 1996) [Wa].

As described in Chapter 19, one relaxation of central government's control of finances has been the facility to transfer responsibility for parking enforcement from the police to local authorities, together with arrangements under which revenues from charges relating to civil offences are retained by the local authority [NIb]. This is important, not only in the context of funding but also because it gives local authorities greater control over parking, as a crucial element of demand management policy. However, the use of revenues from any form of road–use pricing remains a vexed issue. In evidence to the Select Committee on Transport, the then Secretary of State for Transport advised that, since such a charge "...would be a levy as opposed to a charge for a service rendered" it would "...fall under public expenditure and would be treated accordingly" (Transport Committee, 1995). With this interpretation, the revenues would accrue to the Treasury and, hence, become part of general revenue.

This contrasts with the advice that motorway tolls would constitute charges for a service, which would permit retention of at least part of the revenues for re–investment in motorway construction and maintenance (Transport Committee, 1994). However, in his evidence on congestion charging, the Secretary of State also said "...it would be open for a policy to be developed, which would allow money to return to those who had contributed it." (Transport Committee, 1995). This position was explained in the Government response to the Transport Committee's report, which stated '...the view that urban road tolls would be general government revenue, rather than negative expenditure, is not a policy decision but derives from the application of national accounting conventions that are internationally recognised." It went on "..the Government accepts the general principle that financial arrangements should ensure that communities which implemented congestion charging should be economically better off as a result. The Government will consider the detail of the necessary arrangements further as and when necessary" (DOT, 1995b).

This raises the critical issue of the non–hypothecation of tax–revenues, a key element of Treasury policy. This is that no revenues derived from taxation can be committed to specific items, or programmes, of expenditure. This contrasts with other countries, such as the United States, where funds raised through a specific tax, such as the Federal Gas Tax, can be used to pay for particular activities, including highway funding through the Federal Highway Trust Fund. So long as net revenues from urban congestion charging are deemed to be Treasury revenues, it is unlikely that any local authority will choose to implement a congestion charging policy, unless there are complementary improvements to public transport, for which investment funds will be required.

This raises another important issue, that of 'additionality'. Even if it were possible to devise arrangements under which the net revenues from congestion charging were retained or returned for expenditure locally (whether directly or, for example, through increased Government funding), it would be necessary to ensure that these funds were in addition to, rather than in substitution for, those which Government would otherwise have allocated through its various channels. Ensuring additionality of funding is likely to prove a crucial test, locally, in determining the acceptability of congestion charging. The City of London Corporation commissioned a series of studies which have examined innovative approaches to the funding of transport projects, including the use of congestion charging revenues (LBS, 1993).

Goodwin has proposed that the revenues from congestion charging should be allocated according to a 'rule of three', with one third being allocated to highway improvements, one third to public transport and one third to either general tax relief or increased general expenditure (Goodwin, 1989). Subsequently, Small has suggested an alternative allocation, with one third allocated to reimburse travellers as a group, one third for new transport services and one third to substitute for general taxes currently used to pay for transport services (Small, 1992).

21.15 The Political Dimension

While the principles of demand management may be sound, and the need amply warranted, the implementation of measures to achieve a significant reduction in vehicular traffic levels is fraught with risk. There are risks about the degree of success of such measures; risks about the nature and extent of the impacts on the local economy, particularly in relation to other competing towns and cities; and risks about the responses of the electorate to measures which will limit the extent to which they can use their cars.

Politicians tend to avoid potentially risky policies. If significant demand management programmes are to be implemented, political leaders (both national and local) will need to be satisfied that there is sufficiently strong public support and to have confidence in the effectiveness of the plans. To achieve these requirements, they would need to be satisfied that a programme can be implemented in such a way as to minimise the risks, either of failure or of impacts which prove unacceptable to the electorate.

21.16 Public Attitudes

Research suggests that there is a growing awareness of the adverse effects of motor traffic in general, and congestion in particular, on the quality of life and on the environment in urban areas. However, although an increasing proportion of car-users would be prepared to consider switching some trips to public transport, there is little evidence of a widespread willingness to forego the perceived mobility benefits of the private car. Measures designed to reduce the use of cars, through improvements to public transport or restricting the use of cars in city centres, appear to have more popular support than fiscal measures such as congestion charging or increases in fuel duty, although some of the concerns about congestion charging can probably be mitigated, if it forms part of an integrated package of measures (Jones, 1995).

Even if congestion charging can be shown to have beneficial effects on congestion and the environment, successful implementation would require its acceptance by the public as a reasonable measure. This is likely to require the provision of good information and extensive consultation. But, most importantly, the role of congestion charging must be within a broad, well-founded, urban transport policy coupled with a long-term commitment to use a substantial part of the net revenues to improve public transport and the environment, over and above normal public expenditure.

21.17 The Assessment of Demand Management

The assessment of plans and policies is addressed in Chapter 9. Although the basic principles apply equally to the assessment of demand management measures, some particular aspects require more detailed consideration. The first is assessment of the equity of impacts arising from the incidence of benefits and costs. There are two forms of equity, 'horizontal' and 'vertical'.

Horizontal equity is concerned with the distribution of the costs and the benefits between different groups of transport users; for example, whether city centre car–commuters incur net disbenefits due to traffic restraint, while other commuters benefit, or whether peak–hour travellers suffer net disbenefits while travellers at other times of day enjoy net benefits, and so on.

Vertical equity is concerned with the distribution of the costs and benefits between different sectors of the community, for example, by income class or some measure of ability to absorb the consequences. There is concern that congestion charging would be 'regressive', with the less well–off incurring net disbenefits while the better–off enjoy net benefits. In fact, both the Hong Kong studies and the London Congestion Charging Research Programme (DOT, 1995a) showed that, on average, the less well–off benefited, mainly because of the consequential improvements to bus services. But that conclusion would not necessarily apply in all locations or with all types of charging structure. Vertical equity is also concerned with the impacts on particular social groups, such as people with impaired mobility and those with particular travel needs.

Demand management is likely to have differing impacts in the short– and longer–term. In the short–term, an individual response might be to accommodate the charge in some way, continuing with the same basic travel patterns, albeit by another mode or at a different time of day. However, as opportunities arise for more radical change, such as changing one's job or relocating a business, so more extensive changes might take place. Ideally, both should be assessed, although the tools for assessing longer term change, including location decisions, are not as well developed as those for assessing shorter term change.

This issue of short– and long–term change is particularly pertinent in the context of understanding the possible impacts of demand management on the local (urban) economy and the distribution of those impacts by locality and by sector. These are likely to prove of importance in decision–making. As there is little real understanding about the detailed workings of urban economies, this is a particularly difficult topic, in which judgement, informed by research, must play a key role.

21.18 References

Barrett G (1995) 'Transport Emissions and Travel Behaviour: A Critical Review of Recent European Union and UK Policy Initiatives'. Transportation, 22(3).

Blair B (1994) 'Road Charging in the 1990s, An Overview and Guide to the Literature'. British Library, London.

Bly PH, Hunt PB, Maycock G, Mitchell CGB, Porter J, and Allsop RE (1995) 'Future Scenarios for Inland Surface Transport', TRL Report 130, TRL.

Blythe PT and Hills PJ (1995) 'The use of Smart–cards in Road–tolling and Road–use pricing systems' IEE Colloquium London (March), IEE.

CBI (1989) 'Trade Routes to the Future', Confederation of British Industry.

CIT (1990) 'Paying for Progress', Chartered Institute of Transport.

CIT (1992) 'Paying for Progress: Supplementary Report', Chartered Institute of Transport.

CIT (1996) 'Better Public Transport for Cities', Chartered Institute of Transport.

Dasgupta M, Oldfield R, Sharman K, and Webster V (1994) 'Impact of transport policies in five cities' Project Report 107, TRL.

DOE (1976) 'The Control of Private Non–Residential Parking: Consultation Paper', DOE.

DOE (1996) 'The National Air Quality Strategy', DOE.

DOE/DOT (1994) 'Planning Policy Guidance Note 13 Transport', (PPG13), Stationery Office [Sa] [Wb].

DOT (1964) 'Road Pricing: The Economic and Technical Possibilities', Stationery Office

DOT (1967) 'Better Towns with Less Traffic', Stationery Office.

DOT (1995a) 'The London Congestion Charging Research Programme, Final Report' Stationery Office. (Note: a report of the Principal

Findings is also available from Stationery Office, and a summary of the Final Report is available in a series of six papers published in Traffic Engineering + Control between February and August 1996.)

DOT (1995b)   'Urban Road Pricing. The Government's Response to the Third Report 1994 – 1995 of the Transport Select Committee', Stationery Office.

Downs A (1992)   'Stuck in Traffic: Coping with Peak–Hour Congestion', The Brookings Institute, Washington DC.

Gallup (1989)   'Report on Survey in Five West London Boroughs', Metropolitan Transport Research Unit.

Gomez–Ibanez JA and Small KA (1994)   'Road Pricing for Congestion Management: A Survey of International Practice' NCHRP Synthesis of Highway Practice 210, Transportation Research Board, National Academy Press, Washington DC.

Goodwin PB (1989)   'The 'Rule of Three: a Possible Solution to the Political Problem of Competing Objectives for Road Pricing' Traffic Engineering + Control 30(10).

Gray M, Hodson N and Gordon G (1993)   'Teleworking Explained', John Wiley, Chichester.

Hamer R, Kroes E, and van Ooststroom H (1991)   'Teleworking in The Netherlands: an Evaluation of Changes in Behaviour'. Transportation, 18(4).

Hewitt P, Johansson B and Mattsson L–G (1995)   'Road Pricing: Theory, Empirical assessment and Policy', Kluwer Academic Publishers, Dordrecht.

Holland EP and Watson PL (1978)   'Traffic Restraint in Singapore'. Traffic Engineering + Control 19(1).

ICE (1989)   'Congestion', The Institution of Civil Engineers, Thomas Telford.

IHT (1996)   'Guidelines on Developing Urban Transport Strategies', The Institution of Highways & Transportation.

Jones PJ (1992)   'Review of Available Evidence on Public Reactions to Road Pricing' Polytechnic of Central London for the Department of Transport.

Jones PJ (1995)   'Road Pricing: The Public Viewpoint, in Road Pricing: Theory, Empirical Assessment and Policy' (Editors: Johansson B and Mattsson L–G), Kluwer Academic Publishers, Dordrecht.

Jones PJ and Hervik A (1992)   'Restraining Car Traffic in European Cities: an Emerging Role for Road Pricing', Transportation Research A, 26(2).

Koenig BE, Henderson DK and Mokhtarian PL (1996)   'The Travel and Emissions Impacts of Telecommuting for the State of California Telecommuting Pilot Project', Transportation Research C, 4(1).

Koh WTH and Lee DKC (1994)   'The Vehicle Quota System in Singapore: an Assessment' Transportation Research A, 28(1).

LTA (1996)   'White Paper on Land Transport', Land Transport Authority, Singapore.

LBS (1993)   'The City Research Project: Meeting the Transport Needs of the City', London Business School for the City of London Corporation.

May AD (1975)   'Supplementary Licensing: An Evaluation', Traffic Engineering + Control, 16(4).

May AD (1994)   'Potential of Next–Generation Technology in Curbing Gridlock: Peak–Period Fees to Relieve Traffic Congestion', Transportation Research Board, National Academy Press, Washington DC.

Menon G, Lam S–H and Fam HSL (1993) 'Singapore's Road Pricing System: Its Past Present and Future', ITE Journal December 1993. Washington DC.

MVA (1991) 'Edinburgh Joint Authorities Transportation and Environmental Study, Strategies Study' Final Report, The MVA Consultancy.

NEDO (1991) 'A Road User Charge? – Londoners' Views', National Economic Development Office.

Olszewski P and Turner DJ (1993) 'New Methods of Controlling Vehicle Ownership and Usage in Singapore', Transportation, 20(4).

Pendyala R, Goulias L and Kitamura K (1991) 'Impact of telecommuting on spatial and temporal patterns of household travel', Transportation, 18(4).

RCEP (1994) Royal Commission on Environmental Pollution Eighteenth Report, 'Transport and the Environment', Stationery Office.

Sanderson J (1994) 'A Matrix Approach to Setting Parking Standards', Transport Planning Systems, 2(1).

Shoup DC and Wilson RC (1992) 'Commuting, Congestion, and Pollution: The Employer–Paid Parking Contribution', papers Presented at the Congestion Pricing Symposium, June 10–12 1993. US Department of Transportation, Washington DC.

Shrouds JM (1995) 'Challenges and Opportunities for Transportation: Implementation of the Clean Air Act Amendments of 1990 and the Intermodal Transportation Efficiency Act of 1991' Transportation, 22(3).

Small KA (1992) 'Using Revenues from Congestion Pricing', Transportation 19 (4).

Topp H and Pharoah T (1994) 'Car–Free City Centres' Transportation 21(3).

Transport Committee (1994) 'Charging for the Use of Motorways' House of Commons, Stationery Office.

Transport Committee (1995) 'Urban Road Pricing' House of Commons, Stationery Office.

TRB (1994) 'Curbing Gridlock: Peak Period Fees to Relieve Traffic Congestion' Volumes 1 and 2. Transport Research Board, National Academy Press, Washington DC.

Turner DJ and Olszewski P (1993) 'A review of Road Pricing Policies in Singapore', Institution of Transportation Engineers (ITE) 1993 Annual Meeting, The Hague. ITE, Washington DC.

UITP (1991) 'European Attitudes Towards Urban Traffic Problems and Public Transport', UITP, Brussels.

UN (1987) 'Our Common Future (The Bruntland Report)', Report of the United Nations Committee on Environment and Development.

University of Westminster (1996) 'Public Attitudes to Transport Policy and the Environment: An In–Depth Exploratory Study', University of Westminster, Transport Studies Group.

Vickrey W (1959) 'Statement to the Joint Committee on Washington Metropolitan Problems', US Congress

Vincent RA and Layfield RE (1978) 'Nottingham Zones and Collar Scheme – the overall assessment' Report 805, TRL.

Wachs M (1990) 'Regulating Traffic by Controlling Land Use: The Southern California Experience' Transportation, Vol 16, No 3.

World Bank (1996) 'Sustainable Transport: Priorities for Policy Reform', The World Bank, Washington DC.

Chapter 22 Pedestrians

22.1 Introduction

Walking is an indispensable part of the transport system in every urban area. Over 80% of trips under one mile long are made on foot and the proportion is even greater for shorter distances. Even among journeys over a mile, 10% are on foot, although this represents only 2.4% of the distance travelled. In total, about one-third of trips in urban areas are made entirely on foot (although this depends crucially on how 'a trip' is defined). For education and shopping trips, walking represents over 20% and 10% respectively for journeys over a mile, with higher proportions for shorter journeys. Everyone needs to walk – for work, shopping, education or leisure. For those with a choice of mode, more could be done to encourage people to choose to walk. Among the factors which favour walking are its cheapness, its healthiness and its flexibility.

In urban areas, people should be able to walk in reasonable comfort and safety, as walking is an essential part of a wide variety of activities. People walk in order to get to specific destinations but walking–around is also an integral part of shopping, leisure or sight–seeing. Indeed, the freedom with which a person can walk about and look around is a useful guide to the civilised quality of an urban area (Buchanan *et al*, 1963).

It is vital to the environmental quality of urban areas to provide a high standard of pedestrian facilities, recognising the vulnerability of all pedestrians and the special needs of the young, the elderly and people with disabilities (DOT, 1997) [Sa].

22.2 Vulnerability of Pedestrians

Pedestrians are particularly susceptible to risks posed by other road–users, although individuals may under– or over–estimate the actual level of risk in any given situation. However, the perception of risk influences many social activities and, in extreme cases, even community relationships and identity. The mobility of vulnerable groups, including children and the elderly, is especially affected by the perceived risk from traffic volume and speed. However, this perception is well grounded in the high proportion of pedestrian accidents that result in death or injury (see Chapter 16) and the injurious effect on health of the emissions from vehicle exhausts (see Chapter 17).

22.3 Strategies to Provide for Walking

In many urban areas, the needs of vehicular traffic have taken precedence over the needs of pedestrians and it appears that the needs of pedestrians have not been given the attention they deserve. This is not only inefficient but also results in a poor environment. A complete transportation strategy would include the development and maintenance of a comprehensive, safe, well–signed and well–lit network of pedestrian routes, providing easy access to major attractions.

The development of a robust urban transportation strategy must include an analysis of pedestrian needs. The resulting plan should provide a balance between the requirements of private vehicles, public transport, pedestrians and cyclists. Edinburgh, for example, has established specific targets for an increase in walking compared with other modes of transport. Policies now recognise the vulnerability of pedestrians and the need to discriminate in their favour (Davies, 1992). Analysis should determine the nature and preferred routes of walking trips. For example, pedestrians when shopping exhibit a more random and diverse pattern of movements and, therefore, need more space than pedestrians walking between a public transport interchange and, say, an office complex, who are likely to seek the shortest and quickest route.

The level of pedestrian activity is a useful measure of the vitality and commercial viability of a town. Counts should be taken at different locations, both within a town centre and elsewhere in the urban area, and at different times of the day and evening. The counts should be taken in the same locations at the same time each year to monitor trends, especially in retail activity. Surveys of this kind are used to assess the impact of activities, such as special promotions in the town centre and the opening of new retail centres, both within and outside the urban area and to provide a rationale for a footway improvement strategy.

Pedestrians are concerned about the condition of footways and footpaths, including unevenness, raised edges, slipperiness, broken paving slabs, gaps and poor quality repairs (May *et al*, 1991). There is a demonstrable need for a comprehensive strategy for inspection and maintenance of footways and other pedestrian facilities. All pedestrian areas should be inspected on a regular basis and a record made of any footway or footpath in a condition worse than a pre-determined threshold standard. The strategy should allow for a maintenance response which balances the efficient use of resources and preserves acceptably safe surfaces. Generally, pedestrians seem to prefer elemental paving to blacktop and prefer sand or brick colours. The strategy should identify materials which are both economical and appropriate to the location and use of the pavement.

22.4 Developing Pedestrian Networks

Pedestrian networks should be planned carefully and implemented incrementally. They should be related to cycling (Chapter 23) and should be incorporated with town centre strategies (Chapter 12). Walking is both the slowest and most flexible form of transport but may, nevertheless, be the quickest means of making short trips. In order to decide which parts of a pedestrian network require improvement, the designer needs to have a clear understanding of the patterns of pedestrian activity. However, pedestrians, unlike vehicles, do not confine themselves to specific routes but rather follow the shortest and most direct path between their origin and destination. Surveys can be undertaken by a variety of techniques, using interviews, filming and observations. Generally, a combination of survey techniques should be used so as to cross-validate data.

The National Travel Survey (NTS) indicates a significant reduction in the distance walked by children aged 5 to 15 between 1975/76 and 1989/91. One of the factors involved is the trend towards taking children to school by car and the under-lying reasons for this need to be understood. Of especial importance are the perceived hazards of walking in urban areas. Designers should plan safer networks of walking routes for everyone but routes to and from schools should be given priority.

Footways and footpaths should be aligned as directly as possible between the main trip origins and destinations. Pedestrians prefer to see the place to which they are heading. Whilst gentle curves will probably be followed by pedestrians, sharp curves will not be followed readily unless physical barriers deter the taking of short-cuts. All pedestrian footways and footpaths should have a minimum width of 1800 mm but should be wider wherever possible.

Most pedestrian journeys begin and end in buildings or transport interchanges. The relationship between the entrances to buildings and the pedestrian network is of particular significance. Changes in level should be avoided but, where a difference in level is inevitable, the needs of people with impaired mobility must be considered. Bridges, high level walkways and subways should be avoided, unless they relate naturally to the main entrances of nearby buildings.

The quality of a street scene is particularly important to pedestrians. Routes should be planned so as to allow both close and distant views of features of interest. The boundary to the footway should be of a consistently high quality. Hard and soft landscaping should be provided and maintained. Pedestrians enjoy animated and lively street scenes and the presence of a modest flow of vehicles (say, up to 500 vehs/h) is generally acceptable where pedestrian flows are light.

Pedestrians tend to be concerned about personal security. Routes should be developed that will be used by reasonably substantial and predictable flows of people. Corners and angles of buildings or structures, where individuals might not be visible to others, should be avoided where possible.

Local authorities are responsible for the naming and numbering of streets and for ensuring that these are properly displayed. A clear and consistent system of street name plates should be adopted. Key pedestrian destinations, and the quickest route to them, should be signed by a carefully devised area-wide system. Several authorities have adopted a particular style, such as finger-post signs with gold lettering embossed on a black background.

22.5 Dropped Crossings

Dropped crossings may be provided either:
   ❑ to allow vehicles to gain access across footways into buildings or onto land; or
   ❑ to assist pedestrians, especially those with mobility impairment, including those with prams or push–chairs, when crossing a carriageway.

The former should be kept to an absolute minimum and their provision and construction should be controlled by the Highway Authority. In some circumstances, planning permission may be necessary. There may be a vertical face of up to 25 mm to the upstand of a dropped kerb at a vehicle crossover, to ensure that surface water is retained on the carriageway. A problem with dropped crossings is that they can encourage cyclists to opt to use the

Photograph 22.1: A Toucan crossing with an L—shaped tactile area.

footway and, to avoid this, specific provision for cyclists should be made wherever possible (see Chapter 23).

At locations where significant numbers of pedestrians are likely to want to cross a carriageway, the kerbs should be dropped to facilitate crossing with prams or pushchairs and by people in wheelchairs. There should be no vertical face on the upstand of a dropped kerb at a pedestrian crossing so that wheelchair users are not delayed in regaining the

footway. The gradient of ramps at all crossing places where kerbs are dropped should not be greater than eight per cent (1:12) but a gradient of five per cent (1:20) is preferred.

Care should be taken to assist people with visual impairment at appropriate crossing points and reference should be made to the most recent DOT guidance. Tactile surfaces should be used to identify the presence of a dropped kerb (DOT 1991a). Only 'modified blister paving' should be used, comprising rows of flat—topped 'domes' 5 mm (± 0.5 mm) high. It should be noted that DOT advice on tactile surfaces is being reviewed and the DOT Mobility Unit should be contacted to ascertain the latest position.

The layout and colour of the surface will depend on the type of crossing. Full details can be obtained from the Department of Transport's Mobility Unit (see Section 22.14).

Two examples of layouts are:
❑ an 'L' shaped area (see Photograph 22.1), leading pedestrians to the push—button box at a Toucan crossing; and
❑ a 'T' shaped area (see Photograph 22.2), which leads pedestrians to the centre of the crossing.

Crossing places at side—roads should ideally be located beyond the tangent point of the kerb radius. The raised kerb radius should be continued to give positive guidance to drivers turning at the junction and should enable pedestrians with visual

Photograph 22.2: An example of a Pelican crossing with a T—shaped tactile area and guard railing. Courtesy: David Nicholls.

impairment to locate the straight section of dropped kerb. If it is either impractical, due to footway width, or undesirable, since pedestrians might not be seen by drivers, to locate the crossing with dropped kerb in the side–road, the crossing point will have to be located on the pedestrians' line of travel. At all locations, the crossing point on one side of the road should be directly opposite that on the other side.

Consideration should be given to waiting restrictions where vehicles habitually park across dropped kerb crossing places. Where restrictions are not justified, an advisory white line carriageway marking, parallel with the dropped kerb, might be effective.

22.6 Pedestrian Refuges

Refuge islands are a relatively inexpensive method of improving crossing facilities for pedestrians. The width of the island is important. Whilst current standards allow an absolute minimum width of 1.2 m, this is inadequate for more than occasional individual pedestrians. The effective area,and hence the standing capacity, of the refuge should be related to its actual use at peak periods of pedestrian flows. Where people with pushchairs or in wheelchairs are likely to cross, the island should be at least 2.0m wide.

The residual carriageway width should be sufficient to allow vehicles to pass without tracking too close to pedestrians waiting on the refuge. Special consideration should be given to the needs of cyclists, with special provision made if necessary. Greater lane–widths should be allowed on bends and particular care should be taken where refuge islands are incorporated into traffic–calming measures, such as flat–topped road humps.

Refuge islands are usually formed by kerbs or prefabricated steel, 'D'–shaped in plan. The width of the crossing for pedestrians should be similar on both footways and on the refuge island. Dropped kerbs, to carriageway level at the island, should be provided (see Section 22.10 for further details) and tactile paving may also be appropriate (the DOT's Mobility Unit can advise) [Sc].

Refuges should be sited where a majority of pedestrians actually want to cross. If, for overriding safety reasons, this is not possible, then short lengths of pedestrian guard–railing should be installed to guide pedestrians to the provided crossing point. Refuges should not be sited where vehicle drivers' and pedestrians' views of each other are likely to be obstructed by parked vehicles. If there is no practical alternative site, the imposition of waiting restrictions

should be considered. Care should be taken when siting refuges near bus stops.

The probability of traffic queues that would extend across the refuge should be examined. Pedestrians should not be faced with having to squeeze between queueing vehicles in order to use a refuge. The dangers are especially marked for vulnerable people and particularly those with impaired mobility.

Refuge islands should be marked by internally illuminated bollards (HMG, 1994) [NIa]. On roads not subject to a 30 miles/h limit, or where drivers may have difficulty in judging the presence or size of a refuge, supplementary lighting can be added to the island with illuminated 'KEEP LEFT' signs fixed to the lighting column (IPLE, 1982a).

22.7 At–Grade Pedestrian Crossings

A full pedestrian crossing may be justified where pedestrians experience significant delay or danger in crossing a road. Pedestrian crossing provision at signal–controlled junctions is discussed in Section 40.8. The Department of Transport now recommends the use of an explicit procedure, based on a site assessment record and an assessment framework (DOT, 1995a) [Sb]. The purpose of the procedure is to ensure that all relevant information is collected and that the grounds for decisions, and their consequences, are made clear.

The boundaries of the site assessment should extend approximately 50m on either side of the site of the intended pedestrian crossing. However, the exact length depends on the existence of side–roads and major entrances to buildings across the footway.

Factors which should be recorded are:
- ❑ carriageway and footway types and widths;
- ❑ the nature and form of any existing pedestrian crossing;
- ❑ existing road lighting standards;
- ❑ minimum visibility–distances for pedestrians and drivers;
- ❑ waiting and loading restrictions;
- ❑ public transport stopping points;
- ❑ locations of nearby junctions;
- ❑ other major pedestrian crossings or school crossing patrols;
- ❑ skid resistance of the carriageway(s);
- ❑ surroundings affecting pedestrian movement;
- ❑ flow and composition of pedestrians;
- ❑ average time taken and difficulty experienced in crossing the road;

□ vehicular flow, composition and speed; and
□ records of recent crashes and casualties in the vicinity.

The difficulty of crossing can be determined by one of three methods. These are: acceptable gap analysis; data–logger method; or by the judgement of an experienced engineer (DOT, 1995a) [Sb]. A *précis* of the information, recorded in the site assessment, is then included in the assessment framework. The assessment framework considers all reasonable pedestrian crossing options against the more important factors. The options should include refuge islands, Zebra crossings or signal–controlled crossings. Each of these should be compared with the 'do–nothing' option. The most likely factors to influence the decision are:
□ pedestrians' current difficulty in crossing;
□ local accident trends;
□ vehicle–delays in the peak period;
□ vehicle–speeds;
□ local representations;
□ installation costs; and
□ the present value of operating costs.

The assessment framework should annotate clearly the effects of each option. Whilst the incidence of crashes and casualties is important, it is difficult to predict accurately the consequences of introducing a particular type of pedestrian crossing.

General Siting Requirements

Various requirements should be met in the siting of both uncontrolled and controlled pedestrian crossings (DOT, 1995b) [Sb]. They should be located well away from potential conflict points at uncontrolled road junctions. Drivers need adequate time to see and react to pedestrian crossings. Signalled–controlled crossings on a major road should be a minimum of 20m, and Zebras an absolute minimum of 5m, away

Photograph 22.3: A Pelican crossing with built–out kerb and guard–rail.

from the give–way line on a side–road. Those on side–roads should be sited well away from the give–way line.

Special care should be taken in siting pedestrian crossings near roundabouts. Zebras are preferred but, if signal–controlled crossings are necessary, then a staggered island arrangement should be used. The detailed choice of site should have regard to the pedestrian desire–lines and flows, vehicle–speeds, visibility, vehicle–flows, size of the roundabout and the width of the crossing.

Drivers' ability to see pedestrians waiting to cross is crucial. Visibility must not be obscured by street furniture or parked vehicles. For 85 percentile speeds of 30 and 40 miles/h, the desirable minimum visibility is 65m and 100m respectively. One method of improving visibility is to build out the kerb (Photograph 22.3). Any equipment such as guardrailing, signal posts or control cabinets, must be sited with care, so as to achieve maximum visibility and not to cause difficulties for people with impaired mobility (see Section 22.10). Whilst there should be no surface water lying at a crossing point, gully gratings and any statutory utility boxes should all be sited away from the crossing or vice–versa. Similarly, crossings should not be sited immediately adjacent to bus stops but, if this is unavoidable, bus stops should always be beyond (ie downstream of) the crossing.

High skid–resistance surfacing should be laid on both vehicle approaches to the crossing. The length of the carriageway surfacing will depend on the approach speeds and collision potential of the site. Similarly, the need for advance warning signs depends on the 85 percentile speeds: 50 miles/h, or greater, for a signal–controlled crossing and 30 miles/h for a Zebra crossing.

Zebra Crossings

Zebra crossings have the advantage of relatively low cost but must be installed only where they are the most appropriate type of crossing (HMG, 1990) [NIb]. To exercise priority over traffic, pedestrians have actually to be on the Zebra crossing markings. This form of crossing is not ideal, therefore, where traffic speeds or volumes are high. Zebra crossings may also be unsuitable where pedestrian flows are so high that pedestrians are likely to dominate the crossing and cause long delays to vehicular traffic.

Road markings and details of studs and materials are set out in the Traffic Signs Manual (DOT, 1991c). Zebra crossings are characterised by flashing amber globes, the construction and performance of which are given in BS.873. Zig–zag markings are laid on both the approaches and the exits to the crossing and on both sides of any central refuge island. Zig–zags ban waiting or parking, prohibit vehicles from overtaking each other and warn pedestrians of the increased risk of crossing in the zig–zag area.

Whilst it is inadvisable to install Zebra crossings in the vicinity of speed humps, they can be installed on flat–topped road humps (DOT, 1991b) [Sc]. This provides a pedestrian crossing of the carriageway at footway level but drivers must be given adequate warning of the arrangement. Where the width of the carriageway exceeds 11.0 m, a refuge island should be constructed and must include two flashing amber beacons.

After dark, the safe and satisfactory functioning of pedestrian crossings relies on the approaching drivers' ability to see pedestrians clearly. Roads with significant numbers of pedestrians should be lit in accordance with BS5489 (BSI, 1992). This should normally provide sufficient illumination for a pedestrian crossing. However, it is essential to inspect the site after dark and to assess the lighting, in the context of the street and the traffic speed, especially in shopping streets where there may be other lighting that affects visibility. If there is still difficulty in seeing pedestrians waiting at the kerb or on the crossing, supplementary lighting can be installed. A number of new lighting techniques have been assessed, as reported by the Institution of Lighting Engineers in 1997. If supplementary lighting is installed, it should be positioned with care so as not to cause glare to drivers or pedestrians. The use of vandal–resistant lanterns, mounted at least 3m above the ground but not obscuring the beacon globes, is recommended (DOT, 1995b) [Sb]. Experienced lighting engineers should be involved in the design of illumination at all pedestrian crossings.

Photograph 22.4: A Puffin crossing with an L–shaped tactile area, dropped kerbs and guardrailing.

Signal–Controlled Crossings

The incorporation of pedestrian facilities at signal–controlled traffic junctions is dealt with in Chapter 40.

There are currently three types of independent signal–controlled pedestrian crossings: Pelicans; Puffins; and Toucans. The operational cycles and timings for Pelicans, Puffins and Toucans are set by the Department of Transport (DOT, 1995b) [Sb]. Briefly, the differences between the three are that:

❏ 'Pelicans' (Photograph 22.3) use far–side pedestrian signal heads with a green–man aspect, are demanded by a pedestrian push–button and have a fixed duration of flashing amber to traffic, concurrent with flashing green to pedestrians (HMG 1987) [NIc];

❏ 'Puffins' (Photograph 22.4) use near–side pedestrian signal heads, with an extendable all–red crossing period which is demanded by both kerbside and on–crossing pedestrian detectors (to cancel demands which are no longer required) (DOT, 1993c); and

❏ 'Toucans' (Photograph 22.1) use far–side pedestrian and cycle signal heads and the same on–crossing detection as the Puffin and are used by both pedestrians and cyclists (DOT, 1993a). These are likely to be replaced, as the standard form of Toucan crossing, with a near–side mounted signal similar to Puffin pedestrian crossings.

The general siting requirements are the same for all three. The Puffin has been developed, using kerbside and on–crossing detectors, to provide an efficient crossing with advantages to both drivers and pedestrians. The Toucan provides a crossing for cyclists and pedestrians. Cyclists are not permitted to use Zebra, Pelican and Puffin crossings.

All signal–controlled crossings must use approved equipment and must comply with current regulations regarding position and mounting height of the signal–heads and road–markings (DOT, 1981a and 1981b). Drivers, either when approaching or waiting at the stop–line, must be able to see at least one signal–head clearly. As one signal may be masked by parked vehicles or other obstructions, it is normal to align at least two signals to be seen on each approach. At most pedestrian crossings, these objectives can normally be achieved with one primary and one secondary signal, the latter mounted at either the centre or the off–side of the road. The use of 'primary' visors (which are cut away) on the secondary signal heads is normally recommended, to improve the visibility of the signal from the stop–line. However, if the road is particularly wide or the approach alignment is poor, it may be necessary to install additional signals. In appropriate circumstances, tall posts or even overhead mounting can be used. On roads with an 85 percentile speed of more than 35 miles/h, additional primary signals should be provided. Whatever layout is chosen, access for maintenance should be a key consideration and trees and other vegetation should be trimmed to keep sight–lines clear. The pedestrian push–button boxes should be mounted so that the push–button is between 1.0m and 1.1m above the ground and on the right–hand side from a pedestrian's view (see Section 22.10 for details of the requirements for people with impaired mobility). For Toucan crossings, push–buttons are provided on both the left and right sides, whilst at any crossing that horse–riders may use, the mounting height will have to be chosen accordingly. For Puffin crossings, the box should be mounted to the right of the pedestrian and at the kerb–edge nearest to the approaching traffic. This may mean that additional boxes are needed for staggered crossings.

The choice of site and layout of any refuge island should allow sufficient space for the expected numbers of pedestrians waiting to cross the carriageway, whilst still allowing sufficient space for those passing by. Wherever a refuge is provided, a staggered crossing arrangement, with two independent crossings over each half of the carriageway, must be installed. A staggered refuge layout is optional where the carriageway is between 11m and 15m wide, but is essential for carriageways greater than 15m wide. The layout of the stagger should be such that pedestrians on the island are facing on–coming traffic. If this orientation is impossible, then a straight crossing is recommended. Staggered crossings are also not recommended for one–way streets. If necessary, additional advisory signs should be used to aid pedestrians but formal authorisation for their use is required. Staggered crossings should have an absolute minimum of three metres between crossing limits and, due to the need to install guardrails on both sides of the refuge island, the island should be a minimum three metres width overall. It is important to ensure that there is sufficient capacity on the island for all pedestrians likely to accumulate there. Dropped kerbs and tactile pavings should be provided, subject to current DOT advice.

At most crossings, approaching vehicles should be detected by either inductive loops or microwave sensors. However, vehicle actuation may not be necessary where the pedestrian crossing is linked to other adjacent traffic signals or is part of an urban traffic control scheme. In a 30 miles/h limit or where the 85 percentile speed is less than 35 miles/h, fixed–time operation is an option. In such circumstances, a pedestrian pushing the button will actuate the operational cycle, which will normally commence with a 20– to 30–second vehicle–precedence

stage. At a busy crossing, this period could be extended, so that vehicle delays are minimised. The 'steady green–man' time should take into account the overall kerb–to–kerb distance that pedestrians have to walk.

Before installation, it is essential to consult with the police and to give notice to the public. The Secretary of State must also be formally notified [NId]. The new crossing should be well publicised and the Road Safety Officer should offer instruction to any local schools or old persons' homes. Leaflets to aid people's understanding of Puffin and Toucan crossings are available (DOT, 1993a and b).

School Crossing Patrols

Children are particularly vulnerable when crossing roads. Indeed, there is some evidence to suggest that drivers are less likely to give way to children than to adults at pedestrian crossings [NIe]. The appropriate local authority can, with police agreement, operate a school crossing patrol. The Road Traffic Regulation Act 1984 (HMG, 1984) defines the nature and limitations of such patrols [NIf]. The decision to provide school patrols rests with the Local Authority and they should develop consistent assessment procedures for vetting requests for patrols. These should include such matters as:

❑ traffic volumes, composition and speed;
❑ main routes to and from school;
❑ the complexity of adjacent junctions;
❑ the volume and average age of child pedestrians;
❑ the availability of safe waiting places;
❑ sight lines and visibility distances;
❑ street lighting and signing; and
❑ traffic fumes.

The sites themselves should be safe to operate and should be justified by these criteria. Consultation with all interested parties is essential. Once a positive decision has been made, adequate publicity information should be given to potential users, who should also be advised that responsibility for the safety of their children remains with parents.

Careful selection and appropriate training in the operation of patrols is essential, particularly if there is a signal–controlled crossing at the site. Advanced warning signs should be erected using the standard red triangle 'Children' symbol, with a supplementary 'Patrol' plate. Flashing amber warning lights should be added when the 85 percentile speed of vehicles is over 35 miles/h or where the forward visibility of drivers to the patrol is less than 100 m.

22.8 Grade–Separation

In urban areas, pedestrians are particularly at risk

when crossing roads with heavy traffic flows. For this reason alone, pedestrians are safer when physically separated from traffic. However, pedestrians are often concerned about their own security and sometimes resent the inconvenience of longer routes or apparently unnecessary steps or slopes. Grade–separation should look natural, in terms of topography and the manner in which it fits into the grain of the surrounding built environment. In many town and city centres, separation can be achieved by removing traffic from certain streets (see Section 22.9). However, pedestrians inevitably have to cross major urban roads. In these situations, designers should investigate the feasibility of separating pedestrians from road traffic vertically. Grade–separation which is confined within the highway is often inconvenient for pedestrians. A safe segregated road crossing should not involve a much longer walk, exposed to the elements, and create any anxiety for personal security. Successful grade–separation, either by footbridges or subways, gives pedestrians the feeling of remaining on the level and on their natural desire–line, whilst vehicles undergo the changes in grade and level.

Footbridges

The design and layout of footbridges should accommodate the likely pedestrian flows and movements and should cater for the needs of people with disabilities and people with prams or pushchairs (DOT, 1987a). The widths and layout have to be varied if they are also to be used by cyclists or equestrians (DOT, 1986) [Sb]. It is important to ensure that vehicles cannot gain access to footbridges, without restricting access for people in wheelchairs or children in prams or pushchairs.

The width of a footbridge should not be less than 1.8m. A minimum of 2.0m is appropriate where cyclists and pedestrians are not separated. Parapets should be at least 1.15m high but increased to 1.4m if cyclists are expected to use the bridge. In locations exposed to wind and weather, it might be desirable to cover the footbridge and, in these circumstances, the minimum internal headroom should be 2.3m. Similarly, where objects might be thrown onto the carriageway below, or where the bridge is unusually high above the carriageway, high parapets of unbreakable transparent material might be appropriate.

The appearance of a bridge is important to both pedestrians and to vehicle occupants passing beneath. It can be fabricated from steel, reinforced or pre–stressed concrete, timber or aluminium alloy, although this last material is not recommended where vandalism or theft is prevalent. On aesthetic matters,

the advice of the Royal Fine Arts Commission should be sought. The appearance of a bridge, especially when crossing a dual–carriageway, is enhanced by having no central support. If one is necessary, then it will need to be protected from possible collision impacts of vehicles. The design should also incorporate drainage, have waterproofed and non–slip surfaces and be well–lit. The lighting should relate to that of the approaches and ground–level mounted columns, columns on the bridge itself or parapet fittings can be used.

The approach to the bridge should be by ramps of a gradient no greater than five percent but stairs should also be incorporated with horizontal landings at regular intervals. The accesses should be as short and direct as possible and should follow the main pedestrian desire–lines. All ramps and stairs should be provided with handrails on both sides and appropriate provision made for people with impaired mobility (see Section 22.10).

Subways

Whereas complete separation of pedestrians from vehicular traffic should eliminate the risk of pedestrian casualties, some people have an aversion to going 'underground'. Wherever grade– separation, by way of a subway, is considered, the layout and design should promote the illusion that the highway has been elevated to cross the natural pedestrian route. The desired effect can be achieved by wide approaches, good 'through' visibility and the maximum possible subway width. A feeling of personal security will be fostered if the subway is in constant view of other people and if there are no places where a felon might be concealed. Great care should be given to the detailed specification of wall and ceiling finishes. Materials should be used which are designed to reflect light, deaden sound, be vandal–resistant and yet be easy to clean and maintain. Vehicles should not be able to enter a subway other than for maintenance and servicing. The aim is to produce a welcoming and pleasant environment.

Factors to consider in justifying the construction of a subway include:
- ❏ the pedestrian and cycle flows likely to use the facility;
- ❏ the use by children, and other vulnerable people, who might experience difficulty or excessive risk in crossing the highway at grade;
- ❏ the type and width of the road to be crossed;
- ❏ vehicle speeds, flow and the proportion of heavy goods vehicles; and
- ❏ the capital cost and present value of future maintenance costs.

The predictions of potential use should have particular regard to:

❑ whether or not the subway would form an access route to a school, play area or other amenity;

❑ the location, convenience and safety of alternative routes which pedestrians might take; and

❑ any likely changes in land–use in the vicinity over the next 15 years.

The subway should be sited as close to the major desire–lines of potential users as practicable and should be as short as possible. Its construction should preferably involve elevating the carriageway, although existing underground services may be too costly to divert.

Whilst the objective of good design is to provide generous dimensions, excessive costs and other practical difficulties may limit the available options (DOT, 1993d). Nevertheless, at all changes in direction, there should be adequate sight distances. This can be achieved by corners of 4.6m minimum radius. However, if cycles share the subway with pedestrians, different criteria will apply (see Chapter 23).

Subways should be well lit, in recognition of users' concerns about underground passages and fear of crime. Lighting should be incorporated at the early stages of design (IPLE, 1982b) and should aim to achieve:

❑ a bright appearance of the interior of the subway, to encourage confident use and to discourage abuse;

❑ a high level of lighting on stairs, ramps and approaches, for pedestrians' safety;

❑ avoidance of deep shadows;

❑ luminaires resistant to attack by vandals;

❑ ease of maintenance with reliable, long lamp–life;

❑ good colour rendering of surfaces, to give a sense of spaciousness; and

❑ emergency lighting in the event of failure of the mains power supply.

The visual problem when approaching a subway is to see sufficiently far into it, whilst there is daylight outside, yet achieving the reverse effect after dark. One way to achieve this is to use higher levels of illumination for, say, the first 6 metres. After dark, this initial 'threshold' lighting level could be reduced to match that of the exterior lighting.

22.9 Pedestrian Priority Areas

The pedestrian environment in an existing street can be improved in many ways. Most techniques involve the restriction of traffic using the street but do not necessarily offer pedestrians legal priority over any residual vehicles (DOT, 1987b) [Sb]. Such improvements are most applicable to streets incorporating leisure activities, where the pattern of pedestrian movements tends to be random.

The value of pedestrianisation schemes in improving the attractiveness and commercial success of town centres has been demonstrated widely, especially for retail shopping streets specialising in comparison rather than convenience goods (see Chapter 12). The prime quantifiable benefits come from reductions in accidents and pollution. Studies show that most schemes improve retail turnover but sometimes not during an initial one– or two–year period. 'Fringe' shops and convenience stores can suffer a net loss of trade (IHT, 1989b and ICE, 1993).

The quality of the design of schemes can have a marked effect on their success. Imaginative designs that create a distinct sense of place and avoid uniformity can enhance the vitality of town and city centres. The opportunities afforded by area–wide refurbishment and redevelopment of centres are generally best seized by schemes with mixed land–uses that include the addition of more homes and community facilities. For real success, pedestrianised areas should not become deserted in the evenings. Where this may happen, consideration could be given to allowing vehicles to enter and park in such areas outside the normal working day.

There are two statutory means of improving the pedestrian environment in an existing street. The Road Traffic Regulation Act 1984 (HMG, 1984) permits the closure of roads to traffic but generally for no more than eight hours in any 24–hour period [NIg]. Subject to consultation with the public and police, the highway closure Order may be permanent. However, the facility for traffic to use the carriageway outside the times of closure must remain. Apart from variations to the surface treatment, only limited changes can be made. The Road Traffic Regulation Act 1994 also allows for closures to be made on an experimental basis (see Section 13.2) [NIh].

Frontagers to the street are entitled to compensation if they are adversely affected by the removal of vehicular rights of access [NIi]. Because the rights–of–way for vehicles are removed permanently, the width, surface and layout of the street can be changed. However, reasonable access to underground services should be maintained. The emergency services must be consulted on their needs for access to frontage premises and to the street itself.

The conception and design of any scheme to restrict or remove vehicles from existing streets requires care and sensitivity by the designer. Special consideration should be given to whether or not to exclude buses, so as to ensure continued ease of access to the street and to the buildings in it. Whilst proposals may generally be welcomed, especially in busy shopping areas, the changes may also have an adverse impact on certain businesses. Full and early consultation on proposals is essential. Among those who should be involved from the outset are the police, businesses based within the vehicle restriction area, transport firms who deliver there and groups representing local residents and people with impaired mobility. Adequate access has to be maintained for goods deliveries and especially for the collection of cash from banks, building societies and large shops. There may well need to be compromises between the conflicting wishes of different interest groups.

Special consideration must be given to public transport operators, especially if any bus routes have to be diverted. In any scheme, access by bus should remain at least as good as that from the nearest car park. Taxi operators may object to potential loss of trade and longer journey– lengths for their customers, so taxi stands should be located so as to minimise inconvenience. If cyclists are excluded from the street, then the impact of their displacement to alternative routes should be assessed. People with impaired mobility can be assisted by a 'Shopmobility' scheme, which involves the free loan of wheelchairs, sometimes with power assistance, from reserved parking areas close–by. Whereas all vehicles, except those relating to emergency services, statutory undertakers, street cleansing, funerals and security carriers, would normally be prohibited from using the street or area, it is possible to offer exceptions (see Section 13.2).

If vehicles are permitted to enter the street, the design of the scheme should be such as to maintain a clear distinction between footways and carriageways. Signing of the restrictions must be clear. A comprehensive review of pedestrian direction signs should be undertaken to ensure that people can continue to find the most convenient route to all major facilities. Similarly, local traffic direction signs may well need to be amended, so as to minimise drivers' confusion, particularly for those unfamiliar with the area.

Any scheme should be carefully monitored and necessary adjustments or improvements made. Assessments should be made of the effects on pedestrians, vehicle movements and parking. The

difficulties and dangers for pedestrians crossing roads to which traffic may have been diverted should be monitored and appropriate remedial measures implemented.

22.10 Facilities for People with Disabilities

A significant proportion of people who live in urban areas have some degree of impaired mobility. To help them to achieve a reasonable quality of life, their needs must be understood and accommodated. The advice and assistance of such people who actually live in the area should be sought before changes are made to the layout of pedestrian facilities, especially pedestrian crossings (TRL, 1991).

The use of dropped kerbs and tactile surfaces at pedestrian crossings has been described in Section 22.5. Where a cycletrack runs alongside a footway or a footpath, an appropriate method of delineation should be used (DOT, 1990) [Sa]. One option is to use a tactile surface to enable blind and partially sighted people to position themselves on the correct side of such a shared route. Advice on this can be obtained from the DOT's Mobility Unit (see Section 22.14).

Blind or partially–sighted people can usually follow kerb lines or the facades of buildings. However, they can experience particular problems in finding their way in pedestrianised streets or urban squares. Different surface textures can provide a valuable aid. Alternatively, 'directional guidance' paving with a series of flat–topped, round–ended ribs can be used to provide a guide. The paving is laid with the ribs indicating the direction of travel (TRL, 1992). Further advice on this can be obtained from the DOT's Mobility Unit (see Section 22.14). In conservation and other historic areas, it is often difficult to meet the needs of people with physical disabilities or with visual impairment and the advice of the Civic Trust and English Heritage should be sought.

Maintenance works in urban areas should be undertaken with particular regard to people with disabilities. Special care should be taken wherever maintenance works interfere with facilities provided for such people. The New Roads and Street Works Act (HMG, 1991) requires that facilities provided for people with disabilities, such as tactile surfaces, are reinstated in full by public utilities after street works have been undertaken [Nij].

Street furniture, including street–lighting columns and barriers, can prove a real hazard to people with disabilities (IHT, 1989a). Street furniture should be carefully positioned so as not to be on the natural routes taken by blind or partially–sighted people. The presence of necessary street furniture can be marked by a surround of granite setts or an obstruction–warning material, which has a soft or springy feel.

Changes in level should avoid the use of steps wherever possible. If steps are unavoidable, the top and bottom of flights of steps should have warning surfaces of 'corduroy paving' and step–cages should be painted white or another contrasting colour. The stairs should be divided by handrails into flights not wider than 1800mm. There should be a maximum of 16 risers in any flight, with landings at top and bottom. Generally, ramps are preferred. A maximum gradient of five per cent should be provided. Where this is impracticable then the gradient should be no greater than eight percent, with level rest–areas.

Signal–controlled crossings remain one of the biggest issues of concern, especially to elderly people with disabilities. The Traffic Signs Regulations and General Directions 1994 (HMG, 1994) [NIa] require that all tactile or audible signals provided to assist visually impaired people at pedestrian crossings must be of a type approved by the Secretary of State for Transport [NIk]. Two types of audible signal are available. The standard unit, located in the pedestrian push–button box, produces a series of bleeps when activated and can be used at crossings of single carriageways. It should not be used in the vicinity of another crossing or on a staggered crossing of a dual–carriageway, in case there is confusion about which crossing has been activated. At staggered crossings, the alternative audible signal should be used. Known as 'bleep–and–sweep', it emits a distinctive sound of four beeps followed by a rising tone. The volume can initially be adjusted manually but is then automatically adjusted to the ambient background noise level, so as to be heard by someone close to the loudspeaker but not by anyone waiting at the other crossing. The audible signal, in both cases, operates while the steady green–man pedestrian signal is lit (DOT, 1995b) [Sb].

The standard tactile signal consists of a small rotating cone protruding from underneath the push–button box. Although tactile signals are not generally so useful to people with visual impairment as audible signals, they are essential to pedestrians who also have impaired hearing and are also helpful where audible signals are not provided or are switched off at night. Great care should be exercised in siting audible or tactile signals. The equipment must not be capable of activation if the red lights to traffic fail. Push–button boxes must be easily accessible to pedestrians waiting to cross and there should be a clear path without obstructions, such as guard–rails.

If the sound of the audible signal is likely to cause a nuisance to nearby residents, then either the sound output can be reduced or a time–switch incorporated to cut out the sound altogether at night.

Puffin crossings can provide assurance to pedestrians with disabilities, and especially to the elderly, that they have time to cross. Not only do Puffins incorporate pedestrian detection at the kerb–side but also infra–red detectors to extend the red time to vehicles, if a pedestrian is still crossing the road when the flashing green ends.

Possibilities for helping visually–impaired people to distinguish the type of crossing include a design of push–button box of a controlled crossing which emits a continuous low–pitch ticking sound, and a raised letter Z attached to Zebra crossing poles.

22.11 Guard–Rails

The installation of pedestrian guardrails should be considered only where there are real risks of accidents should pedestrians walk onto the carriageway. Guardrails are intrusive and unsightly. Their purpose is to restrict people's freedom of movement. This will be resented unless their installation is self–evidently necessary. The use of guardrails should be avoided unless there is no practical alternative in terms of pedestrian safety.

Guardrails are a continuous safety fence placed on the footway with sufficient clearance (500 mm minimum) from the kerb–face (BSI, 1976). They therefore narrow the footway, which causes a reduction in footway capacity and this should be considered before installation. There should be no gaps, through which a small child could squeeze, at any breaks in the guardrailing, such as at trees, signs or similar obstructions. Guardrails should comply with BS.3049 and can be either painted or anodized. Several proprietary makes are available and careful selection of a type which is both easily erected and repaired is worthwhile. Decoration of guardrails can help to relieve their inherent monotony. The ability of drivers to see pedestrians crossing at the end of a length of guardrailing should be checked carefully. Special guardrailing that provides increased visibility is available.

Where guardrail is installed in streets with retail or commercial premises that do not have rear service facilities, real difficulties can be experienced in loading or unloading goods. It is possible to install gates in the guardrail but these should be located with extreme care and nominated frontagers should accept responsibility for their closure. A gate left open, after the servicing of premises has been completed, can cause a real hazard to pedestrians and can negate the benefits of the guardrailing.

22.12 Personal Security

Whilst the fear of personal crime is out of proportion to its reality, nevertheless the risks should be considered carefully in the layout and design of urban public areas. Statistics show that most street crimes, including physical assault and abuse, are the work of opportunists when circumstances appear to be in their favour. Indeed, surveys of such acts after dark indicate that young men are far more likely to be attacked than women. In a few areas, however, women are more vulnerable. Evidence of anxiety can be seen in the way many women avoid using subways during the hours of darkness. Experience in the UK and abroad relates such anxiety, and the opportunity for crime, to those locations which are not 'watched' naturally by the community. To combat the problem, the priority should be to create conditions where there is always a mix of different ages and groups of people and where the physical layout does not include places where threatening groups can gather. A good urban design produces natural surveillance of public places and streets from adjacent buildings.

Studies in America indicate that streets taken over by heavy flows of traffic tend to lose their community activity and hence the incidental surveillance of the footways. On the other hand, traffic–free areas without adequate levels of activity can also promote anxiety in pedestrians. Indeed, many private shopping precincts are locked to the public outside opening hours. The feeling of being watched can be enhanced in two ways. Good lighting is a significant deterrent to crime and enhances a feeling of personal security. Also, an increasing number of local authorities, working in close conjunction with the relevant police authorities, are installing closed circuit television (CCTV). From a central control room, operators are able to see, at any time of the day or night, whole street–scenes, car parks or other public areas. Modern high–definition cameras are able to focus–in on fine details, such as individual people's faces or vehicles' number plates. In the event of any unsocial or potentially criminal activity being recorded, the police are advised and the video–tape may be accepted, later, in evidence.

22.13 References

BSI (1976) BS 3049 'Pedestrian Guard Rails (Metal)', British Standards Institution.

BSI (1992) BS 5489 'Road Lighting', British Standards Institution.

Buchanan CD et al (1963) The Buchanan Report 'Traffic in Towns', Stationery Office.

Davis A (1992) 'Liveable Streets and Perceived Accident Risk : Quality of Life Issues for Residents and Vulnerable Road Users', Traffic Engineering + Control. 33(6).

DOT (1981a) TA13/81 (DMRB 8.1) 'Requirements for the installation of Traffic Signals and Associated Control Equipment', Stationery Office.

DOT (1981b) TA14/82 (DMRB 8.1) 'Procedures for the Installation of Traffic Signals and Associated Control Equipment', Stationery Office.

DOT (1986) LTN 2/86 'Shared Use by Cyclists and Pedestrians', DOT [Sb].

DOT (1987a) BD 29/87 (DMRB 2.2.8) 'Design Criteria for Footbridges', DOT.

DOT (1987b) LTN 1/87 'Getting the Right Balance – Guidance on Vehicle Restriction in Pedestrian Zones', DOT [Sb].

DOT (1990) TA 4/90 'Tactile Markings for Segregated Share Use by Cyclists and Pedestrians', DOT (subject to revision – contact DOT's Mobility Unit for advice) [Sc].

DOT (1991a) Circular DU 1/91 'The Use of Dropped Kerbs and Tactile Surfaces at Pedestrian Crossing Points', DOT (subject to revision– contact DOT's Mobility Unit for advice) [Sc].

DOT (1991b) TA Leaflet 3/91 'Speed–Control Humps', DOT [Sc].

DOT (1991c) 'Traffic Signs Manual', DOT.

DOT (1993a) TA 10/93 'Toucan' An Unsegregated Crossing for Pedestrians and Cyclists', DOT.

DOT (1993b)   Pamphlet 'How to Use a Puffin Crossing', DOT.

DOT (1993c)   'The Use of Puffin Pedestrian Crossings' DOT.

DOT (1993d)   TD 36/93 (DMRB 6.3.1) 'Subways for Pedestrians and Pedal Cyclists Layout and Dimensions', Stationery Office.

DOT (1995a)   LTN 1/95 'The Assessment of Pedestrian Crossings', DOT [Sb].

DOT (1995b)   LTN 2/95 'The Design of Pedestrian Crossings', DOT [Sb].

DOT (1997)   Mobility Unit '1996/97 Overview', DOT [Sa].

HMG (1984)   'Road Traffic Regulation Act 1984', Stationery Office.

HMG (1987)   'The 'Pelican' Crossings Regulations and General Directions 1987', Statutory Instrument 1987 No. 16, Stationery Office.

HMG (1990)   'The 'Zebra' Pedestrian Crossings Regulations' Statutory Instrument 1971 No. 1514 updated by Amendment Regulations 1990 Statutory Instrument No. 1828, Stationery Office.

HMG (1991)   'New Roads and Streetworks Act', Stationery Office.

HMG (1994)   'Traffic Signs Regulations and General Directions 1994', Stationery Office.

ICE (1993)   'Tomorrow's Towns: An Urban Environment Initiative', Thomas Telford.

IHT (1989a)   'Reducing Mobility Handicaps : Towards a Barrier Free Environment', The Institution of Highways & Transportation.

IHT (1989b)   'Guidelines on Pedestrianisation', The Institution of Highways & Transportation.

IPLE (1982a)   Technical Report No. 12 'Lighting of Pedestrian Crossings', Institute of Public Lighting Engineers.

IPLE (1982b)   Technical Report 13 'Lighting of Pedestrian Subways', Institute of Public Lighting Engineers.

May AD et al (1991)   'Pedestrians preference for footway maintenance and design', HT.

TRL (1991)   Contractor Report 184 'An Ergonomic Study of Pedestrian Areas for Disabled People', TRL.

TRL (1992)   Contractor Report 317 'Tactile Surfaces in the Pedestrian Environment –Experiments in Wolverhampton', TRL.

22.14 Further Information

ACC (1991)   'Towards a Sustainable Transport Policy', Association of County Councils.

DOE (1994)   Planning Research Programme – URBED – 'Vital and Viable Town Centres – Meeting the Challenge', Stationery Office.

DOE (1995)   'PPG.13: A Guide to Better Practice', Stationery Office. [Sd] [Wa].

Department of Transport, Mobility Unit, 1/11 Great Minster House, 76 Marsham Street, London SW1P 4DR. Tel: 0171 271 5252. Fax 0171 271 5253.

Jacobs J (1993)   'The Death and Life of Great American Cities', Penguin Books.

Rigby J (1991)   'Developing and Implementing York's Environmentally Sustainable Transport Strategy', RIBA Conference, Royal Institute of British Architects.

Chapter 23 Cycling

23.1 Introduction

Cycling is an increasingly important element in transportation strategies to achieve sustainable development (DOT, 1996a). It offers health, environmental, economic and other benefits. It is suitable for many local journeys and can be used in combination with public transport for longer trips. The total number of journeys by bicycle in the UK is equivalent to those made on British Railways and London Underground combined (Morgan, 1991). In some towns, such as Cambridge, York and Oxford, cycling accounts for around 20% of journeys to work. Most cycle journeys (51%) are for commuting, business or education, with leisure accounting for 31% (Figure 23.1). By comparison, 31% of journeys by all modes are for commuting, business or education with 32% for leisure (DOT, 1994a). The overall level of cycle-use in the UK is low, being only two percent of all journeys, compared with some other European countries. The UK Government recognises a potential to transfer some short journeys from car to bicycle and has set a national target of doubling the number of journeys by bicycle by 2002, compared to the 1996 base, with a further doubling by 2012. The National Cycling Strategy sets out the framework to achieve this target (DOT, 1996b).

During the 1980s, a number of experimental and innovative cycle schemes were undertaken in the UK (Harland et al, 1993) and abroad. The results indicate that integrated physical and policy measures, including engineering, education, encouragement and enforcement, are required in order to increase cycle-use and to improve safety for cyclists (EFTE, 1994). For cycling to replace trips by car, complementary demand management measures are also necessary. Investment in measures to assist cyclists is rising (Davies et al, 1995). It is important, therefore, that cycle–schemes meet their objectives and are cost–effective.

23.2 Objectives and Strategies

The principal objectives of providing for cyclists are:
- ❑ to maintain and, preferably, increase the level of cycle–use; and
- ❑ to reduce danger to cyclists, thereby reducing the accident rate per distance cycled and the perceived risk .

These objectives should be set in the context of wider sustainable transportation goals, including reducing the growth of car–use and promoting alternatives, such as those set out in PPG13 (DOE/DOT, 1994) (see also Chapter 6) [NIa] [Sa] [Wa].

To achieve these objectives, positive consideration of the needs of cyclists is required in all highway, traffic management, safety and maintenance programmes. Providing cycle facilities should not be an objective in itself. Cyclists require roads or cycle tracks that are safe, convenient and pleasant to use. Detailed guidance on making the existing road infrastructure more suitable for cyclists, and on developing new facilities, is provided in Cycle–friendly Infrastructure: Guidelines for Planning and Design, a collaborative project by the IHT, DOT and cycling organisations (IHT et al, 1996). In this, a hierarchical approach to improving conditions for cyclists is recommended.

Hierarchy of measures
Traffic reduction. Can traffic volumes be reduced sufficiently to achieve the desired improvements in attractiveness and safety for cyclists? Can heavy lorries be restricted or diverted?

Traffic calming. Can speeds be reduced and drivers' behaviour modified to achieve the desired improvements?

Junction treatment and traffic management. Can the problems that cyclists encounter, particularly accident–locations, be treated by specific junction treatment or other traffic management solutions, such as contra–flow cycle lanes?

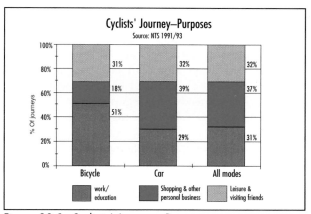

Figure 23.1: Cyclists' Journey—Purposes.

Photograph 23.1: College Green, Bristol — now closed to motor vehicles Courtesy: Mike Ginger.

Re–allocation of the carriageway–space. Can the carriageway space be re–allocated to give more space to cyclists, perhaps in conjunction with buses?

Cycle–lanes and cycle–tracks. Having considered and, where possible, implemented the above, what specific cycle lanes or tracks are now necessary?

The specific measures which are adopted depend on the overall transportation strategy for the area and the local conditions. It is necessary to consider the intended function of the roads in the network, their physical form and their actual use. The design solution may involve adjusting one or more of these factors. For example, the appropriate design solution for a road that is used as a short–cut by through traffic may be to make a short length of the street one–way with a contra–flow cycle–lane, thus modifying both form and use.

There is no single correct solution to providing suitable infrastructure for cycling: much will depend on the broader traffic, environmental and planning objectives and on the available funds. Measures are likely to be more easily funded and implemented if they benefit the wider community, not just cyclists. Strategies that emphasise traffic restraint, speed reduction and promotion of environmentally–friendly modes will tend to benefit cyclists.

Cycle audit procedures are recommended, to ensure that opportunities to benefit cyclists are properly considered in all highway and traffic scheme design. A cycle audit is not the same as a safety audit. Cycle audits seek opportunities to improve cycling conditions, whereas safety audits seek to avoid dangerous design for all users, including cyclists. Further guidance can be found in the IHT guidelines (IHT *et al*, 1996).

314

23.3 Cycle Networks

The purpose of providing a cycle network is to concentrate resources to enable cyclists, of a wide range of abilities and experience, to move more safely and conveniently between all points in a town and also to reach the surrounding countryside. The basis of the cycle network is the existing road network, augmented by special facilities where appropriate.

A good cycle network will have the following features:

❑ 'coherence' – the cycling infrastructure should form a coherent entity, linking major trip origins and destinations; routes should be continuous and consistent in quality;

❑ 'directness' – routes should be as direct as possible, based on desire–lines, because detours and delays will deter use;

❑ 'attractiveness' – routes should be attractive to cyclists on subjective as well as objective criteria – good lighting, personal safety, aesthetics and integration with the surrounding area are important;

❑ 'safety' – designs should minimise casualties and perceived danger for cyclists and other road–users; and

❑ 'comfort – cyclists need smooth, well–maintained surfaces, with regular sweeping and gentle gradients and routes must be convenient to use, avoiding complicated manoeuvres and interruptions.

Segregation of cyclists from motor vehicles is not essential as an objective. Broadly speaking, cyclists can mix safely with vehicular traffic of all kinds at speeds below 20 miles/h. They can also mix safely with vehicular traffic at speeds between 20 miles/h and 30 miles/h, unless volumes are high or there are significant numbers of Heavy Goods Vehicles (HGVs). Additional lane–width, or possibly segregation, is desirable where traffic flows are heavy. Where speeds are between 30 miles/h and 40 miles/h, some form of segregation or additional lane–width is preferable. Above 40 miles/h, segregation is necessary for the majority of cyclists. However, local circumstances, such as kerbside parking, lane–widths, side–road junctions, driveways and available space are crucial and consideration should first be given to reducing motor vehicle speeds. Segregation will rarely be appropriate on low–flow rural roads.

Cycle networks should be planned on the basis of cyclists' trip origins, destinations and desire–lines. Information on actual and suppressed demand, including leisure trips, should be collected. This can be obtained from the National Census journey–to–work data, classified traffic counts,

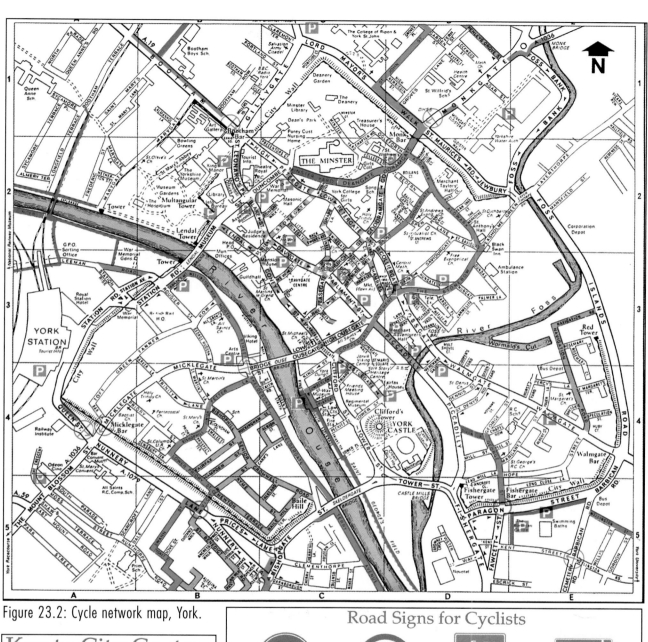

Figure 23.2: Cycle network map, York.

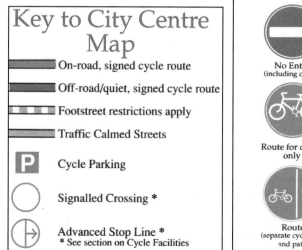

Key to City Centre Map

━━━ On-road, signed cycle route

━━━ Off-road/quiet, signed cycle route

┅┅┅ Footstreet restrictions apply

▨▨▨ Traffic Calmed Streets

P Cycle Parking

◯ Signalled Crossing *

⊕ Advanced Stop Line *
  * See section on Cycle Facilities

Road Signs for Cyclists

No Entry (including cycles)

No Cycling

Recommended Cycle Route

Lane Reserved for Cycles Only

Route for cycles only

No Motor Vehicles (cycles allowed)

P / **Cycle Parking**

Contra-flow Cycle Lane

Route shared with pedestrians (separate cycle track and path)

(No separation)

Acomb / **Directions on recommended route**

Bus and Cycle Lane

Figures 23.2: Key to city centre map and road signs for cyclists.

TRANSPORT IN THE URBAN ENVIRONMENT

315

specific surveys and consultation with local cyclists. Traffic models are available that can predict potential levels of cycle trips and assign these to the local road network, based on population and socio–economic data (Rickman, 1995). Accident–location plots can also be useful in identifying routes used by cyclists and sites requiring treatment.

Equipment exists for the collection of cycle flows on dedicated cycle–tracks. Automatic equipment for counting pedal cycles in mixed traffic streams has proved problematic but improvements are expected, due to new techniques and increased computing power. Manual classified counts should record pedal cycles but staff must be properly briefed or the data may be unreliable, particularly where cycles are a small proportion of the total traffic. Manual cycle–only counts are more reliable and can record additional factors, such as gender and age range. Inductive loops, used to detect bicycles at signals, can be linked to dataloggers to monitor the numbers of cyclists. Infra–red or microwave detectors can also be used. Other monitoring systems, using piezo–electric sensors, are also in use. The sensor is located in the surface of the cycle track, usually within a concrete pad or smooth tarmacadam surface. The detectors pass the information to data–recorders, which store it for future analysis.

Route–choice criteria must be taken into account. Cyclists will usually choose the quickest route for most journeys. They are reluctant to accept detours, unless there are significant compensating advantages. Cyclists will avoid routes that are hilly, perceived as dangerous or have bad riding surfaces (Hopkinson et al, 1989).

Once the pattern of demand has been established, opportunities for traffic management or construction measures should be assessed. In practice, this will be an iterative process. It is important that physical opportunities alone do not determine which measures come forward, in isolation from knowledge about cyclists' desire–lines.

A network proposal plan should be produced, that shows speed limits, traffic calmed routes/areas, traffic management and accident remedial schemes, cycle–lanes/tracks and cycle–parking locations. Maps and publicity are valuable in raising public and professional awareness of cycle–routes and cycling in general. An example is shown in Figure 23.2.

23.4 Construction and Maintenance

Relatively minor defects in road or cycle–track surfaces can be an accident hazard for cyclists, whereas for drivers they may be merely an inconvenience. A good quality riding surface is essential for comfort and safety and can also affect the speed at which cyclists can travel. Ideally, that part of the highway used by cyclists should be smooth and maintained to a higher standard than the remainder of the carriageway, otherwise cyclists may avoid it.

The surface of a cycleway should have an even profile. Defects in either longitudinal or transverse profile can cause loss of control. The surface should have a smooth macro–texture, to give a comfortable ride, and a harsh micro–texture, to ensure good skid–resistance for cornering and braking (IHT et al, 1996). Asphalt gives the best riding surface, provided it is properly laid on a good foundation. It is relatively cheap to lay and maintain, with laser–guided machines to achieve a sufficiently even surface. Hand–laid asphalt is usually uncomfortable at speeds above 10 miles/h. Dense Bitumen Macadam (DBM) is porous and may require an asphalt base–course to maintain the integrity of the construction. Materials with gravel content should be avoided, as these become polished and slippery in wet conditions. Concrete can provide an acceptable riding surface over short stretches and is almost maintenance free. Block paving can be acceptable for limited stretches, provided it is well laid and meets the criteria of evenness and texture. However, block paving tends to trap glass and other debris and can be uncomfortable. Paving slabs, or flags, tend to result in an uneven surface and often have poor skid–resistance.

Distinguishing different areas of carriageway by surface colour is useful. Red has been commonly used to indicate cycle–lanes or cycle–tracks, although green is now used more frequently. Coloured surfaces should be used consistently, so that road users know what to expect. It is recommended that coloured surfaces should be used for areas of the carriageway to discourage others vehicles from entering, such as at positioning lanes at junctions and bus/cycle lanes. Surface treatment is usually achieved by a pigmented slurry seal. It should be borne in mind that the application of slurry seal greatly reduces texture depth and can result in surfaces becoming slippery in the wet. A slightly more costly, but generally more satisfactory, alternative would be a 'Macamit'–type aggregate surface dressing, 12–15 mm thick. A more suitable material for use on carriageways would be a coloured surface dressing, using a small aggregate (3–6mm) and an epoxy resin binder.

The transition from cycle–track to carriageway is critical for cyclists' comfort and safety. The transition should be as smooth as possible with no upstand. This is best achieved by using full radius kerbs, minimum radius 2.0m, or dropped kerbs that are fully flush with the cycle track and the carriageway. An alternative, though less satisfactory arrangement, is to use channel squares or bull–nose kerbs laid flush. Dropped kerbs used upside down can give a reasonably smooth transition.

Where the road has a pronounced camber, reconstruction may be necessary to avoid an uneven transition from cycle–track to carriageway. As cyclists tend to use the edges of the carriageway, efficient drainage is very important; so, too, are well–constructed and maintained gullies and gratings. On minor roads, cyclists are normally able to cycle away from the road–edge. However, on busy roads, particularly those with sub–standard lane–widths, cyclists will be less able to avoid gullies. On such roads, it is recommended that the gully openings should be in the kerb–face, not in the carriageway surface. Improvements of this type should be phased–in with structural maintenance or other programmes.

Reinstatement of the paved surface, immediately after roadworks, should match existing levels. The New Roads and Streetworks Act (HMG, 1991) sets out standards for reinstatement by statutory undertakers [NIb]. Computerised maintenance management programmes can be used to set priorities. It is recommended that routes of importance to cyclists should be identified for priority maintenance. Provision should be made for cyclists at road–works, with appropriate signing and diversion routeing. Delays and detours for cyclists should be minimised. Routes used by cyclists require regular sweeping to remove glass, loose gravel, litter and other detritus. Cycle–tracks and cycle–lanes require more frequent sweeping than all–purpose roads, as they are not routinely 'swept' by motor vehicles. The costs and arrangements for adequate sweeping should be fully considered at the planning stage.

23.5 Signing and Road–Marking

Consistent and high quality signing will assist cyclists with route–finding and advertise the presence of cyclists to other road–users. Cycle destination signs should normally include two destinations, the next destination and the major destination. It is helpful if primary routes are also identified by a name or number. Signs should conform to the current regulations (HMG, 1994 and DOT, 1994b) [NIc]. It is essential that signs should be made secure, so that they are not easily removed or turned around. Poles

on cycle–tracks should be positioned at least 0.5m back from the edge, so as to maintain the effective path–width and they should not be located in the cycle–track. Signing should indicate route continuity and, where appropriate, cycle–route priority. 'Cyclists Dismount' signs should be used only where absolutely essential and the use of 'End' markings should be minimal. 'Give Way' markings may be more appropriate, supported by markings and upright signs. 'Cyclists Rejoin Carriageway', for which special authorisation is required, may be helpful to explain route continuity. The 'Except Cycles' plate should be added to 'No Through Road' signs where cyclists have a through route. Where a cycle–route passes through an unlit park or similar area, signs to indicate 'Alternative Route After Dark' may be indicated but need special authorisation.

Raised–rib road–markings consist of a continuous line marking, with ribs across the line at regular intervals. Concern that these would cause discomfort and possible accident problems for cyclists led to the development of an alternative design for use on all–purpose roads. They should not be used where cyclists are likely to cross the markings, including those locations where cyclists are likely to cross when riding parallel to the markings (DOT, 1995d) [Sb].

23.6 Safety for Cyclists

Cyclists present little danger to other road–users, other than pedestrians on shared facilities, but are particularly vulnerable to injury in collisions with motor vehicles. About eight percent of all reported road casualties are cyclists. In accidents involving an adult cyclist and a motor vehicle, only 17% were found to be the fault of the cyclist (Mills, 1989). Moreover, approximately three–quarters of cyclists' accidents are not reported to the police and therefore do not appear in the road traffic accident statistics, normally quoted. Accidents not involving a motor vehicle, and those occurring off the carriageway, such as on cycle–tracks, are also rarely reported.

The reported cyclist casualty–rate per distance travelled is twice that for a pedestrian and 16 times that for a car–occupant and the cyclist casualty–rate per journey stage is four times that for a car–occupant. However, comparisons between modes are not straightforward: one third of cyclist casualties are children under 16, yet the estimates of cycle traffic, on which the rates are calculated, exclude children's play in the street. The average cycle journey–length is also only one quarter of the length of the average car journey (O'Donoghue, 1993) (see also Chapter 16).

Casualty-rates per distance travelled for motor vehicle passengers have declined steadily, whereas the cyclist casualty-rate has remained static or increased. Fatality-rates, however, have fallen. More hazardous conditions have led to less cycling, which tends to increase the dangers to remaining cyclists as, with fewer of them, drivers are less alert to their presence. Cyclists' safety in the UK is about the same as in Germany but inferior to that in Denmark and the Netherlands.

Any significant increase in the amount of cycling may be expected to increase the number of cyclists injured, unless this can be offset by safety improvements. Indeed, if safety for cyclists is not improved it is unlikely that cycling will increase significantly. Evidence from Europe suggests that, where cycling increases, the accident-rate per distance cycled declines. In the UK, York City Council has encouraged cycling, maintained high levels of cycle-use and succeeded in meeting national casualty-reduction targets. If cycling were to increase, as a result of transfers from private cars, risks to other road-users should also diminish. The British Medical Association (BMA, 1992) points to the net gains in health and fitness that would arise through an increase in cycling. Evidence also suggests that, if the reductions in risk to other road-users are taken into account, more cycling need not lead to increased total casualties, so long as measures are implemented to provide for cyclists on the transport network.

Reducing the danger perceived by cyclists is also important if cycling is to increase. Perceived danger may be a good indicator of actual danger for cyclists (Sissons *et al*, 1993). Traditional procedures for selecting local safety schemes can overlook 'treatable' groups of cyclist accidents, because they are less numerous and appear more dispersed than motor vehicle accidents. Cyclist accidents tend to cluster along certain routes, as well as at junctions. It is necessary to consider each road-user group separately and to address their individual needs within an integrated framework (Hall *et al*, 1989). Appropriate methodologies are proposed by Hall (1993).

23.7 Training, Publicity and Promotion

Cycle Proficiency Training

Good cycle proficiency training for children is essential to enable them to cycle safely and independently. The most effective training schemes are those which involve stages, completed at different ages, with each stage conducted over several weeks (rather than intensively over a shorter period) and including on-road training (Savill *et al* 1996). Integrating cycle training with lessons on science, technology, environment and physical education will help to make it more relevant and memorable. Cycle training in schools is recommended for children aged nine to 11, although in Denmark and elsewhere it starts at a younger age. A cycle training code of practice was produced jointly in 1994 by the DOT, the Road Safety Officers Association, the Cyclists Touring Club (CTC) and the Royal Society for the Prevention of Accidents (RoSPA). Recognised training courses includes RoSPA's 'Rightrack'. Many local authority Road Safety Officers have developed or adapted their own cycle training schemes, often in conjunction with the police, local schools and cyclists' organisations.

There is some demand for adult cycle training, principally from those who have not ridden since childhood or who lack confidence to cycle in today's traffic. Successful schemes have been provided, involving riding skills, the Highway Code, route planning and cycle maintenance. Detailed advice for adult cyclists is provided by Franklin (1988).

Safety, Education and Publicity

Improving the status of cyclists is important to improving road safety and encouraging cycling. Safety education and publicity material should portray cyclists as legitimate and valued road-users, undertaking everyday journeys. Material aimed at drivers should emphasise the need for drivers to exercise appropriate care, particularly regarding speed, in the vicinity of cyclists. This approach is set out in the DOT's strategy for improving the safety and freedom of child pedestrians (DOT, 1996d) and is highly relevant to other vulnerable road-users, such as cyclists. For campaigns aimed at cyclists, material that has a high information content, such as cycle route maps and technical advice on equipment, is particularly valuable and likely to be of genuine interest. Danger should not be exaggerated (Davies *et al*, 1997). Factual information on cycle helmets – the types, their uses and limitations – can be helpful. If an accident occurs, wearing a helmet may prevent or reduce the extend of injury (Royles, 1994). However, the advantages of helmet-wearing should not be exaggerated. In Australia, where cycle helmet use became compulsory in 1991, the casualty reduction effect appears to have been slight. The main effect has been to reduce the amount of cycling (Robinson, 1996). The safety benefits of helmet-wearing may be outweighed by the loss of health benefits (Hillman, 1993).

Promoting Cycling

Organisational, financial and attitudinal factors can

havea major influence on people's willingness to cycle. In order to encourage cycling, policy and promotional initiatives are required, in addition to infrastructure and safety measures. Attitudes to cycling are influenced by peer presure and the culture of employers and society (Davies *et al*, 1997). The Department of Transport's Cycle Challenge fund has supported innovative schemes which address these issues, including cycle–friendly employer schemes, school cycling projects and cycle centres, which provide comprehensive back–up services. Cycling is also being encouraged as an integral part of 'green' commuter plans and travel awareness campaigns. In these contexts, cycling is seen as part of the solution to wider objectives, rather than as a single issue campaign.

23.8 Traffic Calming

Traffic calming can assist cyclists and other road–users, by reducing motor vehicle speeds and encouraging drivers to pay greater attention to vulnerable road–users. As well as reducing casualties, well–designed area–wide traffic calming can help to increase levels of cycle–use (Hass–Klau *et al*, 1990). Consideration of cyclists in traffic–calming programmes may help to determine priorities: for example, schemes that provide safe routes to schools may be given higher priority. Bus routes and routes followed by emergency vehicles should be taken into account, when developing area–wide traffic–calming schemes. The definition of a hierarchy of routes can assist this process. Some routes may be important through routes for cyclists and these should be designed accordingly (see also Chapters 13 and 20).

Most traffic–calming schemes originate as safety or environmental schemes, in which cycling is only one consideration. The specific needs of cyclists should always be considered from the outset, so that traffic calming can improve routes for cyclists and so reduce accidents and promote cycling. Badly–designed traffic–calming measures can increase dangers to cyclists and cause them to divert to other routes, which might also be unsatisfactory.

The IHT guidelines provide advice on good design for cyclists in traffic–calming schemes (IHT *et al*, 1996). Examples of provision for cyclists are illustrated in the County Surveyors' Society report (CSS, 1994).

Recommended general principles are:
- ❏ that traffic–calming schemes should be seen as an opportunity to encourage and facilitate cycling, as a means of transport, and the specific needs of riders should be considered from the outset. This

Photograph 23.2: Road closure with cycle–gap, Oxford.

Photograph 23.3: Rumble strip with cycle–gap, Kensington. Courtesy: TRL.

Photograph 23.4: Cycle by–pass at traffic throttle, Oxford.

may require special features, such as cycle–bypasses, ie short stretches of segregated route, allowing cyclists to bypass the traffic–calming features, or it may simply mean paying attention to the design and construction details of standard traffic–calming features;

❑ that designs in which the presence of cyclists becomes the principal speed–reducing feature should be avoided – for example, a road narrowing that leaves insufficient width for drivers to pass a cyclist. Even if all drivers behave considerately, some cyclists will feel intimidated in these situations, particularly by large vehicles;

❑ that features that endanger the stability of cyclists, such as rumble strips and upstands on turning manoeuvres, should not be used unless a satisfactory alternative is provided for cyclists;

❑ that designs should take account of likely obstructions, particularly illegally parked vehicles, and maintenance operations, which may limit the use of cycle by–passes;

❑ that surface materials should be skid–resistant, particularly in wet weather, and obstructions in the carriageway, including all ramp–faces, should be clearly visible after dark;

❑ that access restrictions imposed on motor traffic, such as banned turns, one–way streets and road closures, should provide an exemption for cyclists, unless there are overriding safety reasons which prevent this (see Photograph 23.2); and

❑ that local cyclists and cycling groups should be consulted at an early stage on the appropriateness and design of all proposed traffic–calming schemes.

Specific design recommendations can be made, in relation to particular features, as follows:

❑ 'road humps' – transitions and gradients should be gentle, with no upstands; 75mm high round–top humps, or road humps with cycle by–pass facilities, such as speed–cushions, are preferred in asphalt. Humps with a sinusoidal profile have been installed in Edinburgh and have been more widely used on the Continent (CROW, 1993);

❑ 'rumble strips'– a gap 0.75m–1.5m wide should be provided for cyclists and positioning should take account of cyclists' desire lines and any provision for parked cars (see Photograph 23.3);

❑ 'horizontal deflections' – at pinch points, traffic islands and chicanes, a cycle by–pass or a shared running lane 4.5m wide is recommended. The latter is likely to have little or no traffic–calming effect but may still provide a satisfactory pedestrian crossing refuge. If neither is possible, a shared running lane of three metres or less, in which overtaking is not possible, may be preferable, provided that vehicle speeds are low.

This may require a prior speed–reducing feature. Where cyclists' safety would be compromised by traffic islands or central refuges, alternative measures should be sought. Chicanes should be designed so that the paths of cyclists and motor vehicles do not conflict on exit; and

❑ 'cycle by–passes' – these should be 1.0m – 1.5m wide between faces. They should be straight through and as short as possible. Where they are more than a few metres in length, or are kinked, the full width will be more important. They should remain at carriageway, not footway, height. Where cycle flows are heavy, 1.8 m width, or an alternative design, may be necessary (see Photograph 23.4).

23.9 Cycle–Lanes

Cycle–lanes, bus/cycle–lanes and wide nearside lanes are all useful techniques for assisting cyclists in appropriate circumstances. Cycle–lanes help to alert drivers to the presence of cyclists and give cyclists greater confidence. They help cyclists to pass queueing traffic and lead cyclists to special facilities at junctions, such as advanced stop–lines. They are most useful where there are few side–roads and no parking or loading requirements. Cycle–lanes can be used to narrow the carriageway visually, particularly where the objective is to reduce the number of running lanes. However, they do not necessarily induce drivers to give cyclists greater clearance when overtaking. They are not kept clear of debris by the passage of motor vehicles, so additional sweeping is required. If used in unsuitable locations or have a substandard width, cycle–lanes can lead to increased accident–rates (Wegman *et al*, 1992). They are unnecessary on roads with low vehicular traffic flows

Photograph 23.5 Bus/cycle–lane, Shrewsbury. Courtesy: Peter Foster.

and on roads with 20 miles/h speed limits. Cycle–lanes should be a minimum of 1.5m wide and preferably two metres. If 1.5m cannot be accommodated, 1.2m may be acceptable for very short sections, such as at the approach to advanced stop–lines.

Mandatory cycle–lanes, marked by a solid white line and backed by a traffic regulation Order, prohibiting motor vehicles from using the lane, are better respected by drivers than advisory cycle–lanes, marked only by a broken white line. Coloured surfacing helps to keep motor vehicles out of cycle–lanes (see IHT *et al*, 1996 and DOT, 1989). Mandatory cycle–lanes must be discontinued at side–road junctions. Discontinuing advisory lanes, where they cross side–road junctions, encourages motor vehicles turning left into the minor road to do so at the junction and not encroach into the cycle–lane. However, it may also lead to abrupt turning movements by vehicles, rather than a gradual merge with bicycle traffic. Other options worth considering are: continuing an advisory lane through the junction (emphasised by a coloured surface); or allowing motor vehicles to merge with the cycle lane, in advance of the junction (Wilkinson *et al*, 1994).

Cyclists should be permitted to use bus–lanes unless there are overriding safety reasons to exclude them. On busy roads in urban areas, a bus–lane will give cyclists greater separation from general traffic than would a cycle–lane on its own. It may also be easier to justify a combined bus/cycle–lane than a lane exclusively for one mode (see Photograph 23.5). Introducing and enforcing parking restrictions in bus–lanes is likely to be more successful than in cycle–lanes. Where there is adequate carriageway width, bus–lanes should be 4.25m – 4.6m wide to

Photograph 23.7 Staggered stop–lines, Cambridge.

allow safe passing within the lane. However, narrower widths can work satisfactorily, depending on the flows of buses and cycles and traffic in the adjacent lane. Instead of peak–hour arrangements 12–hour or 24–hour bus–lanes are preferable. Permitting motor–cyclists to use bus–lanes is not recommended.

Wide nearside lanes allow large vehicles to pass cyclists in relative safety and comfort. They are useful on roads where there is occasional parking or loading or where there are significant numbers of heavy goods vehicles or buses. They are usually inexpensive to install, particularly if carried out when resurfacing or repainting carriageway markings. No traffic regulation Orders or additional signs are required. They should be 4.25m wide. If narrower, cyclists will have insufficient clearance from passing traffic and, if wider, traffic may form two lanes. Additional width can usually be taken from outer lanes. However, wide nearside lanes do not have the same attraction for

Photograph 23.6 Advanced stop–line for cyclists, Edinburgh. Courtesy Philip Noble and Edinburgh City Council

Photograph 23.8 Bypass to traffic signals and entry restriction.

'new' cyclists as cycle–lanes and they do not channel traffic.

23.10 Signal–Controlled Junctions

Signal–controlled junctions offer designers various possibilities for installing features to assist cyclists (see also Chapter 40).

'Advanced stop–lines' for cyclists enable them to position themselves, more safely, ahead of motor vehicles, thus reducing conflicts between left–turning vehicles and straight–ahead or right–turning cyclists (see Photograph 23.6). The position of the approach cycle–lane (nearside or central) needs to be carefully considered. There is no evidence that advanced stop–lines reduce the capacity of the junction. On roads with three or more lanes, a two–stage, 'jug handle' turn will assist less–confident cyclists to turn right (DOT, 1993a and 1996c) [Sb].

'Staggered stop–lines', where the cycle–lane is continued one or two metres ahead of the main stop–line but without a widened reservoir, can also be beneficial to cyclists. These help to place the cyclists in the driver's view. Staggered stop–lines may be appropriate where the right turn is not available or where, for some local reason, a standard advanced stop–line cannot be accommodated (see Photograph 23.7).

'Cycle by–passes' may also be incorporated into signal–controlled junctions to enable cyclists to bypass the signals, particularly for cyclists turning left or going straight ahead at T–junctions. Cyclists' speed and manoeuvres should be considered when determining signal–phasing, cycle times and linking of sets of signals. The length of the intergreen on staggered junctions is particularly important. Signal–controlled junctions are generally preferred to roundabouts by cyclists, for safety reasons and because their rights of way are better respected (see Photograph 23.8).

23.11 Roundabouts

Some types of roundabout can be particularly hazardous and intimidating for cyclists. Problems tend to arise due to uncertainty, as to whether or not drivers will give way to cyclists, and also because of the high speeds of motor vehicles through some types of roundabout. Over 50% of cyclist accidents at roundabouts are due to an entering vehicle striking a circulating cyclist. Turning right on large roundabouts is particularly difficult for cyclists. Roundabouts have a substantially worse accident–rate than other junction–types for cyclists – about four times the rate for traffic signals. However, the accident–rate for motor vehicles (other than motor cycles) at roundabouts is generally good. Large roundabouts cause the most concern, although small roundabouts with flared entries and inadequate deflection also have a poor record for cyclist accidents (Layfield et al, 1986). Segregated left–turn lanes are also unsafe for cyclists. Conventional and signalised roundabouts have a better cyclist accident record. In some circumstances, mini roundabouts can form useful features in traffic–calming schemes, although they are not always 'cycle–friendly' (see also Chapters 16 and 39).

Experience of unsegregated roundabouts in Continental Europe suggests that roundabouts can be a reasonably safe junction–type for cyclists, if they have:
- ❑ a 24m – 32m external diameter;
- ❑ a circulatory carriageway that prevents overtaking, ie less than 8m wide; and
- ❑ radial entry arms, slightly curved towards the centre island (Balsiger, 1995).

These designs can accommodate flows of 2000 motor vehicles per hour and substantial numbers of cyclists. This suggests that it is the design of roundabouts, rather than roundabouts per se, that determines the risk to cyclists. Other Continental roundabout layouts give circulating cyclists priority over entering and exiting motor vehicles. These work best where there is a high flow of cyclists, to achieve drivers' observance of the cyclists' priority. Their use in the UK would need to be accompanied by intensive and effective driver–information.

Alternative techniques to improve the safety of existing roundabouts for cyclists have been well established in the UK (Allott & Lomax, 1993) and include:
- ❑ signalising the roundabout (Lines, 1995) and providing advanced stop–lines;
- ❑ altering the geometry, particularly reducing entry–widths, increasing deflection and narrowing the circulatory carriageway;
- ❑ reducing vehicle–speeds on entry;
- ❑ altering sight–lines and conspicuity; and
- ❑ attending to road markings and signs, particularly circulatory lane–markings.

Peripheral cycle tracks, segregated from the carriageway, can be useful at some large roundabouts. However, they may introduce delays at crossing points with each approach arm such that cyclists do not use them. Signal–controlled 'Toucan' crossings of the arms may be necessary but should be located carefully to prevent blocking of the junction exit (see Section 22.7). Cycle–lanes on the outer edge of the carriageway encourage cyclists to ride in the area that entering drivers find most difficult to see. Therefore, continuing cycle–lanes onto roundabouts is not recommended.

23.12 Grade–Separated Junctions

Crossing the mouth of an entry, or exit, slip–road at a grade–separated junction is particularly hazardous, as motor vehicles will often cross a cyclist's path at high speeds. Junction designs that reduce entry speeds and increase deflection on entry will help to reduce risks to cyclists. Advisory dog–leg cycle–crossings can be provided at existing junctions (DOT, 1988) [Sb], although this is not entirely satisfactory. Alternative routes that avoid such junctions should be made available wherever possible (see also Chapter 43).

23.13 Priority Junctions

Cycle–lanes on the minor road should normally be discontinued well before a junction with a major road, with the distance depending on the turning movements at the junction. Traffic islands and bollards can protect right–turning cyclists. Cyclists should normally be exempted from prohibited turns (see Photograph 23.9). Converting three–arm priority junctions to mini–roundabouts can lead to an increased accident–risk to cyclists. It seems that junctions with clear priorities are preferable for cyclists (Summersgill,1989) (see also Chapter 38).

Photograph 23.9 Exemption from right–turn restriction, Birmingham.

23.14 Cycle–Tracks

Cycle–tracks are most useful where the volume and/or speed of motor traffic is high and where there are few side–roads or other interruptions. Route continuity and safe and convenient crossings of side–roads are crucial. Cycle–tracks tend to reduce accidents on links but increase accidents at junctions, particularly in urban areas and where intervisibility is poor (Wegman *et al*, 1992). Good quality riding–surfaces and frequent sweeping are essential.

A cycle–track alongside the carriageway should be 2m – 3m wide. Where provided, they should be installed on both sides of the road for safety and convenience. One–way cycle–tracks are safer than two–way, as accidents tend to occur between traffic turning into side–roads and cyclists travelling contra–flow on the cycle–track. Where two–way use is likely, the design should accommodate this (DOT 1986b and DOT 1989) [Sc].

Cycle–tracks should be given priority at junctions with minor side–roads, wherever this can be achieved safely. Where space permits, the cycle–track should be bent gradually away from the main carriageway by 4m – 8m. Priority can be emphasised by the use of a raised crossing. Where there is sufficient space to accord priority, by 'bending out' the cycle–track, it may be preferable to maintain priority by merging cyclists back onto the carriageway into an advisory cycle–lane prior to the side road. If cyclists are not accorded priority, 'Give Way' signs should be used. 'Cyclists Dismount' signs should not be used (see Section 23.5). Dropped kerbs should be flush, with no upstand. Guidance on side–road crossings can be found in the IHT guidelines (IHT *et al*, 1996).

Footways converted to shared use in urban areas rarely provide a good quality cycle–facility and may inconvenience pedestrians. Space should first be sought within the carriageway. If footways are converted they should have light pedestrian flows, few driveways or minor road crossings and good intervisibility.

Cycle–tracks away from the carriageway, such as those created on disused railway lines, will have different characteristics but should still conform to high standards of safety and design, particularly regarding sight–lines, personal security and maintenance (Sustrans, 1994). Where they are intended as commuter routes, lighting is desirable. This should normally be provided to footway lighting standards and BS 5489.

23.15 Road Crossings

The Department of Transport provides advice on the range of facilities that can be provided to enable cyclists and pedestrians to cross roads (DOT 1986a, 1995a and 1995b) [Sc]. These include unsignalled and signal–controlled crossings, such as the Toucan, which is an unsegregated crossing for cyclists and pedestrians. Useful layout details for a number of crossing arrangements are provided in the National Cycle Network guidelines (Arup and Sustrans, 1997). To meet the needs of cyclists, dropped kerbs should be fully flush, with no upstand (DOT, 1991) [Sd].

23.16 Cycling and Pedestrians

Cyclists and pedestrians have many common characteristics and interests. Both groups, in their contribution to transport, are environment–friendly, cause few accidents to others but are vulnerable to injury from motor vehicles. Cycling and walking are highly efficient modes for local trips and can form an important component of longer journeys. However, despite common interests, they cannot be regarded as a single mode (CTC *et al*, 1995).

Provision for cyclists should normally be made within the carriageway. If no satisfactory on–carriageway solution can be found, it may be appropriate to consider converting the footway to shared use or to seek an off–road alternative. Consultation with local organisations is important, particularly with those representing people with mobility difficulties, such as blind and partially–sighted people. Guidance is provided by the DOT (1986b and 1990) [Sc] [Sb] . Segregation requirements tend to be dependent on the types of user and the flow levels. Where new facilities are created, such as cycle–paths away from roads, unsegregated use is more likely to be acceptable but personal security issues should be considered carefully.

Where streets are pedestrianised, cyclists should not necessarily be excluded unless there are good reasons and suitable alternative routes (DOT 1993b) [Sb]. When pedestrian densities are high, cyclists tend to modify their behaviour (Trevelyan *et al*, 1993). Segregating cyclists and pedestrians in pedestrianised areas will not always be necessary or desirable. Where it is desirable, cycle–movements can be combined with selected motor vehicles, such as buses and service vehicles, permitted at particular times of day or channelled by defined paths. Admitting cyclists to pedestrian zones can help to maintain their vitality outside shopping hours.

23.17 Grade–Separated Crossings

Where high quality grade–separation can be achieved, this will be superior to crossing at grade. Detours, gradients, accident–risks and personal security all need to be considered.

For new subways, the DOT standard for trunk road construction (DOT, 1993c) indicates suitable dimensions and layout. Straight approaches, straight–through visibility, flared entries and good lighting will improve their safety and acceptability. Converting existing pedestrian subways to shared use is recommended only where a useful and high quality outcome can be achieved. Guidelines on dimensions are recommended by the DOT (DOT, 1986a) [Sc].

Bridges can be useful to cyclists, depending on the on–ramp gradients and detours. Parapet heights should be 1.4m–1.5m and adequate forward visibility on entry and exit is also important (see also DOT, 1995c).

23.18 Cycle–Parking

Good quality cycle–parking can encourage cycle–use, particularly at workplaces, at railway stations and in town centres (see Photograph 23.10). Theft is a major problem; some 200,000 pedal cycles are reported stolen each year and the total number stolen each year is estimated at 600,000 (DOT, 1996e) and usually less than 10% are recovered. This is a burden for the police and a deterrent to cycling.

Cycle–parking standards should be included in development control guidance, issued by local planning authorities. Commuted payments may be

Photograph 23.10: Covered cycle–parking, Aston University in Birmingham.

MINIMUM NUMBER OF PARKED CYCLES THAT SHOULD BE ACCOMMODATED

SAMPLE CITY (1991 population)	Bristol (370,300)	Cambridge (101,000)	Oxford (109,000)	York* (100,600)
% journeys to work by cycle (1991 Census)	4%	28%	18%	19%
TYPE OF LAND USE per				
Shops, Services (A1/A2) 100 m²; (staff) 100 m²	2 2	4	4	1 2
Restaurants, Cafes, Public Houses (A3) (bar area) 50 m²; (dining area) 50 m²; staff	4	30 5	15 1 per 3	1 1 per 4
Business offices (B1) 100 m²; 1,000 m²; (staff) 1,000 m²	1 5	3	4	20
Industry (B2) 200 m²; 1,000 m²; 5,000 m²	4 4 12	5 25 125	5 25 125	4 20 100
Warehouses (B8) 200 m²; 1,000 m²; 5,000 m²	4 4 6	5 25 125	5 25 125	2 10 50
Hotels, Guesthouses (C1) 20 beds; 100 beds; staff	2 10	1 per 2	1 per 2	1 per 4
Hospitals, Nursing Homes (C2) 100 beds; staff	10 (in above)	1 per 2	33 1 per 2	5 1 per 4
Clinics, Health Centres (D1) treatment room; staff	2 (in above)	2	2 2	
Secondary Schools (D1) 500 students; staff	100	300	300	166 1 per 6
Colleges, Universities (D1) 500 students; staff	100	500	500	166 1 per 6
Halls of Residence (C2) student			1	
Other dwellings, Flats (C3) unit; bedroom	1	1	1	1
Libraries, Museums (D1) 200 m²; staff	2	25	6 1 per 2	1 per 20 seats 1 per 4
Theatres, Cinemas (D2) 100 seats; staff	2		25	5 1 per 4
Sports, Leisure centres (D2) staff	1 per 10 players	1 per 15 seats	use dependent 1 per 2	use dependent 1 per 4

*figures slightly lower in outer areas

NOTES:
1. Where figures were given for square feet, these have been translated to the nearest value for square metres.
2. Not all figures submitted by the local authorities have been included here; some have provided more details.
3. Specific land-use circumstances may alter some of the figures given.
4. Staff numbers given are for non-residential staff.

Fig 23.3: Examples of Local Authority cycle–parking standards for new developments. Source: Cyclists' Touring Club.

appropriate where on–site provision cannot be made in full (see also Chapter 27). 'Sheffield' parking stands are generally adequate for short–term parking; lockers and supervised cycle–parks provide better security and weather protection for medium to long–term parking. Guidance on development control standards and technical details is published by the Cyclists' Touring Club (CTC, 1993) and the London Cycling Campaign (LCC, 1995).

23.19 Bike–and–Ride

The combination of cycling and public transport, particularly trains, can be very effective and can increase, substantially, the catchment area of railway stations. 'Bicycle stations', where cycles can be parked, hired, repaired and bought are increasingly common in continental Europe. At smaller stations, cycle lockers are often provided for secure commuter parking. Considerable scope exists for improving access and provision for cyclists at railway stations in the UK.

23.20 Legislation

Legislative references regarding cyclists, and provision for cyclists, in England and Wales are provided in Cycle–friendly Infrastructure (IHT *et al*, 1996). Guidelines for Scotland are contained in the Scottish Office Cycling Advice Note 1/90 Making Way for Cyclists (SODD, 1990).

23.21 References

Allott & Lomax (1993)	'Cyclists and Roundabouts – Update Report', Cyclists' Touring Club, Godalming.
Arup & Sustrans (1997)	'The National Cycle Network – Guidelines and practical details', Sustrans, Bristol.
Balsiger O (1995)	'Cycling at roundabouts: safety aspects', Conference Papers, Velo–City Conference '95, Basel.
BMA (1992)	British Medical Association 'Cycling: towards health and safety', Oxford University Press.
CROW (1993)	Centre for Research and Contract Standardisation in Civil Engineering 'Sign up for the bike – Design manual for a cycle–friendly infrastructure', CROW The Netherlands.
CSS (1994)	'Traffic Calming in Practice', County Surveyors' Society/Landor Publishing London.
CTC (1993)	Cyclists' Touring Club 'Cycle Parking', CTC Godalming.
CTC and PA (1995)	Cyclists' Touring Club and Pedestrians Association 'Joint Statement on providing for walking and cycling as transport and travel', CTC Godalming.
Davies DG and Young, HL (1995)	'Investing in the Cycling Revolution: A review of Transport Policies and Programmes with regard to cycling', Cyclists' Public Affairs Group, Godalming.
Davies D Halliday M, Mayes M and Pocock R (1997)	'Attitudes to Cycling: a a qualitative study and conceptual framework', TRL Report 266, Crowthorne.
DOE/ DOT (1994)	Planning Policy Guidance note PPG13 'Transport' Stationery Office [Sa] [Wa].
DOT (1986a)	Local Transport Note 1/86 'Cyclists at road crossings and junctions', Stationery Office [Sc].
DOT (1986b)	Local Transport Note 2/86 'Shared use by cyclists and pedestrians', Stationery Office [Sc].
DOT (1988)	Traffic Advisory Leaflet 1/88 'Provision for Cyclists at grade separated junctions', DOT [Sb].
DOT (1989)	Local Transport Note 1/89 'Making way for cyclists', Stationery Office [Sc].
DOT (1990)	Traffic Advisory Leaflet 4/90 'Tactile Markings for Segregated Shared Use by Cyclists and Pedestrians', DOT [Sb].
DOT (1991)	Disability Unit Circular 1/91 'The Use of dropped kerbs and tactile surfaces at pedestrian

crossing points', DOT (subject to revision – contact DOT's Mobility Unit for advice) [Sd].

DOT (1993a)   Traffic Advisory Leaflet 8/93 'Advanced stop lines for cyclists', DOT [Sb].

DOT (1993b)   Traffic Advisory Leaflet 9/93 'Cycling in pedestrian areas', DOT [Sb].

DOT (1993c)   TD 36/93 (DMRB 6.3.1) 'Subways for Pedestrians and Pedal Cyclists: Layout and Dimensions', Stationery Office.

DOT (1994a)   'National Travel Survey 1991/93', Stationery Office.

DOT (1994b)   'Traffic Signs, Signals and Road Markings Bibliography', DOT.

DOT (1995a)   Local Transport Note 1/95 'The assessment of pedestrian crossings', Stationery Office [Sc].

DOT (1995b)   Local Transport Note 2/95 'The design of pedestrian crossings', Stationery Office [Sc].

DOT (1995c)   TA 67/95 (DMRB 5.2.4) 'Providing for Cyclists', Stationery Office.

DOT (1995d)   Traffic Advisory Leaflet 2/95 'Raised Rib Markings', DOT [Sb].

DOT (1996a)   'Transport – The Way Forward'. The Government's response to the Transport Debate CM3234, Stationery Office.

DOT (1996b)   'The National Cycling Strategy', DOT.

DOT (1996c)   Traffic Advisory Leaflet 5/96: 'Further Development of Advanced Stop Lines', DOT [Sb].

DOT (1996d)   'Child pedestrian safety in the United Kingdom', DOT.

DOT (1996e)   Transport Statistics Report 'Cycling in Great Britain', Stationery Office.

EFTE (1994)   European Federation for Transport and the Environment, 'Greening Urban Transport – Pedestrian and cycling policy' T&E 94/6 and Annex 94/6A, Brussels.

Franklin J (1988)   'Cyclecraft –skilled cycling techniques for adults', Unwin, London.

Hall R, Harrison J, McDonald M and Harland D (1989)   'Accident analysis methodologies and remedial measures with particular regard to cyclists', Contractors Report 164, TRL.

Hall R (1993)   Confeence papers 'Velo–City Conference', McClintock, H, (Ed), Nottinghamshire County Council.

Harland G, and Gercans R, (1993)   'Cycle Routes' Project Report 42, TRL.

Hass–Klau C, and Crampton G, (1990)   'Cycle safety: a comparison between British and (West) German cities', PTRC.

Hillman M (1993)   'Cycle helmets – the case for and against', Policy Studies Institute, London.

HMG (1991)   'New Roads and Streetworks Act 1991', Stationery Office.

HMG (1994)   SI No 1519 1994 'The Traffic Signs Regulations and General Directions' Stationery Office.

HMG (1996)   'Transport Statistics Report – Cycling in Great Britain', Stationery Office.

Hopkinson PG, Tight M and Carsten O (1989)   'Review of literature on pedestrian and cyclist route choice', ITS Leeds University.

IHT, DOT, CTC, and BA (1996)   'Cycle–Friendly Infrastructure – Guidelines for Planning and Design', IHT.

Layfield R and Maycock G (1986)   'Pedal Cyclists at Roundabouts' Traffic Engineering + Control, 28(6).

LCC (1995)   'Cycle Parking Equipment and

Installation Standard', London Cycling Campaign, Cyclists Touring Club.

Lines CJ (1995) 'Cycle accidents at signalised roundabouts', Traffic Engineering + Control, 36(5).

Mills P (1989) 'Pedal cyclist accidents – a hospital based study', Research Report 220, TRL.

Morgan JM (1991) 'Cycling in Safety?' Proceedings of Safety '91 Conference, TRL.

O'Donoghue J (1993) 'How much cycling is there? How safe is it?' Papers to, Velo–City Conference. McClintock, H, (Ed), Nottinghamshire County Council.

Rickman M (1995) 'Effective cycle route planning: the application of modelling techniques to the UK', Papers to, Velo–City Conference. IG Velo, Basel.

Robinson D (1996) 'Head injuries and bicycle laws', Acident Analysis and Prevention 28(4), pp 463–475.

Royles M (1994) 'International Literature and Review of Cycle Helmets', TRL Report PR 76.

Savill T, Bryan–Brown K, and Harland G (1996) 'The effectiveness of child cycle training schemes', TRL Report 214.

Sissons Joshi M, Senior V and Smith, GP (1993) 'A survey of risk perception of cyclists and other road users' Papers to, Velo–City, Conference McClintock H, (Ed), Nottinghamshire County Council.

SODD (1990) Cycling Advice Note 1/90 'Making Way for Cyclists' Scottish Office Development Department.

Summersgill I (1989) 'Accidents at mini roundabouts', Proceedings of PTRC Summer Annual Meeting, London.

Sustrans (1994) 'Making Ways for the Bicycle: a guide to traffic–free path construction', Bicycle Association, Coventry.

Trevelyan P and Morgan, J (1993) 'Cycling in Pedestrian Areas', Report 15, TRL.

Wegman F and Dijkstra, A (1992) 'Safety effects of bicycle facilities: the Dutch experience', in 'Still more bikes behind the dykes', pp 93–102, CROW, The Netherlands.

Wilkinson W, Clarke A, Epperson B and Knoblauch, R (1994) 'Selecting roadway design treatments to accommodate bicycles', Federal Highway Administration, Report No. FHWA–RD–92–073, Washington DC

23.22 Further Information

DOT (1995) Traffic Advisory Leaflet 9/95, 'Cycling Bibliography', DOT.

Mathew D (1995) 'More Bikes – Policy into best practice', Cyclists' Touring Club, Godalming.

McClintock H (1996) 'Bicycle Planning: A Comprehensive Bibliography', (4th Ed), Vol 1: 'United Kingdom'; Vol 2: 'International'. Department of Urban Planning, University of Nottingham.

Royal Borough of Kingston upon Thames, London (1996) 'London Cycle Network – Design Guide', Borough of Kingston upon Thames.

SODD (1996) 'Cycling into the Future – The Scottish Office Policy on Cycling' The Scottish Office Development Department.

The proceedings of the international Velo–City cycle–planning conferences, normally held every two years, are a valuable source of information and best practice, for example, Velo–City '93 in Nottingham and Velo–City '95 in Basel.

Chapter 24 Measures to Assist Public Transport

24.1 Introduction

High quality, road–based public transport services are vital to achieve maximum effectiveness from the road network and to offer an acceptable alternative to non-essential use of private cars. Policies and measures should assist effective operation and enable a high quality service to be provided but, since 1985, the major responsibility for providing and operating bus services in the UK rests with the private commercial sector [Wa].

Bus systems

Most urban public transport is provided by buses. Buses can transport large numbers of people while occupying relatively little road-space, thus offering a highly cost-effective use of resources. Buses also, crucially, provide mobility to those who do not have the use of a car. Specially equipped public transport vehicles can also provide accessible transport for people whose mobility is impaired.

Buses can be flexible in operation and can respond rapidly to changing patterns and levels of demand but are adversely affected by urban traffic congestion. If the inherent advantages of bus systems are to be realised, buses must have a good on–street operating environment. There is often a strong case for providing buses with priority over, or complete segregation from, other road vehicles, to protect bus services from the effects of traffic congestion and to improve route–frequencies, speeds and reliability. Speed and reliability of a bus service is also affected by ticketing arrangements and bus design. Vehicle and staff availability, bus route–planning and good on–the–road management of the service further influence reliability.

Guided Buses, Trams and LRT Systems

Conventional buses can be protected from the effects of traffic congestion by their segregation on sections of carriageway or track but it is possible also for buses to be automatically 'guided' on the track. Guided buses retain some, if not all, of the flexibility advantages of normal bus systems, whilst promoting a technologically–advanced image. They provide a 'bridge' between ordinary buses and tracked forms of public transport.

Trams or light rapid transit (LRT) (see Chapter 34) can transport large numbers of people, although the need for a fixed track means that network–wide passenger accessibility, without interchange, will be at a lower level than for a bus system. Trams/LRT are perceived by the public generally as modern, high quality and environmentally–acceptable modes of transport. Essentially, trams/LRT face the same on–street operating problems as buses but, unlike buses, tram/LRT systems introduced into British cities have been provided with a high degree of segregation and priority over other traffic. This has been achieved either through construction of purpose–built track and bridges, as in Sheffield, or through the use of redundant railway lines, as in Manchester, in combination with preferential traffic management and control [Wb].

24.2 Legislative Framework and Responsibilities

The Transport Act 1985 (HMG, 1985) [Sa] established a competitive market for the provision of bus services outside London, with road service licensing and a deregulated system based on bus service registration [NIa]. Within Greater London, the majority of bus services are privately supplied under contract with London Transport Buses, which plans and regulates the network of services.

While the intention of the 1985 Act was to promote competition, the Act included, *inter alia*, provisions to ensure safe and acceptable standards of operation (so called, 'quality licensing'). The Act enables local authorities to apply to the Traffic Commissioner for Traffic Regulation Conditions, to prevent danger to road–users and/or to reduce severe traffic congestion, by limiting the number of buses using particular roads. Highway authorities still retain wide powers under the Road Traffic Regulation Act 1984 (HMG, 1984) to control the use of individual roads and routeing of all classes of vehicles [NIb]. However, if they consider that undue constraints are imposed on their operations, bus operators can, under certain circumstances, appeal.

In planning and designing measures to assist buses, the Traffic Regulation Orders (TROs) and signs should recognise that, from 31 December 1996, buses are defined as:

❑ motor vehicles constructed or adapted to carry

more than eight passengers; and
❑ local buses not so constructed or adapted (HMG, 1994) [NIc].

A 'bus' is a public service vehicle used for the provision of a local service as defined in the Transport Act 1985 (HMG, 1985) [Sa] and not being an excursion or tour bus [NId].

Traffic signs with the word 'bus', or the bus symbol, thus apply to this very broad definition. Permitted variants allow for the word 'local' to be added to appropriate signs, which then refer to the more specific definition of 'local bus'.

Outside Greater London, responsibility for the bus system and the infrastructure on which it operates rests with:
❑ 'bus operators', who operate commercial services and/or provide non–commercial services under contract to local authorities;
❑ 'Traffic Commissioners', who license operators, register services and enforce standards;
❑ 'Passenger Transport Executives (PTEs)', where they exist;
❑ 'The Vehicle Inspectorate Executive Agency', which is responsible for the annual testing and spot testing of buses [NIe];
❑ 'the Department of Transport and Government Offices', who provide grants to assist funding of bus priority and other measures [NIf];
❑ 'the police', who enforce most of the associated traffic regulation measures; and
❑ 'local authorities', whose responsibilities are:
  ❑ to provide, maintain and manage the road and traffic systems on which buses operate;
  ❑ to promote on–street parking regulations, which heavily influence many bus operations;
  ❑ to apply to the Traffic Commissioner for traffic regulatory conditions on bus services, to ensure safety and/or to reduce severe bus–based congestion; and
  ❑ to contract 'socially–required' bus services not provided commercially;

Although bus services in London are operated under contract by private companies, London Transport Buses administers a system of service licensing. Local Borough councils in London also have extended powers to manage on–street parking and loading.

Two legislative regimes exist in relation to services using guided buses. These are based either on a TRO or a Transport and Works Order (TWO), as follows:
❑ a scheme may be introduced under a Traffic Regulation Order where:

❑ the guided busway is built on land forming part of an existing public highway;
❑ the buses are not electrically–powered from an external source; and
❑ the operation can be controlled adequately by normal traffic signs and signals; or
❑ a Transport and Works Order (England/Wales) or private legislation (Scotland) is needed, where a guided busway involves [NIg]:
  ❑ equipment which restricts public rights of passage;
  ❑ bus operation with electrical power derived from an external source;
  ❑ operation outside the existing highway limits; and
  ❑ planning permission.

The legislative background to trams/LRT systems is similar to that for Guided Buses (see Chapter 34).

24.3 Government Policies and Guidance

The Government supports improvements to all forms of urban transport through Transport Policies and Programmes (TPPs), which are submitted annually by local highway authorities (see Chapter 4) [NIh]. TPP guidelines recognise the scope for encouraging a shift in travel–demand between modes, such as from cars to public transport, and may provide resources for a wide range of urban public transport related facilities. Bus–priority measures, for example, are eligible for Department of Transport (DOT)/Government Office funding, as part of bids for Supplementary Credit Approval (SCA) [NIe].

Planning Policy Guidance Note PPG13 (DOE/DOT, 1994) [Sb] emphasises the importance of bus– priority schemes and is supported by PPG6 (DOE, 1996) [Sc], which stresses the importance of a high level of bus service in relation to new developments [NIi] (see also Part IV).

24.4 General Approach to Bus–Priority

Bus–priority measures have several aims, as follows:
❑ to reduce delays to buses arising from traffic congestion and thus save bus operating costs, passengers' travel–time costs and bus–fleet requirements;
❑ to improve the reliability of bus services, so as to make bus travel more attractive;
❑ to increase mobility for those members of the community who do not own or have use of a car;

❑ to increase accessibility to major traffic generators, like shopping centres and inter–modal transport interchanges; and

❑ to make a contribution to traffic restraint and the management of congestion, by the provision of efficient and high quality alternative services.

Bus–priority measures vary in scale, from simple traffic management measures, such as exemptions for buses from a manoeuvre prohibited to other traffic, through to area– or route–based schemes, where buses are provided with priority over complete routes, using a comprehensive package of traffic management and control measures [Wc].

Planning and Design of Bus–Priority Measures

The basic approach to bus–priority scheme planning, design and evaluation is described in the DOT guide LTN 1/91 [Sd] Keeping Buses Moving (DOT, 1991). The design process involves a standard approach, consisting of a feasibility study (including 'before' surveys), consultation, detailed design, implementation and 'after' surveys/monitoring. A typical study project brief is described in the London Bus–Priority Network Design Brief (LTB, 1994).

As part of the feasibility study, bus–priority measures should be subject to operational and economic evaluations. The operational evaluation should determine that the proposed scheme can function safely and effectively and will include consideration of layout, junction capacity, bus stop design and the loading/unloading needs of frontage premises. The economic evaluation should determine the benefits to bus operators and passengers, any disbenefits which may arise to other road–users and the capital costs of the scheme. The evaluation should also take account of wider issues, such as increased bus regularity (and thus reduced passenger waiting time), environmental impacts and policy considerations, such as when the transportation strategy is to encourage transfer of passengers from car to bus (see also Chapter 9).

Road–based public transport is supplied by a variety of vehicle–types but bus–priority measures can be used by all vehicles defined as a 'bus' or by 'local buses' only. Other classes of vehicle permitted to use bus–priority measures can be specified in the relevant Traffic Regulation Order. Bus–priority schemes should concentrate on assisting buses but it may be appropriate to permit some other categories of vehicle to use the priority measures, provided that:

❑ road safety is not jeopardised;

❑ effective and efficient operation of the bus–priority measure is not compromised;

❑ the legal definitions of the vehicle classes are clear; and

❑ the other vehicles are sufficiently distinctive for unequivocal enforcement.

Typically, the other vehicle categories which may be permitted to use bus–priority measures are:

❑ emergency vehicles (police, fire and ambulance);

❑ pedal cycles; and

❑ taxis.

Vehicle categories not generally permitted to use bus priority measures include:

❑ private cars;

❑ motor cycles ;

❑ goods vehicles; and

❑ high occupancy cars (see Section 24.13).

However, there are examples where HGVs and motor cycles are also permitted to use bus–priority measures (eg in the access–control scheme for Newcastle's central area).

24.5 Comprehensive Route–Length Bus–Priorities

Bus–priority measures, combining physical traffic management measures, such as bus–lanes and bus–advance areas, and traffic control systems, such as active bus–priority at signals, are most successful when implemented along bus–route corridors and linked to other improvements, such as passenger information at bus stops, improved waiting facilities, more frequent services, a review of waiting and loading requirements, bus stop clearways and cages, easily accessible buses and park–and–ride facilities. In combination, the measures not only improve bus operations but also the image and public perception of the service, in a way that could encourage higher patronage and hence a transfer from other modes.

Bus–priority measures, particularly when linked along a route, may form an important part of an overall strategy for dealing with urban congestion. Examples include projects along the Wilmslow Road in Manchester, in South and West London and Uxbridge Road in London, as well as long established schemes in Oxford. Linked bus–priority measures, ie a comprehensive approach to bus–priority along a route, have been shown to be highly cost–effective. Packages of measures have been shown to result in First Year Rates of Return (FYRR) in excess of 100%.

London's Priority (Red) Routes

In London, the Road Traffic Act 1991 (HMG, 1991) designated a network of roads as Priority 'Red'

Routes (see Section 13.14). These routes are subject to special parking controls, which are applied on an end–to–end to basis, traffic management and bus–priority measures with clear objectives:

❏ to improve the movement of all classes of traffic on the Priority Red Route Network, so that people and goods can reach their destinations in London more easily, reliably and safely;

❏ to provide special help for the efficient movement of buses;

❏ to reduce the impacts of congestion;

❏ to improve the local environment;

❏ to provide better conditions for pedestrians and cyclists; and

❏ to discourage car–commuting into central London and traffic from crossing the central area.

Park–and–Ride Schemes

City centres provide the focus for a wide variety of trip purposes from surrounding areas and many trips will commence by car because of the widespread distribution of origins. Park–and–ride can be an effective policy to assist in reducing central area traffic congestion, by intercepting these car trips and encouraging people to complete their journey by public transport. Out–of–town park–and–ride schemes using bus services to the central areas are operated in many cities including Oxford, York, Chester, Norwich, Exeter, Shrewsbury and Bristol (EHTF, 1993) and are planned in many more. Similarly, park–and–ride may be operated with tram, LRT and local train systems (Noble *et al*, 1993). For example, the Metro system on Tyneside has four major park–and–ride interchanges [Wd].

Oxford provides an example of a successful park–and–ride system, introduced as part of an integrated city centre traffic policy. Typically, in 1992, over 3,600 cars (about 4,500 people) entered four parking sites daily with 4,500 cars (about 8,500 people) on Saturdays. However, traffic flows into the city centre remained broadly constant, as trips which transferred to buses tend to be replaced by other car trips. While some of the bus services required revenue support, other services were commercially registered and claimed to be profitable.

Key criteria for a park–and–ride scheme are that the site should be:

❏ close to an interchange with a major highway to provide easy and safe access for car–users;

❏ near the edge of a built–up area and beyond the usual limits of congestion;

❏ capable of offering a direct bus, tram, LRT or train service to the city centre, with priority or segregation where necessary;

❏ capable of accommodating about 500 parking spaces, the minimum needed to support a financially viable dedicated bus service;

❏ accessible by regular bus, tram, LRT, or train services, if special park–and–ride services are not operated all day;

❏ of compact layout, to limit the walking distance (especially when it is raining) from parked cars to the public transport stop;

❏ furnished with relevant, up–to–date information, attractively displayed;

❏ equipped with good lighting and good surfacing; and

❏ designed to provide a high degree of personal and vehicle security.

Enforcement

Bus–priority measures and parking regulations are liable to violation by other drivers and require rigorous enforcement. New methods, such as the use of camera technology, to improve enforcement of bus–priority and compliance with traffic regulations can reduce the need for intense police effort (TDL, 1995).

Roadworks

Bus services tend to suffer disproportionately from roadworks, with services often disrupted along route–lengths where other vehicles find alternative routes. Highway engineers and police can assist in minimising problems for bus passengers, by ensuring that bus services are given special consideration and by adopting temporary bus–priority measures, wherever feasible (LTB, 1996).

24.6 With–flow Bus–Lanes

A with–flow bus lane (see Photograph 24.1) is an area of carriageway reserved for the use of buses and, occasionally, other permitted vehicles for all or part of

Photograph 24.1: With–flow bus–lane in London.

TRANSPORT IN THE URBAN ENVIRONMENT

the day, in which the buses operate in the same direction as the general traffic flow.

With–flow bus lanes enable buses to bypass traffic queues, usually on the approaches to signal–controlled junctions or roundabouts. With–flow bus lanes:

❑ are usually located at the kerbside, in order to serve bus stops, but some off–side bus lanes exist, for example, to assist right–turning buses;

❑ give buses priority at the locations and times most needed;

❑ minimise disruption to normal traffic patterns;

❑ need only be part–time, thus allowing reasonable access to frontage properties; and

❑ are relatively inexpensive to implement, with the capital cost often repaid by benefits in less than one year.

Times of Operation

Bus–lane operating periods should be determined, primarily, by the times and duration of traffic congestion. Thus, bus–lanes may operate during peak periods, am or pm or both, or all day or weekday or all week but times and days of operation in any one urban area should be standardised to avoid confusion to road–users. Bus–lanes which operate all day, say 07:00 hours to 19:00 hours or 24 hours, are more readily understood by other road–users and are consistent with a general policy of encouraging public transport. However, all day lanes materially affect frontage access for loading and off–loading and, where frontage loading requirements are intense, peak–period only bus–lanes may be unavoidable.

Where all day bus–lanes exist, the loading issue may be resolved by:

❑ servicing frontage premises from nearby side–streets; or

❑ loading 'out of (07:00–19:00) hours'; or

❑ direct frontage service–access, notwithstanding the all day bus–lane.

Permitted Use by Other Traffic

Pedal cyclists are usually permitted to use with–flow bus–lanes for safety reasons, since otherwise they would be required to ride in the main traffic stream outside the bus–lane.

Taxis are sometimes permitted to use with–flow bus–lanes, on the grounds that they perform a public transport service, provided that:

❑ taxi volumes and set–down/pick–up behaviour does not interfere with bus operations (off–line taxi–stop bays may be possible); and

❑ taxi–use does not encourage infringement of bus–lane regulations by other vehicles, ie taxis

should be easily identifiable vehicles, such as London 'black' cabs, or should carry a prominent taxi sign.

Motorcycles are not normally permitted to use bus–lanes, as they travel at the same speed as general traffic and should not be encouraged to weave or overtake on the inside of a queue, by incursion into a bus–lane. However, there are examples where motorcycles are permitted to use bus–lanes.

Layout

The location and design of the start and finish of with–flow bus–lanes are crucial. Lanes must start upstream of the end of the predicted traffic queue and the bus/other traffic diverge at that point should be carefully designed to ensure a safe distance for non–priority vehicles to merge. With–flow bus–lanes should normally be at least 3.0m wide, but, where there are significant numbers of cyclists, a width of 4.25m – 4.6m is preferable. Above 4.25m, a designated cycle–lane (1.0m) may also be provided alongside the kerb by carriageway marking.

Most with–flow bus lanes are terminated, ie 'set–back', before the traffic signal stop–line of the junction they approach. The set–back ensures that the full width of the stop–line is available to all vehicular traffic during the green signal period and thus the capacity of the junction is maintained and left turns made possible. The length of the set–back should be such that buses entering from the bus–lane can clear the traffic signal stop–line on the first available green phase. As a general guide, the set–back length, in metres, should normally be twice the green time, in seconds. A shorter set–back can be used, if the junction approached is not the constraint on the capacity of the route or if the bus–lane continues downstream of the junction. In these cases, a short set–back will allow 'left turns and buses only'.

A with–flow bus–lane may be extended right up to the signal stop–line under four conditions:

❑ if a reduction in the traffic capacity of the junction is acceptable, as part of an overall traffic restraint strategy for the area;

❑ if the junction is not the critical constraint on the capacity of the route;

❑ if safe provision can be made for left–turning traffic; and

❑ if right–turning traffic can be accommodated in such a way that it does not restrict flow in the other non–priority lane(s).

Signing and Road–Marking

Signs and road markings must convey sufficient information to drivers to enable them to obey the

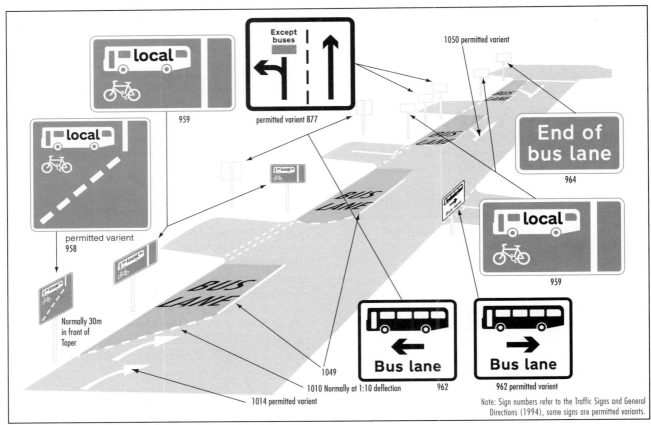

Figure 24.1: Schematic layout of a with–flow bus lane in a one–way street.

regulations applying to the scheme. Signs should be in accordance with current regulations (HMG, 1994) [NIj] and a typical signing layout is shown in Figure 24.1.

Bus and other traffic lanes are separated by carriageway markings comprising a solid white line 250mm–300mm wide. The application of coloured road surfacing also assists with compliance. Bus–lane throughput is a function of bus flow, the number of

Photograph 24.3: A contra–flow bus–lane in London.

bus stops and passenger demands at stops but bus flows and passenger demands do not usually impose capacity constraints on the design of bus–lanes. Research has shown that a single–lane bus–lane, with 'normal' passenger demand at stops, can cater for about 120 buses/hour, without special measures (NATO, 1976). Above this level, special measures, such as provision for overtaking at stops through the use of bus stop bays or variation in the bus–lane width, are likely to be necessary (Photograph 24.2).

Photograph 24.2: Extra bus–lane width for overtaking.

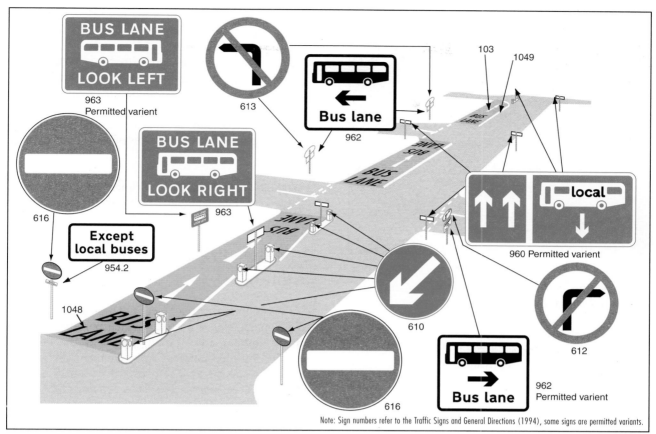

Figure 24.2: Schematic layout of a contra—flow bus lane.

24.7 Contra—flow Bus—Lanes

A contra—flow bus—lane (see Photograph 24.3) is a traffic lane reserved for the use of buses travelling in the direction opposed to the general traffic flow. Contra—flow lanes usually operate on a 24—hour basis.

Contra—flow bus—lanes are usually introduced in area—wide one—way traffic systems, where the effect is to create a two—way road with 'buses only' allowed, in one direction, and all types of vehicle including buses, in the other. By allowing buses to travel against the direction of traffic flow, contra—flow lanes enable

buses to avoid unnecessary diversions and thus save both journey distance and time and often improve access for buses to passengers' destinations. The main characteristics of contra—flow bus lanes are:
- ❑ that buses follow the same route on outward and return journeys in one—way systems, resulting in convenience and benefits to passengers;
- ❑ that savings are achieved in bus—kms and bus—hours;
- ❑ that reductions occur in bus—passengers' walk—times to main destinations; and
- ❑ that, if well—signed, they are easily understood and respected by other drivers.

Permitted Use by Other Traffic
Pedal cyclists may be permitted to use contra—flow bus—lanes, where a minimum lane—width of 4.25m can be provided. However, cyclists can experience difficulties at entry/exit points and at side—roads, where traffic crosses the lane.

Layout
Contra—flow bus lanes should not normally be less than 3.0m wide. Pedestrians' safety may be an issue and the design of pedestrian crossing facilities and pedestrian protection, such as short lengths of guardrailing to channel pedestrians to suitable crossing facilities, should receive special attention.

Signing and Road—Marking
Signing and road—markings should be in accordance with the current regulations (HMG 1994) [NIj]. Physical separation, either a continuous island or a series of long islands, is normally used. While ensuring that other vehicles do not enter the lane, this may introduce potential difficulties, such as:
- ❑ reducing the perception by pedestrians of 'two—way operation';
- ❑ causing tracking damage if the lane is narrow;
- ❑ creating difficulties for buses having to take avoiding action in emergency or breakdown; and

Photograph 24.4 A bus–advance lane.

❑ creating difficulties for loading/unloading to frontage premises.

Contra–flow lanes may be delineated, instead, by a solid white line, 250mm–300mm wide, supplemented by traffic islands and/or double white lines with hatching between them. Coloured surfacing reinforces the special nature of the lane.

At the entry, a 'No Entry Except Local Buses' sign should be used wherever possible (HMG, 1994) [NIj]. However, if cyclists are permitted to use a contra–flow bus–lane, then current regulations do not permit the signing of an exemption for cyclists to 'No Entry' signs. Thus, either a separate 'cycle–gate' must be provided or all motor vehicles could be prohibited with an exemption for buses and cycles, although the general level of compliance by other vehicles may be reduced. Figure 24.2 shows typical entry–signing.

Loading and Unloading
If traffic flow in the opposing direction is heavy, and if loading is allowed in the lane, it may be difficult and unsafe for buses to overtake stationary vehicles.

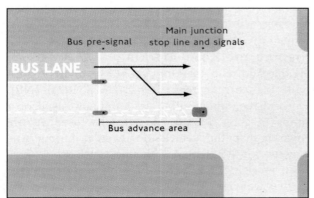

Figure 24.3: Layout of advance–area for buses, with pre–signals.

In these circumstances, possible remedies include:
❑ a double width contra–flow bus–lane;
❑ servicing from nearby side–streets;
❑ delivery vehicles permitted to park, whilst loading/unloading, along the off–side kerb of the bus–lane, where it is physically segregated or partly–segregated; and
❑ the provision of loading–bays, within the curtileges of buildings fronting the contra–flow lane.

Particular care is needed at pick–up points on a contra–flow bus–lane and special traffic islands may be required.

24.8 Bus–Priority Using Traffic Signals

A 'Bus–Advance Area' (see Photograph 24.4) is a traffic management measure which permits buses to advance into an area of road, clear of traffic, before a signal–controlled junction. Pre–signals, in advance of the junction, always control traffic entry to the advanced area, with a bus–lane provided up to the pre–signals (see Figure 24.3). The objective of the pre–signals and advance area is to re–order vehicles, so that buses may be given priority to reach the junction first. The maximum traffic throughput of the junction is unchanged.

'Traffic metering', also termed 'queue relocation' or 'gating', involves a bus–lane running right up to the upstream stop–line of a congested junction or section of road and alters, by adjusted signal–timings, the volume of traffic which can enter the congested section. The objective of traffic metering is to control the flow of traffic at the upstream junction by reducing capacity at the metered junction, so that it, rather than the downstream junction, becomes the critical junction in the network. The bus–lane enables buses to by–pass the relocated traffic queue.

Combined bus–advance areas and traffic metering can be installed, using pre–signals to manage and relocate queues to areas where bus–lanes will still allow buses to be protected from congestion.

Design of Advance–Areas for Buses
Conventional with–flow bus–lanes generally terminate with a 'set–back' from the signal stop–line and are designed to ensure that buses clear the stop–line on the first green, where the buses are mixed with the platoon of general traffic within the set–back area. Greater priority may be given to buses by the use of pre–signals, to create a bus–advance area between the pre–signal and the junction.

Pre–signals control traffic upstream of the junction. A bus–lane taken up to the location of the pre–signal, enables buses to overtake the traffic queue. When the pre–signal is red for other traffic, buses may proceed to the main junction signal, taking their preferred lane in the advance area. They may either proceed into the advance area with no control or be subject to a 'give–way' where the road narrows or be under signal–control throughout. Pedestrian crossings can be incorporated with a separate bus–lane stop–line, although this is not generally favoured. Specific site circumstances will determine the best form of layout. Thus, the pre–signals do not always control bus movements, whereas they always control non–bus traffic movements.

Shortly before the junction signal turns green, general traffic is released from the pre–signal and enters the advance area (if there is space) to make full use of the green–time at the junction. Detailed monitoring in London and elsewhere indicates that the benefits of schemes of this type are high and scheme costs can often be recovered in less than one year (Astrop et al, 1994). The concept can be extended to provide a segregated lane for buses right up to the signal stop–line of the junction, if space permits or can be created. Buses can then be given a separate signal stage or an early start from the main signals. Generally, overall junction capacity will be reduced. The bus–lane must also be long enough to enable buses to enter the lane freely and so overtake the whole traffic queue. If their entry to the bus–lane is blocked, total throughput at the junction may be impaired.

Traffic Metering

Traffic metering on a main route requires the linking of two or more sets of traffic signals and a system for measuring congestion in the critical section of road between those signals. Traffic signal–timings are adjusted at the upstream signals, to meter traffic flow to the level which can be accommodated by the downstream road section. A bus–lane is provided to enable buses to overtake the traffic queues on the approach to the upstream traffic signals. Traffic metering provides journey time and reliability benefits for buses over a congested route section, where it may not be possible, for operational or physical reasons, to provide bus–only lanes.

Examples of traffic–metering schemes can be found at Bitterne Road, Southampton (DOE, 1970 to 1976) and Dewsbury Road, Leeds (Fox et al, 1995).

The technique of combining bus–advance areas and pre–signals with traffic metering is particularly applicable on approaches to town centres where,

Photograph 24.5: Uxbridge Road/Park View Road, Ealing.

because of constraints such as narrow road widths and loading requirements, bus–lanes cannot be introduced. The Uxbridge Road/Park View Road (Southall) scheme in Ealing (see Photograph 24.5) has shown a significant reduction in bus journey times throughout all periods of the day. The technique showed these benefits on all seven days of the week, with overall savings to general traffic as well (LBE, 1995).

Exemptions from Prohibited Turns

Allowing buses to make turns prohibited to other traffic can give buses a considerable advantage, as journey distance can then be shorter than for other traffic. Clear, well–located signs are necessary to prevent other vehicles making the turns intended for buses only. Any scheme involving selective turns for buses must take into account the number of buses involved, their occupancy and the implications for junction capacity and road safety.

24.9 Junction and Network Bus–Priority at Traffic Signals

Traffic signal bus–priority can utilise Selective Vehicle Detection (SVD) within various traffic control strategies, such as vehicle–actuation, fixed–time Urban Traffic Control (UTC) and SCOOT (see Chapter 41) to provide 'active' bus–priority. Alternatively, bus volumes and passenger numbers, plus bus stop dwell– and cruise–times, can be used as inputs into traffic signal–timing calculations, with the aim of minimising delays and stops to passengers rather than vehicles, and thus provide 'passive' priority. Bus–priority at signals is relatively inexpensive to implement, with the capital cost generally balanced by benefits in months rather than years. Moreover, it is complementary to other bus–priority traffic management measures, such as bus–lanes and advanced areas.

Buses can be given priority at traffic signals by making signals respond to the arrival of a bus utilising an SVD system. Buses fitted with transponders, or other types of electronic device, are able to communicate with the traffic signal controller. As buses approach the signals, they are detected and the traffic signal–timings can be altered in their favour. The transponder is interrogated either via a roadside beacon or detector–loop buried in the road and a coded signal is sent to the signal controller, which then alters the traffic signal–settings in one of two ways. Either the green–time for the approaching bus is extended (extension) or, if the bus is approaching lights which are red, other green phases in the signal–cycle are shortened or omitted to bring forward the next green phase for the bus (recall). In the latter case, the time lost to other phases may be compensated during the next signal–cycle. Where bus flows are heavy, an 'inhibit' facility can also be set. Where buses are turning right, communication via a beacon or a detector loop can be used to call the next stage, enabling that bus to make the turn.

Photograph 24.6: An example of a buses–only road link.

SVD for buses has been applied, for example, in Oxford, Swansea (linked to SCOOT) and widely in London, where, by 1996, around 350 vehicle–actuated signal junctions had been installed outside the UTC/SCOOT controlled area and about 4,500 buses had been fitted with transponders. The system has paid for itself, typically, in about 15 months, with bus delays at most junctions reduced by around one–third and with a reduction in variability of around one–fifth.

In most large urban areas in Britain, traffic signals are controlled by some form of computer–based UTC system. Active and passive bus–priority can be provided within UTC systems. Active bus–priority, giving an extension or recall, has been incorporated into Version 3.1 of the traffic responsive SCOOT UTC system (termed BUS SCOOT) (Bowen et al, 1994) and is also available within fixed time UTC, using the SPRINT (Selective Priority Network Technique) algorithm. Results from the PROMPT trial in Camden Town, London, indicated that BUS SCOOT gave bus delay–savings that averaged 22% (ie five seconds per bus per junction). Overall, the benefits repay about 72% of system costs within the first year (Hounsell et al, 1995). Passive bus–priority can be provided within SCOOT and fixed–time UTC (BUS TRANSYT).

24.10 Bus–Only Roads (or Links) and Bus–Only Streets in Town Centres

A road or link restricted to bus–use usually allows buses to take a more direct route than other vehicles, for example between a new housing area and the existing road network, or to by–pass congested junctions (Photograph 24.6). The use of a bus–only street in a town centre is, typically, restricted solely to buses, although limited access by other categories of vehicle, such as taxis, may be allowed or the street may be accessed during limited time–periods for servicing.

Bus–only roads or links enable buses to maintain their route–patterns and to avoid needless detours where road systems have changed; thus, services can continue to provide long–established access for passengers to business and shopping areas, where such access may be denied to other vehicles. They also improve the environment of shopping streets by restricting traffic while, at the same time, maintaining accessibility.

Bus–only roads or links usually require a 'bus–gate' at the point(s) of access, to ensure compliance by other

TRANSPORT IN THE URBAN ENVIRONMENT

vehicles. These could be traffic signals, actuated by the buses, or physical barriers, surmountable only by buses, or signs, such as 'No Entry Except Local Buses', often coupled with local road–narrowing.

A bus–only street is often used by pedestrians and, thus, the bus 'track' should be emphasised by the use of different running levels or materials or colours to increase pedestrians' awareness and safety. Their alignment should discourage high speeds (James *et al*, 1991). Kerbs are not always necessary but may be considered (at a minimum height), to facilitate drainage on curved alignments and, at stops, to prevent buses overrunning and to help passengers boarding and alighting.

Bus–only roads or links should not cause loading problems, since they are generally purpose–built without frontage access. For bus–only streets, servicing is a key planning constraint, as the streets are generally located in existing shopping areas. The conventional solutions apply, ie access limited to certain specified times of day and provision of facilities in nearby side–roads or at the rear of premises.

24.11 Bus–Stopping Places

Siting

Bus stops must be sited to allow passengers to board and alight, safely and conveniently, with minimum disruption to other road–users. Stop locations should be convenient for main shopping and business areas, right beside stations and as close to other main passenger origins and destinations as possible. The needs of elderly and disabled people should be recognised. Provision of new bus stops, or re–siting of existing stops, occurs when bus services change or new developments open. Wherever new bus stops are proposed, or an existing stop is to be moved, discussions should be held between the bus operators or PTE (in London, LT Buses), the local Highway Authority and the police, in order to determine the most suitable location. The criteria for new bus stops are that, ideally, they should be located:

❏ near pedestrian routes to and from the main generation points of bus trips;
❏ close to pedestrian crossing facilities;
❏ close to main junctions, to facilitate passenger interchange with other buses, but without interfering with junction capacity or compromising road safety;
❏ to minimise walking distances between interchange stops and cross roads;
❏ 'tail–to–tail', where possible on opposite sides of the road for safety reasons and allowing sufficient

space between the rear–ends of bus stop markings for other vehicles to pass;
❏ away from residential and other sensitive frontages, where noise and disturbance are undesirable; and
❏ never between a signal detector and a stop–line, where Selective Vehicle Detection (SVD) is in use.

In practice, these criteria may not all be achievable, in every instance, in which case safety considerations must dominate.

Spacing

Typical bus stop frequency is between two and three stops/kilometre. In densely populated areas, town centres and residential developments, bus stops should preferably be no more than 300m apart. Stops may be split so that buses on different routes, but using the same street, stop at different points, because of high frequency (more than 25 buses/hour) and/or lengthy stop–occupancies. A balance should be sought between the advantages of splitting stops, to reduce bus–on–bus delays and traffic congestion, and the disadvantages of reduced convenience for passengers. Bus routes with common destinations should share the same stop.

Layout

Buses should be able to approach and leave stops without delay or obstruction. Vehicles parked close to or at bus stops prevent buses from reaching the kerbside and force buses to stop in the carriageway. This causes difficulties for passengers trying to board or alight, especially for elderly or disabled people and people with children or shopping who have to walk on the road and negotiate a higher step onto the bus. Preferred bus stop layouts are shown in Figures 24.4(a) to 24.4(e). The layouts apply to urban conditions, ie roads with speed–limits up to 40 miles/h, and for 12m buses with doors at both front and centre. If other buses are used, the designs may have to be adjusted. The overall aim is to permit buses to stop within 200mm of the kerbside, without overhanging or over–running the footway. Other bus stop and bus–bay designs have been developed (LBPNSG, 1995).

Most stops in urban areas will be conventional kerbside stops. Figure 24.4 provides examples. However, buses often experience difficulty in manoeuvring to the kerbside, due to parked or loading vehicles. Bus stop 'boarders' help to resolve this problem. Boarders require less kerb–length than conventional bus stops located between otherwise continuous parked cars. They provide an effective deterrent to kerbside parking at the stop itself and

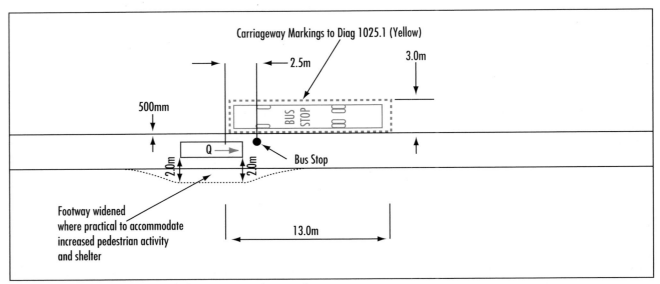

Figure 24.4(a): Typical kerb–side bus stop, unobstructed.

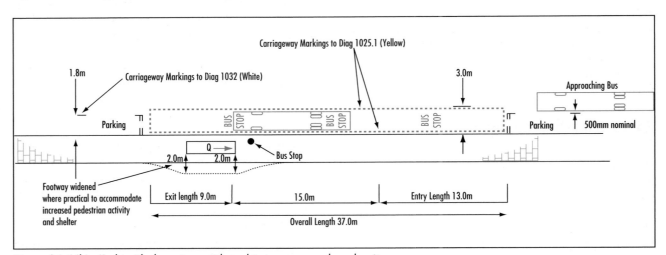

Figure 24.4(b): Kerb–side bus stop, with parking on approach and exit.

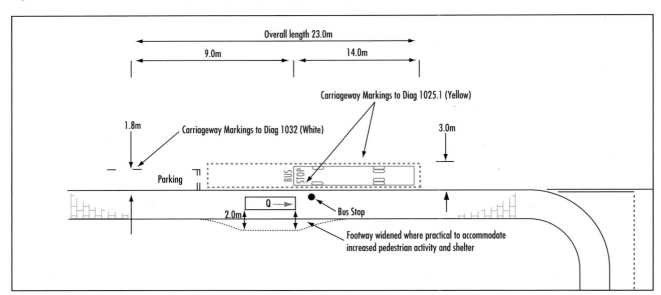

Figure 24.4(c): Kerb–side bus stop on the side of a junction.

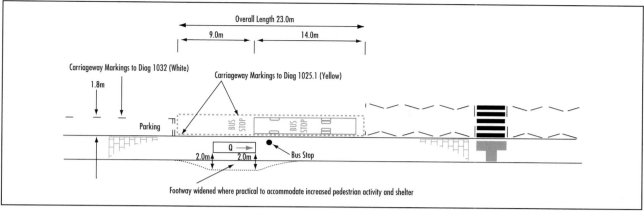

Figure 24.4(d): Kerb–side bus stop on the exit of a pedestrian exit.

they define clearly the parking areas up– and down–stream. They enable buses to align with the kerb, create passenger waiting–areas, without conflicting with general pedestrian flows, and ease bus boarding, by allowing the height of the kerb/platform area to be raised. Figures 24.5(a) to 24.5(c) show arrangements for these bus–boarders. Bus bays do not assist buses in the same way, because bus drivers can experience difficulty in re–entering a traffic stream when leaving a bus bay and the bays can attract illegal parking. The design of many existing bus bays is unsatisfactory, particularly where their geometry does not enable buses to stop close to the kerb. Preferred layouts are shown in Figures 24.6(a) and 24.6(b).

Where kerbside parking and loading is a problem in the vicinity of a stop, bus stop clearways and 'cages' should be provided (HMG, 1994). On the London Priority (Red) Route Network, 24–hour, seven days per week 'no–stopping–except–buses' arrangements are provided at all bus stops. On the London Bus–Priority Network (LBPN), in addition to the 24–hour provision, other standard hours are 07.00–midnight. At certain locations, it may be necessary to restrict the use of a stop to local buses only.

Footway Treatment

Passenger–waiting areas should be attractive, convenient and well–lit. The needs of the elderly and people with impaired mobility should be considered in the design. Where possible, bus stops should be sited on footways that are sufficiently wide to avoid obstruction to pedestrians by waiting bus passengers. Where footways are narrow, bus–boarders should be considered to enable bus passengers to wait away from pedestrian paths. If a 2.0m, full–width bus–boarder is feasible, the kerb and bus–boarder may be raised to between 160 and 180 mm at the kerb and sloped back to meet the existing kerbline. This reduces step–height to buses without impeding pedestrians on the footway. The addition of 'Kassel' kerbs (see Photograph 24.7) allows buses to stop within a few millimetres of the kerb without any damage to tyres. Bus shelters are beneficial at stops, where space permits, and high quality shelters should be used to improve passengers' comfort and convenience.

At terminus stops, at major commercial developments, at LRT/Metro interchanges and similar locations, buses may stand for some time and

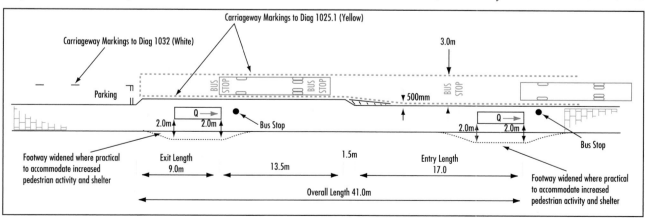

Figure 24.4(e): Multiple bus–stops, including one with a slight build–out.

TRANSPORT IN THE URBAN ENVIRONMENT

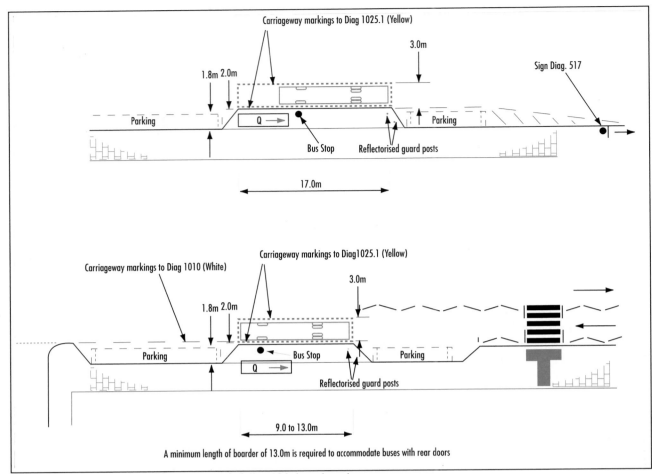

Figure 24.5(a): Examples of bus stops with full–width bus–boarders.

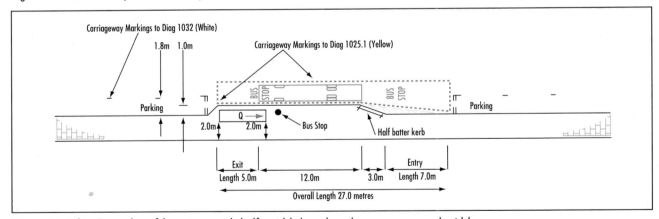

Figure 24.5(b): Examples of bus stops with half–width bus–boarder – narrow road width.

will usually need to turn round. Clearway regulations may be necessary at some stands to keep them free of other traffic. It is preferable to provide a turning–area off the highway, unless a suitable roundabout or gyratory exists nearby.

Bus Stations

Bus stations assist buses to provide good accessibility to town centres or major developments. In smaller, concentrated town centres, a well–sited bus station will be able to serve the majority of passengers' objectives. In larger town centres, a single bus station may only be able to serve a minority of passenger–trips and could impose unnecessary constraints and costs on bus operators. Bus stations should:

❑ provide a focal point for passenger–journeys to and from a town centre and allow good accessibility to town centre facilities;

❑ allow easy interchange for passengers between bus services;

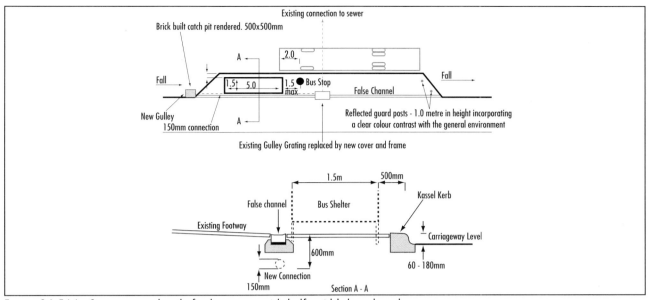

Figure 24.5(c): Construction details for bus stops with half–width bus–boarder.

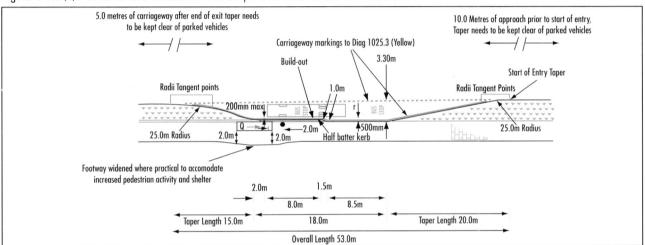

Above: Figure 24.6(a): General arrangement of bus bay.   Below: Figure 24.6(b): Low cost amendment to an existing bus bay.

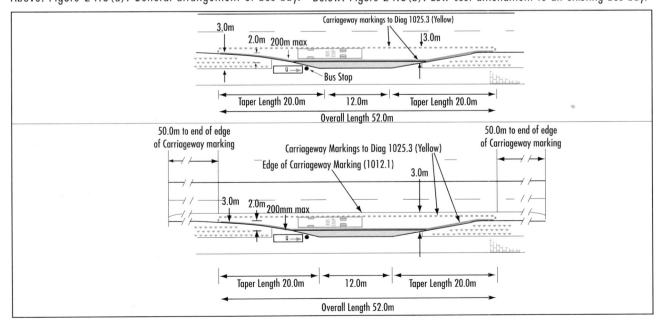

Photograph 24.7: Bus–boarder and Kassel kerbs.

❑ provide a high quality passenger environment, including shelters, information, security and other facilities;
❑ permit operators to improve the efficiency of management of bus services, by the provision of bus–stacking and lay–over areas and crew–change facilities; and
❑ allow buses to enter and leave the bus station site unobstructed by traffic congestion and without introducing costly diversions.

Bus station design depends on the expected movement patterns of passengers and buses, as well as bus standing and lay–over requirements. Basic design objectives are:
❑ to encourage maximum passenger use;
❑ to ensure the quality, safety and security of passengers;
❑ to minimise the potential for conflict between passengers and vehicle–movements;
❑ to minimise walking distances between the main passenger origins and destinations and the bus station;
❑ to minimise the number and distance of bus movements within the station; and
❑ to provide for an efficient sequence of setting–down, standing, waiting and boarding.

Bus station design (LTB, 1995; and BRPT, 1981) depends on specific site–factors, such as available land–area, site–shape, topography, local road–pattern and access, passengers origins and destinations, the peak numbers of buses and passengers and the manoeuvring capability of the largest bus in service. Four basic layouts of bus station are shown in Figure 24.7, as follows:

❑ 'island layouts' – which are compact but have operational problems for buses and for passengers' safety, as frequent crossing of lanes is necessary;
❑ 'perimeter layouts' – where a single main passenger destination is served, such as a shopping centre or railway station. These are not recommended where main passenger destinations exist on more than one boundary, as jaywalking can occur and bus circulation between setting–down, standing and boarding requires additional empty–running kilometres for buses;
❑ 'central concourse layouts' – these are preferred to the narrow island layouts for restricted sites, since stops can be organized to ensure an efficient setting–down, standing and boarding sequence and passengers can be provided with high quality facilities and shelter in a single area; and
❑ 'reversing layouts' – which save space but are appropriate only for low frequency services, as buses cannot queue in the circulation space while waiting for a designated free bay and, for safety reasons, supervised reversing is usually necessary.

In addition to interchange at bus stations, many other facilities involve interchange between buses and other modes, including railway stations, park–and–ride facilities, kiss–and–ride facilities, ferries and airports. Design considerations similar to those for bus stations apply but, additionally, the potential for transfer between modes should be maximized. This can be achieved with:
❑ good accessibility for vehicles of all kinds (for example, combining kiss–and ride with park–and ride); and
❑ a high level of convenience (eg short walking distances) and comfort (eg weather protection) to encourage passenger interchange.

Timetables and count–down information should be displayed at all public transport stops. Local authorities and PTEs play an important role in providing timetable information at stops, for passengers, as a first priority, and also for local residents. The image of public transport and passenger services can be enhanced by the use of real–time displays based on automatic vehicle location (AVL) systems. These show the expected time to the arrival of the next bus, tram or LRT vehicle and its service number (see Photograph 24.8) (see also Chapter 15).

24.12 Busway Transit and LRT

Busway transit is defined as a public transport system which utilises buses, operating on exclusive rights of way, termed busways. Busways, provided over

Photograph 24.8: A 'countdown' display at a bus stop based on AVL

significant but not necessarily all route–sections, protect buses from the effects of traffic congestion and enable a rapid and reliable service to be offered. Busways should be segregated physically from general traffic, to the maximum extent feasible, in order to minimise violation of their right of way. Buses may be guided or non–guided. As with LRT, a properly planned busway transit system should not just consist of a busway track, the system should also incorporate high quality buses, bus stop information services, passenger facilities and 'smartcard' fare–collection systems.

Mass transit based on busways has many advantages over fixed–track systems, including:

❑ 'performance' – buses operating on segregated busways can provide a reliable and regular service at high operating speeds. With traffic management and signal–priority and even, in selected critical locations, grade–separation, service performance can approach that of LRT with similar segregation;
❑ 'flexibility in operation' – buses can join and leave a busway at intermediate points along its length. Thus, buses serving many areas of a city may use part, or all, of a busway and passengers from a wide catchment area benefit from services, without the interchange required by rail systems;
❑ 'flexibility in implementation' – as with any segregated track system, busways will affect other traffic, particularly loading and unloading to frontage properties. However, unlike LRT, a segregated busway may be discontinued for short lengths and some less rigid form of priority provided;
❑ 'lower capital costs' – costs of infrastructure and rolling stock are substantially less than for fixed–track systems of equivalent capacity, along the same route and with the same degree of segregation;
❑ 'scope for rapid and incremental development' – busways may be introduced and used effectively over short sections, so a system may be enhanced and expanded, as demand grows and resources permit; and
❑ 'passenger capacity' – busways are unlikely to offer greater capacity than LRT with equivalent degrees of segregation from other traffic. However, high capacities (ie 25,000 to 35,000 passengers/h) are not often needed in Britain. Even so, segregated busways have been recorded with 200 buses/h carrying 20,000 passengers/h at uniform headway and a constant speed of about 19 km/h (Cracknell et al, 1992).

In Britain, there are few purpose–designed busways, although a segregated contra–flow lane may be regarded as a basic form of busway. Examples of purpose–built schemes are the Runcorn segregated busway and Redditch, which has a part–segregated busway. In other countries, examples of busways operating include: Ottawa, with 20km of segregated bus track; Seattle, with a two kilometre bus–only tunnel, with metro–type stations and dual–powered buses; Pittsburgh, with two segregated busways, one used jointly by buses and trams; Curitiba Brazil, with the most extensive system–wide busway network in the world, having schemes on five major corridors serving a city of near two million population; Liege, with busways introduced into the existing road network, partly using ex–tram rights of way; and France, with busways in Paris, Envreaux, Montpelier and Nice.

Busways can be operated with guided buses and examples are schemes in Ipswich and in Leeds (Photograph 24.9). Various forms of bus guidance exist. The most usual form involves lateral guide–wheels, mounted on conventional buses, which

| Island | Concourse | Perimeter | Reversing |

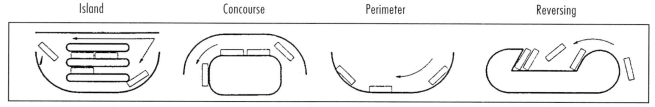

Figure 24.7: Bus station layout.

run along guide–rails or kerbs on both sides of the busway track. This system operates in Leeds, Adelaide (Australia), Essen and Mannheim (Germany). Other prototype schemes have been developed, involving buses guided by signals from cables buried in the road surface and buses guided by an arm lowered from each bus into a slot and rail cut into the road surface. The advantages of guided buses are: the reduced width required for the busway track, since it need only be wide enough to accommodate a bus and its attached guidance system – usually about 2.6m kerb–to–kerb (although the structural width of the guidance kerbs and safety clearances must be added); the smooth train–like ride (a consequence of the high quality track which forms the busway); and the enhanced public perception of a guided system. Against these factors must be set the capital costs, lack of flexibility (only specially equipped buses can use the guideway) and problems which may occur in the event of a bus–breakdown on the guided way.

Safety and ease of enforcement favour rail–based systems. Moreover, the physical presence of rails, when street–running is used, appears to deter illegal parking. The main advantage of LRT systems is the quality image they project, which may assist in attracting car–users to public transport as an acceptable alternative mode in the context of car restraint policies. This must be set against the high capital and operating costs. Design considerations for LRT systems are discussed in detail in Chapter 34.

Stops

Where buses operate on–street, details described in the foregoing paragraphs for conventional bus stops apply. For guided buses, in cases where the guidance wheels on each bus are permanently deployed, clearances and layouts of bus stop bays must take this dimension into consideration. Where stops are within a busway, whether non–guided or guided, the following apply:

❑ for island stops or platforms, suitable arrangements must be made for boarding and alighting passengers to cross adjacent roads to and from the stops; crossing may be combined with junction signals or may use measures such as a Pelican crossing; and

❑ island platforms should incorporate physical measures for safety and ease of use, such as channelising guardrails, a back shelter–wall to protect waiting passengers from the nearside traffic stream, dropped kerbs at the crossing points for access for mobility– impaired people and textured surfaces for the visually impaired. The height of the kerbs and bus stop waiting–area will depend on the type of bus and guidance system which is to be operated.

No comprehensive design standards exist for stops. Nevertheless, generally:

❑ stop–spacing may vary from 250m in densely–developed inner city areas to 1km in suburban areas;

❑ wherever possible, stops should be located on segregated sections of track, to avoid delays to other vehicles and to ensure stops are not obstructed by other road vehicles;

❑ dimensions of stops will reflect the sizes of rolling stock used. Platforms may range from high (915mm) to low (400mm). Most modern systems are likely to adopt low–floor vehicles and thus lower platforms will most often be used; and

❑ platforms should preferably be 3m wide or more (including allowance for guardrails) although, in critical cases, a minimum of 2.5m may be acceptable.

24.13 Other Facilities

High Occupancy Vehicle (HOV) lanes are traffic lanes for use by buses and other vehicles (usually cars) with three or more occupants. HOV lanes do not benefit buses alone nor do they have a great impact on modal choice. The objective is to promote higher car occupancy and, thereby, to improve the efficiency of road–use. Numerous examples of HOV lanes exist in cities in the US and other countries (Kain, 1992).

Where bus flows are too low to justify a lane exclusively for buses, allowing taxis and/or goods vehicles into the lane may justify a combined scheme.

Photograph 24.9: Guided bus in Leeds.

The introduction of disability discrimination legislation has stimulated the provision of public transport which is accessible to people with impaired mobility, including those who travel in wheelchairs. For some of these people, the journey between their home and a bus stop will be unmanageable, even if buses are accessible. Weather conditions, state of health, time of day and other factors will determine their ability to use public transport or their need instead to rely on a door–to–door services, such as community buses, dial–a–ride and accessible taxis.

To achieve a genuine door–to–door service, the vehicles should be able to get as close to the pick–up point as possible and to drop passengers off as close as possible to their final destination. The value of the services will be entirely lost if the vehicles are unable to gain access into a shopping precinct or pedestrian area, in order to drop off an elderly or mobility–impaired person, in safety and within a walking distance that can be managed (for some, this may as little as 50m). Factors necessary for a good service include: dedicated drop–off points; tactile surfaces; dropped kerbs; and, as minibuses with rear–loading passenger lifts are likely to be used, sufficient clear space behind the vehicle for wheelchair–users to manoeuvre safely.

24.14 References

Astrop AJ, and Balcombe RJ, (1994) — 'Performance of Bus Priority Measures in Shepherd's Bush', DOT.

Bowen GT, Bretherton RD, Landles JR, and Cook DJ, (1994) — 'Active Bus Priority in SCOOT', Institution of Electrical Engineers.

BRPT (1981) — 'Urban Planning and Design for Road Public Transport', British Road Passenger Transport.

Cracknell JA, Cornwell PR and Gardner G (1992) — 'The Performance of Busway Transit in Developing Cities' TRL.

DOE (1970 to 1976) — 'Working Group on Bus Demonstration Projects, Summary Reports 1–9', DOE.

DOE (1996) — Planning Policy Guidance Note 6 'Town Centres and Retail Developments', DOE [Sc] [We].

DOE/DOT (1994) — Planning Policy Guidance Note 13 'Transport', DOE/DOT [Sb].

DOT (1991) — Local Transport Note 1/91, 'Keeping Buses Moving', DOT [Sd].

EHTF (1993) — 'Bus–based Park–and–Ride – a Good Practice Guide', English Historic Town Forum.

Fox K, Balmforth C, Franklin P, Montgomery F and Siu YL (1995) — 'Integrated ATT strategies for urban arterials: DRIVE II PRIMAVERA – 3. The Dewsbury Experiment', Traffic Engineering + Control, 36(7/8).

HMG (1984) — 'Road Traffic Regulation Act 1984', Stationery Office.

HMG (1985) — 'Transport Act 1985', Stationery Office [Sa].

HMG (1991) — 'Road Traffic Act 1991', Stationery Office.

HMG (1994) — 'Traffic Signs Regulation and General Directions,' Stationery Office.

Hounsell N, and McLeod F, (1995) — 'Field Trial Implementation and Evaluation, PROMPT (V2049) Deliverable No. 19' University of Southampton for DRIVE II Project, PROMPT.

James N, Jessop A, and Roberts J (1991) — 'Space Sharing' Report No. 92, Transport and Environmental Studies (TEST).

Kain JF (1992) — 'Increasing the Productivity of the Nations Urban Transportation Infrastructure – Measures to Increase Transit Use and Car Pooling', Department of Transportation, US.

LBE (1995) — 'Park View Road Pre–Signals – Before and After Study', London Borough of Ealing.

LBPNSG (1995) — London Bus Priority Network Steering Group 'Guidelines for the design of bus bays and bus stops to accommodate the European standard (12m) length bus', LBPNSG.

LTB (1994)	'London Bus Priority Network – Design Brief, London Borough of Bromley', London Boroughs/London Transport Buses.
LTB (1995)	'Bus station design guide', London Transport Buses.
LTB (1996)	'Guidelines for treatment of buses at roadworks', London Transport Buses.
NATO (1976)	NATO CCMS Report No. 45 – 'Bus Priority Systems', TRL.
Niblett R and Palmer D (1993)	'Park and Ride in London and the South–East' Highways & Transportation 40(2).
TDL (1995)	'Cameras to Catch Bus Lane Cheaters', Traffic Director for London, Press Notice 15 December 1995.

24.15 Further Information

DOT (1989)	'Provisional Guidance on the Highway and Vehicle Engineering Aspects of Street Running of Light Rapid Transit Systems', DOT.
DOT (1995a)	'Guided Bus', A Briefing Note by the Bus and Taxi Division of the DOT, DOT.
DOT (1995b)	'Light Rapid Transit (and Related) Systems'. A Briefing Note by the Bus and Taxis Division of the DOT, DOT.
Hounsell N (1993)	'Bus Priority in SCOOT areas', TCSU.
Robertson D I and Vincent R A, (1975)	'Bus Priority in a network of fixed time signals', TRL LR 666, DOT.
TRL (1997)	'Bus TRANSYT – A User's Guide', SR 266. DOT/TRL.
Wood, K and Baker R T, (1992)	'Using SCOOT weighting to benefit strategic routes', Traffic Engineering + Control 33(4).

Chapter 25 Management of Heavy Goods Vehicles

25.1 Introduction

An efficient system of road freight transport is essential to the national economy. Industry and commerce, shops and offices all depend on a complex flow of raw materials, components and finished goods. Yet lorries often have a negative image and can cause nuisance such as noise, environmental intrusion and damage to infrastructure. This has led to the development of a wide range of lorry management techniques, aimed at mitigating such nuisance.

It is now generally recognised that it is physically and economically impossible to build new roads to solve all the environmental and operational problems now facing transport. Nor is it possible to reduce, significantly, the number of lorries on our roads by shifting freight to alternative modes, notably railways. However, in many cases, lorry problems can be alleviated by relatively low cost management schemes. The approach to lorry management tends to vary from place to place, in recognition of the local economic situation and quality of the environment. While this is to be expected, consistency of approach is important, in terms of problem-identification and assessment of possible solutions, in a manner which recognises both the impact of lorries on the community and the needs of commerce and the freight transport industry.

Lorries are now quieter, safer and less polluting than ever before, because:
- ❑ engine noise has been reduced to half the levels of ten years ago;
- ❑ air brake silencers are now standard equipment on most new lorries and recent EC legislation has reduced permitted compressed air noise even further;
- ❑ better brakes have improved safety and manoeuvrability;
- ❑ more sophisticated suspensions have reduced roadwear and, coupled with better body design, lowered body noise levels;
- ❑ lorry exhaust emissions have been radically reduced – and further improvements are on the way; and
- ❑ vehicle manufacturers and operators are striving continually to improve safety and reduce the environmental impact of lorries. Their achievements have been impressive but they recognise that lorries must be made even safer and more environment friendly.

25.2 Legislative Framework and Responsibilities

Lorry weights and sizes are regulated by the Road Vehicles (Construction and Use) Regulations 1986 (HMG, 1986) [NIa]. The maximum permitted weight for general use on roads in the UK is 38 tonnes. However, since March 1994, lorries up to 44 tonnes, with six axles and road friendly suspension, have been permitted for use in combined transport operations, ie moving freight to or from a railhead prior to or after transport by rail. In December 1996, the Government published a consultation document seeking views on proposals to permit 44 tonne lorries for general use in the UK (DOT, 1996). A fully laden 44 tonne vehicle, on six axles, with road friendly suspension, with a 10.5 tonne axle weight limit and with adequate minimum axle spacings would cause no more road or bridge wear than most 38 tonne, five axle vehicles already allowed on UK roads, and considerably less wear than the 40 tonne, five axle vehicles which will be allowed on the roads from 1 January 1999 under EC Directive 85(3).

Many, but not all, of the measures which can reduce the environmental impact of lorries require legal procedures to be followed for their implementation. The relevant Acts of Parliament and Statutory Instruments should always be consulted for precise detail. For example, sections 6 and 1 of the Road Traffic Regulation Act 1984 (HMG, 1984) deal with traffic regulations in Greater London and elsewhere, respectively. Section 2 provides for the types of 'amenity' control on goods vehicles, which a local authority may introduce [NIb]. These include:
- ❑ control of 'through' routes used by heavy commercial vehicles; and
- ❑ prohibitions or restrictions on the use of heavy commercial vehicles (subject to specified exceptions) in such zones or roads "...as may be considered expedient for preserving or improving the amenities of their area or of some part or parts of their area".

A 'heavy commercial vehicle' (HGV) is defined in section 138 of the Act as any goods vehicle with an operating weight (ie gross permitted weight) exceeding 7.5 tonnes [NIc]. Section 138 should be consulted, if necessary, for further definitions of 'operating weight' in relation to articulated vehicles, solo articulated tractor units and drawbar–trailer combinations (ie a rigid lorry drawing a separate trailer).

Control of through routes relates to 'positive' routeing; ie specifying mandatory through routes for lorries. However, experience has shown that mandatory positive routeing networks are not very practical; signing requirements are extremely complex and drivers' understanding and compliance are likely to be poor. For this reason, lorry control Orders now invariably rely on 'negative' routeing; ie specifying roads and zones which are prohibited to lorries but subject to exemptions, such as for loading or unloading. Also, the designation of 'advisory routeing' for lorries is now common–place (see Section 25.9).

Section 3 of the Act provides that an Order must not prevent access to premises by vehicles for more than eight hours in 24, subject to certain exclusions [NId]. These include Orders where the objective is to preserve or to improve the amenities of an area, by prohibiting or restricting the use of heavy commercial vehicles on a road or roads in the area.

When exercising their powers to control lorries, local authorities must have regard to their duties under section 122 of the Act [NIb]. These duties are:
   ❏ to exercise their powers (so far as is practicable, having regard to the matters specified below) to secure the expeditious, convenient and safe movement of vehicular and other traffic (including pedestrians) and the provision of suitable and adequate parking facilities on and off the highway.

Due regard must be paid:
   ❏ to the desirability of securing and maintaining reasonable access to premises; and
   ❏ to the effect on the amenities of any locality affected and the importance of regulating and restricting the use of roads by heavy commercial vehicles, so as to preserve or improve the amenities of the area through which the road runs.

Orders may provide for the prohibition or restriction of waiting, loading and unloading.

Section 61 of the Act provides for the designation of land, which is not part of a highway, as a 'loading area' [NId]. This can usefully be employed when off–street delivery or service areas are obstructed by unauthorised parking. However, an Order requires consent by the owner and occupier of the land and this may not always be readily forthcoming because of existing use.

Section 9 of the Act enables authorities to implement experimental Traffic Orders, for a maximum period of 18 months [NIe]. In general, an experimental Order can make any provision for any purpose specified in relation to permanent Orders under sections 1 and 6. Schedule 9 includes provisions which require consent by the Secretary of State for certain Orders, including Orders (except in Greater London) preventing access to premises for more than eight hours in 24.

The Local Authorities' Traffic Orders (Procedure) (England and Wales) Regulations 1996 (HMG, 1996e) [Sa] describes the statutory procedures to be observed by authorities when exercising their powers to make Traffic Orders (see also Chapter 13) [NIf]. The regulations are complex but the key requirements are [NIg]:
   ❏ for proposed Orders, authorities must consult with organisations representing persons who use any road to which the Order relates or are otherwise likely to be affected. The Freight Transport Association and the Road Haulage Association are among the consultees who must be consulted in relation to all proposed traffic Orders (see Chapter 13) [NIh];
   ❏ details of proposed Orders must be published in the London Gazette [NId] and local newspapers;
   ❏ all objections must be considered;
   ❏ if Orders are made despite objections, the objectors must be notified in writing, with the reasons why their objections have been over–ruled; and
   ❏ a public inquiry must be held if there are unwithdrawn objections to an Order which would prevent access to premises or loading or unloading, unless wholly within peak hours [NIi]: ie 07.00 to 10.00 hrs and 16.00 to 19.00 hrs.

Many of the above requirements are relaxed for minor or experimental Orders. The regulations should be consulted for further information.

The Traffic Signs Regulations and General Directions 1994 (HMG, 1994) authorise a range of signs including those for lorry controls [NIj]. For amenity controls, the 7.5 tonne sign to diagram 622.1A, as specified in HMG (1994), is most commonly used. The only permitted variant is 17T. Sign 622.2 should be used to indicate the end of a lorry control indicated by a 622.1A sign. Signs to diagram 618.2 in HMG (1994) can be used to indicate loading exemptions for goods vehicles in pedestrianised areas. Diagrams 629, 629.1 and 629.2 indicate, respectively, signs for width, length and height restrictions.

Diagram 625.2 indicates a 'weak bridge' weight limit and 627.1 the supplementary plate, 'except empty vehicles'. The Department of Transport recommends that authorities use the empty vehicle exemption for all weak bridges with a capacity of 17T and above (DOT, 1994a). The permitted variants for 'weak bridge' signs are: 3T, 7.5T, 10T, 13T, 17T, 25T and 33T, as set out in the General Directions (HMG, 1994). Signs to diagram 665 and 666 indicate the entrance to, and exit from, a controlled lorry parking zone.

25.3 Lorries and Operator–Licensing

The Goods Vehicles (Licensing of Operators) Act 1995 (HMG, 1995) provides for a comprehensive system of goods vehicle operator–licensing, founded on safety and environmental considerations [NIk].

An 'operating centre' is the place where a lorry is normally kept when not in use. References to a particular operator's operating centre(s) mean the place or places, specified in the operator's licence, where the vehicles specified on it are normally kept. An operator's licence must be held by the operator of any goods vehicle over 3.5 tonnes maximum gross weight, unless exempt. In practice, the 'operator' is usually the person or company who employs the driver.

Once a licence has been granted, its renewal is necessary only if the operator wishes to increase the number of vehicles authorised or to alter any conditions, on it. Traffic Commissioners can discipline operators, at any time, if they fail to observe any of the statutory requirements (eg regarding maintenance) or breach conditions placed on a licence. Local residents, who are adversely affected by an operating centre, can make representations to the Traffic Commissioners [Sb] at any time [NIl]. Unless representations relate to alleged breaches of statutory requirements or licence conditions, they are considered at a five–yearly review of the licence. If significant environmental difficulties emerge, the Traffic Commissioners [Sb] may convene a Public Inquiry.

Statutes prescribe a number of bodies, including local authorities, who can object to the granting of a licence, on specified grounds.

25.4 Planning Conditions

Goods vehicle operator–licensing is not related to planning law. The two processes are entirely separate. A Traffic Commissioner [Sb] can only control licensed vehicles. He cannot control vehicles smaller than 3.5 tonnes gross nor can he control vehicles which merely visit a site.

It has been common practice for many authorities, sometimes by mutual agreement with an operator, to impose routeing conditions or accept undertakings, relating to planning consent for new developments – typically new or extended quarries. The legality of such conditions has always been unclear. However, in 1994, the Secretary of State issued an important direction, in which he ruled as 'inappropriate': "...the establishment of an approved route in respect of heavy goods vehicles associated with the operation of the quarry and with the removal of minerals from the land ...neither conditions nor planning obligations can lawfully control the right of passage over public highways" (DOE, 1994).

25.5 Lorries: Benefits and Environmental Effects

The overwhelming reliance on lorries for the movement of goods is hardly surprising, given the door–to–door flexible and cost–effective service offered by lorries, coupled with the relatively short distance of the average freight trip in the UK (89km in 1995).

While car traffic in urban areas continues to increase, total lorry traffic remains largely unchanged. During the period 1983–93, car traffic on urban roads increased by 12% from 57.9 to 65.0 billion vehicle–km. During the same period, lorry traffic remained constant at 3.8 billion vehicle–km. However, the mix of urban lorry traffic has changed substantially. Development of 'just–in–time' distribution systems by industry, with consolidation of suppliers' deliveries at out–of–town regional distribution centres, has eliminated many town centre lorry trips by individual suppliers. However, this reduction in urban lorry traffic has been off–set by a substantial increase in small– and medium–sized goods vehicles, reflecting economic growth in the service industries.

Although the total number of heavy lorries registered in the UK has declined gradually for many years, their average permissible size and weight has increased and there is continuing public concern about their effect on the environment, on people and on local communities. Complaints are made about noise and exhaust emissions and pedestrians, cyclists and car drivers feel that their safety is threatened. These problems are perceived at their worst when lorries use unsuitable roads, often because there is no adequate alternative.

Industry itself is well aware of the environmental effects of lorries and much effort is devoted by many companies to minimising those effects in their operating practices. Vehicle manufacturers are continually improving vehicle–design, to make lorries safer and more environment friendly. Improvements in tyres, suspension, brakes, fuel–technology, exhaust emissions, engine–noise and body–design have made a major contribution to reducing the environmental impact of lorries. The Road Vehicles (Construction and Use) Regulations 1986 (as amended) (HMG, 1986) prescribe stringent design and in–use requirements to limit environmental nuisance [NIa]. The statutory operator licensing system extends to the environmental as well as the safety aspects of lorry operation.

However, there are still opportunities to manage the movement of lorries by using regulatory controls, traffic management techniques and advisory and voluntary routeing agreements. These measures are described below.

25.6 Evaluation of Lorry Controls

An objective cost–benefit and environmental assessment should always be undertaken, at the planning stages, for all but the smallest lorry control schemes. The assessment should take account of both the net environmental benefits and the additional operating costs incurred by industry. More detailed advice is available in the following works:
   ❏ Lorries in the Community (DOT et al, 1990a);
   ❏ Lorries and Traffic Management (DOT et al, 1990b);
   ❏ Lorry Management Schemes (IHT, 1981); and
   ❏ Code of Practice for Highways and Traffic Management in London (ALBES et al, 1985).

Whenever lorry management measures are being planned, early consultation with affected operators and their representative bodies (notably the Freight Transport Association and the Road Haulage Association) is vital. The wider effects of all schemes should be assessed beforehand and monitored afterwards, to ensure that localised problems have been alleviated and not merely displaced elsewhere.

Further information on economic and environmental assessment techniques is given in Chapters 9 and 17.

25.7 Strategic Lorry Plans

The proportion of lorries in a large conurbation that are on through trips is generally small. Consequently, strategic area controls of the type suitable for smaller towns are not usually effective in reducing overall environmental disturbance in conurbations, because through vehicles will simply re–route into adjoining areas. A ban on vehicles passing through the whole conurbation would affect only a very small proportion of total flows.

Successful strategic lorry–routeing controls depend upon the existence of a network of purpose–designed roads which are able to accommodate increased lorry flows, ie lorries displaced from environmentally sensitive roads. Experience suggests that strategies which concentrate lorry nuisance on a small number of roads, some of which are sensitive, may not be acceptable. The roads identified for an urban lorry network are likely to be busy roads with poor existing environmental conditions.

25.8 Positive Assistance for Lorries

Lorry management is not only about regulating, restricting or banning lorries. It is also about positive initiatives to assist lorry access to, and circulation within, urban areas for essential freight movements. Generally, the more efficiently industry can operate its lorries, the fewer are needed. Conversely, the inefficiency and unreliability of many urban road networks, coupled with over–restrictive control measures, can generate increased lorry traffic as industry strives to maintain service to customers. Positive assistance to lorries will usually yield environmental as well as economic benefits. For example, where obstruction is caused by lorries loading or unloading, in unsuitable places, positive measures might include:
   ❏ the provision of off–street servicing (potentially costly but should not be ignored by default as a possible option);
   ❏ the designation of laybys, or localised widening using the loading bay sign (Figure 25.1), which also prohibits parking by Orange Badge holders;
   ❏ revised regulations for the times of waiting or loading; and
   ❏ enforcement of kerbside car parking regulations, violations of which often cause delivery drivers to double park.

Low cost improvements can often reduce or eliminate, difficulties caused by lorries negotiating a sub–standard junction (see Figure 25.2). For example, Newcastle City Council has introduced a shared

Figure 25.1: Loading bay sign.

TRANSPORT IN THE URBAN ENVIRONMENT

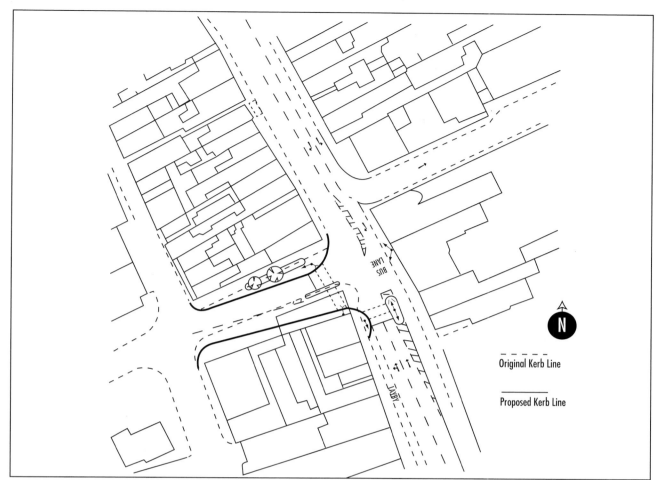

Original Kerb Line

Proposed Kerb Line

Figure 25.2: Example of low–cost junction improvement to assist lorry operation.

bus/lorry priority (no–car) lane (at Barras Bridge in the city centre) to facilitate essential passenger and goods movements (see Chapter 24). The lane functions well and others are planned on major radial routes. In London, no–car lanes are under consideration for certain locations on the developing priority (red) route network. Where bus usage is likely to be insufficient to justify an exclusive bus lane, a shared–priority lane for buses and lorries can sometimes provide a viable alternative (see Photograph 25.1).

25.9 Non–Regulatory Controls

The national primary route network is clearly the national advisory route network for use by lorries and other long–distance traffic. The DOT leaflet Commercial Vehicle Drivers: Know Your Traffic Signs (DOT, 1995) contains the following advice:

"Away from motorways, the best routes for commercial vehicles are the primary routes. These form a national network between towns and can easily be recognised by their distinctive green signs

...on a long journey, away from the motorway (network), use the green signs".

Traffic management on primary routes, therefore, should be consistent with their role as commercial vehicle routes.

Off the primary route network, the signing of preferred routes for lorries, using the 'white lorry' sign, may provide a simple solution, particularly where the problem is of environmental intrusion by long–distance traffic passing through built–up areas. If the majority of lorries passing through are engaged on local journeys, an unacceptable time/distance penalty may be incurred by diverting them on to the alternative route and compliance will be poor.

A number of authorities have introduced local advisory routeing schemes, to identify preferred routes for lorries servicing town centres, industrial estates, retail parks and other significant generators of lorry movement (see Photograph 25.2). Typically, the area concerned is divided into zones, with the preferred routes serving each zone codified by letter, number, symbol or colour and signed accordingly. For

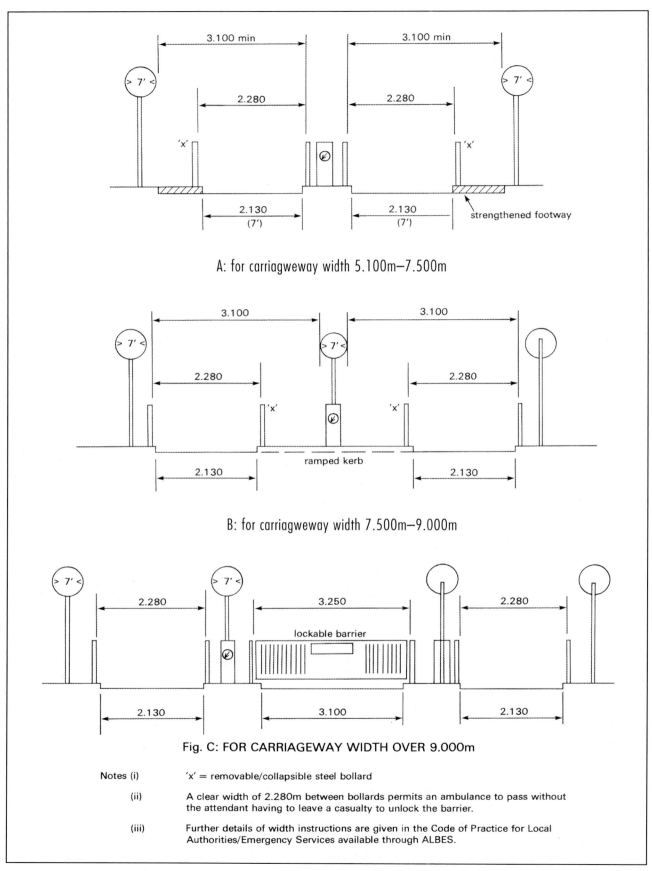

A: for carriagweway width 5.100m–7.500m

B: for carriagweway width 7.500m–9.000m

Fig. C: FOR CARRIAGEWAY WIDTH OVER 9.000m

Notes (i)   'x' = removable/collapsible steel bollard

(ii)   A clear width of 2.280m between bollards permits an ambulance to pass without the attendant having to leave a casualty to unlock the barrier.

(iii)   Further details of width instructions are given in the Code of Practice for Local Authorities/Emergency Services available through ALBES.

Figure 25.3: Typical layouts for width restrictions.   Source: ALBES *et al* (1985).

Photograph 25.1: A priority lane shared by buses and goods vehicles in Newcastle upon Tyne.

Photograph 25.2: Local advisory signs for lorries.

maximum benefit, such schemes need good publicity in the form of maps and leaflets and full co-operation by local firms in advising their suppliers.

Although physical barriers are non-regulatory, in the sense that they are self-enforcing, they require a supporting Traffic Regulation Order (TRO) if placed on the highway, because they have the effect of preventing the passage of vehicles otherwise lawfully allowed (see Chapter 13).

The simplest physical measure is a height-limiting barrier, although this is normally used only to give advance warning of physical restrictions ahead (eg a low bridge). It can, however, be useful when placed at the entrance to a car park or other off-street location, to prevent access by oversized vehicles. Sometimes, an automatic warning system is provided at low bridges, with a sign activated by a height-detector unit.

Physical width-restrictions, authorised by Orders under the Road Traffic Regulation Act 1984 (HMG, 1984) may be used to narrow the carriageway to less than 2.5m at selected points, to prevent the passage of large vehicles (a 2.28m clear width between posts is often used) [NIm]. Figure 25.3 shows a number of typical layouts. The method of construction, using, for example, bollards or planting boxes, needs to be particularly robust to withstand a certain amount of abuse. Especially in the early stages of introduction, drivers who have not seen, or have chosen to ignore, the associated warning signs may try to squeeze through. It is usually necessary to locate such a width-restriction adjacent to an existing junction, which can then be used as a turning-head for those drivers who have not complied with the warning signs (Photograph 25.3). Width-restrictions can

effectively create culs-de-sac without turning space, so that large vehicles can become trapped and may have to reverse for long distances. It is therefore important that they are located intelligently and signed well in advance.

Other problems are associated with physical width-restrictions. They are inappropriate for bus routes and can be detrimental to normal servicing requirements, particularly of refuse vehicles, furniture removal vans and fuel deliveries. Although ambulances will normally be able to pass a width-restriction, fire appliances will usually be prevented and the additional distance for emergency vehicles may have important consequences. It is possible to provide a lockable barrier or removable bollards for fire appliances but this does increase journey time unless it is operated automatically. Whenever width-restrictions are contemplated discussions should be held with the emergency services at the earliest stage.

The Traffic Calming Act 1992 (HMG, 1992) amended the Highways Act 1980 (HMG, 1980) [Sc] with new provisions, allowing works to be carried out '...for the purposes of promoting safety and preserving or improving the environment' [NIn]. A wide range of physical measures is now available to reduce the environmental impact of traffic, including chicanes, humps, speed-control tables and carriageway-narrowings. Authorities should note that the Highways (Traffic Calming) Regulations 1993 (HMG, 1993) [Sd] state specifically that [NIc]:

"No traffic-calming work shall be constructed or maintained in a carriageway, so as to prevent the passage of any vehicle, unless the passage of that

Photograph 25.3: A typical width–restriction with a gate for use by emergency vehicles.

vehicle is otherwise lawfully prohibited."

Therefore, if the intention is physically to prevent the passage of lorries, a Traffic Regulation Order is required. The DOT has published advice for authorities wishing to implement traffic–calming schemes (DOT, 1993a [Se], 1993b [Sf] and 1994b) (see also Chapters 13 and 20).

25.10 Regulatory Controls

The Road Traffic Regulation Act 1984 (HMG, 1984) contains wide–ranging powers for local authorities to regulate traffic, either generally or in relation to particular classes of vehicle [NIm]. For many years, local authorities have taken powers under the Act to protect sensitive or weak parts of the highway infrastructure from the effects of heavy lorries, for example, prohibiting them from using weak or low bridges, using both regulatory and advance warning signs.

In 1987, the Government commenced a fifteen year programme to assess and, where necessary, strengthen bridges, in view of the impending introduction of 40 tonne lorries in the UK from January 1999 (under EC Directive 85(3)) and to ensure the network's continuing ability to accommodate, safely, the increasing demands of traffic growth. However, the programme is substantially behind schedule, on both trunk and local roads and many local authorities have expressed concern about the economic and environmental implications, if widespread weight restrictions have to be imposed on bridges, pending adequate funding for their strengthening.

Each year there are around 1,000 bridge strikes by overheight vehicles or loads. While many such strikes cause little damage, some cause great damage to bridge structures, posing serious hazards to rail and road users

and imposing heavy costs of delay and disruption. New legislation is likely to be introduced in 1997, which will require all lorries to carry a prominent notice in the cab indicating the overall travelling height of the vehicle (or, where relevant, its load) if over 3 metres. Special warning devices will be required on vehicles fitted with power–operated high–level equipment, such as lorry mounted cranes, skip carriers and tower wagons.

In 1973, the legislation was extended by the Heavy Commercial Vehicles (Controls and Regulations) Act 1973 (HMG, 1973) (which amended the Traffic Regulation Act 1967 – now RTRA 1984) to include environmental considerations. Regulatory signs are available to limit the weight, width and (less commonly) the length of vehicles in unsuitable areas (see Section 25.2).

While a Traffic Regulation Order must be used to reinforce physical measures, the aim should always be to achieve compliance by good design and drivers' understanding. Relying solely on statutory controls can impose a heavy burden on police manpower. The simplest form of restriction to affect heavy vehicles is a prohibition of on–street loading and unloading, as well as parking, introduced, usually, to maintain traffic flow on important routes. Where there are consequences for servicing frontage premises, such restrictions are usually introduced only in peak traffic hours. A high proportion of urban deliveries and collections take place in the morning, so it is important that loading and unloading bans are not unduly restrictive. They should be the minimum necessary to avoid peak–hour traffic disruption. Wherever possible, loading and unloading should not be prevented before 08.00, as the 07.00 hrs to 08.00 hrs period is critical for many urban delivery operations.

Overnight on–street parking of lorries is generally undesirable, particularly in residential areas. Under goods vehicle operator–licensing requirements, a licence must specify the operating centre(s) where vehicles specified in the licence are normally kept (see Section 25.3) [NIp]. Therefore, lorries returning to base at night should normally be parked off–street in their operating centre(s). However, lorries away from base overnight need alternative facilities.

Where authorities perceive a local problem with on–street lorry parking, some basic research should reveal whether the vehicles concerned are locally or remotely based. An informal approach to locally–based operators should disclose why the relevant vehicle(s) are not being parked at their operating centre. For away–from–base vehicles, provision of off–street parking facilities might be necessary. Such facilities could be developed either specifically for the purpose or, perhaps, as an adaptation of existing facilities such as surface car-parks. The basic requirements are good hard-standing, ease of manoeuvring for larger vehicles, good lighting and sufficient security to discourage vandalism and petty theft.

An alternative approach is to designate roads in appropriate non-residential areas for on-street lorry parking. Roads in industrial and commercial areas and trading and industrial estates can provide opportunities for this purpose.

In many older shopping streets, where rear servicing facilities are not available, conflict often occurs between delivery vehicles and other activities on the street, especially pedestrian movements. This has encouraged many local authorities to restrict access to such streets during certain times of the day, usually mid-morning to late afternoon, to coincide with the highest levels of pedestrian activity. Others have introduced full or partial pedestrianisation (see Chapter 22). The amenity benefits of pedestrianisation are self-evident, with the reduction in vehicular intrusion, noise and emission levels and a safer environment.

Studies have shown that pedestrianisation can increase trade and the perceived benefits of this increased trade, within a traffic-free environment, are often a key factor in the decision to pedestrianise a high street or town centre. However, the linkage between pedestrianisation and increased trade is not axiomatic and authorities should take particular care to assess any assumed economic benefits of pedestrianisation before implementation (CELRC, 1990). It is, in any case, vital to consider the implications for deliveries and other essential servicing activities as part of any scheme appraisal. For all but the smallest schemes, comparative surveys of pedestrian and delivery activity should be undertaken. These may show that deliveries can be allowed until mid-morning or even later, without significant vehicular/pedestrian conflict. Prohibiting deliveries earlier than this (Saturday excepted) is then unnecessarily restrictive, because a high proportion of deliveries may be disrupted. For such schemes to be successful, full consultation should take place with both traders and the operators of delivery vehicles (FTA, 1983).

Some existing premises can be adapted, to provide loading bays and service yards, and rear service-ways can sometimes be created with the consent and co-operation of occupiers and owners. For new industrial and town centre developments, see Chapters 30 and 12.

Intrusion by lorries is likely to be perceived as being at its worst in residential areas, where people might reasonably expect peace and quiet (see Chapter 17 on environmental management). Particular aggravation can occur at night or at weekends.

Removing heavy vehicles from residential areas may be achieved in a number of ways, the simplest being to prohibit them from a short length of a street, by reference to 7.5 tonne or 17 tonne maximum gross vehicle weight. This acts as a 'plug' and forces drivers to divert to a more suitable route (see Figures 25.4 and 25.5). While this technique can alleviate the problem of through traffic, using short-cuts through residential areas, it can be very disruptive to vehicles on local servicing activities, such as fuel deliveries and furniture removals. Sufficient advanced warning should be given, by erecting signs some distance from the 'plug' and showing the restrictions on advance direction signs. Controls which incorporate exemptions now usually specify 'except for loading', as this is easier to enforce than 'except for access'.

It has already been emphasised that lorry management controls should achieve compliance by good design and drivers' understanding. Poor design and drivers' failure to understand will result in poor compliance. Area-wide controls must therefore be

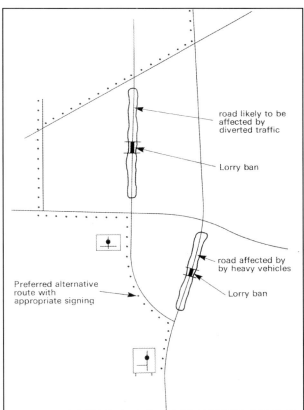

Figure 25.4: Layout of lorry controls to protect an individual road or a small area.

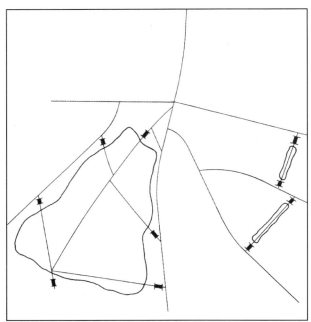

Figure 25.5: Layout of lorry controls to protect an environmental area and adjacent roads.

signed comprehensively to warn drivers sufficiently early for them to take sensible alternative route–decisions. For example, 17 tonne two–way 'plugs' could be located on all access routes to a sensitive town, thus forming a protective cordon. A permit scheme can then be introduced, to allow

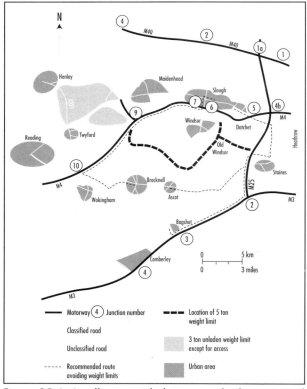

Figure 25.6: An all area–wide lorry control scheme around Windsor.

passage across the cordon for local firms and their suppliers. This measure can be effective in preventing through traffic and can reduce the exposure to heavy vehicle noise for those living in the area, while maintaining reasonable facilities for access and deliveries. A pioneering scheme at Windsor is illustrated in Figure 25.6 (IHT, 1981 and DOE *et al*, 1978). The diversion and enforcement costs of such a cordon approach may be significant and authorities should consider the level of environmental gain in relation to cost before embarking on such projects.

Superficially, the use of permits as a basis for administering lorry control–schemes is attractive and a number are in operation. However, this approach must be treated with caution, since it is administratively cumbersome, vulnerable to misuse and, if widely adopted, would create serious difficulty for operators having to acquire and display a multitude of permits for use in different areas. The original objective, in amending the Road Traffic Regulation Act 1984 to provide for exemption permits, was to cater for particular circumstances, where a limited number of specific vehicles required exemption from a general control. Permit schemes should be used accordingly. Developments in access–control systems using electronic tags may make permit–enforcement easier (see Chapter 18).

Where measures are introduced which have no element of self–enforcement, then it is probable that regular police enforcement will be needed to ensure effectiveness. Other priorities for police resources have meant that such measures tend, increasingly, to be abused. Consequently, in recent years, there has been much interest in the use of cameras to provide a higher level of enforcement presence. Automatic video camera–based schemes have been introduced successfully to enforce weight–limits, for example, at Barnard Castle and at Tilehurst in Berkshire.

25.11 Rail Freight and Inter–Modal Distribution

Rail freight has played a vital role in Britain's economy since the industrial revolution. Although overall traffic volumes have declined steadily, with the development of road freight services, the railways have retained a core of business consisting mainly of bulk raw materials, fuels, semi–finished products and containers, mostly related to imports and exports. In the financial year 1994/95, Britain's railways carried 97.3 million tonnes of freight. This would have generated in excess of four million extra lorry–journeys, if it had all been carried by road. Rail–freight volumes, although only accounting for about 6.5% of total freight movements, are nonetheless still significant, especially in terms of the heavy and bulky nature of the goods that the railways carry.

As emphasised in the introduction, it is not practical to reduce the overall number of lorries on the roads significantly by shifting freight to alternative modes, not even to the railways. The diffuse nature and relatively short distance of many freight movements make road transport the only practical option, particularly in urban areas. However, there is a general consensus, shared by industry, that all practicable means to move more freight by rail should be pursued. Transferring particular traffic to railways can have substantial local effects on the level of lorry traffic. The Rail–Freight Facilities Grant Scheme provides for government grants towards the capital cost of new or modernised facilities, particularly private sidings, handling equipment and railway wagons. These grants may help to make rail freight more competitive with road transport and could lead to a substantial reduction in lorry traffic locally but to only a marginal reduction overall. To qualify for a grant, it is necessary to demonstrate that a proposal would yield environmental benefits, would not otherwise be commercially viable and that, without the grant, the goods would go by road.

Rail freight can only be increased, sustainably, by services being developed which meet industry's needs and compete effectively both in price and in performance with other means of transport. This requires that the needs and expectations of customers are identified by rail–freight service providers and given a positive response. It also requires a stable business structure, free from political uncertainty (FTA, 1995).

There is some potential for inter–modal (combined) transport. However, in Britain, the potential for combined transport is limited by the large proportion of relatively short trip–lengths. Combined transport could be viable for relatively long north–south movements within Britain and, through the Channel

Tunnel, to more distant places on the Continent. To establish inter–modal freight distribution in urban areas, an entirely new infrastructure would be needed.

25.12 Abnormal Loads

The road layout and structure in most urban areas has been designed, or evolved, to cater for a certain maximum size and weight of vehicle. The normal size and weight limits are contained in the Road Vehicles (Construction and Use) Regulations 1986 (HMG, 1986) and it is unlawful to exceed these [NIa]. However, some loads are very large or heavy and cannot reasonably be divided up in either size or weight for the purposes of transport. These loads, therefore, may have to be transported on special vehicles, which are larger and heavier than normal lorries (see Photograph 25.4).

The Motor Vehicles (Authorisation of Special Types) General Order 1979 (the 'STGO') (HMG, 1979) authorises the use on the roads of specific kinds of abnormally large and/or heavy vehicles which, either laden or unladen, exceed the Construction and Use Regulations limits [NIa]. Such vehicles include abnormal load carriers, mobile cranes and large tipper lorries.

Under the Construction and Use Regulations and the STGO, advance notice must be sent to the police, who may consult the Local Highway Authority for advice on appropriate routes, for the movement by road of the following:
   ❑ a load projecting more than 305mm on either side of the vehicle;
   ❑ a vehicle or load exceeding 2.9m in width;
   ❑ a rigid vehicle or load exceeding 18.3m in length;
   ❑ a motor vehicle, trailer and load together exceeding 25.9m in length;
   ❑ a load projecting more than 3.05m over the front or rear of the vehicle; and
   ❑ a vehicle or vehicle–and–load exceeding 76,200 kg (75 tonnes) gross weight.

The STGO authorises use on the roads of specific kinds of vehicle which, either laden or unladen, exceeds the Construction and Use axle or gross weight limits, provided that all highway and bridge authorities along the route:
   ❑ have been notified in advance of details of the vehicle, its load and the route it intends to take; and
   ❑ are indemnified against any damage caused thereby and that the vehicle travels at a reduced speed on ordinary roads (not applicable on motorways) [NIr].

Photograph 25.4: An abnormal indivisible load.

Recommended routes for abnormal loads will not necessarily be the same for each load, as the choice will be constrained by physical features, such as bridges under and over the carriageway, constricted junctions and built–in street furniture. For especially large loads, the route may be extremely circuitous, in order to avoid these features, but local authorities need to be aware of the most appropriate routes for different types and sizes of load. These routes may be designated as especially appropriate for abnormal loads and opportunities may be taken to replace existing street furniture with designs which are easily demountable and, in the longer term, it may be beneficial to amend layouts which regularly present problems.

The informal status of a route set aside for abnormal loads should also be taken into account and not compromised when any junction improvements or maintenance schemes are designed. Loads requiring this type of consideration, although important for local industry, are unlikely to be frequent enough to create any particular environmental or road safety problems.

25.13 Hazardous Loads

Modern industrial processes increasingly require the transport of hazardous loads, not only within and between industrial areas but also to warehouses, stores and hospitals. The United Nations has a comprehensive system of classifying all dangerous goods under headings such as flammable, explosive, corrosive, toxic and radioactive. The relevant regulations governing the transport of such goods by road are listed under references HMG (1996a, b, c and d) [NIs].

A UK Hazard Information System (UKHIS) has been developed for dangerous goods tankers and tank containers. This is designed to give the emergency services all the necessary information in the event of an accident involving spillage. An important part of this is the HAZCHEM 'emergency action code', incorporated on the orange hazard–warning panels displayed on the sides and rear of lorries carrying such materials, which informs the emergency services of the nature of the hazards that could arise in the event of an incident, so that the appropriate action is taken promptly to minimise any possible effects (see Figure 25.7) (HMG, 1996a and 1996b).

Vehicles carrying dangerous goods in packages which are subject to the regulations (HMG, 1996a and 1996c) should display a plain orange plate to front and rear and the packages should also carry appropriate danger–warning signs. The circumstances in which this requirement applies are described in the Regulations (HMG, 1996c and 1996d) and in approved requirements and guidance provided by the Health and Safety Executive. If the vehicle is empty, the sign should either be removed or covered up.

The accident–rate for road–tank vehicles carrying hazardous loads is significantly less than that for ordinary goods vehicles (Gandham *et al*, 1982), although the potential for damage and environmental impact, if and when a serious accident does occur, is clearly much greater.

Department of Transport Circular Roads 1/87 (DOT, 1987) describes procedures which highway authorities should adopt to combat the dangers which may arise following a road accident involving hazardous substances. If an accident occurs to a vehicle carrying dangerous goods, the police and fire

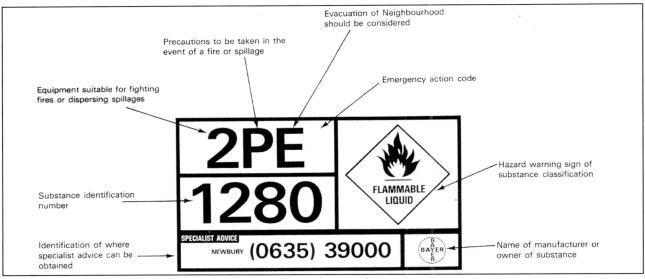

Figure 25.7: HAZCHEM plate for use on lorries carrying hazardous goods.

TRANSPORT IN THE URBAN ENVIRONMENT

brigade will normally be informed by the driver or, if he is unable to do so, by someone else at the scene. For bulk loads, the emergency services will obtain technical advice or assistance, when necessary, from the specialist advice telephone number shown in the HAZCHEM orange hazard information panel (see Figure 25.7). For packaged goods, the name and address of the manufacturer, importer, wholesaler or supplier of the substance will appear in the transport document and the driver is required to carry information, in writing, concerning the nature of the hazards created by the substance (if released) and the actions to be taken in an emergency.

Once the immediate emergency is over, a highway authority may be called upon to remove dangerous substances from the highway. The Highway Authority has a statutory duty under section 22 of the Control of Pollution Act 1974 (HMG, 1974) to remove any spillages from the highway, where such removal is necessary for the maintenance of the highway or for the safety of traffic [NIt]. Under section 140 of the Highways Act 1980 (HMG, 1980) [Sc], a highway authority also has power to remove anything deposited on a highway which constitutes a nuisance, danger or substantial inconvenience to users of the highway and to recover the cost of removing it from the person or firm responsible [NIu].

If a dangerous substance has to be removed from a highway and there is any doubt about the method appropriate and the safeguards to be taken, the Highway Authority should seek technical advice from the company concerned. In all cases, the emergency services should be consulted and informed of the actions to be taken, including (where necessary) the evacuation of people living in the vicinity while the highway is being cleared.

25.14 References

ALBES, DOT, and Metropolitan Police (1985)	'Code of Practice for Highways and Traffic Management in London', Association of London Boroughs Engineers and Surveyors.
CELRC (1990)	'How to Get Pedestrian Rental Growth', Colliers Erdman Lewis Research Consultancy.
DOE (1994)	EMP 1030/219/22 and EMP 1000/529/10, 'Application by Salvesen Brick Ltd, Mouselow Quarry, Glossop, Derbyshire', DOE.
DOE/DOT/TRL (1978)	Report SR 458 'Effects of Lorry Controls in the Windsor Area', DOE.
DOT (1987)	Circular Roads 1/87 (WO 7/88) 'Spillages of Hazardous Substances on the Highway', DOT.
DOT (1993a)	Circular Roads 2/93 (WO46/93) 'The Highways (Traffic Calming) Regulations 1993', DOT [Sa].
DOT (1993b)	Traffic Advisory Leaflet 7/93 'Traffic Calming Regulations', DOT [Sf].
DOT (1994a)	Circular Roads 4/94, Annex F 'Signing of Weak Bridges', DOT.
DOT (1994b)	Brochure 'Safer by Design – A Guide to Road Safety Engineering', DOT.
DOT (1995)	Leaflet 'Heavy Commercial Vehicle Drivers – Know Your Traffic Signs', DOT.
DOT (1996)	'Lorry Weights: a Consultation Document', DOT.
DOT et al (1990a)	'Lorries in the Community', DOT, Civic Trust, and County Surveyors' Society
DOT et al (1990b)	'Lorries and Traffic Management', DOT, Civic Trust, and County Surveyors' Society
FTA (1983)	'Planning for Lorries', Freight Transport Association.
FTA (1995)	'The Rail Freight Challenge – increasing rail freight by meeting customers' needs', Freight Transport Association.
Gandham B and Hills PJ (1982)	'Monitoring the movements of hazardous freight by road', Research Report 45, Transport Operations Reasearch Group, University of Newcastle upon Tyne.
HMG (1973)	'Heavy Commercial Vehicles (Control and Regulation) Act 1973', Stationery Office.

HMG (1974)   'The Control of Pollution Act 1974', Stationery Office.

HMG (1979)   'The Motor Vehicles (Authorisation of Special Types) General Order 1979', Stationery Office.

HMG (1980)   'Highways Act 1980 (and amendments)', Stationery Office [Sc].

HMG (1984)   'The Road Traffic Regulation Act', Stationery Office.

HMG (1986)   'The Road Vehicles (Construction and Use) Regulations' (as amended), Stationery Office.

HMG (1992)   'Traffic Calming Act 1992', Stationery Office.

HMG (1993)   'Highways (Traffic Calming) Regulations 1993' SI 1993 No. 1849, Stationery Office [Sd].

HMG (1994)   'The Traffic Signs Regulations and General Directions 1994', Stationery Office.

HMG (1995)   'The Goods Vehicles (Licensing of Operators) Act 1995', Stationery Office.

HMG (1996a)   'The Carriage of Dangerous Goods by Road Regulations', Stationery Office.

HMG (1996b)   'The Carriage of Dangerous Goods (Classification, Packaging and Labelling) and Use of Transportable Pressure Receptacles Regulations', Stationery Office.

HMG (1996c)   'The Road Traffic (Carriage of Explosives) Regulations 1996', Stationery Office.

HMG (1996d)   'The Radioactive Substances (Carriage by Road) (Great Britain) Regulations 1996', Stationery Office.

HMG (1996e)   'The Local Authorities Traffic Order (Procedure) (England and Wales) Regulations, 1996', Stationery Office [Sa].

IHT (1981)   'Guidelines: Lorry Management Schemes' The Institution of Highways & Transportation.

25.15 Further Information

DOT (1992)   Local Authorities Circular 5/92 'Traffic in London: Traffic Management and Parking Guidance', Stationery Office.

DOT (1995)   Local Authority Circular 1/95 'Guidance on Decriminalised Parking Enforcement Outside London', Stationery Office.

DOT (annual)   'The Transport of Goods by Road in Great Britain', Stationery Office.

DOT (annual)   'Transport Statistics Great Britain', Stationery Office.

FTA (1995)   Freight Matters 3/95 'Vehicle utilisation, making the most of every journey', Freight Transport Association.

FTA (1996)   Freight Matters 5/96 'Lorries in urban areas, delivering the goods and serving the Community', Freight Transport Association.

IHT (1990a)   'Guidelines for the Safety Audit of Highways', The Institution of Highways & Transportation.

IHT (1990b)   'Guidelines for Urban Safety Management', The Institution of Highways & Transportation.

IHT (1994)   'Guidelines for Traffic Impact Assessment', The Institution of Highways & Transportation.

Part
IV

Highway and Traffic Considerations for Development

Chapter 26 Transport Aspects of New Developments

26.1 Basic Principles

The emphasis on the need to restrain car–use and promote public transport is embodied in Government policy statements. Despite this emphasis, new developments still have an impact on local road traffic and will usually generate more private, rather than public transport, trips. The level of private vehicle trips can create congestion and affect the environment and it is therefore important to focus on the use of road-based transport and to assess the predicted impact of the scheme. In the design of new developments, it is vital to accommodate and encourage alternative transport modes, such as walk, cycle and public transport, but it is the impact of vehicular traffic which often causes most concern. It is sometimes possible to design–out the direct environmental effects of any noise or visual impact from within a development but more difficult to mitigate the external effects of vehicular traffic.

26.2 Planning and Policy Framework

A planning authority considers a planning application within the context of its approved development plan. That plan itself has to be consistent with the hierarchy of adopted policies and plans. These are: [NIa].
- ❏ Planning Policy Guidance;
- ❏ Regional Planning Guidance;
- ❏ Structure Plans; and
- ❏ Local Plans.

For metropolitan and unitary authorities, the Structure and Local Plans are replaced by a single Unitary Development Plan (UDP) [Nla].

Planning Policy Guidance (PPGs) notes are issued by Government as guidance on what local authorities have to take into account when preparing their development plans and are material to the consideration of individual planning applications [Nlb] [Sa].

Within the context of transportation planning in England [Wa], the key PPGs are:
- ❏ PPG1 – General Policy and Principles (DOE, 1992a) encourages developers and Local Authorities to consider the issue of access to the development of land and buildings [Sb];
- ❏ PPG4 – Industrial and Commercial Development and Small Firms (DOE, 1992b) [Sc] emphasises the locational importance of sites in terms of accessibility for staff and products;
- ❏ PPG6 (revised version) – Town Centres and Retail Development (DOE, 1996) [Sd] emphasises that retail developments should be accessible by a choice of modes of transport; this usually means locating them in, or next to, existing town centres or in other locations well served by public transport;
- ❏ PPG12 – Development Plans and Regional Planning Guidance (DOE, 1992c) [Se] includes sections on 'transport infrastructure', 'roads' and 'safeguarding transport routes';
- ❏ PPG13 – Transport (DOE/DOT, 1994) [Sf] [Wa] is concerned with land–use related transport and access policy. The key aims of the policy are:
    - ❏ to reduce growth in the length and number of motorised journeys;
    - ❏ to encourage alternative means of travel which have less environmental impact; and
    - ❏ to reduce reliance on the private car. Refer also to PPG13 – A Guide to Better Practice (DOE/DOT, 1995); and
- ❏ PPG15 – Planning and the Historic Environment (DOE, 1994) refers to the means by which highway authorities can institute measures to enhance and protect the historic environment and would apply to any highway or access schemes associated with new developments [Sg].

Regional Planning Guidance (RPG) is issued by Government in the light of advice prepared by local authorities and addresses strategic issues such as the provision of housing, employment and transport [Sh].

Structure Plans, prepared by County Councils [Sk], and Unitary Development Plans should be compatible with regional planning guidance to provide strategic planning guidance over a wide area. Local Plans, prepared by District Councils [Sl], cover a smaller area and are required to conform to the Structure Plan, provide the detailed planning framework and identify specific proposals and sites for developments. The combination of all of these plans and policies provides the basis upon which decisions are made regarding new development proposals. Since 1991, planning decisions must be in accordance with development plans, unless material considerations indicate otherwise [NIc]. This has established a plan–led system of development control.

Development proposals themselves are dealt with through applications to the Local Planning Authority for consent to develop [NId]. This can either be a consent in principle, with reserved matters to be determined at a later date (outline application), or a consent in detail, which will define the exact scale and nature of the development including, for example, the architectural form of buildings, means of access and details of landscaping (detailed application).

In preparing a planning application, a developer needs to be mindful of all guidance and policies which may affect a scheme and should consider the transport issues specifically. The documents supporting such an application should address these issues explicitly.

As PPG 13 [Wa] makes clear, it is important that local authorities develop their land–use and transportation strategies in parallel, ensuring that they are complementary. The relationship between the changing pattern of land–use, the cumulative impact of development and consequent demand for transport, and the changing pattern of transport provision are critical. Transport implications will be a significant factor in influencing local authorities' decisions on the identification of development sites for different purposes.

The PPGs establish a number of key transport objectives that should be considered in connection with new development. Amongst others, key questions to be addressed are:
- ❏ does the development aim to reduce car travel?;
- ❏ does it promote the use of non–car modes?; and
- ❏ does it promote existing town centres?

Where these issues are relevant, the supporting documents should seek to quantify the impact of the proposal. This may require the consideration of before and after scenarios or the comparison with alternative proposals. In many cases, it will not be possible to produce any detailed quantitative analysis but some form of qualitative assessment would generally be of value.

Whilst the process will continue to be heavily dependent on predictions of the growth and use of private cars, it is essential that new developments are planned to encourage other modes and to accommodate these safely from the beginning.

Consequently, it is as important to design for less–mobile users, pedestrians, cyclists and public transport users, as it is for private cars and service vehicles.

26.3 Transport Impact Assessment

Many planning applications are of a size or type that would generate additional trips on the adjoining transport infrastructure. This additional demand may necessitate changes to be made to the highway layout or to public transport services. Wherever possible, opportunities should be taken to provide direct access to public transport and to pedestrian/cycle infrastructure, thus helping to modify the overall transport impact.

The developer or promoter should provide a full and detailed assessment of how trips to and from the development might affect the highway network and/or public transport facilities. The transport impact assessment should be an impartial description of impacts and should include both positive and negative aspects of the proposed development.

Transport impact assessment addresses two related issues. These are:
- ❏ volume/capacity: what will be the effects of additional traffic on the safety and efficiency of the existing network?; and
- ❏ environment: what will be the effects of additional traffic in terms of noise, pollution and visual intrusion?

Impact assessments are now usually produced by developers in support of a planning application and the primary responsibility rests with the developer and not the Local Authority. In order to establish a common format for assessment, The Institution of Highways & Transportation, with the endorsement of the Department of Transport, has published Guidelines on Traffic Impact Assessment (IHT, 1994). The Guidelines recommend that a TIA should be produced where:
- ❏ traffic to and from the development exceeds ten percent of the average traffic flow over the same period on the adjoining highway; or
- ❏ traffic to and from the development exceeds five percent of the traffic flow on the adjoining highway, where traffic congestion exists, or will exist, within the assessment period, or in other sensitive locations

These thresholds should be applied in the absence of alternative guidance from the Highway Authority [NIe] in the form of an approved or adopted policy.

Based on guidance given in PPG13 [Wa], the Department of Transport requires a more rigid threshold for the trunk road network. The Department of Transport [Si] regard as material any increase in excess of five percent in most cases, although, where the capacity of the road is, or is near to being, exceeded, a smaller percentage increase may well be material and require a TIA to be produced. Where the adjacent road or junction is already at capacity, any increase at all in vehicular traffic will be material.

Traffic problems often relate to peak hours; therefore, the thresholds should be applied to these peak periods. However, it may also be appropriate to consider other time periods, such as all day or the peak periods of traffic generated by the development, if it is thought that the impact for such periods is likely to be of concern, for example Saturday shopping or Sunday tourism.

A TIA may be required even though the conventional threshold tests do not apply. An example might be where the percentage increase in vehicle numbers may be small but where most, if not all, of the additional vehicles are large goods vehicles, such as at a landfill site or quarry (the Department of the Environment issues specific Minerals Planning guidance relating to mineral extraction).

Furthermore, there will be developments so significant in size that TIAs should be undertaken as a matter of course. As a guide, proposals which are likely to attract additional traffic sufficient to warrant a TIA are:
  ❏ residential development in excess of 200 units;
  ❏ business (Land Use Classes B1 and B2) with a Gross Floor Area (GFA) in excess of 5000m² [Sm];
  ❏ warehousing (Class B8) with a GFA in excess of 10000m² [Sm];
  ❏ retail development with a GFA in excess of 1000m²;
  ❏ 100 trips in/out combined in the peak hour; or
  ❏ 100 off–street parking spaces, with a single access to the street network.

The same threshold approach should be used to establish the 'area of influence' of the development. Hence, the study area should include all links and associated junctions, where traffic to and from the development will be likely to exceed 10% of the existing traffic (or five per cent in congested or other sensitive locations) or such other threshold as may have been established by the local Highway or Planning Authority.

Prior to undertaking a full TIA, a scoping study should be carried out by the developer, in conjunction with the Planning and Highway Authority, to agree the key aspects to be addressed by the TIA. The scoping study should set out details of data to be collected, the area of analysis, key junctions to be considered, the methodology to be adopted and the years of assessment. Such a scoping study will provide a basis for assessing the level of resources that will be required to undertake the TIA. This scoping study will be invaluable to all parties involved and should ensure that work is not undertaken unnecessarily and that resources are directed to those aspects requiring most attention.

Approval in principle to establish a particular type of development is often sought by way of outline planning consent, before committing further time and resources to an application for detailed consent. The access arrangements for a site is one area of technical analysis where outline conceptual designs may not be sufficient to determine the practicality or safety of a scheme. An outline design often contains insufficient information to enable a highway authority to enter into an agreement with a developer, relating to the costs and layout of the access, and therefore needs to be treated with caution. If appropriate agreements are not determined at the outline stage, it may not be possible to reach a satisfactory outcome at the detailed application stage. Consequently, even with an outline application, sufficient details of the access should be provided.

It should be noted that the Secretary of State, as the Highway Authority for trunk roads and motorways, has powers to direct the Local Planning Authority to refuse an application on highway grounds, whereas the Local Highway Authority can only advise the Local Planning Authority, leaving to them to decide how to determine the application [Sj].

26.4 Trip Attraction

There are several databases which contain information about the level of traffic likely to be attracted to a development. A database allows a user to select existing developments similar to the proposed development and to examine traffic levels that occur at these sites.

However, there is normally a wide spread of trip–rate values even for similar developments and the reasons are not immediately obvious. The IHT Guidelines make the point that using a median value creates a forecast with a 50% chance of being exceeded. If car park size or junction capacity is to be derived from such values, there could be major risks associated with undersizing or under–designing such facilities.

Consequently, the Guidelines advise that it would be prudent to consider the design elements based on a trip–rate higher than the average. An 85th percentile value is recommended (ie, a value not exceeded by 85% of all values).

When traffic data is to be used for an environmental impact assessment (EIA), it is more appropriate to use the most likely estimate, that is, an average rather than the 85th percentile value.

The Guidelines recommend that assessments should be undertaken at the year of opening and for a year either 10 or 15 years later. Forecasts should relate specifically to the type of road, locality and time period being assessed. In most cases, the use of National Road Traffic Forecasts (NRTF), which are forecasts of the total annual vehicle–mileage in the country, is not likely to be appropriate.

Assessors will, therefore, need to consider local traffic trends, the availability of local forecasts, or applications derived from the National Trip–End Model, which is maintained by the Department of Transport. The National Trip–End Model has advantages over the National Road Traffic Forecasts, in that its forecasts are more locationally–based and provide an opportunity to sub–divide travel patterns by journey purpose.

For trunk roads, the Department of Transport [Wb] normally require an assessment period of 15 years after opening and may require consideration of a range of traffic conditions based on low and high forecasts of vehicular traffic growth.

26.5 Design Considerations

Proposals for new developments will include layouts of access roads, service yards and car parking. These are considered in Chapters 28, 29, and 30. Pedestrian access, facilities for cyclists and the design of public transport infrastructure, such as bus stops and shelters, are examined in Chapters 22, 23, and 24.

Where highway authorities require independent safety audits in support of proposals for new highway works associated with development proposals, these should be undertaken in line with IHT Guidelines for such audits (IHT, 1996) (see also Chapter 16).

The TIA should illustrate access to, and the internal layout of, the site and demonstrate how facilities are to be provided for disabled people, servicing traffic, pedestrian access, cyclists and public transport.

26.6 Environmental Considerations

Guidelines prepared by the Institute of Environmental Assessment (IEA, 1992) establish a range of procedures to evaluate environmental impacts. These techniques follow closely the procedures set out in Chapter 11 of The Design Manual for Roads and Bridges (DMRB). Further consideration of these issues is set out in Chapters 9 and 17.

26.7 Developers' Contributions

The IHT Guidelines on TIAs deliberately separate the issue of assessments from who should pay for infrastructure improvements [NIf]. Sections 27.4 and 27.5 set out in detail the rules covering instances where developer contributions may be sought to fund transport works. There are essentially four elements to this. These are:

❑ a developer should expect to fund works necessary to provide immediate access to the site;

❑ a new development may be forecast to attract traffic which would have a significant impact on local roads, for example, where roads around the development area are already operating at or near capacity. In these circumstances, the Local Authority might indicate that it would refuse planning permission, until such time as necessary works were implemented, leaving the developer to consider whether or not to provide a financial contribution to the work. As PPG 13 [Wa] makes clear, the assessment of what is necessary should include the scope for providing enhanced access by public transport, cycling and walking;

❑ where traffic from a development is likely to compromise the safety or free flow of traffic along the trunk or motorway network, the Secretary of State, as the Highway Authority for the aforementioned, has a responsibility to exercise his development control powers. Broadly, policy is not to allow direct access to motorways and to limit access to the strategic trunk road network, unless improvement works can be realised to offset the traffic effects of the development. Normally, developers are expected to pay for trunk road improvements necessary to ensure that conditions on the road, both at opening date and the appropriate design year (see Section 26.4), are no worse than they would have been if the development had not taken place (DOT, 1996); and

❑ where the development proposal provides less car parking than the Local Authority considers adequate, the developer might pay a commuted sum to the Authority to provide alternative parking and/or enhanced public transport.

It is common practice for authorities to seek some form of contribution to help finance the enhancement of public transport. This can take the form of improved infrastructure, such as bus stops, shelters and additional bus lanes, or contributions to improve the level of provision.

Local authorities can also request 'commuted sums' to pay, for example, for car parking away from the development site itself. It is also common for contributions to be requested to other community benefits. Local transport plans and development briefs are a good way of setting out likely requirements in advance of development applications. However, developers should not be asked to fund works which are not related to the impact of their own development; contributions sought should be necessary, relevant and reasonable.

Local authorities can use their development plan process to set down appropriate criteria governing when and where contributions to improved transport infrastructure might be required. Such criteria must meet the requirements set down by the DOE.

26.8 TIA Contents

The IHT Guidelines (IHT, 1994) set down a suggested format for the presentation of a TIA. Clearly many TIAs will not need to cover all of the items identified within the Guidelines' structure but authors should state clearly why particular issues are not considered to be relevant.

The document should be accompanied by a 'non–technical summary' similar to that required of an Environment Impact Assessment.

Table 26.1, on the following pages, sets out the suggested format of a TIA. This format is not meant to be prescriptive but is illustrative of an approach which has been found to be helpful.

Photograph 26.1: A typical example of new development.

26.9 References

DOE (1992a)	PPG1 Planning Policy Guidance 'General Policy and Principles', DOE.
DOE (1992b)	PPG 4 Planning Policy Guidance 'Industrial and Commercial Development and Small Firms', DOE.
DOE (1992c)	PPG12 Planning Policy Guidance 'Development Plans and Regional Guidance', DOE.
DOE (1994)	PPG15 Planning Policy Guidance 'Planning and the Historic Environment', DOE.
DOE (1996)	PPG 6 Planning Policy Guidance 'Town Centres and Retail Developments', DOE.
DOE/DOT (1994)	PPG13 'Planning Policy Guidance – Transport' DOE/DOT [Wa].
DOE/DOT (1995)	PPG 13 'Guide to Better Practice', DOE/DOT.
DOT (DMRB)	'Design Manual for Roads and Bridges', Stationery Office.
DOT (1996)	'Control of Development Adjacent to Trunk Roads: Guidelines for the Highways Agency', DOT.
IEA (1992)	'Guidelines on the Environmental Impact of Traffic from Developments', Institute of Environmental Assessment.
IHT (1994)	'Guidelines on Traffic Impact Assessments', IHT.
IHT (1996)	'Guidelines for the Safety Audit of Highways', IHT.

26.10 Further Information

Regional Planning Guidance issued by DOE
Structure Plans and Unitary Development Plans.

1 Non–Technical Summary

❏ a brief non–technical resume of the projected traffic impact of development

2 Existing Conditions

❏ description of current transport policies for the area (including DOT, Structure Plan, Local Plan, etc);
❏ quantification of current traffic flows on links and junctions within the affected area;
❏ examination of historic accident records, where appropriate;
❏ quantification of pedestrian flows at critical locations;
❏ identification of critical links and junctions;
❏ identification of committed highway works in the area; and
❏ identification of developments with planning consent but not yet implemented.

3 Proposed Development

❏ description of current planning policies for the site including parking guidelines;
❏ description of current use of the site and its recent usage history;
❏ description of proposed use, including site area and development phasing;
❏ specification of size of the development; and
❏ provision of site plan for proposed development, where available.

4 Modal Choice/Trip Attraction

❏ quantification of current trip attraction of the site;
❏ estimation of projected modal split;
❏ estimation of trip attraction, specified by direction and vehicle type, for:
  ❏ weekday;
  ❏ peak hour; and
  ❏ development peak;
❏ justification of the values used;
❏ identification of times when traffic impact is at its greatest, i.e. the peak combination of network and development traffic;
❏ for multi–purpose sites, provision of details of each significant element;
❏ specification of trip attraction by phase (if appropriate); and
❏ specification of trip attraction by construction period (if appropriate).

5 Trip Distribution

❏ definition of catchment area;
❏ consideration of competing opportunities;
❏ identification of transfer trips, i.e. the trips previously attracted to an alternative site;
❏ identification of non–primary trips, ie 'pass–by' and 'diverted' trips that might already be on the network;
❏ distribution of trips to potential opportunities; and
❏ justification for the methodology adopted.

6 Assignment of Development Traffic

❏ identification of traffic routeing to and from the site;
❏ definition of turning movements at the site entrance; and
❏ provision of modified traffic projections at key links and junctions within the affected area.

7 Assessment Years

❏ estimation of traffic growth over time for:
  ❏ network traffic; and
  ❏ development traffic;
❏ estimation of traffic flows on the adjacent links and at key links and junctions within the affected area for:
  ❏ base year, i.e. first year of full operation; and

❑ base year plus 10 years; or,

    ❑ year of completion of infrastructure plus 15 years, if a new or modified highway infrastructure is required;

❑ inclusion of committed highway and development proposals that affect local traffic conditions; and

❑ possible requirement for additional separate assessments for specific phasing proposals and for construction traffic impacts.

8 Highway Impact

❑ indication of the proposed site access layout;

❑ justification of the design;

❑ traffic assessments on other key links and junctions within the affected area;

❑ identification of reserve capacity and queue lengths, where appropriate;

❑ identification of alternative designs for key links and junctions within the affected area which may be necessitated by the increased traffic movements;

❑ identification of any departure from design standards; and

❑ safety assessment of all designs.

9 Environmental Impacts

❑ identification of the environment impact arising from the traffic consideration of the proposed development;

❑ special consideration required for sensitive and residential areas; and

❑ consideration of measures that might be appropriate to mitigate against any environmental disadvantage.

10 Road Safety

❑ examination of historical data for accident factors, trends and groups, for example, regular occurrence of one type of accident or involvement of one type of road–user; and

❑ preparation of a safety audit on any proposed change to the highway layout.

11 Internal Layout

❑ definition of internal road and circulatory layout, with dimensions and plan;

❑ consideration of service and emergency vehicle routes;

❑ definition of aisle widths, road marking, traffic safety, visibility, etc; and

❑ consideration of vehicle speed–control measures.

12 Parking Provision

❑ determination of level of provision and justification;

❑ consideration of essential operational, visitor, disabled spaces;

❑ specification of bay and aisle dimensions and location of spaces;

❑ verification that vehicles can access each space with adequate turning provision; and

❑ determination of service area requirements.

13 Public Transport

❑ indication of intended public transport provision;

❑ determination of siting of bus stops, routes, etc; and

❑ determination of access to bus/rail facilities.

14 Pedestrian/Cyclists/People with Disabilities

❑ indication of specific provisions;

❑ indication of safety and security provisions; and

❑ indication of facilities for disabled.

Table 26.1: TIA Format. Source IHT, 1994.

Chapter 27 Development Control Procedures

27.1 Introduction

Over recent years, the transport focus has changed from an assumption of unrestrained traffic growth to a realisation of the need for all modes of transport to play their full part. The role of development control is uppermost in ensuring that development proposals comply with prevailing planning guidance and making appropriate provision for access by all transport modes without undue environment impact. The transportation aspects of development control procedures consequently form a key part in the decision-making process regarding any development proposal.

Professionals involved in the process, whether engaged by a developer, a local planning authority, an interested party or an objector, require appropriate technical expertise and need to be familiar with the relevant planning and highway legislation and associated procedures.

27.2 Land–Use and Transport Implications

A change in land-use will often have transport implications, particularly the need to take into account its interaction with neighbouring land–uses and its consequent effect on the adjacent transport system. The role of professional advisers on transportation aspects is to provide a technical and procedural appraisal of the transport implications of the proposed development. The assessment of the traffic impact of any proposed development should identify the potential transport problems and benefits that would result from the development and any associated traffic management measures and highway alterations necessary to alleviate them.

The transportation aspect of development control is usually only one of a number of considerations that relate to a development proposal. It is therefore important for the professional adviser, whoever he or she is representing, to appreciate all the other material issues and to blend the transportation advice to ensure that the development provides a balance of benefits to all the interested and affected parties.

Transportation aspects which need to be considered include:

❏ assessing the viability of altering the access to a particular land-use, or analysing an existing land-use, to identify potential improvements to the access arrangements;
❏ advising on appropriate access arrangements for the proposed land-use, with particular regard to public safety and convenience;
❏ advising on the need for off-site improvements to off-set the effects of the development;
❏ providing a procedural programme for transportation aspects of the development; and
❏ managing the access arrangements of the potential land–use from concept to completion.

In any of these circumstances, the adviser's role is to provide a technical appraisal that is professional, unbiased and presented in a concise and easily interpreted format, together with a realistic procedural programme.

The technical appraisal for a land-use can be considered in three parts. These are:
❏ the overall transport and environment implications;
❏ the impact on the adjoining highway network; and
❏ the internal site access arrangements. (see photograph 27.1)

Photograph 27.1: Internal site arrangement for access.

The IHT Guidelines on Traffic Impact Assessment (IHT, 1994), provide a systematic means of assessing the transport impact of developments that potentially have a significant effect on the highway network. The document recommends minimum thresholds relating to the scale of those developments which ought to be subject to a Transport Impact Assessment (TIA). Carrying out a TIA is an appropriate discipline and procedure for the technical appraisal of development proposals (see Chapter 26).

A particular responsibility that falls on the adviser is to achieve adequate and safe access arrangements, without prejudice to the safety of the users of the adjoining public highway. In this respect, it is suggested that, once a scheme for the development access has been drawn up, it is tested for its user-friendliness to all categories of vehicles, pedestrians, cyclists, public transport users and the mobility impaired, and is subject to a safety audit (see Chapter 16).

An aspect of development traffic impact that may require special attention is construction traffic, which may require temporary access works to facilitate the delivery of materials and/or plant. It may be possible to reduce the scale of access once construction is complete and this should be taken into account in the access design, particularly for schemes in environmentally sensitive areas.

Development schemes in environmentally sensitive areas, such as conservation areas, areas of natural beauty, landscape areas and traditional villages require particular care in respect of materials and environmental aspects of the layout. A survey of the locality will reveal the materials and features that provide the local distinctiveness, which should be considered in the development of the scheme by all the parties concerned with the project.

The development of land is controlled by legislation and the associated national and local policies will indicate whether the development can progress and the manner in which it should proceed (see Chapter 26). The transport supply and access aspects of a development are usually fundamental considerations which need to be resolved, at least in principle, before a developer can be optimistic about his scheme and prior to the submission of any formal planning application. A realistic and successful developer is, therefore, likely to consider engaging an experienced engineer as part of his professional advisory team from the outset. The Local Planning Authority will, through the prescribed planning application consultation process, normally involve the Local Highway Authority [Sa].

For a highway engineer to make a valuable contribution to a developer's advisory team or to the Local Highway Authority [Sa], it is important to understand the legislative framework pertaining to development control.

27.3 Legislative Framework, Roles and Responsibilities

Several Acts of Parliament and associated Statutory Instruments relate directly to development control

and thus provide the legal framework within which developers and local authorities are obliged to perform. These Acts and Statutory Instruments are usually accompanied by Department of the Environment Circulars and Planning Policy Guidance Notes (PPGs) [Sb], as a means of providing additional guidance and interpretation (see Chapter 26) [Sc].

The principal Acts are the Town and Country Planning Act 1990 (HMG, 1990a) [Sd] [NIa], as amended by the Planning and Compensation Act 1991 (HMG, 1991a), which is intended primarily to improve the efficiency of the planning system and to make the compensation code fairer, and the Highways Act 1980 (as amended) (HMG, 1980) [NIb] [Sd]. These Acts of Parliament and the associated Statutory Instruments, the Planning Policy Guidance Notes [Sc] [NIc] and the relevant Circulars, have implications for transport aspects of development proposals which are summarised below.

The Town and Country Planning Act 1990 [Sd] and the Highways Act 1980 [Se] between them define, in Part I, the terms 'Local Planning Authority' and 'Local Highway Authority' [Sa].

In a two-tier county and district council arrangement, both the County Planning Authority and the District Planning Authority can be construed as local planning authorities, except where there is specific provision to the contrary [NId]. The Local Highway Authority for all highways, except trunk roads, is the County Council [NIe] [Sa]. A unitary authority is both the Local Planning Authority and the Local Highway Authority for all highways except trunk roads. The Secretary of State for Transport [Sf] [Wa] is the Highway Authority for trunk roads, which are managed by the Highways Agency [NIe] [Sg] [Wb]. The Highways Agency uses agreements with highway authorities or with private sector commercial companies, for the purposes of trunk road maintenance, improvement and development control (DOT, 1988) [Sh].

The Town and Country Planning Act 1990 [NIa] [Sd]

The Act requires the various planning authorities to prepare Development Plans, Structure Plans [Si], Unitary Development Plans, District-wide Local Plans [Si] and Subject Plans, such as County-wide Waste Disposal Plans and Mineral Plans. Planning applications must be determined in accordance with the Development Plan, unless material considerations indicate otherwise. The Secretary of State has powers under the 1990 Act to 'call in' and determine planning applications for developments that are not in accord with the appropriate Development Plan [Sj]. It is important for developers and their advisers to appreciate the implications of their development proposals and to take into account the procedural consequences, as must the Authorities in their discussions with applicants.

Section 57 of the 1990 Act [NIf] [Sk] states that '...planning permission is required for the carrying out of any development of land' subject to certain specified qualifications. 'Development' is defined in Section 55 but excludes the maintenance or improvement of roads. [Sl].

Section 91 of the Act [NIg] [Sm] specifies that every planning permission granted, or deemed to be granted, shall be subject to a condition that the development must be commenced within five years of the grant of consent.

Section 92 [NIh] [Sn] provides a further restriction with regard to outline planning permission. The application for approved and reserved matters must be made within three years of the grant of outline consent and the development must commence not later than two years after the final approval of reserved matters.

Part VIII of the Act [NIi] [So] imposes, on developers and local planning authorities, obligations regarding the preservation and planting of trees as part of a development scheme. This is particularly relevant to any highway or access scheme that affects existing trees or, as a consequence, provides lands, such as embankments or cuttings, that would be suitable for tree-planting.

Part IX of the Act [NIj] [Sp] allows a local authority to acquire land required for a purpose which is necessary to achieve proper planning of the area. In this circumstance, a developer and his advisers, together with the Local Authority, need to study carefully the associated case law to ensure that this power of purchase can be exercised and that the appropriate procedures and arguments of justification are applied. This provision could be considered where a modest alteration to a public highway is necessary to provide access to a potential development site.

Part X of the Act [NIk] [Sq] provides for the necessary orders and procedures for developments that involve the stopping up or diversion of a public highway . This section of the Act applies where the stopping up or diversion of any highway is necessary to enable development to be carried out, in accordance with planning permission granted under Part III of the Act [NIl] [Sr].

An application for the stopping up or diversion of a footpath or bridleway is usually made to the relevant Local Planning Authority and for other public highways the application is made to the Secretary of State for Transport [NIm] [Sf] [Wc]. As part of the stopping up/diversion procedure, the Local Authorities, in whose areas the particular highway is situated, have to be consulted, together with any affected statutory undertaker. Any resultant objection will cause the Secretary of State to hold a Local Inquiry.

The draft Stopping Up/Diversion Order is advertised and objections considered valid by the Secretary of State, particularly from persons likely to be affected directly, would also cause a Local Inquiry to be held. A provision for the Secretary of State or a local highway authority to acquire, compulsorily, the lands necessary to achieve a public highway stopping up/diversion Order is included in the Act. A developer and his highway adviser ought to establish, as soon as possible, whether stopping up or diversion Orders are necessary, since the procedures can seriously affect the development programme. In this circumstance, it is prudent for any planning application to include all the highway and access details and for the highway adviser to negotiate this aspect of the development scheme with all the parties that could cause a Local Inquiry to be held.

Schedule I of the Act [Ss], ' Distribution of Functions', sets out the various roles and duties of planning authorities within the development control process. In particular, paragraph 1 defines 'County Matters' as being the types of development for which a planning application would be submitted to, and determined by, a county planning authority. Paragraph 1(1) (a) specifies the categories of mineral development that are defined as a County Matter and paragraph 1(1) (j), by virtue of the Planning (Consequential Provisions) Act 1990 (HMG, 1990b), prescribes as County Matters the use of land, or the carrying out of operations, for the deposit of refuse or waste materials or the erection of any building, plant, or machinery for the purposes of disposing of refuse or waste materials. In the two–tier situation, a planning application for development, not defined as a County Matter, is submitted to the District Council [St] [Wd].

The Highways Act 1980 (As Amended) (HMG, 1980) [NIb] [Se]

Various provisions in the Highways Act will influence or facilitate development works associated with the public highway or the creation of a new highway.

Sections 4 to 9 – Agreements Between Authorities [Su].

Section 4 provides for an agreement between the Secretary of State and a local highway authority, where it is proposed to improve a trunk road and the works affect another highway under the Local Highway Authority's control. This procedure needs to be discussed with all the parties concerned and allowed for in any trunk road improvement programme. Section 5 provides for an Agreement to allow a local highway authority to maintain and improve a highway (other than a trunk road) which the Secretary of State has procured. This again may influence a highway improvement scheme programme. Sections 6 to 9 relate to the delegation of certain functions with regard to trunk roads and metropolitan roads and also facilitate agreements between local highway authorities for the transfer of the specified functions for a particular highway [NIe].

Section 38 – Powers of Highway Authority to Adopt by Agreement [NIn] [Sv].

A highway authority may undertake to maintain a highway by agreement with any person liable for the maintenance of that highway. The terms of the Agreement can include for all the Highway Authority's costs to be met and for it to relate to the construction of a new highway or the making up of a private street.

Section 59 Recovery of Expenses due to Extraordinary Traffic [NIo] [Sw].

A highway authority may recover any excess expenses incurred in the maintenance of a highway that is caused by excessive weight passing along the highway or other extraordinary traffic thereon.

Sections 116 to 123 – Stopping Up and Diversion of Highways [NIp] [Sx].

Section 116 allows a highway authority to make an application to a Magistrates' Court for the stopping up or diversion of a highway that is under their jurisdiction (see also Section 27.3 – Part X of the Town and Country Planning Act 1990). The particular highway has to be shown to be unnecessary or to be being diverted to make it more commodious to the public. Section 117 allows a person who desires a highway to be stopped up or diverted to request the Highway Authority to use its powers under Section 116. Section 118 permits a Council to make a Public Path Extinguishment Order for the stopping up of a footpath or bridleway on the ground that it is not needed for public use, which, if objected to, has to be confirmed by the Secretary of State. Where it is necessary to stop up or divert a public highway to

enable development to take place it is appropriate to use the Town and Country Planning Act provisions and procedures. Section 119 is a similar provision for the diversion of a footpath or bridleway, where the owner, lessee or occupier of the land that contains the footpath or bridleway satisfies the Council that the stopping up and diversion are expedient.

Sections 124 to 129 – Stopping Up of Means of Access to a Highway [NIq] [Sy].

A highway authority may make an Order to stop up a private means of access from the highway to any premises if they consider it to cause danger to, or to interfere unreasonably with, traffic on the highway. In certain circumstances, the Order may need to be confirmed by the Secretary of State. Section 127 allows for an agreement between the Highway Authority and the occupier of premises (or any other person with an interest in them) to stop up the private means of access to the premises. These sections of the Act contain specific provisions to protect those affected by such an Order and include for the payment of compensation where appropriate.

Section 184 – Vehicle Crossings Over Footways and Verges [NIr] [Sz].

A highway authority is required by this section to ensure that any vehicular access onto a public highway, and across a verge or footway associated with the highway, provides safe access and egress and facilitates the passage of vehicular traffic on highways, as far as practicable.

Sections 219 to 225 – The Advance Payments Code [NIs] [Saa].

The erection of a building with a frontage onto a private street may result in liability for the payment, to the Street Works Authority, of such a sum of money, or equivalent security, that provides for any potential street-work costs regarding that frontage. The term frontage relates to all the building plot boundaries that abut an existing or proposed street. Section 219 contains a schedule of exemptions which includes the normal circumstance where the existing or proposed private street is the subject of an agreement with the Street Works Authority, under Section 38, for the making up and adoption of the private street. It is, however, important for a developer, or his adviser, to check with the Street Works Authority whether there is likely to be such a liability regarding his building works and whether a Section 38 Agreement or other exemption would apply.

Section 278 – Contribution Towards Highway Works By Person Deriving Special Benefit from them [NIt] [Sab].

A highway authority is permitted to undertake

highway alteration works under an agreement with, and either partially or wholly at the expense of, a person who would derive a special benefit from them. The section specifically permits the Highway Authority to be paid for the maintenance of the works to which the Agreement relates. An agreement under this section is only permitted if the Highway Authority is satisfied that it will be of benefit to the public.

The Planning and Compensation Act 1991 (HMG, 1991a)

Section 12 of the Planning and Compensation Act 1991 [Sac], entitled 'Planning Obligations', amended Section 106 of the 1990 Act to enable a planning obligation to be entered into by means of a unilateral undertaking by a developer, as well as by agreement between a developer and a local planning authority [NIu]. The associated Circular, No. 16/91 (DOE, 1991) ,advises that unilateral undertakings are not intended to replace the use of agreements and expects them to be used principally at planning appeals where there are planning objections, which only a planning obligation can resolve, in respect of which the parties cannot reach agreement. Where a developer offers an undertaking at appeal it will be referred to the Local Planning Authority to seek their views. In such cases, it is particularly important that the developer's technical advisers measure their advice carefully, taking account of the relevant statutory instrument, ie The Town and Country Planning (Modifications and Discharge of Planning Obligations) Regulations 1992 (HMG, 1992a).

The Permitted Development and Procedure Orders

The 1988 General Development Order has been superseded by two Orders, namely the Town and Country Planning (General Permitted Development) Order 1995 (HMG, 1995a) [Sad] and the Town and Country Planning (General Development Procedure) Order 1995 (HMG, 1995b) [Sae]. Guidance is given in the associated Circular 9/95 (DOE, 1995a) [NIv] [Saf].

Procedure Order [Sag]

The procedure order sets out the procedural requirements for planning applications and appeals and planning registers and related matters and defines the roles and obligations imposed on the applicant, the Planning Authority, and the statutory consultees.

Article 1 confirms the principal reserved matters in relation to an outline planning application as siting, design, external appearance, means of access, and landscaping of the site.

Article 3 allows an applicant for outline planning consent to reserve any, or all, of the five reserved matters above and for the Local Planning Authority to grant permission, subject to conditions identifying reserved matters for future submission and approval.

The Planning Authority is permitted to identify any, or all, of the five reserved matters above for consideration as part of the outline application and has one month after the receipt of the application to notify the applicant what further details they require.

Article 10 provides a tabulated schedule of categories of development and the person or authority that needs to be consulted on a particular category of development. The schedule of consultees not only imposes an obligation on the Local Planning Authority and the consultee but also provides an applicant and his advisers with a list of persons or authorities who need to be involved in any discussions concerning the planning application.

Article 15 provides the Secretary of State with the ability to direct a local planning authority to refuse planning permission or impose specified conditions on a grant of planning permission for development affecting certain existing and proposed highways, which includes trunk roads. A local highway authority has no powers of direction and makes recommendations in response to a local planning authority's consultation.

Article 20 provides the time–period within which a local planning authority is obliged to notify the applicant of their decision. This is eight weeks from the date of receipt of the application, unless an extension to the determination period has been agreed with the applicant. This also relates to approvals of reserved matters. In the circumstance where an Environmental Assessment was submitted with the application then the permitted period of determination is extended to sixteen weeks [Sah].

Permitted Development Order [Saf].

The Permitted Development Order sets out a comprehensive schedule of development that is permitted without the need for a planning application. Two overriding preclusions regarding permitted development, which are particularly relevant, are that:

❏ any development for which an Environment Assessment is required does not benefit from permitted development rights; and

❏ the permission granted by the schedule of permitted developments does not authorise any development which requires, or involves the formation, laying out or material widening of, a means of access to an existing highway, which is a trunk or classified road, or creates an obstruction to the view of persons using any highway used by vehicular traffic, so as to be likely to cause danger to such persons. The exceptions to this relate to maintenance or improvement works in a private street, development under local or private Acts or Orders, development by local highway authorities and toll roads.

In addition, Part 13 of the permitted development schedule includes: '...the carrying out by a local highway authority, on land outside but adjoining the boundary of an existing highway, of works required for, or incidental to, the maintenance or improvement of the highway'.

Town and Country Planning (Use Classes) Order 1987 and Circular 13/87 [Sai]

It is necessary for a planning applicant to be aware that the proposed land-use will usually fall within a particular Use Class as defined in the Town and Country Planning (Use Classes) Order 1987 (HMG, 1987a), with guidance in Circular 13/87 (DOE, 1987) [NIw]. The Town and Country Planning Act 1990 excludes from the definition of development any change of use where both the existing and proposed uses fall within one Use Class. Part 3 of Schedule 2 of the Permitted Development Order 1995 also provides a table of change of use between certain Use Classes that are permitted development. It is particularly important for an applicant and his advisers and the Local Planning Authority and its consultees to appreciate this, since a potential development site and its access may be suitable for the intended use but may not be suitable for another use within the relevant Use Class. An obvious example is Class A3 , 'Food and Drink', which includes cafes, restaurants and public houses. A site may be adequate in terms of access and car parking for a cafe but not a public house. In this circumstance, negotiations can be carried out with the Planning Authority to achieve a planning consent for the proposed use that excludes other inappropriate uses within the same Use Class [Saj].

Circular 11/95 The Use of Conditions in Planning Permissions, (DOE, 1995b) [Sak]

The conditions imposed on any grant of planning permission by the Local Planning Authority are important both to the developer and to the Planning Authority and their consultees. The aspects of the advice in the Circular that are particularly relevant to a developer's highway and access considerations are set out below [NIx].

Outline Permissions (Paragraph 43)

An applicant who proposes to carry out building operations may choose to apply either for full planning permission or for outline permission, with one or more of the reserved matters reserved by condition for subsequent approval by the Local Planning Authority. An applicant cannot seek outline planning permission for a change of use alone or for operations other than the proposed building operations.

As a general rule, an outline planning approval establishes the feasibility of the proposed development and, in that respect, an adequate means of access needs to be proven before outline planning permission is granted. It is therefore recommended that an outline planning application includes full access details.

Conditions Affecting Land Outside the Application Site

It is usual for conditions to relate to the lands within the application site or to other lands under the control of the applicant. The Courts have, however, held that there can be circumstances, in particular cases, where only such control is necessary as is required to enable the developer to comply with the condition.

Informal Discussions

Pre-planning application discussions between the Planning Authority and the applicant are encouraged and ought to involve the applicant's professional advisers and the Highway Authority's representative.

Six Tests for Conditions

As a matter of policy, conditions should only be imposed where they satisfy the test as to whether they are: necessary; relevant to planning; relevant to the development to be permitted; enforceable; precise; and reasonable in all other respects. The circular provides detailed guidance on each element of this test. This is an important consideration to a highway authority, when recommending conditions for imposition on a grant of planning consent. The test is also important for a developer's highway adviser, when advising on the implementation of any conditions, as to whether they should be subject to discussions with the Local Planning Authority or the subject of a planning appeal.

Highway Conditions

Paragraphs 64 to 70 of the Circular particularly refer to highway conditions. Paragraph 71 refers to lorry routeing and Paragraph 72 advises that '...conditions may not require the cession of land to other parties, such as the Highway Authority'. The dedication of land to another party, such as the land necessary to provide the widening of an existing public highway, is normally achieved through an agreement with the relevant Highway Authority.

Negative Conditions

Paragraphs 38 to 41 of the Circular, titled 'Conditions Depending on Others Actions', refer specifically to the use of negative conditions (ie a certain stage in the development programme cannot progress until a particular identified aspect of the development proposals has been achieved). For example, 'No other development works shall commence until the access from Bickington Road has been constructed in accordance with the approved plan number XYZ'. Reason: In the interests of public safety.

Annex C of Planning Policy Guidance Note 13 'Transport' (DOE/DOT, 1994) [Sal] [Wf] provides more comprehensive advice regarding conditions requiring works in the highway. It is fundamental that '... planning permission cannot be granted subject to conditions which specifically require works on land outside the application site and outside the control of the applicant'.

Planning Appeal Rules and Regulations

General

An applicant can appeal to the Secretary of State [NIg] in the event of:
- ❏ a planning application not being determined within the statutory period;
- ❏ a refusal of planning permission;
- ❏ a condition, or conditions, imposed on a grant of planning permission that is not acceptable to the applicant;
- ❏ a refusal to approve a submission of a reserved matter application; or
- ❏ a planning enforcement notice.

Appeals are determined using one of three procedures. These are:
- ❏ an exchange of written representations;
- ❏ a Hearing; or
- ❏ a Public Inquiry.

Written representations and Inquiry appeals are governed by statutory rules and regulations as follows:
- ❏ Town and Country Planning (Appeals) (Written Representations Procedure) Regulations 1987 (HMG, 1987b) [Saw];
- ❏ Town and Country Planning (Enforcement Notices & Appeals) Regulations 1991 (HMG, 1991b) [San];
- ❏ Town and Country Planning (Enforcement)(Inquiries Procedure) Rules 1992 (HMG, 1992b) [Sao];
- ❏ Town and Country Planning (Inquiries Procedure) Rules 1992 (HMG, 1992c) [Sap]; and
- ❏ Town and Country Planning Appeals (Determination by Inspectors) (Inquiries Procedure) Rules (HMG, 1992d) [Saq].

Hearings appeals are subject to a non-statutory Code of Practice, set out in Annex 2 to DOE Circular 10/88 (DOE, 1988) [Sar].

Timetable

All concerned with planning appeals should study carefully the relevant rules or regulations, in order to be familiar with the timetable for evidence submission and the necessary lines of communication. The Inquiries Procedure Rules also specify whether a summary is required with a proof of evidence.

Town and Country Planning (Fees for Applications and Deemed Applications) Regulations 1989 (As Amended) (HMG, 1989) [NIz] [Sas]

These regulations provide the basis on which planning application fees are assessed. The fees are payable to the Planning Authority that is to determine the application.

Circular 31/92 (DOE, 1992) [Sat]

Paragraphs 24 and 25 of Circular 31/92 advise that the payment of the planning application fee should accompany the planning application when it is lodged. Although all valid planning applications will be registered, the eight week determination period does not commence until the correct fee has been paid and honoured. Paragraph 31 of the Circular defines the site area, as the area to which the application relates, and is normally shown edged red on the plan accompanying the application. Other land in the same ownership, but not being developed, is normally identified separately by blue edging.

Planning Charters

Although the advisory booklets Development Control: A Charter Guide (DOE, 1993) and Planning: Charter Standards (DOE, 1994) [NIaa] [Sau], published jointly by the National Planning Forum, the Department of the Environment and the Welsh Office, apply principally to local planning authorities, there are implications for statutory consultees, such as the Highway Authority and the applicant/developer and his technical advisers.

Duties and Obligations

The duties imposed on those involved in the highway aspects of development control, as particularly inferred by the legal framework, are set out below.

The professional adviser representing the applicant/developer has a duty:

❑ to provide clear, concise and unbiased technical advice to the applicant/developer at each stage of the procedure and to ensure that the client complies with all relevant legislation. The advice ought to highlight the likely difficulties, disadvantages and consequences of the proposed development in transport terms and propose the means of mitigation to offset identified problems;

❑ to encourage the applicant/developer to provide sufficient design/investigation funding and time, before a planning application is lodged, to allow a proper assessment of the development scheme and its traffic/highway implications. The applicant/developer should also be persuaded to facilitate pre–planning application discussions with the Local Planning Authority, the Local Highway Authority, and any other affected consultee; at this stage the highway/access design brief and Transport Impact Assessment parameters should be agreed;

❑ to represent the client's interests by establishing and developing aspects of the scheme that are acceptable to the Planning and Highway Authorities and comply with the recognised standards, consistent with the codes of practice of the Professional Institutions;

❑ to analyse the highway and access aspects of the scheme throughout the planning application process, taking into account all the various alternatives and responses. In this respect, the adviser needs to investigate and resolve the procedures necessary to achieve the scheme or to optimise the chances of the scheme being successful;

❑ to endeavour to achieve a transport access scheme that meets the ultimate site operator's requirements, both from the point of view of users and employees, with the minimum disruption and maximum benefit to the public at large, and all to a realistic and reasonable extent; and

❑ to discuss all procedural and technical matters with the client and the other members of his team and to negotiate in a fair open-minded manner with the Local Planning Authority, the Highway Authority and the consultees.

The Applicant/Developer has a duty:

❑ to propose development which minimises the impact on the environment and is suitable for its purpose;

❑ to set a realistic programme for achieving each stage of the procedures necessary to achieve the development, in discussion with the technical advisers and the Planning/Highway authority, and

to be willing to revise the programme to accommodate unforeseen delays;

❑ to accept professional advice from the technical advisers and to make any recommended adjustments to the scheme; and

❑ to comply with the relevant legislation and to follow the advocated and recognised development control procedures.

The Planning and Highway Authority [Sa] has a duty:

❑ to act within the procedural recommendations of the Department of the Environment [Sb], as set out in the Circulars, Policy Guidance Notes [Sc] and Charters that are in place;

❑ to consider the development proposals against adopted national and local policies and to provide advice on the interpretation of these policies to the applicant and/or the advisers, preferably in pre-application discussions;

❑ to advise the applicant of any obvious amendments to the proposal or any procedures that could, or should, be adopted that would optimise the proposed scheme and minimise delays;

❑ to obtain, at the earliest opportunity, any decision or other advice that is required to assist in the determination of the application or achievement of the proposed scheme;

❑ to consider the benefits or disbenefits of the proposed scheme to all potential users of the development scheme and the general public, as well as the developer; and

❑ to be fair, reasonable, assistive and courteous to the applicant/developer and the technical advisers and to provide concise, positive and factual advice, throughout the negotiations, to both the applicant and the Statutory Authorities, with the intention of reaching a conclusion, as soon as is realistically possible.

27.4 Agreements with Developers

Local Authority Powers
The powers of Local Authorities for the purpose of making an agreement with a developer for alterations to a highway are Section 106 of the Town and Country Planning Act 1990 (HMG, 1990a) [Sav], as substituted by Section 12 of the Planning and Compensation Act 1991 (HMG, 1991a) [NIu] [Sac], or Section 278 of the Highways Act 1980 [Saw], as substituted by Section 23 of the New Roads and Street Works Act 1991 (HMG, 1991c) [NIt]. Other assistive powers are contained in Section 111 of the Local Government Act 1972 (HMG, 1972) [Sax] and Section 33 of the Local Government (Miscellaneous) Provisions Act 1982 (HMG, 1982). Section 38 of the Highways Act 1980 [Sv] enables highway authorities to achieve

agreement with a developer for the construction, laying out and adoption of a new street. Section 184 of the Highways Act 1980 [Sz] allows authorities to achieve agreement with a developer for the construction of vehicle crossings, footways and verges.

Section 278 of the Highways Act 1980 [NIt] [Sab]

A developer and a highway authority may enter into an agreement under this provision for the execution of highway works that are necessary to achieve access to a development site. The design and construction of the works are usually undertaken by the Highway Authority and, if not, must be to the reasonable satisfaction of the Highway Authority, and are either partially or wholly at the developer's expense.

The Highway Authority is also permitted to acquire the lands necessary to achieve the proposed highway scheme, either by negotiation or compulsory purchase, again either partially or wholly at the developer's expense. It should be noted that the acquisition of any essential land by compulsory purchase does not necessarily avoid the payment of any 'ransom value' attributable to the lands under consideration. There are provisions in the Section for the Highway Authority to recover the costs of the works, which include the power to close the access relating to the Agreement or to put a charge on the developer's lands [NIab].

An agreement under this section may also provide for the making of payments to the Highway Authority by the other party to the Agreement, in respect of the maintenance of the works to which the Agreement relates. Such payments would form part of the Agreement and are normally received as a single commuted sum. This may include specific elements of the scheme, such as traffic–signal operation costs.

The advantage to a highway authority of this form of agreement is that it provides direct control of the design and construction of the highway works, which is particularly important for traffic-sensitive routes in the highway network. As a general rule, the Highways Agency uses this form of agreement for development-associated highway alteration works on trunk roads (DOT, 1991). A disadvantage is that the Highway Authority will be under considerable pressure to achieve the target date for the completion of the works as initially agreed with the developer, notwithstanding that unforeseen difficulties may be encountered.

Section 106 of the Town and Country Planning Act 1990 [NIu] [Sav]

An agreement under this provision may be made with a local planning authority to make alterations to a public highway considered to be necessary to achieve access to a development site. The terms of such an agreement are not restricted to the extent of Section 278 of the Highways Act 1980 [Sab] [We] and a highway authority could allow the developer to use his own consultants and contractor to design and carry out the highway works, albeit with safeguards in the Agreement to ensure that the Highway Authority's duties and obligations, regarding the public highway, are protected.

A Section 106 Agreement [Sav] for highway works is more flexible than a Section 278 Agreement [We] and could address, through its terms, the concerns of both the Highway Authority and the developer, with the exception that it is difficult in practice for a highway authority to achieve direct control over a developer's contractor.

A Section 106 Agreement [Sav] for highway works would normally provide for the full details and specification of the highway works, although it could reserve other matters, such as a security bond, inspection fee and contractor approval, until the commencement of the works is imminent. In certain circumstances, either on account of the nature of the works or in the absence of any steep gradients or differences in level in the vicinity of the works, it may be possible for the Section 106 Agreement to include, initially, the highway works in plan and sample section only, reserving the full sectional information to be agreed before the works commence. This facility can assist with the development programme or the earlier issue of a planning consent notice.

The provisions of the Town and Country Planning Act [Sd] can also be used for the payment to the Highway Authority of commuted sums in respect of the future maintenance of small areas of landscaping associated with the highway, principally for the benefit of the development rather than the wider public. Commuted payments may also be sought in respect of any shortfall in car–parking provision below the Local Authority's required minimum guideline. PPG13 [NIc] [Sg] advises authorities that these guidelines should be set as a range of maximum and operational minimum levels for broad classes of development and location. DOE Circular 1/97 Planning Obligations (paragraph 13.10) (DOE, 1997) [Say] advises on the circumstances where contributions would be appropriate to improve the accessibility of sites inadequately served by transport modes other than private cars.

The accession of private lands to a public authority, such as the ceeding of land to a highway authority to provide a new or widened public highway, cannot be achieved by a planning condition and would therefore be incorporated in the terms of an agreement, associated with the development, by reference to specific statutory powers relating to the Local Authority's function.

Circular 1/97, Planning Obligations (DOE, 1997) [Say] advises that '...the cost of subsequent maintenance and other recurrent expenditure should normally be borne by the body or the Authority in which the asset is to be vested' but identifies the exception of the commuted maintenance payments for highway works allowed for in Section 278 of the Highways Act [Sab]

Section 38 of the Highways Act 1980 [Sv].
This allows for an agreement between a developer and a highway authority for the construction and subsequent adoption, by the Authority, of a proposed highway. The most common use of this provision is to achieve the ultimate adoption of housing and commercial estate roads. With regard to housing estate roads, a nationally available 'Model Agreement' has been drawn up by the House Builders Federation and the Local Authority Associations, representing the Highway Authorities and their agents.

Section 38 [Sv] can also be used, either in isolation or as part of another agreement, to achieve the dedication and ultimate adoption of development access roads and parts of a highway recognised as necessary to achieve adequate access to development, such as a new footway adjacent to an existing publicly–maintained carriageway on the development site frontage. The existence of a Section 38 Agreement for a proposed highway, or an existing private street, allows exemption from the APC (Advance Payments Code) provision of the Highways Act, Section 192, [Saz], that would otherwise apply to the erection of any new building that fronts the private street or proposed highway.

Section 184 of the Highways Act 1980 [NIr] [Sz]
This section of the Act relates to the provision of vehicle crossings over an existing footway or verge, that is part of the existing highway, in order to gain access to premises. It imposes on the Highway Authority the obligation, in determining the prescribed works, to have regard to safe access and egress from premises and the need to facilitate the passage of vehicular traffic on highways. Paragraph 3 of Section 184 refers to land that is to be developed in accordance with a planning permission granted, or

deemed to be granted, under the Town and Country Planning Act 1990 (HMG, 1990a) [Sd] and allows a highway authority, in this circumstance, to require a new, improved or altered footway or verge crossing to provide access to the premises concerned.

There is a reference in Section 184 to the extent of the works that may be prescribed by the Highway Authority in that acceleration and deceleration lanes can be considered as part of the crossing works.

Moreover, the cost of the works is met wholly by the owner of the premises and may include the statutory undertakers' diversion costs, the cost of the works and the Highway Authority's supervision costs. The Highway Authority decides whether the prescribed works are carried out by themselves or the developer's contractor.

Sections 106 [Sav] and 278 [Sab] are normally appropriate for the purpose of an agreement relative to the highway works necessary to achieve access to a development. Alternatively, in some circumstances. Section 184 [Sz] could be utilised.

General
A highway authority should consider which form of agreement is necessary for the particular highway alteration works and access works and should discuss this with the applicant/developer and technical advisers before reaching a conclusion. There may be pre–agreement complications, which also need careful consideration and should be identified for early discussions with the applicant/developer and advisers. For example, the highway alteration and access scheme may be dependent on Traffic Orders, which may require committee approval for advertisement, and then be considered by a later committee for determination, when any objections would be assessed. At the very least, this procedure would delay the Agreement and, at the worst, the highway alteration scheme may not go ahead or may need to be amended.

Another aspect of a planning application that includes alterations to an existing public highway is whether or not the affected part of the public highway needs to be included within the red outline that identifies the application site and whether or not the highway works themselves require planning consent.

Planning permission is always required for highway works promoted by a development scheme. This facilitates public consultation for highway alteration works through the planning application procedure. With regard to whether highway works need to be included in the application site, this is appropriate if the highway works are considered by the Local Planning Authority not to be permitted development.

27.5 Planning Obligations and Planning Gain

Circular 11/95 – The Use of Conditions in Planning Permissions (DOE, 1995b) [Sak]

The Circular, in paragraphs 12 and 13, refers to planning obligations under Section 106 of the Town and Country Planning Act as a means of overcoming a planning objection to a development proposal. It advises, however, that the Secretary of State prefers the use of planning conditions, wherever possible, since the imposition of restrictions by means of a planning obligation deprives the developer of his rights to appeal against the imposition of inappropriate or onerous requirements. Section 106A of the Town and Country Planning Act does, however, allow a developer to apply to a local planning authority to discharge or modify a planning obligation on the expiry of five years after the obligation is entered into. Section 106B allows for an appeal to the Secretary of State if a local planning authority decides that a planning obligation shall continue to have effect without modification.

Circular 1/97 – Planning Obligations (DOE, 1997) [Saz]

The Department of the Environment Circular 1/97 Planning Obligations provides: '...advice on the proper use of planning obligations made under Section 106 of the Town and Country Planning Act 1990 (as substituted by Section 12 of the Planning and Compensation Act 1991) and of similar obligations under other powers, including local legislation'. Paragraph B2 then states that: '...properly used, planning obligations may enhance the quality of development and enable proposals to go ahead which might otherwise be refused. They should, however, be relevant to planning and directly related to the proposed development, if they are to influence a decision on the planning application. In addition, they should only be sought where they are necessary to make a proposal acceptable in land–use planning terms'. The Circular also sets out the circumstances in which certain types of benefit can reasonably be sought in connection with a grant of planning permission and provides tests regarding validity, reasonableness and scale.

Appeal Decisions

A professional adviser, representing a developer, a highway authority or an objector, will need to ascertain the current advice, legislation and Court decisions that relate to the transportation aspects of development control. Apart from the Planning Policy Guidance Notes, the Government Circulars and the legislation, the Journal of Planning and Environment Law and the Encyclopedias of Planning Law and Practise and Highway Law and Practice provide useful sources of up to date information (see Section 27.7)

A number of development–associated appeals to the Secretary of State, the Court of Appeal and the House of Lords, relating to planning obligations and planning gain, have been assessed by David Brock in a Journal of Planning and Environmental Law article (Brock, 1994) and, in the Journal of Planning Law Notes of Cases (July 1995 edition, pages 581 – 604), there is a summary of the relevant appeals. In general terms, the advice put forward in Circular 16/91 [Saz] was confirmed by a House of Lords appeal decision, which related to a case known as the 'Tesco Witney' case. In the light of this advice and these Court decisions, it would appear prudent, from the point of view of both developers and local authorities, to ensure that any planning obligations relate directly to the development under consideration and could not be regarded as an unrelated planning gain.

27.6 References

Brock D (1994)   'Negotiating Planning Agreements after Tesco', Journal of Planning and Environmental Law (August).

DOE (1987)   Circular 13/87 'Town and Country Planning (Uses Classes) Order' (WO 24/87), Stationery Office [Sai].

DOE (1988)   Circular 10/88.'Town and Country Planning (Appeals) Regulations 1987' (W015188), Stationery Office [Sam].

DOE (1992)   Circular 31/92 'The Town and Country Planning (Fees for Applications and Deemed Applications) (Amendment) (No 2) Regulations' (W073/92) DOE [Sas].

DOE (1993)   'Development Control – A Charter Guide', DOE [Sau].

DOE (1994)   'Planning Charter Standards' DOE [Sau].

DOE (1995a) Circular 9/95 'General Development Order Consolidation' (WO 29/95), Stationery Office.

DOE (1995b) Circular 11/95 'The Use of Conditions in Planning Permissions' (WO 29/95), Stationery Office [Sak].

DOE (1997) Circular 1/97 'Planning Obligations', DOE.

DOE/DOT (1994) PPG 13 Planning Policy Guidance Note, 'Transport', DOE/DOT [Sal] [Wf].

DOT (1988) Circular 4/88 (WO 42/88) 'The Control of Development on Trunk Roads', DOT

DOT (1991) Circular 6/91(WO 65/91) 'Development in the Vicinity of Trunk Roads – Agreements Under s278 of the Highways Act 1980', DOT.

HMG (1972) 'Local Government Act.' 1972 Stationery Office [Sbb].

HMG (1980) 'Highways Act 1980 (As Amended)', Stationery Office [Se].

HMG (1982) 'Local Government (Miscellaneous) Provisions Act 1982' Stationery Office [Sax].

HMG (1987a) 'Town and Country Planning (Use Classes) Order 1987' Stationery Office [Sai].

HMG (1987b) 'Town and Country Planning (Appeals) (Written Representations Procedure) 1987 Regulations', Stationery Office [Sba].

HMG (1989) 'Town and Country Planning (Fees for Applications and Deemed Applications) Regulations 1989 (As Amended)', Stationery Office [Sas].

HMG (1990a) 'Town and Country Planning Act 1990' Stationery Office [Sd].

HMG (1990b) 'Planning (Consequential Provisions) Act 1990' Stationery Office.

HMG (1991a) 'Planning and Compensation Act 1991' Stationery Office.

HMG (1991b) 'Town and Country Planning (Enforcement Notices and Appeals) Regulations 1991', Stationery Office [San].

HMG (1991c) 'New Roads and Street Works Act 1991', Stationery Office.

HMG (1992a) 'Town and Country Planning (Modifications and Discharge of Planning Obligations) Regulations 1992', Stationery Office.

HMG (1992b) 'Town and Country Planning Enforcement (Inquiries Procedure) Rules 1992' Stationery Office [Sao].

HMG (1992c) 'Town and County Planning (Inquiries Procedure) Rules 1992', Stationery Office [Sap].

HMG (1992d) 'Town and Country Planning Appeals (Determination by Inspectors) (Inquiries Procedure) Rules', Stationery Office [Saq].

HMG (1995a) 'Town and Country Planning (General Permitted Development) Order', Stationery Office [Sar].

HMG (1995b) 'Town and Country Planning (General Development Procedure) Order', Stationery Office [Sae].

IHT (1994) 'Guidelines on Traffic Impact Assessment', IHT.

27.7 Further Information

Encyclopedia of Planning Law and Practice – Sweet and Maxwell.

Encyclopedia of Highway Law and Practice – Sweet and Maxwell.

Journal of Planning and Environmental Law – Sweet and Maxwell.

Chapter 28 Parking Related to Development

28.1 Introduction

Parking issues are a significant element of transport policy since the availability or not of parking has a major influence on people's choice of mode of transport. It can be argued that all parking is related to development and local authorities use parking policy to influence the demand for travel by private car within their areas.

Such policies cover both parking provided by local government or the private sector and parking provided specifically in association with new development or re–development proposals. Publicly available parking issues are addressed in Chapter 19.

Development projects vary considerably in scale, content and location. Different land–uses are increasingly being provided on the same sites, or in close proximity, and operational economies of scale are leading to larger developments. The resulting demands for car, lorry and motorcycle parking can lead to a scale and nature of parking provision which can be operationally complex and environmentally intrusive. Addressing the balance between operational requirements and environmental impact presents a challenge for designers.

28.2 Planning Policy Guidance

Planning policy guidance related to development has been issued by Government for a number of years through a series of Planning Policy Guidance notes (PPGs) covering a wide range of land–uses and designed to provide a consistency of approach across different regions of the country (see Chapter 26) {NIa} [Sa].

These policy documents have a significant effect on both the location and nature of new developments in the UK, including parking guidelines that should apply. Many local authorities have published parking guidance and apply different requirements depending on the location of the site and its proximity to good public transport. These local guidelines, which should translate national policy into local situations, should always be used when designing a new development.

28.3 Statutory Planning Framework

In general terms, the planning system operates at two levels. The highest level, the Structure Plan, provides strategic planning guidance over a wide area. Local plans provide the detailed planning framework identifying specific proposals and sites for development. The Structure and Local Plans provide the basis upon which decisions are made regarding new development proposals [NIb].

Where a proposed development does not conform to the Structure Plan, the facility exists for the Strategic Planning Authority to 'call in' the application for determination rather than determination being made by the Local Planning Authority. Local Government reorganisation has simplified this situation, in many areas, by combining Strategic and Local planning functions in one new unitary authority.

A full explanation of the planning and development control system is given in Chapters 4, 26 and 27.

28.4 Demand for Parking

The location and nature of developments affect the amount and method of travel and the pattern of development is itself influenced by transport infrastructure and transport policies. Hence, the scale and demand for parking provision is influenced both by the location of the development and by the

Photograph 28.1: Typical parking provision at a new development.

availability of alternative forms of transport. For example, levels of car ownership are generally higher in rural areas where the provision of public transport is more dispersed and less frequent. Conversely, in urban areas, where there may be better public transport alternatives, the ownership and use of a car should be seen as less important. Furthermore, the traffic levels experienced in certain high activity areas, like shopping centres or business areas, may mean that a local authority wishes to discourage the use of private vehicles and may seek to use its parking control powers as an instrument of that policy.

Any development produces different types of parking demand which can be classified in various ways. It is often useful to divide demand into short–stay and long–stay requirements. For example, a single residential unit usually needs at least one permanent parking space for the occupiers. The level of provision will vary, however, depending upon the type and size of property, from zero for, say, sheltered accommodation for old people, to perhaps three spaces for larger units with several occupants. All dwellings need some spaces for visitors' vehicles but, as this is usually a short–term need, each visitor's space serves several individual trips each day.

For commercial development, parking is also often categorised under three headings, which are:
- ❑ operational parking;
- ❑ commuters' parking; and
- ❑ visitors' parking

Operational parking is defined as that which is essential for the operation of the business concerned and can include provision for members of staff who require to use a vehicle to carry out their duties. This parking is essential and the need should be met regardless of the location of the development.

Commuters' parking generally covers parking by employees who drive to and from work but do not use the vehicle for business purposes during working hours. Depending on the nature and location of the development this may or may not be provided close to the site or at all.

Visitors' parking, which can cover visitors to business premises or shoppers visiting retail outlets, can vary in scale, dependent on the nature of the development, and should be accommodated in a manner consistent with the local parking policy. Visitors' parking should include provision for emergency access.

Estimates of parking demand can be made using a

variety of techniques, including analysis from first principles, comparisons with similar developments elsewhere and the use of databases. When estimating demand, facilities for disabled people, cyclists, taxis and public transport users should be considered carefully to provide for changes in travel mode arising from demand management policies.

28.5 Differing Parking Requirements

Perceptions of parking requirements may vary depending on the viewpoint of, for example, a property developer, a potential building occupant, a local authority or a user of the facilities.

Photograph 28.2: Typical surface parking layout.

Different emphases will be placed on different aspects of each development and these are affected by commercial considerations. For example, city centre office developments are generally more easily let or sold if some parking space is available for the exclusive use of staff or visitors and a developer may try to achieve as much parking space as possible in areas where parking is difficult. In other circumstances, the Local Authority may be seeking to ensure that a development includes adequate parking space, so that all its parking needs are met within the site without a detrimental spillover onto surrounding streets. However, in the immediate vicinity of underground stations or other major public transport interchanges, there may be little justification for providing more than a limited amount of off–street parking, unless it is for a 'park and ride' facility developed in co–operation with the Local Authority. Occupants of properties usually wish to have sufficient parking space for their own needs, allowing them the option of using private transport if they wish.

The demands of potential users of facilities can also influence the scale and nature of parking provision.

Category	Type of Area	Offices and Commercial	Residential	Retail
A	City centre location ,with excellent public transport and a strong economy, where severe parking restraint needs to be applied	Operational space only (based upon a tight definition of operational space). Commuted payments would be required to assist financing of public transport, in recognition that most staff would not be provided with parking.	Standards relaxed to allow wherever possible the re–introduction of residential uses into the centre (but limited to one space per dwelling)	Spaces to be provided to meet short–stay demand (when economic price for parking is charged) but should be provided in communal parking areas. Commuted payments in recognition that most staff do not have parking
B	Town centre location with good public transport where parking restraint needs to be applied	Operational space more broadly construed, with some commuted payment to assist public transport, in recognition that most staff would not be provided with parking.		
C	Smaller town centres – areas adjacent to town centres and along strategic public transport corridors, where some parking restraint needs to be applied	Standards based on about 50% of demand, with a contribution to assist public transport in recognition that 50% of staff would not be provided with parking.	Where space allows, up to two spaces per dwelling	Town Centre space to be provided in communal parking areas. Based upon demand standards when economic price for parking is charged. Development sites to be designed, nevertheless, to favour public transport use
D	Other areas, including "out of town" centres	Based on demand standards (economic pricing of parking space should be encouraged).	Full demand standard (ie up to 3 spaces per dwelling)	Based upon demand standards when economic price is charged for parking facilities.

Table 28.1: A parking policy, based upon a four–tiered hierarchy of standards.

For example, convenience shopping is predominantly carried out in bulk and by car, hence requiring a high provision of parking space. In urban areas, this can conflict with overall transport policy and can influence decisions on location and scale of development.

In certain special circumstances, such as where existing property is converted to a new use or is in a conservation area, departures from standards may be justified for practical reasons or on amenity grounds.

28.6 Local Authority Guidelines

Most local authorities publish parking guidance for new developments either as part of a Local Plan or as free–standing documents [NIc]. These documents serve the dual purpose of providing guidance to potential developers and to the Local Planning Authority itself when considering development proposals. The context within which these are published is increasingly one of seeking to contribute towards a balanced transport policy for an area with different levels of provision being required in different areas.

Government is keen that authorities should not seek to impose higher car parking provision than the developer would wish to construct, unless there are significant road safety or traffic management advantages.

Examples of this include lower levels of parking provision for central urban areas where traffic congestion is a problem and public parking spaces are likely to be controlled in number, duration of stay and price. However, if parking policy is to contribute substantially towards a sustainable transport policy, encouragement needs to be given to use public transport rather than private cars in urban areas. This can be achieved, in relation to new developments, through the use of lower parking provision in urban corridors where public transport is being encouraged (see Section 28.7).

What is sought is a range of acceptable parking requirements, defining not only the operational minimum but, importantly, the maximum provision as required in PPG13 (DOE/DOT, 1994) [Sb] [Wa].

One recognised approach is to adopt a hierarchal set of parking requirements. For instance, a four– tiered system could be adopted within a Structure Plan which sets differing guidelines for different types of area. This could take the form set out in Table 28.1. Having established such a strategy at a Structure Plan level, it would then be the responsibility of the Local Planning Authority to define the exact boundaries of each zone. This approach could be extended to a regional level.

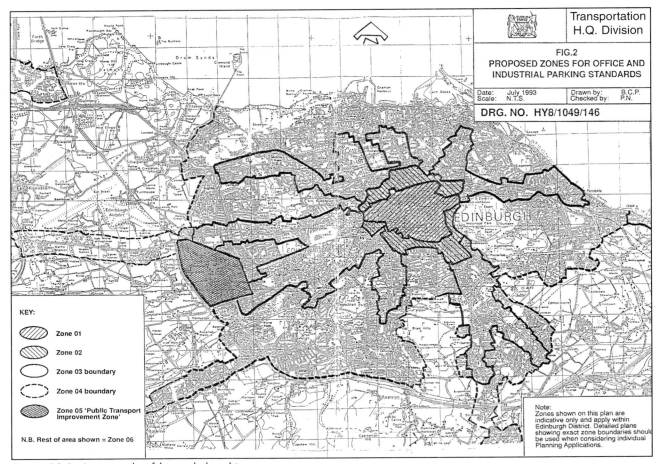

Figure 28.1: An example of hierarchal parking zones.

One of the key benefits of such an approach is that it avoids the temptation for developers to move between towns seeking the "best deal" on parking provision.

Where less than maximum demand provision for car parking is adopted, commuted payments can be sought by the Local Authority from developers as a contribution towards provision of public transport infrastructure. In practice, this is achieved by agreement with the developer either to fund directly a new piece of infrastructure or to contribute a fixed sum to a central fund dedicated to the provision of new facilities. Government guidance on planning obligations makes clear that contributions from developers are to be related directly to the particular development proposal (see Sections 27.4 and 27.5) [Sd].

In smaller towns and rural areas, where access to public transport is limited, congestion is lower and consequently car ownership is higher, it is generally more appropriate to permit higher parking provisions than in larger urban areas. This ensures an adequate operational performance of the development without any risk of overspill onto adjacent highways.

Wherever possible, every effort should be made to match the appropriate parking provision to the development proposals. In some cases, this requires

the use of planning control to avoid the possibility of future changes in operations, resulting in significantly increased parking demand and associated operational problems, for example, a change from a carpet retail warehouse to a DIY store. Control may be achieved through the use of planning agreements under the Town and Country Planning Act 1990 (HMG, 1990) [NId] [Sc], which can be used to specify the permitted type of retail use within the general planning land–use category. A similar agreement may be required in controlling different types of residential use, such as student, elderly or general, which generate significantly different levels of demand for parking (see also Chapter 27).

If parking policy is to contribute fully to an integrated transport strategy, then control of Private Non–Residential (PNR) parking spaces is essential.

28.7 Hierarchical Parking Provision

The 'ABC Location Policy' in the Netherlands

The use of hierarchial standards for parking have been adopted widely in the Netherlands. To combat decentralisation, the Dutch Government has adopted a policy aimed at concentrating employment–intensive land–uses around public transport routes and interchanges. The policy, known as 'the ABC location policy', is based on establishing and then matching accessibility definitions for locations and mobility definitions for businesses (DOE/DOT, 1995).

Accessibility Definitions

'A' locations – main public transport interchanges in town centres, easy access by cycling and walking, fast and frequent rail services to other centres – parking for no more than 20% of the workforce, with supporting park–and–ride facilities on feeder routes.

'B' locations – at a transport interchange in a district centre or a bus interchange in a small town, near a main trunk road or motorway junction, – parking for up to 30% of employees.

'C' locations – no specific public transport requirements and therefore within the immediate vicinity of motorway junctions normally at, or near, the edges of urban areas.

Mobility Definitions

'A' people–intensive land–uses in relation to surface area and a high need for public transport.

'B' commercial services, such as the clothing industry, instruments and optics, sport and recreation, social services, and retail of all kinds.

'C' goods–intensive uses and those dependent on private transport.

To match mobility and accessibility, businesses of Mobility 'A', can only be located in locations with accessibility 'A' and, likewise, for types 'B' and 'C'.

In the Netherlands, parking spaces at offices are allocated according to employment levels usually with a maximum of one space for every two jobs. Often the provision is much less: in Amsterdam, the parking norm for offices in the city centre is zero and elsewhere only 25 places per 100 jobs.

Parking demand is further discouraged by the Travel Demand Management initiative, whereby employers encourage their staff to use the most appropriate mode of travel for journeys to work and other business trips, according to environmental criteria.

In response to the ABC approach, developers in the Netherlands have proposed fewer developments adjacent to motorways and show greater commercial interest in 'A' locations but resistance to promoting sites with severely constrained car parking has continued. Furthermore, developments in the pipeline which had not yet been implemented, and do not adhere to the ABC approach, have proceeded.

Current Practice in the UK

British local authorities are advised to devise 'accessibility profiles' for types of development of particular sizes, in a way that makes clear to developers that the choice is between town centre sites, where parking provision will be restricted since public transport is good, and less central sites, where contributions will be sought from the developer to help fund public transport improvements.

Parking policies should avoid the scope for competitive provision of parking by neighbouring authorities keen to attract new development. The UK Government has proposed maximum levels of off–street parking spaces permitted at new business developments in London (DOE/DOT, 1995). Exceptions to these standards may need to be permitted where justified by the economic importance of particular development proposals. Local authorities are encouraged to seek a balanced approach, which recognises the legitimate need for parking but sets that against the environmentally damaging effects of more traffic. The guidance advises the London Boroughs, when arriving at local standards, to consider the availability of public transport alternatives, forecast traffic levels and the relationship with other forms of parking. Thus, in London, a co–ordinated approach to the establishment of car parking space standards throughout the city is being developed. Using such an approach parking space standards would be relatively high for a development which has poor public transport and low restraint whereas, in central areas, with relatively high restraint and good public transport, the parking space standard would be low.

A particular problem is parking associated with retail stores. Parking guidelines for such sites vary. A common requirement for a food superstore in the UK would be one space per $10m^2$ of gross floorspace but there are no national standards.

Standards are viewed as guidelines rather than fixed rules. On average, food superstores would have 12 spaces per 100 m^2 of sales area; lower levels apply to other shops. Out–of–town sites often have a higher provision of parking space. The majority of town and city centres have existing levels of provision below recommended standards. Furthermore, parking for retailing in town centres is often badly located, poorly designed and ineffectually managed.

By comparison, in the Netherlands, supermarket shops typically have only five places per 100m² out of town and up to three spaces per 100m² elsewhere. Smaller shops generally have even fewer parking spaces. Furthermore, land–use controls severely restrict retail developments outside towns.

If more stringent controls are to be applied to off–street parking, it will certainly become necessary to have complementary on–street controls, if parking is not simply to spill over onto the adjacent streets.

28.8 Design Considerations

Where parking space is to be provided, the following points should be considered when preparing development plan requirements or in assessing individual planning applications.

Accessibility and convenience
The location of parking and loading areas should be sufficiently close to the building or land they serve to reduce the likelihood of drivers parking indiscriminately to avoid walking. Acceptable proximity may be affected by the nature of the walk involved. A longer walk may be acceptable in a safe and pleasant environment with easy gradients and good lighting. As a guide, a 400m walk distance can be taken as a generally accepted maximum.

Disabled persons
Location is particularly important for disabled persons and any allocated spaces should be as close as possible to the destination, sufficiently wide to allow wheel chair access and connected to the destination without steps. Ramps or lifts may be necessary.

Vehicle access and safety
Geometric standards should be applied which allow

reasonably comfortable clearance for the types of vehicles for which the spaces are provided. Special attention will be necessary at turning points and to give adequate headroom and ground clearance on ramps. Good standards of visibility must be maintained at all times and this is particularly important where the car park access joins a main road. It is also necessary to ensure that vehicles waiting for a space do not cause queues to extend back beyond the access road.

Operation and maintenance.
Some form of access control is usually needed so that car parking spaces are used in the way that is planned. In some places, this might extend to fully automatic doors, grills or even cages for individual vehicles to prevent vandalism. The running surface should be resistant to attack by oil or petrol. It may also be necessary to employ attendants to ensure that operational and visitors' spaces are used correctly. Good design can minimise supervision and maintenance. Robust and vandal–proof light fittings and safety barriers may have to be provided.

Impact on the surrounding road network.
The number of spaces provided should relate to the capacity and functions of the surrounding road network and the characteristics of the use of the particular development. These issues are addressed in detail in Chapter 26.

28.9 Sizes and Layout of Spaces

Table 28.2 gives the dimensions of some typical vehicles and the space required for parking areas. These may be used as basic reference values but different layouts, such as parallel, herring–bone and in–line, have slightly different overall space requirements and the detailed layout of parking spaces will be site–specific (see also Chapters 19 and

	Car	Light Van	Coach (60 seats)	Heavy Goods Vehicle	
				Rigid	Articulated
Vehicle Dimensions (m)	3.8 x 1.7* 5.0 x 2.0*	up to 6.0 x 2.1	up to 12.0 x 2.5**	up to 11.0 x 2.5**	up to 15.5 x 2.5**
Recommended Dimensions of Parking Space (m)	4.8 x 2.4***	5.5 x 2.4	14.0 x 3.5	14.0 x 3.5	18.5 x 3.5
Overall Area per Parked Vehicle (m²), including access and manoeuvring space	20 to 25	20 to 30	100 to 150	100 to 150	150 to 200

*   Typical dimensions
**   Maximum dimensions normally permitted in the UK
***   Car parking spaces for disabled persons should have an additional adjacent 1m strip for wheelchair access. A similar provision should be made where spaces are allocated for 'parent and children' bays.

Table 28.2: Typical parking space requirements in surface vehicle parks.

30). Further guidance can be obtained from PPG13 – A Guide to Better Practice (DOE/DOT, 1995) [NIa] and the publication Design Recommendations for multi–storey and underground car parks (IHT/IStructE, 1984).

28.10 References

DOE/DOT (1995)   PPG13 – 'A Guide to Better Practice', DOE/DOT [Wa].

HMG (1990)   'Town and Country Planning Act 1990', Stationery Office.

IHT/IStructE (1984)   'Design Recommendations for multi–storey and underground car parks', IHT/IStructE.

28.11 Further Information

BCC (1995)   'Car Park Design Guide' – Birmingham City Council.

Burt, S and Sparks, L (1991)   'Setting Standards for Car Park Provision: The Case of Retailing', Traffic Engineering + Control, (May).

CROW (1991–1995)   'Parking and Mobility', Centre for Research and Contract Standardisation in Civil and Traffic Engineering, CROW, Series 1991–5.

DOE/DOT (1994)   Planning Policy Guidance Note 13, 'Transport', DOE/DOT.

DOT (1992)   Traffic in London: 'Traffic Management and Parking Guidance', Local Authority Circular 5/92, DOT.

DOT/LAA (1987)   'Traffic and Parking', Tapworth Report, DOT/LAA.

Flint, D (1992)   'A Traffic and Parking Plan for Exeter', Highways & Transportation, (November).

HTOA (1994)   'Car Parking Standards for Hertfordshire', Hertfordshire Technical Officers Association.

LPAC (1993)   'Advice on a Parking Strategy for London', London Planning Advisory Committee.

NEDO (1991)   'Company Car Parking', Traffic Management Systems Working Party, National Economic Development Office.

Trani, C et al (1995)   'Parking Policies', PIARC.

Tutt, P and Adler, D (1979)   'New Metric Handbook – Planning and Design Data', Butterworth Architecture.

Chapter 29 Residential Developments

29.1 Introduction

This chapter deals particularly with new developments on self–contained sites. However, it is recognised that, in many existing urban areas, much development, and re–development, is carried out on a small scale or in locations where existing housing is replaced by new dwellings. Guidance is also needed on design standards for residential roads and footpaths where improvement schemes are being considered but such guidance should not be prescriptive and should encourage flexibility in approaches to design.

The scope for adopting the standards suggested in this chapter may be limited, in these circumstances, but where existing conditions are unsatisfactory, such as congested traffic movements, inadequate safety or a generally poor environment, other techniques may achieve some improvement. These are explained in Part 3, which deals with the management of traffic on the existing road network. Where development proposals would simply perpetuate an undesirable situation, they may well be in conflict with the adopted Structure, Local or Unitary Development Plans and should be refused by the Local Planning Authority [NIa].

The Department of the Environment and the Department of Transport have issued a comprehensive set of recommended principles, for the layout of residential roads, footways and footpaths, in Design Bulletin 32 (DOE/DOT, 1992) to assist with the preparation of development plans and supplementary guidance. The Bulletin emphasises a balance that must be struck between highway design standards and other requirements. Engineers, planners, public utilities' representatives, public transport operators, developers and housing designers need to collaborate, when local standards are being produced for inclusion in Local and Unitary Development Plans, when scheme designs are being prepared and when proposals are being considered during the development control process. Since DB32 was first issued, almost all highway authorities have published new or revised standards for the adoption of roads and most of these have accepted the bulletin's recommendations [NIb].

The second edition of DB32 (DOE/DOT, 1992) was produced following detailed consultation with a wide variety of interests and experience gained in the use of the first edition.

29.2 Roads in New Developments

The majority of developments in urban areas are on small infill and redevelopment sites which will have existing roads on at least one boundary. Figure 29.1 shows a small redevelopment site set within a network of heavily trafficked roads. Access can only be gained from a narrow lane along one side and visibility at the entrance is restricted by existing buildings. The narrow frontage, terraced houses are arranged formally around one shared surface access road. Parking provision is integral with the dwellings and in grouped garages and hardstandings. Restrictions on the use of such roads to give direct access to individual dwellings and parking spaces may determine whether or not a site can be used for housing. There may also be restrictions on the type, spacing and design of junctions which can be provided to connect any new roads required to serve the new development.

Such factors are usually less critical when large sites are being developed. Figure 29.2 shows part of a proposal for a large–scale housing development with a network approach to road layout. Note that the street pattern is characterised by formal squares and crescents, with a rectangular layout of interconnecting roads. The houses are a mixture of

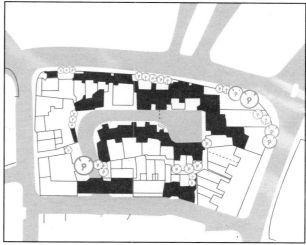

Figure 29.1: A redevelopment within a network of existing urban roads.

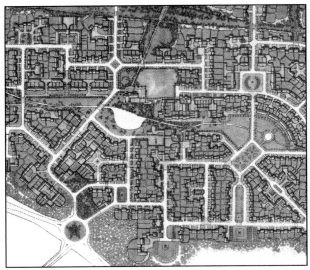

Figure 29.2: an example of a large-scale housing development.

individual frontage dwellings, groups of houses on private drives and courtyards.

An economical layout of new roads can normally be provided within a site to serve individual dwellings. Parking spaces and existing roads can often be extended to provide new points of access. In these circumstances, the principles which govern the layout of new roads are usually at least as important as those which govern access from existing roads.

Roads and traffic issues must be considered as an integral part of the design of residential developments. Roads should be seen as part of the overall urban design and their layout will play an important part in creating surroundings which are safe, convenient, nuisance-free (DOT, 1996b and 1996c), visually attractive and economical to construct and maintain. It is important to incorporate the principles contained in PPG13 (DOE/DOT, 1994) [NIc] [Sa] [Wa] (see also Chapter 26), regarding the need to reduce the environmental effects of transport overall by promoting the use of less environmentally-damaging modes such as walking, cycling and public transport. Careful consideration should also be given to how the needs for access and mobility of disabled people are to be met in the design.

The accommodation to be provided is normally determined by local planning policies and market demands which, together, will determine the details of building density, dwelling design, parking provision, open space, landscaping and other design features. Site requirements, which affect frontage access and parking arrangements, will be important in determining the minimum amount of space in a development that will need to be occupied by roads.

29.3 Highway Adoption Agreements

In England and Wales, new highways are usually adopted as 'highways maintainable at public expense' through agreements between developers and highway authorities, normally under section 38 of the Highways Act 1980 (HMG, 1980) [NId] [Sb] (see also Section 27.4). Such agreements provide a mechanism for developers to have their completed roads adopted, seek to protect house purchasers from future road charges, ensure that access to services is available both in emergencies and for routine maintenance and that powers are available to remove any obstructions that may make access difficult or cause damage to services.

Requests for adoption agreements are normally welcomed by highway authorities, because they can then specify layout and construction standards as a condition for adoption, thereby reducing the likelihood of future demands from residents for the adoption of roads which may be costly to maintain and/or difficult to improve.

Authorities normally wish to adopt the carriageways, verges, footways and other land that is needed for visibility splays at junctions and on bends. This is to ensure that vegetation or other obstructions to visibility can be removed or trimmed, as necessary. Authorities may also be prepared to adopt grouped parking spaces contiguous with carriageways and open spaces, amenity areas and play areas, on payment by the developer of a commuted sum to cover the anticipated costs of future maintenance.

Adoption agreements are normally sought by private sector house–builders because completed agreements assist house sales. In addition, house purchasers seldom wish to be responsible for the maintenance of spaces outside dwelling curtilages once the development is complete. Equally, public sector developers, such as local housing authorities, housing associations and new town development corporations, do not normally wish to maintain roads even where they are prepared to maintain common open spaces and landscaped areas.

Public utilities normally expect their services to be located within the limits of adopted roads and footpaths. This may restrict the numbers of dwellings that can be served by any unadopted roads that have to be provided. Public utilities usually expect, *inter alia*, service strips to be demarcated when underground services have to be laid in adopted verges adjacent to privately maintained gardens (NJUG, 1986 and 1991) [NIe].

29.4 Hierarchy of Roads in Residential Developments

Highway authorities normally set separate standards for the two main categories of roads in the hierarchy which are appropriate in residential development (see also Chapter 11).

Local distributor roads form the links between access roads and district distributor roads. Direct access to individual dwellings and parking spaces is not normally allowed from these in areas of new development. This category includes local roads whose function is to distribute originating and terminating traffic within districts. For the purpose of this chapter, they include roads which distribute access traffic or provide bus routes within large residential developments. Such roads are often also referred to as 'major collector roads' or 'access collector roads' and should have design speeds of 30 miles/h or less (see Chapter 31).

Access roads form the major part of residential road networks and provide direct access to individual dwellings and parking spaces. This category includes roads with traditional layout and cross-section; i.e. with footways separated from the carriageway by kerbs and a change in level. Differences between the numbers of dwellings served by these roads are commonly indicated by the use of terms such as 'major access road' and 'minor access road' but the design speed for all access roads should be less than 30 miles/h. This category also includes roads whose surface is shared by both pedestrians and vehicles (see Photograph 29.1). Such roads are commonly called 'shared surface access roads.' This generic term covers a wide range of different layout arrangements which are sometimes called 'access ways', 'mews courts' and 'housing squares'. When such roads serve only a handful of dwellings, and are not to be adopted by the highway authority, they are often called 'shared private drives' or 'shared driveways'.

Design of shared-surface access roads should have regard to people with mobility impaired. People who are blind or are partially-sighted can experience difficulty where the footways and carriageway are flush, particularly if there is a need to cross that part of the space designated as carriageway. In the circumstances where a shared-surface access road serves more than about 30 dwellings, it may be appropriate to consider using a minimum 25mm kerb upstand to delineate the footway.

29.5 Vehicular Access to Development Sites

When a site is being appraised to see whether it would be suitable for residential, or indeed any, development, it is necessary to consider whether the existing roads have sufficient spare capacity to carry the amount and type of traffic that would be generated by the new development; i.e to check whether such traffic would cause any existing traffic or accident problems to be exacerbated or new ones to be created.

Where alternatives are available, the number and location of points of access to a site may be determined by considerations such as the volume, type and destination of vehicular traffic using the existing roads, the volume of vehicular traffic likely to be generated by the scheme and the main directions which such traffic is likely to take when setting off from the dwellings and travelling to destinations outside the area (see also Section 29.6). In general, development should be avoided which might encourage the use of trunk roads and other through routes for short trips, as this undermines their strategic role. The Design Manual for Roads and Bridges (DMRB) Volume 6, Section 2, Part 7, TD 41/95, 'Vehicular access to all purpose trunk roads' (DOT, 1995a) provides guidance and mandatory requirements with regard to the layout of minor road access to trunk roads, as well as general advice with regard to pedestrians, cyclists and equestrians.

It is also important to consider whether junctions of appropriate capacity and geometry can be created (see Part 5) including, when appropriate, additional connections to existing roundabouts or signal-controlled junctions. Developers and highway

Photograph 29.1: An example of a shared-surface access road in a residential area.

authorities often need to enter into agreements (see also Chapter 27) to finance the improvement of roads or junctions which, although outside a proposed development, are, nevertheless directly affected by it. Developers will need to discuss such constraints and opportunities with local highway and planning authorities before design work is started.

For larger developments, they also need to consult local bus operators about their requirements. The need to make provision for buses varies according to local public transport objectives and policies. Provision might involve amending an existing bus route, relocating existing bus stops or altering footways or footpaths to provide improved access to, and facilities for, bus stops (see Chapter 24). Access for pedestrians and cyclists must also be considered (see Section 29.7).

29.6 The Layout of Roads and Footpaths

Roads, footways and footpaths should be regarded as an integral part of the overall housing layout and should help to create an attractive environment. They should also be designed to create a safe and nuisance–free environment, whilst ensuring the shortest practicable routes for pedestrian access to dwellings and convenient access by service vehicles. Roads and footpaths should be considered as part of wider urban design and account should always be taken of the local character of the area. The extent to which these aims can be achieved will depend partly on the availability and location of points of access to roads serving the site.

Where there is little likelihood of 'through' traffic becoming a problem (for example, because speed restraint measures discourage such traffic), a network configuration of roads may provide the greatest convenience of access (see Figure 29.3). Depending on the number of points at which access can be obtained from the surrounding distributor road network and the extent to which 'through' traffic could become a problem, more restrictive layouts may need to be considered. Some routes through the area may need to be made more tortuous (see Figure 29.4) or, in extreme circumstances, extinguished altogether (see Figures 29.5 and 29.6). Where any 'through' motor vehicle routes are closed, provision should be made so that cyclists can still proceed through any closures.

Depending on the locations of main destinations outside the site and any differential speed restraints on the access roads, the layouts in Figures 29.3 to 29.6 should ensure well-distributed vehicle flows. As the

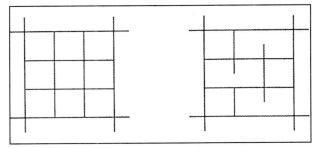

Figure 29.3.   Figure 29.4.

number of points of access to surrounding roads is reduced (see Figures 29.7 to 29.9), flows will clearly become more concentrated at these points and, in larger developments, direct access may have to be limited on the 'stem' roads. Asymmetric layouts (see Figure 29.10) tend to distribute flows less evenly across the network and to require, therefore, a larger area of roadspace per 1000 dwellings for the same level of accessibility by vehicle.

Residential access roads, other than culs–de–sac and other short sections similar in layout, should be adequate to serve all but the largest buses. The views of public transport operators should be sought on the design of those roads to be designated as bus routes. Wherever possible, the opportunity should be taken to provide direct routes for buses between the densest parts of a residential district and the most frequently used destinations. This can be achieved, while at the same time preventing the route from being used as a short–cut by through traffic, by the use of bus–only gates or links. Where a bus link is proposed, the police should be consulted about enforcement. Any bus link or gate should be designed to be self

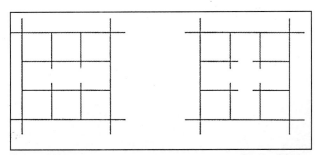

Figure 29.5.   Figure 29.6.

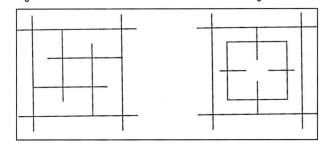

Figure 29.7.   Figure 29.8.

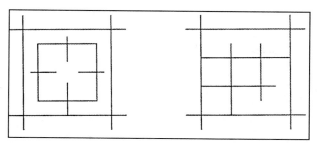

Figure 29.9   Figure 29.10.

enforcing, as far as possible. When an existing bus route needs to be extended into a large site with only one point of access, it is normally necessary to provide a loop road or turning and layover facilities, if a terminus is needed. The layout of the development should be such that no dwelling is more than about 400m from a service bus stop (see also Chapter 24).

The principles set out in DB32 (DOE/DOT, 1992) suggest that roads serving more than around 300 dwellings should not normally give direct access to individual dwellings or parking spaces, although many existing roads of course do not meet this standard. On the other hand, the visual character of the development and the need for security will be enhanced if the fronts of dwellings mostly face stretches of road, except, perhaps, the distributor roads around the edges of the development.

Crime prevention has become an increasingly important consideration in the layout of roads and footpaths in residential developments (NHBC, 1986), particularly in some inner city areas. Careful design to ensure 'natural surveillance' can help to reduce risks. In culs-de-sac serving only a few dwellings, this can be achieved by ensuring that houses overlook each other. In other cases, the greater volume of passing traffic can provide the necessary surveillance provided traffic volumes are not so excessive that residents turn their backs on the road. Risk of assault may be reduced if separated footpath routes are kept to a minimum (see also Section 29.6) and risks of car theft and vandalism may be reduced if parking provision is made mostly within dwelling curtilages or in locations that can be easily supervised from the windows of adjoining houses (see also Section 29.8). Other features, such as good street lighting, an absence of concealed places of refuge and the use of vandal-proof fittings, all help. Developers should consult their local police force about the need for such measures. [Sd]

Although the most important contributions to safety can be achieved by excluding through traffic and reducing traffic volumes, the overall configuration and the geometry of road design can also ensure that

drivers keep to speeds of well below 30 miles/h, especially along access roads. A 20 miles/h standard may well be appropriate to 'design-in' a low speed regime (DOT, 1994d).

On access roads with footways and traditional cross-sections and visibility standards, research suggests that the use of 90° bends with tight radii (see Figure 29.11) and only short lengths of road between junctions or bends can help to bring speeds down to as low as 20 miles/h (Bennet, 1983). The research also showed that minor reductions in carriageway width appeared to make little difference to speeds and even more substantial reductions, like those produced by lines of parked cars, had only a limited effect.

Other effective methods of restraining traffic speeds, if carefully designed and sited, are chicanes (Figure 29.12), speed control islands (DOT, 1995b) (Figure 29.13, and Photograph 29.2) and road humps (DOT, 1996a), which include flat-top road humps, round-top road humps, raised junctions and speed cushions (DOT, 1994c) (Figure 29.14) (see also Chapter 20). Legislation permits new developments to be designed as 20 miles/h zones (DOT, 1990 and1991). Incorporating appropriate features from the start can be cheaper than having to install them later. However, physical traffic-calming features should not be used as an easy or cheap alternative to the principles of good design for lower speeds on residential roads. Their use can result in problems for buses, service vehicles, emergency vehicles and cyclists and should only be incorporated into new developments with great care.

Drivers need to be made aware, on entry and as they traverse a site, that they are in surroundings where the needs of cyclists and pedestrians, including those with impaired mobility, are expected to take precedence over the convenience and free-flow of motor vehicles. Even in large developments, the needs of drivers should not predominate. Motor vehicle movements should be accommodated such that safety is not compromised but every encouragement should be given for trips to be made by more environmentally-friendly modes, such as walking, cycling and public transport.

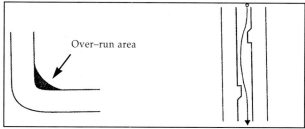

Figure 29.11: 90° Bend   Figure 29.12: Chicane.

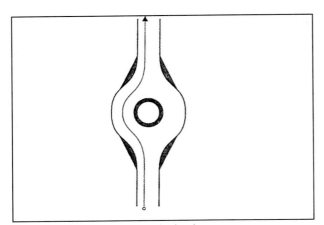

Figure 29.13: Speed–control island.

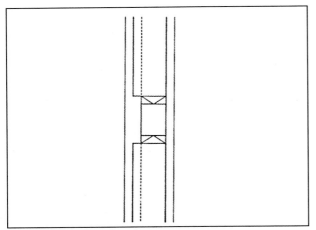

Figure 29.14: Speed cushion.

The aim of giving priority to the needs of pedestrians and cyclists in the design of surroundings in the immediate vicinity of homes can normally be achieved by serving as many of the dwellings as possible from roads which carry low volumes of traffic and where vehicle speeds are low – preferably below 20 miles/h. The layout principles set out in this section should enable this aim to be met (DOT, 1987).

If there are existing trees or shrubs on the site, a survey will be needed to establish their position and condition, so that the layout of roads, footpaths and underground services can allow for as many as possible be retained.

Carefully considered hard and soft landscaping (see Photograph 29.2) is needed to help distinguish shared–surface access roads from other types of road (see also Section 29.6). A flowing alignment of curves and verges with variable widths to accommodate trees and shrubs may be beneficial on the longest roads. It is also desirable to allocate space to accommodate planting in rectilinear road layouts and in all places where the scene would otherwise be dominated by views of back gardens, garages or screen fences.

Photograph 29.2: Carefully considered hard and soft landscaping.

Special care is needed in the selection of pavings, trees and shrubs (DOE, 1973 and La Dell *et al* 1982), so that maintenance is minimised and vegetation does not obscure visibility or cause damage to services underground. Whenever possible, trees and shrubs should be located in places that will be maintained by individual householders, usually in gardens within dwelling curtilages. Elsewhere, maintenance agreements could be sought prior to adoption.

29.7 Provision for Pedestrians and Cyclists

To complement the principles of site access and layout, the design of the development as a whole needs to ensure that footways, footpaths and cycle–tracks are convenient for residents and visitors to use, especially those who have impaired mobility. This can be achieved by the provision of dropped kerbs and tactile surfaces at crossing points. Links between roads should be created when they would provide routes that are significantly shorter than along the residential roads and should avoid the use of busy distributor roads. Such routes also need to be laid out to help strangers to find their way and to be convenient for those who make regular door–to–door collections and deliveries.

The widths and alignment of footways alongside main carriageways need to ensure that pedestrians do not have to step off the footway when passing each other. It is also necessary to ensure that the widths and alignments of both footways and footpaths are adequate to meet public utilities' requirements, wherever services need to be accommodated beneath them.

Footways normally need to be provided along both sides of local distributor roads and pedestrian safety and comfort are improved if footways can be separated from the carriageways by verges along most heavily trafficked local distributor roads. Footways are also normally required along both sides of access roads. Given the emphasis on encouraging walking, footways should be omitted only in very exceptional circumstances where pedestrian safety and convenience would not be compromised. Care should be taken to avoid the provision of narrow verges that are difficult to maintain.

It is normally safe for both footways to be omitted from access road culs–de–sac, thereby creating shared–surface access roads, provided that:
  ❑ only small numbers of dwellings are to be served (DB 32 suggests a maximum of 50);
  ❑ vehicular speeds can be kept very low;
  ❑ drivers are warned at the entrance that they are entering a shared surface (by carriageway narrowing or a ramp) (DOT, 1994a);
  ❑ surface finishes contrast visually with the surfaces of roads with footways [NIe];
  ❑ there are no blind spots masking inter–visibility;
  ❑ grouped parking spaces are clearly demarcated from pedestrian and vehicular routes (DOE, 1977);
  ❑ features, such as lamp–posts, bollards and planting, are located so that they do not cause obstructions or hazards to vehicles, cyclists and people with impaired mobility when moving between their home and parking spaces; and
  ❑ major pedestrian routes do not need to pass through the shared surface area.

Public utilities normally expect markers to be provided to denote the limits of adopted areas, especially when there are services under verges that are contiguous with privately–maintained front gardens [NIe].

Underground services may need to be located beneath carriageways and in some shared–surface access roads, and the public utilities should routinely be consulted about the detailed requirements for the location of their services. Services normally need to be ducted where they must cross, or are located beneath, the carriageway (NJUG, 1986).

Separated footpaths may be wasteful of land and an unnecessary maintenance burden if pedestrians find it more convenient to walk along roads to reach their destinations or prefer to do so for other reasons, such as the fear of assault. When there are clear benefits to be obtained from the provision of footpaths or subways, they need to be convenient to use, well–lit and not provide potential hiding places or traps for litter.

Special provision for cyclists is likely to be required when a significantly more direct route to local facilities or places of work would result. Cycle tracks might be provided either contiguous to a distributor road or as an unsegregated route available to pedestrians and cyclists, though preferably separated from each other by some physical demarcation. Measures to assist pedestrians and cyclists are considered in Chapters 22 and 23 respectively and the management of pedestrians at different types of junction is described in appropriate chapters in Part V.

29.8 Provision for the Movement of Vehicles

The widths and alignment of carriageways provided in a residential development should take account of:
  ❑ the adequacy of off–street parking provision (see also Section 29.8);
  ❑ the expected speeds and volumes of traffic;
  ❑ the frequency with which cars may need to pass one another; and
  ❑ the need for provision for service vehicles to pass one another.

On safety grounds, it is also important to provide intervisibility between pedestrians and moving vehicles and to ensure that visibility on bends and at junctions is sufficient to enable drivers to slow down, or stop if necessary, to avoid collision.

Widening of carriageways on bends is usually required when the combined 'swept–paths' of vehicles likely to pass one another is greater than the width of the straight carriageway. Overrun areas can be effective in discouraging higher speeds by smaller vehicles but allowing extra space for larger vehicles to manoeuvre. Care needs to be taken that they are not located where pedestrians might cross (DOT, 1993).

The spacing, layout and dimensions of turning areas should be designed to cater for the sizes of vehicles normally expected to use them. This avoids the need for vehicles to reverse over long distances (ie not greater than 30 m). The siting should also aim to minimise the likelihood that vehicles may use turning bays for parking.

Turning areas at the heads of culs–de–sac serving more than a small number of dwellings should be large enough for a refuse vehicle to turn. Unwanted parking in turning areas is most easily discouraged if the turning areas provide access directly into parking spaces within the curtilages of adjacent dwellings. It may also be possible to design junctions which can accommodate turning vehicles along access roads carrying low volumes of traffic.

The spacing, layout and dimensions of junctions should take account of the types and numbers of vehicles that will be likely to use the junction, the likely directions of turning movements at the junction and the extent to which danger and/or delays may be caused by conflicting vehicular movements. Further information on the characteristics of different types of junction and their design requirements is given in Part V.

Priority junctions are appropriate for most of the busier locations within residential developments but , where a local distributor road serving a large development joins the existing road network, some form of control may be necessary, such as a roundabout or traffic signals. The design of priority junctions is dealt with in Chapter 38, roundabouts in Chapter 39, and traffic signals in Chapter 40.

Visibility at junctions should always be sufficient to enable drivers to slow down or stop, if necessary, to avoid collision and to enable drivers turning into a more heavily trafficked road, such as a local distributor, to make turning their movements safely.

Requirements for the spacing of adjacent and opposed junctions, kerb radii and 'x' and 'y' sight–line dimensions at junctions (see Chapter 38) depend on the numbers of dwellings being served by the major and minor roads at individual junctions, and the expected speeds of the vehicles using them, but minimum standards for safety are always required in accordance with local authority specification.

29.9 Provision for Parking

The numbers of parking spaces to be provided, and their location, can make a substantial difference to the resulting widths of carriageways required to avoid damage from vehicles parking on footways and verges and to the standard of visibility achieved, in practice, on the carriageway. For these reasons, the design of roads and parking provision in a residential area should always be considered together.

Planning authorities usually provide guidelines for the parking provision appropriate to different types of development in their area (see Chapter 28). These guidelines usually take account of projected levels of car ownership and the mixture of dwelling types and sizes, as well as the authorities' transport policies for the area in which the site is located.

Planning Policy Guidance Note 13 'Transport' (DOE/DOT, 1994) [Sc] [Wa] states that: '...a certain level of off–street parking provision may be necessary for a development to proceed without causing traffic problems but, in order to realise the potential of locational policies and to avoid disadvantaging urban areas through added congestion or because of their poorer level of car access, local planning authorities should:
  ❏ be flexible in the requirements for off–street residential parking space and reduce or waive them, where necessary, in order to provide quality and affordable high density development in areas of good access to other means of travel;
  ❏ ensure parking requirements in general are kept to the operational minimum; and
  ❏ not require developers to provide more spaces than they themselves wish, unless there are significant road safety or traffic management implications.'

The aim is to make each group of dwellings self–sufficient with regard to its parking provision except, perhaps, in inner city developments. Experience suggests that few drivers are prepared to use parking spaces located more than a short distance away from their destinations. Thus, to help avoid indiscriminate on–street parking:
  ❏ sufficient off–street parking spaces should be provided for the expected numbers of residents' and visitors' cars and long–stay parking by service vehicles;
  ❏ some spaces for short–stay parking should be provided on or alongside carriageways giving direct access to dwellings; and
  ❏ routes between off–street parking spaces and dwelling entrances should be short enough to discourage long–stay parking on carriageways.

To achieve these aims, parking spaces for residents and their visitors normally need to be located no further from dwelling entrances than the distances required for service vehicles and should be conveniently served by footpaths and footways. In sheltered housing developments, or in special accommodation for the disabled, consideration should always be given to the need for ambulances, or specially adapted vehicles, to gain access. For convenience, and to minimise risks of car theft and vandalism, as many residents' and visitors' parking spaces as possible need to be located within dwelling curtilages or at least in proximity to, and within sight of, the dwellings they serve. Photograph 29.3 shows a high density housing scheme, with a mixture of garages and communal spaces,and Photograph 29.4 shows a high density housing scheme with all spaces communal. Local authorities often set guidelines for

Photograph 29.3: High density housing with a mixture of garages and communal parking spaces.

maximum walking distances to parking areas and typical figures are given in Table 29.1.

A smaller number of parking spaces per dwelling is normally required when provision is made in the form of unassigned grouped hardstandings; i.e when none of the spaces is provided for the sole use of an individual household. In this situation, the needs of individual households with above average car ownership can be balanced against those with below average car ownership and the total number of spaces required for visitors' and service vehicles is also minimised.

When one or more spaces per dwelling are to be assigned for the sole use of individual households (i.e when the spaces are to be provided within dwelling curtilages or in assigned hardstandings or garages outside curtilages), some additional unassigned spaces are required for visitors' and service vehicles and to meet the needs of households with more cars than can be accommodated within their assigned spaces.

Photograph 29.4: High density housing with all parking spaces communal.

In some high density developments, typically in central area locations, high land costs might justify the construction of a multi-storey or underground car park, controlled by issuing passes or machine-readable tags to residents. If an existing car park is either close by, or incorporated in, the same comprehensive redevelopment, it may be possible for reserved spaces to be allocated to residents on a contracted basis.

29.10 Housing Areas Without Cars

One innovative idea for reducing car dependency is to design new housing developments without allocating any space for car parking. Such settlements enable space that would have been used for parking to be re-allocated for play areas, parks or even more intensive developments. In addition to energy savings, other environmental benefits include reduced noise and air pollution. Car-free settlements must be located where good accessibility by other modes, such as walking, cycling and public transport, is achievable. Clearly, access must still be available for service and emergency vehicles.

Various types of households, such as older people, younger singles or couples and lower income groups,

Typical Recommended Distance	to dwelling entrance	to curtilage entrance
Maximum walking distance from a refuse vehicle (m)	n/a	25
Maximum distance from a refuse vehicle when refuse facilities shared i.e. special handling may be necessary (m)	n/a	9
Maximum walking distance from communal parking spaces (m)	50	n/a
Maximum distance for the approach of a fire appliance (when no floors are greater than 6m above ground) (m)	45	n/a

Table 29.1: Typical maximum access distances for different requirements.

may find car–free settlements attractive. In addition, there are many people who need to use a car only occasionally but would like to have one available at any time. For such people, owning a car may be too much but not having a car available at all would not be enough. In Germany, the concept of Stadt–Auto, a neighbourhood–based car rental scheme for residents in car–free settlements, has been developed (Glotz–Richter, 1995). Residents must sign a covenant not to own a car or to park one near to the car–free settlement but, in return, are able to hire one via the Stadt–Auto scheme. The scheme's system of booking and charging results in low basic (car–ownership) costs but higher driving costs, the opposite of the usual situation. It has been estimated that, in German car–free settlements, one car is sufficient for about 15 users. Thus, for one car–free settlement in Bremen, it was estimated that about 20% of the area of a housing estate could have been freed for purposes other than parking.

Car–free settlements are aimed at encouraging a more rational use of cars and may be particularly appropriate for the 30% of households in Britain who currently lack access to a private car.

29.11 Improvement of Existing Housing Schemes

The advice given in this chapter can be used as a basis for assessing and designing schemes to improve the layout of existing residential roads and footpaths. These might form part of programmes to improve poor quality estates or urban safety management schemes, such as 20 miles/h speed–limit zones. Special considerations apply when planning such schemes. Careful assessment of the existing character of the area is required. The views of residents, and of others who need to make journeys within the area, have to be sought, both at the outset and when alternative solutions are being developed. The existing layout will often be a constraint on developing the ideal solution as will the availability of funds. These matters are covered in more detail below (see also Chapter 17).

A full appraisal of the existing area is required. The views of residents, local service operators and businesses should be sought, both on features which are worth preserving or enhancing and those which are perceived to cause problems. This should be backed up by a thorough survey of the area including:
❑ identification of the potential road hierarchy, sample counts of vehicle, cyclist and pedestrian flows, injury accident statistics and identification of bus routes; and

❑ drawings showing the existing layout, with details such as road widths, driveway positions, areas of carriageway used for on–street parking, positions of drainage gulleys and underground service routes, street furniture and landscape features.

The consultation process during this first stage will identify perceived benefits, such as the more popular pedestrian or cyclist routes, valued meeting or play areas and landscape features, such as existing trees. Problems might include danger, perceived or recorded, noise, severance, parking difficulties and damage to footways and verges due to overrunning or illegal parking.

The general approach to scheme design is similar to that employed for new residential layouts. For example, non–access traffic should be discouraged and vehicular speeds restrained. The important difference is that the nature of the existing layout may require the use of special measures which would not normally be considered for new scheme design. The costs and benefits of different solutions should be compared, including the costs of enforcement and maintenance. The views of residents and others directly affected should be sought on the options identified for consideration. The introduction of experimental measures is a useful way of testing proposals which may prove controversial, provided the costs of installation and subsequent removal, or modification, are reasonable and the local people are made fully aware of the reasons for adopting this approach.

The extent to which non–access traffic should be discouraged must be considered in the context of the overall traffic management strategy and distributor road hierarchy planned for the surrounding area. Selective closure of junctions with distributor roads may be appropriate but care is needed to avoid excessive flows on unsuitable roads as a result. Link or junction closures necessitate the provision of turning areas and a through route should be provided for pedestrians, cyclists (DOT, 1987) and, possibly, emergency services (DOT, 1994b). One–way traffic on residential streets would not normally be desirable because of their tendency to increase traffic speeds.

Where the introduction of road closures is not feasible, or would be insufficient in themselves to restrain speeds and discourage non–access traffic, it may be necessary to consider specific speed–control measures. Islands, refuges and chicanes can be effective in controlling speed but often humps and raised junctions may be necessary. Such measures should be introduced with particular care and only

Typical Recommended Layout Guidelines	Access Roads	Local Distributors
Range of Minimum carriageway widths (m)	5.5/4.8/4.1/3.0	6.7/6.0
Minimum centre–line radii (m)	29/20/10	90/60
Minimum junction spacing: Adjacent (m) Opposite (m)	29/20 15/0	90 40
Minimum kerb radii at junctions (m)	6/4	10

Table 29.2: Typical dimensions for residential roads.

N.B. Variations in the minimum guidelines relate to recommendations by different authorities and also reflect variations in the scale of the road, depending on the volume and type of vehicular traffic it is intended to carry.

after full consultation (see Chapter 20). These measures can also create opportunities for tree and shrub planting or hard landscape features.

Full account should be taken of the needs of cyclists and pedestrians, especially mobility–impaired people (IHT, 1986). Accident risks are a particular concern, especially when crossing or moving along busy roads. In such cases, opportunities should be taken to, for example, widen footways or construct refuges at crossing points. On the other hand, creation of shared–surface access roads may be appropriate, if traffic speeds and flows are very low and the needs of blind and partially sighted pedestrians are recognised.

On–street parking problems need to be considered in relation to the overall traffic management policy in the area and its surroundings. Restraint of parking can cause displacement of parked vehicles and resultant problems elsewhere. It may be possible to use garden land or generous verge–width to create parking areas but this may be regarded as an environmental intrusion. Occasionally, footways may be used but only where proper provision for pedestrians remains. On the other hand, widening of footways may need to be considered at crossing points where parked vehicles would otherwise cause visibility problems.

29.12 Detailed Design Considerations

Developers always need to consult Local Plans, Unitary Development Plans [NIa] and highway authorities' design guides for their requirements on matters of detailed design, such as the widths and alignment of carriageways, footways, verges and footpaths and the provision of adequate visibility on bends and at junctions. However, in considering the detailed design of road and footpath layouts, developers should have regard to wider urban design issues and seek to ensure that the scheme takes account of the local character of the area. Table 29.2 gives an indication of the range of carriageway widths and some other dimensions that might be appropriate, depending on the volume and type of traffic carried.

It is also necessary to consult the Fire Service, about local requirements for access by fire appliances, and the public utilities about local requirements for the widths of footways and verges and the location and demarcation of underground services.

A comprehensive scheme of traffic signs, including street name–plates and road markings, is needed as an integral part of residential road networks, although every effort should be made to keep such features to a minimum to help enhance the visual character of the development and to minimise maintenance costs. Within areas designated as 20 miles/h zones, it is not necessary to sign individual traffic calming measures. Nevertheless, all signs and road markings must comply with the current Traffic Signs Regulations and General Directions (HMG, 1994).

In developments which have to have access directly from district distributor roads subject to higher speed limits, particular attention needs to be given to the erection of speed–restriction signs and it is essential to ensure that any speed limits, or other Order–making procedures, are progressed to coincide with the opening of the road. Any Orders required to control the movement of traffic or parking can only be made by the appropriate authority and most regulatory traffic signs have no legal status without a Traffic Regulation Order (see also Chapters 13 and 27).

Most developments need to incorporate a scheme of street lighting in accordance with the British Standard Code of Practice for Road Lighting (BSI, 1992) and this should be designed to complement the design, style and ambience of the development. Lighting should not create glare nor cast light upwards.

The construction and drainage of carriageways, footways, verges, footpaths and cyclepaths needs to be appropriate to the topography and soil conditions on the site, with forms of edge–restraint and cross–falls that deal adequately with surface water run–off. Local authorities usually provide their own specifications for the construction of roads, which are to be adopted for maintenance at public expense.

To help minimise the risk of damage to footways and verges by vehicles over–running it is normally necessary to ensure that adequate protection is afforded by robustly–constructed kerbs, banks and special features such as bollards.

The differences in carriageway surfacing materials, which are required to demarcate shared surface access roads from other types of road, should not only remain visible and provide a surface safe for vehicular movement, over many years, but should also be economical to maintain and continue to be acceptable environmentally after repairs have been carried out. These matters should be discussed with the Local Authority during the design stages [NIe].

Specially dimensioned and reserved parking spaces usually need to be provided adjacent to all dwellings designed for occupation by old or disabled persons and access to these dwellings should always avoid steps or steep gradients. Requirements for the dimensions of hardstandings, garages and parking forecourts and for the demarcation of hardstandings can sometimes be found in Local Plans and Unitary Development Plans or in design guides issued by planning and highway authorities. National guidance on such matters has also been published (DOE, 1987; IHT, 1991 and Noble, 1983).

Grouped hardstandings may need to be individually demarcated to avoid the waste of space and obstruction that can be caused by random parking. A distinct change of surface material or colour is usually better for this purpose than surface markings. Parking spaces contiguous with carriageways can be demarcated with setts or coloured blocks laid flush with the road surface.

29.13 References

Bennett, G T (1983) 'Speeds in Residential Areas', The Highway Engineer, Vol 30 No 7 (July).

BSI (1992) BS5489'Road Lighting: Part 1: Guide to the general principles. Part 3: Code of Practice for lighting for subsidiary roads and associated pedestrian areas. Part 9: Code of practice for lighting for urban centres and public amenity areas', British Standards Institute.

DOE (1973) Housing Development Note II 'Landscape of new housing'. DOE.

DOE (1977) Housing Development Note VII, 'Parking in new housing schemes: Parts 1 and 2, DOE.

DOE/DOT (1992) Design Bulletin 32, 'Residential roads and footpaths layout considerations', DOE/DOT.

DOE/DOT (1994) PPG13 Planning Policy Guidance Note 13: 'Transport', DOE/DOT [Wa].

DOT (1987) Traffic Advisory Leaflet 1/87, 'Measures to Control Traffic for the Benefit of Residents, Pedestrians & Cyclists', DOT.

DOT (1990) Circular Roads 4/90, '20 miles/h Speed Limit Zones', DOT.

DOT (1991) Traffic Advisory Leaflet 7/91 (Wo 2/91), '20 miles/h Speed Limit Zones',DOT.

DOT (1993) Traffic Advisory Leaflet 12/93, 'Overrun Areas', DOT.

DOT (1994a) Traffic Advisory Leaflet 2/94, 'Entry Treatments', DOT.

DOT (1994b) Traffic Advisory Leaflet 3/94, 'Emergency Services and Traffic Calming: A Code of Practice', DOT.

DOT (1994c) Traffic Advisory Leaflet 4/94, 'Speed Cushions', DOT.

DOT (1994d) Traffic Advisory Leaflet 9/94, 'Horizontal Deflections', DOT.

DOT (1995a) Design Manual for Roads and Bridges, Volume 6, Section 2, Part

7, TD 41/95,' Vehicular access to all purpose trunk roads', Stationery Office.

DOT (1995b)   Traffic Advisory Leaflet 7/95, 'Traffic Calming Islands', DOT.

DOT (1996a)   Traffic Advisory Leaflet 2/96, '75mm High Road Humps', DOT.

DOT (1996b)   Traffic Advisory Leaflet 4/96, 'Traffic Management and Emissions', DOT.

DOT (1996c)   Traffic Advisory Leaflet 6/96, 'Traffic Calming: Traffic and Vehicle Noise', DOT.

Glotz–Richter, M (1995)   'Car free housing and car sharing as steps towards a new urban lifestyle', paper presented at the Future of Transport in Cities conference, Edinburgh 1995.

HMG (1980)   'Highways Acts 1980', Stationery Office.

HMG (1994)   'Traffic Signs Regulations and general Directions', Stationery Office.

IHT (1991)   'Reducing Mobility Handicaps', IHT.

La Dell, T, Parker, I and Cole, R (1982)   'Shrub Planting on Housing Estate Roads'. Institution of Landscape Design (May).

NHBC (1986)   'Guidance on How the Security of New Houses can be Improved', National House Building Council.

NJUG (1986)   'Recommended Position of Utilities' Mains and Plant (Services in New Streets) for New Works', the National Joint Utilities Group Publication No. 7.

NJUG (1991)   'Provision of Mains and Services by Public Utilities on Residential Estates', the National Joint Utilities Group, Publication No 2.

Noble, J (1983)   'Activities and Spaces', The Architectural Press.

29.12 Further Information

Baines, C (1982)   'Land for New Housing: The Builders Manual,' New Homes Marketing Board (May), London.

BSI (1990)   BS5588 Part I (AMD 1): 'Code of Practice for Residential Buildings', British Standards Institution.

DOE (1992)   The Building Regulations 1991, Fire Safety: B5 'Access, and Facilities for the Fire Service', DOE.

IHT (1996)   'Cycle–friendly infrastructure – Guidelines for Planning and Design', IHT.

Lindsay, A and Fieldhouse, K (1990)   'Landscape Design Guide' Volume 1 – Soft Landscape, Volume 2 – Hard Landscape,' PSA Projects. Gower Press.

LCC (1996)   'The Lincolnshire Design Guide for Residential Areas' Lincolnshire County Council.

Local highway authorities' standards for road layout and construction (generally).

Local planning authorities' standards for parking provision and aspects of design, such as landscaping (generally).

Local public utilities' requirements for the provision of services (generally).

Noble, J and Jenks, M (1996)   'Parking Demand and Provision in Private Sector Housing Developments', School of Architecture, Oxford Brookes University.

Chapter 30 Non–Residential Developments

30.1 Introduction

This chapter deals with issues relating to the layout and design of non-residential development and, more particularly, with the larger industrial and commercial developments which regularly attract significant numbers of people and vehicles. The type of development considered here frequently requires the provision of large carparks and, in many cases, an internal road layout designed to cater for the manoeuvring characteristics of light or large goods vehicles. However, careful consideration should be given to accommodating and encouraging public transport, pedestrians and cyclists (DOE/DOT, 1994).

In most cases, the land to be allocated for non-residential uses will have been identified as part of the development plan process. The relationship with other non-residential uses and with adjacent residential areas should be considered as part of this process and should be in the context of national planning policy guidance, in particular PPG4 (DOE, 1992), PPG6 (DOE, 1996), Regional Structure Plans and Local Plans [NIa]. This policy guidance, particularly that relating to transportation issues, takes account of the influence of transport on development location and is discussed in Chapter 26. PPG13 (DOE/DOT, 1994) [NIb] [Wa] aims to reduce the need to travel, through the optimum location of development and associated traffic restraint measures. At the development level, this requires detailed assessment of overall area–wide transport impact. It may be that layout and area–wide impacts are addressed in parallel, as part of studies of development feasibility and justification [Sa].

Many local authorities publish their own design guides to address the issues which arise in planning either free–standing developments or estates. Since these guidance documents are the result of both collaboration between authorities and contacts with developers during the planning and implementation stages, they can often reflect current best practice and should include consideration of pedestrians, cyclists and public transport users, as well as vehicular traffic. Consistency has emerged on many of the main issues, although there remain minor differences due to local characteristics or experience. Collectively, these publications contain valuable information on practical issues, provide a summary of current best

practice and perform a useful role in setting adoptable standards.

The issues discussed in this chapter relate to both free–standing, non–residential, developments and to various estates and development zones in their many shapes and forms. For larger development areas, the hierarchical approach to road network configuration, described in Chapter 11 in relation to the wider transport network, can be applied to each zone or estate thereby providing a localised network of distributor roads and access roads. Consideration should be given to the needs of cyclists and pedestrians and to servicing by public transport, particularly in large developments.

Photograph 30.1: Springkerse development in Stirling.
Supplied by Walker Group (Scotland).

30.2 Types of Development

Irrespective of the type or mix of development being considered, a number of general objectives can be used as guiding principles when approaching the design, which should aim:

❏ to minimise conflict between vehicles, pedestrians and cyclists;

❏ to promote a layout which encourages low vehicular speeds;

❏ to develop a design which considers landscape issues and the relationship between buildings, roads, footpaths and cycle tracks;

❏ to ensure that there is adequate provision for public transport;

❏ to ensure that adequate off–street parking for

cars and cycles is provided to reduce the dangers and inconvenience caused by on–street parking;

❑ to provide on–street parking only where it is safe and suitable to do so; and

❑ to provide safe and convenient routes for pedestrians and cyclists which match–up with those in areas adjacent to the development site.

Applying the above guiding principles should help to promote good road, footpath and cycle route design, thereby ensuring a safe and efficient business environment. In addition, the provision of adequate off–street parking and on–site vehicle facilities will help to minimise damage and intrusion into adjacent streets.

While the term 'non–residential' is self–explanatory, in that it covers all development other than residential, there are some collective types of development which may be more common than others when considering the aspects raised in this chapter. These developments are or several types.

Retail Developments – these include free–standing food superstores, non–food retail warehouses, collections and combinations of these as retail parks, or shopping centres (town centre, edge–of–town or of out–of–town). The increasing emphasis on multi–modal access to these types of development means that access for pedestrians, cyclists, buses and, occasionally, trains requires careful attention. However, the provision of appropriate levels of parking for cars and cycles and facilities for bulk deliveries and servicing is essential for successful operation.

Industrial Developments or Estates – these are, typically, light industry and warehousing, with perhaps some low–key on–site retailing. In these locations, provision of well–designed and appropriate levels of employees' car–parking is important, as well as car–parking for visitors. Access for pedestrians and cyclists should also be provided and linked to adjacent local networks, where appropriate, and cycle parking close to the destination should be included. If considered early in the design process, these facilities can often be incorporated at little cost. An important design constraint is the need to cater for the loading, parking and manoeuvring of the significant number of light and large goods vehicles likely to visit the site. Industrial estates, particularly larger estates, benefit from being served directly by public transport, which may be in the form of regular local bus services or special peak–period services for the workforce and, if practical, train services.

Photograph 30.2:   An example of separated parking and loading facilities for individual commercial units.

Office Developments – these include large free–standing office developments such as regional or corporate headquarters buildings or business parks. In these cases, on–site car–parking is an important design issue, the estimation of overall provision often being significantly influenced by location. Access by cycle or on foot should be encouraged by the provision of suitable pedestrian and cycle facilities, including convenient cycle parking. This, and the provision of public transport services, particularly to business parks, can help to reduce demands on infrastructure in the context both of not overloading network capacity and of limiting the supply of on–site car–parking. Science Parks may also be considered within this general category. These are generally smaller industrial estates with a significantly high technology or research and development content.

Distribution Parks, Freight 'Villages' and Freight Terminals – these may be collections of large and small warehouse buildings and large open areas for handling bulk goods and requiring frequent access by commercial vehicles of all sizes. Operators may include users, such as bulk freight handlers, large retailing operations, mail order companies and courier companies. Parking areas are also required for staff and visitors but larger areas of hard–standing are required for commercial vehicle manoeuvring and layover parking. Recent development initiatives associated with the transfer of freight from road to rail, the Channel Tunnel rail links and rail freight terminals would also be considered within this group. The increasing development of break–bulk facilities requires careful consideration because of the interaction between a relatively small number of large vehicles, or trains, delivering goods and a much larger number of smaller vehicles used for

TRANSPORT IN THE URBAN ENVIRONMENT

distribution, often at 'unsocial' times of day (see Chapter 5).

Exhibition Centres, Stadia, Leisure Theme Parks, etc – these are locations where high levels of car and coach parking provision are required and where, ideally, there should be good local public transport links. A common single characteristic of this development category is the intensity of the peaks in demand which they place on the surrounding transport systems. A high degree of flexibility is required to cope with exceptionally high demands created by timing of events, opening hours, seasonal variations, etc.

Photograph 30.3: Car parking adjacent to offices at a business park.

Due to the nature and scale of the above developments, they should almost always be accessed directly from primary or district distributor roads. This level of service will ensure that large peak volumes of traffic and large goods vehicles are dealt with in the most effective manner and cause minimum inconvenience to surrounding residential areas.

Some types of bulk freight and loads being hauled over long distances can be transported more easily and cheaply by rail. Developments which involve certain industrial processes, or use bulk raw materials, may be attracted to sites with an existing railhead or where new sidings could be provided. With the emphasis on reducing pressures on road infrastructure and the environmental impact of heavy lorries, government assistance and support may be available to encourage more use of rail for heavy bulk freight over longer distances (see Section 25.11).

30.3 Access from Highways Public Transport and Other Modes

In dealing with site–specific issues of development, the first considerations are those relating to access, which forms the interface between the development and the wider transport network or system. In this context, access considerations must address all modes of transport and all kinds of traveller, including people with impaired mobility.

Since much of the traffic activity associated with development is road–based, a primary consideration is usually road access. In the case of new development, this may involve the adaptation of existing site–access arrangements. More commonly, and in the interests of operational efficiency and safety, a new access may need to be formed or an existing access to be upgraded. A highway authority is likely to require a developer to finance these works (see Chapter 27) and the type of junction required will be dependent on traffic estimates for the development. This would normally be addressed as part of the Transport Impact Assessment process (IHT, 1994) (see Chapter 26). Road junctions should be designed in accordance with the advice given in Part V.

Public transport services to new developments are also important. The greater the opportunity created for good public transport access, the more likely that growth in car traffic demand will be minimised on the surrounding road network Preferential access for public transport, such as by the provision of bus gates, may be feasible especially if considered at an early stage of design. Bus bays or turning circles, together with good passenger facilities, such as shelters and timetable information provided within development areas, also make access to public transport easier. The provision of easy access by public transport creates an alternative for those who would otherwise have travelled by car and it opens up access to the development for those who have no car available to them and who would not otherwise go there. From the outset, public transport access arrangements should be discussed with local operators to ensure that the development can be served efficiently; otherwise, facilities which are added on later may not be used.

Facilities for pedestrians and cyclists require special attention, particularly where internal routes are connected to networks outside the site. Routes within the site and on its fringes need to be designed to minimise conflict with motor vehicles. Many car park designs have little to offer car–users by way of safety or convenience, once they leave their cars and have to

Photograph 30.4: Good provision for public transport access planned to reduce the need and attractiveness of cars for shopping trips.

proceed on foot to the buildings where they want to go. This is particularly important for developments where a large number of children can be expected and should be addressed in the design. Personal security is another issue to consider when designing pedestrian and cycle routes and car parks and it is also important to give careful consideration to desire-lines for movement beyond the development. Provision which is inadequately linked to the wider network will not operate satisfactorily. A similar approach should be taken to providing for people with impaired mobility. Although pedestrian and cycle facilities may be small in scale, their design requires considerable effort and research if they are to prove attractive to users.

There may be an opportunity for facilities to be shared by pedestrians and cyclists but these need careful design (see Chapter 23). In addition, secure parking for bicycles should be provided in locations that are overlooked and are close to frequently-used entrances. Access arrangements for service vehicles and emergency services may need special consideration for some types of development, such as food stores, retail parks and certain types of industrial processes, and may require the provision of separate service access.

30.4 Internal Circulation and Traffic Management

At the same time as designing the best access arrangements for the development area or site,

careful and considered design of the internal layout can help to avoid unnecessary difficulties on the highway network beyond the development. This is especially true for larger developments, such as superstores, retail and business parks, and large trading estates.

One example concerns the distance between a major access junction onto the public highway and the first internal junction, such as a roundabout or an entrance/exit to a car parking area, within the development site. This distance needs to be sufficient to avoid any queueing back onto the main public highway. Every effort should be made to make the inbound trip from the public highway uninterrupted.

Likewise, drivers entering a development should not be confused by a lack of direction or be faced with too many route-choices immediately on entering the site, particularly, for example, in large developments where more than one car-park is provided. It is important to ensure that traffic is always guided to where it wishes to go and that all the direction signing and road markings are designed to achieve this. Layout features, street furniture and signs which are conflicting and do not conform to legislation should be avoided. Provision of a layby with a street map display listing companies located within the development may also be helpful.

In the same way that care is exercised in ensuring that entering traffic does not block back onto the public highway, it is also necessary to ensure that traffic queueing to leave the site does not block any of the entry routes. This has implications for the location of different types of uses on the site. For example, petrol filling-stations and 'fast food' outlets should normally be located on exit roads.

In some types of development, it may be beneficial to provide separate access points for customers' or visitors' cars, public transport and service traffic and to segregate these types of traffic at the earliest opportunity. This minimises the conflict between vehicle types and permits largely separate circulatory systems to be designed to the standards required for the different types of vehicle, namely, private cars, buses and large goods vehicles.

On industrial trading estates, the layout of individual factories and depots should be designed to assist their operational functions and should aim to achieve the most efficient layout, commensurate with road safety. Roads and footpaths, whether private or intended for adoption by a highway authority, should provide a safe and efficient means of access for workers, visitors and the range of service and delivery vehicles which might be anticipated when a number of different industrial processes and commercial activities are grouped together. Estate layouts will vary according to the size and topography of the site, the mix and size of factory units and the opportunities for safe access to the external distributor road network. Flexibility should be built into the distributor road network to allow for subsequent growth in the nature and volume of traffic.

On larger estates and on large mixed use developments, such as retail parks, problems can arise which are similar in nature to those experienced on the wider transport network. One of these problems is the risk of accidents due, in particular, to vehicle speeds. Developments designed as a traffic–calmed scheme from the outset can assist in creating a safe environment for its users.

Caution should be exercised when designing measures, such as traffic–calming features, to ensure that they conform with current legislation and advice. Specific measures and their uses are described in Chapter 20.

In large sites and estates, development of the entire area may be phased over a number of years. It is important that future demands are anticipated and that the estate layout is developed as part of a longer–term programme. Where infrastructure is being provided in the first instance to meet short–term needs, adequate space should be set aside for possible future modifications where these might be necessary.

30.5 Layout of Parking Areas

This section gives a brief overview of the main car–parking issues affecting the layout design for non–residential developments. More detailed information is given in Chapters 19 and 28.

Car parking associated with new industrial or commercial developments is generally provided at ground level, principally due to their non–central locations, with lower land values, and to the cost of constructing multi–storey car–parks. Occasionally, the construction of a multi–storey car–park may be justified at edge–of–town sites, for example, where insufficient land is available for surface parking or to reduce walking distances or on commercial grounds where parking revenues are sufficient to finance it.

Access to a car park should ideally be possible from all directions but, where there are restrictions, either to entry or exit, adequate signing is required. The design of the main entry road should help the transition from the higher traffic speeds on the external road network to inside the parking area, where pedestrian activity and low speed manoeuvring should predominate.

The internal distribution system should be designed so that:

❏ a vehicle arriving from any entry point can reach any parking bay and, on exit, can visit any internal facility such as pick–up points or petrol filling stations;
❏ conflicts between different users such as pedestrians, cyclists, cars, buses and delivery vehicles are minimised;
❏ safe, secure, and pleasant walkways are provided from any point in the parking areas to the entrances;
❏ cyclists and pedestrians from external routes have segregated and environmentally–friendly routes to the buildings;
❏ public transport has priority over cars in providing access to the buildings;
❏ signing for cyclists and pedestrians is given as much prominence as signing for vehicles;
❏ special parking provision is available for disabled drivers;
❏ appropriate high standard lighting is provided, particularly on footpaths and cyclepaths;
❏ landscaping reduces the visual scale of parking areas and creates a pleasant environment; and
❏ road markings, to establish the circulation pattern, are always supplemented by appropriate signing because road markings can be obscured when vehicles are travelling close to one another.

Large single–level car parks should be subdivided by walkways and landscaping into smaller areas, each of which should be well signed and given a unique identity to help drivers find their cars on their return. Walkways may be identified by different coloured or textured paving or a small change of level. Gaps for pedestrians should be provided between bays at appropriate intervals and car parks associated with large supermarkets should have areas set aside for customers to deposit shopping trolleys.

Special facilities should be provided for disabled persons or other special needs, such as parents with young children, to minimise travel distances. Parking–bay widths should be sufficient for wheelchair access and ramps or dropped kerbs provided to deal with changes in level.

Traffic aisles provide access to parking bays. Maximum aisle–widths should relate to the function

Photograph 30.5: Well–designed circulation and landscaping within a large parking area.

of the car park with shorter stay car–parks having shorter 'through' aisles. If any aisle forms a cul–de–sac, the length should be reduced to that at which a passing driver can see whether there are any vacant spaces without having to enter the aisle.

Parking bays are usually positioned at 90° to the aisles, although oblique angles may be used to fill awkwardly shaped sites. Landscaping and areas for storing trolleys should be protected by kerbs or other physical barriers.

Local authority design guides specify the various dimensions for traffic aisles and parking bays. Parking bays are generally made significantly wider, when providing for the disabled or parent–and–child, and smaller variations in dimensions may be used to differentiate between long–stay general parking and short–stay spaces. Increased bay-widths permit safer and quicker manoeuvring of vehicles and makes getting in and out of vehicles much easier especially with parcels and shopping. Depending on the shape of the site, it is generally possible to provide 400 to 500 car-parking spaces per hectare. Further information on parking layout is contained in Chapter 28.

30.6 Commercial Vehicle Requirements

The manoeuvrability of large goods vehicles depends on their size, on whether they are rigid–bodied or articulated, on the number of axles and on the skill and judgement of the driver. Although some manufacturing processes employ specialist vehicles with manoeuvring characteristics that can be translated directly into a design, the majority of industrial estates, factories and distribution depots will involve a wide range of vehicle types and sizes.

In most aspects of highway design, it is seldom practical or economical to provide facilities to cope

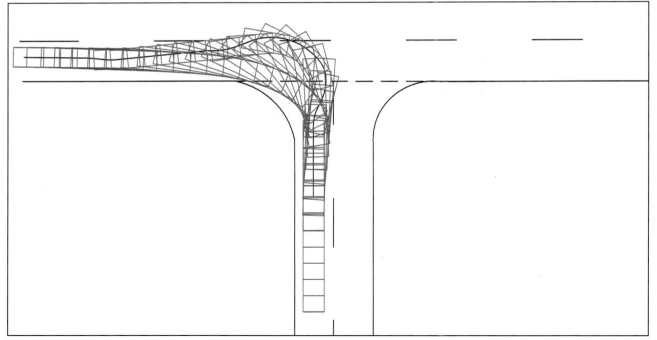

Figure 30.1 – Swept path for a 90° turn by an articulated vehicle, generated from AutoTRACK.

TRANSPORT IN THE URBAN ENVIRONMENT

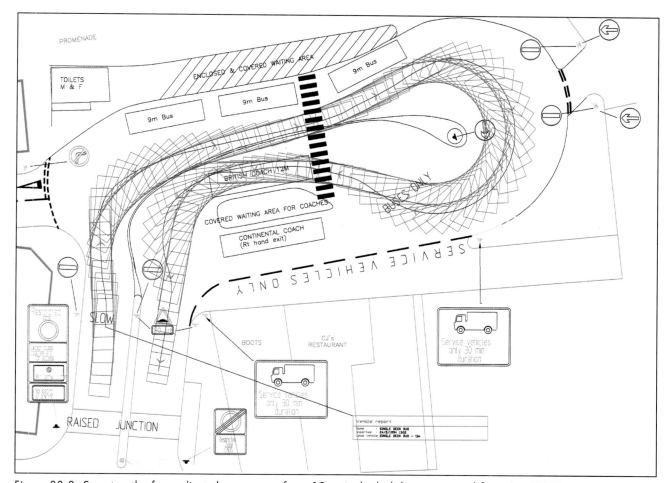

Figure 30.2: Swept path of complicated manoeuvre for a 12m single deck bus, generated from AutoTRACK.

with the worst of all possible operating conditions, such as large numbers of the largest vehicles all choosing to visit a site in the same short space of time. However, designs should be able to cope with commonly occurring circumstances, recognising that a small number of the larger vehicles will occasionally experience some difficulty in manoeuvring and that individual vehicle performance varies greatly with axle–configuration and spacing.

In design practice, the manoeuvring requirements can be checked using two methods. The first is to use standard templates, based on vehicle types and design techniques originally developed by the Freight Transport Association (FTA, 1983). A second method is to use one of the computer software packages, either to generate the swept path for the particular vehicle or to superimpose it on a pre–drafted layout drawing. Figures 30.1 and 30.2 provide examples of this type of application either for generating a swept path for a specified turn (Figure 30.1) or providing a detailed plot of a more tortuous vehicle manoeuvre (Figure 30.2).

Due to the changing nature of distribution and delivery methods, it is not possible to generalise on

the likelihood of finding smaller or larger vehicles in particular locations. In this context, modern computer methods of addressing the manoeuvrability issues in detail are invaluable.

Detailed designs for accommodating large goods vehicles can often be found in the publications produced by local authorities. The standards set out in these documents, which are usually the adopted policy of the authority, are aimed at ensuring that a safe design is applied consistently by all parties. They assist in understanding the implications at the planning stage and ensure that the infrastructure is provided to a standard which is subsequently adoptable by the roads authority.

Whilst these detailed issues are addressed elsewhere (see Chapters 13 and 25), a number of general points are worth noting. Developments which attract significant numbers of large goods vehicles will require areas where they may be turned safely and conveniently, parked securely when not in use and loaded easily and efficiently without disrupting other traffic.

In large new developments, a highway authority will

almost always insist on off–street turning facilities being provided as part of the development. Where culs–de–sac are planned, a turning area should be constructed at the end. Reversing a large goods vehicle is a difficult and hazardous manoeuvre, particularly due to the severely restricted rearward vision, so turning areas for these reversing manoeuvres require careful design if they are to be used safely and effectively.

Failure to provide infrastructure to an appropriate standard will result in inconvenience and difficulty for drivers, the likelihood of damage to kerbs, footways and any adjacent street furniture, structures or landscaping, as well as an increased possibility of accidents. Tyre–scrub on all non–steered axles can also occur when trailers are turned too sharply. From these examples, it can be seen that low standard design, even though it may save construction costs, may well result in increased maintenance costs. Turning areas should always be kept free of obstructions such as parked vehicles, trailers, skips and pallets.

Security and convenience are important factors for any freight or haulage site, where vehicles or trailers are likely to be left for long periods. Site size and shape, together with the variable lengths and manoeuvrability of vehicles, affect the layout of individual parking areas.

Well–designed loading areas located away from the highway will overcome problems associated with loading and un–loading on–street and make these tasks easier and more efficient to perform. Such areas should therefore be incorporated in all new shopping, industrial and freight handling developments. As with car–parking facilities, it is important to encourage their use by making them more attractive and convenient to use than the adjacent highway.

Finally, it is important that adequate loading and parking areas are provided within the curtilage of premises, since many new estates with an open plan layout tend to encourage parking on roadways or turning areas which can interfere with normal traffic movement.

30.7 References

DOE (1992)   PPG4 Planning Policy Guidance Note – 'Industrial and Commercial Development and Small Firms', DOE.

DOE (1996)   PPG6 Planning Policy Guidance Note –'Town Centres and Retail Developments', DOE.

DOE/DOT (1994)   PPG13 Planning Policy Guidance Note – 'Transport' DOE/DOT [NIb] [Wa].

FTA (1983)   'Designing for Deliveries', Freight Transport Association.

IHT (1994)   'Guidelines on Traffic Impact Assessment', The Institution of Highways & Transportation.

30.8 Further Information

All relevant Local Plans and design guidance issued by local authorities.

Part

V

The Development and Design of Highways and Other Infrastructure Schemes

Chapter 31 Design Concepts

31.1 Introduction

This chapter deals with general principles relating to the design of highway networks in urban areas and is, to a large extent, based on the standards and advice provided by the Department of Transport [Sa]. Much of this information has now been compiled into a single multi–volumed document, known as the Design Manual for Roads and Bridges (DOT, DMRB).

When taking decisions on design, it is important to be aware of the overall strategy for the area through which a proposed road will pass. The funding of local transport by the Department of Transport [Sa] encourages proposals to be put forward, which are part of a package of measures for an area and not merely incidental road schemes considered in isolation.

In many instances, the Authority responsible for a road will have a transport strategy which may identify a hierarchy of priorities. These may vary depending on the location. Pedestrians may, for example, be the top priority on routes in the city centre but not necessarily in more remote locations. However, giving priority to one class of road–user often leads to a reduction in priority for other users. For example, cycle–friendly measures, such as widened nearside lanes, specially marked lanes on the carriageway and advanced stoplines, result in a reduction in the lane provision for other road–users. Similarly, the introduction of bus lanes on an existing carriageway is likely to lead to an increase in queue-lengths for the remaining traffic, which is restricted to fewer lanes.

Sometimes, priorities can be achieved for certain classes of vehicle, without the need to allocate roadspace solely for their use. The fitting of transponders to vehicles, coupled with the latest signal technology, can allow priority to be given to vehicles at traffic signals when they are detected in an approaching flow.

Pedestrian and cyclist measures should always be considered, particularly at road junctions. Carriageway widths can often be reduced to provide more footway area and to reduce both the distance and the time required for pedestrians to cross the road.

Local residents and operators of businesses in the vicinity of any road proposals should be given an opportunity to comment on alternatives being considered, in addition to any formal consultations carried out with organisations such as the police, emergency services and public transport authorities. A scheme which can achieve widespread support during the design process is likely to have fewer objections raised at the later stages.

The recommendations in the Design Manual for Roads and Bridges are sometimes difficult to achieve in dense urban areas, where there are many constraints in the form of buildings and public facilities. In applying these standards, it is necessary, therefore, to strike a balance between achieving desirable standards and their impact on the area.

31.2 Quality of Design and Construction

Quality in design and construction covers a range of issues, such as fitness for purpose, technical soundness and the choice of materials, as well as criteria such as aesthetics, durability, safety, ease and cost of maintenance, the effect on the environment and, not least, minimising the whole–life cost. Financial constraints will sometimes limit what can be achieved in terms of quality, although care and skill in both the design and construction processes will ensure the most satisfactory result.

Management of the design and construction process has usually been on an informal basis, with in–house systems of controls and checks aimed at eliminating error and thereby ensuring technical quality. Many organisations demonstrate their commitment to quality by seeking Quality Assurance (QA) accreditation under ISO 9001. The Government has encouraged this move towards QA accreditation, by making Quality Management a prerequisite for the highway works which it is procuring, either directly or through conditions attached to grant–aid. Many suppliers, and all materials laboratories, involved in Government–funded schemes must demonstrate a commitment to Quality Management.

Technical soundness, durability, maintenance and safety issues can be addressed by the application of appropriate standards. Care must be exercised in the application of the individual standards to ensure compatibility of the results in any given set of circumstances. The aim of the design should be a scheme which is fit for its purpose, is aesthetically pleasing, and enhances, or at least does not detract from, the environment.

The strategy of national and local government, to remove through traffic from town centres by the construction of by–passes and traffic management, provides an opportunity to address the quality issues of design of the urban environmen – as it affects all classes of users, including vehicle drivers, cyclists, pedestrians and people with disabilities. Also, in a climate of environmental awareness, sensitive quality treatment of urban areas is expected by those who live and work in them. Utilitarian schemes are no longer acceptable. Manufacturers of construction materials have recognised the trend, and the demand, and offer a wide range of high quality materials for most aspects of highway construction, providing opportunities for innovation in design. Although there is a wide choice available, it is important to avoid the proliferation of products and styles. A co–ordinated approach, based on simplicity and the avoidance of clutter, is an important design principle. Schemes can be enhanced further by the addition of tasteful and sensitive hard and soft landscaping. Consideration should also be given, at the design stage, to future requirements for maintenance (see Chapter 12).

The choice of materials and treatment will require approval by the Planning Authority and, in preparing a design, due regard should be given to the Authority's planning policy or other strategies. Consideration should also be given to the scheme in its setting, as this will significantly influence the acceptability of design and materials. Later difficulties can be avoided by early consultation with the Planning Authority. Above all, quality control must be applied from design through specification and to construction.

31.3 Levels of Service

Levels of service describe the different operating conditions, which can occur on a road at different times, in terms of speed, safety, drivers' comfort and vehicle operating costs. The Design Manual for Roads and Bridges defines a sequence of service level standards in the form of a hierarchy of geometric design criteria related to design speed. A design based solely on maximum peak–hour travel demand and desirable minimum standards may result in a scheme which is unacceptable, because of its cost and

the environmental impact that it may have in an urban area.

It is also vital that the designer considers the concept of level of service for all users, and not only drivers, at the outset and during the design process. The design brief for any urban scheme must embrace the need to accommodate and encourage movement by bus and on foot or cycle, to avoid the high degrees of severance often created in the past. A balance must be achieved in provision for all users which does not advantage one group by disadvantaging others. Generally, as traffic speeds increase, design standards need to be more generous in order to provide a given level of service.

Levels of service are normally deemed to be taken into account in the Department of Transport's [Sa] COBA [Sb] method of economic assessment, where higher levels of service, in terms of higher operating speeds, are traded–off against the additional costs involved. However, COBA was not designed specifically for use in urban areas. A general outline of demand modelling and of economic and environmental appraisal techniques is provided in Chapters 8 and 9.

31.4 Estimation of Future Traffic Demands

Future traffic demand for major urban roads is usually estimated using transportation models. These can take account of underlying growth in demand, assumed changes in land–use, parking strategy, the attractiveness of alternative modes of travel, the capacity of the surrounding road network and the design of the proposed scheme. The models can also estimate any likely increase in traffic demand resulting from the increased capacity provided by the scheme. Models can also be used to estimate future demand for public transport schemes. A fuller explanation of the role and capabilities of transportation models is provided in Chapter 8.

Traffic forecasts are subject to a considerable degree of uncertainty. Consequently, consideration should be given as to how sensitive the design is to variations in the predicted traffic flow. Forecasts may have been made using low and high traffic growth scenarios; in which case, these can be used as a basis for checking the robustness of the design. If a single scenario forecast has been made, it will be sensible to consider the effect that changes in the predictions would have on the design. A balanced view can then be taken about the appropriateness of the design, based on costs, the range of forecasts which the recommended design could accommodate, the consequences of over or under–design and other such factors.

It may also be possible to minimise the consequences of under–estimating demand, by providing flexibility in the design for the provision of increased capacity. This might be achieved by making provision for the future construction of additional carriageway width or alterations to the layout or type of control employed at junctions. Such provision will generally involve some additional initial cost and judgement will have to be exercised as to whether this is justified.

31.5 Design Speed

Drivers regulate their speed in accordance with the road layout and their perception of prevailing road conditions. Apart from the amount of traffic on the road, weather and daylight conditions, the main factors that influence speed are visibility, curvature, road width, surface conditions, the presence of junctions and accesses and speed limits. In urban areas, traffic speeds are also influenced by frontage activities, pedestrians, pedestrian crossings, the presence of cyclists and buses and bus stops.

In selecting a design speed for a new road, account needs to be taken of the wide range of vehicular speeds exhibited in practice. Designing for the fastest drivers would be unnecessarily expensive and often environmentally unacceptable. It would also encourage higher average speeds by ordinary drivers, with a consequential decrease in safety at conflict points. To design for the much slower drivers could be unsafe for the faster drivers. The general practice is, therefore, to design for the 85th percentile speed (the speed below which 85% of vehicles travel, or may be expected to travel, in free–flow wet conditions) (see Chapter 36). The selected design speed should also be appropriate to the area through which the road passes and to the planned role of the route in the local road hierarchy (see Chapter 11).

The selected design speed should then be used as the basis for coordination of all the various elements of geometric design and the design should aim to create an operating environment that encourages drivers to conform to the chosen design speed.

31.6 Design Standards

The design of roads of national significance, ie those that form part of the trunk road network, is carried out in accordance with the requirements of the Department of Transport [Sa] [Wa]. The Department publishes a wide range of Circulars, Technical Standards and Advice Notes, many of which now comprise the Design Manual for Roads and Bridges (DOT, DMRB). The Design Manual covers a wide range of situations and local authorities need to consider, in each case, whether the standards are appropriate for their own roads.

Choice of Road Layout

The choice of layout and lane–provision for a major urban road will be derived from careful consideration of the desired level of provision, in terms of capacity, operating efficiency, safety, the constraints imposed by the urban environment, the effects of the proposals on that environment and the available sources of finance.

The alignment of a new urban road, or major improvement to an existing road, is normally constrained to a 'corridor of opportunity' determined by the extent of the land which can realistically be acquired. Also, the valid needs of particular vehicle-types or user–groups, such as pedestrians, cyclists and bus passengers, may constrain the design of alignment and layout.

Geometric Standards

The alignment for a new urban road should, whenever possible, be so designed that the various geometric elements, such as curvature, superelevation and sight distances, provide at least the minimum values consistent with the chosen design speed of the road. It is equally important that these geometric elements are not excessive, as this can lead to operating speeds, in practice, much higher than the intended design speed.

The severe constraints on road alignment, frequently encountered in urban areas, may mean that it is not possible to achieve the desirable design standards without prohibitive cost or environmental damage (see Photograph 31.1). The Design Manual recognises the need for a flexible approach to design in order to avoid unacceptable consequences, in terms of cost and environmental impact, which could result from rigid application of the desirable minimum standards. The Manual makes provision, therefore, for relaxations to, or departures from, the desirable minimum standard (see also Chapter 36).

Each situation where a relaxation is proposed should be considered on its merits, with regard to cost, environmental impact, safety and the effect on the various geometric parameters. Situations requiring exceptional reduction of standards, beyond those appropriate to 'relaxation', constitute 'departures'.

Photograph 31.1: Inner city one–way circulation–reduced standards on 90° bend.

Whilst relaxations are still considered to conform broadly to acceptable standards and provide a satisfactory level of service, without appreciable reduction in safety, departures can reduce safety significantly and careful consideration and approval is required before their adoption. Some of the adverse consequences of departures may be mitigated by adopting measures to improve safety by introducing, for example, special carriageway surfacing, additional signing, restrictive road marking or lower mandatory speed limits. For trunk roads, all proposed departures must be submitted to the Department of Transport [Sa] [Wb] for assessment and approval.

31.7 Frequency of Intersections

In urban situations, the requirements of access have a major influence on scheme design. It is essential to maintain access to existing land and property. The frequency with which junctions should be provided along a new or improved road depends on its role in the road hierarchy (see Chapter 11). The more important routes, ie primary distributor roads, should ideally have intersections with roads only of the same, or immediately lower, category. Existing roads of even lower categories which cross these roads should be diverted under or over the new road or closed to traffic. Where this is not a practical proposition, because of the constraints imposed by the urban environment, alternative solutions may have to be sought. Often an acceptable compromise can be achieved which will permit a higher frequency of intersections, at–grade, by incorporating a

signalised Urban Traffic Control (UTC) scheme linking the junctions, in order to maximise flow, regulate speed and maintain safety (see Chapter 41).

Type of Junction

The type of junction required will depend on a number of factors, such as the number of roads to be connected, their various roles in the road hierarchy, the volume and composition of traffic on each (on– and off–peak), turning movements and, not least, what can physically be accommodated.

Consideration should be given to all the options; ie grade–separation, signal–controlled gyratory junctions, roundabouts, priority 'T' junctions (with and without signal control) and the linking of junctions through UTC (see Photographs 31.2 and 31.3) (see Chapters 37 to 43).

Photograph 31.2: Complex inner city signal–controlled junction with quality paving materials, architectural features and planters.

Photograph 31.3: Major signal–controlled roundabout with block paved crossings, architectural features and planters.

31.8 Provision for Pedestrians

Safe and convenient paths for pedestrians alongside and across major routes in urban areas are essential and these must be planned during the earliest stages of design (see Chapter 22).

On national routes and primary distributors, the aim should be to segregate pedestrian routes from the carriageway. In providing crossings to the carriageway, consideration should be given to the most appropriate form; ie grade–separation or at–grade. It should be borne in mind that there is, amongst many users, a fear of using subways and, if grade–separation is adopted, it should be made as attractive as possible to encourage use and to discourage pedestrians from crossing at unsuitable or unsafe locations. On district distributor roads in established urban areas, especially those with bus routes, significant pedestrian flows should be anticipated, particularly in the vicinity of bus stops, and the widths of footways should be increased. Elsewhere, for example in sub–urban areas, it may be acceptable to provide only the minimum width of 1.8m. Safety measures, such as short lengths of guardrails to channelise pedestrian movements and signal–controlled crossings at appropriate locations, should be incorporated from the start where safety would be improved. Staggering the two halves of a signal–controlled crossing frequently allows for more pedestrian green time than a straight crossing, although these can produce longer overall walking distances. The balance between pedestrians and other traffic should be assessed and pedestrian facilities should be provided at signal–controlled junctions, other than by exception (see Photograph 31.4).

Wherever possible, the urban environment should be made less intimidating and more accommodating to

Photograph 31.4: Block paved signal–controlled crossing.

disabled people. Guidance has been produced by the Department of Transport Disability Unit (DOT, 1991) and by the Institution of Highways and Transportation (IHT, 1991). Designs to cater for the needs of people with different disabilities can sometimes conflict and designers should liaise closely with the groups representing their interests (see Chapter 22).

31.9 Provision for Cyclists

The Royal Commission on Environmental Pollution (HMG, 1994a) has recommended that comprehensive networks of safe cycle–routes should be developed in all urban areas. Many local authorities have already introduced cycle–priority schemes and more are planned.

Cyclists are legitimate, but highly vulnerable, road - users and should be catered for in ways which promote their safety and, where possible, their comfort and convenience. Highway authorities are increasingly developing strategies and policies to provide better accommodation for cyclists in the urban scene (see Chapter 23). The Highways Agency [Sc] [Wc] and many local authorities have appointed a Cycling Officer, to act as a focus and to provide specialist advice to policy–makers and designers.

Highway and traffic management schemes should be designed to relieve situations which are hazardous and intimidating to cyclists and should avoid creating new ones. Many techniques are available and the Department of Transport [Sa] has published useful guidance (see Chapter 23). This includes the design of cycle lanes, cycle tracks, road crossings and junctions sharing space with pedestrians and the use of signs and surfacing materials (see Photograph 31.5). The aim is to assist those seeking to make the general

highway infrastructure safer and more convenient for cyclists. Dedicated cycle routes are an important part of this, but only a part. The emphasis should be on reducing motor vehicle volumes and speeds and on using traffic management techniques to reduce accidents and to give cyclists a positive advantage. Planning for cyclists should be considered as part of mainstream transportation planning and not as an afterthought (see Chapter 23).

31.10 Provision for Buses

The majority of public transport journeys in urban areas are made by bus. With the introduction of the 'Package Approach' to funding by the Department of

Photograph 31.5: Toucan crossing, cycleway and bus lane highlighted in red calcined bauxite.

Transport [Sd], highway authorities are now giving more priority to schemes which assist buses. A number of innovative measures are available including pre–signals, queue relocation, gap generation, detection of buses stopping and tidal flow arrangements. Signal technology can also be used, together with loop detection or AVL transponders fitted to the buses, to give priority to buses even where there are no physical measures such as bus lanes (see Chapter 24).

Attention should be given to the siting of bus stops and shelters, the provision of adequate footpath widths and to the routes used by pedestrians moving to and from bus stops.

Bus priorities should be seen as part of an overall package of measures to enhance bus services, along with improvements to bus stations and passenger information. A park–and–ride site may be complementary to bus priority measures in order to attract drivers from their cars and onto buses at an earlier stage of their journey.

Public transport operators should be consulted at as early a stage as possible about any facilities that are planned.

31.11 Fixed–Track Systems

The increase in congestion and pollution in urban areas has encouraged renewed interest in the benefits of fixed–track mass transit systems, which have operating characteristics different from normal railways. Such systems are collectively known as 'Light Rapid Transit', sometimes abbreviated to LRT (see Chapter 34).

Design standards for fixed–track systems vary greatly across the range of different systems, which makes any attempt at standardisation difficult. Nevertheless, an at–grade intersection between an LRT system and a road can be treated as a kind of highway junction. The layout of the junction may be priority, roundabout or signalled but the best form of control will depend on the nature of the LRT system and on the traffic and pedestrian flows.

The level of priority which can be given to a fixed–track system will depend on the nature and layout of the junction that is adopted, the traffic flows and degrees of saturation of the approaches, the frequency of the fixed–track service and the direction and densities of all public transport services.

31.12 Minimising Environmental Impact

The impact of new road schemes and road improvement schemes on the environment can be minimised, if environmental assessment forms an integral and iterative part of the planning and design process and especially if care is taken in the design stages. The intention should be to avoid, wherever possible, sensitive locations, for example sites of scientific interest and sites of special scientific interest (SSIs and SSSIs), conservation areas, archaeological sites, housing areas, hospitals and schools and to blend the new/improved road into its surroundings (see Chapter 17).

Particular consideration should be given to the following aspects of design:
❑ selection of alignment and profiles for the new road and its structures which enable it to blend with its surroundings;
❑ sympathetic design of lighting and signing;
❑ sensitive reinstatement of buildings and an appropriate choice of materials and street furniture to match and blend with the existing townscape, including the use of coloured or textured materials to enhance the attractiveness of the area (see Photograph 31.6);
❑ avoiding, wherever possible, splitting buildings and carefully designing the remedial treatment of any truncated structure, such that the final appearance is aesthetically balanced;
❑ softening areas of hard landscaping by the planting of grass, shrubs and trees;
❑ minimising noise impact by the incorporation of sympathetically designed barriers and/or by the use of cuttings. Where possible, landscaped and planted mounds, or a combination of mound and wall, should be used. Walls and barriers should be

TRANSPORT IN THE URBAN ENVIRONMENT

Photograph 31.6: Quality seating, fencing and paving material incorporated in a city centre highway scheme.

constructed of appropriate materials to blend attractively with the surroundings and care should be taken that reflected noise from these barriers does not adversely affect other sensitive areas;

❑ minimising air quality problems in sensitive locations, as far as possible, by utilising traffic management measures to keep traffic flowing smoothly and avoiding congestion and interrupted flow situations;

❑ determining the likely future pedestrian desire–lines once the scheme is completed and ensuring that adequate and suitably–located footways and crossing facilities are provided. Care should be taken to allow for the needs of disabled persons in these provisions;

❑ avoiding undue severance of communities; and

❑ anticipating problems and disruption, which may arise during the construction period, and making provision in the design and the contract to minimise these effects.

Clearly, some of these objectives are more easily achieved with a new road than when improving an existing corridor. However, designers have the responsibility to minimise the environmental impact of their design, whatever the circumstances. With sensitive design, the improvement of an existing corridor can result in an enhanced environment.

31.13 The Management of Construction

Building an urban road scheme will inevitably cause disruption and, usually, the bigger the scheme the greater the disruption. Clearly, this disruption should be kept to a minimum and liaison should be maintained continuously with the local community through the design and construction period.

The management of construction to avoid disruption begins during the design and contract preparation stage, when the designer should give careful consideration to how the scheme can be built. It is important to consider each element of the works individually, and in combination, to ensure that conflicts are avoided. Sufficient provision must be included in the contract to enable supervisory staff, in conjunction with the contractor, to exercise control over the construction. However, strict control to avoid disruption through the provisions of the contract will frequently inhibit efficient working and a balance has to be sought.

Advance contracts and/or temporary works, for example the temporary diversion of traffic, can provide additional capacity to facilitate construction of a scheme. Alternative routes should be provided wherever possible. Temporary signing, temporary traffic control, staged completions and delayed possessions are all means by which disruption can be reduced. It may be necessary to restrict construction to off–peak or night–time; in which case careful consideration of the environmental consequences is required.

Irrespective of the size of the scheme, strict limitations on nuisances caused by noise, vibration, dust etc. should be imposed by the contract. Reference should be made to Health and Safety Executive guidance and, where appropriate, advice should be sought from the local Environmental Health Officer. It is important also to consult and work closely with the police, emergency services, passenger transport and statutory undertakers at all stages of design. Ducts for cables should be considered when improvements are undertaken, in order to minimise future maintenance problems.

The construction industry has one of the worst accident records, with the highest number of fatalities, of any industry. Good management of the construction process is essential to prevent accidents. The Construction (Design and Management) Regulations (HMG, 1994b) is important legislation aimed at making the construction industry safer. The Regulations require the management of health and safety, through risk identification and assessment, and the management of risk at all stages of a scheme up to, and including, demolition at the end of its life to eliminate, avoid, or lessen the foreseeable risks. The Regulations apply to all parties associated with construction projects, ie clients, designers, contractors and workers, and identify the duties and responsibilities of each in the management of health and safety.

31.14 Maintenance Liabilities

The general maintenance of the new or improved highway and of the associated street furniture, traffic signs and street lights should be considered during the design process. Closure of a carriageway or even a traffic lane, to accommodate maintenance vehicles and equipment and to provide a safety margin for workmen, can impose additional operating costs and delays on other road–users. Temporary traffic arrangements, increasing capacity or making provision for maintenance, where the opportunity and budget allow, can provide flexibility for these occasions. Whole life costs, including the delay costs associated with maintenance, should be established when determining minimum carriageway construction thickness. Additional depth of pavement, for a longer design life, may be more cost–effective, in order to defer or obviate the costs involved in strengthening the pavement at a later stage.

The use of a needlessly wide range of materials will make maintenance more difficult and designers should be aware that specially manufactured products or 'one–off' materials may be difficult and/or expensive to replace when they are damaged. Temporary repairs should be avoided, where possible, because they tend to become 'semi–permanent' and an otherwise attractive environment can soon take on an air of neglect. In undertaking a repair, it is important to obtain the best possible match in the replacement materials. Many of the proprietary or specialist materials may have long delivery periods and it is important that a sufficient stock of these materials is available, not only for use by the Authority, but also, for example, by statutory undertakers, to enable them to effect a matching repair.

31.15 References

DOT (DMRB)   'The Design Manual for Roads and Bridges', Stationery Office.

DOT (1991)   Disability Unit Circular 1/91, 'The Use of Dropped Kerbs and Tactile Surfaces at Pedestrian Crossing Points', DOT.

HMG (1994a)   Royal Commission on Environmental Pollution 18th Report – 'Transport and the Environment', Stationery Office.

HMG (1994b)   Statutory Instrument No. 3140, 'The Construction (Design and Management) Regulations 1994', Stationery Office.

IHT (1991)   'Reducing Mobility Handicaps', The Institution of Highways & Transportation.

31.16 Further Information

DOT   'The Manual of Contract Documents for Highway Works', Stationery Office.

Chapter 32 Alternative Concepts of Road Link Capacity

32.1 Introduction

Previous chapters have discussed the growth in the demand for travel and for goods transport and, in particular, the tendency towards growth in private car traffic. Against this background and the need for sustainability in the transport system, the issue of the capacity of a road link is fundamental.

Capacity may be thought of as:
- ❑ the maximum number of vehicles that can pass along a link (vehicular capacity) and this may be refined into flows that will result in varying degrees of congestion (the concept of 'level of service');
- ❑ the maximum number of people that can pass along a road link (person capacity) and this depends on the type of vehicle in which those people travel and on the number of pedestrians and cyclists; and
- ❑ the maximum number of vehicles that may be accommodated without exceeding certain (chosen) environmental standards (environmental capacity).

32.2 Link Capacity

The capacity of a link can be defined as the maximum ability of a length of road to accommodate the passage of people or vehicles. In urban areas, the vehicular capacity of a route is generally governed by that of its major junctions rather than by its links. The maximum ability of an approach to an at–grade junction to cope with vehicular flow is, in turn, governed by the junction design and is limited by the magnitudes of cross–flows and turning movements of vehicular traffic and the presence of pedestrians and cyclists.

The capacity of a link, or of a traffic lane on a link, to accommodate the passage of people depends on the mix of vehicles which is permitted to use the link. An urban traffic lane open to general traffic will rarely accommodate more than 2,000 people per hour, whilst a lane of the same width, reserved for public transport, can have a far higher capacity for person – movement, although this depends on the seating capacities of the buses. As an example, the passenger–flow on a bus lane, three metres wide, on the A23 at Streatham Hill in south London, routinely exceeds 4,000 people per hour in the morning peak period. Furthermore, this is with only 66 buses per hour – far less than the traffic capacity of the lane. Busways of comparable width in South America have been observed to carry as many as 40,000 people per hour, though this necessitates special convoy–control techniques.

Traffic volumes and capacities

The Department of Transport [Sa] gives recommended 'design flows' for urban links to be used as a starting point by those assessing a range of carriageway – width options (DOT, 1985). However, these are not intended to represent maximum capacity but are offered as guides to the practical design flow, providing a reasonable level of service, after accounting for the various obstructions to traffic flow which commonly occur on urban roads. Moreover, a designer rarely has an unconstrained choice from a range of carriageway–width options in an urban area. More commonly, the need is to manage the width between largely fixed highway boundaries so as to satisfy, as far as possible, all the competing needs – such as safety and ease of movement for pedestrians, cyclists, buses and general traffic, servicing of frontages and, perhaps, some on–street parking. Depending on the nature of the road, apportionment of the available space can be used to promote or discourage use by various categories of user. Link–capacity can also be managed by time of day through the use, for example, of part–time bus lanes or tidal flow lanes.

Where a traffic lane carries only 'through' traffic and where practical constraints are largely absent, high vehicle flows can be accommodated. Photographs 32.1 and 32.2 show roads on which single traffic lanes routinely carry between 1,700 and 2,400 passenger car units (pcu) per hour per lane in peak hours. Variable speed controls on the M25 motorway in south–east England are designed to reduce the speed limit to 60 miles/h, when the traffic volume reaches 1,650 vehicles per hour on a lane, and to 50 miles/h at 2,100 vehicles per hour. Their purpose is to reduce the unstable nature of traffic flow at high volumes and hence minimise the risk of flow exceeding capacity, which leads to flow–breakdown, queueing and excessive delays.

The concept that traffic volumes have a direct effect on the achievable vehicle speeds on a road link is now widely accepted with regard to rural roads and motorways. There is, however, less evidence for such straightforward speed/flow relationships on urban road links. An extensive search for such relationships in the 1970s concluded (Buchanan *et al*, 1976) that mean running-speed between junctions was independent of traffic flow. More important was the land-use activity and the associated frontage access alongside a link. These practical limitations on link capacities are discussed below.

Photograph 32.1: Suburban two-lane two-way road capable of carrying 1,700 to 2,000 pcu per hour.

32.3 Practical Limitations on Link Capacity

Junctions
On most urban routes, junction capacity rather than link capacity limits the maximum traffic flow. In general, existing junction approach widths are not much greater than the corresponding carriageway widths between junctions, while flow through the junction is limited by the green time available at traffic signals or by give-way markings at roundabouts.

Lane width and lateral clearance
Narrow traffic lanes and/or obstructions close to the kerb, such as walls or bridge abutments, can reduce link capacity. Although this is rarely an important issue on urban streets, it can be significant at higher speeds on sub-standard dual carriageways. Detailed American advice (TRB, 1985) suggests that, on dual two-lane carriageways, without other practical constraints, the provision of lanes ten feet (3.0m) rather than twelve feet (3.66m) wide reduces capacity

by nine percent. The presence of a tall lateral obstruction one foot (0.3 m) from the kerb could increase this loss to 15%.

Parked and loading vehicles.
Many urban roads are fronted by premises, particularly shops, which require regular delivery of goods. Unless adequate, off-street, service bays exist or there is space in nearby side-streets, which is convenient not only for loading but also for vehicle entry and egress, then kerbside space must be set aside for vehicles to load and unload. This, effectively, precludes use of the nearside lanes by through traffic, unless the lanes happen to be wide enough to allow moving vehicles to pass large, stationary vehicles. Loading can be prohibited during peak periods but this is inconvenient and will generally be resisted by traders (see Chapter 25).

On-street parking is very important in many locations. Traders may regard short-term on-street parking facilities as crucial to their business, although this is seldom borne out in practice. Residents may have no other opportunities to park their cars when car-ownership exceeds the number of garage spaces in an area. Even when waiting or loading is prohibited, illegal parking can adversely affect traffic flow on a link. Rationalising and enforcing the provision of parking can restore previously wasted capacity, as in the Priority (Red) Routes in London (DOT, 1992) (see also Chapter 19).

Photograph 32.2: Urban road where traffic management maintains optimum balance of link capacity.

Vehicles turning right into side–roads or frontage assesses

On a two–way road, a vehicle turning right into a side–road must wait for a suitable gap in oncoming traffic. If no turning pocket is provided, this blocks a lane to straight–ahead traffic until the turning manoeuvre is completed. If the space between the turning vehicle and the kerb is, for example, a designated bus lane or occupied by a parked car, then no straight–ahead traffic can move until the turning vehicle clears.

Vehicles turning right out of side–roads or frontage accesses.

Especially at busy times, vehicles turning right out of side–roads, may, rather than wait for simultaneous gaps in both directions of flow, leave the side road and wait, sideways–on to opposing traffic, while they attempt to force their way into their desired traffic stream. Opposing traffic is then blocked until the manoeuvre is completed.

Where carriageway width permits, it may be possible to provide carriageway markings or kerbed islands which will allow such turning vehicles to wait in the centre of the road without blocking other traffic (see Chapter 37).

Bus stops

A stopped bus obstructs other vehicles from using the nearside traffic lane. If parked vehicles prevent the bus from getting close to the kerb, the adjacent lane will also be obstructed while passengers, including less mobile people, step down from the kerb and up again from the carriageway to board the bus. One–person operation of buses has extended dwell times at stops, though more widespread use of season tickets and travelcards has, to some extent, off–set this effect. Bus bays, often constructed to prevent this obstructive effect, can also delay a bus as it waits to join a stream of overtaking cars (see Chapter 24).

Pedestrian crossings

Zebra crossings can severely restrict traffic flow, especially where they are well–used, such as outside a busy railway station or college. In some locations, this may reflect an appropriate priority for pedestrians over traffic. Elsewhere Pelican, and other light–controlled, crossings reduce this obstructive effect by interrupting the pedestrian flow so that they cross in groups. Pedestrian crossings incorporated within a signalised urban traffic control (UTC) system give 'green man' signals, at times when the least disruption is caused to traffic, but, nevertheless, limit traffic flow to below the capacity of the link.

The resulting limitation on the throughput of an urban link

From the examples referred to in Section 32.1 above, it can be seen that, in ideal circumstances, a traffic lane can have a vehicular capacity of the order of 2,000 pcus per hour per lane which is equivalent to the maximum saturation flow rate on the approach to a traffic signal junction. In practice, few urban links have upstream and downstream junctions which can accommodate a flow of this magnitude. Furthermore, on most urban links, the practical constraints described above limit the traffic flow that could otherwise be accommodated from the geometry of the link. The resulting maximum traffic flow on a link is, therefore, directly dependent on the nature and the severity of the constraints which operate, especially at junctions.

32.4 Environmental Capacities of Links and Areas

The concept of environmental capacity was advanced by Colin Buchanan in the 'Traffic in Towns' report (Buchanan et al, 1963), where it was defined as:

'The capacity of a street or area to accommodate moving and stationary vehicles, having regard to the need to maintain the (chosen) environmental standards.'

In other words, environmental considerations could be used to determine the upper limit of traffic flow and the proportion of heavy vehicles, consistent with a maximum acceptable level of, say, pedestrian delay and noise nuisance previously established as the minimum environmental standards for a particular street. The minimum standard and the most significant environmental factor (eg noise, risk of accident, etc.) might vary from street to street and at different times of day, according to the type of activities taking place and their sensitivity to intrusion by vehicular traffic. Thus, a street with wide footways and predominantly commercial frontages might have a higher figure for environmental capacity (ie a lower environmental standard) than a street with narrow footways fronted by terraced housing. Likewise, the environmental capacity for a residential street, based upon noise nuisance, would be higher during the day than during the night. Also, the degree of priority to be accorded to pedestrians and cyclists will have a bearing on the acceptable maximum level of vehicular traffic. The starting point for such decisions would normally lie in designating a hierarchy of the road network (see Chapter 11) but local factors, such as frontage activity, conservation area designation and the location of schools and hospitals, would also play a part.

Buchanan suggested that the environmental capacity for an access road or local distributor lies, typically, in the range 300–600 vehicles per hour, demonstrating that the maximum traffic flow compatible with a good environment will be substantially lower than the traffic capacity determined merely by the width and alignment of the carriageway or the intersections and other constraints along its length. Thus, to ensure that the environmental capacity of a street or network is not exceeded, design features or traffic management measures are often required to restrict traffic flow and to control the type and speed of vehicles permitted to use certain streets. There are many ways in which this can be achieved and a variety of techniques is explained in Chapter 20.

Deterioration of air quality has received increasing attention as an environmental concern. Under the Environment Act (HMG, 1995), local authorities are required to assess air quality and to identify areas where air pollution exceeds, or is forecast to exceed, thresholds specified in the national air quality strategy. Authorities must design Air Quality Management Areas and must draw up plans to reduce the levels of pollution to below specified thresholds. Such plans may include the management or control of traffic and thus could be fed into the consideration of the environmental capacity of links in the road network.

32.5 Balance of Capacities in a Hierarchical Network

Little purpose is served in providing link capacities on a route which are much greater than the capacities through junctions, although providing space to store queueing traffic may be a consideration. At peak times, some short links may be filled with vehicles queueing to pass through downstream junctions. Generally, though, there will be opportunities to use some of the roadspace between junctions to provide bus or cycle lanes, to widen footways in areas busy with pedestrians or to designate loading or parking spaces at frontages, provided that sufficient roadspace remains to balance the traffic flow with the capacity of the junctions, after allowing for the practical constraints described in Section 32.2. If the overall capacity offered is less than the traffic demand, drivers will tend to seek other routes. This may be acceptable, even beneficial, on an access road or local distributor whose environmental capacity is being exceeded and where less harmful alternative routes are available. On a district or primary distributor route, an excess of traffic demand over capacity will create a potential for the undesirable use of local residential roads by through traffic, ie 'rat–running'.

32.6 Assessing the Impact of Changes in Link Capacity

Whatever criteria are used to define link capacity, any proposed change to the capacity of a link should, like any other investment proposal, be appraised using appropriate methods which encompass all relevant effects. Details of economic and environmental appraisal methods are set out in Chapter 9.

When a road link or junction is subjected to traffic demand above its capacity, congestion ensues, speeds drop, delays are caused to vehicles, journey times become unpredictable, fuel is wasted by queueing traffic and vehicle exhausts pollute the air to a greater extent than would be the case with the same volume of smoothly–flowing traffic.

In respect of accidents, it is not always the case that increasing the capacity of a congested link will improve its safety record. Certainly, the severity of accidents tends to worsen with increased traffic speed and accidents can 'migrate', when traffic patterns are altered. Safety audits of individual schemes are needed to predict the consequences on accident levels (see Chapter 16). Likewise, it is not automatic that an increase in the capacity of an urban link provides overall benefits. When the capacity of a congested link is increased, there may be an immediate reduction in fuel consumption and pollution from vehicle exhausts if traffic flow is smoother. However, if traffic demand has been suppressed by congestion, the extra capacity may lead to traffic growth with consequent increases in fuel usage and pollution. If, on the other hand, the extra capacity is provided as part of an area–wide plan to strengthen the road hierarchy, then the benefits of diverting traffic away from residential environments onto the improved link may be realised without encouraging overall traffic growth.

32.7 References

Buchanan, C et al (1963)	The Buchanan Report 'Traffic in Towns', Stationery Office.
Buchanan, M and Coombe D (1976)	'A note on traffic speeds in London', Traffic Engineering + Control, (June).
DOT (1985)	TD20/85 DMRB 5.1 'Traffic Flows and Carriageway Width Assessment', DOT [Sb].
DOT (1992)	LA Circular 5/92 'Traffic in

London: Traffic
Management and Parking
Guidance', DOT.

HMG (1995)   'The Environment Act',
                    Stationery Office.

TRB (1985)   'Highway Capacity Manual'
                    Transportation Research Board
                    (US).

Chapter 33 Procedures for the Planning and Approval of Transport Infrastructure Schemes

33.1 Introduction

This chapter outlines the planning and statutory procedures required for the procurement of highways and fixed-track public transport facilities, including the powers to acquire land and the costs associated with acquisition. It should be noted that local authorities follow different procedures from the Department of Transport [NIa] [Sa] in the promotion of highway schemes. Consideration is also given to the funding of highway infrastructure by private finance, in association with development and as enabled under the New Roads and Street Works Act 1991 (HMG, 1991b) [NIb].

33.2 Initial Steps

Structure Plans and Part I of Unitary Development Plans (DOE/DOT, 1992) [NIc] [Sb] (see also Chapter 4) should define the primary transport network for an area and should set out general proposals for any major improvements to it, which are intended to commence within the timescale of the plan. Local plans and Part II of Unitary Development Plans should elaborate the proposals for the improvement of the primary network and indicate other improvements to non-primary transport links and services.

Schemes, not already included in the development plan, should be appraised and evaluated against, and be consistent with, the objectives drawn from the plan to ensure that they are properly integrated and compatible with the planning of transport and land–use matters (see Chapter 6). The provision and protection of transport infrastructure proposals in England is explained more fully in Planning Policy Guidance (PPG) Note 13 (DOE/DOT, 1994) [NId] [Sc].

Planning permission will normally be required for transport infrastructure. Applications require notification, consultation and publicity for the proposals comparable with the procedures applied to private applications [Sd].

33.3 Legal Framework

Highways
In respect of the legal procedures required to progress a highway scheme, the two main aspects which need consideration are the procedures involved up to, and including, the grant of planning permission and the procedures involved in the preparation and implementation of Highways Orders and Schemes.

Planning Permission
Planning permission is required under the Town and Country Planning Act 1990 (HMG, 1990) for development as defined in section 55 of that Act [NIe] [Se]. Development consists of either a material change in the use of land or operational development, consisting of building, engineering, mining or other operations in, on, over or under the land. For the purposes of the Act, land includes a building, as provided by section 336 of the 1990 Act [Sf].

Planning permission is required for the construction of new roads, whether these are proposed to be built by the relevant Highway Authority [NIf] for the purposes of the Highways Act (HMG, 1980) [Sg] or by a private developer; for instance, where the road is provided in conjunction with, and as part of, some other development scheme, such as an industrial or business park.

A distinction needs to be drawn between the provision of new roads, which may or may not become highways maintainable at public expense, and the carrying out of ancillary works. The latter fall within two categories. First, under section 55(2)(b) of the 1990 Act [Sh], the carrying out on land within the boundaries of a road, by a local highway authority [Si], of any works required for the maintenance or improvement of the road are deemed not to be development. Secondly, planning permission is granted by Part 13 of Schedule 2 of the Town and Country Planning (General Permitted Development Order) 1995 (HMG, 1995a) [NIf] [Sj] for works required for, or incidental to, the maintenance or improvement of a highway, by a local highway authority [Si], on land outside but adjoining the boundary of that highway. However, the 1995 Order [NIg] [Sj] specifically provides that nothing in that part of the Schedule authorises any development which requires or involves the formation, laying or material widening of a means of access to an existing highway, which is a trunk road or classified road, or creates an obstruction to the view of persons using any highway used by vehicular traffic, so as to be

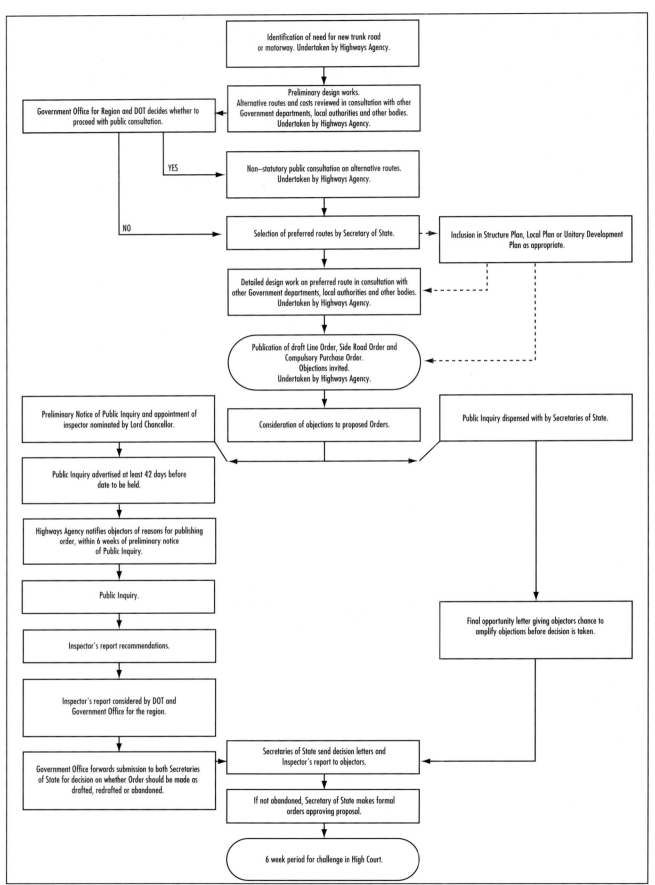

Figure 33.1: Main procedural stages for the planning and approval of trunk roads. Source HMG (1980).

TRANSPORT IN THE URBAN ENVIRONMENT

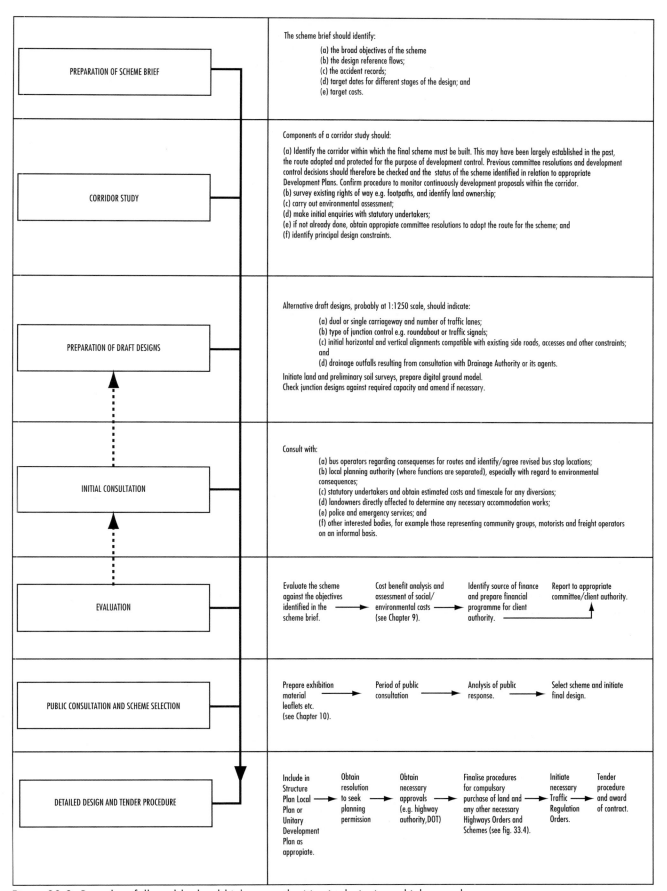

Figure 33.2: Procedure followed by local highway authorities in designing a highway scheme.

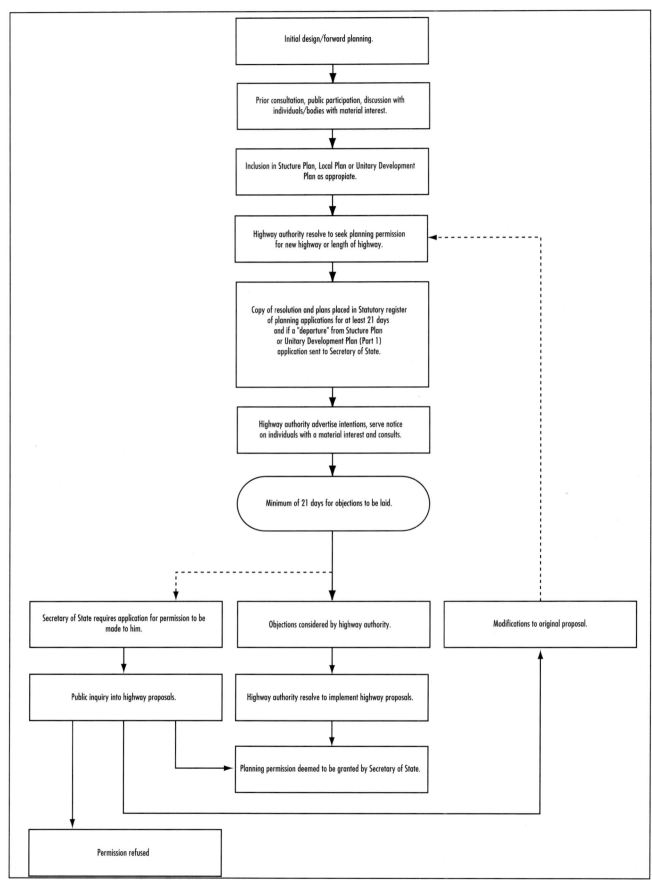

Figure 33.3: Procedure followed by local highway authorities in seeking planning permission for a highway scheme.

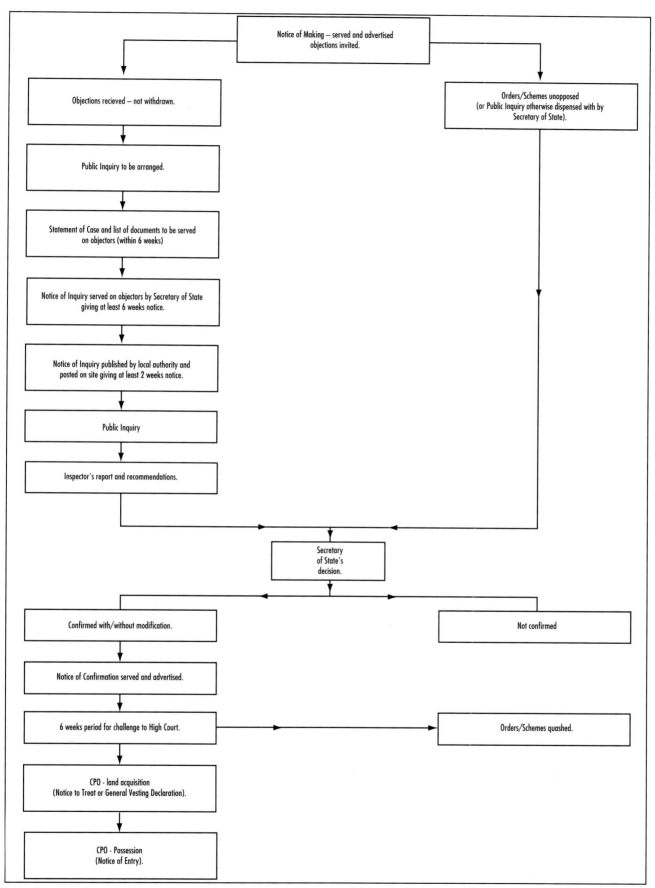

Figure 33.4: Procedure followed by local authorities in dealing with objections (if any) to a highway scheme.

likely to cause danger to such persons.

Once a planning application has been made, then the Local Planning Authority [NIh], which has jurisdiction to deal with the application, proceeds to handle it under Part III of the 1990 Act [NIi] [Sk]. There are a large number of considerations and policies which a local planning authority should consider when deciding whether or not to grant planning permission for a road scheme. In brief, these are:

❑ section 54a of the 1990 Act [Sl] provides that "where, in making any determination under the Planning Acts, regard is to be had to the development plan, the determination shall be made in accordance with the plan unless material considerations indicate otherwise". The development plan may comprise a county Structure Plan and/or a district Local Plan (where the latter exists) or in the case of unitary authorities, a new Unitary Development Plan [NIc] [Sm];

❑ the Department of the Environment in England [Sa], on behalf of the Government, publishes, from time to time, planning policy guidance notes and circulars, which indicate the Government's nationwide planning policy on certain issues. Of particular note is PPG 13: Transport (DOE/DOT, 1994) [NId] [Sc];

❑ emerging development plans, setting out policies which a local planning authority propose to adopt as the development plan in the future, have some weight in the decision-making process and, generally, their weight increases the further they progress through the process of drafting, advertising/consultation and final adoption; and

❑ issues, such as the planning history of the site and of adjacent land, the effects of the proposed scheme in terms of amenity, visual impact, traffic generation and the like, are clearly material considerations in the determination of a planning application. Generally, issues which relate to land–use planning are material considerations in the determination of planning applications and the courts are the ultimate arbiters of whether or not such a matter is a material consideration.

Since there is a duty upon a local planning authority to determine planning applications, section 70 of the 1990 Act [NIj] [Sn] provides that, where an application is made to a local planning authority for planning permission, they may either grant planning permission unconditionally or subject to such conditions as they think fit or they may refuse planning permission. In dealing with an application, a local planning authority shall have regard to the provisions of the development plan, so far as it is material to the application, and to any other material

considerations. The link is made at that point with Section 54a of the 1990 Act [Sl], as described above.

Orders and Schemes

The extent to which legal orders and schemes may be required in support of highway proposals should be established as soon as possible during the early planning stages. Failure to do so may delay the road–building programme and, at worst, could jeopardise the proposals entirely.

A variety of legal Orders and Schemes may be required under powers mainly contained in the Highways Act 1980, (HMG, 1980) [Sg]. These include:

❑ Line Orders – section 10 [NIk] [Sp];
❑ Side Roads Orders – sections 14 and 18 [So];
❑ Special Roads Schemes, including Line Orders – section 16 [NIl] [Sp];
❑ bridges over, or tunnels under, navigable waters Section 106 [NIe] [Sq];
❑ Stopping–up of private accesses – sections 124 and 125 [NIm] [Sr]; and
❑ Compulsory Purchase Orders – sections 239, 240 etc. [NIn] [Ss] and the Acquisition of Land Act 1981 (HMG, 1981) [St].

Full compliance is required with all relevant statutory requirements and advice circulars. In particular, the relevant prescribed forms must be used and the requirements relating to the publication and service of statutory notices must be closely followed.

The confirming Authority is the Secretary of State for Transport [NIo] [Su] [Wa] and, in the event of objections being submitted against the proposals, it is likely that a Public Inquiry will have to be arranged. Here again, it is necessary for there to be strict compliance with the statutory requirement governing the publication, and service, of all necessary statutory notices relating to the Inquiry.

Orders and Schemes, approved by the Secretary of State, become operative when notice of confirmation is first published. There then follows a six–week period, during which the validity of the Orders and Schemes may by challenged on legal grounds by application to the High Court. Any such challenge could involve considerable delay, even if successfully resisted.

The works authorised by the majority of Orders and Schemes can be implemented as soon as the challenge period has been successfully negotiated. In the case of Compulsory Purchase Orders, however, further action is required to secure the acquisition of the necessary land and rights. This is achieved either by serving Notices to Treat or by executing a General Vesting Declaration. It may also be necessary for

Notices of Entry to be served, in order to gain physical possession of the land prior to the commencement of construction works.

A non-exhaustive list of the principal source documents currently in use, in relation to local authority Orders and Schemes, is set out in Section 33.10. Some of the references set out in Section 33.11 are also relevant.

Once a local highway authority, or the Secretary of State in respect of trunk roads, has resolved to approve a highway scheme which affects land outside the highway or which includes specified features, such as additional lanes, elevated carriageways, subways or roundabouts, then that scheme must be declared against Local Search Enquiries. Such enquiries are made of local authorities to determine any possible effects upon land/property within the area of that authority. It is vitally important that approved highway schemes are declared as soon as possible, to minimise the costs of land/property acquisition.

An alternative to approval and declaration of an actual highway scheme is approval and declaration of a highway improvement line. In these circumstances, only the extent and general nature of the highway scheme has to be defined with none of the detail. Procedurally, the establishment of improvement lines can be extremely cumbersome. Hence, this course of action should be followed only where it is not possible to achieve the definition needed for a detailed scheme, but it is a useful way of avoiding blight notices.

Both improvement schemes and improvement lines should be declared against planning applications for the development of land. The objective of this is to minimise abortive investment in development, thereby ensuring, as far as possible, that development accords with future highway improvement proposals whilst, at the same time, minimising the costs of land/property acquisition.

Under the provisions of Article 10 of the Town and Country Planning (General Development Procedure) Order 1995 (HMG, 1995b) [Sx], local planning authorities are required to consult the Local Highway Authority [Si] or the Secretary of State for Transport [Su] [Wb], where development proposals are likely to result in a variety of effects upon a highway as specified in Article 10. Additionally, the Local Planning Authority must consult the Secretary of State for Transport [Su] [Wc], where development proposals affect an existing or proposed highway as specified in Article 15. Local planning authorities are

then required to take account of representations received from the Highway Authorities or their agents. Directions from the Secretaries of State are binding on local planning authorities.

Only the Secretaries of State have the power to direct local planning authorities to refuse or restrict the granting of planning permission for development, so as to protect the opportunity to improve or build a highway. This applies both to highways for which the Local Highway Authority [Si] is responsible and to those which are the responsibility of Secretaries of State. The Secretaries of States' powers of direction under Article 10 of the Town and Country Planning Act (General Development Procedure) Order 1995 are laid out in Articles 14 and 15 (HMG, 1995b) [Sw].

Whilst local planning authorities may seek to influence development, so as to protect highway improvement proposals and minimise blight, they can only actually control development in this respect in two situations: first, where highway schemes have been included in formal development plans, local planning authorities are empowered to control development so as to protect those schemes, under the provisions of the Town and Country Planning Act 1990 (HMG, 1990) [Sx]; and secondly, when directed to do so by the Secretaries of State, as explained above. Good liaison and co-operation between authorities is clearly beneficial to all parties.

It would be wrong to allow land or buildings to be blighted or lie idle for a considerable period because of a potential road scheme. Therefore, safeguarding powers should only be used when a particular road scheme is included in the development plan and, in the case of trunk roads, is also included in the current Roads Programme. If a road proposal, included in a draft or approved development or alteration, gives rise to blight notices, then the Local Highway Authority [Si] or the Department of Transport [Sa] [Wd], whichever is responsible for the scheme, will be the appropriate authority under section 169 of the Town and Country Planning Act (HMG, 1990) [Sw], provided that the proposal has been included in the plan with their consent. If a local planning authority includes a road proposal without the relevant Highway Authority's consent, it may be that the Local Planning Authority will be held to be the appropriate authority under section 169 of the 1990 Act [Sy].

Footways and cycleways
Section 75 of the Highways Act 1980 (HMG, 1980) [Sab] gives authorities powers to widen footways and, generally, to vary the widths of carriageways and footways.

Cycleways or carriageways are either advisory, in which case a Traffic Regulation Order (TRO) is not required, or mandatory, in which case a TRO is required, to prevent use by others and to allow for appropriate signing. This procedure to convert all, or part, of a footway to a cycle track involves 'removing' all or part of the footway under section 66 of the Highways Act and then 'constructing' a new cycle track under section 65 (1) of the Act (see IHT, 1996) [Sac].

33.4 Land Acquisition

The wide variety of land–uses, land–owners and occupiers in urban areas can make the acquisition of land for a road difficult. When land cannot be purchased by agreement between the owner and the Highway Authority, it may be acquired compulsorily, using powers currently provided by the Highways Act 1980, section 239 (HMG, 1980) [NIn] [Sz]. Guidance on the making, and submission for approval, of compulsory purchase Orders under the Act is given in DOE and DOT Circulars (DOE, 1994 and DOT, 1981) [Saa]. These draw attention to the need to provide a planning backing to justify compulsory purchase and to the importance of identifying related matters, such as development plan proposals, planning applications and appeals, on which it may be desirable to arrange for a concurrent, or joint, inquiry with the highway compulsory purchase Order, on the grounds of speed and efficiency. The acquisition of areas of special category land, such as public open space, burial grounds, allotments and common land can involve additional procedures, which usually require replacement land to be provided (HMG, 1981).

33.5 Land Compensation

The Land Compensation Act (HMG, 1973) [NIp] [Sad] provides a right to compensation for depreciation of value of interests in land, caused by physical factors arising from the use of highways, aerodromes and other public works. The Act gives highway authorities [Si] powers and duties to mitigate the injurious effect of such works on their surroundings and amended the law relating to compulsory purchase and planning blight.

As a result of this legislation, designers of road schemes must give careful consideration to the following:
❑ compensation can be claimed (under Part I of the Act) by owners of property, where its value is depreciated by noise and other specified physical factors (including traffic calming measures) arising from the use of a new or altered highway (the period during which claims can be made is six

years, starting one year after the first opening of the new or altered highway). Note that intensification of traffic on an unaltered highway (ie. an existing use) is specifically excluded as a valid reason for making a claim for compensation;
❑ in certain instances, discretionary purchase powers are available to highway authorities [Si] under section 246 of the Highways Act (HMG, 1980) [Sae] introduced in the Land Compensation Act (HMG, 1973) and the 1991 Planning and Compensation Act (HMG, 1991a) [NIq]. The Highways Agency [Saf] [We] issues public guidelines for discretionary purchase under section 246(2A) [Sae];
❑ the Noise Insulation Regulations 1975, as amended by the Noise Insulation (Amendment) Regulations (HMG, 1988), made under section 20 of the Act, require a highway authority [Si] to make offers of noise insulation, or grant, to occupiers of dwellings subjected to additional noise at, or above, the specified level, due to the use of a new highway or a highway to which a new carriageway has been added. The noise insulation package includes the provision of secondary windows, venetian blinds, double or insulated doors and supplementary ventilation – eligible rooms are living rooms and bedrooms only; and
❑ these Regulations give certain powers to a highway authority [Si] to make similar, discretionary, offers of noise insulation, or grant, where dwellings are subjected to additional noise at, or above, the specified level due to the use of a highway altered in vertical or horizontal alignment other than by resurfacing.

There are also powers to enable offers of noise insulation to be made to occupiers of properties adjacent to the site or roadworks, if it is felt that construction of the works will give rise to noise at such a level that it will seriously affect the enjoyment of a dwelling for a substantial period of time.

33.6 Public Consultation

The procedures for the planning and approval of road schemes provide several opportunities for public consultation. These include development plans, planning applications, line orders, side–road orders, traffic regulation orders and compulsory purchase orders. Nowadays, most highway authorities will carry out a large scale, non–statutory, public consultation exercise using the techniques described in Chapter 10.

33.7 Public Inquiries

There are statutory rights under the Highways Act for

objections to be made to line orders, side–roads orders and to any compulsory purchase orders necessary for a road scheme to proceed [Sah]. If objections to the published draft Orders are made, and not subsequently withdrawn, a local public inquiry will be held (HMG, 1980, Schedule 1 para 7). Strictly, if the only objections are to the compulsory purchase order, and relate solely to matters of compensation, an inquiry need not be held but this rarely happens. The procedures for the conduct of public inquiries are described in Chapter 10 (see also Section 33.11).

Trunk Roads

The main procedural stages followed by the Department of Transport [Wf] for the planning and approval of trunk roads under the Highways Act 1980 (HMG, 1980) [Sai] are set out in Figure 33.1. Trunk road proposals should also be included in Structure Plans, Local Plans and Unitary Development Plans [Sm] when they are included in the current roads programme. Where an order under the Highways Act, as specified in section 24 of the Town and Country Planning Act (HMG 1990) [Saj], has already been made on a particular trunk road scheme, the Secretary of State for the Environment [Wg] and the Local Planning Authority have discretionary power to disregard objections to the inclusion of the scheme in the development plan.

Local Authority Roads

The main steps which local highway authorities follow in developing a highway scheme, from initial design brief to tender stage, are set out in Figure 33.2. Figures 33.3 and 33.4 outline the procedures, which must be followed by a local highway authority to obtain planning consent and confirmed orders, to permit a scheme to be built. These differ from the procedures followed by the Department of Transport [Sak] [Wh].

33.8 Privately–Funded Roads

New Roads in New Development

Where new residential or commercial development is to take place, the issue of new publicly–maintainable roads will usually arise. Guidance for new residential roads can be found in Design Bulletin 32: Residential Roads and Footpaths – Layout Considerations (HMG, 1992b) [NIr], as supplemented by local highway design guides produced by local highway authorities [Si]. The standards for road and footpath layouts, which result from applying the advice in these documents, will allow a sensible balance to be struck between planning and highway considerations in the design of most new developments, where the design

should also encourage the use of walking, cycling and other forms of transport. The advice and guidance in the documents is equally applicable to new developments, whose roads are not intended for adoption (see Part 4). The above advice and guidance should be taken into account at the initial stage of a new residential development ie. when preparing and determining planning applications.

Following building regulation approval, the Local Highway Authority [Si] is charged with issuing, to developers, Advanced Payment Code Notices [Sal], which advise the developer of the requirements to place, with the Local Highway Authority [Si], a sum of money estimated to be the cost of completing the areas of new adoptable highway within the development. This money must be deposited with the Local Highway Authority [Si] before development commences.

As a voluntary, and normally used, alternative to the Advanced Payment Code procedure, developers may enter into an adoption agreement with a local highway authority, under the provisions of section 38 of the Highways Act 1980 (HMG, 1980) [NIs] [Sam]. Such agreements, which are in a nationally agreed model form, allow for approval of design and specifications of all aspects of new road construction, dedication of land, arrangements for inspection of construction, staged adoption and the deposit of a bond to cover potential default. The money deposited can be returned to the developer as and when the new roads are completed to the satisfaction of the Local Highway Authority. However, in the event that satisfactory completion is not obtained, the money can be used by the Authority to complete the roads.

Financing of Major Roads

The New Roads and Street Works Act (HMG, 1991b) is "an act to amend the law relating to roads so as to enable new roads to be provided by new means"; ie through private finance. It relates to special roads, as defined by the Highways Act 1980, to which all the provisions contained in the Act apply [NIt]. It enables the Highway Authority, usually the Department of Transport [Wi], to enter into a concession agreement with a concessionaire for the design, construction, maintenance, operation and improvement of a special road, in return for which the concessionaire has the right to charge tolls for the use of the road.

The Act provides for the concessionaire to transfer, to the Highway Authority, any land that he holds that is required for the scheme. It also provides for compensation payments by a highway authority to the concessionaire, and to a highway authority by a concessionaire, in the event of failure, by either party, to fulfil the requirements of the agreement.

Under section 2 of the Act, a concession agreement may authorise a concessionaire to exercise highway functions. Such functions are also one of the matters prescribed by statutory instrument as having to be covered by a 'Concession Statement' published with the Toll Order (see schedule 2–1(3)&(4) of the Act). The Act further defines the requirements placed on both the Highway Authority and the concessionaire on termination of the agreement, either at term or prematurely.

The authority for the charging of tolls, by either a highway authority or a concessionaire, is provided by the making of a toll order. For a special road, the order is made by the Secretary of State. The procedures required for the making of a toll order should be run in parallel with those required under the Highways Act (HMG, 1980). The toll Order and the Orders for a scheme under the 1980 Act cannot be made separately: if either the toll Order or the scheme cannot be confirmed, both fail [HMG, 1980 section 6(5)] [NIu] [San].

The toll Order identifies the period over which tolls will be charged but cannot specify a maximum toll to be charged by a concessionaire, unless the route includes a major crossing for which there is no convenient alternative. The maximum toll to be charged by a highway authority must be specified in every case. Toll Orders can be varied by the Secretary of State but only with the agreement of the concessionaire, if the road is subject to a concession. The Act also covers how provision can be made for the collection of the tolls.

Section 38 of the Highways Act 1980 (HMG, 1980) [NIs] [Sao] has been amended to provide for the adoption and maintenance of privately–constructed special roads.

Private Finance Initiatives and Design Build Finance and Operate (DBFO) Schemes

Although the Government may be responsible for delivering a service and the cost of providing it, the Government need not be responsible for providing the service or the initial finance. The Private Finance Initiative (PFI) is the name given to policies announced by the Chancellor of the Exchequer in 1992. The aim of the PFI is to have services and infrastructure, formerly provided and financed by the public sector, provided instead by the private sector and financed by private capital. This allows the Government to bring forward the benefits of schemes earlier than the Exchequer could afford, against a background of constraint in public spending, but necessarily requires the capital to be repaid.

The PFI in highway procurement is about highway authorities contracting with Design, Build, Finance and Operate (DBFO) companies to provide, manage and maintain roads over, typically, a 30–year period. For the private sector, the PFI offers the opportunity for further business and profit. In return for this opportunity, the Government, in transferring the cost of providing, operating and maintaining highways over an extended period, is also seeking to transfer risk from the public to the private sector, whilst at the same time seeking overall value for money.

Private money is more 'expensive' than public money, because Government can borrow more cheaply. However, PFI is based on the expectation that the private sector will be able to deliver projects more efficiently and cheaply, thus offsetting the extra cost of capital borrowing. A test has been devised to prove such value for money when considering schemes for PFI. Adequate transfer of risk from the public to the private sector must also be demonstrated.

Risk is present at all stages of highway scheme development; feasibility, planning, Orders, public inquiry and construction. But perhaps the greatest risk lies in the accurate forecasting of traffic, on which future income and long–term maintenance costs depend. In the past, the pricing of risk has presented the greatest difficulty for the public sector and now presents a corresponding difficulty for the private sector. While it is reasonable to expect the private sector to bear the cost of the accustomed risks (ie finance, design, construction, operation and maintenance), the public sector might well be better placed to carry other risks, such as planning permissions.

Despite undoubted benefits in bringing forward the construction of new highway schemes, the use of private finance involves long–term commitments. These commitments need to be taken into account by Government, in determining the scale of financial resources available for new infrastructure. Clearly the widespread introduction of projects financed by means of PFI could also result in a variety of different ventures for different parts of the network.

Trunk road procurement through DBFO initiatives is usually paid for by the Department of Transport [Sa] [Wj] in the form of 'shadow' tolls, although PFI can fund the procurement of highway infrastructure through actual tolls as, for example, with the Skye Bridge. There are arguments for and against private tolls. Whilst they clearly generate additional direct revenue, which can ease the burden of repayment, they could lead to the transfer of traffic to non–tolled roads, with the attendant damage to the highway network and the environment.

Government policy is to introduce electronic motorway tolling when the technology has demonstrated its capabilities and Parliament has approved appropriate legislation. All the technical elements exist but there is room for improvement in the standard of enforcement that is achievable.

33.9 Fixed–Track Systems

Legal Background
The construction and operation of railways, tramways, trolley vehicle–systems and other systems of guided transport would normally need statutory authorisation, unless the systems were to be constructed entirely on private land. Until 1992, statutory powers for such schemes in England and Wales were obtained by promoting a private Bill in Parliament. The Transport and Works Act (TWA) (HMG, 1992c) established a new procedure, under which these schemes could be authorised by an Order made by the Secretary of State [NIv] [Sap].

The purpose of a TWA Order is to provide the powers necessary for the implementation of a scheme, including compulsory acquisition powers and powers to interfere with public rights of way, for example, by the placing of equipment in the street. On making a TWA Order, the Secretary of State may give a direction that planning permission be deemed to be granted for development authorised by the Order. Obtaining a TWA Order will also enable an operator to claim a defence of statutory authorisation against actions for nuisance arising from the proper operation of the system.

The procedures under the TWA have many similarities to those which apply to highway schemes and other forms of development, particularly in the arrangements for objections to schemes to be considered at a public local inquiry. A full description of the procedures is provided in the Transport and Works Act (TWA) 1992: A Guide to Procedures (HMG, 1992d). The key features of the process are outlined below.

Promoters are encouraged by the Department of Transport to carry out thorough consultation with relevant statutory bodies, such as local authorities and bodies with environmental responsibilities, and with the general public, well in advance of making the formal application for an Order to the Secretary of State for Transport. This is intended to help to identify and resolve problems before the form of a scheme is finalised, so reducing the potential for objections and lengthy inquiries. Consultation is also an essential part of the environmental assessment process which is required, under the European Community Directive, for all but the smallest of schemes. These matters are the subject of non–statutory guidance. They are backed up by a number of statutory requirements in the TWA Applications Rules (HMG, 1992e) which are designed to ensure that an applicant involves relevant bodies in the preparation of an application.

Application for a TWA Order
A promoter has to provide, along with an Order application, a number of documents which define the scheme for which powers are being sought. These documents include:
- ❑ a draft of the proposed Order;
- ❑ an environmental statement;
- ❑ large–scale plans and sections of the proposed works; and
- ❑ plans and drawings to support a request for deemed planning permission.

The TWA Applications Rules (HMG 1992e) include comprehensive publicity requirements to ensure that the public is made aware of an application. Promoters have, for example, to publish a notice of their application, twice, in at least one local newspaper and also to post notices at the site of proposed works. Copies of application documents have to be served on a variety of statutory bodies and others who would be affected by a scheme and must be made available for public inspection. All owners and occupiers whose land is subject to proposed compulsory acquisition also have to be served with a notice of the application.

Promoters must pay an application fee to cover the Department of Transport's costs in processing the application. The level of fee relates to the area of land required for the proposed works.

Objections
The statutory objection period runs for a minimum of 42 days from the date of application, during which time anybody may make a representation to the Secretary of State about the application. At the end of the objection period the Secretary of State has four weeks, normally, to decide whether to hold a public inquiry, or a hearing, into any objections or whether to consider objections by means of the written representations procedure. The Secretary of State may, however, allow time for negotiations to be carried out before he decides which procedure to

invoke. Under section 11 of the TWA, persons, whose land is subject to proposed compulsory acquisition powers, and local authorities, in whose area works are to be carried out, may require the holding of an inquiry or hearing. If there are no objections to an application, the Secretary of State can proceed to determine the application directly.

Public Inquiries

The Secretary of State will usually decide to hold a public inquiry where the number of objections makes it impracticable to deal with them by an exchange of written representations between the parties. The TWA inquiry procedures (HMG, 1992f) follow, largely, the pattern of those applying to highways and other forms of development. The cost of holding an inquiry, including the provision of an independent inspector, is borne by the promoter. Parties to the inquiry are, nevertheless, expected to meet their own costs in preparing and presenting evidence.

The purpose of an inquiry is to facilitate the examination of a scheme, and its effects, in a public forum. Typically, environmental and planning issues are a major topic of debate at an inquiry but any other relevant matter may be discussed, such as the acceptability of compulsory acquisition proposals and, sometimes, the financial viability of a scheme. Inspectors will normally allow anyone to speak at an inquiry who has something relevant to say.

After the inquiry, the Inspector writes a report on the evidence heard. It will include a recommendation to the Secretary of State as to whether the Order should be made and as to whether any conditions should be attached to the planning direction.

Decisions

The Secretary of State is responsible for deciding whether to make, or to refuse, an Order. The Secretary of State may also decide at this stage to modify an Order, for example in the light of evidence presented to an inquiry. If, however, the Secretary of State considered that a scheme should be significantly altered, the Secretary of State would be likely to ask the promoter to apply for an Amending Order. This would ensure that all those affected by the changes would have a proper opportunity to consider them and to make representations. The Secretary of State's decision on an Order may be challenged in the courts.

Land Compensation

Statutory blight procedures, under the Town and Country Planning Act 1990 (HMG, 1990), apply to schemes promoted under the TWA which involve the compulsory acquisition of land from the date on which the Order application is made. Promoters must, therefore, have funds available to meet these costs which they may incur before an application is determined.

The normal Compensation Code applies to the compulsory acquisition of property under a TWA Order, as to highway schemes and other forms of development. Thus, as well as paying compensation for the market value of property which is compulsorily acquired, a promoter may have to pay compensation for adverse effects caused by the carrying out of works and for depreciation in the value of property affected by physical factors arising from the use of the authorised works. In addition, the Noise Insulation (Railways and Other Guided Transport Systems) Regulations (HMG, 1996) impose a duty on the body responsible for the construction of a new railway line or guided transport system to provide or pay for insulation to nearby dwellings, if noise exceeds certain prescribed levels.

Parliamentary procedures

TWA Orders are not normally subject to any Parliamentary consideration. Where, however, the Secretary of State considers that proposals included in a TWA application are of 'national significance' and makes an announcement to that effect within 56 days of the application being made, the procedures in section 9 of the TWA apply. In these cases, the application is referred to Parliament before an inquiry is arranged. The Secretary of State can only make an Order relating to a scheme of national significance, if both Houses pass a motion approving the application (in principle). Thus, if either House rejected a proposal in principle, the Secretary of State could not progress the application any further.

Section 12 of the TWA also provides that applications shall be subject to Special Parliamentary Procedure in the circumstances specified in sections 18 and 19 of the Acquisition of Land Act 1981, (HMG, 1981). This means that, where the compulsory acquisition of land held inalienably by the National Trust is opposed, or where common land or open space is to be acquired without the provision of exchange land, the TWA Order, once made, has to be referred to Parliament for confirmation.

Other relevant legislation

Fixed–track systems, which employ rails within the carriageway, need to comply with the provisions of the New Roads and Street Works Act 1991 (HMG 1991b). Consideration should be given to designating the route as 'streets with special engineering difficulty'. This effectively prevents any statutory undertaker from carrying out work, which would affect the track, until plans have been agreed with the highway authority.

The operator will also be required to comply with the provisions of the Act when carrying out any maintenance work to the system on the highway, which involves opening up the carriageway or footway or creating a diversion.

Any fixed–track system will be subject to the Railways and Other Transport Systems (Approval of Works, Plant and Equipment) Regulations 1994 (HMG, 1994) requiring the Secretary of State's consent before any new or altered works, plant or equipment, including rolling stock, may be brought into use. This consent is given by Her Majesty's Railway Inspectorate. For kerb–guided buses, a dispensation may be granted by the Inspectorate.

All new railways are subject to the regulatory system, established by the Railways Act 1993 (HMG, 1993), unless exempt from its requirements in the manner provided by the Act. The Act includes provisions for the licensing of train and network operators, the franchising of passenger services, open access to railway networks and for controlling the closure of railways.

33.10 References

DOE (1994)　　Circular 14/94 (WO 4/95) 'Compulsory Purchase Orders: Procedures,' DOE [Saa].

DOE/DOT(1992)　Planning Policy Guidance PPG12 'Development Plans and Regional Guidance', DOE/DOT [Saq] [Wk].

DOE/DOT (1994)　Planning Policy Guidance: PPG13 'Transport', DOE/DOT [Saq] [Wl].

DOT (1981)　　Circular Roads 1/81 (WO 1/81) 'Notes on the Operation of Compulsory Purchase Orders for Highways,' DOT.

HMG (1973)　　'Land Compensation Act 1973', Stationery Office [Sad].

HMG (1980)　　'Highways Act 1980', Stationery Office [Sg].

HMG (1981)　　'Acquisition of Land Act 1981', Stationery Office [St].

HMG (1988)　　'Noise Insulation Regulations 1975 as amended by the Noise Insulation (Amendment Regulations) 1988', Stationery Office [Saq].

HMG (1990)　　'Town and Country Planning Act 1990' (as amended), Stationery Office [Sx].

HMG (1991a)　　'Planning and Compensation Act 1991', Stationery Office.

HMG (1991b)　　'New Roads and Street Works Act 1991', Stationery Office.

HMG (1992a)　　'Town and Country Planning General Regulations 1992' SI 1992 No. 1492, Stationery Office [Sar].

HMG (1992b)　　Design Bulletin 32: 'Residential Roads and Footpaths – Layout Considerations', Stationery Office.

HMG (1992c)　　'Transport and Works Act 1992' Chapter 42, Stationery Office.

HMG (1992d)　　'Transport and Works Act 1992 : A Guide to Procedures' Stationery Office.

HMG (1992e)　　SI No. 1992/2902 'The Transport and Works (Applications and Objections Procedure) Rule 1992', Stationery Office.

HMG (1992f)　　SI No 1992/2817 'The Transport and Works (Inquiries Procedure) Rules 1992', Stationery Office [Sas].

HMG (1993)　　'Railways Act', Stationery Office.

HMG (1994)　　'Railways and Other Transport Systems (Approval of Works, Plant and Equipment) Regulations 1994', Stationery Office.

HMG (1995a)	'Town and Country Planning (General Permitted Development Order) 1995', Stationery Office [Sj].
HMG (1995b)	'Town and Country Planning (General Development Procedure) Order 1995' Stationery Office [Sv].
HMG (1996)	'Noise Insulation (Railways and Other Guided Transport Systems) Regulations', Stationery Office [Sas].

33.11 Further Information

DOE Circulars	1/90 'Compulsory Purchase Inquiries Procedure Rules' 2/90 'Compulsory Purchase Regulations', DOE [Saa].
DOE Pamphlets	[Sad] [Wm] Land Compensation Your Rights Explained: Booklet 1 – 'Your Home and Compulsory Purchase' 1991. Booklet 2 – 'Your Home and Nuisance from Public Development'1992. Booklet 3 – 'Your Business and Public Development' 1992. Booklet 4 – 'The Farmer and Public Development' 1992. Booklet 5 – 'Insulation Against Traffic Noise' 1992. 'Compulsory Purchase Orders A Guide to Procedure', Stationery Office (1978)Addendum (1981) Reprint 1988.
DOT Leaflets	[Sau] [Wn] 'Public Inquiries into Road Proposals – What you need to know', HMG (1995). 'Trunk Road Planning and the Public', HMG (1993). 'Trunk Road Proposals and your Home', HMG (1993).

Statutory Instruments

| HMG (1971) | 'Stopping–up of Access to Premise (Procedure) Regulations' No.1707, Stationery Office. |

HMG (1984)	'Road Traffic Regulation Act 1984' Chapter 27, Stationery Office.
HMG (1988)	'Town and Country Planning (Assessment of Environmental Effects) Regulations' No.1199, Stationery Office [Sav].
HMG (1990a)	'Compulsory Purchase (Vesting Declarations)Regulations' No.497, Stationery Office [Saa].
HMG (1990b)	'The Compulsory Purchase by Non–Ministerial Acquiring Authorities (Inquiries Procedure Rules' No.512, Stationery Office [Saa].
HMG (1992)	'The Transport and Works (Model Clauses for Railways and Tramways) Order' No.3270, Stationery Office [Sas].
HMG (1993)	'The Special Road Schemes and Highway Orders (Procedure) Regulations' No.169, Stationery Office.
HMG (1994a)	'The Compulsory Purchase of Land Regulations' No.2145, Stationery Office [Saa].
HMG (1994b)	'Highways (Inquiries Procedure) Rules' No.3263, Stationery Office [Sal].
HMG (1994)	'The Compulsory Purchase by Ministers)Inquiries Procedure) Rules 1994; No.3264, Stationery Office [Sal].

Chapter 34 Fixed–Track Systems

34.1 Introduction

The importance of high quality public transport, as a component of the urban transport system, is now widely recognised. A substantial proportion of the population rely on public transport as their only means of travel, particularly the young, the elderly and those with some mobility impairment. Good quality public transport is also an essential part of any integrated package approach to transportation planning in medium and large cities, where it is not feasible to meet the total demand for private motorised travel and there is a need to provide an attractive alternative to private cars.

Two essential features of public transport, if it is to be attractive to a wide range of users, are accessibility and reliability. Both these characteristics can be enhanced greatly if the public transport system can be protected from the influences of other road traffic, especially where there is any risk of congestion. A proven method of achieving this protection is the use of segregated fixed–track systems. Moreover, the environmental quality of urban areas, particularly town and city centres, can only be sustained if provision of high capacity transport does not itself impose unacceptable environmental impacts (RCEP, 1994).

This chapter describes the features of fixed–track systems, the legislation applying to their introduction, some key design factors and the integration of tracked systems within the highway. Any system which uses rails within the carriageway is termed a tramway and is subject to the safety principles and guidelines laid down by Her Majesty's Railway Inspectorate (HMRI) (HSE, 1996). Approval to operate any new system must be obtained from the Secretary of State for Transport prior to public service commencing, such approvals being the responsibility of HMRI (HMG, 1994) [NIa].

34.2 Types of Fixed–Track System

System classification

The wide range of fixed–track public transport systems, with differing characteristics of vehicle, track and method of operation, may be considered under three broad headings:
- ❑ partially segregated, bus–based systems;
- ❑ partially segregated, rail–based systems; and
- ❑ fully segregated systems.

Systems which utilise conventional buses, sometimes termed 'busway transit', are described in more detail in Chapter 24. Fully segregated systems, which include conventional railways and metro systems, are not within the scope of this book but are listed for the sake of completeness.

Partially–segregated, bus–based systems
(see Chapter 24)
The characteristics of bus–based systems are determined by the type of vehicle and the type of track. Most examples use conventional buses, powered by diesel or gas engines, or electrically–powered trolleybuses. The track may be either a conventional carriageway built to normal highway standards for heavy commercial vehicles or it may incorporate a method of steering the vehicles automatically. There are few examples of the former, the most extensive being in Runcorn, where a dedicated system was built in the 1960s for bus use. In the case of automatically steered vehicles, there are three primary methods – kerb guidance, central rail guidance and electronic (cable) guidance.

Kerb–guided buses run on a track which has vertical kerbs about 180mm high to form vertical rails, which guide small horizontal solid rubber–tyred wheels attached to the front axle of the bus. Articulated buses may have guidewheels fitted to all axles. Thus, the vehicle is steered automatically, when running on the track, yet can be driven manually on normal roads. A short 'funnel' is negotiated under manual guidance to enter the guided track and short gaps in the track can be allowed for pedestrian crossings (see Photograph 34.1). At road intersections, longer gaps are required and the bus must leave and re–enter the guided sections. The track gauge, ie the distance between the vertical rails, is 2.60m for 2.50m wide buses, which allows 50mm each side for the guide wheels. The wheels protrude by 50mm outside the width of the vehicle body. In Ipswich, a guideway has been designed for 2.40m wide midibuses and the gauge is therefore 2.50m.

Guided bus track may be combined with tramway track to permit buses and trams to operate along the same lane, segregated from other traffic. This is relatively straightforward with metre gauge tram track but more difficult with standard gauge track. Guided buses operate in Essen (Germany) and Adelaide (Australia), with smaller examples in Mannheim (Germany) and in Ipswich and Leeds.

Photograph 34.1: The kerb–guided busway in Kesgrave, Ipswich.

Electronic cable–guided buses follow a buried cable in the carriageway which controls the steering mechanism. They have been demonstrated on test tracks at Fürth and Rastatt in Germany, TRL at Crowthorne and in Newcastle. This system is used in the service tunnel of the Channel Tunnel.

Rail–guided buses have a mechanical device, which uses a form of double–flanged wheel to engage with a central rail in the carriageway for operation in guided mode. The guidewheels are retracted for manual operation off the guideway. The rail is flush with the road surface, as for tramway track, and may therefore be crossed by other road vehicles. Originally developed in the 1930s, the principal design is the GLT (Guided Light Transit) from Belgium, known in France as the TVR (Transport sur Voie Reservée). A 4km test track operated at Rochefort in Belgium and the first commercial application is planned to operate in Caen, France (see Photograph 34.2). GLT vehicles are rubber–tyred, double articulated with diesel and electric power packs. They operate on overhead

Photograph 34.2: A rubber–tyred guided light transit (GLT) vehicle for Caen, France. Courtesy of Bombardier Eurorail.

power supply when in guided mode and on diesel when in manual mode.

Trolleybuses are 'fixed route', by virtue of their overhead mains power supply, but are not 'fixed-track', as they can manoeuvre up to about 5m each side of the overhead lines. They may become fixed-track when operated as guided buses, as in Essen. Trolleybuses have not been used in Britain since 1972. A new prototype trolleybus has been demonstrated in Doncaster. Most trolleybuses have an auxiliary power supply, either traction battery or small internal combustion engine, to enable them to manoeuvre off–wire for depot movements, minor diversions or emergencies.

Photograph 34.3: Guided duo–buses in the median of a dual carriageway in Essen, Germany.

'Duo–buses' have dual power packs, to enable them to operate either in normal passenger service, as trolleybuses with overhead current collection, or off–wire, using the diesel engine. Duo–buses are used

Photograph 34.4: A major signal–controlled intersection between an urban distributor road, a reserved track tramway and an entrance to a bus station – Piccadilly Gardens in central Manchester (before reconstruction of the bus station in 1995).

446

on the guided busways in Essen (see Photograph 34.3) and in normal traffic in Nancy (France).

Partially–segregated, rail–based systems

Rail–based systems, which can operate on a highway, normally use a grooved rail section which is laid flush with the carriageway surface so that it does not present any danger or impediment to normal road traffic. When operating on–street, such systems are known collectively as tramways (see Photograph 34.4).

A 'tramway' is defined in the Transport and Works Act (HMG, 1992) [Sa] as "a system of transport used wholly or mainly for the carriage of passengers and employing parallel rails, which:
❑ provide support and guidance for vehicles carried on flanged wheels; and
❑ are laid wholly or mainly along a street or in any other place to which the public has access (including a place to which the public has access only on making a payment)."

Light Rail is a tramway which has a significant proportion of its route on tracks, segregated from ordinary road traffic, either within the highway or on separate rights–of–way. Where it operates on fully segregated and fenced alignments, it may be classed as a railway but the system as a whole is still termed light rail. These systems now operate in Manchester and Sheffield, a third system (Midland Metro) is under construction in the West Midlands and a fourth in Croydon.

Fully segregated systems

Fixed–track systems that do not share space with road traffic or pedestrians, except at level crossings, are fully segregated. They may run at grade, elevated or in tunnel and may share a highway alignment, for example in the median strip of a dual carriageway.

Light Rapid Transit (LRT) is a term which embraces all the fixed–track modes which have vehicle weight and operating characteristics lighter than normal railways; in particular, they can negotiate steeper gradients and smaller curve radii. Thus, light rail, monorails, guideways, peoplemovers and some metro systems are all forms of LRT. Traditional tramways, guided buses and trolley buses have some of the characteristics of LRT.

The Tyne and Wear Metro in Newcastle upon Tyne and the Docklands Light Railway in London are both fully segregated and both may be classified as light rail systems within the broader classification of light rapid transit. However, Tyne and Wear Metro is, in many ways, indistinguishable from a heavy metro

system, except that the vehicles are lighter and based on a type that is designed for street running (but never implemented) and there is a simplified signalling system. The Docklands Light Railway is automated with no drivers and the tracks are, therefore, fully segregated and fenced throughout. Much of the alignment is elevated, or in tunnel, to achieve this degree of segregation. Part of the Becton extension route is located in the median of a dual carriageway but this does not constitute street–running.

Underground or Metro systems are constructed to heavy railway standards with longer trains and platforms, a larger loading gauge (except for tube railways) and a full block safety signalling system. They are fully segregated throughout, often with a high proportion of route in tunnel. Examples in Britain are the London Underground, Glasgow Underground and the Merseyrail network based on Liverpool. Most metro systems use third rail current collection, while most light rail systems use overhead power supply, which is essential where any section is not fully segregated.

Automated Guideway Transit (AGT) is a form of light rapid transit which has a proprietary track and fully automatic operation. They cannot therefore be used in any street running applications but may follow highway alignments on elevated track. Most AGT systems use medium sized cars, either singly or in short trains, with some form of rubber–tyred wheels. They are usually designed for specific local movements, such as intra–airport travel or as links between car parks and shopping centres or other major traffic demands. Examples include the installations at Gatwick and Stansted airports (see Photograph 34.5) and the monorail at Alton Towers (JIG, 1995).

Peoplemover is a generic term used to describe any transit system which is primarily intended for low

Photograph 34.5: The AEG Peoplemover at Gatwick Airport.

Photograph 34.6: The Parry Peoplemover. Courtesy of JPN Parry and Associates.

speed, short distance journeys, effectively to aid pedestrian movements. It may be used to describe some systems which are also categorised as AGT. Peoplemovers include various small car systems, some cable–hauled and others using conventional rubber–tyred or steel–wheeled traction technology, as well as continuous belt systems such as travelators. A system, which has been proposed for a number of medium sized towns in Britain, uses conventional steel–wheeled technology but with a flywheel energy storage system and low voltage power supply for intermittent charging at stops (see Photograph 34.6).

34.3 Advantages and Disadvantages of Fixed–Track Systems

Speed and reliability
Public transport systems which operate in mixed traffic on the highway are subject to delays due to traffic congestion, which can be severe in urban areas especially at peak times. Extra running times may also be needed because more people board and alight at peak periods. Accordingly, provision for some level of delay normally has to be included within the schedules, possibly with differentials according to time of day and direction of travel. This reduces the average speeds that can be achieved. More importantly, delays create variations in journey times and affect the reliability of service, which is arguably the most important characteristic of any public transport system. Even small delays can result in extensive disruption, particularly on high frequency services, where delays to one vehicle will affect all the vehicles on the route leading to the familiar phenomenon of 'bunches and gaps'.

Bus–priority measures can greatly reduce delays and improve reliability (see Chapter 24). However, interference from other road traffic is still significant and enforcement of measures such as bus lanes is often difficult. Systems on dedicated fixed tracks enable a more positive level of priority to be established, generally with more extensive segregation from road traffic.

A tracked system, by its nature, has to be designed into the highway throughout its length and this demands detailed consideration of all traffic management problems and opportunities at every location. Even where a system is street–running, without any specific priority measures, the location of the tracks must take account of parking, frontage access for loading, turning movements at intersections and so on. Furthermore, because the vehicles are tracked, any potential obstructions must be avoided because the vehicle cannot deviate from its pre–planned path.

Most tracked systems will have a high proportion of segregated track, which allows higher operating speeds to be achieved as well as greater reliability. Where there is any element of street–running, signal priority (as for buses) and other traffic management measures can help to minimise delays and allow consistent journey times to be achieved throughout the day. Hence, systems like the Manchester Metrolink can achieve 99.8% reliability, which is rarely achieved by any bus–based system.

Line capacity and productivity
The capacity of a normal road vehicle is constrained by the Construction and Use Regulations governing vehicle length and weight. The maximum capacity for a bus is about 100 passengers, although this can be increased to about 130 for articulated buses, if a high proportion of standing space is provided.

Most tracked systems allow vehicles to be operated in trains and this, potentially, enables much higher capacities to be obtained. Even single vehicles are not constrained by the normal legal length limits for road vehicles and can therefore have a higher capacity. For example, light rail vehicles up to 34m or more in length can accommodate up to 250 passengers. The maximum length of train that will normally be permitted to operate on–street is 60m, which allows a two–car train of 30m long cars, having a total seating capacity in excess of 400 passengers. At minimum headways of 90 seconds, this gives a theoretical maximum capacity of around 15,000 passengers per hour per track for light rail. In practice, such high capacities are rarely needed and capacities in the range 2,000 to 5,000 passengers per hour are more

Photograph 34.7: Fixed–track systems can offer level boarding. This example is in Grenoble, France.

typical of major urban corridor flows in British cities.

The higher vehicle or train capacity, together with the higher operating speeds, enables higher productivity levels to be achieved than for buses, in terms of passengers carried per driver employed. A light rail system may typically achieve a figure three times that for a bus system. However, it is probably more relevant to consider total staff required per passenger–kilometre and this will reduce the difference between the productivity of buses and fixed–track systems, because of the need for additional staff to maintain track and power supply systems.

Level access

A requirement for any new tracked system is to provide level access through at least one door. A substantial proportion of the population have some mobility impairment, including those in wheelchairs, those with pushchairs or shopping trolleys, luggage, or those who have difficulty in climbing steps (see Photograph 34.7).

A tracked system ensures that the vehicle always follows the same path and makes provision of level access close to the platform relatively easy to achieve. For most systems, it requires a section of straight track through the platform and for a short distance on the approach and exit tracks. While there is a growing number of low floor buses in service, which are easier to access for many people, they cannot provide the same ease of access as tracked systems because of the difficulty of ensuring that the vehicle is always sufficiently close, and parallel to, the kerb from which passengers are boarding. This problem can be solved by introducing some form of local track or automatic guidance, known as 'stop docking'.

Enforcement

Enforcement of traffic regulation Orders is a continual

difficulty for the police and local authorities and, with increasing traffic levels, there is a tendency for infringements to increase. Parking control is particularly difficult and one vehicle stopped for even a short period at a critical location can result in extensive delays to public transport operations. Experience suggests that tracked systems are essentially self–enforcing. Hence, while illegal parking in ordinary bus lanes is commonplace, illegal parking in reserved tram lanes, or guided bus lanes, is rare.

Most forms of tracked system can be constructed with a form of carriageway surfacing that can be used by road traffic or with forms of surfacing that are clearly different to a carriageway surface. Examples of the latter include ballasted track and grassed tracks (see Photograph 34.8). It is, therefore, clearly apparent to drivers of road vehicles that they are not permitted on those sections. Enforcement should be encouraged by the physical design and layout of segregated systems.

Environmental impact

Most fixed–track systems use electricity as their primary power source, with all the environmental benefits associated with electric traction, including low noise and vibration levels and zero emissions (except those associated with power generation). Visual intrusion is a disbenefit for systems requiring overhead power supply, although it is possible to design relatively unobtrusive overhead line equipment.

Where systems are diesel–powered, or part diesel and part electric, there may still be some environmental advantages over buses, through reduction in the frequency of acceleration and braking that a reserved track alignment affords.

The construction of a fixed–track system usually requires reconstruction of the carriageway or adjacent

Photograph 34.8: The tramway in Strasbourg makes extensive use of grassed tracks.

pedestrian areas and this may present an opportunity to carry out environmental improvements, such as pedestrianisation and hard or soft landscaping, which would not otherwise be undertaken.

Capital costs

A major disadvantage of fixed–track systems is their high capital costs. The construction of trackwork, associated highway works, power supply and signalling, together with rolling stock and depot, may typically cost in the range five million pounds to fifteen million pounds per km (1996 prices) depending on the amount of segregated track. While this may be significantly less than an urban motorway with comparable capacity, it still represents a major investment which can only be justified where corridor flows are sufficiently great. A partially segregated system, such as guided bus, can be built for much lower cost, typically in the range one million to four million pounds per km of guideway, again depending on the amount of segregated track. Capital costs for intermediate systems, such as GLT, may be expected to be about 70% to 80% of those for a light rail system.

One of the major costs relates to the diversion of statutory undertakers' utilities. City centre streets and major radial roads are usually well stocked with services. It is normal practice to relocate all, except deep sewers, outside the swept path and this may alone result in costs of two million pounds or three million pounds per km. Systems based on dual mode vehicles, those that can operate on tracks or on a normal carriageway, can avoid the cost of service diversions by accepting that, when access to services is required for maintenance or renewal, vehicles will be diverted off the tracks. However, this will inevitably reduce the reliability of the system.

Maintenance costs

In addition to the normal operating costs associated with any public transport undertaking, such as staff and fuel, a fixed–track system must bear the maintenance costs of its infrastructure, including track, overhead line equipment and signalling and control facilities. Where tracks are located within the highway, an agreement will be needed with the Highway Authority on the boundaries of responsibilities for maintenance, including drainage.

Traffic restraint

To accommodate a fixed–track alignment within the highway, and to provide the necessary degree of segregation and priority, may well mean some reduction of highway capacity for vehicular traffic and may involve loss of on–street parking spaces (see Photograph 34.9). Other changes may be needed, such

Photograph 34.9: The tramway in Strasbourg illustrates the reallocation of highway space to create a tramway reservation and cycleway, leaving a single lane carriageway for road vehicles.

as relocation of bus stops, taxi ranks and loading bays or the introduction of traffic management schemes to divert traffic away from the fixed–track route.

Such measures may result in restraint of traffic through an overall reduction in the vehicular capacity of the network. However, this must be seen in the context of increasing the total capacity for person movements. It is essential to develop a comprehensive package of measures for all modes of transport, to ensure that the maximum benefits are obtained from the investment and that any adverse effects are minimised.

Flexibility

Fixed–track systems have traditionally been regarded as limited, because of their fixed track, and therefore inflexibility; ie unable to respond to changes in patterns of demand and unable to be rerouted at short notice. Bus systems, on the other hand, are almost infinitely flexible, being able to operate anywhere, subject only to the type and condition of the highway. A guided bus is able to combine the characteristics of both fixed–track and flexible routing, as is GLT, but with some restrictions.

Flexibility of operation in terms of ability to deviate from a route may, in some cases, be a disadvantage. Once a vehicle leaves the protected environment of its track, it can be subject to all the vagaries of the highway network, with consequent loss of reliability. It also becomes less easily identified by the passenger – there is never any doubting where a fixed–track system runs or stops. Indeed, just the presence of a track lends a feeling of permanence to the system.

Interchanges

The walk–in catchment for a fixed–track system is limited to a radius of about 1km from each stop or station. The area served can be expanded greatly, if interchange facilities are provided so that passengers can arrive at a stop by feeder bus, car (driver or passenger) or cycle. However, interchanges can be costly to provide, maintain and control and a change of vehicle is always a deterrent compared with a through journey which does not involve a change.

The benefits of interchanging are likely to be that a faster and more frequent service can be provided, for the same price, and that a wider choice of destinations can be offered. It also results in a simpler network which is easier for passengers to understand. These benefits have to be weighed against the time, cost and inconvenience penalties of changing vehicles. Cost penalties can be eliminated by through fares and ticketing and time penalties can be minimised by coordinated scheduling of services. Inconvenience can be reduced by careful design including cross–platform interchange, where possible.

Integrated bus and fixed–track systems are commonplace in major cities abroad but are not easily achieved in a deregulated environment where fixed–track and bus systems must compete. However, there are still potential benefits and there are some excellent examples of interchanges functioning well with deregulated bus services, such as Meadowhall in Sheffield, Bury and Altrincham in Greater Manchester and Heworth in Tyne and Wear.

Timescales

The construction of a fixed–track system is a major project which is likely to take at least two or three years, after perhaps six or seven years in planning, obtaining approvals and tendering. Hence, the total implementation timescale may be around ten years.

Depending on the location and type of system, considerable disruption is probable during the service diversion and construction stages. Constructing systems in tunnel may reduce the degree of disruption on the surface, if bored tunnels are used, but will increase it if cut and cover methods are used. In many cases, cut and cover is needed at stations for either type.

A partially segregated system, such as guided bus, can be implemented in stages over a longer period, as funding allows or traffic levels demand. A complete system does not have to be built initially, as the buses can operate on ordinary roads over the remainder of the route. Thus, an incremental approach can be

adopted. Guided busways can usually be implemented in a timescale shorter than for other fixed–track systems.

34.4 On–street Running with Priority Right of Way

Street running in mixed traffic

Where space is insufficient to provide any form of segregation, and it is not feasible to divert traffic onto an alternative route, tracked systems, in the form of tramways, may share traffic lanes with general traffic. The alignment must be designed carefully to take account of all traffic movements, including turning flows, and frontage servicing. It must also take account of the need for stops, which must be at a platform or kerbside; on any new system, passengers are not permitted to board or alight in the carriageway, as was common practice with old–style tramways.

If tracks are located at the kerbside, a twenty four hour loading prohibition will be needed. Even if the system is only operational for 16 hours per day, access will still be needed during the remaining 8 hours for maintenance purposes. Where there are any conflicts with turning movements, or trams merging into the traffic stream, it will probably be necessary to install traffic signals. These should be linked to minimise delays to trams.

If there is any possibility of congestion along the shared track, consideration should be given to metered signal control, whereby traffic entering the section is constrained to ensure free flow conditions on the tracked section. Trams entering the street running section can have priority on demand at the entry signals but such priority must be authorised by a Road Traffic Regulation Order.

Reserved lanes

Lanes may be reserved for use by fixed–track systems. The general approach to their design, signing and carriageway marking is similar to that for bus lanes (see Chapter 24) but modified as necessary to comply with the guidelines in Section G of the Railway Safety Principles and Guidance (HSE, 1996). With–flow reserved lanes may be at the kerbside, in the centre of the carriageway or on the offside. Contra–flow lanes may be located on the nearside or offside of general traffic lanes but should be physically segregated by normal height kerbs.

Particular attention must be paid to the needs of pedestrians and cyclists. Controlled pedestrian crossings should be provided, where necessary, and cycle lanes should be provided on parallel routes or alongside the tram lane.

Where there is no requirement for general traffic to use a lane except in an emergency, a 'trambaan' may be provided. The tracks and paved area of the swept path are raised 150mm (exceptionally 75mm) above the traffic lane and separated from it by a chamfered kerb. This allows road vehicles to drive over the tracks to avoid an obstruction, such as an illegally parked car. It is a useful facility when the carriageway alongside the tramway is only wide enough for a single traffic lane.

Segregated tracks within the highway

Where the highway is of adequate width, it may be possible to incorporate a reserved track which is segregated from other traffic by some form of physical separator. It is, therefore, only for use by trams and can not be used by any other vehicles. It may be surfaced in a form of paving, such as concrete blocks, if pedestrians have free access, or it may be grassed or ballasted. It should always be paved in a different colour and texture from any adjacent carriageway to deter motorists from attempting to use it (see Photograph 34.9). A segregated section must be signed according to the traffic signs regulations and the design of kerbing, bollards and other street furniture should reinforce the prohibition of road vehicles.

A reserved track may be centrally located, as in the median strip of a dual carriageway, or may be at the side of either a single or dual carriageway. The footways should normally be located at the extremities of the highway in cross section but may, in exceptional cases, be located between the reserved track and the carriageway. Careful attention to the design of intersections is needed and signal control will be needed where there are significant traffic flows or where sightlines are restricted. Pedestrian and cycle facilities, including signal control, should be provided as appropriate.

Tram–only streets

If there is no requirement for general traffic movements along a street or for servicing access, it may be feasible to designate a street for use only by trams. The street may be retained as a conventional carriageway with kerbs or it may be repaved, without normal kerbs, in block paving or some other form of surfacing. If there is any access required for emergency vehicles, then a form of paving suitable for their use must be retained, over at least part of the carriageway.

Photograph 34.10: Pedestrians, cyclists and trams in a pedestrianised area – Kaiserstrasse in Karlsruhe, Germany. Streets may be reserved for use by buses and trams, either using the same lanes for both or with separate lanes for buses and trams. Various combinations can be devised according to the routes to be accommodated, for example two–way trams with one–way buses (see Section 34.6).

Pedestrian precincts

Fixed–track vehicles can be accommodated more easily than road vehicles in pedestrian precincts, because their path is fixed and predictable. Identifying the swept path is a requirement and can be reflected in the paving design to define the swept area, so that it is clear to pedestrians which areas may be traversed by a tram (see Photograph 34.10). The swept path should also be defined by a small level difference (about 25mm), which can be recognised by visually impaired pedestrians and guide dogs.

Guided buses using kerb– guidance are more difficult to accommodate in pedestrian areas, unless they operate as normal buses. However, at low speeds, it may be feasible to include sections of guideway with frequent gaps to allow pedestrians to cross without any obstruction. As with tramways, a small level difference at the edge of the swept path is still required.

Parking and loading

Wherever tracked systems operate on–street, parking and loading bays should be provided, where necessary, to maintain reasonable access to frontage properties. They should be clearly defined so that there is no risk of parked vehicles obstructing the passage of trams. If possible, parking bays should be relocated to adjacent streets not on the fixed–track route.

34.5 Off–street Running on Segregated Right of Way

Segregated alignments alongside highways

A segregated alignment may be located alongside a highway but outside the highway boundary. In this case, the footway should be between the fixed–track alignment and the carriageway. The tracks may be fenced and should be ballasted or grassed. They should not be paved in a surface which could suggest that they are highways. At intersections with side-roads, footpaths or cycleways, the fixed track will need to be treated as if it were a highway for the purposes of junction design and signalling, according to the principles laid down by HMRI (HSE, 1996). It is preferable for the tracks to be separated from the highway by at least 15m at junctions, to create a storage space for vehicles queueing to enter the major road. Alternatively, the junction should be fully signal–controlled (see Photograph 34.4).

Private rights of way

Fixed–track alignments may be located away from highways on any suitable land, such as derelict land, abandoned highway reservations, former railways or canals or public open space. The land will normally be acquired by the promoter under a Transport and Works Act Order [Sb] and remains as private land. Such alignments may be fully fenced and all other pedestrian, cycle, or vehicular traffic prohibited or may be open with some access allowed. If it is likely that the route will prove attractive to pedestrians, or cyclists, then consideration should be given to creating a separate public right of way for them alongside the private right of way.

Railway alignments

A tracked system may be located within an existing railway alignment, if there is sufficient unused space alongside the operational railway tracks. Adequate safety clearances must be allowed and fencing may be required to meet the separation criteria demanded by the incumbent railway track authority. Alternatively, such sections may remain in the ownership and control of the track authority, with the new system given running powers over the tracks. This would probably only be acceptable where the tracked system uses track, signalling and a method of operation broadly compatible with the existing rail operation, for example light rail or metro but not guided bus.

It should be noted that, in the UK, any fixed–track system which operates on fully segregated and fenced alignments may be regarded as a railway for the purposes of safety principles by HMRI. Discussions may be necessary to confirm the status of specific sections.

Mixed running – shared track

In exceptional circumstances, a new fixed–track system may share its track with an existing heavy rail system or with another form of fixed–track system. Examples in operation include Karlsruhe (Germany), where dual voltage light rail trains operate on mainline tracks to outer suburban destinations and also run over street tracks into the city centre, and Essen (Germany), where guided buses share light rail tracks on central reservations and in tunnel (see Photograph 34.3).

Although there are no current examples in the UK, shared track has been extensively researched by British Rail Research on behalf of a number of promoting authorities. Proposals for shared track between heavy rail and light rail have been made in Nottingham and Cardiff and a proposal for shared track between guided busway and freight trains has been made in Bristol. Shared track will feature on the extension of the Tyne and Wear Metro to Sunderland, but this does not involve any form of street running.

Before any form of shared track can be implemented in the UK, a fail–safe method of ensuring separation of the different types of rolling stock will have to be acceptable to HMRI. This could involve some form of automatic train protection or a method of controlling physical access to the shared section, with proof that the section has been cleared. When such methods have been developed, and when suitable rolling stock has been specified, there are a considerable number of potential applications which could increase substantially the role of fixed track systems in the UK and reduce the capital cost of their introduction.

Grade–separation

Where space is insufficient to accommodate a fixed–track alignment at grade, it may be necessary to consider grade–separation, either at a local intersection or along a section of route. The fixed track may be located in an underpass or on a flyover. Examples of both can be found on the Sheffield Supertram, near the city centre.

Longer sections of tunnel, either cut and cover or bored, may be needed to penetrate dense urban centres where surface running is not practicable as, for example, with the Tyne and Wear Metro through the centres of Newcastle upon Tyne and Gateshead. Longer sections of elevated track may be needed, for example, on radial routes where there are frequent road intersections with heavy cross traffic. Grade-separation is essential if fully automatic driverless operation is adopted, as on the Docklands Light Railway in London, or if the track form is incompatible with the highway surface, as for example with the peoplemovers at Gatwick and Stansted airports (see Photograph 34.5).

Any form of grade–separation is likely to increase, considerably, the cost of the alignment. It will also increase the cost of any stops or stations, possibly by a factor of ten to fifteen times the cost of a surface stop. Such stops will almost certainly be more difficult for passengers to reach and may require lifts and escalators, adding further to both capital and maintenance costs. Tunnel portal ramps, if located in the highway, can be a barrier to pedestrian movement and be visually intrusive. For all these reasons, grade–separation should only be considered where there are clear benefits, which justify the additional costs, or where there are specific local topographical features which make grade–separation easy to achieve.

34.6 Mixed Operation with Buses

Bus and tram lanes
Where a fixed–track system operates in a reserved lane within the highway, it may be shared with other classes of traffic, particularly buses. Thus, a tramway may share its lane with buses, or guided buses, or a GLT route could be shared with buses. The locations at which their respective lanes join and diverge must be carefully designed, clearly signed and marked, to reduce the risk of road vehicles following the wrong lane.

Bus and tram only streets
Streets reserved exclusively for buses and trams may either have shared lanes, in one or both directions, or may have separate lanes for buses and trams. The choice will depend on the available width of carriageway, the location of bus and tram stops, passenger objectives to be served and the routeing patterns of buses and trams, including any turning movements.

Stops and stations
It is preferable for any stops or stations for buses and trams to be segregated, ie located away from the shared section of lane, so that stopped buses do not delay trams and vice versa. This may require bus and tram stops to be located alongside each other, in which case special attention must be given to pedestrian movements to, and between, the stops.

Trackwork
Where there is any element of shared use of traffic lanes, the design and construction of the track and the surrounding carriageway structure and surfacing must reflect the loadings that will be imposed by buses or other vehicles. For example, buses constrained to a tracked lane may well cause rutting of the track. Construction of the track must be capable of withstanding such concentrated loading.

If there are points within the trackwork, especially electrically–operated facing points, they should not be located where they may be damaged by heavy movements of road vehicles. Pre–sorting may be used to locate them in more appropriate positions.

Pedestrian and cycle facilities
Wherever there is any form of shared use of lanes, particular attention should be paid to pedestrian facilities and, where appropriate, cycle lanes (see Photograph 34.9). Controlled crossings should be provided where uncontrolled crossings would not offer adequate safety or where sightlines are inadequate. Pedestrian refuges should be provided, where possible, between tracks in opposite directions or between bus and tram lanes, if they are separately signalled.

Driver training
The use of any form of shared track will be a new experience for most drivers and it is desirable to include relevant training to ensure a clear understanding of the procedures to be followed. This is equally necessary for bus and tram drivers. Drivers' eye–view videos and the use of simulators can be helpful in encouraging appreciation of hazardous situations and how to react.

34.7 Interchange Facilities

Multi–modal interchanges
As indicated in Section 34.3, the opportunities for travel by public transport can be greatly expanded by the provision of interchanges, which allow easy movement between different modes of transport or between different services of the same mode. Fixed–track systems can benefit particularly, as interchanges potentially increase their otherwise limited catchment areas around stops. Access to a fixed–track system may be by car, cycle, bus or other fixed–track system such as heavy rail train or underground railway.

The design of an interchange should take account of the approach and departure route of each transport mode, the parking stand or platform facilities required and passenger facilities. Particular attention should be paid to the pedestrian routes taken for any given interchange movement, so that the distances to be traversed, both horizontal and vertical, are minimised. The ideal solution is when the predominant interchange movements can be accommodated on either side of a single platform, ie the so-called 'cross platform' interchange.

Access to shopping centres

An interchange will be most successful where it combines a number of functions. Thus, a bus/tram interchange located in a good town centre, or district centre, location will generally attract more passengers than an interchange which is more remotely located. Ideally, interchanges should be sited close to shopping centres, with easy level pedestrian access, unobstructed by major road traffic flows. Where differences in level cannot be avoided, lifts and escalators should be provided.

Highway access

For the smooth and reliable operation of an interchange, it is essential that vehicles should enter and leave with the minimum of delay and by the shortest practicable route. It is also desirable that modes are segregated, in particular cars queuing to enter a car park should not obstruct access roads for buses or trams. The access roads to the interchange should be designed to facilitate efficient operation and should allow congestion-free movement to and from the major roads used by bus or fixed-track systems. Signal control with priority for public transport movements should be incorporated into the junction designs.

Grade-separation/multi-level interchanges

Where there are large numbers of vehicle and passenger movements to be accommodated, the interchange times, and the size of the interchange site, may be reduced by constructing a multi-level complex. This can also help to segregate different modes, for example buses and rail vehicles, or buses and cars. If there are fully segregated heavy rail services, some vertical separation is essential, if only for passengers to cross the tracks. A multi-level interchange will inevitably be more costly than a simple ground level interchange but may enable more efficient transfers to be achieved. It will be necessary to consider carefully the options available, and the site constraints, before determining the optimum solution.

34.8 Network Information Systems

Passenger information needs (see Chapters 15 and 24).

If passengers are to be attracted to a new system, adequate information about the system must be readily available. Information is needed before a

journey commences, at stops and stations, on the vehicle exterior, inside the vehicle during the journey and at stops when leaving the vehicle. Much of the responsibility for providing this information will lie with the operator but good highway directional signing should be provided along all highway approaches to stops and, where appropriate, along pedestrian or cycle routes.

Most fixed-track systems operate on the 'open station' principle, which means that passengers do not have to pass through any ticket check on entering the vehicle and stops are usually unstaffed. It is therefore important that all the information needs of passengers are met by means other than enquiry to a member of the operator's staff.

Static/passive information display

The most important need for static information is at the station or stop, giving details of routes, departure times or frequencies, first and last departures, fares and general information on the use of the system. Maps or diagrams of the system and the area around the stop should be included. It should be noted that passive information gives details of what is scheduled to operate and not what is actually operating. However, passive displays, usually in printed paper form, are a relatively cheap and effective form of providing information.

Real-time passive information displays

Real-time displays reflect the actual operating conditions at the time. They display actual times and destination for next departure or minutes to next departure. They are particularly useful where lines serve more than one terminal point, where headways are longer than a few minutes or where the service is subject to frequent delays.

The information may be displayed on roller blind, dot-matrix or liquid crystal (LC) displays and should be located so that it is visible from all parts of the platform. It is activated automatically by the vehicle transponders, so that it always provides accurate and updated information.

Interactive off-line systems

Interactive systems allow the passenger to search for specific information. This is most commonly done through enquiry offices and telephone enquiries. Computerised information systems are also available, such as Teletext or Videotext, at travel centres or on home or office computer terminals.

Real-time interactive systems

It is now technically feasible to devise systems where

the user has access to real–time information about the actual position of vehicles and may obtain verbal, visual or printed copy of the required data. Such systems are still in the early stages of development and it is unlikely that their relatively high cost would be justified for fixed–track systems, which are inherently reliable and frequent.

Automatic Vehicle Location (AVL) and Global–Positioning Systems (GPS)

Several different technologies exist for determining the actual positions of vehicles, using AVL systems based on closed circuit data transmission, microwave or radio links. It is also possible to locate vehicles by GPS systems, which offer a relatively low cost and reliable method of monitoring vehicle location. This may be used for operational control purposes, as well as providing real–time passenger information.

34.9 Control and Operations

Principles of control systems

A fixed–track system demands a high level of discipline to achieve efficient, safe, operation. This discipline, usually, is exercised through a control strategy, whereby all aspects of the operation of the system are monitored and controlled from a single location. The primary aim of the control system is to ensure the safe operation of all aspects of the system at all times. The second objective is to maintain as reliable a service as possible, given all the extraneous factors which apply at the time. Usually, the control system is staffed every day of the year, 24 hours per day, whether the system is operating or not. When not in public operation, there are likely to be engineering works, routine maintenance or other activities requiring a control function. The control strategy is documented in operating manuals (the Rule Book and a Book of the Road), which will form a key part of the staff training programme and, together, contain all the required information and procedures for operating the system. A range of communications systems is available to assist in the control function, providing links to equipment, staff and other relevant organisations. On a tramway, safe operation is the responsibility primarily of the driver and any control system is there to facilitate the driver's safe and efficient operation of the system.

Vehicle location, driver communication and Inspectorate control

The Controller will monitor the positions of all vehicles and, in the event of any disturbance to the scheduled positions, will take any appropriate action to restore service to the correct status. Such deviations can be monitored manually or may be computer–controlled. Instructions to drivers are issued by the Controller, using the cab radio, and may also be relayed through roving inspectors. Inspectors may exercise a range of functions in the control of the service, such as revenue control, passenger assistance and on–site control of emergency situations.

Passenger information and publicity

Clear information should be readily available for passengers, so that they can use the system as easily as possible. This must include details of the routes, services and timings, fares systems and emergency facilities. Most fixed–track systems are, by their nature, relatively easy to understand with a simple route–structure and vehicle–paths that are predictable. They will, however, be an unfamiliar form of transport for many intending passengers and for other road–users. It is essential, therefore, that a good level of publicity is provided to show people how to use the system and to highlight safety issues. Particular emphasis should be given to new features of the system, such as the swept path, new traffic signs and signals and pedestrian facilities.

Control centre operations

An Operations Control Centre is the focal point for all communications and information displays related to the operation of the system. These will normally include:
- ❏ the location of vehicles;
- ❏ the state of the signalling system;
- ❏ the state of the power supply system;
- ❏ an internal telephone network;
- ❏ radio communications with staff;
- ❏ closed circuit television monitors;
- ❏ public address systems;
- ❏ passenger information systems;
- ❏ passenger emergency communications(on platforms and in–vehicle);
- ❏ monitoring lifts and escalators;
- ❏ ticket vending machine control; and
- ❏ links to emergency services (police, fire, ambulance).

A person will be appointed as Controller, with full responsibility for the operation of the system on a minute to minute basis. The Controller is responsible for continuous monitoring of the movements of vehicles, movements of passengers on stations and the deployment of drivers, inspectors and other operating staff. Note again that tramways may not have a control centre and the signalling systems for tramways are not usually centrally controlled.

Links with Urban Traffic Control (UTC) Centres, Police and Transport Operators

In addition to the links with the emergency services, communication links should be maintained with the UTC centre (where such exists), the police control centre and the control functions of other transport operators, such as bus or train, where there are any significant interactions between the fixed–track system and other public transport systems. This will assist a rapid response to deal with any incident which may affect the smooth operation the system.

Emergency procedures

Procedures should be included within the operating rules and regulations for dealing with any emergency situation that may arise. Such procedures will need to be agreed with the emergency services, the Highway Authority, where the system is street running, and other operators if there is any shared track. They will also need to satisfy HMRI, which is within the Environmental Agency (EA).

34.10 References

HMG (1992)   'Transport and Works Act 1992', Stationery Office [Sa].

HMG (1994)   SI No. 157 'The Railways and Other Transport Systems (Approval of Works, Plant and Equipment) Regulations 1994', Stationery Office.

HSE (1996)   'Railway Safety Principles and Guidance' Part 2 Section G 'Guidance on Tramways', Highways Safety Executive, Stationery Office.

JIG (1995)   'Jane's Urban Transport Systems 1995–96' Jane's Information Group Ltd. Coulsden (published annually).

RCEP (1994)   'Transport and the Environment', Eighteenth Report of the Royal Commission on Environmental Pollution, Stationery Office.

TAS (1995)   'Rapid Transit Monitor', TAS Publications Preston.

34.11 Further Information

Barry M (1991)   'Through the Cities, The Revolution in Light Rail', Frankfort Press Dublin.

DOT (1992)   'Transport and Works Act 1992 A guide to Procedures for obtaining orders relating to transport systems, inland waterways and works interfering with rights of navigation', Stationery Office [Sa].

DOT (1995a)   'Light Rapid Transit (and related) Systems', A Briefing Note by Buses and Taxis Division of the Department of Transport', DOT.

DOT(1995b)   'Guided Buses. A Briefing Note by Buses and Taxis Division of the Department of Transport', DOT.

Durkin J, Lane P and Peto M (1992)   'Guide to the Transport and Works Act 1992 Planning for Infrastructure Developments', Blackstone Press London [Sa].

ETSU–OPET (1994)   For Commission of the European Communities DG XVII, (1994). 'Energy andEnvironmental Implications of Light Rail Systems', ETSU Harwell.

HMG (1968)   'Transport Act 1968', Stationery Office.

HMG (1985)   'Transport Act 1985', Stationery Office.

HMG (1988)   'Road Traffic Act 1988', Stationery Office.

HMG (1991a)   'New Roads and Streetworks Act 1991', Stationery Office.

HMG (1991b)   House of Commons Transport Committee. 'Fourth Report: Urban Public Transport, The Light Rail Option', Stationery Office.

HMG (1994)	SI No. 1519 'The Traffic Signs Regulations and General Directions 1994', Stationery Office.
ICE (1991)	'Light Transit Systems' BH North (Ed), Thomas. Telford.
UITP (1995)	'Light Rail Trends' Brussels.
Wood C (1994)	'Street Trams for London', CILT Transport Research, London.

Journals

Local Transport Today
Light Rail and Modern Tramway
UITP Revue
Railway Gazette International
International Railway Journal.

Chapter 35 The Highway in Cross–Section

35.1 Introduction

All urban roads have more than one function. As well as acting as routes for pedestrians, cyclists and all forms of motor vehicles, they may provide facilities for parking, loading and unloading of goods and passengers and a conduit for public utility services. This may affect the choice of the appropriate cross section.

For most existing urban roads, the junctions act as the constraint on traffic capacity rather than the links between them (see Chapter 32). For new roads and improvements, it is important to relate the two. Also, it may not be appropriate to design urban roads to allow traffic to flow freely at all times. It may be reasonable and economic to accept lower operating speeds at certain times of the year or day, as part of a planned strategy for the urban area.

Road Type (level in the hierarchy)	Undivided carriageway									Dual carriageways		
	Peak hour flow (veh/h) both directions of flow*					Peak hour flow (veh/h) one direction of flow				Peak hour flow (veh/h) one direction of flow		
	2 lanes					4 lanes			6 lanes	4 lanes		6 lanes
	6.1 (m)	6.75 (m)	7.3 (m)	9 (m)	10 (m)	12.3 (m)	13.5 (m)	14.6 (m)	18 (m)	Dual 6.75+ (m)	Dual 7.3+ (m)	Dual 11.0+ (m)
A Urban motorway (Primary distributor)	-	-	-	-	-	-	-	-	-	-	3600	5700
B All purpose road no frontage crossings no standing vehicles negligible cross traffic (District distributor)	-	-	2000	-	3000	2550	2800	3050	-	2950	3200	4800
C All purpose road frontage development side roads, pedestrian crossings, bus stops, waiting restrictions throughout day, loading restrictions at peak hours. (Local distributor)	1100	1400	1700	2200	2500	1700	1900	2100	2700	-	-	-

*60/40 directional split assumed. + Includes division by line of refuges as well as central reservation; effective carriageway width excluding refuge width is used. Based on TD20/85 Table A (DOT, 1985b).

HEAVY VEHICLE CONTENT: The recommended flows allow for a proportion of heavy vehicles equal to 15%. No allowance will need to be made for lower proportions of heavy vehicles; the peak hourly flows at the year under consideration should be reduced when the expected proportion exceeds 15% by the following amounts (veh/h).

Heavy vehicle content (% total vech/h)	Typical reduction in flow level (veh/h)		
	Motorway and dual carriageway all purpose road	10m wide and above single carriageway road	below 10m wide single carriageway road
	per lane	per carriageway	per carriageway
15%-20 %	100	150	100
20%-25c%	150	225	150

Table 35.1: Design–flows (veh/h) for two–way urban roads.

Based on TD20/85 Table C (DOT, 1985b).

Photograph 35.1: A typical urban dual carriageway.

Photograph 35.2: A typical urban primary distributor.

35.2 Lane Widths and Carriageway Widths

Procedures for estimating carriageway widths

The width of a highway link between junctions must be sufficient for its intended function in the road hierarchy (see Chapter 11) and to achieve acceptable levels of safety and operating efficiency. The Department of Transport's procedures for estimating carriageway width are given in TD20/85 (DOT, 1985b) and TA46/85 (DOT, 1985c). Whilst the accent in these is on rural roads, there is also some discussion of application in urban areas. Where demand is sufficient to justify four or more traffic lanes, dual carriageways may be provided. Typical design flows for different widths and types of carriageway are given in Table 35.1 and illustrations of a variety of urban dual carriageway schemes are shown in Photographs 35.1, and 35.2.

Speed/Flow Relationships

Typical relationships between vehicular flows and associated mean speeds for different carriageway widths are shown in Figure 35.1 (see also DOT, 1985b). In general, the heavier the traffic flow the

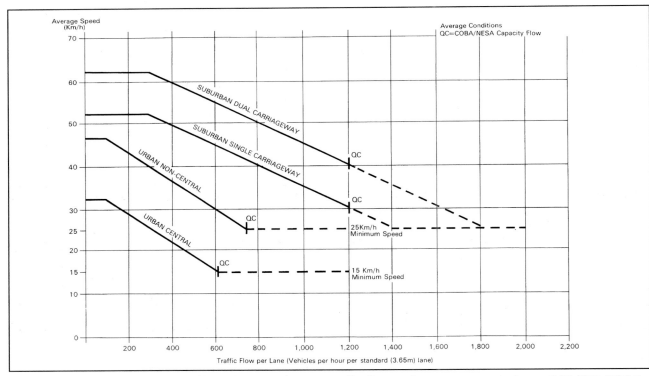

Figure 35.1: Urban and suburban roads average speed against traffic flow per lane.

Source DOT/SDD (1986).

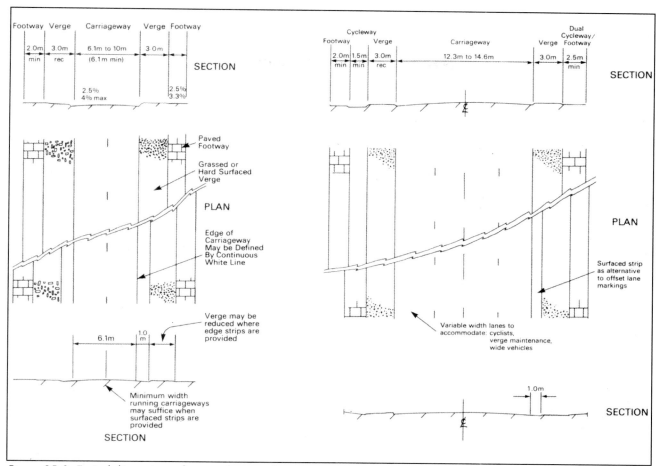

Figure 35.2: Typical dimensions of a two–lane single carriageway. Source DOT (1987).

Figure 35.3: Typical dimensions of a four–lane single carriageway. Source DOT (1987).

lower the operating speed but the lines merely indicate the averaging of a large number of individual cases. Apart from geometric factors, the speed/flow performance of a carriageway is a function of vehicle/driver type, weather and daylight/darkness conditions. Therefore, the graphs can give only a broad indication of the relative performances of differing road types.

Typical Road Widths

Typical cross–sections for single and dual carriageway roads are given in Figures 35.2, 35.3 and 35.4. The designer has more flexibility when selecting verge and central reserve widths, depending on the cost, the availability of land and the function of the facility.

Lane Widths

Lanes on standard single or dual carriageways are normally multiples of 3.65m but narrower lane–widths can be used on local distributor and access roads. A width of 6.75m is often used on two–lane local distributors and widths down to 6.1m, or even 6.0m (Table 1 of TD20/85) for two–way roads,

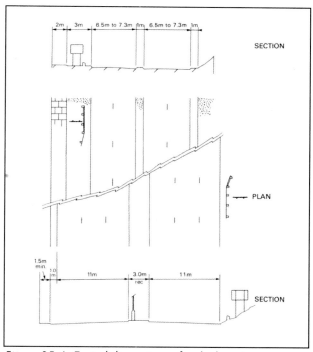

Figure 35.4: Typical dimensions of a dual carriageway. Source DOT (1987).

have been found to operate satisfactorily for local distributor roads, especially if kerb–side parking is restricted.

On primary and district distributor roads, the narrowest widths should only be considered at pinch points, to overcome a particular local problem or where significant cost–savings can be achieved over short distances, such as on bridges or in tunnels. In these circumstances, operating speeds should be kept below 65 km/h by consistent reduction in the other elements of design and this may be reinforced by a regulatory speed limit.

Lane–widths of 3.25m are normally adopted only for those lanes which carry light vehicles; ie those nearest to the central reserve. TD9/93 (DOT, 1993) [Sb] allows the use of lanes down to this width at pinch points and, therefore, by definition, they can only be used over short lengths. This should be a last option, when feasible reductions in width of all other elements in the cross–section have been made. Such reductions are treated as 'departures' (see Chapter 31) and have to be justified, particularly on cost and safety grounds. On major schemes, the CIDEL (Calculation of Incident Delay) program can be used to make an economic assessment of sites with a restricted width (DOT, 1985c) [Sa].

Narrow carriageways are inappropriate where significant numbers of cyclists or large vehicles are anticipated and can result in tracking (ie rutting and uneven wear of the road surface) on narrow straight carriageways. More advice on the needs of cyclists can be found in TA67/95 (DOT, 1995) [Sc], Traffic Advisory Leaflets [Sd] and the IHT Guidelines (IHT, 1996) (see Chapter 23).

Lane–widening is required on small radius curves, as shown in Table 35.2, to allow for the swept path of long vehicles. In urban areas, radii less than 90m may be used particularly on local roads.

Edge–Strips

It is common practice to provide 1.0m edge–strips on non–motorway rural roads. Whilst these may be difficult to provide in urban areas, they do offer significant benefits. They enable stationary vehicles to pull, slightly, out of the traffic flow and they may enable maintenance works to be carried out without reducing the number of traffic lanes or, in certain cases, without a road closure. They may also be used by cyclists, although only if they are kept clear of detritus and, even then, they should not be used in place of more appropriate provision such as proper cycle–lanes.

35.3 Tidal Flow

In situations where traffic flow in opposing directions is unbalanced at different times of the day, consideration can be given to the provision of an odd number of lanes to enable the direction of flow on the centre lane(s) to be altered for fixed periods of time to suit traffic conditions. Care must be taken to ensure that opposing flows are adequately signed and separated (see Photograph 35.3). This is usually achieved by symbolic variable message signs on overhead gantries, often reinforced by the use of different coloured surfacing on the tidal flow lane(s). However, there are a few examples (eg Reading) where short lengths of tidal flow operate successfully without using overhead gantries. Provision for tidal flow may be particularly appropriate for major commuter routes along purpose–built radial roads and at places where the carriageway cannot be widened, such as bridges, tunnels and frontage development; however, the signing equipment can be visually intrusive and safety issues require careful attention.

35.4 Edges of the Carriageway

The edges of carriageways on urban roads will

Lane Width (m)	Radius (m)		
	<150	150-300	300-400
3.65 - 3.53	0.3m per lane	NOT REQUIRED	NOT REQUIRED
<3.53	0.6m per lane	0.5m per lane	0.3m per lane

Table 35.2 Road widening on curves.   Source: TD9/93 (DOT, 1993).

Photograph 35.3: An example of a three-lane tidal flow operation.

normally be kerbed. Kerbs retain the structure of the carriageway, highlight the edge of the road, particularly when driving at night using headlights, protect adjacent areas from encroachment by vehicles and assist drainage. Kerb faces on major roads should be about 100mm high, except, for example, at pedestrian crossing points, where dropped kerbs should be provided to assist prams and wheelchair users. Where footways are provided, or street furniture is placed adjacent to the carriageway, half-battered or bullnosed kerbs are commonly used. Elsewhere, 45° splay kerbs are preferred, as they assist the removal of immobilised vehicles from the carriageway at times of emergency and tend to improve highlighting of the carriageway edge by headlight beams. Where hard shoulders or surfaced edge-strips to the carriageway are provided, extruded asphalt kerbs will form a drainage check, but they are not usually appropriate in built-up areas.

35.5 Footways and Cycleways

Ideally, pedestrian and cycle movements should be segregated from heavy vehicular traffic but this is often not practical in urban areas. For pedestrians, a paved footway at least 2.0m in width should be provided (1.8m is the absolute minimum) to permit prams or wheelchairs to pass in comfort. These are minimum values and should be increased wherever large pedestrian flows are expected or where more space is available, as wider footpaths can provide more distance from passing traffic and give pedestrians a greater feeling of safety (see Chapter 22).

Cycleways should be a minimum of 1.5m wide and preferably 2.0m, with good skid-resistant surfacing. Provision for cyclists should normally be made within the carriageway but segregated from other traffic (see Chapter 23).

35.6 Verges

Where footways are located beside carriageways, they should preferably be separated from the carriageway by a verge. Whilst it may be difficult to provide verges in urban areas, they do offer certain advantages. They provide space to install street furniture, such as pedestrian guardrails, safety barriers, lighting columns and telephone and post boxes, which would otherwise encroach upon available footway space. Public utilities' apparatus should be located in verges, to avoid disruption to carriageways and footways during maintenance work. Thus, the required width of verge will be dictated by constraints of available width and cost, as against the benefits of locating these items in the verge rather than the footway. Where there is no footway, a verge may be necessary to accommodate these items and also to provide sufficient set-back and protection to items at the edge of the highway, such as boundary fences, buildings and structures. For these reasons, no minimum width of verge is specified. Widening to provide clear verges at bus stops and bus laybys is not necessary. Verges should be discontinued at pedestrian crossing points. Verges may be hard surfaced or grassed. A hard surface should be in a different material from the adjacent footway and textured in a way that discourages use by pedestrians (eg cobbles). Grass may be more attractive but the cost implications for maintenance should be considered.

Where structures, such as retaining walls or bridge parapets, run parallel with the carriageway, kerbed marginal strips should be provided between the carriageway and the structure to reduce the potential for collisions. On bridge decks these should be hard surfaced. Widths will depend on the design speed of the road, the crossfall and on whether any street furniture, such as safety barriers or lamp columns, is provided but should always give at least a 0.45m set-back, although TD19/85 (DOT, 1985a) [Sd] refers to set-back being reduced to 0.33m in urban areas, where space is limited (see Figure 35.4; DOT, 1987 and DOT, 1985a).

35.7 Central Reserves

To Improve Safety
On existing multi-lane roads, ie carriageways which have at least four lanes and where space permits,

central reserves may be constructed to segregate opposing flows and to create dual carriageways which offer higher levels of safety. Central reserves would normally be appropriate on multi–lane roads where speed–limits are greater than 40 miles/h. Exceptions may need to be made where tidal flow systems are used or where economic arguments are strong, such as on long or very expensive structures. Notwithstanding the economic arguments, there should always be some form of carriageway marking between opposing traffic flows, such as hatched markings on a coloured surfacing.

To Accommodate Street Furniture

Central reserves provide useful areas to accommodate street furniture, such as traffic signs, street–lighting columns and the legs of sign gantries. Widths must be sufficient to enable clearance to be obtained to these structures and they may need to be protected by safety barriers to reduce the chance, and effects, of vehicle impacts. Such facilities will, however, require routine maintenance, which may involve closing a traffic lane with its potential for creating additional delays (DOT, 1985a and DOT, 1996).

Vehicular Crossing–Points

Vehicular crossing–points should be provided at regular intervals on low speed roads to assist temporary diversions during maintenance works or other incidents. These crossings should be closed by removable barriers to prevent U–turns during normal operation. It should be noted that emergency crossing points are not normally suitable for major maintenance works or on high-speed roads. Further advice on layout and dimensions is available in TA45/85 (DOT, 1985d) [Sc].

The location of vehicle crossing–points will depend on the particular road layout and the configuration of any grade-separated junctions. The most useful locations will probably be, as follows:
□ close to junctions near to the end of a dual carriageway;
□ either side of all other junctions; and
□ approximately mid–way between junctions.

Widths

Narrow central reserves, ie less than 3.0m in width, should be kerbed and hard surfaced, unless they are used as open (French) drains. If pedestrian crossing facilities are provided, a minimum width of 2.5m is required to permit a pedestrian pushing a pram or wheelchair to wait in safety. The normal minimum recommended width of a central reserve is 1.0m. Where a central reserve safety fence is provided then the need to accommodate this, and to provide the necessary set–backs and clearances, will usually dictate the required width.

35.8 Crossfall (or Camber)

Carriageways

Except on curves, where superelevation (see Chapter 36) or elimination of adverse crossfall or camber may be required, carriageways should normally have a crossfall of 2.5% from the crown, or central reservation, outwards towards the side of the road. Excessive crossfall or camber is a source of danger to drivers and cyclists and should be avoided; it may cause loads to be displaced or lead to vehicles slipping sideways in icy conditions. Crossfalls can increase gradually as a result of successive re–surfacing where drainage channels are kept at their original level. Crossfalls should not exceed 5 per cent in urban areas.

Footways

A maximum crossfall of four percent is recommended for footways and verges.

35.9 Other Aspects of Design

Headroom Clearance

There is no general limit on the height of vehicles in Britain, although certain regulations apply in specific instances. For example, the Construction and Use Regulations set a maximum height limit for buses and some semi–trailers. It is generally accepted that an absolute minimum headroom of 5.1m should be provided at the most critical places, after taking account of gradient and crossfall.

To allow for future re–surfacing, or overlays, it would normally be prudent to provide a headroom of 5.3m at construction, unless full–depth reconstruction is planned for local strengthening in the future (see Figure 35.4). The headroom provided under footbridges and gantries is usually 5.7m. (DOT, 1987 and DOT, 1996) [Sc].

Obstructions Beside the Kerb

Lane–capacity can be reduced when obstructions are sited close to the edge of the carriageway, so a minimum set–back must be maintained to allow for overhanging loads and the tilting of vehicles towards the obstruction. Figure 35.4 gives recommended set–backs for a design speed of 50 km/h.

Longitudinal Construction Joints

As far as possible, longitudinal construction joints in the highway pavement should coincide with the division of the carriageway into traffic lanes. This helps to avoid confusion when driving in poor light or when the road surface is wet.

Location of Manholes

Where service manholes for drains or utility plant have to be located within the carriageway, every effort should be made to site manholes away from the wheel–tracks which suffer the worst wear. It is always preferable to locate manholes in verges, footways or on traffic islands, to make maintenance work safer and to reduce delays to traffic. Manholes should preferably not be located in that part of the carriageway used by cyclists. Consideration should also be given, at the design stage, as to how pedestrian movements can be maintained safely when access is required to the manholes.

Laybys and Service Roads

The servicing of land–uses adjacent to a road, including loading and unloading, should be achieved, where possible, off the highway using laybys and service roads.

Bus Laybys

A major consideration in urban highway design is the provision of facilities for public transport. Bus laybys should be located at places with good visibility and discussions should be held with bus operators at an early stage to determine the best way to provide for bus staging and the location of bus stops and laybys. Further information on the location and layout of bus laybys can be found in Chapter 24.

35.10 References

DOT (1985a)   TD19/85 (DMRB 2.2) 'Safety Fences and Barriers', Stationery Office [Se].

DOT (1985b)   TD20/85 (DMRB 5.1) 'Traffic Flows and Carriageway Width Assessment', Stationery Office [Se].

DOT (1985c)   TA46/85 (DMRB 5.1) 'Traffic Flows and Carriageway Width Assessment for Rural Roads', Stationery Office [Se].

DOT (1985d)   TA45/85 (DMRB 2.2) 'Treatment of Gaps in Central Reserve Safety Fences', Stationery Office [Se].

DOT (1987)   'Highway Construction Details', Stationery Office [Se].

DOT (1993)   TD9/93 (DMRB 6.1.1) 'Highway Link Design', Stationery Office [Se].

DOT (1995)   TA67/95(DMRB 5.2.4) 'Providing for Cyclists', Stationery Office [Se].

DOT (1996)   TD27/96 (DMRB 6.1.2) 'Cross–sections and Headroom', Stationery Office [Se].

DOT (various)   'Traffic Advisory Leaflets', DOT [Se].

IHT (1996)   'Cycle Friendly Infrastructure', The Institution of Highways & Transportation.

35.11 Further Information

DOT (1996)   'CoBA 10 Manual' and subsequent amendments, Stationery Office.

Simpson D and ,   'New Highway Design Flow Baker DJ (1987)   Thresholds'. IHT Journal 'Highways & Transportation', 34(5).

Chapter 36 Highway Link Design

36.1 Introduction

Highway link design is concerned with the geometry of road alignment, both horizontally and vertically. Clearly, it is dominated by the operating characteristics of motorised vehicles and the behaviour of drivers. In urban areas, the needs of other road–users, including pedestrians and cyclists, must be recognised and accommodated in the overall design but do not affect the geometric conditions of link design.

The Department of Transport's Technical Design Standard TD9/93 (DOT, 1993) [Sa] provides guidance, primarily for trunk roads, on both rural and urban highway link design. The prime objective is to design a highway link that achieves value for money and an acceptable standard of safety. It is for local highway authorities to decide the standard required for the roads for which they are responsible. Careful consideration should be given to the circumstances in which the Department's standards should be applied to local roads. TD9/93 gives designers the necessary flexibility and guidance to achieve such a design by allowing the use of values, below desirable minimum standards (see Table 36.1), known as 'relaxations'. In extreme cases, values known as 'departures' may be considered but these may require special approval procedures by the Local Highway Authority [Sb] before they can be used on local roads. It is particularly important, in urban areas, that designers are aware of, and make use of, this flexibility (see Chapter 31).

Design Parameter	Design Speed (km/h)					
	120	100	85	70	60	50
A STOPPING SIGHT DISTANCE (m)						
A1 Desirable Minimum	295	215	160	120	90	70
A2 Absolute Minimum	214	160	120	90	70	50
B HORIZONTAL RADII (m)						
B1 Minimum R *without elimination of Adverse Camber and Transitions	2880	2040	1440	1020	720	510
B2 Minimum r *with Superelevation of 2.5%	2040	1440	1020	720	510	360
B3 Minimum R *with Superelevation of 3.5%	1440	1020	720	510	360	255
B4 Desirable Minimum R *with Superelevation of 15%	1020	720	510	360	255	180
B5 Absolute Minimum R *with Super elevation of 7%	720	510	360	255	180	127
B6 Limiting Radius *with Superelevation of 7% at sites of special difficulty(Category B Design Speeds only)	510	360	255	180	127	90
C VERTICAL CURVATURE (m)						
C1 FOSD Overtaking Crest K Value	*	400	285	200	142	100
C2 Desirable Minimum *Crest K Value	182	100	55	30	17	10
C3 Absolute Minimum *Crest K Value	100	55	30	17	10	6.5
C4 Absolute Minimum Sag K Value	37	26	20	20	13	9
D OVERTAKING SIGHT DISTANCE (m)						
D1 Full Overtaking Sight Distance FOSD	*	580	490	410	345	290

Table 36.1: Recommended geometric design standards.   Source: TD9/93 Table 3 (DOT, 1993).

*Not recommended for use in the design of single carriageways.

In urban areas, the strict application of desirable minimum values may lead to disproportionately high construction costs and/or severe environmental impacts upon people or properties. In these circumstances, it may be possible to use lower values (ie relaxations) which will maintain an acceptable level of service without significantly affecting safety and, at the same time, offer significant cost savings and/or environmental benefits. Such values may also provide alignments which mould with the scale and grain of the urban fabric.

Whilst desirable minimum standards should always be the aim, TD9/93 gives a range of relaxations which are considered to conform to standard. Guidance on the assessment of the implications of the use of relaxations is given in Chapter 1 of TD9/93 (DOT, 1993) [Sa].

In extreme situations, significant cost and/or environmental benefits might be secured by using values which are lower than those for relaxations (ie departures) and still not have a significant adverse effect on safety. Whilst the use of departures does require careful consideration and assessment, designers should not dismiss their use out-of-hand but the approval of the Local Highway Authority [Sb] (and the Department of Transport [Sc] [Wa], where grants are involved) will be required before they can be used.

36.2 Design Speed

In urban areas, physical limitations may make it difficult to achieve even desirable design speeds. The role of a road in the hierarchy and a wide approach to safety management are relevant in deciding upon an appropriate design speed. Vehicles' operating speeds are, in any case, affected by the frequency of accesses and junctions, the presence of parked vehicles and pedestrians crossings and the mix of vehicle types.

Mandatory speed-limits are not, in themselves, a useful practical basis for determining an appropriate design speed. For existing urban roads, a guide to an appropriate design speed may be obtained by measuring the existing 85th percentile speed, when traffic is flowing freely in wet conditions (see Chapter 31).

A road alignment should be designed to ensure that the curvature, visibility and super-elevation are all consistent with the chosen design speed of the road. The design speed represents the 85th percentile speed of traffic when the road is first opened to traffic. This speed varies according to the impression of constraint that the road alignment and layout impart to the driver. It is necessary, therefore, to adjust the alignment and layout iteratively until curvature, visibility and superelevation are appropriate for the actual design speed. This involves estimating the alignment constraint (Ac) and the layout constraint (Lc) as described in TD9/93. It is the responsibility of the designer to create a road alignment where the chosen 85th percentile speed of traffic would be achieved without artificial constraints.

When the design principles for a new road have been established, appropriate standards for curvature, superelevation and visibility can be determined to suit the selected design speed of the road. Further advice on the choice of design speed is given in TD9/93 (DOT, 1993) [Sa].

36.3 Stopping Sight Distance

The stopping sight distance (SSD) is the theoretical sight distance required by a driver to stop a vehicle, when faced with an unexpected obstruction in the carriageway. The SSD has two elements:
  ❑ the perception–reaction distance, which is the distance travelled from the time the driver sees an obstruction to the time the brakes are applied; and
  ❑ the braking distance, which is the distance travelled while the vehicle decelerates from the assumed design speed to stop just short of the obstruction.

Drivers' perception and reaction times vary greatly and are affected by age, fatigue and conflicting stimuli or distractions, such as bright lights or noise. The braking distance will depend upon the condition and type of vehicle and the condition of the road surface. Desirable minimum values for design purposes are given in Table 36.1. These are based on a combined perception/reaction time of two seconds, for comfort and safety, and a deceleration of 0.25g (gravitational constant) for comfort.

Height of Vision
A driver's forward visibility requires an uninterrupted view between the driver's eye and any obstruction in the carriageway. Ninety–five per cent of drivers of private vehicles have eye heights of 1.05m or more above the road surface and this figure is used as the lower limit for design purposes. The upper limit for eye height is taken as 2.0m to represent drivers of large vehicles. The height of an object viewed is assumed to be between 0.26m (low object height) and 2.0m. Forward visibility should be provided in both horizontal and vertical planes between points in the centre of the lane nearest to the inside of the curve (see Figure 36.1).

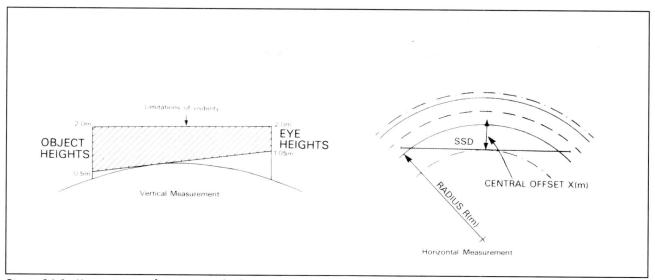

Figure 36.1: Measurement of stopping sight distance.

Source DOT (1993).

36.4 Full Overtaking Sight Distance

On single carriageway roads, overtaking vehicles need to use the opposing traffic lane and, to do so safely, they require adequate sight distances. TD9/93 provides a detailed design methodology for the coordinated design of single carriageway roads. This includes the concept of using full overtaking sight distance (FOSD) to divide the route into overtaking and non-overtaking sections. Minimum overtaking values, which express the length of the overtaking sections as a percentage of the length of the whole route, are prescribed.

Within urban areas, opportunities to provide FOSD may be rare and this will result in both an overtaking value considerably less than the minimum recommended by TD9/93 and lengths where overtaking may be dangerous. It is necessary that drivers are able to decide if it is safe to overtake and both the horizontal and vertical alignment of the road are important factors in making that decision. For example, certain ranges of values of horizontal radius are not recommended for use on single carriageway roads, because the associated forward visibility is such that it may create uncertainty in the mind of the driver (Figure 24 of TD9/93). The use of appropriate carriageway markings may assist. Detailed information is given in TD9/93 but a brief definition of FOSD is given as the 'sight distance required to permit drivers to complete a normal overtaking manoeuvre in the face of an oncoming vehicle'. It consists of four elements (see Figure 36.2):

❑ the perception/reaction distance, which is the distance travelled by the vehicle while the driver decides whether or not to overtake;
❑ the overtaking distance, which is the distance travelled by the vehicle to complete the overtaking manoeuvre (D1);
❑ the closing distance, which is the distance travelled by the oncoming vehicle while the actual overtaking manoeuvre is taking place (D2); and
❑ the safety distance, which is the distance required for clearance between the overtaking vehicle and the oncoming vehicle at the instant the overtaking vehicle has returned to its own lane (D3).

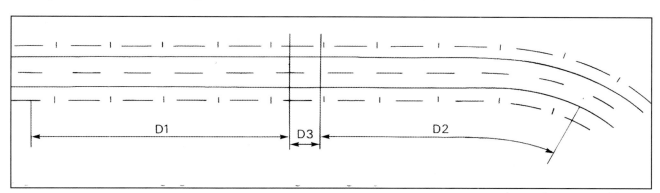

Figure 36.2: Full overtaking sight distance on a single carriageway road.

Source DOT (1984).

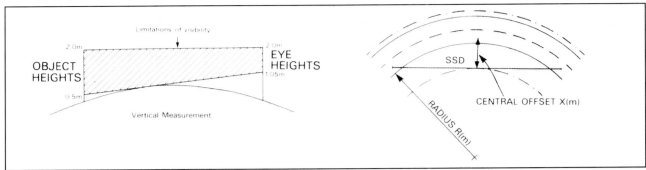

Figure 36.3: Measurement of full overtaking sight distance.

Source DOT (1993).

The time taken to complete an overtaking manoeuvre depends on the relative speeds of the vehicles involved. FOSD does not represent a safe overtaking visibility for all manoeuvres but will provide a reasonable degree of safe overtaking for 85% of traffic, although generally not in peak hours when heavy traffic prevails. Table 36.1 indicates typical values for FOSD.

In order to provide an uninterrupted forward view between the overtaking driver's eye, the vehicle to be overtaken, and the oncoming vehicle, FOSD should be available between points both 1.05m and 2.00 m above the centre of the carriageway, forming an envelope of visibility as shown in Figure 36.3. FOSD should be checked in both the horizontal and vertical planes.

36.5 Obstructions to Visibility

This section refers to visibility requirements for the design of links. The visibility requirements at junctions are covered in Chapters 37, 38 and 39 and the visibility requirements for pedestrians and cyclists are covered in Chapters 22 and 23, respectively.

The available forward visibility is a function of horizontal and/or vertical radius, carriageway width and verge and/or footway width. At lower horizontal radii, the line of sight to achieve desirable minimum SSD/FOSD will fall outside the carriageway and across the central reserve or footway/verge. In the case of the central reserve, a safety fence may obstruct visibility to the low object height of SSD, although full visibility to a higher object, usually taken to be 1.05m, will be available over the top of the safety fence. Urban road alignments and widths are often constrained such that the necessary widening of the central reserve verge or footway, to provide desirable minimum visibility requirements, can only be achieved at a very high cost in financial and/or environmental terms. In such circumstances, the use of relaxations and, in extreme cases, departures

should be considered (DOT, 1993) [Sa]. Obstruction to visibility caused by street furniture, particularly road signs and pedestrian guard rails, should be avoided. This requires attention to detail at early stages of design and should also be addressed in the safety audit.

36.6 Horizontal Alignment

The geometric parameters used in design are normally related to design speed (see Chapter 31). Table 36.1 shows typical desirable minimum values for horizontal curvature, corresponding to a superelevation of five percent, and radii below desirable minimum with a superelevation of seven percent. In urban areas, superelevation is usually limited to five per cent. However, the alignment may be so constrained that to provide desirable minimum radii will result in a very high cost in financial or environmental terms and, again, the use of relaxations or, in extreme cases, departures may be justified.

Superelevation

As a vehicle moves round a curve it is subject to centrifugal force, which causes it to try to slide outwards or to overturn. Side friction between the wheels and the road surface is generated to counter this force but the carriageway can also be superelevated to reduce the overturning effect when V^2/R is greater than 7.0 (where 'V' = velocity and 'R' = radius of curvature).

The percentage superelevation (or crossfall) 'S' required, is found from the equation:

$$S = V^2/2.828\ R$$

where V is the design speed (in km/h) and R is the curvature radius (in metres).

On horizontal curves, adverse camber should be replaced by a favourable crossfall of 2.5%, when the value V^2 is greater than 5.0 and less than 7.0.

Type of Road	Desirable Maximum Gradient
Urban motorways	3%
All Purpose dual carriageways	4%
All purpose single carriageways	6%

Table 36.2: Desirable maximum gradients for various types of road.

Studies of drivers' behaviour on curves have shown that, whilst superelevation provides a desirable contribution to comfort and safety, it need not be applied too rigidly. Thus, for sharp curves in urban areas with at–grade junctions and side accesses, superelevation should be limited to a maximum of five per cent (Southampton University, 1982).

Transition Curves

Superelevation, or removal of adverse camber, should be achieved progressively over the transition curve which introduces the horizontal curvature.

The basic length of transition curve 'L' can be derived from the equation:

$$L = V^3/46.7\ qR$$

where L = length of transition curve (m)

V = design speed (km/h)

q = rate of increase of radial acceleration (m/sec^3)

R = radius of curve (m).

Note that 'q' should normally be less than 0.3m/sec^3 for unrestricted design, although in urban areas it will frequently be necessary to increase it to 0.6m/sec^3 or even higher for sharp curves in tight locations.

The use of too long a transition curve may give the false impression to drivers that the horizontal curve which they are approaching is gentler than it actually is. This is particularly important when the radius of the curve is less than the desirable minimum. In these circumstances, the length of the transition curve should normally be limited to $(24R)^{0.5}$.

36.7 Vertical Alignment

Gradients

Vertical gradients cause vehicles, particularly heavy commercial vehicles, to lose speed and thereby incur additional running costs. Moreover, the likelihood of accidents increases as drivers attempt to overtake slower traffic. Desirable maximum gradients for various types of road are set out Table 36.2. Steeper gradients may be justified in hilly terrain but should be avoided if possible especially where traffic flows are high. A maximum of four percent is suggested for primary distributors. When designing in constrained situations, it should be borne in mind that this is an economic limit rather than one dictated by safety. A gradient greater than 0.5% should be maintained, wherever possible, to assist drainage of the carriageway surface. In flat areas, kerbside drainage channels may have to rise and fall to provide adequate run–off to clear surface water.

Vertical Curves

On the brow of hills, it is necessary to limit the severity of vertical curves because of visibility considerations and for the comfort of vehicle occupants. Ideally, vertical curves should be parabolic, because this form provides a constant rate of change of curvature at constant speed. It is convenient to determine the minimum length of a vertical sag or crest curve using a 'K' value. The minimum length of the curve can be obtained from the equation:

$$L = KA$$

where L = minimum curve length (m) required for

forward visibility and comfort;

K = design speed–related coefficient (chosen from

Table 36.1); and

A = algebraic difference in grades (%).

On crest curves, desirable minimum K is intended to give desirable minimum SSD, where the line of sight falls within the carriageway and is therefore only obstructed by the intervening carriageway. In urban areas, where design speeds may well be at the lower end of the range and where crest K values below desirable minimum are being considered, comfort becomes an issue and designers should be aware of this when using relaxations.

For sag curves, the K values are derived from comfort criteria and SSD becomes an issue only if the sag curve has a low K value and if there is a structure over the carriageway which may obstruct the visibility envelope.

36.8 Co–ordination of Horizontal and Vertical Alignments

It is beneficial to coordinate the horizontal and vertical alignments of a road, as this will avoid 'broken–back' optical illusions on the curvature of bends when the scheme has been constructed. This can be achieved by making all the points, where horizontal and vertical curves change, coincide with one another. Where this is not possible and the curves cannot be separated entirely, the vertical curves should be contained either wholly within, or wholly outside, the horizontal curves. Where curves are allowed to overlap, the resulting optical illusions are damaging to the appearance of the road and may contribute to accidents.

36.9 Climbing Lanes (or Crawler Lanes)

Climbing lanes should be provided to assist heavy commercial traffic in places where gradients are steep and the construction costs can be economically justified. They can be formed by widening a carriageway, or re–allocating the existing width, to provide an additional lane (further details may be found in TD9/93). The design principles in TD9/93 are suitable for urban roads but the economic appraisal methods may not be, particularly when flows are low (see Chapter 9).

36.10 References

DOT (1984)   TA43/84 'Highway Link Design', DOT.

DOT (1993)   TD9/93 (DMRB 6.1.1) 'Highway Link Design', Stationery Office [SA].

Southampton   'The Effect of Road Curvature on
University (1982) Vehicle/Driver Behaviour', DOT.

36.11 Further Information

DOT (1984)   Traffic Advisory Leaflet 43/84 'Highway Link Design', DOT.

Kerman J et al   'Do Vehicles Slow Down at
(1982)   Bends?', report to PTRC Annual Summer Meeting.

Simpson D (1985)   'System Development and Monitoring Designs', paper to ICE Conference (June), Institution of Civil Engineers.

Chapter 37 Junction Design: General Considerations

37.1 Introduction

Junctions are one of the critical elements in a highway transport system as they are the 'pinch points' where delay, accidents and emissions tend to be concentrated.

The optimum design solution for a particular set of conditions is complex and computer programs, specially written to assist with this process, enable a comparison of alternative designs to be made, in terms of the relationships between layout, the patterns of traffic demand and the resulting flows, capacities and delay. These programs can also be used to make assessments of the safety performance of the layouts of certain types of junction. Despite the availability of these aids, designing a junction continues to rely on the knowledge, experience and judgement of the designer.

Within a highway system, the basic purpose of a junction is to facilitate the transfer of traffic streams from one road to another in a safe and efficient manner. These transfers can be between roads of the same or different levels in the hierarchy (see Chapter 11). Ideally, such transfers should only be carried out between adjacent levels in the hierarchy. In other words, for example, transfers should not take place directly between local streets and primary distributors or between district distributors and access roads.

From the definition of its purpose, a junction can be described as an intersection between conflicting traffic streams or between motor vehicles and pedestrians and cyclists. The conflicts occurring at a junction can be categorised into three types:

❏ diverging: traffic streams from a common direction dividing themselves into two or more streams going in different directions;
❏ merging: traffic streams from two or more different directions joining together into a single stream going in one common direction; and
❏ crossing: the intersection of two traffic streams each entering from a different direction and leaving by a different exit.

Within these categories, further conflicts can arise from the different classes of vehicle making the manoeuvres, such as cyclists or buses, and between road vehicles and pedestrians.

These various manoeuvres must be carried out as safely as possible and sufficient capacity must be provided to minimise congestion, delay and fuel consumption. However, these two basic aims of junction design are sometimes in conflict. Thus, the junction designer requires a thorough knowledge of the various aspects and constraints affecting the safety and capacity performance of these important traffic facilities. It is vital, therefore, that designs are subjected to an independent safety audit (see Chapter 16).

37.2 Type of Junction

The various types of junction provide a hierarchy of layouts, which cater for increasing levels of traffic flow. These are:

❏ junctions without any designated priority;
❏ priority junctions (see Photograph 37.1);
❏ priority junctions with channelisation;
❏ roundabouts (see Photograph 37.2);
❏ traffic signal control (see Photograph 37.3); and
❏ grade–separated junctions.

Features of these junction types can sometimes be combined with advantage. For example, in some circumstances, a signal–controlled roundabout can have advantages over either a roundabout or a conventional signal layout. The overall network strategy may also influence the choice of junction type, for example, if positive management or control of traffic is desirable. Figure 37.1 gives an approximate guide to the magnitudes of major and minor road traffic flows that can be accommodated by particular types of junction and further information is available in Department of Transport publications (DOT, 1992 and DOT, 1981) [Sa]. Figure

Photograph 37.1: A typical priority junction.

Photograph 37.2: An urban roundabout.

Photograph 37.3: A signal–controlled junction on a district distribution road.

37.1 provides only a first estimate of the suitability of a particular layout. It does not take into account the pattern of movement through the junction, particularly the proportion of right–turning traffic or variations in the geometry of junction layout. For these reasons, undue reliance should not be placed on this Figure for design purposes.

37.3 Provision and Spacing of Junctions

The frequency and precise locations of junctions to be provided along a new road will depend upon its level in the road hierarchy (see Chapter 11), the proportion of non–local through traffic which it is intended to carry and the nature and presence of intersecting roads. If the number and importance of existing cross–routes would require too many junctions, it may be possible to combine two or more side–roads, before they reach the main road, giving benefits for road safety and junction capacity on the main road.

Junctions should be spaced at regular intervals and the minimum spacing should exceed the stopping sight distance appropriate for the 85th percentile speed of the major road (see Section 31.5). Greater distances should be provided, wherever possible, and particular care will be necessary when providing access to sites which generate large numbers of trips.

37.4 Design Issues and Objectives

The major sources of detrimental impact at junctions can be identified as accidents, congestion and delay, extra fuel consumption, air pollution and noise. The design objectives should set out to minimise these impacts.

Accidents
About two–thirds of personal injury accidents in urban areas occur at or near junctions. Whilst it has been recognised for some time that junctions are a major source of accidents and, although much effort has been put into their reduction, much remains to be done. Generally, the most effective method of reducing accidents at junctions is to separate the conflicting flows as much as possible.

The national average value of preventing a single personal injury accident (PIA) in a built–up area is estimated to be £31,460 at 1994 prices (see also Section 16.5). This value justifies considerable investment in improving the accident record at 'black spot' junctions.

Congestion and Delay
Congestion can spread rapidly when traffic demand exceeds the maximum capacity of a junction and is sensitive to the amount of excess demand and its duration. Thus, there is a continuing need to monitor the capacity at urban junctions. The cost of congestion at even a small urban junction can amount to hundreds of thousands of pounds every year.

Junction delays can be considered as delay due to the

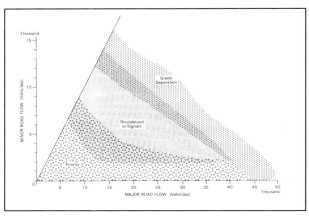

Figure 37.1: Type of junction appropriate for different traffic flows on major/minor roads.

geometric layout and form of control and delay due to congested traffic conditions at the junction. Both sources of delay have to be considered when a particular design solution is being assessed. In general, the geometric source of delay will be predominant at lightly–trafficked junctions, whilst the congestion source is the major contributor to delay at heavily–trafficked junctions.

Junction delay is often only a few seconds per vehicle and rarely more than two minutes at any one location, unless in a particularly congested urban street network. Also, because short delays occur at frequent intervals, drivers tends to bear them without irritation although, when such delays are summed over thousands of journeys in urban areas, the effect on total travel–time can be very substantial.

In considering what level of delay could be acceptable, account should be taken of the overall network strategy. This may seek to minimise delay on some roads, such as bus routes, but accept longer delays elsewhere, as part of a queue–management strategy.

Fuel Consumption
A typical journey in an urban area on a district distributor road with frequent signal–controlled junctions consumes 50% more fuel than the same length of journey by urban motorway. This is caused by lower than optimum speed and greater frequency of stops/starts and speed changes.

Air Pollution and Noise
For a given flow of vehicles, air pollution and engine noise levels increase with congestion and, conversely, reductions in congestion lead to lower levels of traffic pollution. Whilst emissions and noise levels can be determined, a degree of uncertainty remains about their specific impacts on health or on the quality of life. Thus, estimates of overall environmental costs, which are based on generally–accepted monetary values, are the subject of continuing research. With moderate congestion at a small urban junction, it is estimated that approximately 30 tonnes of carbon dioxide (CO_2) are produced per annum. There is also increasing concern about the level of particulates which are emitted mainly from diesel–powered vehicles (see Chapters 9 and 17).

Design Objectives
The major objectives for junction design can be summarised as:
- ❑ minimising accident risk, particularly for vulnerable users;
- ❑ minimising accident severity;

- ❑ providing adequate capacity for vehicular traffic, such that the level of service is compatible with that provided on the approach roads, thus minimising congestion, as measured by the length and duration of queues on the approaches to the junction, and delay to vehicles passing through the junction;
- ❑ providing safe and convenient passage for cyclists and pedestrians, including people with visual or mobility impairment;
- ❑ minimising environmental impacts, such as air pollution and engine noise, by minimising fuel consumption through reductions in the number of speed changes and the number of stops/starts required at the junction;
- ❑ providing an economic solution, so that the cost of implementing the design will be, at least, offset by the economic benefits derived; and
- ❑ minimising conflicts between traffic activity at the junction and existing and planned roadside development in the vicinity.

It is likely that some of these objectives will conflict with each other and a balance will need to be struck. That balance should reflect background policies, such as giving priority to environmental concerns or to vulnerable road–users.

37.5 Design Principles

Principles of good junction design can be considered under four headings:
- ❑ general;
- ❑ geometric and operational requirements of vehicular traffic;
- ❑ safety; and
- ❑ provision for pedestrians and cyclists.

General
The geometric and the control aspects of a junction layout should be considered together to ensure a good design solution to a particular problem. The aim is to provide road–users with layouts that have consistent standards and are not likely to be confusing. On lengths of urban road, sequences of junctions should not, therefore, involve many different types of layout. For example, a length of road containing roundabouts, single–lane dualling, ghost islands, simple priority junctions and grade-separation would create uncertainty for road–users and increase the risk of accidents on that account. Similarly, a signal–controlled pedestrian crossing on one half of a carriageway and an uncontrolled crossing on the other could confuse pedestrians. The most efficient and safest schemes usually contain no surprises for road–users.

Major/minor priority layouts are the most common form of junction control, with the advantage that through traffic on the major road is not delayed. However, high speeds and any possibility of overtaking manoeuvres on the major road should be discouraged at priority junctions.

For more heavily-trafficked junctions, more complex forms of layout and more sophisticated control systems are required. Bearing in mind the uncertainty of traffic forecasting, a designer should always consider whether the layout and control system being designed could readily be converted to a different type of junction, such as signal-controlled, if this should prove necessary in the future.

The consideration of which movement should have priority is important at junctions. At priority junctions and roundabouts, these are fixed but at traffic signals priority is allocated to the different movements during the control cycle-time, which can either be varied for different times of day or be wholly demand-responsive. Also the overall network strategy will affect decisions on priority. Often, it will be appropriate for heavier volumes of traffic to have some measure of priority but care is needed to ensure that the minor flows are not too severely disadvantaged. In some circumstances, public transport vehicles and pedestrians may warrant priority at certain times of day.

Geometric and Operational

Several important principles governing the geometric and operational aspects of a design solution are set out below.

Layout to Suit Traffic Movements and Patterns

The layout should be designed to suit the traffic patterns with the principal movements generally being given the easiest paths. Ease of movement does not have to equate with high speed. Layouts which encourage a smooth, but slower, passage through a junction will be safer for vulnerable road-users. Wherever practicable, the layout should be designed so as to follow the shortest vehicular paths. This improves the smoothness of operation and makes it more readily understood by road-users. Unduly sharp radii, or complex paths involving several changes in direction, should normally be avoided, although in some areas, eg subject to traffic calming, they may be appropriate.

On entering a junction, users should always be able to see quickly, from both the layout and advance traffic signs, the path they should follow and the potential crossing, merging and diverging traffic streams that may be encountered. Drivers should be encouraged to slow down on entry to large roundabouts, so that they have time to see circulating cyclists. To achieve this, the layout, traffic islands, control devices, traffic signs and road markings should all be considered as a single design entity. Uphill approaches to a junction make it difficult for drivers to comprehend the layout and should be avoided wherever possible.

To an increasing extent, computer aided design (CAD) software packages are being used to examine and refine junction layouts. CAD can be used to allow engineers to view layouts from various perspectives and to check sightlines.

Layout to Suit Long Vehicles

Allowance should be made for the swept paths of long vehicles turning in areas where significant numbers of such vehicles can reasonably be expected to use a junction. The turning swept paths normally used are those generated by a 15.5m long articulated or semi-trailer vehicle, with a single axle at the rear of the trailer. However, in the case of staggered priority junctions, the design vehicle is represented by a 18.35m long vehicle with draw-bar trailer (DOT, 1995) [Sb].

Reduction and Separation of Conflicts

The choice of layout will govern the number and type of conflicting manoeuvres that have to be accommodated within a design solution. Roundabout layouts result in the least number of conflict points and eliminate crossing-points entirely, whilst T-junctions and staggered layouts have fewer conflict points than cross-roads. Traffic signal-controlled junctions should, in theory, eliminate most of the conflicting movements, by separating them in time.

Crossing, merging and diverging movements can usefully be separated by physical or painted 'ghost' islands, so that the number of traffic conflicts at any point is reduced. However, layouts which have numerous small traffic islands must be avoided as they are ineffective and confusing. Separation of traffic conflicts means that road-users are faced only with simple choices of direction at any one time. This can lead to greater safety. Nevertheless, for the separation of conflicts to be effective, the junction must be large enough to enable users to identify, in adequate time, those traffic streams that will conflict with their intended path and those that will not. Otherwise, gaps in the priority movement cannot be used efficiently by traffic entering the junction and the flow through the junction will not be optimised.

Visibility

It is important that all road-users have adequate

visibility in each direction to see conflicting traffic movements in sufficient time to permit them to make their manoeuvres safely. This concept applies to all types of junction and to the visibility of pedestrians and cyclists. The specific requirements for each of the junction types is explained in detail in later chapters. Where possible, junctions should be positioned on level ground or in sags rather than at, or near, the crests of hills. As well as having adverse safety implications, poor visibility reduces the flow of turning movements that can be achieved.

Diverging and Merging Lanes

It is important to reduce the speed differences on high speed roads between through and turning traffic and, for this purpose, the incorporation of diverging and merging lanes into the design solution is useful.

Traffic, slowing down on the approach to an intersection in order to turn either left or right into an intersecting road, may impede following vehicles or cycles that are not turning. It is helpful, therefore, where space allows, to permit divergence of the two streams at a small angle by the provision of a diverging lane. Offside 'storage' lanes in the centre of single and dual carriageways are especially useful as they provide a safe space for vehicles waiting to turn right off the major road.

Nearside diverging lanes are useful where there is a heavy left turn from that approach road, especially on higher speed roads and on gradients. Likewise, merging lanes permit turning traffic to accelerate before joining the fast traffic streams on dual carriageway and other high speed roads, where this traffic would otherwise impede flow and be a source of hazard. However, they can also be hazardous when the capacity to absorb merging traffic has been taken up.

Provision for cyclists needs particular consideration, especially where high speed traffic crosses their path as it merges with, or diverges from, a main route. Cyclists are better segregated from these situations (see Chapter 23).

Traffic Signs and Road Markings

The importance of traffic signs and road markings in junction design is often under-rated. They are an integral part of the design process and should be considered from an early stage. No junction design is complete without these features. Advance direction and warning signs should be provided. Designs should be checked to ensure that the proposed layout can be properly signed. Care must be taken with the positioning of signs at the junction itself so that they do not interfere with visibility. Policy and detailed guidance on these aspects are given in the Traffic Signs Manual (HMG, 1994) (see Chapter 15).

Road Lighting

Road lighting will normally be provided at junctions in urban areas, especially when one of the intersecting roads already has lighting. It is recommended that, where road lighting is required at a junction, provision should be in accordance with the British Standard Code of Practice for Road Lighting (BSI, 1992). When an existing junction is being modified, the lighting provision should be checked for suitability with the new arrangement. Any alteration to the lighting should be carried out prior to, or at the same time as, the roadworks.

Safety

Safety considerations are a priority from initial concept to final design and layouts should be subject to safety audits, as described in Section 37.6 (see also Chapter 16).

For the same level and pattern of traffic flows, a major/minor priority junction will usually have more accidents per year than other junction types. These accidents will also be more serious than with other forms of control. For example, in the UK, the average cost of an accident at a priority junction is more than twice that at roundabouts and one and a half times that at traffic signals. The accidents are mainly associated with right-turns and are exacerbated in number and severity by high speeds and the possibility of overtaking manoeuvres on the major road. Key considerations in the attainment of a reduction in both the occurrence and severity of accidents are:
- ❏ a reduction of high vehicle speeds through the junction;
- ❏ the provision of clear visibility for all approaching traffic streams;
- ❏ appropriate geometric standards for the typical design vehicle; and
- ❏ integration of the traffic information and control systems within the junction layout.

Provision for Pedestrians and Cyclists
(see Chapters 22 and 23)

The requirements of pedestrians and cyclists should be carefully considered from the start of the design of junctions, especially in central urban and suburban areas. It may be possible to provide separate routes for pedestrians and cyclists away from the junction, where road widths are less and traffic movements more predictable. This is not always desirable. Where significant numbers of cyclists are expected, consideration should be given to providing for them specifically by, for example, separating out their movements from those of other vehicles but still accommodating them within the carriageway through dedicated lane–space and advanced stop–lines at signals. For pedestrians, and sometimes for cyclists, one of the following facilities should be considered at the junction itself:

- ❏ central island refuges at unmarked crossing places;
- ❏ an unsignalised crossing, with or without central refuges;
- ❏ signal–controlled crossings; or
- ❏ a subway or footbridge.

The type of facility selected will depend upon the expected traffic volumes and the movements of both pedestrians and vehicles and should be designed in accordance with current recommendations and requirements (see Chapters 22 and 23).

At–grade pedestrian crossings on a major road should not be placed over ghost islands where there are no refuges. Defined at–grade pedestrian crossings on a minor road should be beyond the tangent points of the corner radii and should be sited to reduce to a minimum the width to be crossed by pedestrians, provided that they are not involved in lengthy detours from their desired paths. Central refuges should be used where possible. In urban areas, where large numbers of pedestrians are present, short lengths of guardrails should be used to channelise them onto crossings.

In some circumstances, it may be possible to combine facilities for pedestrians and cyclists. Any special provision should be designed to be convenient and easy to use; otherwise, it runs the risk of being ignored.

37.6 Evaluating Alternative Solutions

The uncertainties associated with forecasting future levels of urban traffic flow suggest that it is sensible to test a range of design flows in terms of the resulting alternative junction designs. This process should include varying the proportions of turning traffic, which can influence the type and scale of the junction proposed. This type of analysis can identify a range of proposals, which offer varying levels of service and other strengths and weaknesses.

Where the type of junction to be designed is not predetermined by other factors, assessments of safety performance, operational efficiency and resource costs can be used to assist in the choice of the best design. Evaluation of alternative proposals should cover the identification of which junctions are critical to the operation of the network and should examine the effects of capacity–overloads causing queueing and possible re–routeing using unsuitable roads. Cost–benefit studies will require the use of appropriate values of time for the people and vehicles involved and reference should be made to the Department of Transport's COBA 10 Manual and subsequent amendments (DOT, 1996a) (see Chapter 9) [Sc].

Safety Performance

The junction designer should be concerned with both accident prevention and severity minimisation. The safety performance of a scheme can be evaluated by making assessments of the likely accident–frequency through a safety audit, which is a process that seeks to ensure that highway and traffic schemes are as safe as practicable within the context of the purpose for which they are intended. Considerable research effort has been devoted to the development of models for predicting accident–frequency at junctions and on some kinds of links in urban areas (see Chapter 16).

The research had three objectives:

- ❏ to develop standards for road links and junctions;
- ❏ to build statistical accident–prediction models for junction design; and
- ❏ to design network accident appraisal software for local networks.

Several levels of accident–prediction models have been developed. The simplest models predict total accidents for the junction or link, while the most sophisticated take account of the effects of design geometry and environmental factors on a number of accident categories. Some simple models for pedestrian accidents are available, although not for accidents to pedal cyclists, and the models are insensitive to speed–management policies. The main application is in assessing the safety impacts of traffic engineering schemes, rather than specific measures for speed reduction or the safety of vulnerable road–users.

To be successful, the safety audit process must be well defined and must be systematically applied, at all stages of the planning, design and construction of a scheme. The design standards generally used for highway and traffic schemes take account of safety but, when the various elements of the design are brought together, the resulting scheme may not be the best in terms of safety. In addition, there may be compromises between safety and capacity or departures from standards necessitated by specific site conditions. Safety audits seek to take account of all of these issues and to highlight changes to the scheme design that would optimise safety within the overall scheme objectives (IHT, 1996; DOT, 1994a; and DOT 1994b) (see also Chapter 16) [Sb].

The elements of a junction that should be audited will depend, to some extent, on the nature of the scheme. The audit should identify how the various users would walk, drive or ride through the junction. As a guide, the following should be checked:

❑ geometric design: the type and layout of the junction, including horizontal and vertical alignments and cross–sections;

❑ road markings: including white lines, road studs, raised markings and delineators;

❑ road signs and street furniture: including lighting (intensity and location of posts), all types of signs, islands and bollards, pedestrian guardrails and safety fencing;

❑ road surface: including profile, effect on lighting and skid resistance and storm water provision;

❑ traffic management: including provision for pedestrians, cyclists and the disabled; speed limits and controls, junction control, waiting, loading and parking provision, traffic circulation, one–way streets and banned turns and provision for public transport;

❑ management of incidents: an assessment of the ability of the junction to accommodate emergency traffic management in the event of an accident; and

❑ road works and maintenance: including temporary working during construction, maintenance of schemes, and the signing and operation of road works on the existing network.

Operational Efficiency

Evaluation of the operational efficiency of a junction comprises three major components, namely:

❑ capacity analysis;

❑ determination of queueing characteristics; and

❑ determination of vehicular delay.

Capacity analysis is concerned with determining the ability of the junction to accommodate all the various movements. In the case of priority junctions and roundabouts, this will depend primarily on the arrival patterns of the priority movements, as well as on the junction layout, type of control and gap–acceptance characteristics. The case of signal–controlled junctions is different, in that the arrival patterns of the movements are considered simultaneously and there is no fixed priority, since it changes during the signal cycle.

The capacity of a road intersection for traffic making a particular movement is frequently specified, in terms of its throughput of passenger car units (pcu) per hour. Pcus are introduced to allow for differences in the amount of interference to other traffic by the addition of one extra vehicle to the traffic, according to the type of vehicle. For example, a large lorry is longer, wider and slower than the average car and, therefore, has a considerably greater effect on other vehicles by making it more difficult for them to overtake and by slowing down those which are forced to follow. On any particular section of road, under particular traffic conditions, the addition of one vehicle of a particular type, per hour, will reduce the average speed of the remaining vehicles by the same amount as the addition of, say, x cars of average size per hour. Under these conditions, one vehicle of this type is said to be equivalent to x passenger car units.

In the case of a bottleneck, in particular at an intersection, one can arrive at a slightly different definition, which is, however, equivalent to applying the one given to maximise flow conditions. This definition is that, if a particular type of vehicle, under saturated conditions, requires x times as much time at the intersection than is required by an average car then that type is equivalent to x passenger car units.

For priority junctions and roundabouts, it is normal to assume that heavy vehicles and buses are equivalent to two pcus. However, for saturation flow determination at traffic signals, it is normal to adopt the following values:

❑ car or light goods vehicle (LGV) = 1.0 pcu;
❑ medium goods vehicle = 1.5 pcu;
❑ heavy goods vehicle (HGV) = 2.3 pcu;
❑ bus/coach = 2.0 pcu;
❑ motor cycle = 0.4 pcu; and
❑ bicycle = 0.2 pcu.

The output of the capacity analysis is the servicing pattern, ie the ability of the junction to cater for the through movements of non–priority traffic. If the servicing pattern is input into a queueing model, together with the arrival pattern of the non–priority movements and the queue discipline regime (normally first–in–first–out), estimates can be made of the queue–lengths. If the queueing model is time–dependent, then the vehicular delay characteristics can be determined by integrating the queue–length distribution over time, using appropriate software.

The value of delay that emanates from the queueing model is 'operational' or 'congestion' delay and is due to other traffic. This has to be combined with 'geometric' delay to determine the total delay. It should be noted that geometric delay includes all fixed delay elements, due to both the layout of the junction and the control strategies that are applied to it.

The output of the evaluation process, in terms of queue–lengths (presence or absence of congestion) and total delay, is used to determine the operational efficiency of a particular design strategy. The analysis can be extended to estimate its efficiency in terms of fuel consumption.

Resource Costs

The generalised cost of operating a junction will comprise:

❑ time (delays) costs;
❑ vehicle operating costs; and
❑ accidents costs.

Although these are usually studied separately, for different purposes, the true optimisation of junction design should aim to minimise the aggregate of all of these costs, discounted over the anticipated life of the junction, in comparison with the capital cost of improving it (see also Chapter 9). Wider policy considerations may, however, constrain some of the parameters affecting this optimisation.

In addition to time costs arising from delay, operating costs will be incurred for each of the separate manoeuvres performed at a junction. Vehicle operating costs depend on the speed, type of manoeuvre and the distance travelled by each vehicle. These costs include the additional cost of fuel, oil, tyres, maintenance and depreciation caused by the layout of a junction and the changes in speed and direction of the vehicles travelling through it.

Assessment Programs

In assessing junction designs, account needs to be taken of both capacity and safety requirements. These may sometimes be in conflict and some junction types will perform better than others in each respect. A balance needs to be struck for the situation that exists at each junction. Advice is provided in the IHT/TRL leaflet Designing Junctions to Cut Delays and Accidents (IHT/TRL, 1993).

To facilitate the decisions involved in junction choice and design, the Transport Research Laboratory (TRL) has developed three computer programs for assessing isolated junctions. ARCADY 3 (Assessment and Roundabout Capacity and Delay) and PICADY 3 (Priority Intersection Capacity and Delay) deal respectively with roundabouts and major/minor priority junctions. OSCADY 3 (Optimised Signals Capacity and Delay) deals with traffic

signal–controlled junctions. All of the programs operate on the same principles: given demand flows and turning movements for typical peak hours and the junction geometry, they predict where queues will form, how long they will last and when, and for how long, vehicles will be delayed (DOT, 1996b; DOT, 1993a; and DOT 1993b). These programs also make assessments of the frequency of accidents that might occur with a particular design for certain types of junction.

In addition, other programs can assist in the evaluation process. These include RODEL and ROBOSIGN (see Chapter 39), which address roundabout designs, and LINSIG and SIGSIGN (see Chapter 40), which can be used for the assessment of traffic signal designs. RODEL is especially useful, in that it can be used to generate geometric design parameters for a specified level of service.

37.7 References

BSI (1992) BS5489: 'Code of Practice for Lighting', British Standards Institution.

DOT (1981) TA23/81 (DMRB 6.2) 'Junctions and Accesses: Determination of the Size of Roundabouts and Major/Minor Junctions', Stationery Office [Sa].

DOT (1982) TD22/92 (DMRB 6.2.1) and TA 48/92 (DMRB 6.2.2) 'Layout of Grade Separated Junctions', Stationery Office [Sa].

DOT (1993a) AG18 (3rd Edition) 'PICADY 3 User Guide', TRL.

DOT (1993b) AG22 (Issue A) 'OSCADY 3 User Guide', TRL.

DOT (1994a) HA42/94 (DMRB 5.2.3) 'Road Safety Audits', Stationery Office [Sb].

DOT (1994b) HD19/94 (DMRB 5.2.2) 'Road Safety Audits', Stationery Office [Sb].

DOT (1995) TD42/95 (DMRB 6.2.6) 'Geometric Design of Major/Minor Priority Junctions', Stationery Office [Sb].

DOT (1996a)   'COBA 10 Manual' and subsequent amendments (DMRB Volume 13), Stationery Office [Sc].

DOT (1996b)   AG17 'ARCADY 3 User Guide', TRL.

HMG (1994)   SI 1994 No. 1519 'The Traffic Signs Regulations and General Directions 1994', Stationery Office.

IHT/TRL (1993)   'Designing Junctions to Cut Delays and Accidents', TRL.

IHT (1996)   'Guidelines for the Safety Audit of Highways', The Institution of Highways & Transportation.

Chapter 38 Priority Junctions

38.1 Introduction

The majority of junctions in urban areas take the form of some type of priority junction, which is normally appropriate where traffic flows, particularly to and from minor roads, are relatively light. Where flows are heavier or layouts are complex, other types of layout or control, such as roundabouts or traffic signals, are required to reduce accident risks and to balance or improve capacity. The primary advantage of priority junctions is that the main road flow does not normally experience any delay. Movements from the minor road, and right turns into it, are dependent on gaps in the major traffic stream and this influences both safety and capacity. The problem is that, as main road traffic flows increase, gaps between vehicles get smaller and accidents increase as gap–acceptance gets shorter. As in all junction design, the needs of pedestrians and cyclists should be considered from the outset and not added as an afterthought.

The three basic types of priority junction are:
- ❑ 'T' junctions;
- ❑ staggered junctions; and
- ❑ crossroads.

Each of these types can be divided into four forms of layout, which are derived from the characteristics of the major road. These are:
- ❑ simple layouts;
- ❑ 'ghost island' layouts;
- ❑ localised single–lane dualling; and
- ❑ dual carriageways.

Photograph 38.1: A typical major/minor priority junction.

The main source of advice on the choice and design of priority junctions is contained in the publication TD42/95 Geometric Design of Major/Minor Priority Junctions (HMG, 1995) [Sa]. This relates specifically to new and improved trunk roads but it is often applied to other categories of road.

38.2 Choice of Major/Minor Priority Junction

Table 38.1 gives guidance on the major/minor priority junction forms considered suitable for various major road carriageway types in urban situations. This represents the starting point when choosing the most appropriate type of major/minor priority junction to use at any particular site (see Photograph 38.1).

Simple Junctions
TD 42/95 states that new simple junctions are not suitable when design flows exceed about 300 vehicles two–way average annual daily traffic (AADT) on the minor road and 13,000 vehicles two–way AADT on the major road. However, this advice relates specifically to rural junctions. The only advice relating to urban junctions is that consideration should be given to upgrading existing simple junctions, where the minor road flow exceeds 500 vehicles two–way AADT. Minor road flows of 300 and 500 vehicles per day are low for an urban situation, particularly as no reference is made to the major road flow or to the pattern of turning movements.

Where the occurrence of accidents involving right–turning vehicles is evident, or expected, simple layouts may not be appropriate. This is equally the case where major road traffic would be inhibited by right turns into the minor road. Capacity analysis will help determine if this is likely.

Ghost Islands
The provision of a dedicated lane for traffic slowing and waiting to turn right from the major road has significant road safety and capacity benefits. However, problems can arise if overtaking manoeuvres are thereby encouraged. Consequently, ghost islands should not be used where overtaking opportunities on the upstream and downstream links are very restricted. As an alternative, a nearside passing lane or a left–hand diverging lane loop (see

Carriageway Type		Junction Type								
		Simple			Ghost Island			Dualling		
Standard	Location	T junction	Staggered crossroads	Straight crossroads	T junction	Staggered access roads	Straight crossroads	T junction	Staggered crossroads	Straight crossroads
S2	Urban	Yes	Yes	Maybe	Yes	Yes	No	Yes (D1)	Yes (D1)	No
WS2	Urban	No	No	No	Yes	Yes	No	Yes (D1)	Yes (D1)	No
D2	Urban	No	No	No	No	No	No	Yes (D2)	Yes (D2)	No
D3		No	No	No	No	No	No	No	No	No

Table 38.1: Choice of junction type.

Source: TD42/95.

Key:

S2	Single 2–lane carriageway
WS2	Wide single 2–lane carriageway
D1	Dual single–lane carriageway
D2	Dual 2–lane carriageway
D3	Dual 3–lane carriageway

Figures 38.1 and 38.2) can be beneficial, where a normal ghost island layout cannot be achieved.

Single–Lane Dualling

Dualling of single lane roads is normally used on unrestricted rural single carriageway roads and, therefore, its application in urban areas is not commonplace. However, its use can be beneficial in particular cases, for example, where a pedestrian crossing needs to be located on the major road, near to a priority junction, for which a divided crossing is appropriate.

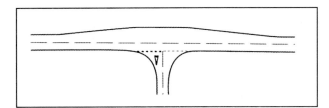

Figure 38.1: A nearside passing lane.

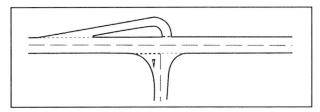

Figure 38.2: A left–hand diverging lane loop.

Dual Carriageways

Major/minor priority junctions may be used on new dual two–lane carriageways but never on new dual–3 all–purpose (D3AP) roads. Such junctions should include the widening of the central reserve to provide an offside diverging lane and waiting space for vehicles turning right from the major road. TD42/95 recommends an upper limit for minor road flows of 3000 vehicles 2–way AADT for D2AP roads in rural areas.

No advice exists specifically for urban roads but capacity analysis will assist in this respect. There may be no preferable alternative to retaining a priority junction on an existing urban D3AP roads but consideration should be given to restricting right–turn movements into and out of the side–road.

Crossroads

It is best to avoid straight crossroads because of their generally poor safety performance. Staggered junctions are better but roundabouts or traffic signal–control may be better still. If crossroads are unavoidable, they should be provided only on single carriageway roads and measures should be introduced to make the priorities clear and to slow traffic down on the minor roads approaches.

Staggered Junctions

Right–left staggers are preferred because, with this layout, traffic moving between the minor roads is less likely to have to wait in the centre of the major road. Also, right–turning movements from the major road will not 'hook' and thereby interfere with each other.

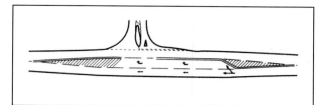

Figure 38.3: Road markings for a protected turning bay.

38.3 Siting of Priority Junctions

If overtaking on single carriageways is likely to be a road safety problem, measures should be included to prevent or discourage it. These may include:
- reduction of the carriageway width, by hatched markings;
- the use of different coloured surfacing materials; and
- the use of double white lines, where the visibility criteria can be met (see Figure 38.3).

In terms of horizontal alignment, priority junctions should not be sited on sharp curves, particularly on the inside of a bend. At skew junctions, careful consideration should be given to measures to ensure that priority is not mistaken.

Priority junctions are best located on level ground or where approach gradients do not exceed plus or minus two percent. Significant downhill major road gradients encourage excessive speed, whilst uphill gradients reduce drivers' appreciation of the junction layout, as they approach it.

On minor roads, downhill approaches can result in over-running at the give-way lines and uphill approaches can reduce both drivers' appreciation and gap-acceptance. The profile of a junction should be such as to enable a driver to see the full width of a junction. This is particularly important at junctions on dual carriageways.

38.4 Road Safety

In urban areas, over half of all personal injury accidents occur at or near major/minor junctions. For the same level and pattern of traffic flow, major/minor priority junctions will usually have more accidents per year than other types of junction. Right turn manoeuvres, both from the major road and from the minor road, are particularly vulnerable and the safety implications of these movements should receive special consideration in junction design.

Road safety at priority junctions can be improved in several ways:

- by protecting pedestrians and making special provision for cyclists;
- by preventing overtaking on the approaches;
- by making crossroads more conspicuous or replacing them with staggered junctions;
- by improving visibility; and
- by preventing right turning, particularly on high speed carriageways.

Road safety audits will provide an overview to ensure a safe combination of design parameters and features for all users (see Chapter 16).

38.5 Road–User Requirements

Cyclists

Major/minor junctions should be designed from the outset with the interests of cyclists (as well as pedestrians) in mind, as 73% of cyclist accidents occur at such junctions.

On the busiest roads, there will be value in providing for cyclists away from the junction, where space permits. This could involve:
- provision of a cycle track away from the carriageway;
- Toucan crossings for shared use by pedestrians and cyclists;
- signposting an alternative cycle route; and
- grade–separation

If cyclists are expected to use the main carriageway then facilities along the length of the carriageway can improve conditions for cyclists generally and raise drivers' awareness of the presence of cyclists, as they approach junctions. The specific facilities to incorporate at a junction are best considered on a site by site basis, according to the dominant turning movements for cyclists at that junction. Every design should emphasise to drivers where they can expect to encounter cyclists at a junction. If the volume of cyclists is significant, or is expected to increase at a particular location, consideration should be given to signalising the whole junction and to providing advanced stop–lines for cyclists. Measures to improve facilities for cyclists are considered in more detail in Chapter 23.

Pedestrians

The needs of pedestrians should be considered throughout the design process. Pedestrian facilities are normally provided in the form of refuge islands, Zebra or signal–controlled crossings. The type of treatment will depend on the anticipated vehicular and pedestrian flows. Chapter 22 deals with pedestrian facilities in more detail.

Defined at–grade Zebra, or signal–controlled, pedestrian crossings on minor roads should normally be a minimum of 15m back from the give–way line and should be sited to minimise crossing widths.

Separation islands are normally situated at the mouth of the side road. Dropped kerbs and tactile surfaces should be used at all pedestrian crossing points for the blind and the partially sighted.

Guardrails should be used only where significant pedestrian activity makes it necessary to channel pedestrians to the appropriate crossing points. Care should be taken to ensure that guardrails do not interrupt visibility for drivers and consideration should be given to the use of types of barrier designed especially for that purpose.

Landscaping

Landscaping can help to define the outline of junctions, to provide reference points and to establish a background for signs. Sensitive use of textured surfaces, choice of street furniture and planting should be used to enhance the general appearance. However, landscaping should not compromise visibility and good maintenance must be provided.

38.6 Geometric Design

Design Speed

Geometric standards are related to the traffic speed of the major road and, for new roads, to the design speed as defined in TD9/93 'Highway Link Design' (DOT, 1993). Advice on measurement of the 85th percentile wet weather speeds on existing roads is contained in TA22/81 (DOT, 1981) (see also Chapter 31) [Sa].

Corner Radii

Where no provision is made specifically for large goods vehicles, the minimum circular corner radius should be six metres. Where provision for heavy vehicles is to be made, a 10m circular corner radius, followed by a taper of 1.5m over a distance of 30m, is recommended. Where a significant number of heavy vehicles is likely to use the minor road, corners should be designed using a three–centred compound curve (HMG, 1995).

Carriageway Widths

Through lanes should normally be a maximum of 3.65m and a minimum of 3.0m wide. A ghost island turning lane should be 3.5m, although relaxation to 3.0m is permissible. In very restricted situations, a minimum width of 2.5m may be used. If it is considered desirable to encourage vehicles turning right from the minor road to execute the turn in two separate manoeuvres, the right–turn lane should be widened to provide a five metre refuge. Carriageway widening is required on tight radii to take account of the overhang and cut–in associated with articulated lorries.

Where significant cycle flows are anticipated, the inside lane should be widened to 4.25m to provide space for lorries to overtake cyclists. Alternatively, additional width, in the form of a 1.5m wide cycle lane, could be provided (see Chapter 23).

Right–Turn Lanes

The layout at right–turn lanes comprises the taper, the deceleration length and the turning/queueing length [see Figures 7/4 and 7/6 in TD 42/95 (HMG, 1995)]. It should be noted that the turning length should be long enough to accommodate predicted peak period queues.

Diverging and Merging Lanes

Diverging and merging lanes are recommended only on roads with design speeds of 85 km/h (53 miles/h) or above. In cases where such lanes are potentially suitable, TD42/95 (HMG, 1995) provides general numerical criteria. Diverging and merging lanes create hazards for cyclists and provision for cyclists should follow the guidelines for grade–separated junctions (DOT, 1988).

Channelisation

Traffic islands can serve several useful purposes:
- ❑ to direct or guide vehicles to take a specific path and to segregate opposing traffic streams;
- ❑ to provide a refuge for pedestrians and waiting traffic;
- ❑ to provide segregated lanes for buses or cyclists in appropriate circumstances; and
- ❑ to provide convenient locations for essential street furniture, such as traffic signs, street lighting, manholes and inspection chambers.

They should be of sufficient size to be clearly visible, otherwise they will be potential accident hazards.

Traffic Signs and Road Markings

The choice and positioning of traffic signs and road markings should be an integral part of the design process. Advance direction and warning signs should be provided and care must be taken with the positioning and the size of these signs. Policy and detailed guidance on these aspects are given in the Traffic Signs Manual (DOT, 1982–86) [Sb].

38.7 Visibility

Drivers emerging from the minor road of a priority junction must have adequate visibility to left and right along a single carriageway major road. Where the major road is a dual carriageway, with a central reserve of adequate width to shelter turning traffic, the standard visibility splay to the left is not required for the minor road but visibility to the left is needed in central reserves. If the major road is a one–way street then, clearly, visibility is only required towards oncoming traffic.

Major Road Distance ('y' distance)

The 'y' distance (see Figure 38.4) is determined by the speed of main road traffic. It must be sufficient to allow side–road traffic to emerge safely and to provide forward visibility to allow major road traffic to stop, if required.

The 'y' distance is determined by the 85th percentile wet weather speed or, if this is not known, by the speed–limit of the road. Tables 38.2 and 38.3 should be used, as appropriate

Minor Road Distance ('x' distance)

The 'x' distance (see Figure 38.4) gives a good field of view for a driver approaching, or stationary at, the give–way line. It also allows oncoming traffic to see emerging side–road traffic. TD 42/95 advises a desirable 'x' distance of nine metres but acknowledges that, in difficult circumstances, this may be relaxed to 4.5m for lightly trafficked simple junctions and, exceptionally, to 2.4m. Annex D of PPG13 Transport (DOE/DOT,1994) [Sc] [Wa] states that a 4.5m 'x' distance will normally be acceptable for less busy simple junctions and busy private accesses. For single dwellings and small culs–de–sac, a minimum of 2.0m is given.

In some circumstances, it is not possible to achieve the above requirements and authorities take a more relaxed view, particularly regarding 'x' distances. A pragmatic and balanced view is sometimes necessary. The following advice from PPG13 highlights this but needs to be used with caution [Sc].

'It is not always practicable to comply fully with visibility standards. Such standards, like all other material considerations in development control, need to be assessed in the light of all the circumstances of each case. However, visibility should not be reduced to such a level that danger is likely to be caused.'

Drivers' Eye–Height

The splay of visibility should be uninterrupted at the typical drivers' eye–height of 1.05m. At junctions near the crests of hills and adjacent to bridges, the effect of vertical alignment on visibility will need careful consideration. Problems are particularly likely when a priority junction occurs near the crest of a hill.

Other Visibility Considerations

As well as visibility between drivers, the visibility of, and between, pedestrians and cyclists using the junction should be considered in the design process. Drivers approaching on the minor road should have an unobstructed view of the junction. TD 42/95 has introduced two requirements in this respect and these are illustrated in Figure 38.4. This ensures that drivers slow down sufficiently to see the junction form clearly. On curved alignments, splays should be made tangential to the nearside edge of the road (see Figure 7/2 in TD 42/95).

Vehicles turning right from the major road need good visibility towards oncoming traffic and this should never be less than the desirable minimum stopping sight distance (DMSSD). TD 9/93 Highway Link Design (DOT, 1993) [Sd] states that DMSSD must be achieved on the immediate approach to a junction. The immediate approach is defined as:

Major road speed (85%ile)	miles/h	75	62.50	53	44	37.5	31	25	19
	(km/h)	(120)	(100)	(85)	(70)	(60)	(50)	(40)	(30)
Major road (y) distance	m	295	215	160	120	90	70	45	33

Table 38.2: Visibility distance based upon 85th percentile wet weather speed.

Speed Limit on Major Road	miles/h	70	60	50	40	30	20
	(km/h)	(113)	(96)	(80)	(64)	(48)	(32)
Major road (y) distance	m	295	215	160	120	90*	45*

* includes an allowance for vehicles travelling 10 km/h above the speed limit.

Table 38.3: Visibility distance based upon prevailing speed–limit.

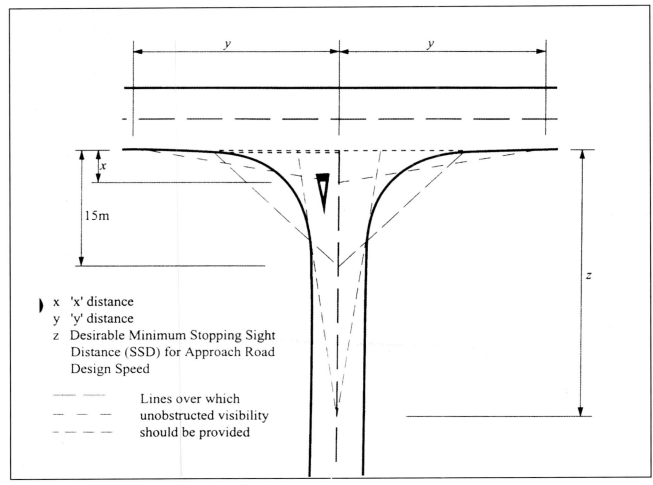

Figure 38.4: Visibility distances on major road and side–road.

□ minor road – 1.5 times the DMSSD upstream of the give–way line; and
□ major road – 1.5 times the DMSSD from the centre line of the minor road

Visibility splays should preferably lie within the curtilage of the highway to ensure that they are not obstructed. Planning controls can be used to prevent any future obstructions and section 79 of the Highways Act 1980 (HMG, 1980) [Se] provides highway authorities with the powers to pursue measures to improve or safeguard visibility.

As outlined earlier, 'x' and 'y' distances are measured respectively along the centre line of the minor road and along the kerbline of the major road. In reality, these are not the points from which drivers observe oncoming traffic or the positions of approaching vehicles. In some circumstances, where decisions on visibility standards are marginal, it may be beneficial to consider the visibility distances that are achieved in reality. Such an approach can assist pragmatic decision–making.

38.8 Traffic Throughput and Delay
Maximum Throughput

Prior to determining whether a design for a priority junction is a satisfactory solution, it is necessary to test the proposed layout using the relevant 'design reference' flows. The maximum throughput of the non–priority movements can be determined by two factors, which are:

□ the number and length of gaps, occurring in the major traffic streams, that can be used by traffic entering or leaving the minor road(s); and
□ the characteristics of the junction layout, such as lane widths, flare lengths and visibility distances.

A quick and simple assessment of the throughput of a particular traffic stream can be obtained by using the following formula:

$$C = 700 - 0.33M$$

where C is the throughput (veh/h) of the non–priority stream(s) and M is the total flow (veh/h) opposing priority movement(s).

TRANSPORT IN THE URBAN ENVIRONMENT

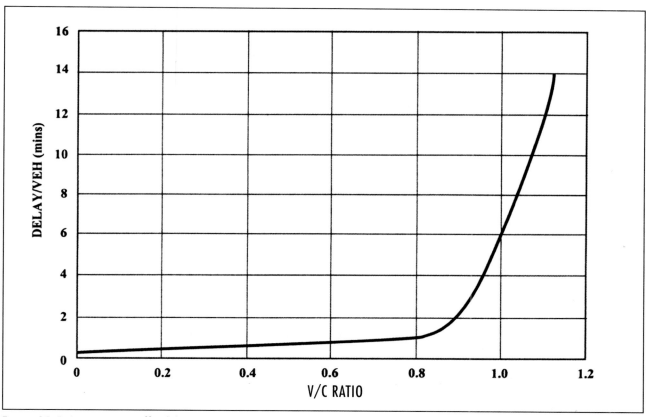

Figure 38.5: Increase in traffic delays with increase in the volume ratio to capacity (V/C).

The most critical traffic streams are usually the right–turn movements into, and out of, the minor road. More detailed equations for turning–stream capacities are given in Annex 1 of TD 42/95 (HMG, 1995).

The computer program PICADY 3 (TRL, 1993) has been developed to calculate the queues and delays that are likely with any particular junction layout and design flows. The program is commonly used to determine the capacity of all types of priority junction and it can also predict accident–rates for some types of junction.

Performance Indicators

The acceptability of any junction design depends upon the performance indicator used. TD 42/95 suggests the use of the ratio of volume to capacity (V/C) as an indicator of likely junction performance and it suggests that a design with a V/C ratio of about 85% is likely to result in a level of provision which will be economically justified in urban areas. Beyond a V/C ratio of around 0.85, in any stream of traffic, delays to that traffic tend to increase disproportionately, as illustrated in Figure 38.5.

Such an approach is simple to apply but should be tempered by examination of the actual demand flows and the significance of delays. Queue–lengths are

only of major significance, if blocking back to adjacent junctions or accesses threatens to impair the performance of the network.

Commonly, junction analysis concentrates on conditions during peak periods, but it may be necessary to assess performance over the range of conditions occurring over a typical week. Trade–offs between what happens in a few peak hours and what happens in the many more off–peak hours can be important in the choice of layout, or even the type of junction control, adopted.

The Use of PICADY 3 (TRL, 1993).
The PICADY 3 User Manual details the use of this program. The following observations relate to some of the complexities that may not be self–evident.

Blocking of Major Road Traffic by Right–Turners
The program can predict the delays to straight–ahead traffic caused by waiting right–turning traffic. However, this is not the case where there are two lanes on the major road and only one is blocked or where blocking occurs only sometimes.

Flaring on Minor Roads
If a minor road has a single–lane approach but is flared close to the junction entry, the traffic on this arm is considered as two streams; one containing the

left–turning traffic and the other containing the right–turning traffic. The queues and delays for the two streams take account not only of the traffic capacity at the give–way line but also of the effect of vehicles being trapped upstream in the single–lane approach.

Pedestrian Crossings

The effect of Zebra crossings close to junctions can be modelled using PICADY 3. This takes into account pedestrian flows at a crossing and the location of the crossing.

However, Zebra crossings cannot be modelled, if they occur at any of the following locations:
- ❏ across a major road between the two minor arms of a staggered junction; or
- ❏ on a flared minor road within the flared two–lane section; or
- ❏ on a major road arm where right–turning traffic blocks the straight–ahead stream.

Furthermore, it is not possible to model the effect of Pelican or Puffin crossings close to junctions.

Site–Specific Corrections

PICADY 3 will not always model a junction with sufficient accuracy. Observations should therefore be made to identify if this is the case. At three–arm junctions, entry capacity can be determined when there is continuous queueing, with corrections then being made to the program. The application of this calibration technique will not be appropriate, if large scale changes in junction layout are planned.

'Marginal effects analysis' predicts the effects of small discrete changes in junction geometry.

The measurement of visibility distances differs from the standards used in TD 42/95 and relates more closely to the distances that are actually achieved.

Prediction of Accident Rates

The prediction of likely accident rates is an important issue. PICADY 3 has the facility to predict accident rates for some types of junction. The increasing range of junctions for which this is possible makes it easier to consider road safety issues, consistently, when evaluating junction design.

38.9 References

DOE/DOT (1994) PPG13 Planning Policy Guidance 'Transport', DOE/DOT [Sa] [Wa].

DOT (1981) TA22/81 (DMRB 5.1) 'Vehicle Speed Measurement on all–purpose Roads', Stationery Office [Sa].

DOT (1982–1986) 'The Traffic Signs Manual', Stationery Office [Sb].

DOT (1988) TA1/88 'Provision for Cyclists at Grade–Separated Junctions', DOT, [Sa].

DOT (1993) TD9/93 (DMRB 6.1.1) 'Highway Link Design', Stationery Office [Sa]

HMG (1980) 'Highways Act 1980', Stationery Office [Sf].

HMG (1995) TD42/95 (DMRB 6.2.6) 'Geometric Design of Major/Minor Priority Junctions', Stationery Office [Sa].

TRL (1993) 'PICADY 3 User Manual', TRL.

38.10 Further Information

DOT (1981) TA23/81 (DMRB 6.2) 'Junctions and Accesses : Determination of the Size of Roundabouts and Major/Minor Junctions', Stationery Office [Sa].

DOT (1995) TD41/95 (DMRB 6.2.7) 'Vehicular Access to All Purpose Trunk Roads', Stationery Office.

TRL (1980) Report SR582 'The Traffic Capacity of Major/Minor Priority Junctions', TRL.

TRL (1982) Report SR724 'The Effect of Zebra Crossings on Junction Entry Capacity', TRL.

Chapter 39 Roundabouts

39.1 Introduction

Roundabouts have a good safety record for vehicular crashes when compared with traffic signals and priority junctions but much depends on the size of the junction, the flow pattern of vehicles and the presence of pedestrians and cyclists, who are particularly vulnerable. However, the risk of accidents increases as the throughput approaches capacity and gap–acceptance times reduce. Roundabouts tend to cause less overall delay to vehicular traffic than signals at low and medium flow levels and are particularly appropriate for large right–turning flows. Roundabouts can give greater priority to flow on minor roads than main roads and this may create queueing on the major road. Signals may consequently be needed on one or more approaches, particularly at peak periods, to deal with uneven demand flows, which can cause congestion or accident problems on individual arms. Introducing signal–control at roundabouts has been found to improve the safety of cyclists, although not necessarily the overall safety of all road–users (Lines, 1995). This is covered in more detail in Chapter 42. Roundabouts are also appropriate at particular locations, where a significant change in road standard occurs, as they have the advantage over other junction types in slowing down all traffic streams and eliminating crossing conflicts but this too can create accident problems when a roundabout is placed at the end of a high–speed dual carriageway.

There are two main types of roundabout, namely, conventional and mini (see Figure 39.1).

A conventional roundabout has a one–way circulatory carriageway around a kerbed island 4m or greater in diameter, usually with flared approaches to allow several vehicles to enter simultaneously. The recommended number of entry–lanes is usually three or four. Roundabouts are not recommended on dual 3–lane roads, because a design to provide sufficient junction capacity would probably require a very large roundabout indeed.

A mini–roundabout has a one–way circulatory carriageway around a flush, or slightly raised, circular marking less than four metres in diameter, with or without flared approaches. Mini–roundabouts can be effective in improving existing urban intersections that suffer overload or accident problems. Their layout should be designed so that drivers are made aware, in good time, that they are approaching a roundabout. They should only be used where all the approaches are subject to a speed–limit of 30 miles/h or less. Technical Directive TD16/93 (DOT, 1993) [Sa] covers the geometric design of roundabouts and various layouts are described in Table 39.1 and illustrated in Figures 39.1 and 39.2.

39.2 Principles of Operation

A roundabout junction operates as a one–way circulatory system around a central island, where entry is controlled by 'Give Way' markings and priority must be given to traffic approaching from the right. The operating efficiency of this type of junction depends on the ability of drivers to respond to safe opportunities to join the stream of circulating vehicles already using the junction.

Roundabouts with more than four arms are generally not desirable, because drivers' comprehension of the layout is affected when roundabouts are large and high circulating speeds may be generated.

Although the initial construction costs may be greater for conventional roundabouts than for other types of junction, because of the larger land-area required, vehicle operating costs are likely to be less because they permit a free flow of traffic when demand is light and they are self–regulating.

The ability of roundabouts to cope with U–turn manoeuvres can be particularly useful, where one or more of the approach carriageways is divided by a continuous central reserve or where U–turns or right–turning traffic would otherwise be dangerous or disruptive.

During uncongested off–peak periods, roundabouts will generally result in less delay than similar junctions with signal–control. However, they are not generally compatible with urban traffic control systems, as they cannot respond to positive control commands. They may also be unsatisfactory where there are cyclists or pedestrians in significant numbers and where special provisions may be required, such as grade– or mode–separation, which can be expensive.

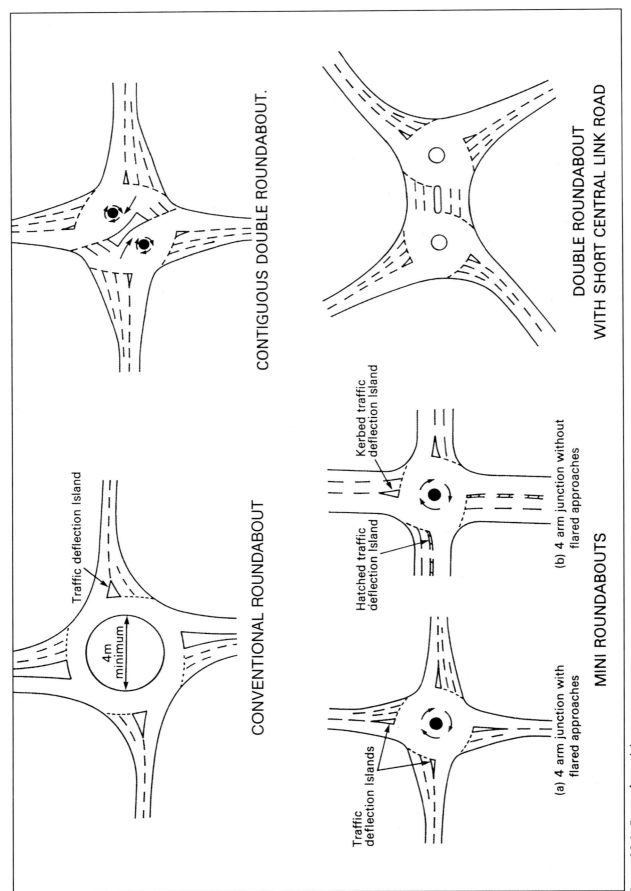

CONTIGUOUS DOUBLE ROUNDABOUT.

DOUBLE ROUNDABOUT
WITH SHORT CENTRAL LINK ROAD

CONVENTIONAL ROUNDABOUT

Traffic deflection Island

4m minimum

Kerbed traffic deflection Island

Hatched traffic deflection Island

(b) 4 arm junction without flared approaches

MINI ROUNDABOUTS

(a) 4 arm junction with flared approaches

Traffic deflection Islands

Figure 39.1: Types of roundabout.

TRANSPORT IN THE URBAN ENVIRONMENT

Type	Description	Typical Use/ Location
Convent-ional	❑ Kerbed central island with diameter greater than or equal to 4m ❑ Flared approaches to allow multiple entry lanes ❑ See Figure 39.1	❑ New developments and construction ❑ Junctions within or at end of dual carriageways ❑ To change direction of a new road at a junction
Mini	❑ Flush or slightly raised central island less than 4m in diameter ❑ Road markings indicate pattern of movement ❑ No street furniture on central island in order to allow long vehicles to overrun ❑ See Figure 39.1	❑ To improve the performance of existing junctions where space is severely constrained ❑ Mainly as conversions from other roundabout and junction types ❑ At sites subject to a 30 miles/h speed-limit
Double	❑ Two conventional or mini roundabouts are placed within the same junction connected by a short link road ❑ See Figure 39.1	❑ For controlling unusual or asymmetric approaches. ❑ At approaches with heavy opposing right–turning movements, staggered approaches and at sites with more than four arms
Grade–separated	❑ At least one traffic movement passes through the junction without interruption, while the remainder are brought to one or more roundabouts at a different level ❑ Compact designs are favoured ❑ For pedestrians and cyclists the roundabout is elevated, to allow easy gradients for pedestrian and cycle network below ❑ See Figure 39.2	❑ On urban motorways and dual carriageways ❑ On high capacity roads and those with high approach speeds of traffic ❑ On new construction where there are high forecast vehicle and pedestrian flows
Ring junctions	❑ A large two-way circulatory system where each approach is provided either with 3-arm roundabouts (normally minis) or with traffic signals ❑ See Figure 39.2	❑ At some special sites to solve particular local problems ❑ For conversion from very large roundabouts which have entry problems ❑ Not recommended for a new facility
Signal–controlled	❑ Traffic entering the roundabout from one or more arms is signal-controlled for all or part of the day ❑ See Figure 39.2	❑ To increase capacity under certain operating conditions
Gyratory systems	❑ Small one-way systems where normal–land use activities can be maintained on the central island	❑ In urban areas, especially town centres ❑ Safe access to the island must be ensured for pedestrians, cyclists and possible maintenance vehicles

Table 39.1: Types of roundabout and their main characteristics.

Photograph 39.1: Segregated cycleway at Cheals Roundabout, Crawley. Courtesy: David Nicholls.

Accident rates and accident severities at roundabouts can be significantly lower than those at signal–controlled junctions of equivalent capacity (see also Chapter 37). The most common problem affecting safety is excessive speed, mainly on entry but also within the roundabout. Cyclists are particularly at risk, especially at conventional roundabouts, where they are over 14 times more likely than a motorised vehicle to be involved in an accident. Decisions as to whether or not roundabouts should be used should, therefore, take account of existing and prospective cycle networks and the amount of cycle–use through the junction. Wherever possible, cyclists should preferably be segregated from other traffic at roundabouts, either vertically or horizontally (see Photograph 39.1).

39.3 Siting of Roundabouts

The decision to provide a roundabout, rather than some other form of intersection, should be based on operational, economic and environmental considerations. Factors to be taken into account at the design stage include the need to reduce speed at certain places for reasons, such as:

❑ to effect a significant change in road standard, say from dual to single carriageway or from grade–separated intersection roads to at–grade intersection roads; or

❑ to emphasise the transition from a rural to an urban or suburban environment; or

❑ to achieve a sharp change in route direction, which could not be achieved by ordinary curves using standard radii.

Roundabouts should preferably be sited on level ground or in sags, rather than at or near the crests of hills, because it is difficult for drivers to appreciate the layout when approaching on an uphill gradient. However, roundabouts on hill tops are not

intrinsically dangerous, if correctly signed and with adequate visibility standards provided on the approaches to the give–way line, ie in excess of the stopping distance for the design speed.

39.4 The Geometric Features of Roundabout Design

The terminology used to describe the geometry of roundabout design is given in Figure 39.3. The main geometric design features are as set out below.

Entry–Path Curvature
Vehicle approach speeds should be moderated to appropriate levels, so as to achieve the desired levels of safety and capacity. This can be achieved by deflecting the vehicle entry–path at the junction approach, using suitably positioned traffic islands, small adjustments to kerb lines and by staggering the entry arms, as shown in Figure 39.4. Entry–path curvature should not exceed 100m radius, otherwise higher accident rates are likely to occur. This, however, is dependent upon the amount of circulating traffic across the arm and the balance of turning movements on that approach.

Where mini roundabouts are being created, at existing junctions where a 30 miles/h speed limit applies, an appropriate entry angle can be achieved by small traffic deflection islands or, less effectively, by road markings. It is sometimes best to experiment first with temporary materials to obtain the optimum entry shape.

Entry Widths
Theoretical capacity is very sensitive to small changes in entry width. It is good practice to add at least one extra lane–width to the entry approach but, as a general rule, not more than two lanes should be added and no entry should be more than four lanes wide. There may be some cases, usually associated with low predicted flows, where increased entry width is not operationally necessary but it is still recommended that at least two entry lanes be provided. This gives added flexibility in the event of breakdown and eases the problem of space for long vehicles turning. Entry widening on the offside by means of sharp reverse curves is not recommended. Segregated left–turning lanes can be beneficial in particular situations (see below).

Flare Length
Theoretical capacity is also very sensitive to small changes in the flare length, which should develop gradually, avoiding any sharp angles, in order to be used effectively.

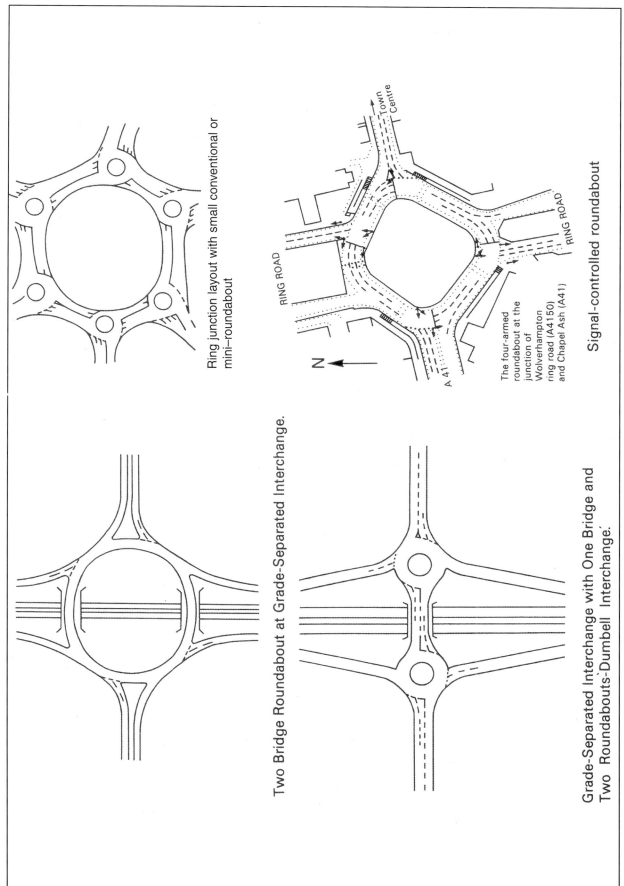

Ring junction layout with small conventional or mini-roundabout

RING ROAD

Town Centre

N

A 41

RING ROAD

RING ROAD

The four-armed roundabout at the junction of Wolverhampton ring road (A4150) and Chapel Ash (A41)

Signal-controlled roundabout

Two Bridge Roundabout at Grade-Separated Interchange.

Grade-Separated Interchange with One Bridge and Two Roundabouts-'Dumbell Interchange.'

Figure 39.2: Types of roundabouts (grade—separated and signal controlled).

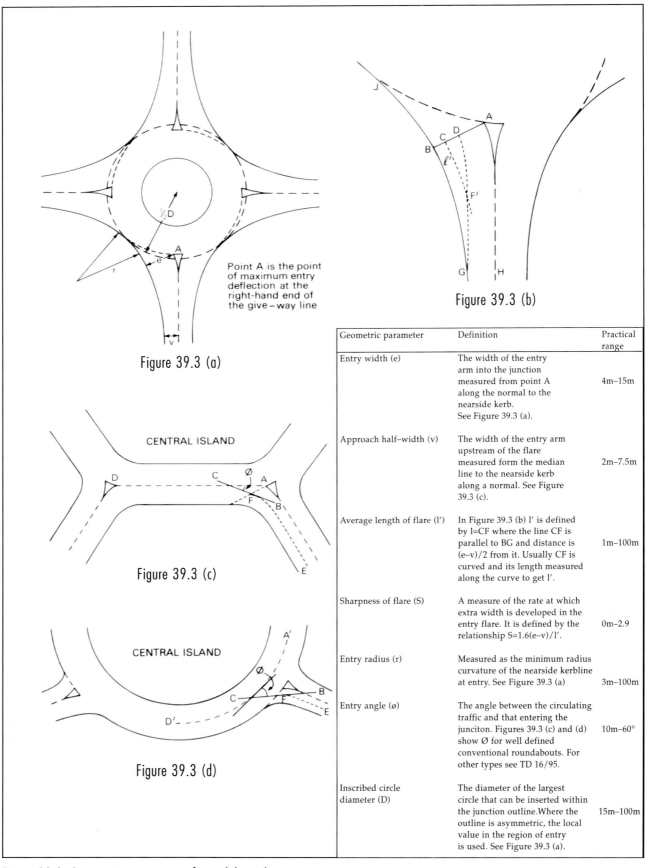

Point A is the point of maximum entry deflection at the right-hand end of the give-way line

Figure 39.3 (a)

Figure 39.3 (b)

CENTRAL ISLAND

Figure 39.3 (c)

CENTRAL ISLAND

Figure 39.3 (d)

Geometric parameter	Definition	Practical range
Entry width (e)	The width of the entry arm into the junction measured from point A along the normal to the nearside kerb. See Figure 39.3 (a).	4m–15m
Approach half-width (v)	The width of the entry arm upstream of the flare measured form the median line to the nearside kerb along a normal. See Figure 39.3 (c).	2m–7.5m
Average length of flare (l')	In Figure 39.3 (b) l' is defined by l=CF where the line CF is parallel to BG and distance is (e–v)/2 from it. Usually CF is curved and its length measured along the curve to get l'.	1m–100m
Sharpness of flare (S)	A measure of the rate at which extra width is developed in the entry flare. It is defined by the relationship S=1.6(e–v)/l'.	0m–2.9
Entry radius (r)	Measured as the minimum radius curvature of the nearside kerbline at entry. See Figure 39.3 (a)	3m–100m
Entry angle (ø)	The angle between the circulating traffic and that entering the junciton. Figures 39.3 (c) and (d) show Ø for well defined conventional roundabouts. For other types see TD 16/95.	10m–60°
Inscribed circle diameter (D)	The diameter of the largest circle that can be inserted within the junction outline.Where the outline is asymmetric, the local value in the region of entry is used. See Figure 39.3 (a).	15m–100m

Figure 39.3: Geometric parameters of roundabout design.

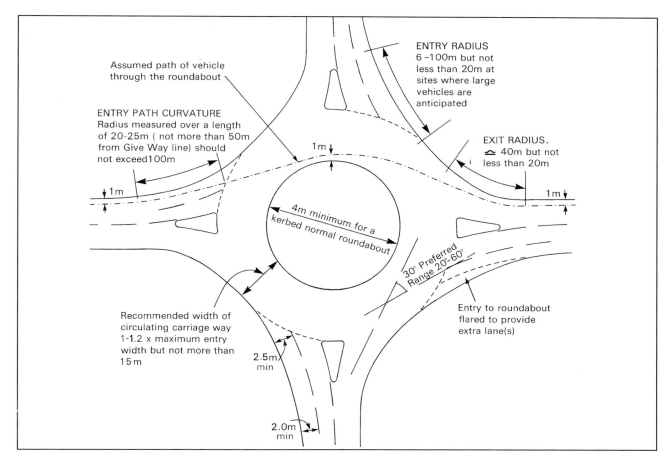

Figure 39.4: Typical layout of a conventional roundabout showing entry details.

Entry Angles and Radii

The entry angle should be considered with effective entry width, to ensure that traffic in the offside lane can enter the circulatory carriageway on a natural path, without conflicting with the central island or traffic entering on its nearside. Entry angles below 20° and above 60° should be avoided, as entry accidents are likely to increase.

The entry radius should always lie between six metres and 100m. A good practical design often lies in the range 20m to 30m. Where a roundabout is designed to cater for long vehicles in particular, the entry radius should not be less than 10m. Increasing the entry radius above 30m provides little (if any) increase in capacity. As values drop below 15m, they produce increasingly severe reductions in capacity and substantially higher operating costs.

Circulatory Carriageway

The circulatory carriageway should not exceed 15m in width and, if possible, should be circular in plan, avoiding deceptively tight bends. The width of the circulatory carriageway should be constant and should lie between 1.0 and 1.2 times the maximum entry width. It is normal practice to avoid short

lengths of reverse curve between entry and adjacent exits, by linking these curves or joining them with straights between the entry radius and the exit radius. One method is to increase the exit radius. However, where there is a considerable distance between the entry and the next exit, as at three–arm roundabouts, reverse curvature may result.

Inscribed Circle Diameter (ICD)

The relationship between the central island diameter and the inscribed circle diameter (ICD) is the most important consideration for the passage of large vehicles. The following advice is based on the turning swept–path generated by a 15.5m long articulated vehicle, with a single axle at the rear of the trailer, which is often adopted as the 'Design Vehicle' for roundabouts. The turning width required by this type of vehicle is greater than that for most other vehicles. The requirements for other vehicles, including a single unit rigid vehicle, such as a 12m long bus or an 18m long drawbar trailer combination, are less onerous.

The smallest ICD for a conventional roundabout that will accommodate the design vehicle is 28m. If this cannot be provided, a mini roundabout should be used. Note that it may be difficult, if not impossible, to meet the entry deflection requirement with conventional roundabouts which have ICDs of about 39m or less. The largest ICD for a mini roundabout should be 28m.

Large Islands

The size of the central islands is in itself not critical but will depend on the requirements for entry deflection, the inscribed circle diameter and the width of the circulatory carriageway. At conventional sites, the island is likely to be circular and will be kerbed.

The presence of a roundabout can be made more obvious to approaching traffic by landscaping and planting. The screening of traffic on the opposite side of a roundabout to the point of entry can, without restricting visibility, avoid distraction and confusion caused by traffic movements of no concern to a driver. Planting can also provide a positive background to chevron signs and direction signs on the central island, while visually uniting the various vertical features and reducing any appearance of clutter. However, good maintenance of landscaped roundabout islands is clearly essential and is likely to be expensive. Generally, the planting of roundabout central islands less than 10m in diameter is inappropriate, as the need to provide visibility leaves only a small central planting area available.

Small Islands

The circular marking (1.0m – 4.0m diameter) of a mini roundabout should be as large as possible in relation to the site and should be domed, up to a maximum height of 125mm at the centre. This doming, in conjunction with the presence of some adverse crossfall, will help to make the roundabout more conspicuous to drivers. No bollards, signs, lighting columns or other street furniture should be placed on the dome or on any central island less than 4.0 m in diameter.

Deflection Islands

Where physical deflection is not possible on the approach, road markings, indicating small traffic islands, should be used to induce some vehicle deflection. These islands should be kept free of all furniture, except the 'Keep Left' bollards and other essential signs.

Segregated Left–turning Lanes

Segregated left–turning lanes are a useful method for giving an improved service to vehicles intending to leave a roundabout at the first exit after the entry. Their use should always be considered when more than 50% of the entry flow, or when more than 300 vehicles/ h in the peak hour, are seeking to turn left

at the first exit. Vehicles are channelled into the left hand lane by lane arrows, supplemented by advance direction lining signs. These vehicles proceed to the first exit without having to give way to others using the roundabout (see Photograph 39.2). Segregation by road markings is more common than by use of additional islands but is less effective because it is subject to abuse. Care needs to be exercised in how vehicles will merge or give way to other vehicles exiting from elsewhere on the roundabout. Segregated left–turn lanes pose particular dangers for cyclists on the relevant approach, who are not turning left, and also when exiting the roundabout with fast moving traffic in the segregated lane on their nearside. Alternative designs which encourage lower speeds may need to be considered.

Pedestrians and Cyclists

Whenever possible, cyclists and pedestrians should be segregated from other traffic at roundabouts. Where cyclists and pedestrians need to be accommodated, particularly where roundabouts are used in traffic–calming schemes, it may be appropriate to consider modifying some of the geometric standards. In these circumstances, it has been found preferable for single–lane approaches without flares, but with deflections, to be used. Multi–lane approaches to roundabouts can mean that circulating cyclists are hidden from the view of drivers approaching the roundabout. Even if the approach cannot be reduced to a single lane, the use of flares should be avoided. Approaches should be deflected towards the centre of the central island, with relatively small radius kerbs, to maximise the conspicuity of circulating cyclists and to discourage

Photograph 39.2: Segregated left–turn lane at Cheals Roundabout, Crawley. Courtesy: David Nicholls.

high entry speeds. The requirement for 'easy exits' should be reviewed, to ensure that circulating cyclists are not compromised and that drivers are not encouraged to accelerate towards any pedestrian crossing which might be located on the exit arm.

39.5 Visibility Requirements

With the increasing tendency to landscape and plant the larger central islands of roundabouts, it is important that visibility requirements are given close attention in order to ensure safe operation. The design has to consider the ultimate height and shape of plants and the operational and cost implications of maintenance.

The following guidelines, based on TD16/93 (DOT, 1993) [Sa], represent good practice concerning the provision of visibility. When these guidelines are not complied with, additional signing is needed to alert drivers of all vehicles to potential hazards. Attention should be given to the visibility of cyclists, taking account of the paths they are likely to take when using the roundabout

Visibility, with the exception of visibility to the right at entry, should be assessed in accordance with an eye–height of 1.05m and an object–height of 0.26m. Visibility to the right at entry should be based upon a driver's eye–height of 1.05m to an object–height of 1.05m.

Visibility on approaches
The forward visibility at the approach to a roundabout should not be less than desirable minimum stopping sight distance for the design speed of the approach measured to the give–way line. In special cases, a departure from standards, to one step below desirable minimum stopping sight distance, may be adopted to avoid severe environmental damage.

Drivers of all vehicles approaching the give–way line should be able to see the full width of the circulatory carriageway to their right, for a distance appropriate to the size of the roundabout (39m to 70m measured along the centre line of the circulatory carriageway). This visibility should be checked from the centre of the offside lane at a distance 15m back from the give-way line. Checks should be made that poor crossfall design or construction does not restrict visibility.

Drivers of all vehicles approaching the give–way line should be able to see the full width of the circulatory carriageway ahead of them for a distance, measured along the centre line of the circulatory carriageway, appropriate to the size of the roundabout (39m – 70m). The visibility should be checked from the centre of the nearside lane, at a distance of 15 m from the give–way line.

Visibility on the circulating carriageway
Drivers of all vehicles circulating on a roundabout should be able to see the full width of the circulatory carriageway ahead of them for a distance appropriate to the size of the roundabout (39 m – 70 m). This visibility should be checked from a point 2m in from the central island.

Visibility of pedestrian crossings
Drivers of all vehicles approaching a pedestrian crossing across an entry should have a minimum distance of visibility to it of desirable minimum stopping sight distance for the design speed of the link. At the give–way line, drivers of all vehicles should be able to see the full width of a pedestrian crossing across the next exit, if the crossing is within 50m of the roundabout. In urban areas, adjacent roadside development may, however, prevent this visibility splay being established fully.

Obstructions to visibility
Signs, street furniture and planting should not obstruct visibility. Infringements by isolated slim obstructions, such as lamp columns, sign supports or bridge columns can be ignored, provided they are less than 550mm wide.

39.6 Safety Considerations in Design

Roundabouts are generally considered to be the safest form of at–grade junction with accidents costing, on average, 50% less than at other junction types and about 70% less than on links. Notwithstanding their good record, care must be taken in the layout design of roundabouts to secure the safety benefits. There are particular difficulties in providing safe operating

Photograph 39.3: Circulating carriageway made two–way and entry controlled by a mini–roundabout.

conditions for cyclists and pedestrians, where excessive speed, both at entry or on the roundabout, is usually the most common problem. Significant factors contributing to high entry and circulating speeds are:
- ❑ inadequate entry deflection;
- ❑ very acute entry angles, which encourage fast merging manoeuvres with circulating traffic;
- ❑ poor visibility to the give–way lines;
- ❑ poorly designed or positioned warning and advance direction signing; and
- ❑ incorrect siting of 'Reduce Speed Now' signs.

Additional safety aspects to be considered in designing a layout are set out below.

Angle between adjacent arms
The accident potential of an entry appears to decrease as the angle, clockwise between its approach arm and the next approach arm, increases. Ideally, entries should be equally–spaced around the perimeter of a roundabout, with a minimum angle of 60° between adjacent arms.

Gradients
Whilst it is normal to flatten approach gradients to about two per cent or less near to the entry, research at a limited number of sites has shown that this has only a small beneficial effect on accident risk.

Visibility to the right at entry
This has comparatively little influence on accident risk, so little is gained by increasing visibility above the recommended level. However, it is important in the immediate vicinity of the entry arm that any cyclists on the circulatory carriageway are clearly visible. As cyclists often hug the inside kerb, it is important that, in this position, they remain in the normal line of vision of drivers approaching and entering the roundabout.

High circulatory speeds
High speeds normally occur on large roundabouts with excessively long and/or wide one–way circulatory carriageways. They can also be caused at smaller roundabouts by inadequate deflection at entries. The solution to high circulatory speed usually has to be fairly drastic, involving signal–control of the problematical entry arms. In extreme cases, the roundabout may have to be converted to a ring junction, in which the circulatory carriageway is made two–way and the entries/exits are controlled by individual mini roundabouts or traffic signals (see Photograph 39.3).

Street Furniture
Care should be taken over the incorporation of signs

Photograph 39.4: Roundabout with spiral markings.
Courtesy: David Nicholls.

and guard rails into the design, since it is important that these features should not obstruct a driver's view on entry to the roundabout.

Two–wheeled vehicles
Although roundabouts have an impressive overall safety record for most types of vehicle, this does not apply to two–wheeled vehicles. Research has shown that, on four–arm roundabouts on Class A roads, injury accidents involving two–wheeled vehicles constitute about half of all those reported. The proportion of accidents involving pedal cyclists is about 15%, although they typically constitute less than two percent of the traffic flow. The accident involvement rates for two–wheeled vehicles, expressed in terms of accidents per road–user movement, are 10–15 times those of cars, with pedal cyclists generally having slightly higher accident rates than motor cyclists.

The provision of pedestrian and cyclist facilities are considered further in Sections 39.8 and 39.9 respectively. Chapters 22 and 23 cover these particular facilities in more detail.

Goods vehicle accidents
The problem of long goods vehicles either overturning or shedding their loads at roundabouts has no obvious solution in relation to layout geometry. Whilst there are only about 60 personal injury accidents a year in this category, there are considerably more damage–only accidents. Load shedding often causes congestion, delay and expense to clear, especially if it occurs at major junctions. Experience suggests that roundabouts where these problems persist usually exhibit one or more of the following design faults:
- ❑ inadequate entry deflection leading to high

entry speeds; or
❑ long straight sections of circulatory carriageway leading into deceptively tight bends; or
❑ sharp turns into exits; or
❑ excessive crossfall changes on the circulatory carriageway.

The computer program ARCADY 3 can be used to estimate accident frequencies for ranges of traffic flows and entry geometry (DOT, 1992).

In general, to reduce accidents, the designer should adopt the following measures:
❑ reduce entry widths;
❑ reduce circulating width (must be between 1.0 and 1.2 x the maximum entry width);
❑ tighten the radius of entry–path curvature (reducing entry width and circulating width can help achieve this);
❑ try to space the arms equally around the roundabout, as this maximises the angle between adjacent arms;
❑ try to ensure that the roundabout itself is fully visible to approaching drivers on all approaches;
❑ try to avoid left hand bends on the approach road within 500m of the roundabout (right hand bends have a better accident record than straight approaches); and
❑ where the number of two–wheeled vehicles is large, consider another junction type or provide other suitable routes to accommodate them.

39.7 Traffic Signs, Road Markings and Street Lighting

Consideration of the need for, and layout of, traffic signs and road markings should be an integral part of the design process. Advice on the use and siting of signs is given in the Traffic Signs Manual (HMG, 1994) [Sb].

Road markings are used to channelise traffic and, where required, to indicate a dedicated lane. Lane indication arrows, to reinforce the map–type advance direction signs on entries, can be beneficial where heavy flows occur in a particular direction. Lane dedication should not be used where entries are less than three lanes wide. Where any particular lane is dedicated, the other lanes should also have arrow markings. This arrangement should always be accompanied by advance direction signing which indicates lane dedication. Lane dedication arrows and markings on the circulatory carriageway will not be necessary in many cases. However, their use, including the use of spiral markings, may be beneficial on roundabouts with unusual operational problems (see Photograph 39.4).

Photograph 39.5: Pedestrian guard–rail maintaining visibility. Courtesy: David Nicholls.

Signing and marking measures found useful in reducing accidents at existing roundabouts with poor safety records include:
❑ the repositioning or reinforcement of warning signs, the provision of map–type advance direction signs, making give–way lines more conspicuous, moving the central island chevron sign further to the left to emphasise the angle of turn, placing another chevron sign above one in the normal position and placing chevron signs in the central reserve in line with the offside lane approach on dual carriageways;
❑ the provision of a ring of black and white paving, laid in a chevron pattern inside the central island perimeter at a gentle slope, to improve the conspicuity of central islands;
❑ the provision of yellow bar markings with decreasing spacing on fast, dual carriageway, approaches;
❑ the reduction of excessive entry widths by hatching or physical means;
❑ the provision of anti–skid road surfacing on entries and on the roundabout itself; and
❑ the erection of 'Reduce Speed Now' signs or count–down markers.

The provision of road lighting at roundabouts is an important safety consideration and should be in accordance with the British Standard Code of Practice for Road Lighting (BSI, 1992) but care has to be taken to minimise the environmental impact. When an existing roundabout intersection is being modified, the lighting layout should be checked for suitability with the new road arrangement and any alterations carried out prior to, or at the same time as, the roadworks.

39.8 Pedestrian Facilities (see also Chapter 22)

Separate pedestrian routes, with crossings, should be located away from the flared entries to roundabouts, where the carriageway widths are less and vehicular traffic movements are more straightforward. However, where this is not practical, the following alternatives should be considered:

❑ an unmarked crossing place (i.e. dropped kerbs), associated with a central refuge wherever possible; or
❑ a Zebra or non–signal–controlled crossing, with or without a central refuge; or
❑ a Pelican or other signal–controlled crossing; or
❑ a subway or footbridge.

The type of facility selected will depend upon the expected volumes and movement patterns of both pedestrians and vehicular traffic and should be designed in accordance with current recommendations and requirements.

If a Zebra, Pelican, Puffin, or Toucan crossing (see Chapter 22) is provided close to the entry/exit points of a roundabout, there will inevitably be consequences for the operation of the roundabout and also, possibly, for safety. Where a crossing is provided within the intersection itself, care will need to be taken with the design to ensure that approach speeds are not excessive and that there is no confusion as to priority. Puffin and Toucan crossings have the advantage that on–crossing detection can be employed to vary the crossing green–times so that, when only a few pedestrians or cyclists are crossing, delays to vehicular traffic is minimised. Roundabout exit radii should not encourage vehicles to accelerate out of the roundabout and hence approach the crossing at high speeds.

In urban areas, where large numbers of pedestrians are present, short lengths of guard rail should be used to prevent indiscriminate crossing of the carriageway. The design of guard railing should not obstruct drivers' visibility. Types of guard rail are available which are designed to maintain drivers visibility to pedestrians through them and vice versa (see Photograph 39.5).

39.9 Cyclists' Facilities (see also Chapter 23)

Roundabouts are a particular hazard for both pedal and motor cyclists. Research is continuing on how to improve the safety of cyclists at roundabouts. The use of peripheral cycle tracks can offer some protection (see Photograph 39.1) but this will normally require the use of Toucan crossings, in order that cyclists can safely cross all the entry/exit roads. Cycle lanes within the circulatory carriageway can help to make drivers aware of the presence of cyclists, and provide a protected area for them. However, the lanes must be swept regularly otherwise they can get covered with debris and cyclists will not use them. Designers should be aware of the following:

❑ conventional roundabouts, with small central islands and flared entries, have accident rates which are about twice as high as those with large central islands and unflared entries. This relationship appears to apply consistently for all types of vehicular road–user;
❑ about 70% of pedal cycle accidents at smaller normal roundabouts involve entry/circulating conflicts. For example, a motor vehicle entering a roundabout collides with a circulating pedal–cycle passing the entry; and
❑ at roundabouts on dual carriageways, the accident rate for cyclists is two to three times greater than that at dual carriageway traffic signals but, for cars, the opposite is true.

It is recommended that where substantial numbers of cyclists are expected the following options should be considered:

❑ a design of roundabout layout with more emphasis on safety than on high capacity; or
❑ an alternative form of intersection, such as traffic signals; or
❑ a signposted alternative cycle route away from the roundabout; or
❑ full grade–separation, incorporating, for example, a combined pedestrian/cyclists' subway system.

Even when cycle–flows are not high, the fact that cyclists are likely to use the roundabout must not be ignored and every effort should be made to protect their safety. Programmes to encourage more cycling are in hand and it is likely that, as a result, more cycling will take place. Sites which presently exhibit little or no cycling activity may well, in the future, experience an increase in this activity. So, it is essential, where future cycling routes can be identified, that account is taken of any likely increase in cycle–flows.

39.10 Capacity and Delay

In evaluating alternative designs, the operational performance of any particular roundabout design needs to be assessed. Several criteria for operational performance have been proposed, including:

□ maximisation of vehicular throughput;
□ maximisation of reserve capacity;
□ optimisation of volume/capacity ratio (V/C); and
□ minimisation of total delays.

Capacity is estimated using Kimber's empirically based equation (Kimber, 1980), which estimates the maximum throughput for each arm of a junction from the flows and six geometric parameters.

$$\text{Entry Capacity } C = k (F - fc\, Qc)$$

where k, F, and fc are geometry–dependent constants and Qc is the circulating flow (pcu/h).

It should be remembered that the throughput estimates have a known standard error of about ±15%, for typical flow values, and significant queues and delays can, therefore, occur before capacity is reached.

Reserve capacity is defined as the difference between the capacity and the demand flows and is often expressed as a percentage of the capacity ie a demand flow of 1,500 veh/h on a capacity of 2,000 veh/h is considered to have a reserve capacity of 500 veh/h or 25%.

For a demand flow of Q, the reserve capacity is calculated, as follows:

$$\text{Reserve capacity (RC)} = (C - Q) \text{ pcu/h.}$$

However, care has to be exercised in using this measure since, as the demand flow rises on all arms, the circulating flow also increases, thereby decreasing throughput.

The ratio of volume to capacity (V/C) serves a similar purpose to the reserve capacity but is expressed differently.

$$\text{V/C ratio} = \text{Volume/Capacity} = Q/C$$

As total in–flows increase, the throughput drops as in the case of reserve capacity. It is common for the Highways Agency [Sc] to recommend that design solutions should have a V/C ratio of 0.85. However, the resulting delays depend on the flows as well as the V/C. Delay, including both traffic and geometric elements, is the best measure of operational performance and is normally measured as average delay per vehicle (see also Chapter 37). Queues are only important when there is a danger of blocking–back to other junctions, thereby causing additional delays.

Queue–length is not a good measure of performance,

because approaches with large flows and capacity can have quite long, but fast–moving, queues with a low average delay per vehicle. Conversely, very short queue–lengths can occur at low flows and capacity, but with quite large delays, as the queue is very slow to disperse.

The interactions between junction layout, capacity, flows and delays are complex and most engineers are advised to make use of computer programs, such as ARCADY (DOT, 1992), RODEL (Crown, 1989) or ROBOSIGN (Kay et al, 1992), which are designed to carry out these calculations. The programs estimate queues and delays and can represent the way in which these vary through time, as occurs through peak periods. The cumulative delay can be converted to cost, using a suitable value of time for comparison with accident and construction costs.

39.11 Safety Evaluation

In addition to the evaluation of operational performance, the safety performance of layouts needs to be assessed in coming to a design solution. Advice is given in Technical Advice note TD16/93 (DOT, 1993) [Sa]. Much of this is based on experience and good practice. The programs ARCADY, RODEL and ROBOSIGN contain forms of an empirical model, which can be used for evaluating the accident probabilities associated with different layouts (Maycock et al, 1984). However there are a number of statistical limitations attached to the use of such models and great care needs to be exercised in interpreting the outputs.

Four main types of vehicular accidents can occur on roundabouts:
□ entry/circulating accidents, involving collisions between an entering vehicle and a circulating vehicle;
□ approaching accidents between vehicles on the approach to the junction, mostly rear–end shunts when one vehicle runs into the back of another, but also including accidents where a vehicle is changing lanes;
□ single–vehicle accidents, involving a vehicle colliding with some part of the junction layout or with street furniture; and
□ other accidents which include pedestrian accidents and a variety of vehicular accidents which occur relatively infrequently, such as circulating vehicles colliding with each other and with other exiting vehicles.

It should be noted that, in the context of accidents on roundabouts, pedal cycles are included in the term 'vehicle'.

The relationship between the geometry and the total number of accidents (all types) is complex. A geometric parameter that may reduce one accident type may increase the incidence of another.

Design layouts should also be subject to an appropriate safety audit, as described in Chapter 16.

39.12 References

BSI (1992)	BS 5489: 'Code of Practice for Road Lighting', British Standards Institute.
Crown BC (1989)	'RODEL – Interactive Roundabout Design', Staffordshire County Council.
DOT (1992)	TA44/92 (DRMB 5.1.1) 'Capacities, Queues, Delays and Accidents at Road Junctions – Computer Programs ARCADY/3 and PICADY 3', Stationery Office [Sa].
DOT (1993)	TD 16/93 (DMRB 6.2.3) 'Geometric Design of Roundabouts', Stationery Office [Sa].
HMG (1994)	SI 1994 No. 1519 'The Traffic Regulations and General Directions 1994', Stationery Office.
Kay WA, Sang L Irani and Katesmark S (1992)	'Advanced Roundabout Design with ROBOSIGN', PTRC.
Kimber RM (1980)	Report LR942 'The Traffic Capacity of Roundabouts', TRL.
Lines CJ (1995)	'Cycle accidents at signalised roundabouts', Traffic Engineering + Control 36(2).
Maycock G and Hall R (1984)	Report LR1120 'Accidents at 4–Arm Roundabouts', TRL.

Chapter 40 Traffic Signal Control

40.1 Introduction

Traffic signals at road intersections allow vehicle movements to be controlled by allocating time intervals, during which separate traffic demands for each arm of the intersection can make use of the available road–space. Signal equipment and control techniques have evolved to cope with a wide range of intersection, layouts and complex traffic demands, including pedestrians, public transport vehicles and cyclists crossing at the intersection. Moreover, traffic signals provide relatively efficient intersection control within limited road–space. Consequently, they are frequently adopted as a means of traffic control at busy urban junctions (see Photograph 40.1). Allocating the amount of time available, as well as the space, to particular movements adds a further dimension to the overall control of traffic.

The use of traffic signals to control traffic movement can bring about major reductions in congestion, improve road safety and enable specific strategies, which regulate the use of the road network, to be introduced. Examples of such strategies might be:

Photograph 40.1: Traffic control at a busy urban junction.

❏ to reinforce the designated route hierarchy;
❏ to give priority to public transport;
❏ to provide crossing facilities for pedestrians and cyclists;
❏ to maximise or limit traffic flow;
❏ to regulate demand and/or manage queueing;
❏ to improve safety; and
❏ to improve throughput at roundabouts which experience problems at peak periods.

Signal–controlled junctions are usually more economical in the use of road–space than conventional roundabouts. They provide equivalent capacity, and can allow more flexibility in layout and land–take to avoid key features, such as historic buildings or public utility equipment.

40.2 Legislative Background and Design Standards

In the United Kingdom, traffic signals are provided under powers contained in the Road Traffic Regulations Act 1984 (HMG, 1984) [NIa] [Sa] [Wa]and they must comply with current directions issued by the Department of Transport [Sa] [Wa]. These include:
❏ regulations covering the details of prescribed traffic signs, which include traffic signals (HMG, 1994);
❏ standards issued by the British Standards Institute;
❏ various specifications relating to traffic signal equipment issued by the Department of Transport and the Welsh Office (DOT, 1983); and
❏ EC standards.

In addition, the Department of Transport and the Welsh Office issues Technical Standards, Advice Notes and Technical Specifications dealing with the operational aspects of signals (DOT, 1973, 1975, 1981a to f, and 1991c) [Sb; Sc; Sd].

40.3 Designing for Safe Use by All
Users (see also Chapter 16)

The design of a traffic signal installation should, from the start, take into account safety for all users and should be subject to a safety audit. There are two elements of safety to be considered in designing a signal installation.

The first element of safety addresses direct conflicts between intersecting or merging streams of traffic, including pedestrians and cyclists. This is normally dealt with by defining the conflict points and calculating the inter–green times. The inter–green should be such that it allows vehicles or pedestrians to clear the conflict–area before the opposing traffic stream can reach it after passing a green signal (DOT, 1981d) [Sd]. Another important factor is the least amount of green–time that a particular approach or movement will receive green, which is known as the minimum green–time (DOT, 1981d) [Sd].

The second element is more difficult to quantify as it is related to users' overall perception of the design and includes the layout and physical design employed to control the intersection. One of the key factors is the location of the signal–heads. They should be positioned so that they are clearly visible only to those drivers, cyclists or pedestrians that they are intended to control and do not mislead other traffic.

The strategy used to control an intersection can influence safety, for example, where vehicles have to negotiate several stop lines in quick succession or where pedestrians arrive at a crossing just as their signal turns to red.

At signal–controlled junctions, there is a relationship between accident occurrence and various design parameters, such as geometric features, stage–sequencing and the proportion of vehicle–types.

40.4 Control Principles

Conflicts occurring between different traffic streams and various categories of road–user decrease the operational efficiency of junctions and increase the likelihood of accidents. Traffic signals can reduce such conflicts by separating movements in time and regulating their position on the road in a way which allows traffic performance to be optimised safely.

However, at many sites in congested urban areas what scope there is for major revisions to junction layout and design will be restricted within existing highway boundaries.

In these instances, it may be impossible to achieve any reserve capacity and designs must aim for the appropriate balance between provision for traffic movement and the needs of pedestrians, cyclists and public transport users. The level of priority afforded to each of these groups of users will be a matter for local judgement and policies.

40.5 Assessing Sites Suitable for Traffic Signal Control

No generally accepted rule or threshold level of traffic flow exists for justifying the installation of traffic signals. Decisions are based on a number of factors considered together, such as:
❑ accident records (numbers and characteristics) and how they relate to local patterns;
❑ the expected traffic speeds and levels of vehicular and pedestrian flows;
❑ the feasibility of alternative types and layouts for the junction, ie priority, roundabout or grade–separated (see Chapter 37);
❑ whether or not signals are the only feasible control (eg due to the limitations of the site); and
❑ whether or not the junction is within an existing or proposed Urban Traffic Control (UTC) area and its proximity to other junctions (see also Chapter 41).

Advantages of traffic signal control include:
❑ minimising the space required, particularly at constrained sites where physical restrictions could make other types of control costly and difficult to provide;
❑ the flexibility to assist traffic using specific approach arms or particular categories of road–user and to respond to a wide range of different traffic conditions;
❑ the ability to make special provision for pedestrians and cyclists;
❑ the ability to link and co–ordinate with other adjacent signal–controlled junctions (see also Chapter 41) to influence the pattern and speed of traffic progression; and
❑ relatively low cost since, for example, capital costs are usually less than for conventional roundabouts or grade–separation.

Disadvantages of traffic signal control include:
❑ increased delays and operating costs, especially in uncongested conditions, such as at off–peak times, when signals may impose more delay and operating costs on traffic than is necessary to resolve conflicts safely;
❑ some increased risk of certain types of traffic accident, such as front–to–rear collisions under braking;
❑ the maintenance costs of signal equipment, with the additional requirement continuously to monitor signal operations and to update signal–settings under fixed time control; and
❑ the limited facility for U–turning manoeuvres.

40.6 Traffic Signal Equipment and Operation

At each individual site a traffic signal installation may require:
- ❏ a traffic signal controller (ie the operating 'brain' of the installation);
- ❏ signal–heads (pedestrian and vehicle heads), poles and any associated regulatory signs;
- ❏ vehicle detectors on each approach;
- ❏ pedestrian and cyclist push–buttons; and
- ❏ ancillary equipment.

It is now common practice to duct and drawpit the cabling runs at traffic signals because of the ease of maintenance and modification.

Traffic Signal Controllers

Controllers are based on microprocessors, which use digital timing and solid–state switching to provide safety, reliability and flexibility. Detailed specifications for micro–processor controlled traffic signals are provided by the Department of Transport (DOT, 1991c).

Controllers are programmed, using proformas defined in the specification (the so–called '141' forms) to operate in a variety of different modes. These are:
- ❏ 'fixed–time operation', in which all the timings run to preset maximum values that may vary by time of day. It is possible to utilise demand–dependent stages, such as a pedestrian stage, within fixed time operation;
- ❏ 'vehicle–actuated' operation, in which the timings vary in response to vehicle demands up to maximum green values;
- ❏ 'cableless linking facility' (CLF), in which pre–determined timings are used to ensure co–ordination with adjacent junctions (the structure of the timings allows for demand–dependent, and full, vehicle–actuation);
- ❏ 'hurry call', in which part of a specific stage is given priority in response to certain vehicles, such as buses, LRT and emergency vehicles, or to respond to queues of traffic. The CLF facility can also be used to control the manner in which the junction operates, for example the normal sequence can be changed; and
- ❏ 'manual operation', for use by police or traffic wardens.

The design of each controller requires that these facilities should be allocated a specified order of priority, with built–in constraints to govern a change of operation, the circumstances under which the method of operation is overridden and other safety features, such as guaranteed minimum green–times. A minimum of four maximum–green sets is usually provided for fixed–time and vehicle–actuated operation.

Controllers can also be used to control adjacent Pelican crossings or to influence the operation of adjacent junction controllers.

Whenever manual control is provided as an option, the controller should be sited so that all approaches to the junction are visible to the operator.

Phases, Cycles and Stages

A traffic signal controller allocates right–of–way, among the various movements at a junction, by showing a green signal to different sets of movements so that conflicting movements do not receive a green signal simultaneously, with some exceptions for turning traffic. The movements are divided into separate sets so that all movements in each set always receive identical signal indications.

The word 'phase' is used to describe a set of movements which can take place simultaneously or the sequence of signal indications received by such a set of movements. In each signal controller there is a normal sequence, in which the various phases receive green. Each repetition of this sequence is called a 'cycle'. A 'stage' is that part of the cycle during which a particular set of phases receives green. These terms are illustrated in Figure 40.1.

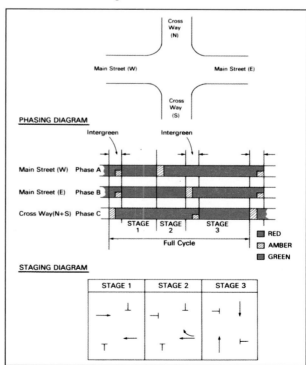

Figure 40.1: A typical phasing and staging diagram.

Note: Phasing is denoted by letters, staging by numbers.

Inter–green and Inter–stage Periods

The period between the end of the green for one phase and the start of the green for another conflicting phase, gaining right–of–way at the same change of stage, is known as the inter–green period between these phases. In microprocessor controllers phase–to–phase inter–greens are specified and this provides great flexibility to achieve the most efficient designs. In the UK this period includes an amber for one phase, which lasts three seconds, and a red–and–amber period for the next phase, which lasts two seconds. These may overlap, run consecutively, or be separated by an all–red period during which both phases receive red signals. The all–red period may be controlled by detectors to maximise efficiency.

The period between stages is known as the inter–stage. The value is based on the inter–greens of the phases which lose, or gain, right–of–way. At complex intersections, with many phases with different inter–greens, inter–stage design that allows phases to run on and start early can achieve throughput benefits.

Vehicle–Actuation

When traffic signals are working in a vehicle–actuated mode, traffic influences the appearance and duration of stages. Vehicle detectors, which are installed on the approaches to the signal, are used to 'demand' phases or extend running phases. For each phase in each stage there will be a minimum green period. This ensures that the stage appears for a safe minimum period.

For each phase showing a red signal vehicles crossing detectors will register a demand for a stage in which that phase has a green signal. This demand will also initiate a preset maximum green time for each running phase. The phase displaying green will continue to do so until there are no more extensions or until the maximum green time is reached. When this occurs a demand for the phase losing right–of–way is automatically generated.

If detectors on an approach already showing a green signal detect a vehicle moving towards it, the green signal will be extended by a preset vehicle–extension period. This can be repeated until the maximum green period has been used up.

In the absence of demands for other stages, the stage which is running will continue indefinitely.

Systems for Vehicle Detection

Vehicle–detection is required for many purposes, such as:

❑ to demand and/or extend the green–time for a phase;
❑ to extend an all–red period;
❑ to demand and/or extend a green–time for a particular category of vehicle, such as buses (see also Chapter 24);
❑ to introduce a 'hurry' call;
❑ to detect the onset and growth of queues;
❑ to change a signal–control strategy; and
❑ to provide data on vehicle flow, density and speed, for use in urban traffic control systems.

Speed assessment and speed discrimination, using detection–based techniques, are required to extend inter–greens on high speed roads in order to improve safety. They extend the green to prevent the signals changing when drivers would find it difficult to stop before the red signal. 'Microprocessor–optimised vehicle actuation' (MOVA) is increasingly being used as an alternative to these techniques (see Section 40.16).

For many years sub–surface inductive loops have been the main form of vehicle–detection. When used in conjunction with self–timing detectors they offer a highly effective, weather–independent detection system. However, the loops are prone to damage by resurfacing or utility works.

Above ground detection (AGD) techniques offer a practical alternative to loop–based detection systems. AGD can be used in poor road surfaces and is unlikely to be damaged by resurfacing or utility works. However, AGD can be influenced by tree branches and by the masking of small vehicles by larger ones. Also, AGD techniques have not replaced loop–detectors where lane–specific detection is required.

Pedestrian crossing facilities are often provided at traffic signals (see also Chapter 22) and can be actuated on demand, using a push–button, or may appear automatically in parallel with defined traffic phases. AGD techniques, associated with, for example, Puffin pedestrian crossings, offer the potential to demand and extend pedestrian phases to provide enhanced pedestrian facilities (DOT, 1993; Dickson et al, 1995).

Signal Aspects

Signal–heads facing drivers have three signal aspects – red, amber and green. A green arrow may be fitted in place of the full green aspect and further green arrows can be added to assist traffic direction and control when used in conjunction with the appropriate regulatory signs and road markings. Unfortunately, drivers behaviour is such that a green arrow tends to be perceived as giving an absolute right of way. Some signs can only be used in conjunction with a Traffic Regulation Order (see the current Traffic Signs Regulations and General Directions) (Chapter 4) [NIb].

The operation of turning arrows is different in Northern Ireland where, under certain traffic flow conditions, it is possible to revert from a full green plus a right–turn filter arrow back to full green, without going through the amber, red, red/amber sequence. When the green arrow is extinguished, it is followed by its own amber signal (see Photograph 40.2). This has the effect of warning drivers that they no longer have priority when turning right, while still permitting them to turn in gaps. Following trials in Northern Ireland, the Traffic Signal Regulations (Northern Ireland) 1979 was amended, in 1992, to take account of this.

Where signal–heads are provided specifically for cyclists, green and amber 'cycle' symbols on a black background should replace the full amber and green aspects, subject to authorisation by the Department of Transport [NIc] [Sa]. Toucan signal–controlled crossings cater simultaneously for cyclists and pedestrians (see Chapters 22 and 23). Signal–heads specifically for equestrian users are also now in use. For pedestrians each signal–head has two aspects – a red and a green person.

Signal–heads are usually mounted on poles but can also be mounted on mast arms, gantries and catenary cables to provide better visibility at difficult sites and on high speed roads. Badly sited or excessive numbers of poles add to urban clutter and create potential hazards for pedestrians.

40.7 The Layout of Signal Controlled Junctions

Location of the Primary Signal
The primary signal is so called because it is usually located behind, but close to, the stop–line and usually at the edge of the nearside footway of the approach.

Arrangements for the layout of traffic signals at existing junctions are generally constrained by features such as buildings and highway boundaries, as well as by the pattern of traffic movements. The aim should be to keep clearance–times between conflicting streams as short as possible, without adversely affecting safety.

Particular issues that affect the location of the

Photograph 40.2: An example of a right–turn filter arrow with its own amber signal.

primary signal are:
- ❑ the needs of pedestrians (DOT 1981c) [Sd];
- ❑ clearance for turning paths of vehicles from other approaches;
- ❑ any unsignalled movements, such as left turns;
- ❑ bus–stopping and bus–lane requirements (see also Chapter 24); and
- ❑ the needs of cyclists (see also Chapter 23).

Examples of various arrangements are shown in Figure 40.2 (a to d).

Duplicate Primary and Secondary Signals
At least two signals should be visible from each approach, usually comprising one primary and one secondary signal. Duplicate primary signals may be required on the off–side of a multi–lane approach or when visibility of the nearside primary signal is restricted. They are recommended on all high speed approaches. On two–way roads, duplicate primary signals should always be placed on a central island and not on the far offside of the carriageway. However, on–one way roads, their placement on the offside of the road is mandatory.

Additional signals, normally sited beyond the junction, are known as secondary signals. They must always display the same information as the primary ones but may give additional information (such as a green arrow aspect) that does not conflict with that shown on the primary. In certain circumstances, it may be undesirable, or impractical, to position the secondary signal beyond the junction (see Figure 40.3). This arrangement is sometimes used to prevent pedestrians relying on watching the traffic signals,

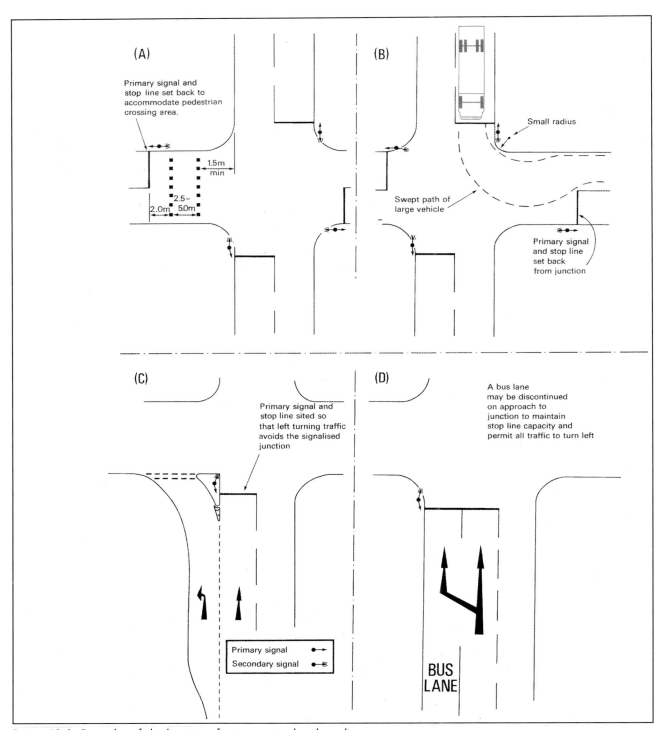

Figure 40.2: Examples of the location of primary signal and stopline.

rather than the traffic, when judging whether it is safe to cross the road and also to provide for the safety of right–turning traffic already in the junction.

Raised signals, placed directly above the carriageway, may be beneficial on multi–lane approaches where there is a risk that low level secondary signals could be obscured by traffic.

Various devices, such as hoods and louvres, can

reduce the possibility of drivers and pedestrians seeing inappropriate signals, for example, where a Pelican crossing is sited only a short distance away from a signalled junction. They can also help signals to be seen in bright sunlight. Secondary signals usually have deeper hoods, so that they are visible only from the relevant approach.

Some examples of layouts showing the location of

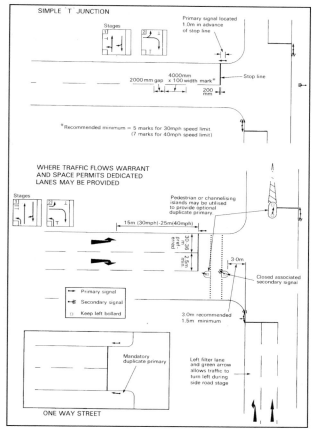

Figure 40.3: Typical layouts at signal controlled T junctions.

traffic signals and their associated carriageway markings are shown in Figures 40.3 and 40.4. Further details on junction layout are provided in the relevant Department of Transport Advice Notes (DOT, 1981c; DOT, 1981d; and; DOT, 1981e) [Sd].

The Siting of Stop-lines
The stop-line on each approach should be located between one and two metres from the primary signal.

40.8 Facilities for Pedestrians and Cyclists

Pedestrians sometimes need exclusive signal-stages. These should be considered where:
❑ the pedestrian flow across any one arm is 300 pedestrians per hour or more; or
❑ the turning traffic into any arm has an average headway of less than five seconds during its green-time and is conflicting with a flow of more than 50 pedestrians per hour; or
❑ there are special circumstances, such as significant numbers of elderly, infirm or disabled pedestrians.

Pedestrian facilities may also be justified on safety

grounds (see Table 40.1, which lists various types of pedestrian facility and DOT, 1981c [Sd]). However, an exclusive pedestrian phase may become counter productive if, in order to provide adequate time for vehicles to clear and pedestrians to cross, the cycle-time becomes so extended that pedestrians are tempted to seek earlier opportunities to cross against the signal. Some examples of alternative ways of locating pedestrian signals are shown in Figures 40.5, 40.6, 40.7 and 40.8 (see also Chapter 22). Information on how cyclists can be accommodated at traffic signals is given in Chapter 23 and in the relevant IHT Guidelines (IHT, 1996).

40.9 Other Applications

WIG–WAG Signals
Flashing signals, known as 'WIG–WAGs', can be used to stop traffic at railway level–crossings, swing–bridges, sites near airfields and at fire and ambulance stations. They consist of two red aspects, arranged horizontally and flashing alternately, and a steady amber aspect placed centrally between the red lamps and displayed before the flashing red lights begin to operate. They remain switched off until control is

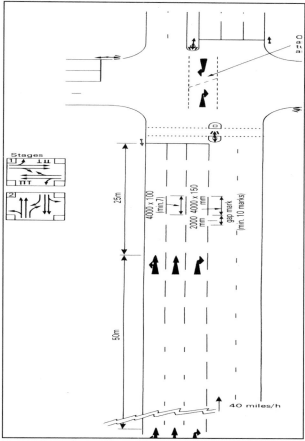

Figure 40.4: Typical layout at a signal controlled cross–roads.

Type of facility	Characteristics
No pedestrian signal	❏ Traffic signals, even without signals for pedestrians, can help pedestrians to crcross by creating gaps in traffic streams. ❏ Especially applicable where there are refugees enabling the conflict in each crossing movement to be with only traffic in one direction and on one-way streets.
Full pedestrian stage	❏ All traffic is stopped. ❏ Demanded from push buttons. ❏ More delay to vehicle than combined vehicle/pedestrian stages. ❏ See Figure 40.5.
Parallel pedestrian stage	❏ Combined vehicle/pedestrian stage often accompanied by banned vehicle movements. ❏ useful across one-way streets. ❏ See Figure 40.6.
Staggered pedestrian facility	❏ Pedestrians cross one half of the carriageway at a time. ❏ Large storage area in the centre of the carriageway required. ❏ See Figure 40.7.
Displaced pedestrian facility	❏ For junctions close to capacity. ❏ The crossing point is situated away from the junction but within 50m. ❏ Normal staging arrangements as above apply. ❏ See Figure 40.8.

Table 40.1: Types of pedestrian facility at signal controlled junctions.

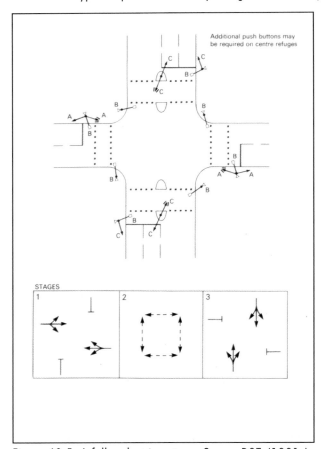

Figure 40.5: A full pedestrian stage. Source DOT (1981c).

required and they may be activated either automatically or manually.

Traffic Signals for Incident and Lane Control

On high capacity roads, such as motorways, special lane control and other informatory signals can be used to stop traffic in the event of an incident, to direct it to change lanes or reduce speed, to warn of hazards or to direct it to take the next exit from a motorway. These signals are sited either at the side of the road or placed above the carriageway on gantries, as lane– specific signals.

Traffic Signals for Ramp Metering

Traffic signals can be provided near the end of the entry slip–roads onto motorways to regulate the volume of traffic entering the main flow, which is uncontrolled. Such traffic signals can improve the overall throughput of traffic on the motorway and can reduce, substantially, the delays caused by 'flow–breakdown' on the main carriageway.

Portable Traffic Signals

Portable signals are often necessary to control traffic at roadworks. Usually, this is to implement a one–way system or to control the use of a single lane past the obstruction. Further information on the use of traffic signals at roadworks is given in DOT (1975) and in DOT (1985) [Se].

40.10 Traffic Signal Design

The aim in designing a traffic signal installation is usually to maximise the throughput of vehicles, cyclists and pedestrians, whilst minimising vehicular delay and waiting time for pedestrians, maintaining a high degree of safety and establishing a balance between the requirements of different streams of traffic. Queue–management techniques may sometimes be necessary to store excess traffic at locations where the required level of capacity is not available or cannot be realised.

The steps to be used as guidelines, when designing a signal–controlled junction, are set out below:

Traffic Flows
It is necessary to determine the demand flows of traffic for all proposed operating conditions, particularly during the four busiest hours of the day. For existing junctions, these data should come from up–to–date classified traffic counts and may be in the form of vehicles or passenger car units (pcu) per hour (TRL, 1963; Wood, 1986; Kimber *et al*, 1986; and Webster *et al*, 1966). For proposed junctions, estimates will be required for which appropriate traffic estimation and assignment techniques should be used (see Chapter 8). As flow data is normally in the form of an average value (eg averaged over the peak hour) fluctuations of vehicle flow–rates within the time–period being considered should be taken into account.

Allocation of Lanes
The layout of a junction should reflect traffic volumes and patterns as well as the needs of pedestrians and cyclists. Traffic lanes should ideally be allocated for each of the vehicular movements allowed but lane–sharing will often be necessary at restricted sites.

An 'approach' is one or more lanes where traffic can be regarded as forming a single queue. A traffic stream comprises traffic on an approach, all of which usually receives right–of–way for the same period. Each approach should, as far as is practicable, be capable of carrying the maximum predicted flow for that approach.

Saturation Flows
The saturation flows need to be determined for each

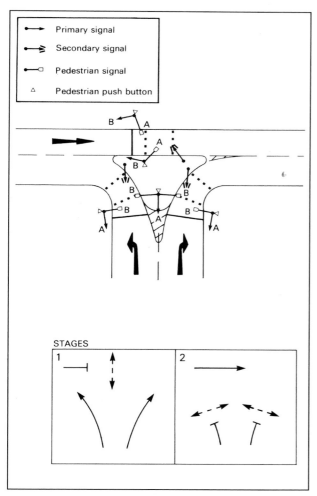

Figure 40.6: A parallel pedestrian stage (one–way street arrangemment). Source DOT (1981c).

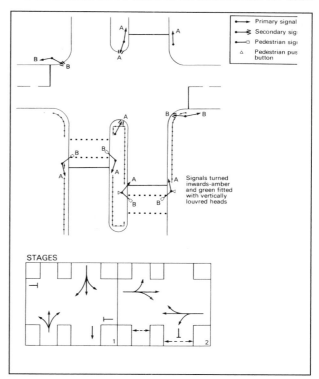

Figure 40.7: A 'staggered' pedestrian facility. Source DOT (1981c).

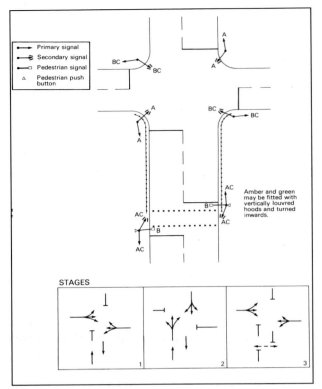

Figure 40.8: A 'displaced' pedestrian facility. Source DOT (1981c).

stream. The saturation flow is the maximum flow-rate that can be sustained by traffic from a queue on the approach used by the stream and depends mainly on:

❑ the number and width of entry and exit lanes available to that stream and the effects of parked vehicles, bus stops etc. on lane width;

❑ the proportion of turning traffic and the radius of turn; and

❑ the gradient of the approach.

Traffic composition also affects saturation flows but is taken into account by calculating flows in passenger car units (pcu) rather than in numbers of vehicles, (ie each vehicle is given an equivalent pcu value (see Section 37.6).

A stream may have more than one saturation flow value, when different movements discharge from it under different signal conditions in the cycle.

For existing junctions, direct observation is always the best method of determining saturation flow (TRL, 1963; Wood 1986). However, in many cases, saturation flow has to be estimated from relationships based on geometric and other characteristics of the site (Kimber *et al*, 1986). Care should be taken in estimating saturation flows at signal-controlled roundabouts (see Chapter 42).

Ordering of Stages

The sets of streams which should run together must be determined, together with the order in which those sets receive green signals. For a particular junction, there will be several options for controlling traffic.

Signal phasing requires cable connections to signal-heads and must be decided before installation. Signal staging, however, is related to the switching of the wired phases and can, therefore, be more easily altered within the controller after installation, provided that the required options have been incorporated.

Conceptually, every vehicle and pedestrian movement can be allocated its own exclusive phase. An examination of these potential phases will indicate which streams should receive green together and which should not. Thus, an optimum number of stages, and the order in which they run, can be chosen, having regard to safety and the maximum green-time for the phases that need it most. Some examples of options are given in Figures 40.9 and 40.10.

Where sufficient space exists, as for example in Figure40.9, consideration should also be given to the introduction of additional signal stop-lines between the component elements of an overall junction. The localised system of signals thus created can enable the throughput of the overall junction to be enhanced, often with reduced cycle-times. This can be achieved

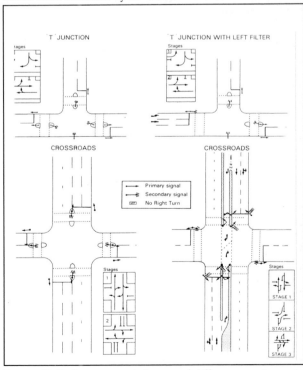

Figure 40.9: Diagrammatic layout of different staging conditions. Source DOT (1981e).

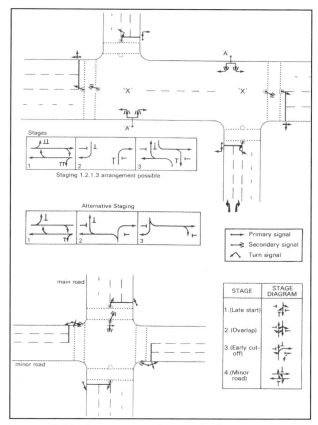

Figure 40.10: Diagrammatic layout of other staging conditions. Source DOT (1981c).

by the effective reduction in lost time which can be afforded by the additional stop-lines (see Hallworth, 1980 and 1983). Such systems of signals can sometimes lend themselves to control by separate sequences of stages (stage-streams) for each component element, rather than utilising a single stage-stream for the junction as a whole (see Figure 40.11).

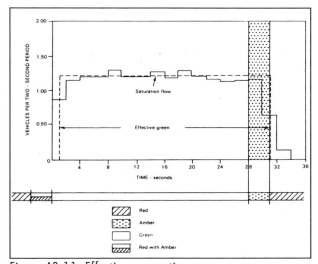

Figure 40.11: Effective green-time.

Signal Timings

Signal timings for installations under isolated control can be calculated using the techniques for manual calculations described in TRL Technical Paper 56 (Webster *et al*, 1966). Alternatively, computer programs are available to perform these calculations (see Section 40.14). The design of UTC systems is dealt with in Chapter 41.

A suggested procedure is as follows:

❑ determine the inter-green times which will provide safe clearance times between conflicting streams;

❑ determine the lost time (see below) for each traffic stream; and

❑ calculate the cycle-time and the duration of the green periods in each phase and thus the duration of each stage, usually to optimise some aspect of traffic performance.

Formulae to calculate cycle-time and green-times by hand are contained in Technical Paper 56 (Webster *et al*, 1966). In these calculations, the combined green and amber period for a stream is treated as effective green-time and 'lost-time'. The 'lost-time' is sometimes known as the starting and stopping losses for each stream. Thus:

$$\text{effective green-time} = (\text{green-time}) + (\text{amber-time}) - (\text{lost-time})$$

This is illustrated in Figure 40.11, from which it can be seen that, in this example, traffic experiences a one second starting delay but runs on through the amber period. However, lost time may also occur at the end of the green+amber period, where the flow ceases more rapidly. The lost time for a stream is best determined from observation but, if this is not possible, a value of two seconds can be assumed. In calculating the cycle-time and the duration of green periods, the total lost time for a junction should be determined from the inter-green times between critical phases and the starting and stopping losses for the traffic streams.

Such calculations are often performed as part of an iterative process of junction design, in which the results for a particular proposal are assessed using a measure of traffic performance (see Section 40.11) and recalculations are then carried out using modified junction layouts or stage-orders.

40.11 Design Techniques

Traffic movements can be arranged in a number of ways to take advantage of site layout or particular traffic demands. Features which should be considered include:

❑ accepting a conflicting move, such as a certain amount of right–turning traffic opposed by oncoming traffic, where the degree of conflict is acceptable and movements can be performed safely;

❑ restricting movements, such as banning right turns or creating one–way streets away from the junction. Care must be taken to ensure that such measures do not create other problems elsewhere;

❑ allowing right–turners unopposed right–of–way, by giving an early cut–off to opposing traffic. This is sometimes proposed as a means of reducing right–turning accidents but its effect on pedestrians should also be considered. An alternative is to give a late start to opposing traffic and these methods are illustrated in Figure 40.10;

❑ choosing signal stages which provide unopposed right–turns, to improve lane–occupancy and to give a definite indication for priority movements;

❑ allowing simultaneous 'non–hooking' right turns;

❑ providing two separate green periods in a cycle (repeated greens) for important movements;

❑ flaring approaches, or increasing the number of entry lanes, by making each lane narrower so long as minimum lane–widths are provided. Both measures increase the number of vehicles that can wait at the stop–line;

❑ providing extra lanes for turning traffic and relating the relevant signal timings to their queue length;

❑ combining the green period for vehicles and pedestrians, where this can be achieved safely;

❑ considering different stage–sequences for different times of the day;

❑ linking to adjacent signals and Pelican crossings; and

❑ linking within an urban traffic control system (see Chapter 41).

See also DOT (1981d) [Sd] on the general principles of control by traffic signals.

40.12 Measures of Performance and Signal Timings

Signal–controlled junctions can be assessed and compared by examining various measures of traffic performance including safety (see Section 40.13).

Degree of Saturation

The ratios of arrival rate to saturation flow–rate for the various traffic streams is important in the assessment of traffic flow performance. For a signal–controlled stream, this ratio is known as the degree of saturation and can be expressed as:

$$x = qc/gs \quad \text{where:}$$

'q' is the average arrival rate;
's' is the saturation flow in the same units as 'q';
'c' is the cycle time; and
'g' is the effective green time in the same units as 'c'.
This expression shows how the degree of saturation for a stream depends on the signal timings.

Traffic Throughput and Reserve Capacity

For any one signal–controlled stream, the maximum traffic throughput of the approach is sg/c but, in practice, it is desirable for the degree of saturation to remain appreciably less than unity. A largest acceptable value may, therefore, be specified and, typically, this is taken as 0.9. With this upper limit, the practical capacity of the approach can be regarded as psg/c, where 'p' is the largest acceptable value. For all approaches at the junction to be operating within their practical capacity, the signal timings must be chosen so that 'q' is less than psg/c for every approach. With any such timings, the amount by which 'q' could be increased without exceeding the practical capacity is called the 'reserve capacity' of the approach. If 'q' exceeds psg/c, the amount by which it does so is called the 'overload' on the approach. The amounts of reserve capacity and overload on different approaches can be influenced by adjusting the signal timings to, for example, equalise the percentage reserve capacity or percentage overload on the most heavily loaded, ie critical, approaches. If this is done there will usually be more spare capacity, or at least less overload, on some other approaches, ie non–critical approaches. One way of deciding what value of 'p' to specify, for design purposes, is to decide on the level of delay that should be exceeded with only a given probability, such as once every 10 cycles.

Vehicular Delay

A commonly used method of assessing performance is to estimate the total amount of vehicular delay incurred by traffic. For streams which have a reserve capacity on the approach, equilibrium values of delay can be estimated using, for example, Webster's formula (Webster *et al*, 1966). For streams on overloaded approaches it is necessary to know how the arrival rate varies over time in order to apply a shared delay formula (Kimber *et al*, 1986).

Two measures of vehicular delay, often used, are average delay per vehicle and rate of accumulation of delay per unit of time. The average delay per vehicle is the difference between the average journey–time through the junction and the average journey–time that would apply if the vehicle were not impeded by the signals. The rate of accumulation of delay is normally expressed as the difference between the average number of vehicles within the zone of influence of the junction at a typical instant and the number that would be there if the same flow of traffic were unimpeded by the signals. Either measure can be applied to an individual stream but only the rate of delay is additive over streams to give a measure of performance for the junction as a whole.

Delay to Passengers

The concept of delay to passengers, instead of just the vehicles, could be used as an alternative measure of performance in urban areas, especially where vehicle–occupancy can vary significantly, such as on heavily patronised public transport routes. Where this is done, estimates should also be made of pedestrian delays, suitably weighted, due to the signal control.

Signal Timings

When a junction is overloaded, the timings that equalise and minimise the percentage overload on the most heavily–loaded approaches will often provide the optimum for vehicle–actuated operation. When a junction has some reserve capacity, timings can be calculated which would minimise the estimated rate of delay for the whole junction, if it were operated on fixed–timings. The resulting green–times provide the optimum for vehicle–actuated operation. In either case, practical constraints on green–times and cycle–times should be respected. In the situation of overload, it may be necessary to have regard to the lengths attained by queues in different streams, especially if there are other junctions upstream which may become blocked by excessive queues. In these circumstances, green–times should be adjusted to limit queueing on the critical approaches.

40.13 Modelling of Accident–Rates

A prediction of the future accident–rate to be expected at a four–arm junction is possible, utilising one of two accident–prediction models available within OSCADY 3, the program for junction capacity calculations. One model (M0) predicts a single value for the total number of injury accidents per annum at the junction using only flow–data and the number of stages. The second, more comprehensive, model (M2), predicts accident frequencies for 11 accident–categories, but requires a wide range of input data, including classified vehicle turning counts,

pedestrian flows and information concerning the physical layout of the junction and the land–uses of the surrounding area.

The two models are based on the results of a study carried out by Southampton University (Hall, 1996). The study examined the accident–records for a sample of 177 four–arm signal–controlled junctions over a four year period and related accident–frequencies by category to functions of traffic and pedestrian flow and a wide range of other junction variables. Further research into accident–rate modelling has been undertaken by TRL, mainly for traffic management applications.

40.14 Computer Programs

The process of calculating timings to minimise vehicular delay, or percentage overload, for given traffic flows at isolated junctions can be carried out, at least approximately, by hand in simple cases. However, phase–based computer programs, like LINSIG (Simmonite, 1985) and SIGSIGN (Sang et al, 1990), are available and can deal equally well with more complicated junctions and can provide better approximations for throughput–maximising timings. LINSIG is a versatile tool for exploring the consequences of different phasing, staging and interstage structures in terms of traffic throughput and delay.

The TRL program OSCADY (Crabtree, 1993) models the operation of a signal–controlled junction as demand varies over the day. Also the TRANSYT program can be used to model a single junction and to optimise its timings, although it is primarily designed to optimise signal networks. The use of computer programs in the design of UTC systems is dealt with in Chapter 41.

40.15 Other Design Considerations

As with priority junctions and roundabouts, the methods of calculation, referred to in Section 40.14, can be used to estimate how capacity, queue–lengths and delay would be affected by the following:
   ❑ changes in layout and sequence of stages in the cycle;
   ❑ changes in the amount of traffic in particular streams; and
   ❑ general increases in traffic over the years, which are likely to affect traffic in all streams.

It must also be borne in mind that traffic conditions

vary continuously and the appraisal of alternative designs should include the considerable scope which exists to adjust signal timings to accommodate a variety of circumstances.

40.16 MOVA

The MOVA (Microprocesser–optimised vehicle actuation) system has been developed for use at isolated, heavily–loaded traffic signal installations. MOVA requires loops to be set in all lanes on all approaches, located approximately 100m and 40m from the stop line depending on the approach speeds. In addition, special loops, to take account of the traffic links and sources, may be required. A MOVA unit is installed in the traffic signal controller, which then communicates with the controller through the UTC interface. A dial–up telephone facility is available to collect data and traffic statistics from the MOVA unit and also to enable faults to be reported automatically (DOT, 1991b) [Sd].

Algorithms for free–flow and congested conditions extend the green–times to values much longer than usual, in order to minimise delays. Average delay reductions in the order of three percent have been observed. MOVA has also been found to be effective in reducing accident rates at installations with high speed approach roads. Due to the longer cycle–times, MOVA is not appropriate for use at locations that have heavy pedestrian flows. Otherwise, all new trunk road installations that are not in a co–ordinated system, should incorporate the MOVA facility.

40.17 References

Crabtree MR and Harrison F (1993)	AG 22 'Users Guide to OSCADY 3', TRL.
Dickson KW and Barker DJ (1995)	'The Puffin pedestrian crossing: pedestrian–behavioural study', Traffic Engineering + Control, 36(9).
DOT (1973)	Circular Roads 5/73 (WO64/73), 'Criteria for traffic signals at junctions', DOT [Sb].
DOT (1975)	Circular Roads 49/75 (WO207/75), 'Portable [Sc] traffic signals for use at roadworks', DOT.
DOT (1981a)	TA13/81 (DMRB 8.1) 'Requirements for the Installation of Traffic Signals and Associated Control Equipment', Stationery Office [Sd].
DOT (1981b)	TA14/81 (DMRB 8.1) 'Procedures for the Installation of Traffic Signals and Associated Control Equipment', Stationery Office [Sd].
DOT (1981c)	TA15/81(DMRB 8.1) 'Pedestrian facilities at traffic signal installations', Stationery Office [Sd].
DOT (1981d)	TA16/81 (DMRB 8.1) 'Junction Principles of control by traffic signals', Stationery Office [Sd].
DOT (1981e)	TA18/81 (DMRB 6.2) 'Junction layout for control by traffic signals', Stationery Office [Sd].
DOT (1981f)	TA12/81 (DMRB 8.1) 'Traffic signals on high speed roads', Stationery Office [Sd].
DOT (1985)	TA 21/85 'Portable traffic signals at roadworks on single carriageway roads', DOT.
DOT (1991a)	'Traffic Signs Manual' – Chapter 8 – DOT [Sf].
DOT (1991b)	TD35/91 (DMRB 8.1.1) 'All purpose Trunk Roads: MOVA System of Traffic Control at Signals', Stationery Office [Sd].
DOT (1991c)	TR0141 'Microprocessor–based traffic signal controller for isolated linked and urban traffic control installations', DOT.
DOT (1993)	The Use of Puffin Pedestrian Crossings' Network Management and Driver Information Division. Network Management Advisory Leaflet, 1993.
Hall RM (1986)	'Accidents at Four Arm Single Carriageway Urban Traffic Signals' Contractors Report, CR65, TRL.
Hallworth MS (1980)	'High capacity signal design' Traffic Engineering + Control, 21(2).

Hallworth MS (1983)

'The Sheepscar signal system; an alternative approach to major intersection design', Traffic Engineering + Control 24(8).

HMG (1984)

'Road Traffic Regulations Act', Stationery Office.

HMG (1994)

Statutory Instrument 1994 No. 1519, 'Traffic Signs Regulations and General Directions', Stationery Office.

IHT (1996)

'Cycle–friendly Infrastructure', The Institution of Highways & Transportation.

Kimber RM and Hollis EM (1970)

Report LR 909, 'Traffic queues and delays at road junctions', TRL.

Kimber RM, MacDonald M and Hounsell N (1986)

Research Report No. 67 'The prediction of saturation flows for road junctions controlled by traffic signals', TRL.

Sang A and Silcock JP (1990)

'SIGSIGN: a phase–based optimisation program for individual signal–controlled junctions', Traffic Engineering + Control, 31(5).

Simmonite BF (1985)

'LINSIG: a program to assist traffic signal design and assessment', Traffic Engineering + Control, 26(6).

TRL (1963)

Road Note 34, 'A method for measuring saturation flow at traffic signals', TRL.

Webster FV and Cobbe FM (1966)

Road Research Technical Paper 56 'Traffic Signals', Stationery Office.

Wood K (1986)

'Measuring saturation flows at traffic signals using a hand held microcomputer', Traffic Engineering + Control Vol 27(4).

40.18 Further Information

Brown, M (1995)

'The Design of Roundabouts', Stationery Office.

DOT/TRL (1982)

Report LR 1063 'Traffic signalled junctions: a track appraisal of conventional and designs' TRL.

Chapter 41. Co-ordinated Signal Systems

41.1 Introduction

Traffic signal installations can be linked together to co-ordinate the time given to traffic at adjacent signal sites in order to control traffic movements over any section of a road network. This is the basis of the majority of present-day urban traffic control (UTC) schemes. By taking account of the available road-space at intersections and balancing the travel-time between successive traffic signals, it is possible to derive widespread advantage in terms of free-flowing traffic and reduced overall journey times. The benefits of co-ordination and the frequency of traffic-signal installations in urban areas have made the use of these techniques commonplace.

Co-ordination between adjacent traffic signals involves designing a plan based on the occurrence and duration of individual signal aspects and the time offsets between them and introducing a system to link the signals together electronically in order to impose the plan. Traffic-responsive systems also require on-line data input from detectors.

Traffic signals are often selected as the preferred means of intersection control in urban areas because of the benefits which can be derived from their co-ordination with other traffic signals both upstream and downstream. It is expected that around 40% of traffic-signal installations in Britain, including pedestrian crossings, will eventually be part of co-ordinated traffic systems.

Detailed advice on the design of traffic signals for individual intersection control is given in Chapter 40. Pedestrian, pedal cyclist and public service vehicle facilities, involving the use of traffic signals, are described in Chapters 22, 23 and 24 respectively.

41.2 Operational Objectives

Co-ordinated signal systems, on their own or in combination with the other network management technologies described in Chapter 18, provide an effective means by which traffic managers can implement a wide variety of flexible strategies for the management of a road network.

The systems can be used to obtain the best traffic performance from a network by reducing delays to vehicles and the number of times they have to stop. Where a network is not congested, this strategy also helps to reduce vehicle noise and pollution. Alternatively, the systems can be used to balance capacity in a network, to attract or deter traffic from particular routes or areas, to give priority to specific categories of road-user or to arrange for queueing to take place in suitable parts of the network; for example, at places where the noise and fumes of waiting vehicles would cause less irritation to passers-by or residents, or where convenient road space exists for queueing or where bus lanes have been provided.

Where co-ordination is achieved by the use of central computers, they can provide the basis for an expanded control system, incorporating such features as variable message signs, including car park information signs, congestion monitoring, priority for public transport and emergency service vehicles and other intervention strategies.

Information from the detectors, used in traffic-responsive systems, can also be processed for use with other network management systems. Similarly, information regarding equipment faults can be collated and used to improve the design and performance of equipment and to manage maintenance work more effectively.

The needs of pedestrians, cyclists and people with impaired mobility should be accommodated. When congestion is reduced and vehicle speeds increase, pedestrians can experience more difficulty crossing a road and Pelican crossings should be linked with signal controlled junctions within co-ordinated systems. Signal cycle-times should be kept as low as possible to provide more opportunities for pedestrians to cross the road safely. This also helps to reduce pedestrian frustration and the consequential risk of accidents. It is also desirable to double- or triple-cycle pedestrian crossings, wherever possible, to reduce pedestrian waiting times.

The Department of Transport has published a framework report for the development of urban traffic management and control (UTMC) systems (Oscar Faber TPA, 1995). This report sets out the requirements for UTMC systems, research needs and key research projects required to develop and utilise such systems.

Photograph 41.1: An example of car park information as part of a UTC system.

41.3 The Benefits of Co-ordination

The potential benefits which can be obtained from the installation of a co-ordinated signal system include:

❏ reduction in passenger or vehicle journey times, number of stops, fuel consumption and environmental pollution;

❏ alleviation of congestion and automatic detection of incidents;

❏ limitation of traffic throughput on selected roads or links;

❏ creation of priorities for buses, LRT, guided buses and bus–only routes (see Chapter 24);

❏ improved facilities for pedestrians and cyclists (see Chapter 22 and 23);

❏ allocation of priority to emergency vehicles responding to incidents and reducing vehicle attendance times, using special signal–timing plans to favour key routes from fire and ambulance stations;

❏ implementation of diversion schemes to deal with emergencies or special events and other control strategies, such as tidal flow schemes;

❏ improved fault monitoring and maintenance of equipment, leading to a reduction in the delays and potential safety hazards caused by faulty equipment;

❏ improved utilisation of car parks and a reduction in the amount of circulating traffic by providing car park information systems as part of UTC (see Photograph 41.1);

❏ the creation of a continually–updated centralised data–bank of information;

❏ interaction with other network management systems, such as route–guidance; and

❏ integration with urban motorway systems.

As an example of what can be achieved, the effectiveness of the SCOOT adaptive UTC system (see Section 41.6) in reducing delay to vehicles has been assessed by major trials in five cities (Robertson *et al*, 1991). The results are summarised in Table 41.1. The trials in Glasgow and Coventry were conducted by the Transport Research Laboratory (TRL) and those in Worcester, Southampton and London by consultants, a university and the local Highway Authority, respectively. In Glasgow, Coventry and Worcester, comparisons were made against a good standard of up–to–date fixed–time plans. Table 41.1 shows that the largest proportinate benefits were achieved in comparison with isolated vehicle actuation but, of course, part of this benefit could be achieved by a good fixed–time system.

The effectiveness of SCOOT varied by area and time of day but, overall, the trials concluded that SCOOT achieves an average saving in delay of about 12% compared with good fixed–time plans, which show up to 20% improvement over isolated vehicle–actuated (VA) controls. Since SCOOT does not 'age' in the way typical of fixed–time plans, it follows that SCOOT should achieve savings, in many practical situations, of 20% or more depending on the quality and age of the previous fixed–time plan and on how rapidly the patterns of flows change.

		Change in excess vehicle–hours through the system		
Location	Previous Control	AM Peak	Off Peak	PM Peak *
Glasgow	Fixed-time	+2%	-14%*	-10%
Coventry	Fixed-time	-23%	-33%*	-22%*
Foleshill		-23%	-33%*	-22%*
Spon End		-8%	0	-4%
Worcester	Fixed-time	-11%	-7%*	-20%*
	Isolated V–A	-32%*	-15%*	-23%*
Southampton	Isolated V–A	-39%*	-1%	-48%*
London	Fixed–time	(average 8% less overall journey time)		

*Results significant at the 95% confidence level

Table 41.1: Proportionate changes in delay resulting from the use of SCOOT.   Source: Robertson *et al* (1991).

On the basis of the surveys and subsequent experience, traffic–adaptive systems are likely to be of most benefit where vehicular flows are heavy, complex and vary unpredictably.

41.4 Suitability of Areas

Adjacent signal–controlled junctions should be considered for co–ordination when the vehicle arrivals are platooned as a result of the control at upstream junctions and when link travel– times are less than 20–30 seconds normally or 60 seconds in particularly free–flowing conditions. Co–ordination can be achieved between as few as two junctions (even between two signal–controlled pedestrian crossings) or on an area–wide network basis. Simple schemes can utilise the co–ordination capabilities of modern traffic–signal controllers. Larger area–wide schemes, making use of a central control computer, become worthwhile when there are about a dozen junctions and pedestrian crossings under signal control within a single locality and the traffic pattern exhibits cyclic downstream platooning at least during peak periods (Wood, 1993).

It is common to divide a UTC network into sub–areas and this should be considered when one or more of the following conditions obtain:
❑ when groups of adjacent signals require different plans or strategies;
❑ where relatively long distances occur between adjacent groups of signals;
❑ where well defined major routes exist, with few significant cross movements;
❑ when queueing space becomes a critical feature at particular junctions;
❑ where complex movements have to be accommodated within a relatively small area; and
❑ when one computer in–station is used to control the UTC systems in several towns.

Co–ordination between sub–areas can be adjusted to meet demands. It is common for individual sub–areas to share a central computer for economy but still operate independently. The Transport Research Laboratory has developed a simple method for deciding which groups of adjacent signals are likely to be worth co–ordinating and the scale of benefits which should result (Robertson *et al*, 1983). A program called COORDBEN is available from TRL and has been used successfully in the UK and overseas.

Critical Nodes
A common cycle–time is needed for a network or sub–area. Cycle–times are often determined by one or more critical nodes (usually junctions) which have the highest degree of saturation (see Chapter 40). In order to limit the degree of saturation of such junctions to reasonable levels (ie around 90%), it is necessary to select an appropriately high cycle–time – thus the critical nodes will dictate the overall network (or

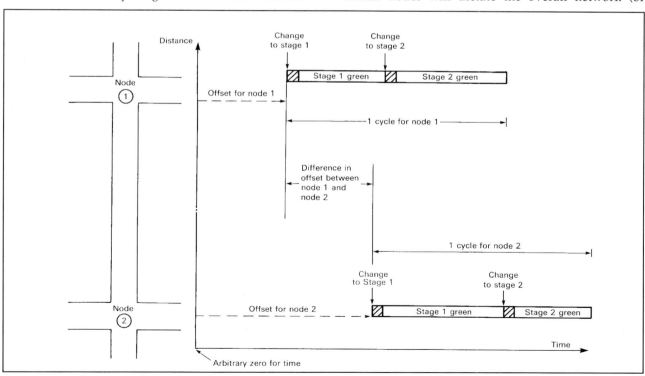

Figure 41.1: Illustration of the term 'offset'.

Source: TRL (1975).

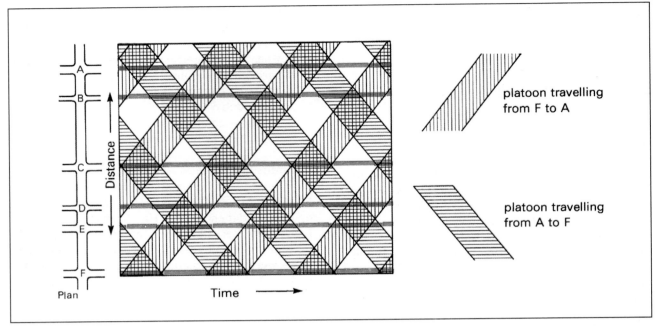

Figure 41.2: A time and distance diagram for linked traffic signals.

Source: HMG (1966).

region) cycle–time. The other nodes, being less critical, can all operate at lower cycle–times and hence may suffer more delay unless they have sufficiently low saturation levels to allow them to 'double cycle' at half the network cycle–time. If only a small number of critical nodes cause a larger number of nodes to suffer such additional delay, localised improvement to the critical nodes may be justified. At junctions, such improvements could include:

❏ adding approach lanes;
❏ prohibiting some of the turning movements; and
❏ making pedestrian facilities 'parallel', in place of 'all red' stages.

In any area–wide design, it is always worthwhile to pay particular attention to detailed layout, phasing and inter–stages at critical nodes.

41.5 Co–ordination Concepts

Signal co–ordination means controlling the starts and durations of the green periods at adjacent sets of signals along a route or within a network.

Common Cycle–times
To maintain signal co–ordination from cycle to cycle, each junction must operate with a common cycle–time or a simple multiple of it. For example, Pelican crossings can often complete two cycles in the time needed for adjacent street junctions to complete one cycle.

Offsets
The green periods occurring at each junction are staggered in relation to each other, by specifying an offset time for each junction with respect to adjacent junctions. The offset is the starting time of a specified stage at the junction to a common time–base of one cycle; this is illustrated in Figure 41.1.

Time–distance Diagrams
Using a time and distance diagram (see Figure 41.2), offsets can be calculated to offer a 'green wave' to the predominant traffic flow and achieve co–ordination of the opposing flow on the same route. In practice, diagrammatic techniques do not always produce the best setting. When more than one or two conflicting traffic streams have to be considered, the problem becomes more complicated and computerised techniques should be employed, as discussed in Section 41.6.

41.6 Alternative Methods of Control

Two basic types of UTC system, currently in use, are based on different control strategies. These are:

❏ fixed–time control systems; and
❏ traffic–responsive control systems.

The latter can be further sub–divided into:
❏ plan–selection systems;
❏ plan–generation systems;
❏ local adaptation;

- ❑ centralised traffic–responsive systems; and
- ❑ traffic–responsive systems with distributed processing.

These systems are described briefly below.

41.7 Fixed–Time Systems

Signal Plans

Fixed–time systems operate with a set of pre–designed signal plans, each of which can be implemented at any time, on receipt of a command from a central control point or using local clocks synchronised by the timings pulses in electrical supply mains. A signal plan is a collection of co–ordinated settings for all the signals in a network and, although it can be calculated by manual methods in simple cases, computer techniques are usually used. The preparation of signal plans involves representing the traffic conditions in the network numerically and producing an index of performance. The signal timings are optimised against various strategy and policy criteria (this procedure is described in the later section on TRANSYT). The signal settings in each plan are fixed in that the green periods and offsets do not vary from cycle to cycle. Thus, fixed–time systems control known patterns of traffic rather than respond to demand. This can be both a strength and a weakness of such systems.

Types of Signal Plan

A typical computer–controlled fixed–time system will have different plans for morning, evening and off–peak weekday conditions and for weekends. It is likely there will also be plans for evening and night–time conditions and for specific occasions, such as processions and sports events. Most modern fixed–time systems have the capability to implement 40 or more plans.

If the network also has traditional vehicle–detection equipment (see Chapter 40), the fixed–time system may also switch to isolated vehicle–actuated (VA) operation during the night, during periods of low flow or if some fault occurs at the central controller. It is also possible to introduce some stages only when there is a demand from a detector (further description of the use of detectors in fixed–time and other systems is given in Section 41.15). However, the cost of maintaining vehicle–detectors is difficult to justify for these purposes alone and the reversion to full isolated–VA operation is becoming less widely used. Where VA operation is not provided, the local controllers switch to cableless linked plans if a fault occurs on the central computer.

Ageing of Signal Plans

The benefits offered by the initial signal plan implemented on–street will depreciate over time as traffic conditions change and the plan becomes less appropriate. It has been estimated that signal plans degrade by about three percent per year, so the initial benefit can be lost typically within five years (Bell *et al*, 1985). The ageing process arises from:
- ❑ any general increase or decrease in traffic over the whole or parts of the network; and/or
- ❑ changes in flows of traffic on different links resulting from re–routeing or altered traffic demands; and/or
- ❑ physical alterations to the street network.

It is also worth remembering that, when a plan is first implemented, vehicles often re–route to take advantage of less congested routes. Consequently, the distribution flows within a network may change in a way that does not then match the assumptions on which the plan was based and this can create a need to update the plan.

Plan Selection

A fixed–time system may, typically, involve between 4 and eight changes of plan during a normal weekday. Because of the day–to–day variations, it is often difficult to decide exactly when to change plan on any particular day but the aim is that changes should be timed to respond to marked variations in traffic flow over the day. Sometimes plans are changed in response to a manual command resulting from visual monitoring of conditions using closed circuit television cameras (CCTV). However, the most common method is to change plans regularly at a particular time each day, determined historically by expected traffic conditions. As this will take place irrespective of prevailing traffic conditions, it may cause some disruption to traffic and reduce the overall performance of the network, whilst adjustments take place. Problems may also arise as a result of the unpredictable non–optimum timings which may occur during the first cycle of the new plans. For these reasons, changing plan too frequently can have a detrimental effect and it is generally better to change plans during off–peak periods.

TRANSYT

In Britain, the most widely–used technique for calculating fixed–time signal setting is the TRANSYT 9 computer program developed by the TRL (Crabtree, 1988). The program models traffic behaviour and carries out optimisation procedures which calculate signal timings giving approximately optimal traffic performance (see Figure 41.3 for the program structure). The program also provides extensive information about the performance of the network, including estimated delays, numbers of stops,

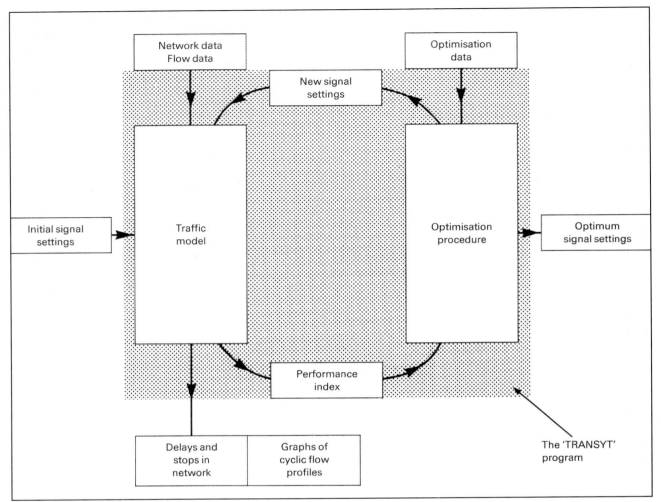

Figure 41.3: The structure of the TRANSYT program.

Source: TRL (1975).

journey speeds and fuel consumption. TRANSYT models traffic behaviour using histograms to represent the arrival patterns of traffic. These are called 'cyclic flow profiles' because they represent the average pattern of traffic flow during one signal cycle (see Figure 41.4). The model produces the best signal settings, consistent with the parameters within the model. Because the model can never reflect reality completely, 0n–site monitoring of timings is essential. It is important to check that the predicted cycle flow–profiles give a reasonably accurate representation of actual traffic behaviour. If not, model parameters within the program must be altered until this is achieved. Even so, some additional fine –tuning of the timings on site may still be required. The signal–optimising part of the program searches for a good fixed–time plan, which will keep down the level of delay by approximately minimising a performance index for the network. The performance index is a weighted combination of the costs of delays and stops on all links. Specific links can also be weighted, so that the optimising process derives more benefit from reducing delays and stops on these links

at the expense of others. These weightings can be used, for example, to give priority to buses, allowing for differences between their movement and that of other traffic along each link to be taken into account.

41.8 Traffic–Responsive Systems

Basic Principles
Traffic–responsive control systems monitor traffic conditions in a network by some form of detection and react to the information received by implementing appropriate signal settings. Thus, systems of this kind adapt themselves to traffic patterns and respond to traffic demands as they occur.

Plan–selection Systems
In this method of responsive control, the information obtained from on–street detection is used to select the most suitable plan from a library of pre–calculated plans. Although this method provides a degree of self–adaptation to traffic conditions, it still requires

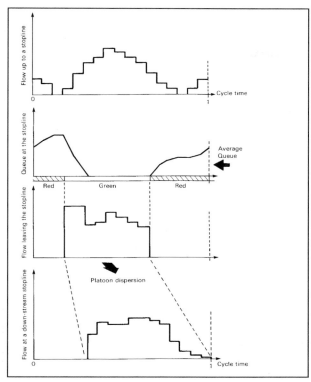

Figure 41.4: Cyclic flow profile in the TRANSYT model.
Source: TRL (1975).

the preparation of fixed–time plans, rather than providing a gradual evolution of signal timings in response to changing traffic conditions. Traffic can be disrupted by frequent changes of plan and usually some restriction is placed on the frequency of such changes. Plan–selection systems are used extensively outside Britain but there is no convincing evidence that systems which change fixed plans on the basis of flows and congestion measurements perform any better than the simpler procedure of changing plans at given times of day.

Plan–generation Systems

Plan–generation systems generate their own fixed–time plans from detector data and implement them. In the past, this technique has been found to give worse control than simple fixed–time plans, because there have been insufficient detectors to provide adequate flow information. The AUT (Automatic Updating of TRANSYT) system in Gothenburg does use detector data to produce new TRANSYT plans. The turning movements are calculated from a combination of detector and historical data but details have not been published. Some systems use a measure of local adaptation at the controllers, to modify the action of centrally imposed fixed–time plans. The best known system of this type is the Australian SCATS system (Laurie, 1982). The basic operation is that an appropriate fixed–time plan is run and the local controllers can omit, or terminate

early, the side–road stage depending on the local demand for the stage in the current cycle. The fixed–time plans are calculated with a particular objective, such as minimum vehicular delay, as for a standard fixed–time system. Local adaptation then increases the main road green–time in some cycles, which should lead to better progressions on the main roads. Because of the emphasis on giving green–time to the main roads, the systems are probably best suited to areas with prominent main roads, such as radial corridors.

Centralised Traffic–responsive Systems

To overcome the problem of plan–preparation and plan–changing, the fully–responsive strategy called SCOOT (Split Cycle Offset Optimisation Technique) was conceived by the TRL and developed by the Department of Transport and British signal manufacturers (DOT, 1995). SCOOT has been introduced in over 130 cities in Britain and overseas and, as described in Section 41.2, has been shown to provide significant benefits over both fixed–time systems, including TRANSYT, and isolated control. The structure of SCOOT is similar to that of TRANSYT, in that both methods use a traffic model of a network which predicts the delay and stops caused by particular signal settings. However, unlike TRANSYT, the SCOOT model is on–line and monitors traffic flows continuously from on–street detectors. SCOOT uses this information to recalculate its traffic model predictions every few seconds and then makes

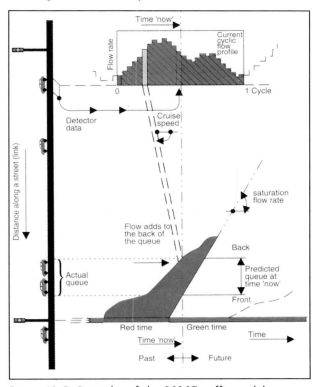

Figure 41.5: Principles of the SCOOT traffic model.

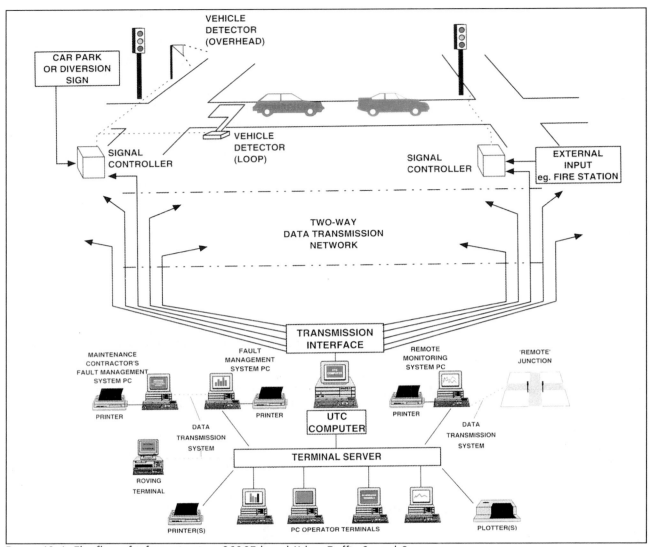

Figure 41.6: The flow of information in a SCOOT based Urban Traffic Control System.

systematic trial alterations to current signal settings, implementing only those alterations which the traffic model predicts will be beneficial.

The structure and principles of SCOOT are illustrated in Figures 41.5 and 41.6.

Advantages of Fully–responsive Systems

The advantages of a fully–responsive strategy, such as SCOOT, over fixed–time systems are:

❏ no need to prepare, or update, fixed–time plans, although the information used to model the network has to be updated periodically;

❏ no sudden changes in signal setting – instead, new plans are continuously evolved;

❏ trends in traffic behaviour can be followed without requiring longer–term predictions of average flows; and

❏ the system will adjust itself to respond to some incidents, such as accidents, and data on the traffic situation is available to operators.

In general, fully–responsive systems are most valuable in areas where congestion is high and flow patterns are complex and variable. However, they do require skilled staff to design and validate the network models. In addition, subsequent changes to the road network, to the uses of land adjoining the highway and to parking and loading activities, do affect traffic responsive systems. The information used to model the network has, therefore, to be reviewed periodically. Where congestion levels are generally low and flow patterns consistent, it is usually best to use fixed–time TRANSYT–based UTC systems.

Traffic–responsive Systems with Distributed Processing

A number of systems are being developed where an appreciable amount of the UTC optimisation is carried out in the local controllers and these are then connected to a central management system by data–links. Examples include the UTOPIA system in Turin (Donati *et al* 1984) and the PRODYN system in Toulouse (Henry *et al*, 1988).

The features and advantages of responsive systems with distributed processing are basically the same as centralised responsive systems. It is possible, however, that distributed processing will result in reductions in the costs of communication between the central and local controllers. This saving may be offset by the lack of real–time information at the central control, which may make it more difficult for the UTC system to respond to inputs from route guidance or traffic information. The benefits and disbenefits of central processing, compared with distributed processing, have therefore to be considered when designing a new system.

41.9 Equipment Requirements

All signal equipment used on public highways in Britain must conform to standards laid down by the Department of Transport (DOT, 1980) [NIa] and type approval, as per Traffic Signs Regulations (TSR) and General Directions (GD).

Traffic signals in a UTC area are usually controlled by a central computer, which sends electronic instructions by telephone–type cables to each junction controller. Local co–ordination may also be achieved either by linking controllers by dedicated cable or by cableless links between microprocessor–based controllers. These operate by having a synchronised time–reference at each junction with co–ordination maintained by regular pulses from the mains supply, so that the need for cable connections between junctions is eliminated (Rudland, 1973).

Although cableless systems do not have the same flexibility as a centrally controlled system, they can be useful as a back–up facility in the event of computer failure or for small groups of signals, in places where the expense of a central computer is not justified. If these systems are connected to a fault–monitoring system, their operation can be modified, or monitored centrally, to ensure that the clocks remain synchronised.

All fixed–time and adaptive UTC systems used in Britain require communication between the central computer and junction controllers on a second–by–second basis. As a consequence, fixed or dial–up data circuits are required and it is not possible, at present, to use data–links, such as cellular radio, where delays in transmission may occur. It may

be possible to use different media as and when devolved adaptive control systems are developed.

Transmission of the data from the central computer is by means of in–station transmission units (ITU). These receive the signals from the computer and transmit the information to the various junctions, by means of time–division or by frequency–shift multiplexing, which accommodates several channels on each data–line by the use of data–concentrators at either end of the line. Alternatively, up to four local controllers can be controlled using one multi–point data–circuit.

Both these techniques can lead to savings in the annual rental costs of the data–lines. However, these savings may be partially offset by increased vulnerability to line–failure, which results from having several controllers dependent on one data–line. It is recommended that detailed comparisons of the installation cost and subsequent annual costs of the various data– communication options should be made, whenever the junctions to be co–ordinated are more than 5km from the central computer.

At each of the junctions controlled by the central system, an out–station transmission unit (OTU) is installed in the controller. This receives the signal from the data–line and interfaces with the local control equipment. The interface is specified in the DOT controller specifications TR 141 (DOT, 1993). The OTUs, which operate at 200 baud or 1200 baud transmission speeds, are specified in DOT Specifications MCE 0312C (DOT 1975) and MCE 0361 (DOT, 1981).

The OTUs use transmission protocols which are specified by the manufacturer of the in–station transmission system and similar units have to be used in any subsequent expansion of the system. Several controller manufacturers now offer OTUs which are an integral part of the controller and which are less expensive than a controller with an interface to a rack–mounted OTU. It is likely, therefore, in subsequent expansion to a UTC system, that economic controllers will only be available from the manufacturer which supplied the in–station transmission equipment. In order to preserve competitive procurement, the use of standard transmission protocols is desirable (Oastler *et al*, 1995).

41.10 Data Requirements

Calculation of signal settings for control strategies requires a considerable amount of data including:
  ❑ a representation of the network, typically as

nodes (junctions) and links (one–way approaches to nodes);
❑ link lengths;
❑ expected peak traffic flows, including turning flows within junctions;
❑ saturation flows for each link;
❑ free–flow journey speeds or times for each link;
❑ details of the cycle of signal operations for each junction, including inter–greens, minimum greens, stage–sequence and appropriate geometric and traffic parameters; and
❑ average queue clearance times.

The suitability of the resulting signal settings for the network depends on the accuracy both of the input data and of the traffic model within the computer program.

41.11 Monitoring of Faults

Improved maintenance of traffic signal equipment can be achieved, at junctions operating under UTC, as they can be monitored continuously and remotely to identify faults, enabling maintenance work to be initiated more swiftly than by conventional methods of periodic checking and reporting. Both on– and off–line computer fault–monitoring and analysis systems have been developed and similar benefits can be achieved through periodic, 'dial–up', monitoring of signal installations which are not on UTC.

41.12 Capability of Software

The local co–ordination software in junction controllers now provides a minimum of 16 cableless linked plans. The main functions of the software, for the UTC systems used in Britain, are specified in the Department of Transport Specification MCE 0360C (DOT, 1983). This software allows the co–ordination of traffic signals using either fixed–time plans or adaptive control under SCOOT. Automatic plan–selection can also be provided by the system suppliers. This flexibility provides engineers with the means to control traffic in towns and cities of every size.

Various other options which can be provided are set out below.

Diversion Sign Control
The system can regulate diversion signs and associated fixed–time plans, either under operator control or remotely by push–button. This facility may be used to close certain streets or areas to traffic, where there is a regular requirement at particular times of the day or week. The system can also be used, under operator control, where streets need to be

closed on a frequent but irregular basis; for example, due to congestion, flood, pollution detection or some other special event.

Car Park Monitoring and Sign Control
The system may be used for car park control and for the signing of car park groups. For monitoring individual car parks, the system collects data on the number of vehicles entering and leaving and controls the signs on approach roads, according to the space available. On a wider scale, the system can also control signs for groups of car parks, ensuring that incoming vehicles are directed to parks where spaces exist. Such signs may be installed strategically around the outskirts of a city (see Photograph 41.1). Information from car parking systems could also be used by the central system to modify traffic signal timings.

Graphic Displays
UTC systems generally include a number of graphic display facilities, which offer a fuller understanding of the traffic situation in the control area. The facilities include diagrams, giving information about queue build–up and dispersal, displays of individual junction operation and time–distance diagrams, to assist in analysing traffic flow and journey times. Facilities may also allow the creation of mimic diagrams, which can be updated in real time to show traffic patterns at a single junction or throughout the traffic network.

Roving Terminal
A roving terminal is a portable lap–top terminal which communicates with the UTC system via a cellular radio link. This provides access to the system from any location, making direct comparisons between the actual traffic and the situation being modelled. Validation displays make it possible to check the performance of the system quickly and effectively under real traffic conditions.

Priority Routes for Emergency Vehicles
'Green waves' can be implemented through UTC systems, to give immediate priority to emergency vehicles travelling through the network. This is especially effective when used in conjunction with automatic detection equipment for these vehicles (Griffin et al, 1980). Fire appliances generally follow predetermined routes to incidents and it is possible to devise special plans to cater for them. The system can be initiated remotely by key–switch at the fire station or introduced by an operator's command. The system usually brings signals to green, along the selected route, about 30 seconds before the emergency vehicle is due to arrive. By this means, not only does the

emergency vehicle receive a green priority at the signals but vehicles are cleared out of its way before the emergency vehicle arrives. It is reported (Williams, 1979) that savings in fire damage costs in Liverpool resulting from the reduction in fire appliance journey–time through this facility, more than justified the costs of installation of the complete UTC system.

Traffic Data–collection

Where traffic detectors are installed in fixed–time UTC systems, traffic counts, detector–occupancy and speed measurements can be transmitted over the data–lines and processed by the central software. This data can be used to give warnings to operators where congestion develops, so that they can introduce special contingency plans. It can also be used to assess changes in daily, weekly, monthly or seasonal traffic flows. Adaptive control systems contain a large amount of useful traffic data derived from the detectors. This includes traffic flows, delays and various detector occupancy–related data, such as traffic density. This information can be assessed from the data messages in the system or from purpose–built databases. A database, called ASTRID, has been developed for storing, processing and displaying SCOOT data (Hounsell et al, 1990). ASTRID stores modelled flow, total delay and congestion data for individual detectors, links, junctions, specified routes, regions and for the whole SCOOT area. It can also calculate average delay per vehicle, average journey–time and speed. The information is available in the form of typical daily profiles, long–term trends and individual daily backup data. The data is available on–line and journey–time information from the on–line system could be provided to traffic information or to route–guidance systems.

Further Developments

Expert computer systems are being developed (Scemama, 1995), which aim to monitor data from UTC detectors and advise the operator on what action to take to alleviate congestion. Moreover, advances in high–speed computing may lead to better modelling and optimisation of co–ordinated traffic signals than is possible with existing algorithms. Present traffic models, such as SATURN or CONTRAM, do not model co–ordinated signal systems fully and do not attempt to model adaptive systems. Improvements in the linking of SATURN models to TRANSYT models and further improvements to the modelling of co–ordinated signals in CONTRAM systems are therefore desirable.

41.13 Priority for Public Transport

Methods of giving priority to isolated traffic signals

are described in Chapter 40. Provision of bus priority, in areas where signal timings are co–ordinated, is more difficult to achieve without increasing overall delay and congestion. However, the potential benefits are considerable, since bus–flows tend to be high in areas where signals are co–ordinated, so long as any additional queues do not cause delays to buses elsewhere in the network.

Three levels of priority can be provided as under.

Passive Priority

For fixed–time systems, bus–stop dwell–times and weighting of links can be input into the BUS TRANSYT program (Pierce et al, 1977). These weightings and dwell–times allow differences between bus movements and that of other traffic along each link to be taken into account. For responsive systems, bus links can similarly be weighted. The benefits of passive priority are limited but the costs of implementation are relatively low. Research on SCOOT indicates possible reductions of between five and eight per cent in delays to buses by using link–weighting facilities. Similar trials of BUS TRANSYT in Glasgow showed about a 16% reduction in delay.

Active Priority

Individual buses are detected on traffic signal approaches and the signal stages are modified appropriately. Within fixed–time UTC systems, the computer may define time–windows, during which the signal timings may be changed to benefit buses. A trial of bus priority in SCOOT took place as part of the EC project PROMPT (Bowen et al, 1994) and bus priority is available in SCOOT Version 3.0.

Bus Tracking

This technique requires a system to track all public transport vehicles that run on–street. The UTC system has to have an interface with a vehicle location system and be able to use this information. The precise arrival–time of the vehicle at the stop–line has to be predicted and the system must have a method of adjusting the signals to give priority to vehicles at their predicted arrival time but only if they are on, or behind, schedule.

41.14 Queue–Management and Gating

Fixed–Time Systems

Congestion on some links of a fixed–time system would have much more serious consequences than it would on other links. The classic example is a circulating link on a gyratory. Under congested conditions, the queue on a gyratory can stretch back beyond the upstream entry and prevent traffic from leaving the gyratory at the corresponding exit. In this situation, the queues increase rapidly and can lock the gyratory totally. TRANSYT includes a feature to monitor the queue–length on critical links, during the off–line optimisation, and to modify the signal timings to prevent queues on critical links blocking upstream junctions. Queue–detectors can be used to identify such queues and to change timings to prevent this blocking–back.

As TRANSYT calculates fixed–time plans off–line, it cannot respond to changes in the behaviour of traffic in the network that lead to congestion. The only facility is to adjust the timings to prevent serious congestion occurring in typical conditions. The robustness of a solution can be tested by additional TRANSYT runs with, say, 20% extra traffic.

Normally, TRANSYT calculates timings to give minimum delay and stops. However, by using the link–weighting facility, it is possible to bias the timings in favour of certain links. If the chosen links are those with the highest capacity, then the resulting signal plans will be biased towards maximising the throughput of each junction but not necessarily to maximising the throughput of the network. There is, generally, no facility to implement, automatically, such timings in congested conditions. The timings are usually prepared in advance and transferred to the UTC system, the operator then implements them manually during congested conditions or they are implemented by timetable at certain times of the day. In the Bitterne Road scheme in Southampton, pre–set traffic restraint plans were selected automatically using strategically located detectors (University of Southampton, 1974). This pioneering scheme has now been incorporated into a SCOOT system.

Adaptive Traffic Control

In the SCOOT adaptive control system, links are assigned a 'congestion importance' factor when the system is set up. This allows SCOOT to operate queue–management in order to reduce the likelihood of queues building back and blocking upstream junctions.

One technique, used to carry out more sophisticated queue management, is known as 'gating'. Gating is used to limit the flow of traffic into a particularly sensitive area. The gating logic allows one or more links to be identified as 'bottleneck' links, where problems are known to occur. Associated gated links are identified, where it is less critical if queues build up (Wood *et al*, 1995). Under normal conditions, no gating action is taken but, when saturation on the

bottleneck link reaches a defined limit, the optimisers will begin to reduce green–time on the gated links in addition to its normal optimisation. An alternative operation of gating is to specify links, downstream of the bottleneck, which will receive increased green time when critical saturation is reached on the bottleneck link.

Traffic limitation strategies can be implemented in SCOOT using split weighting techniques to limit the length of green stages and, on a wider scale, using the gating techniques.

41.15 Detection

Fixed–Time Systems

Where signals continue to operate on a fixed–time basis during the night, drivers often complain that they are stopped by a red signal when there is no other traffic crossing the intersection. This is particularly noticeable at some side–roads. It can be overcome by reverting to full vehicle–actuated operation, during the night, but this requires expensive investment in, and maintenance of, detectors. In this situation, delays to the main road traffic can be overcome by installing presence detectors near the stop–line on side–roads and by introducing the green signal only when a vehicle is actually waiting at the stop–line. Either inductive loops or microwave detectors can be used as presence detectors. Inductive loops are also used to measure traffic volumes, detector–occupancy and, occasionally, speed in traffic–responsive plan systems. Video–processing detection systems are also used for these purposes.

Detectors can be used, on a particular section of road, together with variable message signing (VMS) to measure the speeds of vehicles and to indicate whether they are exceeding statutory and/or recommended safe speed–limits.

Speed measurement with detection can also be used to indicate to road–users the optimum speed of travel to take advantage of linked traffic signal schemes (Teply *et al*, 1990).

Traffic–responsive Control Systems

Inductive detector loops have been used extensively in these systems. Their location is critical to the operation of the system and may be different from those required for other forms of control. The manufacturers detailed recommendations should be followed. Infra–red detectors mounted on lamp posts and video–processing detection systems can also be used to detect vehicles in these systems.

Selective Detection

In order to provide active bus priority within traffic signal systems, buses may have to be located to an accuracy of plus or minus two metres.

A broad range of selective vehicle–detection devices is available and these can be divided into two main categories (Chandler *et al*, 1985). These are:
- ❑ road–side mounted techniques, with no equipment on the vehicle; and
- ❑ vehicle–mounted equipment, using various means of communicating with the roadside.

The first category includes :
- ❑ long detector–loops;
- ❑ detector–loops with signature–processing;
- ❑ microwave or ultrasonic signature–processing ; and
- ❑ video image–processing

The second category includes:
- ❑ infra–red and microwave tags for bus identification;
- ❑ loop–aerial coupling to a transponder unit mounted on the bus (the transponders can either be battery powered or powered from the loop); and
- ❑ bus location, fleet–control systems, using automatic vehicle location (AVL) devices, which interact with roadside equipment.

A disadvantage of this category is the need to equip all of the buses operating in an area with the required devices. They do, however, accurately identify buses which are so equipped as well as providing a two–way communication link between the roadside and travelling buses.

London Transport have fitted 5,000 buses with battery–powered transponders and 300 buses in Leeds have been fitted with transponders powered from loops as part of the European DRIVE 2 project, PRIMAVERA (Fox *et al*, 1995).

These techniques have not yet been developed to a stage where they can identify accurately all selected vehicles in multi–lane congested conditions. The wide variety of bus types now in service also complicates recognition. However ongoing research and development, particularly in video–image processing, may make them viable.

Vehicles as Detectors

Location equipment, fitted for fleet–management or route–guidance purposes, could be used to provide co–ordinated signal systems with information about prevailing traffic conditions. This information can be used either as the basis for completely new co–ordination strategies or to verify the accuracy of existing strategies.

41.16 Future Developments

The Department of Transport has prepared a new specification for urban traffic management and control (UTMC) systems, in consultation with industry, users and the research community generally. The aim of the new specification is to provide a framework for the development of a new generation of UTMC systems, which exploit the potential of modern communications and computer technology to meet a wide range of urban transport policy objectives. The new specification is based on open data–transfer standards and a modular structure. This is intended to encourage competition in development and procurement and to enable authorities to build up their systems in an incremental way, as new applications are developed.

41.17   References

Bell MC and Bretherton RD (1986)	'Ageing of Fixed–Time Traffic Signal Plans'. Proceedings of the Institution of Electrical Engineers, Second International Conference on Road Traffic Control IEE.
Bowen G Bretherton RD	'Active Bus Priority in SCOOT', IEE Conference – IEE.
Landles JR Cook DJ (1994) Chandler MJH	'Traffic Control Studies inLondon Cook DJ (1985) – SCOOT and Bus Detection', Proceedings of PTRC Summer Annual Meeting, University of Sussex.
Crabtree MR (1988)	Report AG8 'TRANSYT 9 – Users Manual', TRL
DOT (1975)	MCE 0312C 'Data Transmission System – Traffic Control', DOT.
DOT (1980)	TD7/80 (DMRB 8.1) 'Type Approval of Traffic Control Equipment,' Stationery Office.
DOT (1981)	MCE 0361 'High Capacity Data Transmission System for Use in UTC', DOT.
DOT (1983)	MCE 0360C 'Urban Town Traffic Control – Functional Specification', DOT.
DOT (1993)	TR 141 'Micro–processor Based Traffic Signal Controller for

Isolated, Limited and Urban Traffic Control Installations', DOT.

DOT (1995) Traffic Advisory Leaflet 4/95 'SCOOT Urban Traffic Control System', DOT.

Donati F, Mauro J, Roncolini G and Vallaurie M (1984) 'A hierarchical decentralised traffic light control system'. The first realisation proceedings IFAC 9th World Congress Vii.

Fox K, Montgomery F, Shepherd S, Smith C, Jones S, and Biora F (1995) 'Integrated ATT Strategies for Urban Arterials'. DRIVE II project PRIMAVERA. 2. Bus Priority in SCOOT and SPOT using TIRIS'. Traffic Engineering + Control June 36(6).

Griffin RM and Johnson D (1980) 'Northampton fire priority demonstration scheme –the before study and EVADE'. Traffic Engineering +Control 21(4).

Henry JJ, Farger JL, and Taffal J (1983) 'The PRODYN Real Time Algorithm', Proceedings 4th IFAC/ICIP/IFORS, Conference Control in Transportation, Baden Baden.

HMG (1966) Road Technical Paper 56 'Traffic Signals', Stationery Office.

Hounsell N and McLeod F (1990) 'ASTRID Automatic SCOOT Traffic Information Database' Contractor Report 235, TRL.

Laurie PR (1982) 'The Sydney Co–Ordinated Adaptive Traffic Systems – Principles Methodology Algorithms'. Proceedings, First Conference in Road Traffic Control, IEE London.

Oastler K and Head J (1995) 'Open Specifications for Traffic Control Equipment'. The Second World Congress on Intelligent Transport Systems, Yokohama.

Oscar Faber TPA (1995) 'Framework Report for the Development of Urban Traffic Management and Control (UTMC) Systems', DOT.

Pierce and Wood (1977) SR 266 'Bus Transyt Users Guide', TRL.

Robertson DI and Bretherton RD 'Optimising Networks of traffic (1991) signals in real time – the SCOOT method'. IEEE Transactions on Vehicular Technology, 40(1).

Robertson DI and Hunt PB (1983) 'Estimating the Benefits of Signal Co–ordination using TRANSYT or SCOOT Optimisation'. Proceedings 53rd ITE Conference, London.

Rudland P (1973) 'Cableless Linking of Traffic Signals', Traffic Engineering + Control 15(1).

Scemama G (1995) 'CLAIRE: Current developments and applications', PTRC Conference on Traffic Engineering for Society', PTRC.

Teply S and Schnableger J (1990) 'Variable speed advisory signals for linking co–ordinated systems in Canada'. Third International Conference on Road Traffic Control IEE.

University of Southampton, (1974) 'An Evaluation of the Bitterne Bus Priority Scheme', TRG Technical report for TRL.

TRL (1975) Report LR888 'User Guide to TRANSYT version 8', TRL.

TRL (1977) SR 266 'Bus Transit Users Guide', TRL.

Williams J (1979) 'Area Traffic Control in Cardiff', Chartered Municipal Engineer.

Wood K (1993) 'Urban Traffic Control Systems Review' Project Report 41, TRL.

Wood K Baker RT and (1995) 'User guide to the 'gating' method of reducing congestion in traffic networks controlled by SCOOT', TRL.

41.18 Further Information

DOT Driver Information and Traffic Management Division, DOT.

Routledge I Kemp S and Radia B (1996) 'UMTC : The Way Forward for Urban Traffic Control' Traffic Engineering + Control, 37(11).

Chapter 42 Signal–Controlled Gyratories

42.1 Introduction

Gyratories are road systems which consist of one–way links connected together, so as make it possible for traffic to circulate along one or more links before exiting. They can take a variety of forms but are most characterised by 'roundabouts', which are usually built as entities rather than configured from a number of existing roads. Whilst roundabouts overwhelmingly constitute the largest class of gyratories, the more general term (gyratories) is used here except where roundabouts specifically need to be identified.

The installation of traffic signals on existing gyratories is becoming common in the UK. There are two major categories: urban gyratories, which tend to be at the smaller end of the size spectrum (typically less than 100m island diameter); and those on roads with speed limits over 40 miles/h, which provide some of the larger examples, including motorway interchanges. The majority of signal–controlled gyratories start out as priority junctions, to which signals are added at a later date, but the fact that some authorities have installed purpose–designed signal–controlled gyratories (Hallworth, 1992) is a reflection of the advantages that this junction arrangement can offer.

Signal–controlled gyratories have many characteristics in common with conventional signal– controlled junctions. However, being multi–node junctions, they present the designer with additional problems, due to the interactions that can take place between the signal–controlled nodes. Consequently, the designer of a layout and control system needs a complete understanding of the operation of the junction (Brown, 1996).

42.2 Effect of Signal Control on Throughput and Safety

Throughput
When the self–regulating nature of a priority–controlled gyratory breaks down, the throughput of particular entries can be restricted by the difficulty of finding gaps in the circulating traffic. The introduction of signals at one or more of the nodes can produce the necessary bias in favour of the affected entries and particularly:

❏ where delays are excessive, due to imbalanced flows, signals can alter the natural priority to give a better distribution of delays by improving the queue–balance;
❏ where throughput is inadequate because of high circulating speeds (rather than high flows), signals can achieve an overall improvement in throughput; and
❏ where it is possible to co–ordinate the gyratory as part of an overall UTC network (usually fixed–time UTC), signals can reduce overall delays.

Installing signals on a gyratory with full–time control often involves making geometric changes, as an integral part of achieving throughput improvements, in ways which might not necessarily be effective or safe on a gyratory with priority control. The scope for changes in geometry with signals includes:
❏ the provision of additional entry–lanes and improvement in the forward visibility for signals, by the reduction or removal of entry–deflection; and
❏ the provision of additional carriageway width, either to complement entry–approach lanes or to accommodate queueing by changing the shape of the gyratory – this often results in less 'circular' and more 'triangular/square' geometry

Safety
The introduction of signals on gyratories can regulate traffic patterns, reduce the need for weaving and merging, remove gap–acceptance problems and reduce speeds. Because of this, they are particularly beneficial for cyclists and can also provide more positive control for pedestrians, when compared with priority roundabouts.

Substantial reductions in accidents between vehicles entering and those circulating are possible on large gyratories. For example, on priority–controlled gyratories accidents often occur at entries where drivers find it difficult to interpret the sizes of gaps. Signal–control can remove this doubt.

A survey involving over 160 signal–controlled roundabouts (CSS, 1997) indicated that 'accident reduction' is cited as the reason for signal–installation in about 18% of roundabouts, with substantial safety benefits achieved in such cases. The survey also suggested that there are net accident–reduction benefits where roundabouts have been equipped with signals for capacity or queue–control reasons.

42.3 Part–time, Partial and Indirect Signal Control

Part–time control

Part–time control of a gyratory is employed to solve throughput problems at certain times of day, allowing priority control to be re–introduced when traffic flows are light. Such signals can be initiated either by queue–detection loop–arrays or by time–of–day control. In some cases, where the traffic flows are strongly tidal, signal control is employed on different nodes at different times of day.

Part–time control often gives lower off–peak delays, but it provides limited scope for making geometric changes as part of the signal installation, since a physical layout that is still suitable for priority control must be retained. Part–time control can also introduce some confusion, due to the need for associated permanent signing and lining, which can lead to an increase in accidents when the signals are not in use. Full–time control should always be considered as the preferred alternative to part–time control, but particularly when any of the following apply:

❑ significant numbers of cyclists;
❑ a requirement for pedestrian–crossing facilities on the gyratory;
❑ a potential benefit from incorporation in a linked system; and
❑ significant queues arising at various times of the day.

Partial control

Partial control of a gyratory is often employed where delays do not occur on all arms and is sometimes combined with part–time control. It can be a useful technique, to avoid having too many closely–spaced signals around a gyratory. Indeed, installing signals at a single entry is sometimes all that is necessary to solve a particular problem, such as queueing back onto a motorway. In such cases, the signal–controlled element can usually be modelled as an isolated junction. Modelling priority junctions can prove more difficult, particularly when there is a succession of them downstream of the signals. Over time, it can become necessary to instal signals at additional entries. The nearer the gyratory tends towards full signal control, the easier it is to predict its operation by modelling (see Section 42.6).

When partial control is used on small gyratories, the lack of positive control on some entries can lead to uncontrolled queueing on downstream circulating links. However, provided uncontrolled entry–flows are small enough, so that such problems can be contained, the omission of signals can be employed to advantage. For example, it can allow a four–entry gyratory to be co–ordinated essentially as a three–entry gyratory (see Section 42.5).

Indirect control

Indirect control at a gyratory is achieved where traffic is controlled by signals some distance upstream from the entry point and is sometimes associated with the provision of pedestrian facilities at the signals. Since circulating traffic will always have priority, the technique can have applications where there is limited queueing space available on the gyratory.

Indirect control is usually applied to that traffic which will contribute most to the circulating flow at the critical node, ie normally to traffic entering at the node(s) immediately upstream of the critical one. In this way, the controlled entry to the gyratory can have its flow–profile reduced, allowing a greater level of flow to enter from the critical approach.

Indirect control of several arms of a gyratory is feasible, provided the signals are synchronised, so that the gaps created from one arm coincide with entry–flow on another. Signals used in this manner can be an effective means of balancing–out the flows on the approaches (Shawaly et al, 1991).

42.4 Innovative Layouts

Signal–controlled gyratories are often provided with signals at each node (on both the entry and associated circulating links). They also usually make use of a single carriageway for circulating traffic. Other arrangements are possible and may be more appropriate to the particular traffic movements and physical constraints which apply at a given site.

Reducing the number of signals/stop–lines

It can be impractical to introduce separate sets of signals on every circulating link, for example, where a gyratory has inadequate separation/storage between signals. In extreme cases, where this applies, it can be advantageous to make the gyratory more elongated, which can make it easier to combine some of the nodes. This may, for example, reduce to near–zero the vehicle storage on one axis of the gyratory but can provide more usable vehicle storage along the other axis. More complex signal–staging is likely to be required at the combined nodes to control the turning movements. Figure 42.1 shows such an elongated gyratory with tram–tracks through the centre.

Such an arrangement can lend itself to partial control, in which the nodes immediately upstream of the

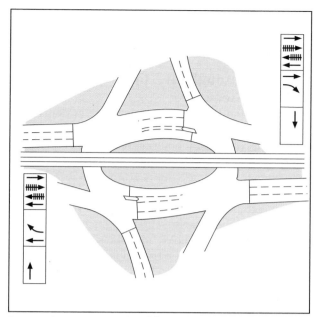

Figure 42.1: An example of four–entry gyratory controlled as two signal–controlled nodes, with tram–tracks through the centre.

elongated links are left with normal priority control of entries. This can reduce some of the lost time inherent in full signal–control of the combined nodes, whilst ensuring maximum storage on the circulating links downstream of the uncontrolled entries. A less common alternative involves the use of give–way control on the circulating links, primarily to control right–turning traffic, but with the retention of signals on the associated entry links (O'Hagan, 1987). The concept of controlling right–turners within the gyratory, with 'priority' rather than 'signal' control, can be scaled down to produce a novel form of signal–controlled junction known as a 'Sig–nabout' (Webb, 1994).

Introducing additional links

By their nature, gyratories force traffic to circulate but sometimes this can create conflicts which can be avoided if a more direct route were provided for particular traffic movements. Cutting straight through the centre of a gyratory with the dominant flow is the extreme case, making it possible to provide higher saturation flows for these movements, since they can usually have a straighter alignment. Such gyratories do exist in the UK but there is general concern that providing direct–links for major traffic movements can lead to confusion as to whether, in the event of loss of signal–control, it is the direct–link or the circulating carriageway which has priority. Simply 'cutting off' a corner of a gyratory with a loop carrying a relatively minor flow can be a safer alternative for reducing traffic conflicts (see Figure 42.2). Ideally, such loops should be long enough to

store the entire traffic platoon which enters each cycle so that the co–ordination of other links on the gyratory need not become compromised. Also, to cater for signal–failure, it is desirable to provide good visibility where such loops intersect the circulating carriageway.

Security of control

With the more complex gyratory designs, such as those involving direct–links or loops, their safety during signal–failure must be considered at an early stage in the design. Manual closure of such links at times of signal–failure may be impractical, so consideration needs to be given to automatic signing or barrier control which can divert the traffic, or indicate the priority, in the event of loss of control. Such equipment will still need to function in the event of mains failure and particularly, where mains power supplies are unreliable, the signals themselves may need to be provided with battery back–up. Standby controllers may need to be installed, which can be brought into service with the minimum of down–time.

42.5 Design Considerations

Queueing restrictions

Compared to conventional signal–controlled junctions, gyratories that have all or most of their entries subject to signal–control usually have a throughput advantage, due to their circulating links. These links can:

❑ allow more than one stream of traffic to flow

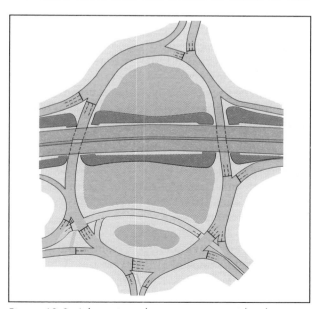

Figure 42.2: A large interchange gyratory, with a loop across one corner. Courtesy: Leeds City Council & Highways Agency.

onto the gyratory at the same time without conflicting; and

❏ provide space in which vehicles can stop and queue until opposing traffic has cleared the conflict area.

In principle, it is better to rely on the former alone without recourse to the latter; but, in practice, it will often be necessary for some turning traffic, including U-turns, to stop and queue on circulating links (see Section 42.10). Unfortunately, this kind of queueing can present problems not found with conventional signal-controlled junctions since, if the queues are large enough, they can present a hazard to circulating traffic and can lead to a reduction in saturation flows on entry approaches.

Furthermore, queueing within small gyratories can present particular problems. As well as the limited storage, there is also the problem that the curvature of small gyratories can result in queued vehicles straddling several lanes rather than filling them uniformly. This can reduce the available queueing length to less than the theoretical value, requiring an increase in green time in order to clear the queue before other traffic streams can safely enter the link.

Cycle-times

Large gyratories, with long link-lengths, have few queueing restrictions and, theoretically, can benefit from long cycle-times (in excess of 90 seconds). However, long cycle-times result in long delays in unsaturated conditions and can be inappropriate for gyratories with clear visibility, where such delays can lead to frustration for drivers.

Many gyratories benefit from having short cycle-times, because of the prevalence of flared approaches. However, small gyratories are often limited to the use of short cycle-times because of their short link-lengths. In particular, those gyratories where traffic queues on circulating links will usually be limited to short cycle-times, because these lead to less red-time and hence smaller queues, which can be more easily contained within the links. Because of such short-link restrictions, there will usually be a particular cycle-time for which the gyratory throughput is optimised.

In practice, an individual signal-controlled gyratory will need to be operated at the cycle-time which best suites its characteristics. This is likely to fall within a range of 'short' cycle-times, typically from 30 to 90 seconds. The upper end of this range may be more applicable when the entry-greens are particularly imbalanced and the lower-end when they are well balanced. Very short cycle-times can sometimes be appropriate to small gyratories but it should be noted that increased frequency of signal-cycling can lead to higher accident-rates at traffic signals.

Gyratory throughput – the effects of size and entry-sequence

In general, the throughput of gyratory systems with all, or most, of the arms signal-controlled is governed both by the individual node characteristics and by the inter-node spacing. The largest gyratories are limited only by the former. As gyratory size reduces, co-ordination between nodes (which is invariably advantageous, in terms of consistency and delay reduction) becomes an essential factor in maximising throughput, since it helps to reduce the interaction between nodes. Co-ordination requires the nodes to be controlled at a common (or occasionally multiple) cycle-time. This dictates that entry-greens must be given in a particular sequence in each cycle.

An important class of gyratories owe their generally good performance to the fact that their entries are sequenced in the opposite direction to the circulation of traffic (i.e. anti-clockwise in the UK). The most notable member of this class, which is an exception to the normal rule relating throughput to size, is the three-entry gyratory. These can yield high throughput (Hallworth, 1980), even when relatively small (less than 100m between nodes), because the efficient two-phase nodes of which they are comprised can be ideally co-ordinated, so that no queueing (other than U-turns) takes place between them. This is illustrated by Figure 42.3, which is a Time-Distance diagram (see also Section 42.6) for such a gyratory operating at its optimum (or natural) cycle-time C_o. When such a system has balanced greens, then $C_o = 2T$, where T is the sum of the three individual cruise-times from each entry signal to the next downstream signal.

Three-entry gyratories are very efficient, in that they can yield considerable total entry-green (G) (where G $= \Sigma gn$, the sum of the individual effective entry-greens) and often more green than there is time available in the cycle (C). The ratio of G/C is a measure of gyratory efficiency – this dimensionless parameter is related to throughput – but conveniently independent of the saturation flow (S) of each entry (efficiency $= (\Sigma gn)/C$, whereas throughput $= (\Sigma gnS)/C$). The gyratory in Figure 42.3 has a G/C ratio of 1.3, which is in marked contrast to conventional three-phase junctions, which are invariably less efficient at dealing with turning movements (excluding left-turns) and which have G/C ratios typically less than 0.8. This is further illustrated by Figure 42.4, which gives the G/C ratio

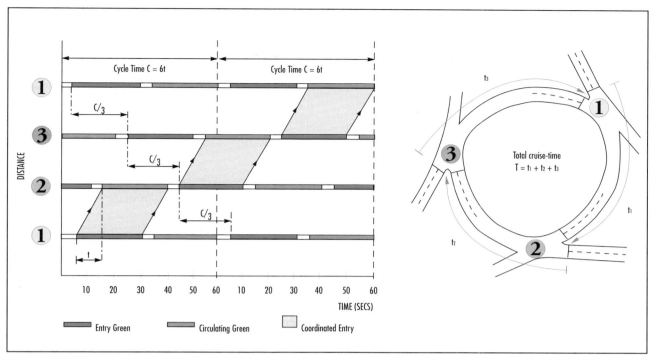

Figure 42.3: Time-Distance diagram for a three-entry gyratory (shown at the optimum cycle time C_0, with the three balanced entry–greens ideally co–ordinated).

for gyratories of different sizes (ie different values of T). When T is equal to 3l (where l is the mean lost time between conflicting phases), the efficiency is constant at 1.0 but, above this value of T, there are distinct 'optimum' efficiencies. All practical values of T shown in Figure 42.4 give greater efficiency (and potentially more throughput) than a conventional three–phase junction (see line for T=0).

The more efficient the three–entry gyratory, the more marked is the reduction in efficiency, when greens become imbalanced, but the efficiency will rarely fall below 1.0, even with large green–imbalance. Such gyratories are insensitive to imbalance in link cruise–time and hence link–length.

Gyratories with more than three entries can also be sequenced in opposition to the traffic circulation (Hallworth, 1992) but it becomes more difficult to achieve full co–ordination between entry and circulating greens. The table in Figure 42.5 demonstrates that, as the number of fully co–ordinated entries increases (from three through four to 'n'), so the efficiency falls (different equations are given for cycle–times (C) above and below the optimum Co). In practice, for more than three entries, full co–ordination is difficult to achieve. Such gyratories are likely to be operated below their optimum cycle–time (since Co=(n–1)T, which will tend to be large for n > 3) and, if so, it will tend to be the circulating links furthest downstream that become

un–co–ordinated. Retaining priority control on selected low–flow entries can reduce the linking constraints of such gyratories, making it easier fully to co–ordinate the remaining signal–controlled entries.

Where opposite arms of a four–entry gyratory have relatively high flows, an alternative sequence, in which these arms enter the gyratory at approximately the same time, may be a better option. However, this will usually require some right–turning vehicles to queue and the implications of this need to be considered carefully, (see Section 42.6). Whilst all gyratory systems need to be co–ordinated and sequenced specifically to the flows, this becomes more important as the number of entries increases. It is difficult to generalise on systems with large numbers of entries, the only option is to examine them in depth. Better still, ways should be explored of reducing the number of entries to a maximum of four or five on any one gyratory.

Saturation flows and lane–usage

Great care should be exercised when applying conventional saturation flow calculations (Kimber *et al*, 1986) to signal–controlled gyratories. This is because, for example, approaches to gyratories often exceed three–lanes, when such calculations are inappropriate. Full account needs to be taken of the site–specific factors which are characteristic of gyratories – such as entry and circulating radii which

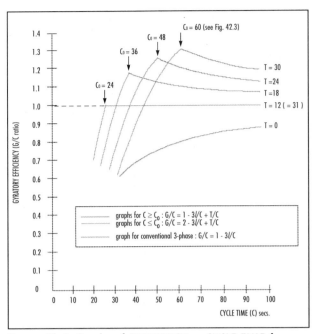

Figure 42.4: Graphs of EFFICIENCY vs CYCLE TIME for three–entry gyratories of different sizes (based on an assumed mean lost time (l) between phases of four seconds).

will reduce saturation flow and any approach flares which may increase it. Also, when considering signal–control at an existing gyratory, vehicle–users' behaviour should be carefully examined. The overall 'saturation flow' of gyratory approaches can often be significantly less than the values appropriate to conventional signals, due mainly to uneven lane–usage, because of the tendency for users to favour the particular entry lane appropriate to their exit.

Where user behaviour is unknown, it is advisable to err on the cautious side when determining saturation flows. The more robust designs will usually make some allowance for unequal use of lanes within their models, principally by making the assumption that drivers travel through the entire system without needing to merge with streams in other lanes.

Circulating links are particularly susceptible to reduced saturation flow through poor lane–usage (Hallworth, 1992), due to the fundamentals of co–ordination and link–length. Well–co–ordinated inflow occurs when vehicles arrive at a green signal, without queuing, and the saturation flow/lane usage cannot then exceed that of the upstream link, however

Number of entries N:		3	4	n
Optimum Cycle time C_o:		2T	3T	(n–1)T
Gyratory Efficiency (ratio of total entry green [G] to the cycle time)	for $C \geq C_o$	$1 - 3l/C + T/C$	$1 - 4l/C + T/C$	$1 - nl/C + T/C$
	for $C \leq C_o$	$2 - 3l/C - T/C$	$2 - 4l/C - 2T/C$	$2 - nl/C - (n-2)T/C$

Figure 42.5: The variation of Gyratory Efficiency (G/C ratio) with number of entries, for the class of signalised gyratories sequenced in opposition to the circulation of traffic.

In the above: T is the sum of the individual link travel time (T = Σt). C is the *actual* cycle time, G is the total entry green (balanced greens being assumed here), and *l* is the mean lost time between conflicting phases (typically four seconds).

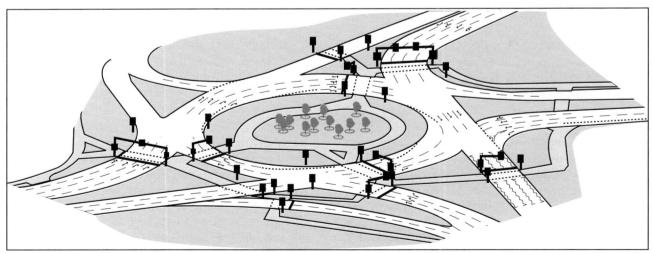

Figure 42.5: Spiral lane markings.

many extra lanes there are. Dis–co–ordinated inflow occurs when vehicles arrive at a red signal and have to queue. This can enable saturation flow/lane usage to be maximised. Long links, where traffic queues for more than one cycle, can facilitate this but short links may not.

One of the ways of accounting for poor lane–usage in modelling is by reducing the overall saturation flow of the approaches. A more realistic alternative is to calculate saturation flows on an individual lane basis, without downgrading for lane–usage, and then to account for lane–usage directly, by assigning particular traffic movements to particular lanes. This requires a realistic prediction of the entry lane which each traffic movement is likely to choose. A full 'origin–and–destination' matrix is required for the gyratory but, even so, the exercise will often prove difficult for multi–lane gyratories with more than four entries. Adoption of this procedure means that the various lanes modelled on an approach will often have significantly different degrees of saturation, as can be the case in reality.

Flared Approaches

Flared approaches can be useful in boosting the saturation flows of entries, particularly at short cycle times. In taking account of flares, it should be noted that they are not always fully utilised, because they can be cut–off by traffic queueing in adjacent lanes. Even where this does not occur, their usage will depend on particular site factors – notably the destination exit that the flared lane indicates to drivers (Lines *et al*, 1990). Version 10 of the TRANSYT program (see Section 42.6) provides a means of investigating the contribution which particular flares can make to throughput (Crabtree *et al*, 1996).

Intergreen periods

The conflict–distances at the signal–controlled nodes of gyratories are often well balanced compared with conventional junctions; also, conflict–angles are usually low and visibility is often good. This can result in intergreen values of 4 seconds being used more commonly than at conventional junctions (Jourdain, 1988). Care should be exercised in using such low values, for example on fast down–gradients at motorway gyratories. One practical solution is to set–back entry stop–lines by a few metres, so that the conflict distances are unequal – this makes it more sensible to adopt a shorter value of intergreen after the circulating green than that after the entry green.

42.6 Design Methodology

The design of signal–controlled gyratory systems can be complex, involving a good deal of iteration. This is because of the difficulty of reconciling the need for high throughput with the achievement of good comprehension by drivers (see Section 42.10) The methodology outlined below involves the use of the TRANSYT program (Vincent *et al*, 1980), as an aid to design, and is suitable for gyratories with most, or all, arms signalised. Whilst more experienced practitioners are able to undertake designs without making use of this tool, it is likely that most will use TRANSYT at some stage in gyratory design. A methodology exists which utilises TRANSYT by making use of weighting factors and queue limits (Lines *et al*, 1990) and this may sometimes be an appropriate alternative to the methodology used here – in any case, many features of the two methods will be complementary. The approach outlined here, and illustrated in a Worked Example designed to be complementary to this Chaprter (Hallworth, 1997), assumes that the designer will wish to spend some time gaining a greater understanding of the factors limiting the design and to influence the signal–timings and impose constraints on the TRANSYT optimisation. More detailed methodologies which use this hybrid approach are also available (Jones, 1992).

Step 1 – Establishing an initial design and TRANSYT model

This involves drawing out an initial layout for testing and collating various data to input into a TRANSYT model. The sequence of activities is as follows:

❏ determine the numbers of signal–controlled nodes and give–way nodes and the approximate positions of each stop–line;

❏ establish the lanes on each approach, ensuring that the number of lanes which directly input flow to a downstream gyratory link are equal to, or exceeded, by the number of lanes on that link;

❏ allocate, provisionally, particular traffic movements to lanes on entry approaches, trying to equalise the flow in each lane. Follow this allocation through to circulating links, ensuring that no merging need take place within the gyratory, or immediately downstream, if necessary re–allocate entry flows in order to achieve this. Until the co–ordination is known, allocate to inflows the same number of lanes on circulatory links as allocated on upstream links;

❏ determine the saturation flow of each lane, or of streams comprising adjacent lanes, if the usage of the lanes concerned is clearly likely to be similar;

❏ determine the intergreens between conflicting entry and circulating links;

❏ set up a link diagram and a TRANSYT file based on the flow, saturation flow and intergreen parameters. Also input the cruise times or distance between stop–lines and the anticipated cruise speeds. The link diagram can be kept relatively simple, although the use of such facilities as dummy links may prove helpful, where there are four or more entries; and

❑ configure each node separately, unless there are some obvious groupings of nodes which suggest inputting these as a combined set of serial stages. Use a single TRANSYT link for each lane or stream which has been allocated its own saturation flow.

Step 2 – Initial Modelling and Design Modifications using TRANSYT.

TRANSYT can be used as the sole method of modelling at this stage. The sequence of activities is, as follows:

❑ decide on a test cycle–time – 60 seconds is a practical starting point because, for typical layouts, it provides a sensible balance between the available vehicle storage and the green–time available in the cycle – and run TRANSYT, allowing it to optimise all nodes; and

❑ examine the TRANSYT output for the degree of saturation on the various links. Any node, with links approaching or exceeding 90% saturation, requires one or more of the following design–modifications:

❑ checking the basic co–ordination indicated by TRANSYT, in terms of green starts and ends. Where co–ordination on a circulating link is poor, it may be possible to reduce the green time on the cirulating links, to the benefit of entry throughput, by distributing the inflow over more lanes;

❑ re–allocating some flows to different entry lanes. But this should only be done if it does not require traffic from different lanes to merge downstream or if downstream lanes can be added to obviate such merging;

❑ allocating left–turn entry movements into dedicated lanes for traffic which will not circulate beyond the next node of the gyratory. This can reduce the flow in other lanes and hence the requirement for green–time. In some cases, it may be appropriate to provide left–filter signals or, in extreme cases, left–turn flows can be allocated free–flow left slips;

❑ adding extra entry lanes. This will, in general, reduce the required green–time but may also require the addition of downstream lanes so as to avoid merging; and

❑ flaring circulating links, relative to upstream links. This can sometimes reduce the green time requirement – on poorly co–ordinated links, the inflow can be distributed over a greater number of lanes.

The TRANSYT model should be updated and re–run repeatedly, by considering the above design modifications either singularly or in combination. After some iteration, it should become apparent whether the proposed gyratory layout is likely to prove adequate within land and budget constraints. Once this appears possible, proceed with fine–tuning of the design.

Step 3 – Fine Tuning the design using TRANSYT and Time–Distance diagrams

At this stage, it is important to obtain a clear picture of how the gyratory co–ordinates. TRANSYT has limitations when modelling short links, because it models vehicles as a 'vertical' queue rather than a queue extending back upstream and, more importantly, it does not provide the designer with a pictorial representation of the co–ordination. It is therefore useful to complement TRANSYT with the use of Time–Distance diagrams (as shown in Figure 42.3). By this means, a designer can visualise the relationship between the timings of the various nodes. The procedure is as follows:

❑ transfer the green start– and end–times from the optimised TRANSYT model (Step 2) onto a Time–Distance diagram. Examine in detail the green times given to each approach and, in particular, how entry links co–ordinate with downstream circulating links;

❑ decide whether the basic co–ordination is acceptable, in terms of drivers' comprehension (see Section 42.10). The co–ordination may be some way from the ideal but, at this stage, it should be possible to form a view as to what the practical gyratory co–ordination should be. Some iteration between Time–Distance diagrams and TRANSYT will be necessary to achieve this;

❑ by inspection of the Time–Distance diagram, derive a simple relationship, involving cycle–time or multiples thereof, which ensures achievement of one or more elements of ideal co–ordination. This will indicate whether a change in cycle–time is likely to improve the overall operation of the gyratory. Running TRANSYT at various cycle–times can assist in identifying the practical range. Usually, the lower cycle–time limit will be set by one parameter (eg lost–time between phases) and the upper limit by another (eg queues on circulating links);

❑ achieve the desired co–ordination by reducing the entry green times (to give entry saturation levels of around 90%). This will provide more green time on circulating links, thus making it easier to achieve good co–ordination within the gyratory. It also has the advantage of ensuring that any increases in flow cause excess queues only on entries. This can be shown by testing with flows scaled up by 20%. On motorway gyratories, this may be unacceptable and a lower entry saturation flow may be necessary, because of the potential for queueing back onto the main carriageway;

❑ whilst striving for acceptable co-ordination, ascertain that the Mean Maximum Queue (MMQ) on each circulating link does not exceed the link length. In practice, because queues can be longer than the MMQ, a value of 60% to 75% of the link-length is suggested (Lines *et al*, 1990). The lower value is appropriate where there are large flow variations in each cycle and the upper value where variations are small. Where such lengths are exceeded, check that this is not as a result of TRANSYT modelling a 'momentary' queue (Jones, 1992), as can occur due to upstream saturation flows exceeding those downstream. The queue graphs within TRANSYT can be helpful in showing this; and

❑ achieve acceptable co-ordination, assisted by the fact that the speed of the front-end vehicles in a platoon travelling towards a red signal, which turns green before the platoon stops, can be significantly lower than the tail-end speed (Jones, 1992 and Davies *et al*, 1980). Allowance can be made for this in the Time-Distance diagrams. If there is a queue already on the approach, co-ordination will need to be estimated in relation to the back of the queue rather than the downstream stop-line.

If, at any stage during fine-tuning of the design, the degrees of saturation of the various links, the co-ordination, or the circulating queues depart from acceptable values, it will be necessary to review the design by reverting to Step 2 above. Parts of the procedure will, in any case, need to be repeated in order to derive the timings appropriate to different times of day.

42.7 Control Strategies

A fundamental decision, when designing the control of a signal-controlled gyratory, is whether to utilise a single set of stages controlling all nodes in a totally serial fashion or whether to allocate a separate set of stages for each node, to allow a greater degree of independent (parallel) control (Jones, 1992).

Totally serial control
This is the most appropriate control for small gyratories, since it has the advantage that the relationship between green times at adjacent nodes can be fixed. With totally serial control, it is possible to fix the duration of certain stages (eg clearance stages), whilst allowing controlling stages to be varied by time-of-day or in response to traffic flow. This form of control can provide the restrictions

necessary for safe operation but it can be inflexible since, once the staging has been fixed, some useful phase combinations may have been precluded.

As the number of nodes being controlled increases, the duration of controlling stages diminishes, ultimately to the point where the intergreen times between stages are required to overlap, effectively giving a negative stage time. Some controllers, built to DOT specification TR0141 (DOT, 1991), can provide this degree of flexibility by, for example, allowing one inter-stage to commence whilst another is still in progress (a 'ripple change').

Parallel stage-streams
Since, theoretically, the number of required stages doubles for each node added, totally serial control can become impractical for gyratories with four or more nodes. A solution is to utilise an individual set of stages, contained in a 'parallel stage-stream', for each node. This offers full flexibility but it can be difficult to provide the restrictions which are often necessary. For isolated control, this can be overcome by making the stage changes, within each stream, dependent on cross-linking conditions generated from stages in a different stream. An alternative, which is particularly useful at signal-commissioning, is to constrain the stage streams using the cable-less linking facility. Generally, however, it is best to limit the overall number of nodes on the gyratory.

UTC control
Control of signal-controlled gyratories using UTC (see Chapter 41) becomes sensible when they are part of a larger signal-controlled network. Most UTC-controlled gyratories are operated using fixed time-of-day plans, since this provides the rigidity of control necessary for safe and consistent operation.

In some cases, the dynamic system SCOOT (Hunt, *et al* 1981) is employed to control gyratories. This has the advantage that timings can be varied to suit different traffic conditions (Knight, 1996). However, care should be exercised when attempting to use SCOOT on any but the largest gyratories, since it is necessary to ensure that sufficient restrictions are placed on the optimisation to avoid unwanted sequences occurring whilst travelling around the gyratory (see Section 42.10).

42.8 Designing for Pedestrians and Cyclists (see also Chapters 22 and 23)

Pedestrian facilities on entry approaches
One of the advantages of a signal-controlled gyratory is that it is often possible to introduce formal pedestrian facilities into the signalling arrangement.

The provision of signals for traffic on the entry approaches of a gyratory provides a natural focal point for pedestrians and it is relatively easy to equip with a pedestrian phase. The route provided for pedestrians across the rest of the gyratory needs to be related to the their desire lines and will affect the signal system that is adopted. Figure 42.5 shows a variety of routing possibilities.

Providing a pedestrian crossing on the exiting side of a gyratory entry will normally require it to be equipped with signals for vehicles and with an associated stop–line. Such crossings can be controlled by a separate stage–stream within the main gyratory controller. Crossings at exits need to be located sufficiently far away from the gyratory to minimise the likelihood of queueing vehicles interfering with circulating traffic. This will depend on traffic flows, overall gyratory sequence and the length of the stop–time for traffic at the pedestrian signals.

Pedestrian facilities on a central island

Routeing pedestrians across the central island of a gyratory can provide a more direct a path. In that case, there may be no need for additional signals, as these may already be present on the circulating approaches; so, in principle, it is a simple matter to introduce a pedestrian phase at that point. However, care must be exercised where the associated entry green time is relatively short, since the time needed for pedestrians to cross can become a limiting factor.

On larger gyratories, speeds are likely to be higher on circulating sections than on exits, since there is usually some physical deflection on the latter. This should be taken into account when designing safe pedestrian routes.

Facilities for Cyclists

One of the advantages of full–time signal–controlled gyratories, as distinct from un–signalled or part–time variants, is that they reduce the dangers for cyclists, (see also Chapter 16). Full–time signalling is now seen as a good technique for making roundabouts safer for cyclists, because the primary conflicts between entering and circulating traffic are resolved by time–sharing. Where roundabouts are equipped with full–time signals on all or some arms, cycle–accidents at entries have fallen by up to 66% (Lines, 1995). Lesser conflicts, caused by the need for cyclists and other traffic to weave across lanes on the gyratory, can also be relieved by the introduction of advanced stop–lines for cyclists on approaches where cycle–flows are highest.

The larger gyratories, where vehicular speeds can be high, are likely to be intimidating to cyclists. So, care

should be taken to design signal co–ordination, so as to encourage reduction of vehicular speed wherever possible. This suggests favouring co–ordination which causes traffic platoons to slow down (see Section 42.10). Signal–controlled gyratories also provide the possibility of routeing cyclists across the traffic flows, along with pedestrians, and Toucan crossings can be used for this.

42.9 Signing and Signalling

Direction signs

One of the more important aids to safety at gyratories is the provision of unambiguous direction signing. With a clear idea of the layout of the gyratory and where their exit is before they enter the system, users can concentrate on moving safely through it. For this, 'map type' advanced direction signs can be helpful. On larger gyratories, it may be necessary to use supplementary gantry signing within the gyratory. But this needs to be kept as simple as possible, to avoid distracting users, since accident problems can arise concerning the visibility of signals where overhead direction signing is provided.

Regulatory/warning signs

On larger gyratories, where there are substantial lengths of carriageway between nodes, both the entering and circulating traffic will have only one direction of travel after passing their signal. In such cases, the use of regulatory 'no right turn' and 'no left turn' signs respectively, might be appropriate, to reinforce the turning prohibitions. These could be supplemented by green arrows at the signals or, since it is an offence to violate a green arrow, replaced by them (HMG, 1994) [NIa]. On smaller gyratories, however, the use of such signs or green arrows can be confusing, since the exiting movement from a circulating link can also be viewed as being a 'left turn'. In such cases, it may be necessary to use standard 'full' green signals and to rely on the use of 'no entry' signs to deter unwanted turns into one–way carriageways. This is a difficult area on which to generalise, the solution adopted should depend on the individual circumstances and always be as simple and unambiguous as possible.

Signal–controlled gyratories require appropriate advance warning signs and, where signals control motorway slip–roads, it may be necessary to provide a sign upstream to warn of any likely queues. In some cases, dynamic signs are provided to direct motorway traffic out of queueing lanes.

Carriageway markings

White lining, consistent around a gyratory, is often

Photograph 42.1: Dot markings to guide drivers into their appropriate lane.

the key to efficient operation. Lane markings which 'spiral' around the gyratory (see Figure 42.5) can be helpful as they tend to reflect natural traffic patterns, providing better information to users and helping to minimise the need to change lane. Their effectiveness depends, in part, on the angle between the entry paths of vehicles and the circulating spiral markings. The application of 'dot' markings to guide drivers may reduce the problems sometimes posed by small angles (McCann, 1996) (see also Photograph 42.1). The use of in–carriageway indicator lights could (in the future) help to overcome such problems which, ideally, require different indications during the cycle.

Lane direction–arrows or abbreviated destination markings on the carriageway can reduce or eliminate the need to merge, which is important for safety, and can encourage a more equal use of lanes which is important for throughput. A problem with such markings is that they are often insufficiently visible to users in heavy traffic and this can limit their

usefulness. To overcome this, they are sometimes reinforced by signs located in the verge (see Photograph 42.2). It is advisable to be flexible and to be prepared to alter the markings in the light of usage. One approach is not to provide any destination lane–markings, until the traffic behaviour has settled down, and then to apply markings only sparingly to resolve particular difficulties or to enhance throughput.

Whilst gyratory signal–timings should ideally be set so that excessive queues will only form on entries, yellow box markings can sometimes be added for good measure. This can be particularly appropriate either where the link–length is little more than the modelled mean maximum queue or where certain entries are uncontrolled. In both cases, there will inevitably be some occasions when the queue exceeds the link–length available.

Locations for signals

When siting signals on a gyratory, the same principles apply as those for conventional signal–controlled junctions but the practicalities can differ. Often, with more than two lanes on approaches to gyratories, it is necessary to consider the use of overhead signals to provide adequate signal–visibility from the central lanes (see Figure 42.5). Also, many of the signals are located at the end of fast slip–roads, such as at motorway junctions. In more urban situations, particularly where existing entry deflection is retained, the need for overhead signals is reduced. On entry approaches, good visibility can often be provided by closely associated secondary signals, sited conventionally on the offside or on the central island of the gyratory. On circulating approaches, good visibility can often be assured by secondary signals mounted on the nearside and, in such cases, it is suggested that offside secondaries are also provided.

42.10 User Comprehension

When equipping gyratories with signals, care must be taken that they do not become too complex for road–users to comprehend, because they frequently involve closely–spaced signals which may not always operate as expected and at which a decision sometimes needs to be made regarding a change of direction. Every effort should, therefore, be made to avoid confusion and to maximise drivers' comprehension of the layout of the gyratory and the way in which it works.

Closely–spaced signals

Signals changing to green at a particular stop–line can

Photograph 42.2: Additional signing located in the verge. Courtesy: Edinburgh City Council.

easily be misinterpreted by drivers waiting in stationary vehicles at a nearby upstream stop–line. If such sequences cannot be avoided, it will be necessary to consider the use of louvres on the downstream green signals, to make these aspects less visible, and also the use of additional secondary signals at the upstream node to help reinforce the controlling red signal. Louvring the downstream amber signals may also be desirable but can present a hazard at the termination of green, without adequate clearance time between the relevant phases.

The proper distance between stop–lines is clearly important in avoiding users' confusion. The distance between stop–lines should normally be at least 50m, particularly where there are more than two lanes on an approach. The achievement of this minimum distance can be aided by setting back entry stop–lines, with intergreens adjusted accordingly.

Ideal gyratory co–ordination

The existence of closely–spaced stop–lines can be misleading to moving, as well as stationary, traffic. Having started from rest at a green signal, users may have an expectation of travelling freely for several seconds and will not necessarily be prepared for signals immediately downstream to turn to red, as they approach them.

The ideal is for all downstream signals to be co–ordinated with upstream entry signals, so that no stopping on the gyratory is required. The ideal co–ordination suggests two extreme conditions relating to the front–end and tail–end of traffic platoons. These are that:

❑ signals turn green before the front–end of each platoon arrives; and
❑ signals remain green until after the tail–end of each platoon clears.

Achieving one or other of these extremes is more feasible than achieving both – which extreme is possible will depend on a number of factors, including the size of gyratory and the cycle–time. Those sequences, consistent only with the latter, can be used to slow traffic down by running the fronts of platoons towards red signals (Jones, 1992). Less desirably, sequences consistent only with the former can encourage higher speeds and can then lead to fast–moving traffic towards the end of the platoon being 'caught' by a red light.

Practical gyratory co–ordination

Ideal co–ordination is not likely to be achieved at many gyratories, since stopping and queueing vehicles will often be necessary for the achievement of maximum throughput. So, practical departures may be needed, such as:

❑ drivers making right–turn movements are likely to have some expectation of stopping; so, given a choice of which movements to stop, minority movements should be chosen;
❑ stopping right–turn movements at the furthest downstream node will help to dissipate the resulting queue and will give drivers more time to adjust to the gyratory. Speeds can, however, increase around the gyratory, so the alternative of stopping them at an earlier node, such as one where the radius is lower, may need to be considered;
❑ stopping right–turn movements at an earlier downstream node, to await the clearance of a major opposing inflow may be unavoidable; and
❑ although it is undesirable to stop vehicles from one approach at the first downstream signal, it is sometimes unavoidable. In which case, it is preferable to choose an entry with low and predominantly local traffic flow, since these drivers are likely to become more familiar with the co–ordination. It is also better to choose an entry with a relatively long distance to the downstream signal. On larger gyratories, entry movements tend to be 'left turns' rather than 'straight across' movements and, as such, vehicles are likely to be travelling slower and drivers more prepared to stop.

In considering the above departures from the ideal co–ordination at the time of design (see Section 42.6), care should be exercised to ensure that the overall co–ordination is likely to be understood by drivers.

Consistency of co–ordination

Even small changes in signal–timings can cause different patterns of co–ordination to be encountered when driving around a gyratory, which may lead to confusion. Care should be taken to ensure that confusing patterns do not occur. If it is necessary to vary the co–ordination significantly to suit different traffic conditions throughout the day, efforts should be made to restrict changes to the peak periods, leaving the off–peak times with patterns that are as consistent as possible.

42.11 References

Brown M (1996)   'The Design of Roundabouts State of the Art Review', DOT.

Crabtree MR, Vincent RA and Harrison S (1996)   'TRANSYT 10 User Guide', TRL Application Guide No 28, TRL.

CSS (1997) 'A review of signal–controlled roundabouts'. Environment Committee, Traffic Management Working Group. County Surveyors' Society.

Davies P, Jamieson B and Reid, DA (1980) 'Traffic Signal control of roundabouts', Traffic Engineering + Control, 21(7).

DOT, (1991) 'Microprocessor–based traffic signal controller for isolated, linked and Urban Traffic Control installations', TR0141 Issue A, DOT.

Hallworth MS (1980) 'High–capacity signal design' Traffic Engineering + Control 21(2)

Hallworth MS (1992) 'Signalling roundabouts Part 1: Circular arguments', Traffic Engineering + Control 33(6).

Hallworth MS (1997) 'Four–entry Signalled Controlled Gyratories – A Worked Example', Traffic Engineering + Control 38(6).

HMG (1994) 'The Traffic Signs Regulations and General Directions 1994', Statutory Instrument No. 1519, Stationary Office.

Hunt PB, Robertson DI, and Bretherton RD (1981) 'SCOOT – a traffic– responsive method of co–ordinating signals', Report LR 1014, TRL.

Winton, RI (1981) Jones SE (1992) 'Signalling roundabouts Part 2: Controlling the revolution', Traffic Engineering + Control, 33(11).

Jourdain S (1988) 'Intergreens at signal–controlled roundabouts', Traffic Engineering + Control, 29(9).

Kimber RM McDonald M and Hounsell NB (1986) 'The prediction of saturation flows for road junctions controlled by signals', Research Report 67. TRL.

Knight P (1996) 'M40 Longbridge interchange junction improvements'

Highways & Transportation, 43(7/8).

Lines CJ and Crabtree MR (1990) 'The use of TRANSYT at signal–controlled roundabouts'. Research Report 274. TRL.

Lines CJ (1995) 'Cycle accidents at signal–controlled roundabouts' Traffic Engineering + Control, 36(2).

McCann V (1996) 'Spiral lane markings at roundabouts – a different angle', Traffic Engineering + Control, 37(7/8).

O'Hagan D (1987) 'The optimisation of traffic management' Highways & Transportation, 34(8/9).

Shawaly EAA Li CWW and Ashworth R (1991) 'Effects of entry signals on the capacity of roundabout entries: A case–study of Moore Street roundabout in Sheffield', Traffic Engineering + Control, 32(6).

Vincent RA Mitchell AI and Robertson DI (1980) 'User Guide to TRANSYT Version 8'. TRL Report LR888, TRL.

Webb PJ (1994) 'Sig–nabout – the development of a novel junction design', Proceedings of the Seventh international conference on Road Traffic Monitoring and Control, IEE conference publication 391 (April).

42.12 Further Information

Barnes AJ (1984) 'Flexible control – Granville Square, Sheffield, as an example' IEE Colloquium on UK Developments in Road Traffic Signalling, Digest No: 1984/5.

Belcher M (1984) 'Roundabout control is signal success', The Surveyor (February).

Bull P and Dunne GM (1983) 'Traffic signal control of Park Square roundabout, Sheffield', PTRC summer annual meeting, University of Sussex.

Chang SH (1994)   'Overcoming unbalanced flow problems at a roundabout by use of part–time metering of signals'. Civil Engineering working paper. 94/T2. (January).

CSS (1993)   'Accidents at signal–controlled roundabouts'. Environment Committee, Accident Reduction Working Group ENV/1–93 (February), County Surveyors' Society.

Crabtree MR (1992)   'The use of traffic signals to improve the performance of roundabouts'. Giratoires 92. Actes du seminaire international, Nantes, France. (October). pp 229–238.

Flanagan TB and Salter RJ (1983)   'Signal–controlled roundabouts' PTRC summer annual meeting, University of Sussex.

Davies P 1980   'Capacity of three–arm roundabouts under signal control', 12th Annual Universities Transport Studies Group Conference, University of Newcastle–upon–Tyne. (January).

Hallworth MS (1987)   'Compound Signalled Junctions – are they really simple'?, Traffic Engineering + Control, 28(11).

Huddart K (1983)   'Signalling of Hyde Park Corner, Elephant and Castle and other roundabouts', PTRC summer annual meeting, University of Sussex.

Lines CJ and Crabtree MR (1988)   'The use of TRANSYT at signal–controlled roundabouts', Traffic Engineering + Control 29(6).

Wright PT and Semmens MC (1984)   'An assessment of the Denham roundabout conversion', Traffic Engineering + Control, 23(9).

Zielinski P (1993)   'Beating Congestion', Highways. 63(3).

Chapter 43 Grade–Separated Junctions

43.1 General Principles

Congestion arising from the conflict between traffic movements at the junction of two or more major roads can be reduced by providing for traffic on different levels. This 'grade separation' should allow the heaviest traffic stream to pass unhindered through the junction, whilst the turning movements take place above or below. This can reduce the area of the junction, increase safety and reduce delays and pollution.

The main carriageway(s) through the junction should maintain the same design standards as the roads they connect. Connector roads should be designed for slower, turning, traffic entering or leaving the main stream. These movements will usually be handled at one or more at–grade junctions (see Chapters 37 to 40). Where there are physical constraints, lower design standards can be adopted through a junction than is the case on the approach road but drivers need to be warned by appropriate speed–limits or signs. Skidding resistance may need to be provided and maintained to a higher level, particularly on areas of carriageway where braking is most frequent.

In free–flow 'braided' interchanges, connector roads provide direct connections for each turning movement separately. Three or even four different levels may be required. Major interchanges of this kind are unlikely to feature in urban areas but further details can be found in TD39/94 (DOT, 1994a) [Sa].

Nevertheless, because of its inherent traffic efficiency and safety, grade–separation should be considered whenever it can be justified economically and local environmental constraints can be satisfied.

43.2 Alternative layouts

There are three standards of grade–separation. These are:

❏ full grade–separation, usually on purpose–built roads with no at–grade junctions (DOT, 1992a and 1992b) [Sa];

❏ limited grade–separation, on a road with, otherwise, at–grade junctions (DOT, 1992b and 1995) [Sa]; and

❏ compact grade–separation, where side–road traffic volumes are less than 10% of the mainline flows and where side–road design speeds can be 30 km/h (DOT, 1994b) [Sa]. Compact grade–separation can also be used on single carriageways with single lane dualling.

The two extremes are shown in Figure 43.1. However, junctions to the different standards should not be mixed along a route, unless the character of the route changes in a way which is made obvious to drivers.

Limited grade–separation can involve the use of left in/left out junctions (see Figure 43.2), which may incorporate merging and diverging tapers, as for priority junctions (see Chapter 38). The junctions operate at lower speeds and the direction signing has

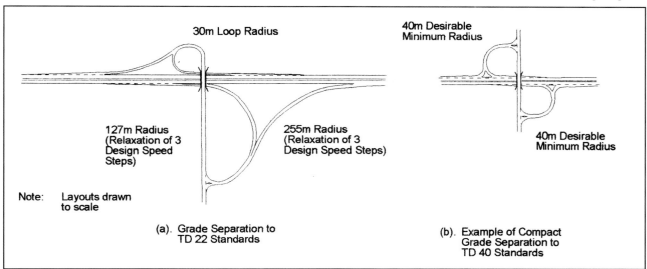

Figure 43.1: Examples illustrating the range of Grade–Separation.   Source: DOT (1994b).

to indicate a turn (sideways T sign) rather than a diverge (Y sign) (DOT, 1995) [Sa].

Compact grade–separation uses similar left in/left out junctions but with even lower speeds and more restricted radii (see Figure 43.3 and DOT, 1994b) [Sa]. The layout of grade–separation for any particular junction will depend on the amount and shape of land available and the expected future overall traffic volumes and turning flows.

The various layouts for full grade–separation for increasing levels of traffic are shown diagrammatically in Figure 43.4 and can be summarised as:
- ❑ a diamond and half cloverleaf, with either priority or signal–control of minor road flows;
- ❑ a 'dumbbell' roundabout, with a single bridge over the main carriageway;
- ❑ a two–bridge roundabout;
- ❑ a three–level roundabout; and
- ❑ a free–flow 'braided' interchange.

Substandard underpasses with only 2.5m headroom, which are used in France, could be considered but safety requirements would show little saving over full provision of headroom. They would clearly only be appropriate where traffic is limited to cars and light vans. Highway authorities should consider the legal implications of this approach before committing themselves to a design with restricted headroom.

Diamond

A diamond layout is the simplest form of grade–separation, requiring less space and only one bridge. The major road may pass over, but preferably under, the minor road. Turning–movements take place where the slip–roads meet the minor road. These at–grade junctions may be priority or

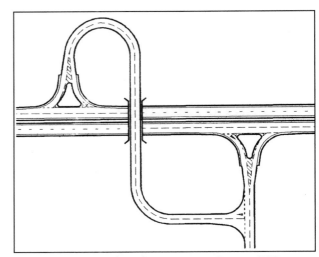

Figure 43.2: Limited grade separation. Source: DOT (1995).

signal–controlled. If demand for certain turning–movements is low, one or more slip roads may be omitted provided an alternative route is available. The junction can also be split to connect with two consecutive minor road crossings.

Half Cloverleaf

A half cloverleaf layout might be convenient when sufficient land is not available in all quadrants. The junction is not as compact as a diamond junction and the curves which form two of the slip–roads may have a lower capacity and less driver comfort where full relaxation of radii is used (see Chapter 36 and DOT, 1993) [Sa].

Dumb–bell Roundabout

The dumb–bell layout is a diamond junction with roundabouts where the slip–roads meet the minor road. It shares the same advantage as a diamond junction, in that only one bridge is required over the main carriageways.

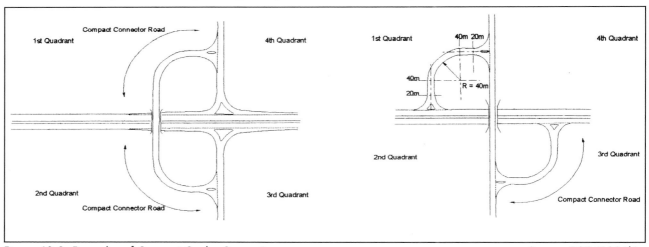

Figure 43.3: Examples of Compact Grade–Separation.

Source: DOT (1994b).

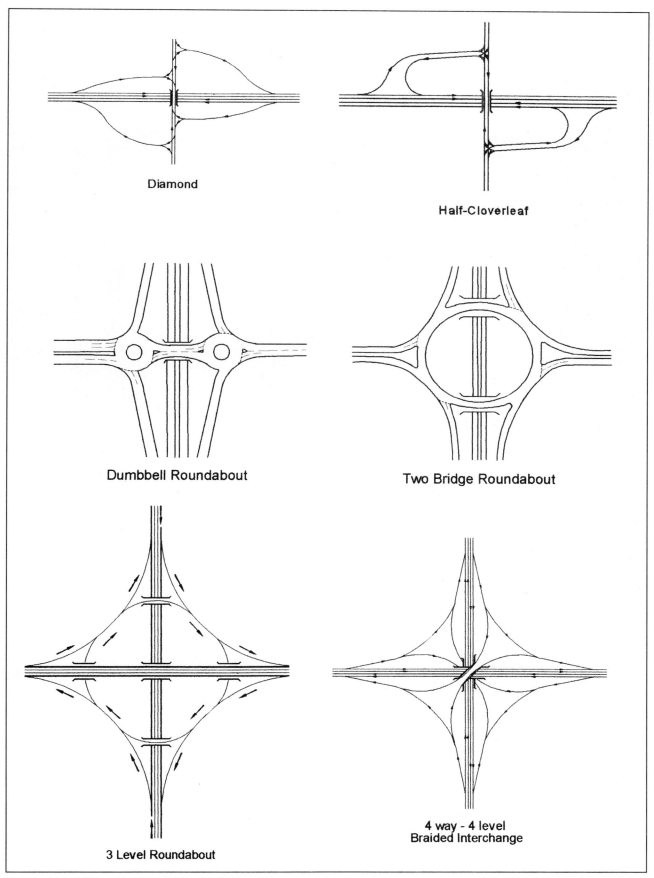

Diamond

Half-Cloverleaf

Dumbbell Roundabout

Two Bridge Roundabout

3 Level Roundabout

4 way - 4 level
Braided Interchange

Figure 43.4: Different layouts for Grade–Separated Junctions.   Source DOT (1992b).

Photograph 43.1: A typical urban grade–separated junction.

Two–bridge Roundabout

This requires a considerable amount of land and the construction of two bridges. If the roundabout is large, long detours can occur which may result in increased vehicle operating costs. Also, vehicles speeds can build up on the longer sides of the roundabout and make it difficult for traffic to enter the circulating carriageway. Two–bridge roundabouts are essential where more than two roads intersect. Some operational difficulties can be overcome if they are designed as signalised gyratories from the start and considerable capacity for turning traffic can be achieved (see Chapter 42).

Three–level Roundabout

A three–level junction might be considered where two major roads of equal importance intersect. The roundabout would normally be at ground level with one major road above and the other below. Through traffic on both major roads would be uninterrupted, while turning–movements would use slip–roads and the roundabout.

Free–Flow or Fully–braided Interchange

This configuration provides for uninterrupted movement for vehicles transferring from one main road to another, by the use of interchange links with a succession of diverging and merging manoeuvres. They are usually only considered for junctions between major roads, such as motorways, outside built–up areas.

Limited Grade–separation

Layouts involving partial or limited grade–separation are more likely to be 'customised' for the location. For example, local differences in ground–levels sometimes enable to be achieved on a compact layout. A typical example is shown in Figure 43.2.

Compact Grade–separation

Examples of more compact grade–separation are shown in Figure 43.3. Photograph 43.1 shows a typical urban grade–separated junction where the slip–roads are kept close to the retaining walls of the underpass to reduce land–take. The main road passes under, to reduce environmental impact, and a separate cycle track is provided.

43.3 Justification of a Grade–Separated Design

Grade–separation usually involves a high capital cost, although this depends on topography. Disbenefits arise from the greater costs of construction and the environmental consequences of land–take and visual impact, which can be substantial in built–up areas.

Nevertheless, by removing traffic conflicts, grade–separation can reduce delay and pollution and improve road safety substantially.

Full standard grade–separation could be justifiable where:

❑ expected traffic flows are too great for at–grade alternatives to operate satisfactorily (ie substantially greater than 30000 vehicles per day average annual daily traffic (AADT); or

❑ a significant proportion of the major road traffic is non–local 'through' traffic; or

❑ all other junctions on the main route are grade–separated; or

❑ the site conditions and constraints preclude an at–grade junction of sufficient size or capacity.

Limited grade–separation can be justified down to 20,000 vechs/day AADT (DOT, 1992b) [Sa]. Compact grade–separation may be justified down to 12500 vehicles AADT (DOT, 1994b) [Sa]. Where the level of future traffic flow is uncertain, there may be a case for staged construction with provision made for later upgrading. Alternatively, it may be possible to induce some of the traffic flow away from the junction by improving suitable alternative routes and so avoid the need for grade–separation altogether.

43.4 Designing for All Users

In many urban areas, the existing road network and the land–use pattern are fairly fixed, with only limited opportunities for alteration or expansion. The design should aim for economic optimisation whereby the benefits of reduced delays, pollution and accidents are weighed against the capital cost.

The design of a grade–separated junction is undertaken in three stages. These are:

❑ planning various options for the junction;
❑ detail design of the junction elements; and
❑ safety audit of the chosen design.

Planning the junction follows on from assessing the nature of the intersection problem, by identifying alternative layouts of the junction to accommodate the demand for vehicular traffic, checking lane–balance and examining routes for pedestrians and cyclists. Wherever possible, pedestrians and cyclists should have segregated routes, which are visible from the vehicular routes for reasons of personal security. Pedestrians and cyclists should not be expected to undertake long detours or to climb severe gradients. However, where this is not feasible, schemes should be developed to guide and assist pedestrians and cyclists, in areas where they are

likely to be most vulnerable, such as in the merge and diverge areas (see Chapters 22 and 23; DOT, 1986 [Sb]; DOT, 1988; SODD, 1989; SODD, 1990; and IHT, 1996a).

The requirements for bus–stopping places, and their relationship with both the pedestrian routes and the vehicular paths and activities, need to be considered. Bus stops may be located in laybys on the slip–roads, preferably on the diverges where vehicles are already slowing down. Adequate stopping sight distance to the laybys needs to be provided, related to the speed of approaching traffic. Curved slip–roads may be unsuitable for the location of bus stops, due to the need for visibility for oncoming traffic and also because bus drivers may have difficulty in seeing other traffic to the rear and thus be reluctant to enter the layby (see Chapter 24).

Direction–signing should also be considered, allowing both for the time taken by drivers to absorb the information and the distance covered at the design speed. It is important to separate the driver's task of route–selection, or confirmation, from that of taking action in traffic. This aids drivers' comfort, therefore safety, and could reduce delays (see Chapter 15).

The compactness of a junction also depends on the requirements for stopping sight–distance. It is assumed that drivers travel at the design speed and, when they see road signs, take appropriate action. Visibility is almost always considered as forward visibility. In the case of drivers approaching a junction, it is normally assumed that they will cross into the exit lane at the design speed of the main carriageway. Where the exit slip–road is short, and road signs indicate the presence of traffic signals or a 'give–way', it can be assumed that drivers will start slowing down when they see the signs. Where this inhibits through traffic, it may be necessary to extend the diverge, so that leaving traffic can move across to the near–side earlier (DOT, 1992a) [Sa].

The design of the junction also has to meet the requirements for maintenance and incident management (see Chapter 37, 6).

Ideally, the heaviest traffic flow should be at the lowest physical level and thus should pass under the minor flow so that:

❑ with the heaviest flow between cutting–slopes or walls, the visual impact and the effect of noise on the local community are minimised;

❑ high speed traffic, leaving the main carriageways, is slowed naturally by the up–grade of the exit slip–roads;

❑ the acceleration of traffic joining the major road

is assisted by the down grade of the entry slip–roads; and

❑ the bridge structures will normally be less costly, as they will be shorter and narrower.

The main disadvantages of constructing roads below ground level include drainage problems and disruption to services. However, such construction may still be justified for the operational advantages listed above.

43.5 Design Standards

Detail Design

For detail design, the elements of the junction can be divided into four categories. These are:

❑ the at–grade junctions with the minor road;

❑ the connector roads, ie the slip–roads and interchange links;

❑ the speed–change areas linking with the major through route, ie merges and diverges; and

❑ sections of carriageway where weaving occurs, ie where one traffic stream has to cross another at a narrow angle.

Advice on the design of at–grade junctions is given in Chapters 37–42. The design should ensure that queueing vehicles do not tail back onto the major route. In the case of dumbbell junctions, a queue from one roundabout may extend back along the minor road and interfere with the other roundabout, thus causing queueing back onto the major road. Computer simulation programs are available which can estimate the probability of certain queue–lengths occurring under different flow conditions.

The design of the alignment of connector roads should follow the recommendations in Chapter 36 (see DOT, 1992a) [Sa]. The design speed is 60 km/h for slip–roads and 70 km/h for interchange links. Where slip–roads exceed 0.75 km in length, they should be designed as interchange links. The choice of design standard may be constrained, requiring speed–limits to be applied. Relaxations in design standards are possible (see Chapter 36 and DOT, 1993) [Sa]. Photograph 43.2 shows a grade–separated junction with relaxation of design standards. This may reduce drivers' comfort but, in constrained situations, is preferable to not providing the facility at all.

'Clover–leaf' loops are a special form of curve, turning through approximately 270°, whose form should be clear to the driver. Where these are part of a grade–separated junction, it may be necessary to introduce advisory speed–limits to warn drivers of the safe negotiating speed. The minimum internal loop radius within an urban motorway junction should be 75m but, for all–purpose roads, this

Photograph 43.2: A typical grade–separated junction with relaxation of Design Standards.

minimum radius may be reduced to 50m for vehicles leaving major roads and 30m for vehicles joining (DOT, 1992a) [Sa].

TD 22/92 (DOT, 1992a) [Sa] favours a maximum gradient of six percent on motorway interchange links and slip–roads, but this can be increased to eight per cent where low turning–traffic volumes and speeds are anticipated. At ten percent, in icy conditions, some stationary vehicles can start to slide.

Design of Merges and Diverges in Full Grade–Separation.

In merges, traffic should be guided to join the through carriageway in an orderly way. In diverges, traffic should be able to leave quickly and unhindered. In the estimation of peak turning flows, sensitivity testing is essential, in case proportions vary by direction. Permitting development to take place up to the highway boundary may prevent later adjustment to the design of the junction.

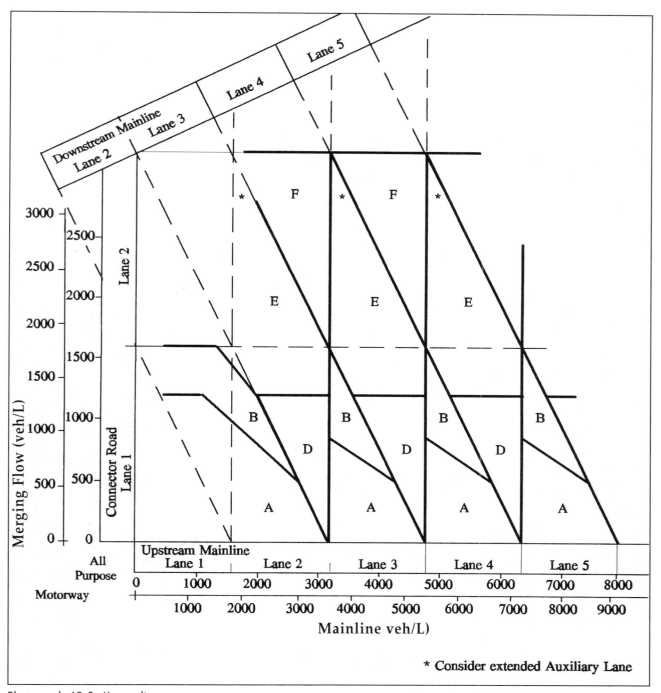

Photograph 43.5: Merge diagram.

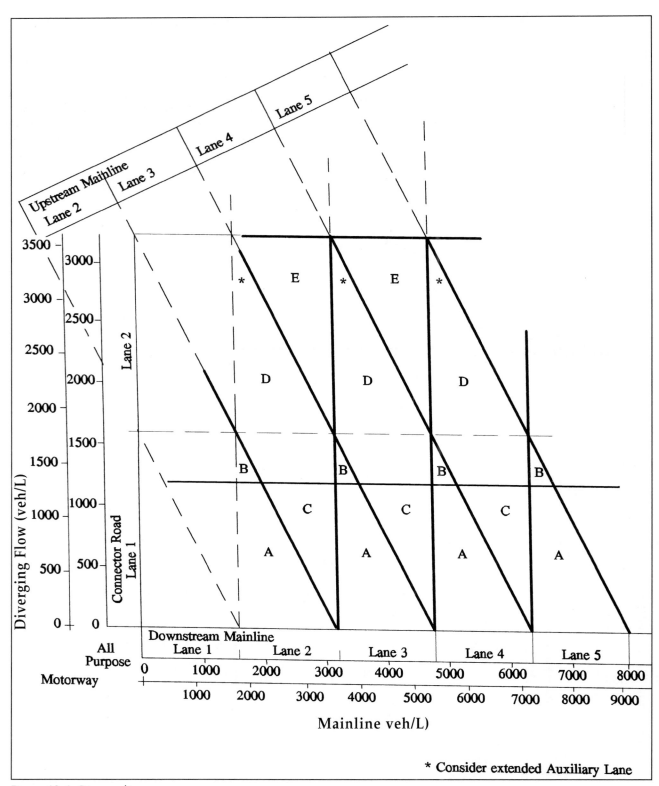

Figure 43.6: Diverge diagram.

The detailed geometric layout for merges and diverges depends on the peak hour traffic flow. The procedure, when designing merges and diverges, should be as follows:

❑ to select, as the design traffic flow, the 30th

highest combination of predicted hourly flows expected in the 15th year of operation or, where entering traffic is constrained by the surrounding network, the most likely peak hourly flow;

❑ to adjust the predicted hourly flows to allow for gradients and large goods vehicles (DOT, 1992a)

TRANSPORT IN THE URBAN ENVIRONMENT

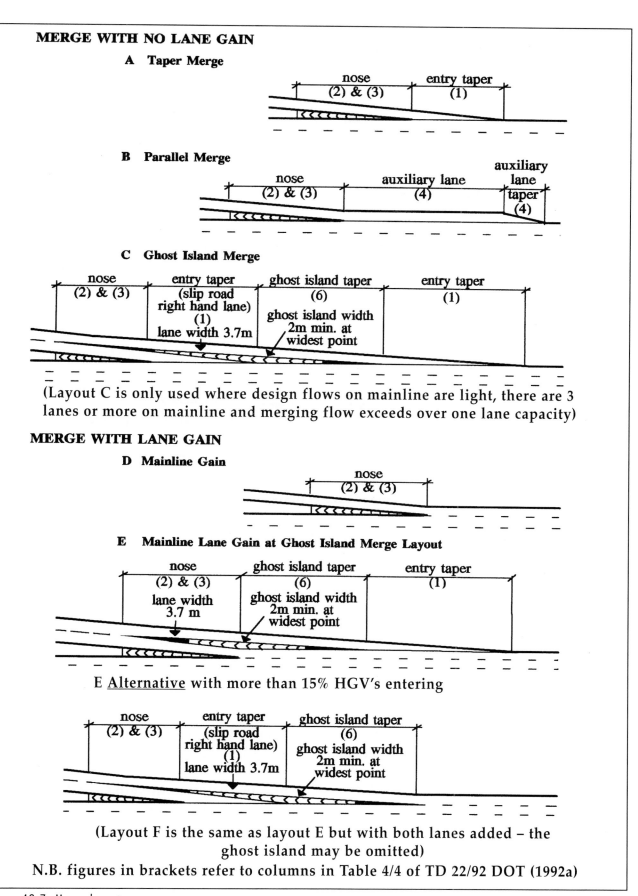

MERGE WITH NO LANE GAIN

A   Taper Merge

B   Parallel Merge

C   Ghost Island Merge

(Layout C is only used where design flows on mainline are light, there are 3 lanes or more on mainline and merging flow exceeds over one lane capacity)

MERGE WITH LANE GAIN

D   Mainline Gain

E   Mainline Lane Gain at Ghost Island Merge Layout

E <u>Alternative</u> with more than 15% HGV's entering

(Layout F is the same as layout E but with both lanes added – the ghost island may be omitted)
N.B. figures in brackets refer to columns in Table 4/4 of TD 22/92 DOT (1992a)

Figure 43.7: Merge layouts.

DIVERGE WITH NO LANE DROP

A   Taper Diverge

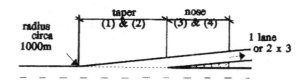

B   Parallel Diverge

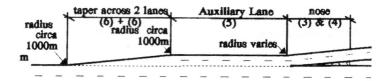

DIVERGE WITH LANE DROP

C   Mainline Lane Drop at Taper Diverge

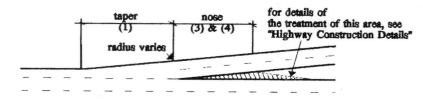

D   Mainline Lane Drop at Parallel Diverge

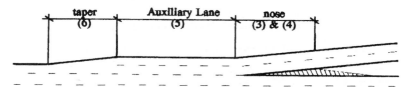

(Layout E is layout D with both lanes off)

N.B Figures in brackets refer to columns in Table 4/5 of TD 22/92

Figure 43.8: Diverge layouts.   Source DOT (1992a)

[Sa]. The average main road gradient should be measured over 1 km (0.5 km from either side of the nose tip). The merge connector gradient should be based on the 0.5km distance upstream of the nose tip, and the diverge connector gradient on the 0.5 km distance downstream of the nose tip; and

❑ to use the adjusted predicted hourly flows to select the merge type from Figure 43.5 or the diverge type from Figure 43.6. Merge types are illustrated in Figure 43.7 and diverge types in Figure 43.8.

The geometric parameters applicable to merges and diverges are specified in TD22/92, (DOT, 1992a) [Sa].

Spacing between Merging and Diverging Sections in Full Grade–separation

The minimum spacing, within a junction, between the noses of successive merges or diverges, or a diverge followed by a merge, should be 3.75 V metres (where V is the design speed in km/h of the through carriageway), subject to the minimum distance necessary for effective signing, which may be greater.

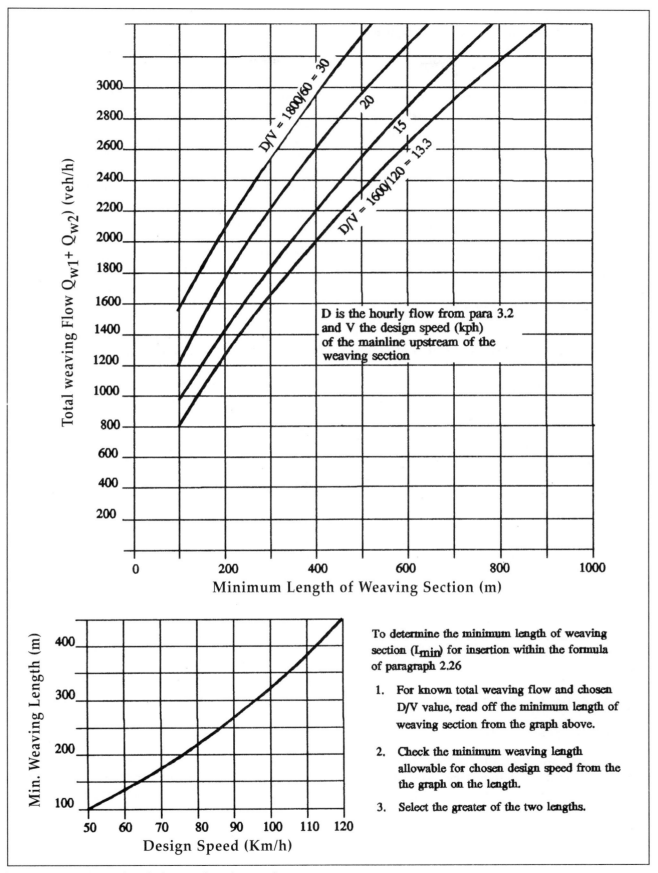

Figure 43.9: Weaving length diagram for urban roads.

Design of Merge and Diverge Tapers in Limited and Compact Grade–separation.

These tapers are the same as for the left–in/left–out junctions of priority junctions (see Chapter 38).

Design of Connector Roads in Compact Grade–separation.

The design methods are the same as for other connector roads (see Chapter 37). As the permitted design speed (at 30 km/h) is below the levels used in that chapter, the geometric parameters given in TD 40/94 (DOT, 1994b) [Sa] should be used. Superelevation should be limited to a maximum of five percent. Gradients should be eight percent or maximum ten percent as a relaxation, only where there is little likelihood of icy conditions. In fixing the gradient, cyclists should be considered. The width of the compact connector road should be from 6.6m to 7.9m, including a 0.6m central hatch–mark before curve–widening (DOT, 1994b) [Sa].

Design of Weaving Sections (DOT, 1992a and DOT 1992b).

The procedure should be as follows:
- ❑ to select the design traffic flow, as for merges and diverges;
- ❑ to adjust the total weaving flow and the associated non–weaving flows, to allow for gradients and large goods vehicles. Appropriate correction factors can be found in TD 22/92 (DOT, 1992a) [Sa]. The average main road gradient is measured over 1 m, being 0.5 km from either side of the merge nose;.
- ❑ to determine the minimum length of weaving section (Lmin), from Figure 43.9;
- ❑ to compare Lmin with the absolute minimum length related to the design speed, in Figure 43.9, with the greater of the two lengths being taken as Lmin, subject to the needs for signing; and then
- ❑ to estimate the number of traffic lanes required, using the relationship recommended in TD22/92 (DOT, 1992a) and TA 48/92 (DOT, 1992b) [Sa].

43.6 Safety Audit (including Drivability/Usability)

Safety audits are normally carried out at the preliminary and detail design stages, as well as after construction and prior to opening (IHT, 1996b). As part of these, it is possible to assess the driveability of a grade–separated junction. This process can also aid in the selection between alternatives, particularly where relaxations or departures have been allowed. Basically, each movement through the junction is assessed in terms of the predicted hourly traffic flow, the decision points passed, the merges and diverges encountered and the design speed steps, by way of relaxations and departures, from the desirable standard. Drivers should not be surprised by unusual or inconsistent features (see Chapter 16). Examples are given in TD39/94 (DOT, 1994b) [Sa].

The same process can be used to assess the useability for pedestrians and cyclists, this time in terms of steps, or up gradients, and conflicts with vehicular traffic for each of the movements and the numbers of users involved, or anticipated, per hour.

43.7 References

DOT (1986)   Advisory Leaflet – Local Transport Note 1/86 'Cyclists at Road Crossings and Junctions', DOT [Sa].

DOT (1988)   Advisory Leaflet 1/88 – Traffic Advisory Unit (TAU) 1/88 'Provision for Cyclists at Grade–separated Junctions', DOT.

DOT (1992a)   TD22/92 (DMRB 6.2.1) 'Layout of Grade–separated Junctions', Stationery Office [Sa].

DOT (1992b)   TA 48/92 (DMRB 6.2.2) 'Layout of Grade–separated Junctions', Stationery Office [Sa].

DOT (1993)   TD9/93 (DMRB 6.1.1) 'Highway Link Design', Stationery Office [Sa].

DOT (1994a)   TD39/94 (DMRB 6.2.4) 'The Design of Major Interchanges', Stationery Office [Sa].

DOT (1994b)   TD40/94 (DMRB 6.2.5) 'The Layout of Compact Grade–separated Junctions', Stationery Office [Sa].

DOT (1995)   TD42/95 (DMRB 6.2.6) 'Geometric Design of Major/Minor Priority Junctions', Stationery Office [Sa].

IHT (1996a)   'Cycle Friendly Infrastructure – Guidelines for Planning and Design', The Institution of Highways & Transportation.

IHT (1996b)   'The Safety Audit of Highways'
                     The Institution of Highways &
                     Transportation.

SODD (1989)   Cycling Advice Note 1/89 –
                     'Providing for the Cyclist',
                     Scottish Office Development
                     Department.

SODD (1990)   Cycling Advice Note 1/90
                     'Making Way for Cyclists',
                     Scottish Office Development
                     Department.

43.8 Further Information

TRL (1996)   CR322 'Operation and Capacity
                     of Weaving Sections', TRL.

Application in Northern Ireland, Scotland and Wales

44 Application in Northern Ireland

44.1 Introduction

The general principles described in this manual concerning the consideration of problems and the development and design of alternative highway and traffic management solutions are appropriate to practice in Northern ireland, but there are a number of differences in the legislative framework.

The Department of the Environment for Northern Ireland has provided a detailed annotation of the text indicating where particular legislation, responsibilities or procedures differ or do not apply in Northern Ireland.

These annotations are shown for each chapter as [NIa] [NIb] etc and an outline of what is appropriate for Northern Ireland is given in 44.2 below. More detailed information can be obtained through direct contact with:

> Roads Service
> Clarence Court
> 10–18 Adelaide Street
> Belfast BT2 8GB
> Tel: 01232 540540
> Fax: 01232 540024
> E-mail: "roads.service.dir.doe@nics.gov.uk"

The most common annotations are:
- ❑ 1a "Transportation in Northern Ireland: The Way Forward",
- ❑ 1c Not applicable to Northern Ireland,
- ❑ 3a Planning Policy Statements,
- ❑ 4c The Road Service being the sole roads authority in Northern Ireland

Most of the other references relate to individual items in the main text.

The following is a list of Acts, Regulations and Orders with important considerations for roads and traffic in Northern Ireland:

Land Clauses Consolidation Act 1845
>Transport Act (NI) 1967
>Planning (NI) Order 1972
>Land Acquisition and Compensation (NI) Order 1973
>Traffic Signs Regulations (NI) Order 1979 as amended
>Private Streets (NI) Order 1980 as amended
>Planning Blight (Compensation) (NI) Order 1981
>Land Compensation (NI) Order 1982
>Access to the Countryside (NI) Order 1983
>Planning (Use Classes) Order (NI) 1989
>Planning (Assessment of Environmental Effects) Regulations (NI) 1989
>Planning (Development Plan) Regulations (NI) 1991
>Planning (NI) Order 1991
>Roads (NI) Order 1993 as amended
>Planning (General Development) Order (NI) 1993
>Planning (Fees) Regulations (NI) 1995 as amended
>Road Traffic (NI) Order 1995
>Street Works (NI) Order 1995
>Noise Insulation Regulations (NI) 1995
>Road Traffic Regulation (NI) Order 1997

Annotation	Comment for Northern Ireland	Chapter Reference

Chapter 1 Transport in the Urban Environment

1a	The publication on the 3 October 1995 by the Department of the Environment for Northern Ireland entitled 'Transportation in Northern Ireland: The Way Forward' sets the framework for the development of Transportation Policy in Northern Ireland	1.2,1.4
1b	In Northern Ireland, Roads Service, an agency within the Department of Environment for Northern Ireland, is the sole authority in dealing with road matters There is therefore no equivalent to the Transport Policy and Programme (TTP) submissions from local highway authorities	1.4
1c	Not application in Northern Ireland	1.5

Chapter 2 Travel Patterns in Urban Areas

No Northern Ireland references

Chapter 3 Transport Policy

3a	The Northern Ireland equivalent of Planning Policy Guidance notes (PPGs) are Planning Policy Statements (PPS). The key Planning Policy Statements in Northern Ireland are: PPS 3 Development Control: Road Considerations PPS 5 Retailing and Town Centres	3.1,3.3
3b	See 1a	3.1,3.3,3.6
3c	See 1b	3.3
3d	There is no corresponding Northern Ireland legislation but the Northern Ireland position has been fully represented in the preparation of the National Strategy	3.3
3e	Responsibility for both Transport and Land Use Policy lie with the Department of the Environment for Northern Ireland	3.3
3f	Structures and Unitary Development Plans are not used in Northern Ireland. Development Plans are prepared by the Department of the Environment for Northern Ireland in accordance with The Planning (NI) Order 1991 and the Planning (Development Plan) Regulations (NI) 1991	

Chapter 4 Roles, Responsibilities and Powers

4a	The Department of the Environment for Northern Ireland is both the Roads and Planning Authority	4.2
4b	As a result of the structure of local government in Northern Ireland since the local government reforms in 1973, the Department of the Environment for Northern Ireland has a wide range of policy, executive and regulatory responsibilities The Department has responsibility for delivering some functions which fall to local government in Great Britain, as well as central government functions	4.3
	In the mid–1990s the Department, as a result of the Government Reform Initiatives which aimed to achieve greater value for money and increased quality of service, created eleven Agencies to deal with its various functions	

	Roads Service on behalf of the Department by virtue of Article 128 of the Roads (NI) Order 1993 is required where it appears necessary to consult with district councils. The Department also as a matter of policy consults with the local authorities on relevant issues within their respective areas	
4c	In Northern Ireland, Roads Service, an agency within the Department of Environment for Northern Ireland, is the sole authority in dealing with road matters	4.3
	While procedures generally follow the basic principles described in the text, the operation of these procedures is carried out under different legislation. The main items of legislation are given in 44.1 above	
	Further advice on roads can be obtained from:	
	Roads Service Clarence Court 10-18 Adelaide Street Belfast BT2 8GB	
4d	Although the Northern Ireland system is broadly similar to that of Great Britain, Northern Ireland has built up its own independent system over the years with differences suited to local requirements	4.4
	In the publicly owned sector operating responsibility for the individual services lies with Northern Ireland Railways, Ulsterbus and Citybus which are subsidiaries of the Northern Ireland Transport Holding Company	
	In 1996 the management boards of these subsidiaries merged to form a common board covering bus and rail operations in the province. A single managing director is now responsible for Group Operations and the combined structure has adopted the name Translink. The constituent companies remain independent bodies under the Translink banner	
	The holding company while managing the companies' property assets, also monitors their performance against corporate plans and public expenditure allocations and reports its findings to the Department of the Environment for Northern Ireland	
	The Department is responsible for setting of financial and other performance targets in conjunction with the holding company	
	The governing legislation is contained in the Transport Act (NI) 1967 and the Transport (NI) Order 1977	
4e	Article 4 of the Road Traffic Regulation (NI) Order 1997	4.4,4.9
4f	Northern Ireland Railways is the sole operator of the railway network in Northern Ireland and may be considered for grant aid from the Department of the Environment for Northern Ireland under the provisions contained within the Transport Act (NI) 1967	4.4,4.8
4g	In Northern Ireland, Planning Service, an agency within the Department of Environment for Northern Ireland, is the sole authority in dealing with planning matters	4.5

While procedures generally follow the basic principles described in the text, the operation of these procedures is carried out under different legislation. The main items of legislation are given in 44.1 above

Further advice on planning matters can be obtained from:

Planning Service
Clarence Court
10-18 Adelaide Street
Belfast
BT2 8GB

4h	In its response to the NI Affairs Committee Report into "The Planning System in Northern Ireland", NC 53 the government undertook to give development plans prime importance in determining planning applications	4.5
4i	See 3a	4.5
4j	See 3f	4.6
4k	Article 128 of the Roads (NI) Order 1993 also see 10.2c	4.6
4l	Roads Service, on behalf of the Department of the Environment for Northern Ireland, as a matter of policy, consult with the police and local frontagers	4.6
4m	See 1b	4.7
4n	**Funding of Roads**	4.8

Funds for the range of Road Service functions are voted directly by Parliament in Annual Appropriation legislation. In estimating the funds necessary for roads, account is taken of other public expenditure priorities within Northern Ireland

Roads Service finance is administered by a central headquarters which disburses money in accordance with agreed policy objectives. The primary aim is to maintain the existing road network to prevent unnecessary deterioration which would be costly to remedy at a later date. The maintenance programme is complemented by substantial investment in major and minor capital projects. This ensures the continued development of the road network in response to both local need and the strategic requirement of the province as a whole

Other Sources of Finance

1 European Community assistance is obtained for selected transport projects under various regulations. Generally, the revenue has not public expenditure implications and does not impact upon the roads vote

2 Developers who construct new residential and industrial/retail estate roads and/or agree to finance highway related works on roads outside their development but adversely affected by it

3 Other revenue such as parking charges and tolls

4o	Articles 23 to 27 of the Roads (NI) Order 1993	4.8
4p	Guidance on Development Contributions is given in PPS 3 Development Control: Road Considerations. See also 3.1a	4.8

Annotation	Comment for Northern Ireland	Chapter Reference
4q	All bridges carrying public roads are the responsibility of a Roads Service, an Agency within the Department of the Environment for Northern Ireland. Road bridges over railways are dealt with under Article 12 of the Roads (NI) Order 1993	4.8
4r	Roads (NI) Order 1993 and Private Streets (NI) Order 1980 as amended	4.9
4s	See 1c	4.9
4t	Within Northern Ireland roads development is 'Crown development' and as such is not bound by the Planning Order. The Department of the Environment for Northern Ireland has however undertaken to go through the planning process for road development	4.9
4u	Northern Ireland Railways, the sole operator of the railway network in Northern Ireland, may construct a new railway or carry out improvements to an existing railway under the provisions contained within the Transport Act (NI) 1967. Anyone other than Northern Ireland Railways seeking to construct a railway would have to do so by means of Private Bill in Parliament	4.9
4v	Land Compensation (NI) Order 1982	4.9
4w	Part IX of the Roads (NI) Order 1993	4.9
4x	Street Works (NI) Order 1995	4.9
4y	Articles 28 to 42 of the Roads (NI) Order 1993	4.9
4z	Article 8 of the Roads (NI) Order 1993 places a duty on the Department of the Environment for Northern Ireland to maintain roads	4.9
4aa	See 1c	4.9
4ab	Part IV of the Road Traffic Regulation (NI) Order 1997	4.9
4ac	Traffic Calming Regulations (NI) 1995	4.9
4ad	The Road Humps Regulations (NI) 1992	4.9
4ae	Article 65 of the Roads (NI) Order 1993	4.9

Chapter 5 Urban Transport into the 21st Century

No Northern Ireland references

Chapter 6 Transport Policy Components

6a	See 1a	6.1
6b	See 3a	6.3
6c	See 1b	6.9

Chapter 7 Data Collection

7a	General information on traffic in Northern Ireland is contained in 2 reports which are available to the public through the Library Service in Northern Ireland or direct from the Roads Service. They are:	7.3
	1 "Annual Traffic Census Report" This gives volumetric	

traffic information for approximately 180 sites on the main routes in Northern Ireland

2 "Annual Report of Vehicle Kilometres of Travel Survey of Northern Ireland". This gives estimates of the amount of travel on the various road classes by class of vehicle. It also includes estimates of classified vehicle counts at approximately 114 sites across Northern Ireland

7b	Northern Ireland Transport Statistics are available in the annual statistical Department of the Environment for Northern Ireland publication "Transport Statistics"	7.3
7c	In Northern Ireland injury accidents information is recorded by the Royal Ulster Constabulary on T1 forms. These forms are modelled on the STATS 19 form. Road traffic injury accident information is passed to the Department of Transport on request	7.4
7d	Accident investigation in Northern Ireland is undertaken in accordance with the Road Safety Engineering Manual published by the Royal Society for the Prevention of Accidents (RoSPA). The procedures are broadly in line with that described in the text	7.4
7e	The Department of Transport Core Census does not extend to Northern Ireland. The Vehicle Kilometres of Travel Survey of Northern Ireland gives estimates of the amount of travel on the various road classes by class of vehicle. See 7.3a	7.5

Chapter 8 Estimating Travellers' Responses to Changes in the Transport System

8a	See 7e	8.4

Chapter 9 Economic and Environmental Appraisal

9a	See 1b	9.4
9b	Planning (Assessment of Environmental Effects) Regulations (NI) 1989 and Article 67 of the Roads Order (NI) Order 1993	9.11
9c	See 4d	9.12
9d	See 4f	9.12

Chapter 10 Involving the Public

10a	See 4l	10.1
10b	See 4c	10.2
10c	Schedule 1,2,3,4 & 5 of the Road Traffic Regulation (NI) Order 1997	10.2
10d	The holding of an inquiry is at the discretion of the Department of the Environment for Northern Ireland	10.2
10e	Procedures for the holding of public inquiries are set out in Article 130 of the Roads (NI) Order 1993 and Schedule 6 of the Road Traffic Regulation (NI) Order 1997	10.2
10f	See 4u	10.2
10g	See 3e	10.2

Annotation	Comment for Northern Ireland	Chapter Reference
10h	See 3f	10.2
10i	See 1b	10.2
10j	See 4t	10.2
10k	See 4ae	10.2
10l	The consultation procedures for development plans are similar in Northern Ireland. The procedures are laid down in the Planning (Development Plans) Regulations (NI) 1991	10.3
10m	See 4b	10.6

Chapter 11 Managing Use of the Road System

11a	Street Works (NI) Order 1995	11.3
11b	The Way Forward published on 3 October 1995 contains the Department of the Environment's new proposals for transportation issues in Northern Ireland	11.4
11c	In Northern Ireland the Roads Service, an agency within the Department of Environment for Northern Ireland, is the sole authority in dealing with road matters. While procedures generally follow the basic principles described in the text, the operation of these procedures is carried out under different legislation. The main items of legislation are given in the Introduction (Section 44.1)	11.4, 11.6
11d	Structure and Unitary Development Plans are not used in Northern Ireland. Development Plans are prepared by the Department of the Environment for Northern Ireland in accordance with The Planning (NI) Order 1991 and the Planning (Development Plan) Regulations (NI) 1991	11.4
11e	Within Northern Ireland the following would apply: Road Traffic (NI) Order 1981 Planning (NI) Order 1991 Roads (NI) Order 1993 Road Traffic (NI) Order 1995 Street Works (NI) Order 1995 Proposed Road Traffic Regulation (NI) Order 1997	11.6
11f	Finding of Roads Funds for the range of Road Service functions are voted directly by Parliament in Annual Appropriation legislation. In estimating the funds necessary for roads, account is taken of other public expenditure priorities within Northern Ireland	11.6

Roads Service finance is administered by a central headquarters which disburses money in accordance with agreed policy objectives. The primary aim is to maintain the existing road network to prevent unnecessary deterioration which would be costly to remedy at a later date. The maintenance programme is complemented by substantial investment in major and minor capital projects. This ensures the continued development of the road network in response to both local need and the strategic requirement of the province as a whole

Other Sources of Finance

1. European Community assistance is obtained for selected transport projects under various regulations. Generally,

the revenue has no public expenditure implications and does not impact upon the roads vote

2. Developers who construct new residential and industrial/retail estate roads and/or agree to finance highway related works on roads outside their development but adversely affected by it

3. Other revenue such as parking charges and tolls

11g	The Road Humps Regulations (NI) 1992	11.10
11h	No application in Northern Ireland	11.11
11i	The Road Safety Plan for Northern Ireland is produced by the Department of the Environment for N Ireland in conjunction with the Department of Health and Social Services and the Royal Ulster Constabulary	11.11
11j	While the making of a proposed Traffic Regulation Order is subject to objection/representation there is no statutory requirement to formally consult	11.15
11k	Statutory Changes Register held by the Land Registers of Northern Ireland, an agency within the Department of the Environment	11.17
11l	Statutory definitions for terminology:	11.7

Road means a public road, that is to say a road which is maintainable by the Department and includes:

a) a road over which the public have a right of way on foot only, not being a footway
b) any part of a road; and
c) any bridge or tunnel over or through which a road passes;
and a special road and a trunk road shall be construed accordingly.
(The Roads (Northern Ireland) Order 1993, Article 2)

Carriageway means a way constituted or comprised in a road being a way over which the public have right of way for the passage of vehicles.
(The Roads (Northern Ireland) Order 1993, Article 2)

Footway means a way comprised in a road which also comprises a carriageway, being a way over which the public have a right of way on foot only.
(The Road Traffic (Northern Ireland) Order 1995, Article 2)

A footpath means a way over which the public have right of way on foot only, not being a footway or part of a road.
(The Road Traffic (Northern Ireland) Order 1995, Article 2)

A street means the whole or any part of the following, irrespective of whether it is a thoroughfare
a) road, highway, lane, footway, alley, passage
b) square, court and any
c) land laid out as a way whether it is for the time being formed as a way or not
(The Street Works (Northern Ireland) Order 1995 , Article 3)

11m	Classification of roads according to their status as traffic routes:	11.7

Trunk road means a road designated as a trunk road under Article 14 of The Roads (Northern Ireland) Order 1993. The Department may designate:
a) any existing road;
b) any road in the course of construction; or
c) any road proposed to be constructed as a trunk road

A classified road is a road classified in accordance with Article 13 of The Roads (Northern Ireland) Order 1993. The Department may by order classify roads in such a manner as appears to it to be expedient and may classify roads as being first–class roads, second–class road, or third–class roads, or as being of any other class

A primary route is a route that is designated by the Department as the most satisfactory route for through traffic between two or more places of traffic importance

Chapter 12 Town Centres

12a	See 11b	12.1
12b	The N Ireland equivalent of Planning Policy Guidance notes (PPGs) are Planning Policy Statements (PPS). The key Planning Policy Statements in Northern Ireland are: PPS3 Development Control: Road Considerations PPS5 Retailing and Town Centres	12.2

Chapter 13 Procedures for Implementing Traffic Management Measures

13a	See 11c	13.1, 13.6
13b	Procedures for making Orders in N Ireland are set out in the Road Traffic Regulation (NI) Order 1997 and the Roads (NI) Order 1993	13.1, 13.6
13c	See 11.15j	13.1, 13.2
13d	Traffic Signs Regulations (NI) 1979 as amended	13.1, 13.4, 13.5
13e	Roads Service, on behalf of the Department of the Environment, as a matter of policy, consult with the police and local frontagers while notifying relevant District Councils	13.2
13f	Article 4, Schedule 1 of the Road Traffic Regulation (NI) Order 1997	13.2
13g	The holding of an inquiry is at the discretion of the Department	13.2
13h	Within N Ireland signs are authorised and equipment approved by the Department of the Environment for N Ireland.	13.2
13i	Within N Ireland Article 5, Schedule 2 of the Road Traffic Regulation (NI) Order 1997 requires that an experimental traffic control scheme be advertised and invite objections/representations before the making of a scheme.	13.2, 13.4
13j	Article 7 of the Road Traffic Regulation (NI) Order 1997	13.2
13k	See 11h	13.2, 13.4, 13.5, 13.6, 13.8, 13.16, 13.17
13l	Planning (NI) Order 1991	13.4, 13.10
13m	The Highway Code for Northern Ireland	13.5
13n	Part IV of the Road Traffic Regulation (NI) Order 1997	13.7
13o	Section 14 of the Chronically Sick and Disabled Persons (NI) Act 1978 and Articles 55 to 57 of the Road Traffic Regulation (NI) Order 1997.	13.8
13p	The Disabled Persons (Badges for Motor Vehicles) Regulations (NI) 1993 as amended by the Disabled Persons (Badges for	13.8, 13.9

Annotation	Comment for Northern Ireland	Chapter Reference
	Motor Vehicles) (Amendments) Regulations (NI) 1995	
13q	Article 68 of the Roads (NI) Order 1993	13.9
13r	Part XI of the Planning (NI) Order 1991	13.9
13s	Article 4 of the Road Traffic Regulation (NI) Order 1997	13.10, 13.13, 13.15
13t	Article 3 of the Roads (NI) Order 1993	13.11
13u	Article 4 of the Roads (NI) Order 1993	13.11, 13.13
13v	The Motor Vehicles (Speed Limits) Regulations (NI) 1989	13.15, 13.16
13w	Article 37 of the Road Traffic Regulation (NI) Order 1997	13.16
13x	Article 36 (2) of the Road Traffic Regulation (NI) Order 1997 allows for the increase or reduction of speed limits.	13.16, 13.17
13y	Article 65 of the Roads (NI) Order 1993	13.17
13z	See 11g	13.17
13aa	Traffic Calming Regulations (NI) 1995	13.17

Chapter 14 Enforcement

14a	See 11b	14.2
14b	See 16d	14.2
14c	See 11c	14.2
14d	See 11h	14.2, 14.5, 14.8
14e	Article 54 of the Road Traffic Regulation (NI) Order 1997	14.5
14f	DVLNI in Coleraine	14.5
14g	Article 15 of the Road Traffic Regulation (NI) Order 1997	14.5

Chapter 15 Information for Transport Users

15a	Article 28 of the Road Traffic Regulation (NI) Order 1997	15.3
15b	See 13d	15.3, 15.6
15c	Schedule I Part I & II of the Traffic Signs Regulations (NI)	15.4
15d	Schedule I Part III of the Traffic Signs Regulations (NI) 1979 as amended	15.4
15e	Schedule I Part IV–VIII of the Traffic Signs Regulations (NI) 1979 as amended	15.4
15f	Within Northern Ireland the criteria for signing from motorways has been set at 75,000 visitors per year	15.5
15g	See 13h	15.6
15h	See 11h	15.7, 15.10
15i	Within N Ireland the Department of the Environment using Article 31(2) of the Road Traffic Regulation (NI) Order 1997 issues Directions that signs and devices shall comply with Chapter 8	15.9
15j	See 11a	15.9

Chapter 16 Road Safety

16a	Roads (NI) Order 1993, Article 8(1)	16.3
16b	Road Traffic (NI) Order 1995, Article 52	16.3
16c	Roads (NI) Order 1993, Article 8(1)	16.3

Annotation	Comment for Northern Ireland	Chapter Reference
16d	In 1989 the Department of the Environment for Northern Ireland set a target of reducing fatal and serious casualties by one third, based on the 1981–85 average, by the year 2000	16.4
16e	See 11i	16.4
16f	The Department of the Environment for Northern Ireland has primary responsibility for road safety engineering and education. A strategic approach to road safety is provided by the Road Safety Review Group	16.5
16g	Within Northern Ireland current policy requires that all schemes projected to cost most than £100,000 should be safety audited	16.11

Chapter 17 Environmental Management

17a	See 12b	17.3
17b	See 11d	17.3

Chapter 18 Technology for Network Management

18a	See 13d	18.6
18b	See 13h	18.6

Chapter 19 Parking

19a	See 12b	19.1
19b	See 11b	19.1
19c	See 11d	19.3
19d	In Northern Ireland the Planning Service, an Agency within the Department of the Environment for Northern Ireland is the sole authority in dealing with planning matters	19.3
	While procedures generally follow the basic principles described in the text, the operation of these procedures is carried out under different legislation. The main items of legislation are given in reference NI 44.1 above	
	Further advice on planning matters can be obtained from:	
	Planning Service Clarence Court 10–18 Adelaide Street Belfast BT2 8GB	
19e	Guidance on car parking standards in Northern Ireland will be issued on behalf of the Department of the Environment through the Roads and Planning Service Agencies	19.3
19f	See 13n	19.4
19g	See 11h	19.4,19.5,19.8,19.12
19h	See 13d	19.5
19i	See 13t and 13u	19.8
19j	Within N Ireland footway parking bands on clearways are made under the powers contained in the Road Traffic Regulation (NI) Order 1997	19.8

Annotation	Comment for Northern Ireland	Chapter Reference
Chapter 20 Traffic Calming and the Control of Speed		
20a	See 13y	20.2
20b	See 13aa	20.2
20c	See 11g	20.2
20d	See 11c	20.2
20e	See 11j	20.5
20f	See 11b	20.5
20g	See 12b	20.5
20h	See 13d	20.8
20i	See 13x	20.9
20j	See 13v	20.9
20k	See 13b	20.9
20l	See 11h	20.10,20.11
20m	Article 36 of the Road Traffic Regulation (NI) Order 1997	20.11
20n	The Road Traffic Offenders Order (NI) 1997	20.12
Chapter 21 Demand Management		
21a	See 12b21.9	
21b	See 11h	21.14
Chapter 22 Pedestrians		
22a	See 13d	22.6,22.10
22b	The 'Zebra' Pedestrian Crossing Regulations (NI) 1974	22.7
22c	The 'Pelican' Crossing Regulations (NI) 1989	22.7
22d	See 13e	22.7
22e	Within N Ireland The Education and Library Boards are responsible for the provision of school crossing patrols	22.7
22f	Article 60 of the Road Traffic Regulation (NI) Order 1997	22.7
22g	See 13i and 13s	22.9
22h	Article 4 of the Road Traffic Regulation (NI) Order 1997	22.9
22i	There is no entitlement to compensation with regard to the removal of vehicular access contained in the Road Traffic Regulation (NI) Order 1997	22.9
22j	Article 30 of The Street Works (NI) Order 1995	22.10
22k	See 13h	22.10
Chapter 23 Cycling		
23a	See 12b	23.2
23b	The Street Works (NI) Order 1995	23.4
23c	See 13d	23.5
Chapter 24 Measures to Assist Public Transport		
24a	Public Transport in Northern Ireland is provided by both state owned and private companies. Although the Northern Ireland	24.2

system is broadly similar to that of Great Britain, Northern
Ireland has built up its own independent system over the years
with differences suited to local requirements

In the publicly owned sector operating responsibility for the
individual services lies with a number of separate companies –
Northern Ireland Railways, Ulsterbus and Citybus – which are
subsidiaries of the Northern Ireland Transport Holding
Company

The Department of the Environment for Northern Ireland is
responsible for co–ordination while permitting the separate
companies to act in a commercial manner and to concentrate in
their own field of operation

The governing legislation is contained in the Transport Act
(NI) 1967 and the Transport (NI) Order 1977

Annotation	Comment for Northern Ireland	Chapter Reference
24b	See 13s	24.2
24c	See 11h	24.2
24d	Motor Vehicles (Construction and Use) Regulations (NI) 1989	24.2
24e	In Northern Ireland the Driver and Vehicle Testing Agency an Agency within the Department of Environment for Northern Ireland is responsible for the annual testing and spot testing of buses	24.2, 24.3
24f	See 11c	24.2
24g	Within N Ireland there is no equivalent to the Transport and Works Act a Private Bill in Parliament would be required	24.2
24h	See 11b	24.3
24i	See 12b	24.3
24j	See 13d	24.6

Chapter 25 Management of Heavy Goods Vehicles

Annotation	Comment for Northern Ireland	Chapter Reference
25a	See 24d	25.2, 25.5, 25.12
25b	See 13s	25.2
25c	Article 31 of the Road Traffic (NI) Order 1995	25.2
25d	See 11h	25.2
25e	Article 5 of the Road Traffic Regulation (NI) Order 1997	25.2
25f	See 13b	25.2
25g	See 13f	25.2
25h	See 11j	25.2
25i	See 13g	25.2
25j	See 13d	25.2
25k	The Transport Act (NI) 1967 provides for good vehicle operator-licensing within Northern Ireland but does not include environmental considerations	25.3
25l	The Department of the Environment fulfils the role of Traffic Commissioners within Northern Ireland	25.3
25m	The Road Traffic Regulation (NI) Order 1997	25.9, 25.10

Annotation	Comment for Northern Ireland	Chapter Reference
25n	See 13y	25.9
25o	See 13aa	25.9
25p	Within Northern Ireland a condition specified on the operators icence is that no overnight parking will be permitted on the public road	25.10
25q	Motor Vehicles (Authorisation of Special Types) General Order (NI) 1968 as amended	25.12
25r	Article 24 of the Motor Vehicles (Authorisation of Special Types) General Order (NI) 1968 as amended gives details of loads which require advance notice to be forwarded to Roads Service for movement on roads (including motorways)	25.12
25s	Similar legislation is proposed for 1997 by the following: – Northern Ireland Office in the case of Explosives – The Department of Environments, Environment and Heritage Service in the case of Radio active Substances – The Department of Economic Development Health and Safety Division inthe case of Carriage of Dangerous Goods as well as Classification, Packaging etc	25.13
25t	Article 7 of the Litter (NI) Order 1994	25.13
25u	Article 94 of the Roads (NI) Order 1993	25.13

Chapter 26 Transport Aspects of New Developments

26a	Structure Plans and Unitary Development Plans are not used in Northern Ireland. Development Plans are prepared by the Department of the Environment for Northern Ireland in accordance with The Planning (NI) Order 1991 and the Planning (Development Plan) Regulations (NI) 1991	26.2
26b	The N Ireland equivalent of Planning Policy Guidance notes (PPGs) are Planning Policy Statements (PPS). The key Planning Policy Statements in Northern Ireland are: PPS 3 Development Control: Road Considerations PPS 5 Retailing and Town Centres	26.2
26c	In its response to the NI Affairs Committee Report into 'The Planning System in Northern Ireland', H.C. 53, the Government undertook to give development plans prime importance in determining planning applications.	26.2
26d	In Northern Ireland the Planning Service, an agency within the Department of Environment for Northern Ireland is the sole authority in dealing with planning matters. While procedures generally follow the basic principles described in the text, the operation of these procedures is carried out under different legislation. The main items of legislation are given in Section 44.1.	26.2

Annotation	Comment for Northern Ireland	Chapter Reference
	Further advice on planning matters can be obtained from: Planning Service Clarence Court 10–18 Adelaide Street Belfast BT2 8GB	
26e	In Northern Ireland the Roads Service, an agency within the Department of Environment for Northern Ireland is the sole authority in dealing with road matters While procedures generally follow the basic principles described in the text, the operation of these procedures is carried out under different legislation. The main items of legislation are given in Section 44.1 Further advice on roads can be obtained from: Roads Service Clarence Court 10–18 Adelaide Street Belfast BT2 8GB	26.3
26f	Within Northern Ireland guidance on Developer Contributions is given in PPS 3 Development Control: Road Considerations	26.7

Chapter 27 Development Control Procedures

27a	Planning (NI) Order 1991	27.3
27b	Roads (NI) Order 1993 and Private Streets (NI) Order 1980 as amended	27.3
27c	See 26.b 27.3, 27.4	
27d	See 26.d	27.3
27e	See 26.e	27.3
27f	Article 12 of the Planning (NI) Order 1991	27.3
27g	Article 34 of the Planning (NI) Order 1991	27.3
27h	Article 35 of the Planning (NI) Order 1991	27.3
27i	Part V of the Planning (NI) Order 1991	27.3
27j	Part VII of the Planning (NI) Order 1991	27.3
27k	Part IX of the Planning (NI) Order 1991	27.3
27l	Part IV of the Planning (NI) Order 1991	27.3
27m	Article 16 of the Access to the Countryside (NI) Order 1983 allows the Department of the Environment to make a public path extinguishment order	27.3
27n	Article 32 of the Private Streets (NI) Order 1980 as amended	27.3
27o	Article 11 of the Roads (NI) Order 1993	27.3
27p	Article 68 of the Roads (NI) Order 1993	27.3
27q	Article 69 of the Roads (NI) Order 1993	27.3
27r	Article 78 of the Roads (NI) Order 1993	27.3, 27.4

Annotation	Comment for Northern Ireland	Chapter Reference
27s	Articles 24 – 33 of the Private Streets (NI) Order 1980 as amended	27.3
27t	Article 122 of the Roads (NI) Order 1993	27.3
27u	Article 40 of the Planning (NI) Order 1991 allows the Department of the Enviornment to enter into agreements facilitating regulating or restricting development or the use of land. The changes made in GB by the Planning and Compensation Act 1991 have not to date been replicated in Northern Ireland	27.3
27v	The Planning (General Development) Order (NI) 1993 set out both the procedures and permitted developments.	27.3
27w	The Planning (Use Classes) Order (NI) 1993 defines Use Classes for Northern Ireland.	27.3
27x	See 26.7f	27.3
27y	Within Northern Ireland appeals are made to an Independent Planning Appeals Commission	27.3
27z	Planning (Fees) Regulations (NI) 1995	27.3
27aa	Planning Service 'Charter Standards Statement'	27.3
27ab	Article 110 of the Roads (NI) Order 1993	27.3

Chapter 28 Parking Related to Development

28a	See 26b	28.2, 28.9
28b	See 26a	28.3
28c	Guidance on car parking standards in Northern Ireland will be issued on behalf of the Department of the Environment through the Roads and Planning Service Agencies	28.6
28d	See 27a	

Chapter 29 Residential Developments

29a	See 26a	29.1, 29.12
29b	Within Northern Ireland guidance is contained in the Department of the Environment's Layout of Housing Roads – Design Guide 1988 or revisions	21.19
29c	See 26b	29.2
29d	See 27n	29.3
29e	Not applicable in Northern Ireland	29.3, 29.7, 29.12

Chapter 30 Non–Residential Developments

30a	See 26a	30.1
30b	See 26b	30.1

Chapter 31 Design Concepts

No Northern Ireland references

Chapter 32 Alternative Concepts of Road Link Capacity

No Northern Ireland references

Annotation	Comment for Northern Ireland	Chapter Reference

Chapter 33 Procedures for the Planning and Approval of Transport Infrastructure Schemes

Annotation	Comment for Northern Ireland	Chapter Reference
33a	See 26e	33.1
33b	Private Street (NI) Order 1980 as amended	33.1
33c	See 26a	33.2
33d	See 26b	33.2
33e	Article 11 of the Planning (NI) Order 1991	33.3
33f	Within N Ireland roads development is 'Crown development' and as such is not bound by the Planning Order. The Department has however undertaken to go through the planning process for road development	33.3
33g	See 27v	33.3
33h	See 26d	33.3
33i	See 27l	33.3
33j	Article 25 of the Planning (NI) Order 1991	33.3
33k	Article 15 of the Roads (NI) Order 1993	33.3
33l	Article 4 of the Roads (NI) Order 1993	33.3
33m	Article 69 of the Roads (NI) Order 1993	33.3
33n	Part IX of the Roads (NI) Order 1993	33.3
33o	The Department of the Environment for N Ireland is both the Roads and Planning Authority in N Ireland.	33.3
33p	Land Compensation (NI) Order 1982	33.3
33q	Article 112 of the Roads (NI) Order 1993	33.3
33r	See 29b	33.3
33s	See 27n	33.3
33t	Articles 23 to 27 of the Roads (NI) Order 1993	33.3
33u	Articles 28 to 42 of the Road (NI) Order 1993	33.3
33v	Within N Ireland there is no equivalent to the Transport and Works Act. Northern Ireland Railways, the sole operator of the railway network in N Ireland, may construct a new railway or carry out improvements to an existing railway under the provisions contained within the Transport Act (NI) 1967. Anyone other than NIR seeking to construct a railway would have to do so by means of a Private Bill in Parliament	33.3

Chapter 34 Fixed–Track System

Annotation	Comment for Northern Ireland	Chapter Reference
34a	Within N Ireland approval to operate any new system must be obtained from the Department of the Environment which is advised by HMRI	34.1

Chapter 35 The Highway in Cross Section

No Northern Ireland references

45 Application in Scotland

45.1 Introduction

The principles and techniques described in this manual are, generally, equally appropriate to the consideration of problems and the development and design of highway and traffic management schemes in Scotland. However, legislative powers and duties are set down under Scottish law which though often similar in principle to that which, holds in England and Wales, may differ in detail or sometimes more substantially. Terminology may also be different.

In particular the term 'highway' is used throughout the manual. This term is not legally recognised in Scotland and such references should be read as 'road'.

The Scottish Office Development Department (SODD) have provided a detailed annotation to the text of the manual, indicating where differences for Scotland should be noted. These are shown in the form [Sa] [Sb] etc in the text for each Chapter and the equivalent information for use in Scotland is provided in Section 45.2 below.

Where more detailed information is required direct contact should be made with:

The Chief Road Engineer
The Scottish Office Development Department
National Roads Directorate
Victoria Quay
EDINBURGH
EH6 6QQ

Tel: 0131 479 3141
Fax: 0131 479 3142

The SODD have also provided details of equivalent Scottish Circulars, Advice Notes and Design Standards and these are annotated in a similar way in the lists of references and sources of information provided at the end of each chapter.

Annotation	Comment for Scotland	Chapter Reference
	Chapter 1 Transport in the Urban Environment	
1a	'Keeping Scotland Moving' A Scottish Transport Green Paper Scottish Office, February 1997	1.2,1.8
1b	The Scottish Office Development Department – European Funds Division	1.7
	Chapter 2 Travel Patterns in Urban Areas	
2a	Scottish Transport Statistics – The Scottish Office 1996	2.10
2b	PPG13: Transport – Scottish equivalent in draft form only at present – NPPG: Transport and Planning	2.10
	Chapter 3 Transport Policy	
3a	PPG13: Transport – Scottish equivalent in draft form only at present – NPPG: Transport and Planning	3.1
3b	'Keeping Scotland Moving' A Scottish Transport Green Paper Scottish Office, February 1997	3.1,3.3

Annotation	Comment for Scotland	Chapter Reference
3c	National Policy Planning Guidance (NPPG's)	3.3
3d	NPPG8, Retailing gives similar guidance in Scotland	3.3
3e	The TPP process does not apply in Scotland	3.3

Chapter 4 Roles and Responsibilities and Powers

4a	'Keeping Scotland Moving' A Scottish Transport Green Paper, Scottish Office, February 1997 provides similar guidance	4.1, 4.8
4b	A Single Tier system of 29 unitary councils and three island councils came into effect on 1 April 1996 under the Local Government etc (Scotland) Act 1994	4.2
4c	In Scotland Central Government's role is carried out by the Scottish Office	4.3
4d	Town and Country Planning (Scotland) Act 1972	4.5
4e	National Policy Planning Guidance (NPPG's)	4.5
4f	Scottish equivalent in draft form only at present – NPPG: Transport and Planning	4.5, 4.7
4g	NPPG – Retailing	4.5
4h	Section 3 of Roads (Scotland) Act 1984	4.8
4i	Part I of Roads (Scotland) Act 1984	4.8
4j	Section 50 of Town and Country Planning (Scotland) Act 1972	4.8
4k	Maintenance and Repair of Highway Structures defined in DMRB Volume 3 – Highway Structures: Inspection and Maintenance	4.8
4l	Roads (Scotland) Act 1984	4.9
4m	Light Railways Act 1896 (HMG 1894) still applicable in Scotland	4.9
4n	Sections 109 and 110 of Roads (Scotland) Act 1984	4.9
4o	Sections 36 to 40 of Roads (Scotland) Act 1984	4.9

Chapter 5 Urban Transport into the 21st Century
No Scottish references

Chapter 6 Transport in the Urban Environment

6a	'Keeping Scotland Moving' A Scottish Transport Green Paper Scottish Office, February 1997	6.1
6b	PPG13: Transport – Scottish equivalent in draft form only at present – NPPG: Transport and Planning	6.2,6.3,6.4, 6.5,6.6

Chapter 7 Data Collection

7a	In Scotland COBA may be used as a PC based alternative to NESA subject to Scottish Office approval	7.7

Chapter 8 Estimating Travellers' Responses to changes in the Transport System
No Scottish references

Annotation	Comment for Scotland	Chapter Reference

Chapter 9 Economic and Environmental Appraisal

9a	In Scotland NESA (Network Evaluation from Surveys and Assignments) is the primary tool for rural schemes where fixed trip matrix analysis is considered appropriate. Advice on its use is given in the NESA Manual a new version of which is to be published in May 1997 (DMRB Volume 15) and in SH1/97, replacing SH2/91, (DMRB Volume 5). This new advice will also replace SH2/86, STEAM and SH3/89. For urban/peri urban schemes, SH1/97 will point to the use of proprietary capacity restrained assignment models in conjunction with an appropriate variable trip matrix economic evaluation program. COBA may be used as a PC based alternative to NESA subject to Scottish Office approval. The appraisal guidance in DMRB Volume 12 should be followed for all schemes	9.6,9.8,9.9

Chapter 10 Involving the Public

10a	The Scottish Office National Roads Directorate exercises similar powers on behalf of the Secretary of State for Scotland	10.2
10b	Similar provision in Roads (Scotland) Act 1984	10.2
10c	'Keeping Scotland Moving' A Scottish Transport Green Paper Scottish Office, February 1997 provides similar guidance	10.2
10d	In Scotland published in the Edinburgh Gazette	10.2,10.3
10e	Roads (Scotland) Act 1985, Section 139 defines the criteria for Public Inquiries being held in Scotland	10.2
10f	Reporter in Scotland usually selected by The Scottish Office Reporters Unit (SOIRU)	10.2,10.3
10g	Secretary of State for Scotland	10.2
10h	Light Railways Act 1896 (HMG 1896) still operates in Scotland	10.2
10i	Scottish equivalent in draft form only at present – NPPG: Transport and Planning	
10j	Town and Country Planning (General) (Scotland) Regulations 1976	10.2
10k	Section 48 of Town and Country Planning (Scotland) Act 1972 allows for appeal to an independent tribunal	10.3
10l	Town and Country Planning – Code of Practice for Local Plan Inquiries, SODD Circular	10.8
10m	The timescales given do not apply to Scotland. Scottish procedures are currently under review	10.2
10n	Scottish procedures are currently under review	10.2
10o	The Court of Session	10.2

Chapter 11 Managing Use of the Road System

11a	The TPP process does not apply in Scotland	11.4
11b	Policy Guidance is porvided by the Scottish Office	11.4

Development Department

11c	Roads (Scotland) Act 1984	11.6
11d	Town and County Planning (Scotland) Act 1972	11.6
11e	Transport Supplementary Grant does not exist in Scotland Since 1996–97 Councils have received a block allocation for capital expenditure on roads and transport and a number of other services outwith the Housing Revenue Account. It is entirely a matter for each Council to decide the priority they should accord to roads and transport projects within their block allocation taking account of local needs and circumstances	11.6
11f	Road Humps (Scotland) Regulations) 1990	11.10
11g	The Scottish Office Development Department Transport and Local Roads Division should be contacted to confirm whether a Local Transport Note is applicable in Scotland	11.12
11h	New Roads and Streetworks Act 1991	11.17
11i	Table 11.2 Statutory Definitions for Teminology (England and Wales) does not apply in Scotland. Table 11.2 Statutory definitions for Scotland (Ref 11i)	

Definition	Additional Information
A Road means any way (other than a waterway) over which there is a public right of passage (by whatever means) and includes the road's verge and any Bridge (whether permanent or temporary) over which, or tunnel through which, the road passes; and any reference to road includes a part thereof.	On some roads this right of passage may only be exercised on foot, on horseback or by specific classes of vehicles (as described by any order which may be applicable). Public roads, refers to those roads which are maintained at public expense (ie by the roads authority). Roads not maintainable at public expense are none the less roads.
A Carriageway is a road which includes a public right of passage by	These rights may be restricted by the implementation of a traffic regulation, speed vehicles (other than pedal cycles only). limit or other orders. The right of way for vehicles does not detract from the established right of pedestrians to cross the carriageway or to pass along it in the absence of a footway.
A Footpath is a road not associated with a carriageway over which there is a public right of passage on foot only.	The essential difference between a footway and a footpath is that the former is adjacent to a carriageway. A way which is exclusively for passage on foot is a footpath. Some footpaths may also include a right of way on pedal cycle (Countryside (Scotland) Act 1967 section 47).
A Bridleway is a public right of way on foot and on horseback.	The right of way may also apply to leading horses or driving animals.
A Cycle Track is a road over which there is a public right of passage by pedal cycle or by pedal cycle and foot only.	

Annotation	Comment for Scotland	Chapter Reference
	Chapter 12 Town Centres	
12a	PPG6: The equivalent is NPPG8: Retailing issued by the Scottish Office Development Department in 1996	12.2
12b	PPG13: Transport – Scottish equivalent in draft form only at present – NPPG: Transport and Planning	12.2
12c	In Scotland National Planning Policy Guidelines (NPPG) are issued by the Scottish Office Development Department	12.3
12d	Traffic Advisory Leaflets 9/96: Cycling Bibliography and 10/96: Traffic Calming Bibliography are issued in Scotland by the Scottish Office Development Department	12.4
12e	In Scotland, enquiries about Traffic Advisory Leaflets should be directed to the Scottish Office Development Department	12.4
12f	Cycling into the Future – Scottish Office Development Department 1995	12.6
12g	There is no Scottish equivalent at present	12.6
12h	Planning Policy Guidance Note 23: Planning Pollution Control – Scottish equivalent(s) are NPPG 10 Planning and Waste Management and Planning Advice Note 51 Planning and Environmental Protection, Scottish Office Development Department	12.6
12i	Planning Policy Guidance Note 24: Planning and Noise – Scottish equivalent(s) are Planning Advice Note 50 Controlling the Environmental Effects of Surface Mineral Workings and PAN 51 Planning and Environmental Protection, Scottish Office Development Department	12.6
12j	Planning Policy Guidance Note 15: Planning and the Historic Environment – Scottish equivalent is NPPG 5 Archaeology and Planning, Scottish Office Development Department	
12k	Planning Policy Guidance Note 1: General Policy and Principles – Scottish equivalent is NPPG1: The Planning System. The Scottish Office Development Department is now responsible for the issue of NPPGs	

Chapter 13 Procedures for Implementing Traffic Management Measures

Annotation	Comment for Scotland	Chapter Reference
13a	Roads (Scotland) Act 1984	13.1
13b	Issued in Scotland by The Stationery Office	13.1
13c	SI 1987 No. 2245, The Local Roads Authorities Traffic Orders (Procedures) (Scotland) Regulations, 1987	13.2
13d	The Secretary of States' Traffic Orders (Procedures) (Scotland) Regulations 1987	13.2
13e	Scottish procedures are currently being reviewed to bring them into line with those in England and Wales.	13.2
13f	Scottish Office Development Department	13.2
13g	See 13e	13.2
13h	Order making powers are conferred on the Secretary of State for Scotland under Section 43 and Schedule 3 of the Road Traffic Act 1991	13.2

Annotation	Comment for Scotland	Chapter Reference
13i	Similar powers are available under Section 201 of the Town and Country Planning (Scotland) Act 1972	13.4
13j	The Local Authorities Traffic Orders (Exemptions for Disabled Persons) (Scotland) (Amendment) Regulations 1991	13.8
13k	The terms of reference to Local Authority Circular 3/91 is still to be confirmed by the SODD/DOT	13.8
13l	Section 68 of the Roads (Scotland) Act 1984 provides Roads Authorities with Order–making powers to stop up or divert a highway if it is considered to have become dangerous or because the road is or is likely to be unnecessary. Section 9 refers to the stopping up or diversion of a road to allow construction of a special road.	13.9
13m	Section 198 of the Town and Country Planning (Scotland) Act 1972	13.9
13o	Section 205 of the Town and Country Planning (Scotland) Act 1972	13.9
13p	Section 200 of the Town and Country Planning (Scotland) Act 1972	13.9
13q	Sections 199 and 203 of the Town and Country Planning (Scotland) Act 1972.	13.9
13r	Section 201 of the Town and Country Planning (Scotland) Act 1972	13.10
13s	Section 129(5) of the Roads (Scotland) Act 1984	13.11
13t	See 13a	13.13
13u	The Transport and Works Act 1992 does not apply in Scotland	13.13
13v	The terms of reference to DOT publication 'Lorries in the Community' is still to be confirmed by the SODD/DOT	13.15
13w	The terms of reference to DOT Circular Roads 4/90 is still to be confirmd by the SODD/DOT	13.16
13x	Road Humps (Scotland) Regulations 1990	13.17
13y	Local Roads Authority	13.17
13z	The terms of reference to DOT Circular 2/92 is still to be confirmed by the SODD/DOT	13.17
13aa	The terms of reference to DOT Circular Roads 4/96 is still to be confirmed by the SODD/DOT	13.17
13ab	Some Traffic Advisory Leaflets are jointly issued by the Department of Transport and the Scottish Office Development Department.	13.17
13ac	Consultation requirements are set out in Section 37 of The Roads (Scotland) Act 1984	13.17
13ad	The Traffic Calming Act 1992 amended The Roads (Scotland) Act 1984 by the addition of sections 39A, 39B and 39C which allow works to be carried out "for the purpose of promoting safety or improving the environment..."	13.17
13ae	The Roads (Traffic Calming) (Scotland) Regulations 1994 – see also Traffic Advisory Leaflet 11/04: Traffic Calming Regulations (Scotland)	13.17
13af	Issued by The Scottish Office Development Department in Scotland	13.18
13ag	Parallel circular issued by The Scottish Office Industry Department	13.18
13ah	The terms of reference to DOT Cricular 5/92 is still to be confirmed by the SODD/DOT	13.18
13ai	The terms of reference to DOT Circular 5/96 is still to be confirmed by the SODD/DOT	13.18

Annotation	Comment for Scotland	Chapter Reference
13aj	SI 1987 No. 2245, Local Roads Authorities Traffic Orders Procedure (Scotland)	13.18
13ak	Town and Country Planning (Scotland) Act 1972)	13.18
13al	SODD (1975): Orders for Resident's Parking Schemes, Circular R347	13.19
13am	SODD (1975): Car Parking and the Medical Profession, Circular 348	13.19
13an	SODD (1980): Speed Limits, Circular 18/80	13.19
13ao	SODD (1985): Speed Limits: Techhnical Guidance Letter 1985	13.19
13ap	The terms of reference to DOT Circular 2/93 is still to be confirmed by the SODD/DOT.	13.19
13aq	Local Transport Notes may be jointly issued by the Department of Transport and the Scottish Office Development Department.	13.19

Chapter 14 Enforcement

14a	PPG13: Transport – Scottish equivalent in draft form only at present – NPPG: Transport and Planning	14.2
14b	The Scottish Office Development Department	14.2
14c	The terms of reference of DOT Circular 1/95 Guidance on Decriminalised Parking Enforcement Outside London are still to be confirmed by the SODD/DOT	14.5

Chapter 15 Information for Transport Users

15a	In Scotland enquiries on, and copies of, Local Transport Notes should be directed to The Scottish Office Development Department, Transport and Local Roads Division	15.3
15b	SODD Planning Circular 10/1984 Town and Country Planning (Control of Advertisements) (Scotland) Regulations 1984	15.3
15c	The Scottish Office Development Department Transport and Local Roads Division	15.6

Chapter 16 Road Safety

16a	Roads (Scotland) Act 1984	16.3
16b	Section 1 of the Roads (Scotland) Act 1984	16.3
16c	There is no Scottish equivalent at present	16.3

Chapter 17 Environmental Management

17a	PPG13: Transport – Scottish equivalent in draft form only at present – NPPG: Transport and Planning	17.3
17b	PPG6: The equivalent is NPPG8: Retailing issued by The Scottish Office Development Department in 1996	17.3
17c	In Scotland enquiries about Traffic AdvisoryLeaflets should be directed to The Scottish Office Development Department	17.5
17d	Planning Policy Guidance Note 23: Planning Pollution Control – Scottish equivalent(s) are Planning Advice Note 51 Planning and Enviornmental Protection, 1997 and NPPG10 Planning and Waste Management, 1996, Scottish Office Development Department	17.7
17e	Planning Policy Guidance Note 24: Planning and Noise – Scottish equivalent(s) are Planning Advice Note 50 Controlling the	17.7

Annotation	Comment for Scotland	Chapter Reference
	Environmental Effects of Surface Mineral Workings, 1996 and PAN 51 Planning and Environmental Protection 1997, Scottish Office Development Department.	
17f	SI 1221 Environmental Assessment (Scotland) Regulations 1988 Part 6	17.8

Chapter 18 Technology for Network Management

No Scottish References

Chapter 19 Parking

19a	PPG6: The equivalent is NPPG8: Retailing by The Scottish Office Development Department in 1996	19.1
19b	PPG 13: Transport – Scottish equivalent in draft form only at present – NPPG: Transport and Planning	19.1
19c	There is no Scottish equivalent at present	19.1
19d	The terms of reference to DOT Circular 1/95 'Guidance on Decriminalised Parking Enforcement outside London' is still to be confirmed by the SODD/DOT.	19.4
19e	The Scottish Office Development Department – Roads Directorate	19.5
19f	SODD (1975): Car Parking for the Medical Profession, Circular R348	19.5
19g	SODD (1974): Orders for Residents Parking Schemes, Circular R347	19.6
19h	SODD (1984): Orange Badge Scheme of Parking Concessions for Disabled and Blind People, Circular 30/84 and letter of 31 May 1984	19.7
19i	SODD(1982): Orange Badge Scheme of Parking Concessions for Disabled and Blind People, Circular 38/82	19.7
19j	SODD (1984): Orange Badge Scheme of Parking Concessions for Disabled and Blind People, Circular 30/84 and letter of 31 May 1984	19.7
19k	The Scottish Office should be contacted to confirm whether a Traffic Advisory Leaflet is applicable in Scotland	19.7

Chapter 20 Traffic Calming and the Management of Speed

20a	Roads (Scotland) Act 1984	20.2,20.5
20b	Roads (Traffic Calming) (Scotland) Regulations 1994	20.2,20.5,20.8
20c	Road Humps (Scotland) Regulations 1990	20.2,20.5
20d	In Scotland enquiries on and copies of Traffic Advisory Leaflets should be directed to The Scottish Office Development Department – Transport and Local Roads Division	20.5,29.0,20.10
20e	PPG13: Transport – Scottish equivalent in draft form only at present – NPGG: Transport and Planning	20.5
20f	SI 1987 No. 2245, The Local Roads Authorities Traffic Orders (Procedure) (Scotland) Regulations 1987	20.9

Annotation	Comment for Scotland	Chapter Reference
20g	The terms of reference of DOT Circular Roads 1/93 which incorporates the Road Traffic Regulation Act 1984 still has to be confirmed by the SODD/DOT	20.9,20.11
20h	The terms of reference of Home Office Circular 38/1992 still has to be confirmed by the SODD/DOT.	20.12
20i	Scottish Office Industry Department (SOID) Circular 7/92	20.12
20j	Scottish Office Industry Department (SOID) Circular 7/93	20.13

Chapter 21 Demand Management

21a	PPG13: Transport – Scottish equivalent in draft form only at present – NPPG: Transport and Planning	21.9

Chapter 22 Pedestrians

22a	In Scotland enquiries on and copies of Mobility Unit Leaflets should be directed to The Scottish Office Development Department	22.1
22b	In Scotland enquiries on and copies of Local Transport Note should be directed to The Scottish Office Development Department	22.7
22c	In Scotland enquiries on and copies of Traffic Advisory Leaflets should be directed to The Scottish Office Development Department	22.7
22d	PPG13: Transport – Scottish equivalent in draft form only at present – NPPG: Transport and Planning	22.14

Chapter 23 Cycling

23a	PPG13: Transport – Scottish equivalent in draft form only at present – NPPG: Transport and Planning	23.2
23b	The Scottish Office should be contacted to confirm whether a Traffic Advisory Leaflet is applicable in Scotland	23.5,23.9, 23.11, 23.16
23c	The Scottish Office should be contacted to confirm whether a Local Transport Note is applicable in Scotland	23.13,23.14,23.15 23.16,23.17
23d	The Scottish Office should be contacted to confirm whether a Disability Unit Circular is applicable in Scotland	23.14,23.15

Chapter 24 Measures to Assist Public Transport

24a	Scottish Development Department (1985); Transport Act 1985 Circular 2 32/85	24.2
24b	PPG13: Transport – Scottish equivalent in draft form only at present – NPPG: Transport and Planning	24.3
24c	PPG6: The equivalent is NPPG8: Retailing issued by The Scottish Office Development Department in 1996	24.3
24d	The Scottish Office should be contacted to confirm whether a Local Transport Note is applicable in Scotland	24.4

Chapter 25 Management of Heavy Goods Vehicles

25a	SI 1987 No. 2245, The Local Roads Authorities Traffic Orders (Procedure) (Scotland) Regulations 1987	25.2

Annotation	Comment for Scotland	Chapter Reference
26b	Traffic Commissioner for Scotland	25.3,25.4
26c	Roads (Scotland) Act 1984	25.9,25.13
26d	Roads (Traffic Calming) (Scotland) Regulations 1994 SI 1994 No. 2488	25.9
25e	The terms of reference to DOT Circular Roads 2/93 is still be confirmed by the SODD/DOT.	25.9
25f	The Scottish Office should be contacted to confirm whether a Traffic Advisory Leaflet is applicable in Scotland	25.9

Chapter 26 Transport Aspects of New Developments

26a	National Planning Policy Guidelines in Scotland (NPPG's)	26.2
26b	PPG1: General Policy and Principles – Scottish Equivalent NPPG1: The Planning System	26.2
26c	PPG4: Industrial and Commercial Development and Small Firms – Scottish Equivalent NPPG2: Business and Industry.	26.2
26d	PPG6: Town Centres and Retail Developments – Scottish Equivalent NPPG8: Retailing	26.6
26e	PPG12: Development Plans and Regional Planning Guidance – Scottish Equivalent in draft form only at present – NPPG: Transport and Planning; though no equivalent of Regional Planning Guidance in Scotland	26.2
26f	PPG13: Transport – Scottish Equivalent in draft form only at present – NPPG: Transport and Planning	26.2
26g	PPG15: Planning and the Historic Environment – Scottish Equivalent NPPG5: Archaeology and Planning covers part of the subject and draft NPPG on Built Heritage in preparation	26.2
26h	RPG not in existence in Scotland, NPPG's normally include such information	26.2
26i	Scottish Office Development Department	26.3
26j	In Scotland, the Secretary of State may recommend refusal and if the Planning Authority favours granting planning permission, the application must be referred to the Secretary of State for his consideration. The Secretary of State, as the overseeing planning authority, has 28 days (which may be extended) to consider whether to convene a public local inquiry	26.3
26k	and by Councils in Scotland, in these cases acting in groups through joint committees,	26.2
26l	and by all Councils in Scotland	26.2
26m	B1/B2 is Class 4 in Scotland B8 is Class 11 in Scotland – Town and Country Planning (Use Classes) (Scotland) Order 1989	26.3

Chapter 27 Development Control Procedures

27a	Local Roads Authority	27.2, 27.3
27b	The Scottish Office Development Department Circulars	27.3
27c	National Planning Policy Guidelines (NPPG's)	27.3

Annotation	Comment for Scotland	Chapter Reference
27d	Town and Country Planning (Scotland) Act 1972	27.3
27e	Roads (Scotland) Act 1984	27.3
27f	Similar powers exist for the Scottish Office National Roads Directorate	27.3
27g	The Scottish Office Development Department National Roads Directorate	27.3
27h	This whole paragraph does not apply in Scotland where all 32 Councils are planning authorities for all purposes and there are no 'local' planning authorities	27.3
27i	There are no UDP's in Scotland and Local Plans are not necessarily District–wide	27.3
27j	Similar powers exist for the Secretary of State for Scotland, to call in planning applications that are covered by various Notification Directions, including those significantly contrary to the Structure Plan	27.3
27k	Section 20 of Town and Country Planning (Scotland) Act 1972	27.3
27l	Section 19 of Town and Country Planing (Scotland) Act 1972. Defines 'Development' and 'New Development'	27.3
27m	Section 38 of Town and Country Planning (Scotland) Act 1972. Also specifies within five years of the grant of consent.	27.3
27n	Section 39 of Town and Country Planning (Scotland) Act 1972 Also specified three years of the grant of outline consent for the application of approved and reserved matters and the development must commence not later than two years after the final approval of reserved matters.	27.3
27o	Part IV, sections 57 to 60 of Town and Country Planning (Scotland) Act 1972	27.3
27p	Part VI of Town and Country Planning (Scotland) Act 1972	27.3
27q	Part X, sections 198 to 200 of Town and Country Planning (Scotland) Act 1972	27.3
27r	Part III of Town and Country Planning (Scotland) Act 1972	27.3
27s	Part 1 of Town and Country Planning (Scotland) Act 1972	27.3
27t	This paragraph does not apply in Scotland	27.3
27u	Section 4 of Roads (Scotland) Act 1984	27.3
27v	Section 16 of Roads (Scotland) Act 1984	27.3
27w	Section 96 of Roads (Scotland) Act 1984	27.3
27x	Section 68 of Roads (Scotland) Act 1984	27.3
27y	Section 68 of Roads (Scotland) Act 1984	27.3
27z	Section 63 of Roads (Scotland) Act 1984	27.3
27aa	Section 17 of Roads (Scotland) Act 1984	27.3
27ab	Section 48 of Roads (Scotland) Act 1984	27.3
27ac	Section 12 in Part 1 of The Planning and Compensation Act 1991 refers to England and Wales and is not included in Part II of the Act Town and Country Planning – Scotland	27.3
27ad	Town and Country Planning (General Permitted Development) (Scotland) Order 1992)	27.3
27ae	Town and Country Planning (General Permitted	27.3

Annotation	Comment for Scotland	Chapter Reference
	Development) (Scotland) Order 1992)	
27af	SODD Circular 5/1992	27.3
27ag	Equivalent Articles in SODD Circular 6/1992, similar time periods apply in Scotland to notify the applicant of the local planning authority decision on the application.	27.3
27ah	In Scotland for 'eight weeks' substitute '2 months' and for 'sixteen weeks' substitute '4 months'	27.3
27ai	Town and Country Planning (Uses Classes) (Scotland) Order 1989 and SODD Circular 6/1989	27.3
27aj	This example does not apply in Scotland because the Use Classes are different	27.3
27ak	SODD Circular 18/1986	27.3
27al	Scottish Equivalent in draft form only at present – NPPG Transport and Planning	27.3
27am	The Town and Country Planning (Appeals) (Written Submissions Procedure) (Scotland) Regulations 1990 (SI 507/1990) and SODD Circular 7/1990	27.3
27an	Town and Country Planning (Enforcement of Control) (No2) (Scotland) Regulations 1992	27.3
27ao	There is no Scottish equivalent as Enforcement Appeal Inquiries are conducted under the 1980 Inquiries Procedure Rules	27.3
27ap	Town and Country Planning (Inquiries Procedure) (Scotland) Rules 1980 (SI1676)	27.3
27aq	Town and Country Planning (Determination of Appeals Appointed Persons) (Prescribed Classes) (Scotland) Regulations 1980 (SODD 1980 SI 1675)	27.3
27ar	SODD 32/1996 Code of Practice for Local Planning Inquiries	27.3
27as	SODD Circular 1/1997 Town and Country Planning (Fees for Applications and Deemed Applications) (Scotland) Regulations 1997	27.3
27at	SODD Circular 1/1997	27.3
27au	Scottish Office Planning Charter Standards Statement, 1996	27.3
27av	Section 50 of the Town and Country Planning (Scotland) Act 1972	27.4
27aw	Section 48 of the Roads (Scotland) Act 1984 also substituted by Section 23 of the New Roads and Street Works Act 1991 (HMG, 1991c)	27.4
27ax	Local Government and Planning (Scotland) Act 1982: Planning Provisions	27.4
27ay	SODD Circular 22/91	27.4
27az	No equivalent in Roads (Scotland) Act 1984	27.4
27ba	Town and Country Planning (Appeals) (Written Submissions Procedure) (Scotland) Regulations 1990	27.6
27bb	Local Government etc. (Scotland) Act 1994	27.6

Annotation	Comment for Scotland	Chapter Reference

Chapter 28 Parking Related to Development

28a	National Planning Policy Guidelines in Scotland (NPPG's)	28.2
28b	PPG13: Transport – Scottish Equivalent in draft form only at present – NPPG: Transport and Planning	28.6
28c	Town and Country Planning (Scotland) Act 1972	28.6
28d	The Scottish Office Development Department Circular 12/1996 – on planning agreements and obligations	28.6

Chapter 29 Residential Developments

29a	PPG13 Transport – Scottish Equivalent Draft PPG Transport and Planning	29.2
29b	Section 16 of the Roads (Scotland) Act 1984	29.3
29c	PPG 13 Transport – Scottish Equivalent Draft NPPG Transport and Planning – Car Parking policies should support the overall locational policies in the development plan.	29.9
29d	Planning Advice Note 46 'Planning for Crime Prevention' The Scottish Office Environment Department, October 1994	29.6

Chapter 30 Non–Residential Developments

30a	In Scotland similar guidance is provided by:	30.1
	PPG4: Industrial and Commercial Development and Small Firms – Scottish Equivalent NPPG2: Business and Industry.	
	PPG6: Town Centres and Retail Developments – Scottish Equivalent NPPG8: Retailing	
	PPG13: Transport – Scottish Equivalent draft NPPG: Transport and Planning	

Chapter 31 Design Concepts

31a	The Scottish Office Development Department.	31.1
31b	In Scotland NESA (Network Evaluation from Surveys and Assignments) is the primary tool for rural schemes where fixed trip matrix analysis is considered appropriate. Advice on its use is given in the NESA Manual a new version of which is to be published in May 1997 (DMRB Volume 13, probably Section 3) and SH2/97 (replacing SH2/91). This new advice will also replace SH2/86, STEAM and SH3/89. For urban/peri urban schemes, SH2/97 will point to the use of proprietary capacity restrained assignment models in conjunction with an appropriate variable trip matrix economic evaluation programme. COBA may be used as a PC based alternative to NESA. The appraisal guidance in DMRB Volume 12 should be followed for all schemes	31.3
31c	The Scottish Office Development Department National Roads Directorate	31.9
31d	The TPP process is not applicable in Scotland	31.10

Annotation	Comment for Scotland	Chapter Reference
	Chapter 32 Alternative Concepts of Road Link Capacity	
32a	The recommended "design flows" are given in Design Manual for Roads and Bridges 5.1, The Scottish Office Development Department	32.1
32b	DOT Departmental Standard TD20/85 and a Scottish Addendum identifying amendments applicable in Scotland are contained in Design Manual for Roads and Bridges 5.1. The Scottish Office Development Department	32.6
	Chapter 33 Procedures for the Planning and Approval of Transport Infrastructure Schemes	
33a	The Scottish Office Development Department – see 31a	33.1
33b	In Scotland Structure Plans and Local Plans are prepared by Planning Authorities as required by the Town and Country Planning (Scotland) Act 1972	33.2
33c	PPG13: Transport – Scottish equivalent in draft form only at present – NPPG: Transport and Planning	33.2
33d	In Scotland, 'Notice of Intention to Develop' procedures apply to development undertaken by a Council	33.2
33e	Planning permission is required under the Town and Country Planning (Scotland) Act 1972 for development as defined in Section 19 of that Act	33.3
33f	Section 275 of the Town and Country Planning (Scotland) Act 1972	33.3
33g	Roads (Scotland) Act 1984	33.3
33h	The equivalent provision is Section 19 of the Town and Country Planning (Scotland) Act 1972	33.3
33i	Local Roads Authorities	33.3
33j	The equivalent provision is contained in the Town and Country Planning (General Permitted Development) (Scotland) Order 1992.	33.3
33k	Part III of the Town and Country Planning (Scotland) Act 1972	33.3
33l	Section 18A is the Scottish equivalent of Section 54A	33.3
33m	Development plans prepared by Planning Authorities in Scotland take the form of Structure Plans and Local Plans	33.3
33n	The equivalent provision is made in Section 26 of the Town and Country Planning (Scotland) Act 1972	33.3
33o	Section 12 of the Roads (Scotland) Act 1984	33.3
33p	Sections 7 to 9 of the Roads (Scotland) Act 1984	33.3
33q	Sections 75 and 76 of the Roads (Scotland) Act 1984	33.3
33r	Sections 69 to 72 of the Roads (Scotland) Act 1984	33.3
33s	Sections 103 to 111 of the Roads (Scotland) Act 1984	33.3
33t	Acquisition of Land (Authorisation Procedure) (Scotland) Act 1947	33.3
33u	Secretary of State for Scotland	33.3
33v	Town and Country Planning (General Development Procedure) (Scotland) Order 1992. Article 15	33.3

Annotation	Comment for Scotland	Chapter Reference
33w	Town and Country Planning (General Development Procedure) (Scotland) Order 1992. Articles 16 and 17	33.3
33x	Town and Country Planning (Scotland) Act 1972	33.3
33y	Section 49 of the Town and Country Planning (Scotland) Act 1972	33.3
33z	Sections 104 of the Roads (Scotland) Act 1984	33.4
33aa	SODD Circular 21/1991: Planning and Compensation Act 1991 Land Compensation and Compulsory Purchase	33.4
33ab	Section 1 of the Roads (Scotland) Act 1984 provides Local Roads authorities with the general power to widen or improve public roads, and Section 25 of the 1984 Act refers to the need to provide 'proper and sufficient' footways on public roads	33.5
33ac	Statutory powers to convert a footway into a cycle track are contained in the Roads (Scotland) Act 1984	33.5
33ad	Land Compensation (Scotland) Act 1973	33.5
33ae	Part IV of Land Compensation (Scotland) Act 1973	33.5
33af	The Scottish Office Development Department National Roads Directorate	33.5
33ag	Section 18 of Land Compensation (Scotland) Act 1973	33.5
33ah	Section 12 of the Roads (Scotland) Act 1984 provides the right of objection up to 28 days form the first public notification	33.7
33ai	A similar process is followed by The Scottish Office Development Department under the Roads (Scotland) Act 1984	33.7
33aj	Section 14 of the Town and Country Planning (Scotland) Act 1984	33.7
33ak	A similar process is followed by Local Roads Authorities under the Roads (Scotland) Act 1972	33.7
33al	Section 56 of the Roads (Scotland) Act 1984	33.8
33am	Under the Roads (Scotland) Act 1984, the Local Roads Authority must adopt any new road completed in accordance with the Construction Consent if asked to by the developer	33.8
33an	Section 27 of the New Roads and Street Works Act 1991 states that Part IIA, paragraphs 15 and 18 of Part III and Part IV of Schedule 1 to the Roads (Scotland) Act 1984 apply to the making or confirmation of a toll order, and that Schedule 2 to the 1984 Act applies to a toll order in terms of its validity and date of operation. As far as is practical, the proceedings required by Part IIA of Schedule 1 of the 1984 Act are to be taken concurrently with authorisation procedures for the provision of the special road scheme in question	33.8
33ao	Part II of Roads (Scotland) Act 1984	33.8
33ap	The Order–making procedure to authorise rail–based and other fixed route public transport systems introduced in England and Wales under the Transport and Works Act 1992 does not apply in Scotland. The Private Legislation	33.9

Procedures (Scotland) Act 1936 remains in force. If an objection is received to a draft Order submitted under this procedure, a public inquiry is held before Parliamentary Commissioners before a Confirmation Act for the scheme can be sought in Parliament

Annotation	Comment for Scotland	Chapter Reference
33aq	PPG12: Development Plans and Regional Planning Guidance – no Scottish equivalent PPG13: Transport – Scottish equivalent for PPG13 is in draft form only at present – NPPG: Transport and Planning	33.10
33ar	SODD Circular 74/1976: Town and Country Planning (General) (Scotland) Regulations 1976	33.10
33as	The Transport Works Act 1992 is not applicable in Scotland. Reference should be made to the Light Railways Act 1896 and or the Private Legislation Procedures (Scotland) Act 1936	33.10
33at	The Land Compensation (Additional Development) (Forms) (Scotland) (No. 2) Direction 1994	33.11
33au	SODD Circular 14/1975: Public Inquiry Procedures	33.11
33av	Town and Country Planning (Scotland) 1988 1221 (S122)	33.11

Chapter 34 Fixed–Track Systems

Annotation	Comment for Scotland	Chapter Reference
34a	The Transport and Works Act 1992 is not applicable in Scotland refer to Light Railway Act of 1896	34.2
34b	Land acquisition procedures will be a component of a private Bill promoted in Parliament as the Transport and Works Act 1992 is not applicable in Scotland	34.5

Chapter 35 The Highway in Cross–Section

Annotation	Comment for Scotland	Chapter Reference
35a	Scottish Office Development Department TD20/85 and TA46/85 each have a Scottish Addendum noting differences applicable in Scotland.	35.2, 35.10
35b	In Scotland the Scottish Office Development Department is responsible for TD9/93	35.2
35c	The Scottish Office Development Department	35.4
35d	The Scottish Office Development Department TD19/85 has a Scottish Addendum noting differences applicable in Scotland	35.6, 35.10
35e	Highway Construction Details from Volume 3 of the Manual of Contract Documents for Highway Works, in Scotland now the responsibility of the Scottish Office Development Department	35.9, 35.10
35f	Some Traffic Advisory Leaflets are published jointly by the Department of Transport and the Scottish Office Development Department and identify differences applicable in Scotland	35.9
35g	Roads and Bridges are the responsibility of the Scottish Office Development Department	35.10
35h	COBA may be used as a PC based alternative to NESA For urban/peri urban schemes, SH2/97 (which will replace SH2/91) will point to the use of proprietary capacity restrained assignment models in conjunction with an	35.11

Annotation	Comment for Scotland	Chapter Reference
	appropriate variable trip matrix economic evaluation program. The appraisal guidance in DMRB Volume 12 should be followed for all schemes	

Chapter 36 Highway Link Design

36a	In Scotland, Advice Notes and Technical Design Standards in the Design Manual for Roads and Bridges are the responsibility of the Scottish Office Development Department	36.1, 36.10
36b	Local Roads Authorities	36.1
36c	The Scottish Office Development Department	36.1

Chapter 37 Junction Design: General Considerations

37a	In Scotland, Advice Notes and Technical Design Standards in the Design Manual for Roads and Bridges are the responsibility of the Scottish Office Development Department. TA23/81 has a Scottish Addendum noting differences applicable in Scotland	37.2
37b	In Scotland, Standards and Advice Notes in the Design Manual for Roads and Bridges are the responsibility of the Scottish Office Development Department	37.5
37c	COBA may be used as a PC based alternative to NESA For urban/peri urban schemes, SH2/97 (which will replace SH2/91) will point to the use of proprietary capacity restrained assignment models in conjunction with an appropriate variable trip matrix economic evaluation program. The appraisal guidance in DMRB Volume 12 should be followed for all schemes.	37.6

Chapter 38 Priority Junctions

38a	In Scotland, Standards, Technical Design Standards and Advice Notes in the Design Manual for Roads and Bridges are the responsibility of the Scottish Office Development Department	35.10, 36.1 38.1
38b	The Scottish Development Department (now The Scottish Office Development Department) was jointly responsible for preparation of the Traffic Signs Manual. The Manual should be used in conjunction with the Traffic Signs Regulations and General Directions 1994 and subsequent amendments.	38.6
38c	PPG13: Transport – Scottish equivalent in draft form only at present – NPPG: Transport and Planning.	38.7
38d	The Scottish Office Development Department	38.7
38e	Section 83 of the Roads (Scotland) Act 1984 provides powers for a Road Authority to serve an 'obstruction notice' imposing restrictions where it is considered that danger arises from obstruction of the view of road users.	38.7
38f	Roads (Scotland) Act 1984	38.9

Annotation	Comment for Scotland	Chapter Reference

Chapter 39 Roundabouts

39a	In Scotland, Standards, Technical Design Standards and Advice Notes in the Design Manual for Roads and Bridges are the responsibility of the Scottish Office Development Department	35.10, 36.1 38.1, 39.1
39b	The Scottish Development Department (now The Scottish Office Development Department) was jointly responsible for preparation of the Traffic Signs Manual. The Manual should be used in conjunction with the Traffic Signs Regulations and General Directions 1994 and subsequent amendments	38.6, 39.7
39c	The Scottish Office National Roads Directorate	39.10

Chapter 40 Traffic Signal Control

40a	The Scottish Office Development Department	40.2
40b	The Design Manual for Roads and Bridges (8.1) contains Scottish Development Department (now Scottish Office Development Department) Technical Memorandum SH6/73: Criteria for Traffic Light Signals at Junctions. Scottish Development Department Circular R287: Criteria for Traffic Light Signals is included as an appendix to SH6/73	40.2
40c	Scottish Development Department (now the Scottish Office Development Department) Circular R362 – Portable Traffic Signals for use at Roundabouts	40.2
40d	In Scotland, Advice Notes and Technical Design Standards in the Design Manual for Roads and Bridges are the responsibility of the Scottish Office Development Department	40.2
40e	Scottish Development Department (now the Scottish Office Development Department) (1976): Circular R362: Portable Traffic Signals for use at Roadworks	40.9
40f	Jointly issued with the Scottish Development Department (now the Scottish Office Development Department)	40.17

Chapter 41 Co–ordinated Signal Systems

No Scottish References

Chapter 42 Signal–Controlled Gyratories

No Scottish references

Chapter 43 Grade–Separated Junctions

43a	In Scotland, Advice Notes and Technical Design Standards in the Design Manual for Roads and Bridges are the responsibility of the Scottish Office Development Department.	43.1
43b	In Scotland, Local Transport Notes are the responsibility of the Scottish Office Development Department	43.2

46 Application in Wales

46.1 Introduction

The general principles described in this manual and most of the legislative and procedural arrangements are applicable in Wales. However, it is the usual practice for the Welsh Office to issue its own, often equivalent, Circulars and there are, in addition, certain matters where practice or standards in Wales are different.

These points are indicated by the annotation [Wa [Wb] etc in the text for each Chapter and advice on practice for Wales is given in 46.2 below.

Inquiries for more detailed information should be addressed to:

Welsh Office Highways Directorate
Phase 1
Government Buildings
Ty Glas Road
Llanishen
Cardiff
CF4 5PL
Tel: 01222 753271
Fax: 01222 823792

Annotation	Comment for Wales	Chapter Reference
	Chapter 1 Transport in the Urban Environment	
1a	The production of Transport Policies and Programmes (TPPs) is not a requirement in Wales. They are used in England to support bids for local authority transport funding. In Wales authorities are only required to bid for projects costing over five million pounds - under the Transport Grant arrangements for road schemes and under Section 56 of the Transport Act 1968 for public transport projects. Other capital projects, including transport packages and local safety schemes, are funded at individual authorities' own discretion from annual borrowing approvals which are allocated by formula and are not earmarked for particular services	1.4
	Chapter 2 Travel Patterns in Urban Areas	
2a	In Wales, the main body of individual Planning Policy Guidance Notes which formerly applied to both England and Wales have been replaced by two notes: "Planning Guidance (Wales) Planning Policy" Welsh Office, May 1996* "Planning Guidance (Wales) Unitary Development Plans" Welsh Office, April 1996* These two documents are in the process of being supplemented by a series of topic based Technical Advice Notes (Wales) (TAN(W)s). *Eight TAN(W)s were issued in December 1996 including TAN(W)4 "Retailing and Town Centres" Further TAN(W)s, including one on "Transport", are projected. *Available as priced publications from Oriel Bookshop, Stationery Office, The Friary, Cardiff CF1 4AA (01222 395548)	2.10

Annotation	Comment for Wales	Chapter Reference
	Chapter 3 Transport Policy	
3a	See 1a	3.2
3b	See 1a	3.3
3c	See 2a	3.3
3d	In Wales, the Welsh Office	3.3
3e	See 1a	3.3
3f	See 2a	3.8

Chapter 4 Roles and Responsibilities and Powers

4a	As a consequence of the Local Government (Wales) Act 1994, there are 22 unitary authorities in Wales responsible for local government functions including local roads Motorways and trunk roads are the responsibility of the Secretary of State for Wales who has appointed eight agent authorities to maintain the network. The Welsh Office, as highway authority for trunk roads and motorways, has similar responsibilities to those of the Department of Transport in England: it is a statutory consultee on planning applications affecting trunk roads and motorways and has powers of direction over local planning authority decisions on the control of development alongside trunk roads and motorways. The remainder of the highway network is the responsibility of the 22 individual unitary authorities. The unitary authorities, together with the 3 Welsh National Park Authorities, are the local planning authorities responsible for day-to-day planning matters including mineral planning	4.2
4b	See 2a	4.5
4c	Welsh Office Highways Directorate	4.6
4d	See 1a	4.7,4.8
4e	Welsh Office Highways Directorate	4.8
4f	Welsh Office Highways Directorate	4.8
4g	See 2a	4.10

Chapter 5 Urban Transport into the 21st Century

No Welsh references

Chapter 6 Transport in the Urban Environment

6a	See 1a	6.9
6b	See 2a	6.11

Chapter 7 Data Collection

7a	Welsh Transport Statistics and Road Accidents: Wales are published annually by the Welsh Office	7.3

Chapter 8 Estimating Travellers' Responses to changes in the Transport System

No Welsh references

Annotation	Comment for Wales	Chapter Reference

Chapter 9 Economic and Environmental Appraisal

No Welsh references

Chapter 10 Involving the Public

10a	In Wales, there is no Highways Agency and the Secretary of State for Wales operates directly through his Highway Directorate in making trunk road proposals (including motorways)	10.2
10b	See 1a	10.2
10c	In Wales, the Secretary of State will appoint the Inspector	10.2
10d	Secretary of State for Wales	10.2
10e	See 1a	10.2
10f	Secretary of State for Wales	10.2
10g	Welsh Office Highways Directorate	10.4
10h	Secretary of State for Wales	10.4
10i	See 2a	10.2

Chapter 11 Managing the Use of the Road System

11a	See 1a	11.4
11b	Welsh Office	11.4
11c	See 1a	11.6
11d	See 1a	11.11

Chapter 12 Town Centres

| 12a | In Wales, the main body of individual Planning Policy Guidance Notes which formerly applied to both England and Wales have been replaced by two notes: Planning Guidance (Wales) Planning Policy Welsh Office, May 1996, and Planning Guidance (Wales) Unitary Development Plans Welsh Office, April 1996 These two documents are in the process of being supplemented by a series of topic based Technical Advice Notes (Wales) (TNA(W)s). *Eight TAN(W)s were issued in December 1996 including TAN(W)4 Retailing and Town Centres. Further TAN(W)s, including one on Transport, are projected *Available as priced publications from Oriel Bookshop, Stationery Office, The Friary, Cardiff CF1 4AA (01222 395548). | 12.8 |
| 12b | Not applicable: advice for Wales is contained in Welsh Office Circular 61/96 | 12.8 |

Chapter 13 Procedures for Implementing Traffic Management Measures

13a	Secretary of State for Wales	13.9
13b	Secretary of State for Wales	13.9
13c	Secretary of State for Wales	13.9

Chapter 14 Enforcement

| 14a | See 12a | 14.10 |

Annotation	Comment for Wales	Chapter Reference

Chapter 15 Information for Transport Users

15a	For Wales, working drawings are available from the Welsh Office Highways Directorate	15.3
15b	Different criteria apply in Wales	15.5
15c	Welsh Office Highways Directorate	15.5
15d	In Wales, application should be made to the Transport Policy Division of the Welsh Office	15.6
15e	Welsh Office Highways Directorate	15.6
15f	For Wales, working drawings are available from the Welsh Office Highways Directorate	15.9
15g	For Wales, working drawings are available from the Welsh Office Highways Directorate	15.10

Chapter 16 Road Safety

16a	In Wales, the corresponding Protocol for Investment in Health Gain: Injuries was launched in 1992	16.5
16b	The Welsh Office provides funds for the services of the Traffic Education Officer for Wales, who liaises with education authorities on a range of road safety activities for schools and teacher training colleges, particularly through the medium of Welsh	16.12

Chapter 17 Environmental Management

17a	See 12a	17.7, 17.8

Chapter 18 Technology for Network Management

No References for Wales

Chapter 19 Parking

19a	See 2a	19.13

Chapter 20 Traffic Calming and the Management of Speed

20a	Welsh Office Highways Directorate	20.2
20b	Welsh Office Transport Policy Division	20.10
20c	Welsh Office Transport Policy Division	20.10
20d	Welsh Office Transport Policy Division	20.11
20e	See 2a	20.15

Chapter 21 Demand Management

21a	See 1a	21.14
21b	See 2a	21.18

Chapter 22 Pedestrians

22a	See 2a	22.14

Chapter 23 Cycling

23a	See 2a	23.20

| Annotation | **Comment for Wales** | Chapter Reference |

24a   The Welsh Office works closely with the Department of   24.1
Transport and others, including unitary authorities, to
develop and implement public transport policies in Wales.
Public transport services in Wales including local, regional
and national rail services and scheduled coach and bus
services, together with some less conventional community
and utility services

The Government's transport deregulation and privatisation
reforms have increased choice for passengers and greater
competition in the provision of services. Greater distances
are being covered by buses and coaches, more passengers
are using airports in Wales, and even the number of railway
stations has grown. For example, there are today some 30
more railway stations in Wales than 10 years ago, and every
reason to expect more refurbishments and openings in the
years to come

Nevertheless, transport growths in recent years has
essentially related to increasing use of the roads, and
although measures have been taken to improve the
competitiveness and attractiveness of public transport
alternatives, it is recognised that an efficient road network
remains an essential element of Wales' transport
infrastructure. At the same time, the effects of traffic
growth are causing increasing concern

Financial support
The Welsh Office's general policy is that public transport
services in Wales should wherever possible be provided on
a fully commercial basis, although local authorities can
support socially necessary but financially unattractive
services. Public transport support by local authorities in
Wales in 1995–96 was nine million pounds, unchanged from the
1994-95 level, in real terms

Under Section 56 of the Transport Act 1968, the Secretary
of State for Wales has the power to make grants towards
the capital costs of providing, improving or developing
facilities for public passenger transport. Projects should not
require continuing subsidy, and must be justified primarily
in terms of non-user benefits – such as a reduction in road
traffic congestion – which must exceed the grant contribution

Financial support is not limited to Section 56 projects. For
example, the Welsh Office has approved 100%
Supplementary Credit Approval support of over £12m to
local authorities through the Strategic Development Scheme
between 1994 and 1999 for track, resignalling and other
infrastructure improvements on the railway link between
Cardiff and the south Wales valleys. In addition, for 1997–
1998, local authority bids approved under the new Welsh
Capital Challenge include grant-related support of £1.1m
over two years for a joint public and private sector scheme

to develop a tramway system for Llandudno. A further two million pounds was approved under the Welsh Capital Challenge for 1997–98 to improve Cardiff Central Square, which incorporates Cardiff Central Railway Station

Concessionary travel

The Welsh Office encourages local authorities to operate concessionary fares schemes to assist the mobility of certain groups, including the elderly and people with disabilities. Most unitary authorities in Wales choose to operate concessionary fares schemes and their expenditure in 1995-96 amounted to nine million pounds, most of which was for the benefit of retirement pensioners and disabled people. Around 330,000 people benefit from these schemes each year

Accessibility

The Government is committed to achieving fully accessible public transport systems for people with mobility difficulties. In recent years, considerable progress has been made in this area – for example, with the introduction of low-floor buses. For the future, the Government is developing the framework for the introduction of regulations under the Disability Discrimination Act 1995 on access standards for public transport vehicles

Bus Priority Scheme

The Welsh Bus Priority Scheme was introduced in 1992 and, since then, has allocated Supplementary Credit Approvals in support of 19 projects throughout Wales. One of the most successful is a dedicated bus-only route between Singleton Hospital and University College Swansea, which has resulted in a 30% increase in passenger levels

For 1997–1998, five projects were approved, including the provision of a range of bus priority detection equipment, kerb improvements and passenger facilities in Neath Port Talbot, the extension of bus priority detection measures in Rhondda Cynon Taff, the adaptation of roadsides for low-floor buses in Swansea

Buses and coaches

The deregulation of local bus services in 1986 led to the development of services better tailored to the needs of the travelling public. There has since been a steady increase in the supply of bus services in the Principality, with a number of vehicle kilometres for local services increasing by around 30% between 1986-87 and 1995-96. This reflects increasing competition as well as the trend towards higher-frequency services using smaller buses

In September 1996, the Department of Transport published 'Better Information for Bus Passengers – a Guide to Good Practice'. The Guide, which is aimed at improving the quality and range of information available to bus passengers, was produced at the initiative of the Bus Working Group, which comprised representatives from local authority associations, bus operators and bus users

Railway

The rail network plays a key role in the transport system in Wales, which has expanded over the past 10 years. There are today more route kilometres and some 30 new stations, several with park and ride facilities. A number of these have been supported by local authorities with resources provided by the Welsh Office

New franchised sector operators are bringing new ideas and new investment to the network. Great Western Holdings secured the franchise for the service between south Wales and London, and has already introduced a number of service improvements. Other franchise holders covering services in Wales, such as Prism Rail and Virgin, are committed to improving services for passengers across the railway network

Conclusion

Effective and efficient transport links are vital for the survival of communities. The Welsh Office's priority for public transport in Wales is to create the conditions for fair competition between and within the various public transport modes, so that the travelling public have more and better choices for their mode of transport

Although public transport services should wherever possible be provided on a fully commercial basis, there are instances where socially necessary but financially unattractive public transport requires revenue support. This is made available to local authorities by the Welsh Office

24b	See 24a	24.1
24c	See 24a	24.3
24d	See 24a	24.5
24e	See 2a	24.14

Chapter 25 Management of Heavy Goods Vehicles

No Welsh references

Chapter 26 Transport Aspects of New Developments

26a	See Ref 12a	26.2, 26.9
26b	Welsh Office Highways Directorate	

Chapter 27 Development Control Procedures

27a	Secretary of State for Wales	27.3
27b	Welsh Office Highways Directorate	27.3
27c	Secretary of State for Wales	27.3
27d	As a consequence of the Local Government (Wales) Act 1994, there are 22 unitary authorities in Wales responsible for local government functions including local roads. Motorways and trunk roads are the responsibility of the Secretary of State for Wales who has appointed 8 agent authorities to maintain the network. The Welsh Office, as highway authority for trunk roads and motorways,	27.3

has similar responsibilities to those of the Department of Transport in England: it is a statutory consultee on planning applications affecting trunk roads and motorways and has powers of direction over local planning authority decisions on the control of development alongside trunk roads and motorways. The remainder of the highway network is the responsibility of the 22 individual unitary authorities. The unitary authorities, together with the three Welsh National Park Authorities, are the local planning authorities responsible for day–to–day planning matters including mineral planning

Annotation	Comment for Wales	Chapter Reference
27e	In Wales, both forms of Agreement may run in parallel	27.4
27f	See Ref 12a	27.6

Chapter 28 Parking Relation to Development

28a	See Ref 12a	28.10, 28.11

Chapter 29 Residential Developments

29a	See Ref 12a	29.13

Chapter 30 Non–Residential Developments

30a	See Ref	30.1, 30.7

Chapter 31 Design Concepts

31a	Welsh Office Highways Directorate	31.6
31b	Welsh Office Highways Directorate	31.6
31c	Welsh Office Highways Directorate	31.9

Chapter 32 Alternative Concepts of Road Link Capacity
No Welsh references

Chapter 33 Procedures for the Planning and Approval of Transport Infrastructure Schemes

33a	Secretary of State for Wales	33.3
33b	Secretary of State for Wales	33.3
33c	Secretary of State for Wales	33.3
33d	Welsh Office Highways Directorate	33.3
33e	Welsh Office Highways Directorate	33.5
33f	Welsh Office Highways Directorate	33.7
33g	Secretary of State for Wales	33.7
33h	Welsh Office Highways Directorate	33.7
33i	Welsh Office Highways Directorate	33.8
33j	Welsh Office Highways Directorate	33.8
33k	See Ref 12a	33.10
33l	See Ref 12a	33.10
33m	These pamphlets are available in Welsh from the Welsh Office Highways Directorate	33.11

Glossary of Initials and Acronyms used in Transport in the Urban Environment

A

AA	Automobile Association
AADF	Average Annual Daily Flow (see Chapter 7)
AADT	Average Annual Daily Traffic (see Chapters 7 and 38)
AAHF	Average Annual Hourly Flow (see Chapter 7)
AAWF	Average Annual Weekly Flow (see Chapter 7)
Ac	Alignment constraint (see Chapter 36)
AC	Audit Commission (see Chapter 14)
ACC	Association of County Councils (now part of the LGA)
ACG	Annual Capital Guidelines (see Chapters 4 and 6)
ACLA	Audit Commission for Local Authorities
ACTRA	Advisory Committee on Trunk Road Assessment (the Leitch Committee)
ADC	Association of District Councils (now part of the LGA)
ADS	Advance Direction Sign (see Chapter 15)
ADS	Automatic Debiting System (see Chapter 18)
AFC	Automatic Fee Collection (see Chapter 21)
AGD	Above Ground Detection (see Chapter 40)
AGT	Automated Guided Transit (see Chapter 34)
AID	Automatic Incident Detection (see Chapter 18)
ALBES	Association of London Boroughs Engineers and Surveyors
ALS	Area Licensing System (see Chapter 21)
AMA	Association of Metropolitan Authorities (now part of the LGA)
AP	All Purpose (roads) (see Chapter 38)
APC	Advance Payments Code (see Chapter 27)
ARCADY	Assessment of Roundabout CApacity and DelaY (see Chapter 37)
ARG	Autonomous Route Guidance (see Chapter 15)ATG Auto Teilet Genossenschaft (see Chapter 21)
ATT	Advanced Transport Telematics (see Chapter 21)
AUT	Automatic Upgrading of TRANSYT (see Chapter 41)
AVI	Automatic Vehicle Identification (see Chapter 18)
AVL	Automatic Vehicle Location (see Chapter 18)

B

BPA	British Parking Association
BS	British Standard
BSI	British Standards Institute

C

CAD	Computer Aided Design
CAPI	Computer Aided Personal Interviewing (see Chapter 7)
CATI	Computer Assisted Telephone Interviewing (see Chapter 7)
CBI	Confederation of British Industry

CCD	Charge Coupled Device (see Chapter 18)
CCT	Compulsory Competitive Tendering (see Chapter 16)
CCTV	Closed Circuit TeleVision (see Chapters 14 and 20)
CEC	Commission of European Communities
CIDEL	Calculation of Incident DELay (see Chapter 35)
CIT	Chartered Institute of Transport
CLF	Cableless Linking Facility (see Chapter 40)
CNG	Compressed Natural Gas (see Chapter 5)
CO2	Carbon diOxide
COBA	COst Benefit Analysis (see Chapter 9)
COE	Certificate Of Entitlement (to own a vehicle in Singapore) (see Chapter 21)
CPT	Confederation of Public Transport
CPZ	Controlled Parking Zone (see Chapter 19)
CSS	County Surveyors' Society (now called CSS)
CT	Civic Trust
CTC	Cyclists' Touring Club (see Chapter 23)

D

D1	Dual single–lane carriageways (see Chapter 38)
D2	Dual two–lane carriageways (see Chapter 38)
D3	Dual three–lane carriageways (see Chapter 38)
DB 32	Design Bulletin (see Chapter 29)
DAB	Digital Audio Broadcasting (see Chapter 15)
DBFO	Design, Build, Finance and Operate (see Chapter 33)
DBOM	Design, Build, Operate and Maintain
DMRB	Design Manual for Roads and Bridges
DMSSD	Desirable Minimum Stopping Sight Distance (see Chapter 38)
DOE	Department Of the Environment
DOH	Department Of Health
DOT	Department Of Transport
DRG	Dynamic Route Guidance (see Chapters 15 and 18)
DRIVE	Dedicated Road Infrastructure for Vehicle Safety in Europe (see Chapter 18)
DS	Direction Signs (see Chapter 15)
DVLA	Driver and Vehicle Licensing Agency (see Chapters 14 and 20)

E

EA	Environmental Agency (see Chapter 34)
EC	European Commission (strictly, the CEC)
ECMT	European Conference of Ministers of Transport (see Chapter 17)
ECN	Excess Charge Notice (see Chapter 14)
EFTE	European Federation for Transport and the Environment (see Chapter 23)
EH	English Heritage
EHTF	English Historic Towns Forum
EIA	Environmental Impact Assessment (see Chapters 9 and 26)
EIB	European Investment Bank
EIP	Examination In Public (see Chapters 4 and 10)

EMA	Environmental Management Area (see Chapter 17)
ERDF	European Regional Development Fund (see Chapter 4)
ERP	Electronic Road–use Pricing (see Chapter 21)
ERTICO	European Road Transport Informatics Co–ordination Office
ES	Environmental Statement (see Chapter 9)
ETC	Electronic Toll Collection (see Chapter 18)
ETSU	Energy Technology Study Unit, Harwell (see Chapter 34)
EU	European Union

F

FOSD	Forward Overtaking Sight Distance (see Chapter 36)
FPN	Fixed Penalty Notice (see Chapters 14 and 19)
FTA	Freight Transport Association
FYRR	First Year Rate of Return (see Chapters 9 and 16)

G

GD	General Directions
GDP	Gross Domestic Product (see Chapter 1)
GFA	Gross Floor Area (see Chapter 26)
GIS	Geographic Information System (see Chapters 7 and 18)
GLT	Guided Light Transit (see Chapter 34)
GOL	Government Office for London (see Chapter 3)
GOR	Government Offices for the Regions (see Chapter 4)
GPS	Global Positioning System (see Chapters 15 and 34)
GSM	Global System for Mobile Communications (see Chapter 15)
GVW	Gross Vehicle Weight

H

HA	Highways Agency
HAZMAT	HAZardous MATerial (see Chapter 25)
HETA	Highways, Economics and Traffic Appraisal (Division of DOT) (see Chapter 7)
HGV	Heavy Goods Vehicle
HMG	Her Majesty's Government
HMRI	Her Majesty's Railway Inspectorate (see Chapter 34)
HO	Home Office
HOV	High Occupancy Vehicle (see Chapters 6 and 21)
HSE	Health and Safety Executive (see Chapters 4 and 34)

I

ICD	Inscribed Circle Diameter (see Chapter 39)
ICE	Institution of Civil Engineers
IEA	Institute of Environmental Assessment
IEE	Institution of Electrical Engineers
IHT	Institution of Highways & Transportation
INSET	In–SErvice Training (see Chapter 16)
IPR	Intellectual Property Rights (see Chapter 18)
IRR	Internal Rate of Return (see Chapter 9)

ISGLUTI	International Study Group on Land–Use/Transport Interaction (see Chapter 8)
ISO	International Standards Organisation (see Chpater 31)
ITS	Intelligent Transport Systems (see Chapters 18 and 21)
ITU	In–station Transmission Units (see Chapter 41)
IVU	In–Vehicle Unit (see Chapter 21)

J and K

None

L

LA 21	Local Agenda 21 (see Chapter 17)
LAA	Local Authorities Association (see Chapter 16) (now the Local Government Association – LGA)
LAM	London Area Model (see Chapter 8)
LBPN	London Bus Priority Network (see Chapter 24)
Lc	Layout constraint (see Chapter 36)
LC	Liquid Crystal (see Chapter 34)
LCD	Liquid Crystal display (see Chapter 15)
LED	Light–Emitting Diode (see Chapter 15)
LGMB	Local Government Management Board (see Chapter 17)
LGA	Local Government Association (incorporating the ACC, the ADC, and the AMA)
LGV	Light Goods Vehicle
LPA	Local Processing Authority (for STATS 19 forms) (see Chapter 7)
LPAC	London Planning Advisory Committee (see Chapter 28)
LRT	Light Rapid Transit (see Chapter 34)
LTA	Land Transport Authority (Singapore) (see Chapter 21)
LUTI	Land–Use/Transport Interaction (see Chapter 8)
LUTS	Land–Use/Transportation Study (see Chapter 8)

M

MEP	Members of the European Parliament (see Chapter 10)
MMQ	Mean Maximum Queue (see Chapter 43)
MOVA	Microprocessor Optimised Vehicle Actuation (see Chapters 5 and 40)
MP	Member of Parliament (UK) (see Chapter 10)

N

NEDO	National Economic Development Office (see Chapters 21 and 28)
NHBC	National House Building Council (see Chapter 29)
NHS	National Health Service (see Chapter 11)
NI	Northern Ireland
NIP	Notice of Intent to Prosecute (see Chapters 14 and 19)
NIS	National Information System (see Chapter 7)
NJUG	National Joint Utilities Group (see Chapter 29)
NPV	Net Present Value (see Chapter 9)
NRTF	National Road Traffic Forecasts (see Chapters 2 and 26)
NTC	National Traffic Census (see Chapter 7)
NTS	National Travel Survey (see Chapter 2)

O

O – D	Origin – Destination (see Chapter 7)
OECD	Organisation for European Co–operation and Development
OPRAF	Offices of Passenger RAil Franchising (see Chapter 4)
ORR	Office of the Rail Regulator (see Chapter 4)
OSCADY	Optimised Signals CApacity and DelaYs (see Chapter 37)
OSGR	Ordnance Survey Grid Reference (see Chapter 7)
OST	Office of Science and Technology (see Chapter 1)
OTU	Out–station Transmission Unit (see Chapter 41)

P

PA	Pedestrians' Association (see Chapter 22)
PAR	Potential for Accident Reduction (see Chapter 7)
PACTS	Parliamentary Advisory Council for Transport Safety (see Chapter 16)
PCfL	Parking Committee for London (see Chapters 14 and 19)
PCN	Penalty Charge Notice (see Chapters 14 and 19)
PCO	Parking Control Officer (see Chapter 19)
pcu	passenger car unit
PFI	Private Finance Initiative (see Chapter 33)
PHF	Peak Hour Flow (see Chapter 7)
PICADY	Priority Intersections CApacity and DelaY (see Chapters 37 and 38)
PIA	Personal Injury Accidents (see Chapters 16 and 37)
PLI	Public Local Inquiry (see Chapter 10)
PNR	Private Non–Residential (parking) (see Chapter 19)
POST	Parliamentary Office of Science and Technology (see Chapter 1)
PPA	Permitted Parking Area (see Chapter 19)
PPG	Planning Policy Guidance
PSV	Public Service Vehicle
PTA	Passenger Transport Authority
PTE	Passenger Transport Executive
PTEG	Passenger Transport Executive Group

Q

QA	Quality Assurance (see Chapter 31)

R

RAC	Royal Automobile Club
RC	Reserve Capacity (see Chapter 39)
RCEP	Royal Commission on Environmental Pollution (see Chapters 1 and 12)
RDS	Radio Data System (see Chapter 15)
RDS–TMC	Radio Data System – Traffic Message Channel (see Chapter 15)
RHA	Road Haulage Association (see Chapter 25)
ROSCO	ROlling Stock COmpany (see Chapter 4)
RoSPA	Royal Society for the Prevention of Accidents (see Chapter 16)
RPG	Regional Planning Guidance (see Chapter 26)
RSG	Revenue Support Grant (see Chapter 4)

RTA	Road Traffic Act (see Chapter 13)
RTRA	Road Traffic Regulation Act (see Chapter 13)
RTUA	'Roads and Traffic in Urban Areas'

S

S	Scotland
S2	Single Two–lane carriageway (see Chapter 38)
SACTRA	Standing Advisory Committee on Trunk Road Assessment
SCA	Supplementary Credit Approval (see Chapters 4, 6 and 11)
SCOOT	Split Cycle Offset Optimisation Technique (see Chapters 5, 18 and 41)

SI	Statutory Instrument
SODD	Scottish Office Development Department
SP	Stated Preference (surveys) (see Chapter 7)
SPA	Special Parking Area (see Chapter 19)
SPRINT	Selective PRIority Network Technique (see Chapter 24)

SR	Severity Ratio (accidents) (see Chapter 7)
SRB	Single Regeneration Budget (see Chapter 6)
SSD	Stopping Sight Distance (see Chapter 36)
SSI	Site of Scientific Interest (see Chapter 31)
SSSI	Site of Special Scientific Interest (see Chapter 31)

STGO	Special Types, General Order (motor vehicle authorisation) (see Chapter 25)
STM	Strategic Transportation Model (see Chapter 8)
SVD	Selective Vehicle Detection (see Chapters 18 and 24)
SWOT	Strengths, Weaknesses, Opportunities and Threats (see Chapter 12)

T

TAM	Traffic Appraisal Manual (see Chapter 7)
TEN	Trans–European Network (see Chapter 4)
TIA	Traffic/Transport Impact Analysis (see Chapter 26)
TMC	Traffic Message Channel (see Chapter 15)
TPP	Transport Policies and Programme (see Chapters 1, 10 and 11)

TRB	Transportation Research Board (US)
TRL	Transport Research Laboratory
TRO	Traffic Regulation Order (see Chapters 4, 13, 15 and 27)
TSG	Transport Supplementary Grant (see Chapters 1, 6 and 11)
TSM	Traffic signs Manual (see Chapter 15)

TSR	Traffic Signs Regulations (see Chapter 15)
TSRGD	Traffic Signs Regulations and General Directions (see Chapters 13 and 15)
TV	Tele–Vision
TVR	Transport sur Voie Reservée (see Chapter 34)
TWA	Transport and Works Act (1992) (see Chapter 33)
TWO	Transport and Works Order (see Chapter 24)

U

UDA	Unitary Development Authority (see Chapter 4)
UDP	Unitary Development Plan (see Chapters 4, 10 and 26)
UITP	Union Internationale de Transports Publique (see Chapter 34)

UK	United Kingdom (of Great Britain and Northern Ireland)
UKHIS	United Kingdom Hazard Information System (see Chapter 25)
UN	United Nations
UNCED	United Nations Commission on Environment and Development (see Chapters 17)
UNDP	United Nations Development Programme
UNECE	United Nations Economic Commission for Europe (see Chapter 3)
URECA	URban EConomic Appraisal (see Chapter 9)
USM	Urban Safety Management (see Chapter 16)
UTC	Urban Traffic Control (see Chapter 40)
UTMC	Urban Traffic Management and Control (see Chapter 41)

V

VA	Vehicle–Actuation (of traffic signals) (see Chapter 41)
V/C	Ratio of Volume to Capacity (see Chapters 18, 38 and 39)
VED	Vehicle Excise Duty
VMS	Variable Message Sign (see Chapters 15 and 18)

W

W	Wales
WCED	World Commission on Environment and Development (see Chapter 1)
WHO	World Health Organisation (see Chapter 18)
WO	Welsh Office
WS2	Wide Single Two–lane carriageway (see Chapter 38)

XYZ

None

Index

Introduction

Index entries are to section numbers; reference to Figures, Photographs and Tables is as designated. Alphabetical arrangement is word–by–word, where a group of letters followed by a space is filed before the same group of letters followed by a letter, eg 'bus stops' will appear before 'business journeys'.

Limited use of the words 'transport', 'transportation' and 'urban' as entry words has been made, and references should be sought under more specific terms. The index is not exhaustive, and no responsibility is taken for any errors or omissions.

Subject	Section	Subject	Section

Subject	Section

Subject	Section

E

Subject	Section

M

Q

Subject	Section
screenline surveys	7.9
Seattle	
busways	24.12
Secretary of State for Transport	27.3
Secretary of State's Traffic Orders (Procedure)	
(England and Wales) Regulations 1990	13.2
Section 56 grants see Transport Act 1968	
security	
objectives	3.6
pedestrians' personal	22.4, 22.12
town centres	12.4
seed oil fuels	
twenty–first century projections	5.3
segregation	
buses	24.1
light rapid transit	24.1
selective detection	
co–ordinated traffic signal control	41.15
Selective Priority Network Technique (SPRINT)	24.9
Selective Vehicle Detection (SVD)	18.4, 24.9
self–completion surveys	7.10
service derivation	8.10
service information	
information provision policy measures	6.6
service levels	
appraisals	9.12
highway design concepts	31.3
service roads	35.9
servicing pattern, junction design	37.6
set–backs	35.9
bus lanes	24.6
settlement patterns	2.6
severance of communities	1.2, 17.5
severity ratio, accidents	7.4
shadow network methods	8.8
shared delay formula	40.12
shared driveways	29.4, photo 29.1
shared surfaces	20.2, photo 20.2
access roads	29.4, photo 29.1
shared–use roads	20.8
sheltered parking	photo 20.3, 20.8
shopping centres	
fixed–track system interchanges	34.7
shopping journeys	2.5
shopping streets	11.3, photo 11.2
shops, town centres	12.3
short–stay car parks see parking	
Shrewsbury	
park–and–ride	24.5
shrubs, residential developments	29.6
side roads, pedestrian crossings	22.5
sig–nabout	42.4
signal control see traffic signal control	
signal–controlled gyratories see gyratories	
signal–controlled pedestrian crossings see	
traffic signal–controlled pedestrian crossings	
signal plans, co–ordinated traffic signal control	41.7
signals see traffic signal control	
signs	
pedestrians	22.4
traffic see traffic signs	
SIGSIGN	37.6, 40.14
simple junctions	38.2
simple tabular methods	
appraisal	9.8
simplified demand models	8.8

Subject	Section
Singapore	
Area Licensing Scheme (ALS)	21.3
Electronic Road Pricing (ERP)	21.3, 21.6
vehicle ownership restraint	21.6
single carriageways	35.2, figs 35.1, 35.3–35.4
single lane dualling, priority junctions	38.2
Single Regeneration Budget (SRB)	4.8, 6.9
single site	
accident analysis and investigation	16.8
safety schemes	table 16.4, 16.10
smaller towns, parking	28.6
smart cards	18.13
congestion charging	21.3
integrated ticketing	18.15
ticketing	21.10, 21.12
twenty–first century projections	5.3, 5.4
Smeed Report	21.3
snow	18.9
social changes	
travel–generating	2.6
twenty–first century projections	5.3
social sustainability	21.2
socially–required bus services	24.2
software	
co–ordinated traffic signal control	41.12
trip planning	15.13
Southampton	
ROMANSE project	18.15, photo 18.11
traffic metering	24.8
SP see Stated Preference	
SPA (Special Parking Areas)	13.2, 19.4
space mean–speed	7.8
space–sharing between pedestrians and vehicles	
town centres	12.4, photos 12.6–12.7
Special Parking Areas (SPA)	13.2, 19.4
special roads	4.8, 33.8
definition	table 11.3
speed	
accidents, significant factor in	20.1
assessment and discrimination	
traffic signal control	40.6
control	20.3
islands, residential developments	fig 29.13, photo 29.2
road safety	16.13
design see design speed	
fixed–track systems	34.3
highway design concepts	31.5
management	20.1
measurement	7.8
roundabouts	39.6
traffic calming see traffic calming	
traffic volume effects	32.2
speed cameras	20.8, 20.12
signing	20.13
speed cushions	20.7, 20.8, fig 20.4
residential developments	fig 29.14
speed/flow relationships	35.2, fig 35.1
speed limitation methods	8.8
speed–limiters	16.13
speed–limits	
20 miles/h zones	20.10
carriageway markings, speed cameras	20.12
changes	20.9

Subject	Section	Subject	Section
sustainability	21.2	20 miles/h zones, alternative routes	20.10
indicators, environmental management	17.3	exclusion from town centres	12.6
objectives	3.6	throughput see traffic throughput	
sustainable development		thumps	20.8
definition	1.3	TIA see transport impact assessments	
environmental management	17.3	tidal flow	35.3, photo 35.3
issues	1.3	time–based charging	21.3
policies	4.7	time–distance diagrams	
policy packages	1.5	co-ordinated traffic signal control	41.5
safety issues	1.6	gyratories	fig 42.3, 42.6
traffic management	11.4	time mean–speed	7.8
SVD (Selective Vehicle Detection)	18.4, 24.9	time–of–day choice	8.6
Swansea		time–of–day control, gyratories	42.3
Selective Vehicle Detection (SVD)	24.9	time–periods, modelling	8.9, 8.10
SWOT analyses, town centre management	12.5	timetable information	
		information provision policy measures	6.6

T

Subject	Section	Subject	Section
		TMC (Traffic Message Channel)	15.11
		TOC (train operating companies)	4.4
T junctions see priority junctions		Toll Orders	33.8
tactile signals		toll–rings, pricing policy measures	6.7
signal–controlled pedestrian crossings	22.10	tolls	4.8, 18.13
tactile surfaces		Toucan crossings	13.11, 22.7, photo 22.1,
pedestrian crossings	22.5,		23.15, photo 31.5
	photos 22.1–22.2, 22.6	gyratories	42.8
signal–controlled pedestrian crossings	22.7	priority junctions	38.5
talking signs, bus passenger information	15.14	roundabouts	23.11, 39.8, 39.9
tankers and tank containers		signal heads	40.6
dangerous goods	25.13	tourist attractions, direction signs	15.5
targets		tourist management strategies, town centres	12.3
environmental management	17.4	Town and Country Planning Act 1990	
policy formulation	3.6–3.7	developers' contributions	4.8
transportation strategy	6.10	development control procedures	27.3, 27.4
taxis		development plans	4.5
lanes, enforcement	14.6	legal challenges	10.3
pedestrian areas	22.9	highway schemes	33.3
role in demand management	21.10	stopping–up of highways	13.9
using bus priority measures	24.4, 24.6	traffic management	11.6
TD9/93 'Highway Link Design'	36.1–36.9, 38.6	vehicular rights, extinguishment	13.10
TD16/93 'Geometric Design of Roundabouts'	39.5, 39.11	Town and Country Planning	
TD22/92 'Layout of Grade–separated		(Assessment of Environmental	
Junctions'	fig 43.1, 43.5	Effects) Regulations 1988	9.11
TD39/94 'The Design of Major Interchanges'	43.1, 43.6	Town and Country Planning	
TD40/94 'The Layout of		(General Permitted Development) Order 1995	27.3
Compact Grade–separated Junctions'	fig.43.1, 43.5	consultation	33.3
TD42/95 'Geometric Design of		planning permission	33.3
Major/Minor Priority Junctions'	38.1–38.8	Town and Country Planning General	
technical appraisal, land–use	27.2	Regulations 1992	
Technical Design Standards		planning permission	10.2
see entries beginning with TD		Town and Country Planning (Use Classes)	
tele–shopping and tele–working		Order 1987	27.3
twenty–first century projections	5.3	Town Centre Management	
telecommunications services		definition	12.5
information provision policy measures	6.6	town centres	12.1
telecommuting	2.6, 21.11	access	12.2
telematics see transport telematics		managing	12.6
telephone surveys	7.10	priority hierarchy	12.6
temporary Orders	13.2	restrictions enforcement	12.5
temporary traffic signs	15.4	time–based restrictions	photo 12.12
TEN (Trans–European Networks)	4.8	bollards	12.4, photo 12.9
terminals, infrastructure policy measures	6.4	buses	12.4, 12.6
theme parks	30.2	care	12.4
thermal infra–red mapping	18.9	categories of roads and spaces	12.6, photo 12.15
three–level roundabouts	43.2	compressed natural gas (CNG), vehicles	
through routes, use for short trips	29.5	powered by	12.6, photo 12.16
through traffic	11.2	Conservation Areas	12.4
		consultation	12.7, photo 12.18

Subject	Section

U

Subject	Section
U–turns	
roundabouts	39.2
UDP see Unitary Development Plans	
UK Hazard Information System (UKHIS)	25.13
UN–ECE (United Nations Economic Commission for Europe)	3.3
UNCED (United Nations Conference on Environment and Development)	1.3, 3.3, 17.3, 17.4
underground services	11.3, 22.9, 29.7
underground systems	34.2
unilateral undertakings by developers	27.3
unitary authorities	27.3
powers and responsibilities	4.2
Unitary Development Plans (UDP)	4.5, 26.2, 27.3, 33.2, 33.3, 33.7
environmental management	17.3
policy formulation	3.3
public involvement	10.2
United Nations Conference on Environment and Development (UNCED)	1.3, 3.3, 17.3, 17.4
United Nations Economic Commission for Europe (UN–ECE)	3.3
unloading	photo 1.2
control	13.5
link capacity effects	32.3
management	19.5
restrictions	25.2, 25.10
traffic management	13.5
Urban and Economic Development Group (URBED)	12.3
urban clearways	13.5
Urban Economic Appraisal (URECA)	9.6, 9.8, figs 9.1–9.2
Urban Regeneration Agency	4.3
urban safety management	16.7, fig 16.1
urban streets	photo 1.1
urban traffic	2.3
urban traffic conditions surveys	7.8
Urban Traffic Control (UTC)	
bus priority	24.9
co–ordinated signal systems see traffic signal control	
fixed–track systems links	34.9
gyratories	42.7
integration with motorway control systems	18.15
junctions	31.7
link capacity effects	32.3
management policy measures	6.5
urban traffic management and control (UTMC)	5.3, 41.2, 41.16
URBED (Urban and Economic Development Group)	12.3
URECA (Urban Economic Appraisal)	9.6, 9.8, figs 9.1–9.2
use classes, development control procedures	27.3
UTC see Urban Traffic Control	
UTMC (urban traffic management and control)	5.3, 41.2, 41.16
UTOPIA	41.8

V

Subject	Section
VA (vehicle–actuated) operation	40.6, 41.7
variable message signs (VMS)	15.2, 15.6, 41.15
car parks	19.11, photos 19.3–19.4
electro–mechanical	18.6
information provision policy measures	6.6
light emitting signs	18.6
parking guidance	photo 18.2
parking management systems	18.6
reflective flip–disk	18.6
route–guidance	18.12
speed control	photo 18.3
technology	18.6
traffic management	11.13
twenty–first century projections	5.3
type approval	15.6
variable speed–limits	20.11
VED (vehicle excise duty)	6.7
vehicle–actuated (VA) operation	40.6, 41.7
vehicle building and servicing industry employment in	1.1
vehicle–detection systems, traffic signal control	40.6
vehicle emissions	
standards	3.3
traffic calming effects on	20.5
vehicle excise duty (VED)	6.7
Vehicle Inspectorate Executive Agency responsibilities	24.2
vehicle location monitoring, fixed–track systems	34.9
vehicle operating costs	
junction design	37.6
savings appraisal	9.8
vehicle ownership	
restraint	21.6
taxes, pricing policy measures	6.7
vehicle path delineation, town centres	12.4
vehicle safety developments	16.14
vehicle technologies	
twenty–first century projections	5.3
vehicle–type categories	fig 7.1
vehicle use regulation	21.5
vehicles	
see also under individual names and types	
parking access and safety	28.8
recycling at end of life	1.3
residential developments	
access to sites	29.5
provision for movement	29.8
travel patterns	2.2
vehicular capacity	32.1–32.3
vehicular crossing points in central reserves	35.7
vehicular delay, traffic signal control	40.12
vehicular rights, extinguishment	13.10
verge–located signs, gyratories	42.9, photo 42.3
verges	35.6
vehicle crossings over	27.3, 27.4
vertical alignment	36.7
coordination with horizontal alignment	36.8
vertical curves	36.7
viability of town centres	12.3, photo 12.1
vibration	17.5
video cameras	
bus lane enforcement	14.6, 14.8
level crossing enforcement	14.6